Hyperbolic Problems

Theory, Numerics and Applications

Volume 2

Series in Contemporary Applied Mathematics CAM

Honorary Editor: Chaohao Gu (*Fudan University*)
Editors: P. G. Ciarlet (*City University of Hong Kong*), Tatsien Li (*Fudan University*)

---

1 Jiongmin Yong, Rama Cont, eds., *Mathematical Finance: Theory and Practice* (2000)
2 F. Dubois, Huamo Wu, eds., *New Advances in Computational Fluid Dynamics: Theory, Methods and Applications* (2001)
3 Hanji Shang, Alain Tosseti, eds., *Actuarial Science: Theory and Practice* (2002)
4 Alexandre Ern, Weiping Liu, eds., *Mathematical Problems in Environmental Science and Engineering* (2002)
5 Haïm Brezis, Tatsien Li, eds., *Ginzburg-Landau Vortices* (2005)
6 Tatsien Li, Pingwen Zhang, eds., *Frontiers and Prospects of Contemporary Applied Mathematics* (2006)
7 Deguan Wang, Christian Duquennoi, Alexandre Ern, eds., *Mathematical Methods for Surface and Subsurface Hydrosystems* (2007)
8 Rolf Jeltsch, Tatsien Li, Ian H. Sloan, eds., *Some Topics in Industrial and Applied Mathematics* (2007)
9 Philippe G. Ciarlet, Tatsien Li, eds., *Differential Geometry: Theory and Applications* (2008)
10 Tatsien Li, Pingwen Zhang, eds., *Industrial and Applied Mathematics in China* (2009)
11 Zhien Ma, Yicang Zhou, Jianhong Wu, eds., *Modeling and Dynamics of Infectious Diseases* (2009)
12 Tomas Y. Hou, Chun Liu, Jianguo Liu, eds., *Multi-Scale Phenomena in Complex Fluids: Modeling, Analysis and Numerical Simulation* (2010)
13 Guiqiang Chen, Tatsien Li, Chun Liu, eds., *Nonlinear Conservation Laws, Fluid Systems and Related Topics* (2010)
14 A. Damlamian, S. Jaffard, eds., *Wavelet Methods in Mathematical Analysis and Engineering* (2010)
15 Tatsien Li, Yuejun Peng, Bopeng Rao, eds., *Some Problems on Nonlinear Hyperbolic Equations and Applications* (2010)
16 A. Damlamian, B. Miara, Tatsien Li, eds., *Multiscale Problems: Theory, Numerical Approximation and Applications* (2011)
17 Tatsien Li, Song Jiang, eds., *Hyperbolic Problems: Theory, Numerics and Applications*, Vol. 1 (2012)
18 Tatsien Li, Song Jiang, eds., *Hyperbolic Problems: Theory, Numerics and Applications*, Vol. 2 (2012)

Series in Contemporary Applied Mathematics CAM 18

Hyperbolic Problems

Theory, Numerics and Applications

Volume 2

editors

Tatsien Li
Fudan University, China

Song Jiang
IAPCM, China

Higher Education Press

World Scientific

NEW JERSEY • LONDON • SINGAPORE • BEIJING • SHANGHAI • HONG KONG • TAIPEI • CHENNAI

Tatsien Li
School of Mathematical Sciences
Fudan University
Shanghai 200433, China
Email: dqli@fudan.edu.cn

Song Jiang
Institute of Applied Physics and
Computational Mathematics
Beijing 100088, China
Email: jiang@iapcm.ac.cn

Editorial Assistant: Chunlian Zhou

Higher Education Press Limited Company
4 Dewai Dajie, Beijing 100120, P. R. China and
World Scientific Publishing Co Pte Ltd
5 Toh Tuch Link, Singapore 596224

ISBN: 978-981-4417-06-8 (set)
ISBN: 978-981-4417-08-2 (v2)

Printed in P. R. China

Scientific Committee

Yann Brenier (France)
Alberto Bressan (USA)
Constantine Dafermos (USA)
Xiaqi Ding (China)
Ling Hsiao (China)
Rolf Jeltsch (Switzerland)
Song Jiang (China)
Kenneth Karlsen (Norway)
Shuichi Kawashima (Japan)
Dietmar Kroener (Germany)
Randall LeVeque (USA)
Tatsien Li (China)
Taiping Liu (USA)
H. J. Nussenzveig Lopes (Brazil)
Pierangelo Marcati (Italy)
Denis Serre(France)
Chi-Wang Shu (USA)
Eitan Tadmor (USA)
Zhouping Xin (China)

Local Organizing Committee

Tatsien Li (Co-Chair)
Song Jiang(Co-Chair)

Feimin Huang
Hailiang Li
Yaguang Wang
Huijiang Zhao
Changjiang Zhu

Preface

The 13th International Conference on Hyperbolic Problems: Theory, Numerics and Applications (HYP 2010) was held in Beijing, China, from June 15 to 19, 2010. Over 200 participants attended the conference and 162 among them came from abroad. There were 10 plenary lectures, 18 invited talks and around 140 contributed talks in parallel sessions.

The objective of the conference is to bring together experts, researchers and students with interest in theoretical analysis, numerical simulations, and applications of hyperbolic partial differential equations and related mathematical models appearing in applied sciences. The conference keeps the traditional balance of the HYP series, blending theory, numerics and applications, and an emphasis is put on nonlinear problems and applications in various fields such as fluid mechanics, elasticity, astrophysics, biomathematics, traffic flow, etc. As has been done in the past, a special effort has been made it possible for young scientists to attend this conference and to promote their interaction with the more senior researchers.

The biannual HYP series of international conferences on Hyperbolic Problems was initiated by C. Carasso, P.-A. Raviart and D. Serre with the first conference held in Saint-Étienne, France in 1986. Since then it has been organized in Aachen (Germany, 1988), Uppsala (Sweden, 1990), Taormina (Italy, 1992), Stony Brook (USA, 1994), Hong Kong (China, 1996), Zurich (Swizerland, 1998), Magdeburg (Germany, 2000), Pasadena (USA, 2002), Osaka (Japan, 2004), Lyon (France, 2006), and Maryland (USA, 2008). Throughout these years, it has become one of the highest quality and most successful conference series in applied mathematics.

Hyperbolic problems, which are probably originated from Euler's study on the acoustic wave in 1755, not only have a long history but also have extremely rich physical background. The development is highly stimulated by their applications to Physics, Biology, and Engineering Sciences, in particular by the design of effective numerical algorithms. Due to recent rapid development of computers, more and more scientists use hyperbolic partial differential equations and related evolutionary equations as basic tools when propose new mathematical models of various phenomena and related numerical algorithms. We believe that various fields in science and engineering will bring us further into future interests in hyperbolic problems.

The scientific committee of this conference consists of Yann Brenier (France), Alberto Bressan (USA), Constantine Dafermos (USA), Xiaqi Ding (China), Ling Hsiao (China), Rolf Jeltsch (Switzerland), Song Jiang (China), Kenneth Karlsen (Norway), Shuichi Kawashima (Japan), Dietmar Kroener (Germany), Randall LeVeque (USA), Tatsien Li (China), Tai-Ping Liu (USA), Helena J. Nussenzveig Lopes (Brazil), Pierangelo Marcati (Italy), Denis Serre (France), Chi-Wang Shu (USA), Eitan Tadmor (USA) and Zhouping Xin (China). We thank all members of the scientific committee for recommending plenary and invited speakers and promoting this conference.

This conference is supported by the Academy of Mathematics and Systems Science, the Institute of Applied Physics and Computational Mathematics, Capital Normal University, Central China Normal University, Fudan University, Shanghai Jiaotong University, and Wuhan University. It is also sponsored by Shanghai Key Laboratory for Contemporary Applied Mathematics, Wuhan Institute of Physics and Mathematics, Chinese Academy of Sciences, National Natural Science Foundation of China, and Tianyuan Foundation for Mathematics.

These two volumes contain 80 original research and review papers which are written by leading researchers and promising young scientists, and cover a diverse range of multi-disciplinary topics addressing theoretical, modeling and computational issues arising under the umbrella of "Hyperbolic Partial Differential Equations". All the articles are peer-reviewed. We would like to thank all the referees for their efforts done for reading and judging the submitted articles. Special thanks go to Dr. Tao Wang for his laborious job arranging the final layout of articles.

Finally, we are extremely thankful to our colleagues Professors Feimin Huang, Hailiang Li, Yaguang Wang, Huijiang Zhao and Changjiang Zhu, and Dr. Xiaoyun Zhai, Dr. Yi Wang, and many graduate students from the Academy of Mathematics and Systems Science, the Institute of Applied Physics and Computational Mathematics, Capital Normal University, Central China Normal University, Fudan University, Shanghai Jiaotong University and Wuhan University, for their assistance in making this conference successful and comfortable.

June 2011
Tatsien Li, Song Jiang

Contents

Volume 2

Contributed Talks

Volume 1

Plenary Talks

Invited Talks

Convergence Analysis of the Particle Method for the Camassa-Holm Equation*

Alina Chertock
Department of Mathematics, North Carolina State University
Raleigh, NC 27695, USA
E-mail: chertock@math.ncsu.edu

Jian-Guo Liu
Department of Physics and Mathematics, Duke University
Duke University, Durham, NC 27708, USA
E-mail: Jian-Guo.Liu@duke.edu

Terrance Pendleton
Department of Mathematics, North Carolina State University
Raleigh, NC 27695, USA
E-mail: tlpendle@ncsu.edu

Abstract

The purpose of this paper is to establish a new method for proving the convergence of the particle method applied to the Camassa-Holm (CH) equation. The CH equation is a strongly nonlinear, bi-Hamiltonian, completely integrable model in the context of shallow water waves. The equation admits solutions that are nonlinear superpositions of traveling waves that have a discontinuity in the first derivative at their peaks and therefore are called peakons. This behavior admits several diverse scientific applications, but introduce difficult numerical challenges. To accurately capture these solutions, one may apply the particle method to the CH equation. Using the concept of space-time bounded variation, we show that the particle solution converges to a global weak solution of the CH equation for positive Radon measure initial data.

*The work of A. Chertock and T. Pendleton was supported in part by the NSF Grant DMS-0712898; the work of J.-G. Liu was supported in part by the NSF Grant DMS 10-11738.

1 Introduction

The purpose of this paper is to establish convergence results for the particle method applied to the Camassa-Holm (CH) equation, given as

$$m_t + (um)_x + u_x m = 0 \quad \text{with} \quad m = u - \alpha^2 u_{xx}, \tag{1.1}$$

which is considered subject to the initial condition

$$m(x,0) = m_0(x). \tag{1.2}$$

Here the momentum m and velocity u are functions of the time variable t and the spatial variable x, and α is a length scale. Equation (1.1) arises in diverse scientific applications and, for instance, can be described as a bi-Hamiltonian model for waves in shallow water [3]. Equation (1.1) can also be used to quantify growth and other changes in shape, such as occurs in a beating heart, by providing the transformative mathematical path between the two shapes, (see, e.g, [8]).

The CH equation exhibits some remarkable properties. Of notable interest is the fact that the equation is completely integrable and yields peakon solutions which are solitons (whose identity is preserved through nonlinear interactions) with a sharp peak. Mathematically, this sharp peak is characterized by a discontinuity at the peak in the wave slope, and therefore are called peakons, [3].

Peakons may be accurately captured by applying particle methods to the CH equation as shown in, e.g., [4, 5, 6, 8]. In these methods, the solution is sought as a linear combination of Dirac distributions, whose positions and coefficients represent locations and weights of the particles, respectively. The solution is then found by following the time evolution of the locations and the weights of these particles according to a system of ODEs obtained by considering a weak formulation of the problem. The main advantage of particle methods is their (extremely) low numerical diffusion that allows one to capture a variety of nonlinear waves with high resolution, see, e.g., [9] and references therein.

In this paper, we apply the particle method for numerical solution of the CH equation. We begin with a brief overview of the particle method and some of its main features which are relevant to our discussion. The main analytical results we provide is the convergence proof of the particle method. While previous convergence results have been established for this equation (e.g. see [1, 2, 4, 7, 10]), we propose a new self-contained method for showing the convergence by establishing BV estimates for the particle solution. We then verify that both the particle solution and its limit are weak solutions to the CH equation to complete our study on the convergence analysis.

2 Description of a particle method

To solve (1.1) by the particle method, we follow the method introduced in [6]. That is, we look for a weak solution in the form of a linear combination of Dirac delta distributions,

$$m^N(x,t) = \sum_{i=1}^{N} p_i(t)\delta(x - x_i(t))). \tag{2.1}$$

Here, $x_i(t)$ and $p_i(t)$ represent the location of the i-th particle and its weight, and N denotes the total number of particles. The solution is then found by following the time evolution of the locations and the weights of the particles according to the following system of ODEs [6]:

$$\begin{cases} \dfrac{dx_i(t)}{dt} = u^N(x_i(t),t), \\ \dfrac{dp_i(t)}{dt} + u_x^N(x_i(t),t)p_i(t) = 0. \end{cases} \tag{2.2}$$

Using the special relationship between m and u given in (1.1), one can directly compute the velocity u and its derivative, by the convolution $u^N = G * m^N$, where G is the Green's function

$$G(|x-y|) = \frac{1}{2\alpha}e^{-|x-y|/\alpha}, \tag{2.3}$$

associated with the one dimensional Helmholtz operator in (1.1). Thus we have the following exact expressions for both $u(x,t)$ and (by direct computation) $u_x(x,t)$:

$$u^N(x,t) = \frac{1}{2\alpha}\sum_{i=1}^{N} p_i(t)e^{-|x-x_i(t)|/\alpha}, \tag{2.4}$$

$$u_x^N(x,t) = -\frac{1}{2\alpha^2}\sum_{i=1}^{N} p_i(t)\mathrm{sgn}(x - x_i(t))e^{-|x-x_i(t)|/\alpha}. \tag{2.5}$$

In practice, except for very special cases, the functions $x_i(t)$ and $p_i(t)$, $i = 1, \ldots, N$ have to be determined numerically and the system (2.2) must be integrated by an appropriate ODE solver. In order to initiate the time integration, one should choose the initial positions of particles, x_i^0, and the weights, p_j^0, so that (2.1) represents a high-order approximation to the initial data $m_0(x)$ in (1.2), as it is shown in [6, 9].

The system (2.2) may be derived in one of two ways. Following [6], we may consider a weak formulation of the problem and make a suitable substitution to derive (2.2) or one may follow [4, 5] by considering the Hamiltonian structure of (1.1). The latter property of the particle system and its complete integrability allows one to establish the global existence results for the solution of (2.2) and to show that for a relatively wide class of initial data there are no particle collisions in finite times [5, Proposition 2.3].

3 Convergence results

In this section,we show that the particle method given by (2.1) and (2.2) converges to a weak solution of the CH equation (1.1). While others have shown this result in a variety of ways, we propose a self-contained concise way of showing convergence by establishing BV estimates for u^N and u_x^N. Once these estimates are established, we use the compactness result, associated with BV functions, to pass to the limit and show that the particle solution associated with the CH equation converges to a weak solution of (1.1). Throughout this section, we shall assume that the initial momenta $p_i(0)$ are positive and that there are no particle collisions in finite time. It should also be noted that the total momentum of the particle system is conserved as it was shown in [6].

3.1 Space and time BV estimates

In what follows, we recall the definition of the total variation of a function and prove that the total variations of both $u^N(x,t)$ and $u_x^N(x,t)$ are bounded.

Definition 3.1. Consider a (possibly unbounded) interval $J \subseteq \mathbb{R}$ and a function $u : J \to \mathbb{R}$. The total variation of u is defined as

$$\text{Tot.Var.}\,\{u\} \equiv \sup \left\{ \sum_{j=1}^{N} |u(x_j) - u(x_{j-1})| \right\}, \tag{3.1}$$

where the supremum is taken over all $N \geq 1$ and all $(N+1)$-tuples of points $x_j \in J$ such that $x_0 < x_1 < \cdots < x_N$. If the right hand side of (3.1) is bounded, then we say that u has bounded variation, and write $u \in BV(\mathbb{R})$.

The following theorem establishes the necessary space and time BV estimates.

Theorem 3.2. *Let $u^N(x,t)$ and $u_x^N(x,t)$ be functions defined in* (2.4) *and* (2.5)*, respectively. Then, both $u^N(x,t)$ and $u_x^N(x,t)$ are BV functions in the two variables x, t.*

Proof. We begin with showing that Tot. Var. $\{u^N(\cdot,t)\}$ and Tot. Var. $\{u_x^N(\cdot,t)\}$ are bounded. Indeed, from (2.3) we have Tot.Var.$\{G(x)\} = 1/\alpha$ and Tot.Var.$\{G_x(x)\} = 2/\alpha^2$. Using the fact that the total momentum of the particle system is conserved, we obtain

$$\text{Tot.Var.}\{u^N(\cdot,t)\} \le \sum_{j=1}^N p_j(t)\text{Tot.Var.}\{G(x)\} = \frac{1}{\alpha}\sum_{j=1}^N p_j(t) = \frac{1}{\alpha}|m_0|, \tag{3.2}$$

$$\text{Tot.Var.}\{u_x^N(\cdot,t)\} \le \sum_{j=1}^N p_j(t)\text{Tot.Var.}\{G_x(x)\} = \frac{2}{\alpha^2}\sum_{j=1}^N p_j(t) = \frac{2}{\alpha^2}|m_0|. \tag{3.3}$$

In order to prove that $u^N(x,t)$ and $u_x^N(x,t)$ have bounded variation with respect to t as well, it now suffices to show that u^N and u_x^N are both Lipschitz continuous in time in L^1, [1, Theorem 2.6].

We first consider expression (2.4) for u^N to have

$$\begin{aligned}
&\int_{-\infty}^{\infty} |u^N(x,t) - u^N(x,s)|\, dx \\
&\le \frac{1}{2\alpha}\int_{-\infty}^{\infty} \sum_{i=1}^N p_i(t)\left|e^{-|x-x_i(t)|/\alpha} - e^{-|x-x_i(s)|/\alpha}\right| dx \\
&\quad + \frac{1}{2\alpha}\int_{-\infty}^{\infty} \sum_{i=1}^N e^{-|x-x_i(s)|/\alpha}|p_i(t)-p_i(s)|\, dx.
\end{aligned}$$

Simple calculations show that

$$\int_{-\infty}^{\infty} \left|e^{-|x-x_i(t)|/\alpha} - e^{-|x-x_i(s)|/\alpha}\right| dx \le 4|x_i(t) - x_i(s)|,$$

and

$$\int_{-\infty}^{\infty} e^{-|x-x_i(t)|/\alpha}\, dx \le 2\alpha.$$

Hence, we have

$$\begin{aligned}
&\int_{-\infty}^{\infty} |u^N(x,t) - u^N(x,s)|\, dx \\
&\le \frac{2}{\alpha}\sum_{i=1}^N p_i(t)|x_i(t)-x_i(s)| + \sum_{i=1}^N |p_i(t)-p_i(s)|.
\end{aligned} \tag{3.4}$$

Using the ODE system (2.2), it now follows from (3.4) that

$$\int_{-\infty}^{\infty} |u^N(x,t) - u^N(x,s)|\, dx \le \frac{3}{2\alpha^2}|m_0|^2\,(t-s)\,,$$

proving that u^N is Lipshitz continuous in time in L^1 and thus has bounded variation with respect to both x and t.

Similarly, for u_x^N we have:

$$\int_{-\infty}^{\infty} |u_x^N(x,t) - u_x^N(x,s)|\, dx \le \frac{3}{2\alpha^3}|m_0|^2(t-s),$$

which together with (3.3) proves that $u_x^N(x,t)$ is a BV function in x, t and the statement of the theorem. □

3.2 Compactness and convergence

In this section, we use compactness properties associated with BV functions to prove convergence of the particle method. To this end, we first give a definition of a weak solution of the CH equation (1.1) and show that the particle solution (m^N, u^N) obtained from (2.1), (2.2) is indeed a weak solution of the CH equation.

Definition 3.3. $u(x,t) \in C(0,T;H^1(\mathbb{R})), m(x,t) = u(x,t) - \alpha^2 u_{xx}(x,t)$ is said to be a weak solution to (1.1) if

$$\begin{aligned}
&\int_{-\infty}^{\infty} \phi(x,0)m(x,0)\,dx + \int_0^{\infty}\int_{-\infty}^{\infty} \left[\phi_t(x,t) - \alpha^2\phi_{txx}(x,t)\right] u(x,t)\,dxdt \\
&+ \int_0^{\infty}\int_{-\infty}^{\infty} \left[\frac{3}{2}\phi_x(x,t) - \frac{\alpha^2}{2}\phi_{xxx}(x,t)\right] u^2(x,t)\,dxdt \\
&+ \int_0^{\infty}\int_{-\infty}^{\infty} \frac{\alpha^2}{2}\phi_x(x,t)u_x^2(x,t)\,dxdt = 0 \qquad (3.5)
\end{aligned}$$

for all $\phi \in C_0^\infty(\mathbb{R}\times\mathbb{R}_+)$.

Proposition 3.4. *Assume that $m_0 \in \mathcal{M}(\mathbb{R})$, then the particle solution $(m^N(x,t), u^N(x,t))$ given by (2.1), (2.2) is a weak solution of the problem (1.1), (1.2).*

Proof. Let $m^N(x,0), m^N(x,t)$ and $u^N(x,t), u_x^N(x,t)$ be given by formulae (2.1) and (2.2), respectively, and $\phi \in C_0^\infty(\mathbb{R}\times\mathbb{R}_+)$ be a test function.

Then, the following relations can be easily established by direct substitutions:

$$\langle m^N, \phi_t \rangle = \langle u^N, \phi_t - \alpha^2 \phi_{txx} \rangle, \tag{3.6}$$

$$\langle m^N u^N, \phi_x \rangle = \left\langle (u^N)^2, \phi_x - \frac{\alpha^2}{2}\phi_{xxx} \right\rangle + \alpha^2 \left\langle (u_x^N)^2, \phi_x \right\rangle, \tag{3.7}$$

$$\langle m^N u_x^N, \phi \rangle = \left\langle \frac{\alpha^2 (u_x^N)^2 - (u^N)^2}{2}, \phi_x \right\rangle. \tag{3.8}$$

Using (3.6)–(3.8) and substituting $m^N(x,t)$ as defined by (2.1) into (3.5), yields

$$\begin{aligned} &\sum_{i=1}^N p_i(0)\phi(x_i(0),0) + \int_0^\infty \sum_{i=1}^N p_i(t)\phi_t\left(x_i(t),t\right) dt \\ &+ \int_0^\infty \sum_{i=1}^N p_i(t) u^N(x_i(t),t) \phi_x\left(x_i(t),t\right) dt \\ &- \int_0^\infty \sum_{i=1}^N p_i(t) u_x^N\left(x_i(t),t\right) \phi\left(x_i(t),t\right) dt = 0. \end{aligned} \tag{3.9}$$

We now add and subtract $\sum_{i=1}^N \int_0^\infty p_i(t)\dfrac{dx_i}{dt}\phi_x(x_i(t),t)\,dt$ into the last equation, use the fact that

$$\frac{d\phi(x_i(t),t)}{dt} = \phi_x(x_i(t),t)\frac{dx_i(t)}{dt} + \phi_t(x_i(t),t)$$

and rewrite (3.9) as follows:

$$\begin{aligned} &\sum_{i=1}^N p_i(0)\phi(x_i(0),0) + \int_0^\infty \sum_{i=1}^N p_i(t)\frac{d\phi(x_i(t),t)}{dt}\,dt \\ &\int_0^\infty \sum_{i=1}^N p_i(t)\left[u^N(x_i(t),t) - \frac{dx_i(t)}{dt}\right]\phi_x\left(x_i(t),t\right) dt \\ &- \int_0^\infty \sum_{i=1}^N p_i(t) u_x^N\left(x_i(t),t\right)\phi\left(x_i(t),t\right) dt = 0. \end{aligned} \tag{3.10}$$

Integrating by parts the second term in the first row in (3.10), and rearranging other terms, we finally obtain:

$$\begin{aligned} &\int_0^\infty \sum_{i=1}^N p_i(t)\left[\frac{dx_i(t)}{dt} - u^N(x_i(t),t)\right]\phi_x\left(x_i(t),t\right) dt \\ &+ \int_0^\infty \sum_{i=1}^N \left[\frac{dp_i(t)}{dt} + p_i(t)u_x^N\left(x_i(t),t\right)\right]\phi(x_i(t),t)\,dt = 0. \end{aligned} \tag{3.11}$$

Since the functions $x_i(t)$ and $p_i(t)$ satisfy the system (2.2), the last equation holds for any ϕ implying that m^N, u^N defined by (2.1), (2.4) is a weak solution of (1.1), (1.2). This completes the proof. □

We are now in a position to establish a convergence result for the particle method. Using the BV estimates for $u^N(x,t)$ and $u^N_x(x,t)$, and the fact that the particle solution is a weak solution to the CH equation, we may establish the following convergence result.

Theorem 3.5. *Suppose that $(m^N(x,t), u^N(x,t))$ is a particle solution of (2.1), (2.2) with initial approximation $m^N(\cdot,0) \overset{*}{\rightharpoonup} m_0$, $m_0 \in \mathcal{M}_+(\mathbb{R})$. Then there exist functions $u(x,t) \in BV(\mathbb{R}\times\mathbb{R}_+)$ and $m(x,t) \in \mathcal{M}_+(\mathbb{R}\times\mathbb{R}_+)$ such that $m^N(x,t)$ and $u^N(x,t)$ converge to $m(x,t)$ and $u(x,t)$, respectively in the sense of distributions as $N \to \infty$. Furthermore, the limit (u,m) is the unique weak solution to the CH equation (1.1) with regularity $u \in C(0,T;H^1(\mathbb{R}))$, $u_x \in BV(\mathbb{R}\times\mathbb{R}_+)$.*

Proof. Using BV estimates for $u^N(x,t)$ and $u^N_x(x,t)$, we refer to the compactness property in [1, Theorem 2.4] and conclude that there exist functions u and u_x and a subsequence (still labeled as $u^N(x,t)$) such that

$$||u^N - u||_{L^1_{\text{loc}}(\mathbb{R}\times\mathbb{R}_+)} \to 0, \qquad ||u^N_x - u_x||_{L^1_{\text{loc}}(\mathbb{R}\times\mathbb{R}_+)} \to 0 \tag{3.12}$$

as $N \to \infty$.

From Proposition 3.4, we know that (m^N, u^N) is a weak solution and thus satisfy equation (3.5). To complete the proof, we need to show that each terms in (3.5) converges to that of the limit solution (m,u). Indeed, by the construction of the initial approximation, one has

$$\lim_{N\to\infty} \int_{-\infty}^{\infty} \phi(x,0)m^N(x,0)\,dx = \int_{-\infty}^{\infty} \phi(x,0)m(x,0)\,dx. \tag{3.13}$$

Furthermore, for any $\phi \in C_0^\infty(\mathbb{R}\times\mathbb{R}_+)$, we have

$$\Big|\iint \phi\left((u^N)^2 - (u)^2\right)dxdt\Big| = \Big|\iint \phi(u^N+u)(u^N-u)dxdt\Big|$$

$$\le \|\phi\|_{L^\infty}(\|u^N\|_{L^\infty} + \|u\|_{L^\infty})\iint_{(x,t)\in \text{supt of } \phi} \big|u^N(x,t) - u(x,t)\big|\,dxdt \to 0,$$

and thus

$$\iint \phi(u^N)^2 dxdt \to \iint \phi u^2 dxdt,$$

as as $N \to \infty$. Similarly, we obtain

$$\iint \phi(u^N_x)^2 dxdt \to \iint \phi(u_x)^2 dxdt.$$

This shows that the limit (m,u) is indeed a weak solution to the CH equation. □

4 Conclusion

In this paper, we have provided a new way of showing that for positive Radon measure initial data, the particle method applied to the CH equation will converge to the unique global weak solution. To this extent, we have only highlighted some of the main results obtained from our study. A full version of this paper will be published in the future and will provide additional details and some verification of results that were omitted above. Furthermore, numerical experiments will be performed that illustrates the no-crossing behavior of the solutions to the CH equation as well as the soliton behavior obtained by the complete integrability of the CH equation.

References

[1] A. Bressan, Hyperbolic Systems of Conservation Laws, The One-Dimensional Cauchy Problem. Oxford Lecture Series in Mathematics and its Applications, vol. 20, Oxford University Press, Oxford, 2000.

[2] A. Bressan and A. Constantin, Global conservative solutions of the Camassa-Holm equation. Arch. Ration. Mech. Anal. 183(2) (2007), 215–239.

[3] R. Camassa and D. D. Holm, An integrable shallow water equation with peaked solitons. Phys. Rev. Lett. 71(11) (1993), 1661–1664.

[4] R. Camassa, J. Huang and L. Lee, On a completely integrable numerical scheme for a nonlinear shallow-water wave equation. J. Nonlinear Math. Phys. 12(1) (2005), 146–162.

[5] ———, Integral and integrable algorithms for a nonlinear shallow-water wave equation. J. Comput. Phys. 216(2) (2006), 547–572.

[6] A. Chertock, J. Marsden and P. Du Toit, Integration of the EPDiff equation by particle methods. Submitted.

[7] A. Constantin and J. Escher, Global existence and blow-up for a shallow water equation. Annali Scuola Norm. Sup. Pisa 26 (1998), 303–328.

[8] D. D. Holm, T. Schmah and C. Stoica, Geometric Mechanics and Symmetry. Oxford Texts in Applied and Engineering Mathematics, vol. 12, Oxford University Press, Oxford, 2009.

[9] P. A. Raviart, An analysis of particle methods, Numerical methods in fluid dynamics (Como, 1983). Lecture Notes in Math., vol. 1127, Springer, Berlin, 1985, 243–324.

[10] Z. Xin and P. Zhang, On the uniqueness and large time behavior of the weak solutions to a shallow water equation. Comm. Partial Differential Equations 27(9-10) (2002), 1815–1844.

Stable Numerical Scheme for the Magnetic Induction Equation with Hall Effect

Paolo Corti
Seminar for Applied Mathematics, ETH Zurich
CH-8092, Switzerland
E-mail: paolo.corti@sam.math.ethz.ch

Abstract

Fast magnetic reconnection can be modeled by Hall MHD equations. We consider a sub-model: the Hall induction equations and design stable finite difference schemes to approximate it. Numerical examples are provided to verify the robustness of the scheme.

1 Introduction

Magnetic reconnection, a widely studied phenomena in plasma physics, is a change of topology of the magnetic field lines that permits a fast change of the magnetic energy into thermal and kinetic energy. One of popular models for fast reconnection [1], are the equations of the form:

$$\frac{\partial \rho}{\partial t} = -\nabla \cdot (\rho \mathbf{u}), \tag{1.1}$$

$$\frac{\partial (\rho \mathbf{u})}{\partial t} = -\nabla \left\{ \rho \mathbf{u} \otimes \mathbf{u} + \left(p + \frac{|\mathbf{B}|^2}{2} \right) \mathbf{I}_{3\times 3} - \mathbf{B} \otimes \mathbf{B} \right\}, \tag{1.2}$$

$$\frac{\partial \mathcal{E}}{\partial t} = -\nabla \left\{ \left(\mathcal{E} + p + \frac{|\mathbf{B}|^2}{2} \right) \mathbf{u} + \mathbf{E} \times \mathbf{B} \right\}, \tag{1.3}$$

$$\frac{\partial \mathbf{B}}{\partial t} = -\nabla \times \mathbf{E}. \tag{1.4}$$

Here ρ, $\mathbf{u}$, p are the gas density, velocity and pressure respectively. $\mathbf{E}$ and $\mathbf{B}$ are the electric and magnetic fields. The total energy $\mathcal{E}$ is given by the equation of state, i.e.,

$$\mathcal{E} = \frac{p}{\gamma - 1} + \frac{\rho |\mathbf{u}|^2}{2} + \frac{|\mathbf{B}|^2}{2}. \tag{1.5}$$

Here γ is the gas constant. Equations from (1.1) to (1.3) represent the conservation of mass, momentum and energy; the last one (1.4) describes the evolution of the magnetic field.

The equations have to obey the divergence constraint:

$$\nabla \cdot \mathbf{B} = 0. \tag{1.6}$$

For ideal MHD, the electric field is given by

$$\mathbf{E} = -\mathbf{u} \times \mathbf{B}. \tag{1.7}$$

However, no reconnection is possible with this model. In order to model fast reconnection, we use a generalized Ohm's law [2], [3]

$$\mathbf{E} = -\mathbf{u} \times \mathbf{B} + \eta \mathbf{J} + \frac{\delta_i}{L_0} \frac{\mathbf{J} \times \mathbf{B}}{\rho} + \left(\frac{\delta_e}{L_0}\right)^2 \frac{1}{\rho} \left[\frac{\partial \mathbf{J}}{\partial t} + (\mathbf{u} \cdot \nabla)\,\mathbf{J}\right]. \tag{1.8}$$

Here L_0 is the normalizing length unit, and δ_e and δ_i denote electron and ion inertia respectively; they are related to electron-ion mass ratio by $(\frac{\delta_e}{\delta_i})^2 = \frac{m_e}{m_i}$.

Using the Ampère's law we can write the electric current $\mathbf{J}$ as

$$\mathbf{J} = \nabla \times \mathbf{B}. \tag{1.9}$$

The Hall MHD equations are non-linear and complicated. A sub-model is the Hall induction equation given by

$$\begin{aligned} \frac{\partial}{\partial t}\left[\mathbf{B} + \left(\frac{\delta_e}{L_0}\right)^2 \nabla \times (\nabla \times \mathbf{B})\right] &= \nabla \times (\mathbf{u} \times \mathbf{B}) - \eta \nabla \times (\nabla \times \mathbf{B}) \\ &- \left(\frac{\delta_e}{L_0}\right)^2 \frac{1}{\rho} \nabla \times ((\mathbf{u} \cdot \nabla)(\nabla \times \mathbf{B})) - \frac{\delta_i}{L_0}\frac{1}{\rho} \nabla \times ((\nabla \times \mathbf{B}) \times \mathbf{B}) \end{aligned} \tag{1.10}$$

with $\mathbf{u}$ being a given velocity field.

For the remaining part of this paper, we will focus on the Hall induction equations (1.10) and onto the design stable numerical scheme for it.

2 Theoretical analysis

We rewrite the advection term in (1.10) using a standard vector identity resulting in

$$\nabla \times (\mathbf{u} \times \mathbf{B}) = (\mathbf{B} \cdot \nabla)\mathbf{u} - \mathbf{B}(\nabla \cdot \mathbf{u}) + \mathbf{u}(\nabla \cdot \mathbf{B}) - (\mathbf{u} \cdot \nabla)\mathbf{B}. \tag{2.1}$$

We note that the term that leads to a lack of symmetry is $\mathbf{u}(\nabla \cdot \mathbf{B})$. For divergence free data (1.6) this term vanishes and the remaining equations are in symmetric form:

$$\frac{\partial}{\partial t}\left[\mathbf{B}+\left(\frac{\delta_e}{L_0}\right)^2\nabla\times(\nabla\times\mathbf{B})\right]=(\mathbf{B}\cdot\nabla)\mathbf{u}-\mathbf{B}(\nabla\cdot\mathbf{u})-(\mathbf{u}\cdot\nabla)\mathbf{B}$$
$$-\eta\nabla\times(\nabla\times\mathbf{B})-\left(\frac{\delta_e}{L_0}\right)^2\frac{1}{\rho}\nabla\times((\mathbf{u}\cdot\nabla)(\nabla\times\mathbf{B}))$$
$$-\frac{\delta_i}{L_0}\frac{1}{\rho}\nabla\times((\nabla\times\mathbf{B})\times\mathbf{B}). \tag{2.2}$$

We have the following theorem:

Theorem 2.1. *Let* $\mathbf{u}\in C^2(\mathbb{R}^3)$ *decays to zero sufficiently fast. Furthermore, assume that the solution of (2.2) goes to zero at infinity, then following apriori estimates hold:*

$$\frac{d}{dt}\left(\|\mathbf{B}\|^2_{L^2(\mathbb{R}^3)}+\left(\frac{\delta_e}{L_0}\right)^2\frac{1}{\rho}\|\nabla\times\mathbf{B}\|^2_{L^2(\mathbb{R}^3)}\right)$$
$$\leq C_1\left(\|\mathbf{B}\|^2_{L^2(\mathbb{R}^3)}+\left(\frac{\delta_e}{L_0}\right)^2\frac{1}{\rho}\|\nabla\times\mathbf{B}\|^2_{L^2(\mathbb{R}^3)}\right) \tag{2.3}$$
$$\frac{d}{dt}\|\nabla\cdot\mathbf{B}\|_{L^2(\mathbb{R}^3)}\leq C_2\|\nabla\cdot\mathbf{B}\|_{L^2(\mathbb{R}^3)} \tag{2.4}$$

with C_1 *and* C_2 *being constants that depend on* $\mathbf{u}$ *and its derivatives only. The above estimates imply that* $\mathbf{B}\in H^1_{loc}(\mathbb{R}^3)$.

Proof. For the first inequality we multiply the equation with $\mathbf{B}$ and then integrate over $\mathbb{R}^3$ resulting in

$$\int_{\mathbb{R}^3}\frac{1}{2}\frac{\partial\mathbf{B}^2}{\partial t}+\left(\frac{\delta_e}{L_0}\right)^2\frac{1}{\rho}\mathbf{B}\nabla\times(\nabla\times\frac{\partial\mathbf{B}}{\partial t})dx$$
$$=\int_{\mathbb{R}^3}\Bigg[\mathbf{B}(\mathbf{B}\cdot\nabla)\mathbf{u}-\mathbf{B}^2(\nabla\cdot\mathbf{u})-\frac{1}{2}(\mathbf{u}\cdot\nabla)\mathbf{B}^2-\eta\mathbf{B}\nabla\times(\nabla\times\mathbf{B})$$
$$-\left(\frac{\delta_e}{L_0}\right)^2\frac{1}{\rho}\mathbf{B}\nabla\times((\mathbf{u}\cdot\nabla)(\nabla\times\mathbf{B}))-\frac{\delta_i}{L_0}\frac{1}{\rho}\mathbf{B}\nabla\times((\nabla\times\mathbf{B})\times\mathbf{B})\Bigg]dx.$$

Partial integration yields

$$\frac{1}{2}\frac{d}{dt}\left(\|\mathbf{B}\|^2_{L^2(\mathbb{R}^3)}+\left(\frac{\delta_e}{L_0}\right)^2\|\nabla\times\mathbf{B}\|^2_{L^2(\mathbb{R}^3)}\right)$$
$$=\int_{\mathbb{R}^3}\Bigg[\mathbf{B}(\mathbf{B}\cdot\nabla)\mathbf{u}-\frac{1}{2}\mathbf{B}^2(\nabla\cdot\mathbf{u})-\eta(\nabla\times\mathbf{B})^2$$
$$+\frac{1}{2}\left(\frac{\delta_e}{L_0}\right)^2\frac{1}{\rho}(\nabla\cdot\mathbf{u})(\nabla\times\mathbf{B})^2-\frac{\delta_i}{L_0}\frac{1}{\rho}\underbrace{(\nabla\times\mathbf{B})((\nabla\times\mathbf{B})\times\mathbf{B})}_{=0}\Bigg]dx.$$

Using the smoothness of u in the above identity leads to

$$\frac{d}{dt}\left(\|\mathbf{B}\|^2_{L^2(\mathbb{R}^3)}+\left(\frac{\delta_e}{L_0}\right)^2\|\nabla\times\mathbf{B}\|^2_{L^2(\mathbb{R}^3)}\right)$$
$$\leq C_A\|\mathbf{B}\|^2_{L^2(\mathbb{R}^3)}+C_B\left(\frac{\delta_e}{L_0}\right)^2\|\nabla\times\mathbf{B}\|^2_{L^2(\mathbb{R}^3)}$$

here $C_A = \max_{k=\{x,y,z\}}\left(\|\frac{\partial(u_1+u_2+u_3)}{\partial k}\|_{L^\infty(\mathbb{R}^3)}\right)$ and $C_B = \|\nabla\mathbf{u}\|_{L^\infty(\mathbb{R}^3)}$.

Applying divergence operator on (2.2), we obtain

$$\frac{\partial\nabla\cdot\mathbf{B}}{\partial t} = -\nabla(\mathbf{u}(\nabla\cdot\mathbf{B})).$$

Integrating over $\mathbb{R}^3$ and then integration by parts, we obtain the estimate (2.4) by setting $C_2 = \|\nabla\mathbf{u}\|_{L^\infty(V)}$. □

3 Numerical scheme

We subdivide the computational domain using a uniform Cartesian mesh with mesh width Δx,Δy and Δz. $\hat{\mathbf{B}}_{i,j,k}(t)$ and $\hat{\mathbf{u}}_{i,j,k}(t)$ are approximations of $\mathbf{B}(\mathbf{x},t)$ and $\mathbf{u}(\mathbf{x},t)$ at point (x_i,y_j,z_k). We also define discrete derivatives $\mathbf{D}=(D_x,D_y,D_z)^\top$ using central differences:

$$\begin{pmatrix} D_x \\ D_y \\ D_z \end{pmatrix} a_{i,j,k} = \begin{pmatrix} \dfrac{a_{i+1,j,k}-a_{i-1,j,k}}{2\Delta x} \\ \dfrac{a_{i,j+1,k}-a_{i,j-1,k}}{2\Delta y} \\ \dfrac{a_{i,j,k+1}-a_{i,j,k-1}}{2\Delta z} \end{pmatrix} \tag{3.1}$$

where $a_{i,j,k}$ is an arbitrary function defined on the mesh. For central difference operators we have the following lemmas:

Lemma 3.1 (Summation by parts). *Let $a_{i,j,k}$ and $b_{i,j,k}$ be grid functions, such that $|a_{i,j,k}|,|b_{i,j,k}|\to 0$ for $i,j,k\to\infty$ then*

$$\sum_{i,j,k} a_{i,j,k}D_x b_{i,j,k} = -\sum_{i,j,k} b_{i,j,k}D_x a_{i,j,k}. \tag{3.2}$$

Proof. This follow directly by a change of index in the sum. □

Lemma 3.2 (Discrete chain rule). *For every finite difference operator D that approximates the first derivative, there exists an averaging operator*

A such that for every $a_{i,j,k} = a(x_i, y_j, z_k)$ *with* $a \in C^2$ *and every* $b_{i,j,k}$ *defined on the mesh,*

$$D(a_{i,j,k}b_{i,j,k}) = a_{i,j,k}D(b_{i,j,k}) + A(b_{i,j,k})D(a_{i,j,k}) + \tilde{a}_{i,j,k} \tag{3.3}$$

holds. If $b_{i,j,k} \in l^2$, *then the residual* $\tilde{a}$ *is bounded i.e.,* $\|\tilde{a}\| \leq Ch\|b\|$ *for a generic mesh size* h *and some constant* $C > 0$.

Proof. For the proof of this lemma, see [4] lemma 3.3. □

For approximating (2.2) we use the following semi-discrete numerical scheme

$$\begin{aligned}\frac{\partial}{\partial t}\left[\hat{\mathbf{B}}_{i,j,k} + \left(\frac{\delta_e}{L_0}\right)^2 \mathbf{D}\times(\mathbf{D}\times\hat{\mathbf{B}}_{i,j,k})\right] &= \bar{\mathbf{A}}\left(\hat{\mathbf{B}}_{i,j,k}\cdot\mathbf{D}\right)\hat{\mathbf{u}}_{i,j,k}\\ -\mathbf{A}\left(\hat{\mathbf{B}}_{i,j,k}(\mathbf{D}\cdot\hat{\mathbf{u}}_{i,j,k})\right) &- (\hat{\mathbf{u}}_{i,j,k}\cdot\mathbf{D})\hat{\mathbf{B}}_{i,j,k} - \eta\mathbf{D}\times(\mathbf{D}\times\hat{\mathbf{B}}_{i,j,k})\\ -\left(\frac{\delta_e}{L_0}\right)^2\frac{1}{\rho}\mathbf{D}\times((\hat{\mathbf{u}}_{i,j,k}\cdot\mathbf{D})\hat{\mathbf{B}}_{i,j,k}) &- \frac{\delta_i}{L_0}\frac{1}{\rho}\mathbf{D}\times\left((\mathbf{D}\times\hat{\mathbf{B}}_{i,j,k})\times\hat{\mathbf{B}}_{i,j,k}\right).\end{aligned} \tag{3.4}$$

Note that t is suppressed for notational convenience. We denote

$$\bar{\mathbf{A}}(\mathbf{B}_{i,j,k}\cdot\mathbf{D}) = A_x(B^1_{i,j,k})D_x + A_y(B^2_{i,j,k})D_y + A_z(B^3_{i,j,k})D_z \tag{3.5}$$

and

$$\begin{aligned}\mathbf{A}\left(\mathbf{B}_{i,j,k}(\mathbf{D}\cdot\mathbf{u}_{i,j,k})\right)^i =& A_x(B^i_{i,j,k})D_x u^1_{i,j,k}\\ &+ A_y(B^i_{i,j,k})D_y u^2_{i,j,k} + A_z(B^i_{i,j,k})D_z u^3_{i,j,k}\end{aligned} \tag{3.6}$$

for $i = 1, 2, 3$. A being the averaging operator defined in previous lemma.

We can show that the following holds:

Theorem 3.3. *Let* $\hat{\mathbf{u}}_{i,j,k} = \mathbf{u}(x_i, y_j, z_k)$ *be the point evaluation of a function* $u \in C^2$ *and let the solutions of (3.4) go to zero at infinity, then the following estimates hold*

$$\begin{aligned}&\frac{d}{dt}\left(\|\hat{\mathbf{B}}\|^2_{l^2(\mathbb{R}^3)} + \left(\frac{\delta_e}{L_0}\right)^2\frac{1}{\rho}\|\mathbf{D}\times\hat{\mathbf{B}}\|^2_{l^2(\mathbb{R}^3)}\right)\\ &\leq C_1\left(\|\hat{\mathbf{B}}\|^2_{l^2(\mathbb{R}^3)} + \left(\frac{\delta_e}{L_0}\right)^2\frac{1}{\rho}\|\mathbf{D}\times\hat{\mathbf{B}}\|^2_{l^2(\mathbb{R}^3)}\right),\end{aligned} \tag{3.7}$$

$$\frac{d}{dt}\|\mathbf{D}\cdot\hat{\mathbf{B}}\|^2_{l^2(\mathbb{R}^3)} \leq C_2\|\mathbf{D}\cdot\hat{\mathbf{B}}\|^2_{l^2(\mathbb{R}^3)} + C_3\max(\Delta x, \Delta y, \Delta z) \tag{3.8}$$

with C_1, C_2 *and* C_3 *constant that depend on* $\mathbf{u}$ *and its derivative only.*

Proof. The proof of this theorem uses the two lemmas 3.1 and 3.2 to mimic the proof of the continuous version of this theorem (Thm. 2.1). A detailed proof will be provided in [5]. □

The scheme (3.4) is semi-discrete and needs to be coupled with a suitable numerical time-integration routine. We have chosen to use a second-order SSP Runge-Kutta method [6].

Remark 3.4. A fourth order version of this scheme is derived by replacing the central difference operator by corresponding fourth-order central difference, e.g.,

$$D_x^{(4)} a_{i,j,k} = \frac{2}{3}\frac{a_{i+1,j,k} - a_{i-1,j,k}}{\Delta x} - \frac{1}{12}\frac{a_{i+2,j,k} - a_{i-2,j,k}}{\Delta x}. \tag{3.9}$$

4 Numerical experiments

We tested the numerical scheme for a 2-d version of the general induction equations(2.2) with the following initial data

$$\mathbf{B}_0(x,y) = 4\begin{pmatrix} -y \\ x-\frac{1}{2} \\ 0 \end{pmatrix} e^{-20((x-\frac{1}{2})^2+y^2)} \tag{4.1}$$

and $\mathbf{u} = (-y, x, 0)^\top$. An exact solution of this problem can be calculated in the pure advection case, i.e. if $\eta = \delta_i = \delta_e = 0$. The solution is given by

$$\mathbf{B}(x,y,t) = \mathbf{R}(t)\mathbf{B}_0(\mathbf{R}(-t)(x,y)) \tag{4.2}$$

where $\mathbf{R}(t)$ is a rotation matrix on the z axis with angular velocity t.

We ran two different tests on the domain $[-2.5, 2.5] \times [-2.5, 2.5]$ with Dirichlet boundary conditions.

Test 1 We test convergence of the scheme for two different central difference operators. One of second order and other of order four.

In absence of a known analytical solution in presence of Hall effect, we have modified the problem. We add known analytical source term to the induction equation; this term is computed so that (4.2) is the solution of the forced version of (2.2).

In Figure 4.1 we show l^2 errors after a time $t = 2\pi$ for different mesh size $N = N_x = N_y$. The theoretical orders of convergence are obtained.

Test 2 As second test we compare the solutions for advection problem and full problem at time $t = \pi$ (Figure 4.2). We note that that the resistivity and the Hall term diffuse the solution and also induce a creation of a small third component in the field.

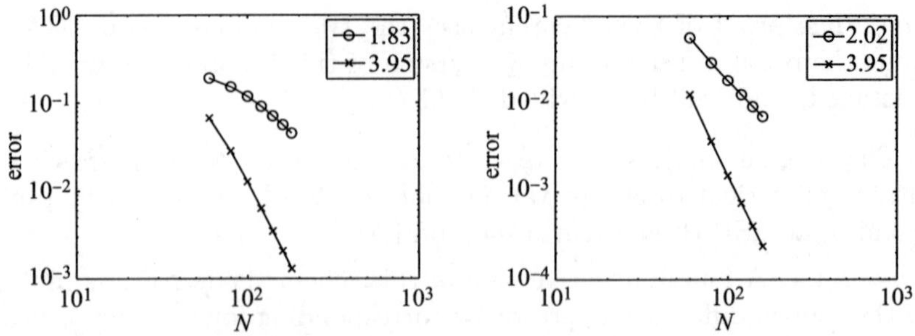

Figure 4.1 l^2 convergence analysis. On the left we have $\eta = \delta_i = \delta_e = 0$, and on the right the forced problem for $L_0 = \rho = 1$, $\eta = 0.01$, $\delta_i = 0.1$ and $\delta_e = 4.5 \times 10^{-2}$. In the legend we show the slope of the lines

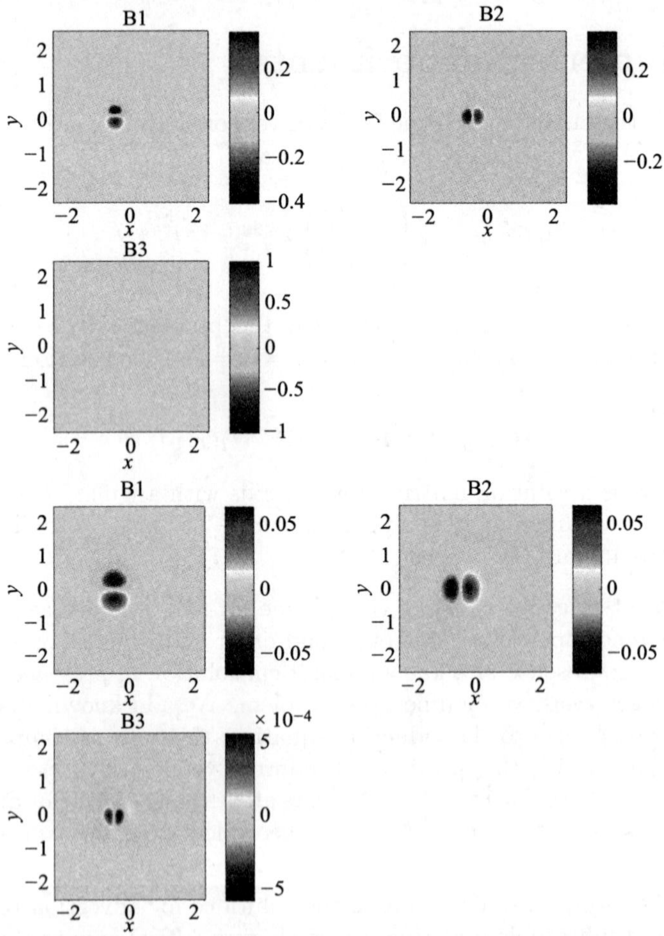

Figure 4.2 Solution after $T = \pi$. On the top we have$\eta = \delta_i = \delta_e = 0$, and on the bottom we have $L_0 = \rho = 1$, $\eta = 0.01$, $\delta_i = 0.1$ and $\delta_e = 4.5 \times 10^{-2}$.

5 Conclusion

The symmetric form of the general induction equations (2.2) posses some energy and divergence estimates. These estimates can be used to build a stable numerical scheme.

The presence of a time-derivative of the current in (2.2) implies that a matrix inversion has to be performed at every time step. Currently, we use a direct solver to invert the matrix. However, the matrix is ill conditioned and suitable pre-conditioners need to be devised to stabilize and accelerate the inversion algorithms. The design of such pre-conditioner is a topic of ongoing research and they will be presented in forthcoming papers.

References

[1] X. H. Deng and H. Matsumoto, Rapid magnetic reconnection in the Earth's magnetosphere mediated by whistler waves. Nature 410 (2001), 557–560.

[2] Z. W. Ma and A. Bhattacharjee, Hall magnetohydrodynamic reconnection: The geospace environment modeling challenge. J. Geophys. Res. 106(A3) (2001), 3773–3782.

[3] X. Qian, J. Balbás, A. Bhattacharjee and H. Yang, A Numerical Study of Magnetic Reconnection: A central scheme for hall MHD. "Hyperbolic Partial Differential Equations, Theory, Numerics and Applications". Proceedings of the 12th international conference held at the University of Maryland, To appear.

[4] S. Mishra and M. Svärd, On stability of numerical schemes via frozen coefficient and the magnetic induction equations. BIT Numerical Mathematics 50 (2010), 85–108.

[5] P. Corti and S. Mishra, Stable finite difference scheme for the magnetic induction equation with hall effect. in preparation.

[6] S. Gottleib, C. W. Shu and E. Tadmor, High order time discretisation with strong stability property. SIAM Review. 43 (2001), 89–112.

A Local Discontinuous Galerkin Discretization for Hydrostatic Free Surface Flows Using σ-transformed Coordinates

Andreas Dedner
Mathematics Institute, University of Warwick
Coventry CV4 7AL, United Kingdom
E-mail: a.s.dedner@warwick.ac.uk

Christoph Gersbacher
Abteilung für Angewandte Mathematik, Universität Freiburg
Hermann-Herder-Str. 10, D-79108 Freiburg, Germany
E-mail: christoph.gersbacher@mathematik.uni-freiburg.de

Abstract

In this paper, we present a Local Discontinuous Galerkin discretization for a class of shallow imcompressible free surface flows. The classical σ-transformation is applied to overcome the difficulty of a time-dependent domain. The transformed system is in conservative form and can be closed in terms of the transformed horizontal momentum and the water height function. We give details on the implementation and show numerical results.

1 Introduction

The most common model for incompressible flows are the incompressible Navier-Stokes equations. The motion of the simulated fluid is described in terms of the multidimensional velocity field u (the density is assumed to be constant and set equal to one), the pressure p, and the evolution in time of the computational domain $\Omega_t \subset \mathbb{R}^d$, $d = 2, 3$, itself. The domain Ω_t represents the fluid region and has (at least) a free upper boundary.

However, numerical simulations of the full Navier-Stokes equations are still very demanding. In the situation of shallow flows a number of simplifications to the equations may be possible: the pressure distribution may assumed to be hydrostatic (see [6]); the depth-integrated Shallow-Water equations represent a even further simplification. In this paper, we will be interested in the numerical simulation of d-dimensional hydrostatic flows by a Local Discontinuous Galerkin (LDG) discretization. In particular, the system of PDEs is still defined in a time-dependent domain.

The LDG-method is a class of Finite-Element methods with piecewise continuous base functions. These methods have become popular in the field of Computational Fluid Dynamics for their conservation properties, higher order approximation and parallelization capabilities.

The application of the LDG-method to three dimensional hydrostatic free surface flow is studied in [1, 3]. The authors propose to adapt the numerical mesh to the motion of the fluid region. This process has to be chosen carefully in order to preserve mass and the convergence rate and to ensure the stability of the scheme.

The basic idea of the discretization presented here will be the transformation in σ-coordinates. The time-dependent domain is mapped onto a fixed reference domain. Yet, the resulting system of PDEs can be written in conservative form. There is no need for grid adaptation, and the discrete water height is discontinuous throughout the whole computation.

This paper is organized as the follows. In section 2 the models used for hydrostatic flows with a free surface are presented. We summarize the σ-transformation as carried out, e.g., in [9]. In section 3 we present our discretization ansatz of limiting to a subset of the unknowns. In section 4 we will give details on the implementation and show numerical results.

2 Model description

In this section, we present the mathematical model used for the simulation of free-surface flows (2.1a)–(2.1g). Then, we apply the so-called σ-transformation which maps the time-dependent domain onto a fixed reference domain.

2.1 Hydrostatic model for free-surface flows

We consider an incompressible hydrostatic flow with a free surface. By $\Omega_t \subset \mathbb{R}^d$, $d = 2, 3$, we denote the domain occupied by the fluid at time t. We assume, that for all times $t \in [0, T]$ it holds

$$\Omega_t = \left\{(x, z) \in \mathbb{R}^{d-1} \times \mathbb{R} \mid x \in \Omega_x,\ b(x) < z < s(x, t)\right\},$$

i.e., both the bottom topography and the position of the free surface are given through smooth functions b and s respectively (see Figure 2.1). By $h = s - b$ we denote the water height, and we assume that $h > 0$.

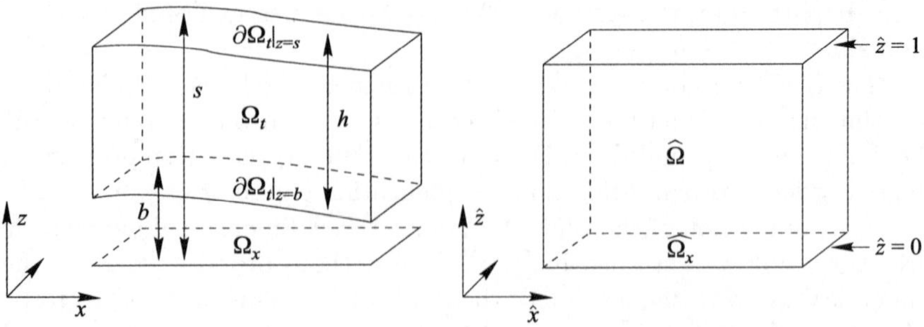

Figure 2.1 The computational domain for the hydrostatic system and after σ-transformation.

The d-dimensional velocity field is $u = (u_x, u_z)$ where u_z is the vertical velocity. The hydrostatic system takes into account conservation of horizontal momentum, the divergence constraint, and the hydrostatic pressure assumption

$$\partial_t u_x + (u \cdot \nabla) u_x + \nabla_x p = \mu \partial_z^2 u_x \qquad \text{in } \Omega_t, \tag{2.1a}$$
$$\nabla \cdot u = 0 \qquad \text{in } \Omega_t, \tag{2.1b}$$
$$\partial_z p = -g \qquad \text{in } \Omega_t, \tag{2.1c}$$

the kinematic boundary conditions

$$\partial_t s + u_x \cdot \nabla_x s = u_z \qquad \text{on } \partial\Omega_t|_{z=s}, \tag{2.1d}$$
$$u_x \cdot \nabla_x b = u_z \qquad \text{on } \partial\Omega_t|_{z=b}, \tag{2.1e}$$

and further boundary conditions

$$\mu \partial_z u_x = 0 \qquad \text{on } \partial\Omega_t|_{z=s}, \tag{2.1f}$$
$$\mu \partial_z u_x = \kappa u_x \qquad \text{on } \partial\Omega_t|_{z=b}, \tag{2.1g}$$

where $\kappa \geq 0$ is interpreted as bottom friction. On the lateral boundaries, no-penetration conditions are prescribed.

We note that the above system (2.1a)–(2.1g) was derived via scaling arguments for shallow water flows as carried out in [6, 4, 7].

The water height can be computed from the following typical evolution equation for shallow water flows

$$\partial_t h + \nabla_x \cdot \left(\int_b^s u_x \, dz \right) = 0 \qquad \text{in } \Omega_x. \tag{2.2}$$

Remark 2.1. In the zero-friction case, the system (2.1a)–(2.1g) may have solutions with the horizontal velocity u_x being independent of the vertical coordinate component z. Then, equations (2.1a), (2.2) simplify to $(d = 2)$

$$\partial_t h + \partial_x(hu_x) = 0 \qquad \text{in } \Omega_x, \tag{2.3a}$$

$$\partial_t(hu_x) + \partial_x(hu_x^2 + \tfrac{1}{2}gh^2) = -gh\partial_x b \qquad \text{in } \Omega_x, \tag{2.3b}$$

i. e., the hydrostatic system conincides with the classical depth-averaged Shallow Water equations (see [6]).

2.2 σ-Transformation and transformed model

In this paragraph, we will present the basic idea of the σ-transformation as it is presented in [9]. Under the assumption of a strictly positive water height, we may perform a coordinate transformation $\hat{t} = t$, $\hat{x} = x$, $\hat{z} = \sigma(x, z, t)$, where

$$\sigma(x, z, t) = \frac{z - b(x)}{h(x, t)}.$$

Then, for all $t \in [0, T]$, it holds

$$\widehat{\Omega_t} := \{(\hat{x}, \hat{z}) \mid (x, z) \in \Omega_t\} = \widehat{\Omega_x} \times (0, 1), \tag{2.4}$$

i.e., the transformed domain is fixed in time. We will therefor write $\widehat{\Omega} = \widehat{\Omega_t}$. The hydrostatic system (2.1a)–(2.1g) is expressed in transformed quantities $\widehat{u}_x(\hat{x}, \hat{z}, \hat{t}) := u_x(x, z, t)$, $\widehat{h}(\hat{x}, \hat{t}) := h(x, t)$, etc. By introducing a new function

$$\widehat{\omega} := \frac{1}{\widehat{h}} \left[\widehat{u}_z - \widehat{u}_x \cdot \nabla_{\hat{x}} \widehat{b} - \hat{z}(\partial_{\hat{t}} \widehat{h} + \widehat{u}_x \cdot \nabla_{\hat{x}} \widehat{h}) \right], \tag{2.5}$$

the transformed system can be written in conservative form: from the divergence condition (2.1b), we get

$$\partial_{\hat{t}} \widehat{h} + \nabla \cdot (\widehat{hu_x}, \widehat{h\omega}) = 0 \qquad \text{in } \widehat{\Omega} \times [0, T], \tag{2.6a}$$

and the momentum equation (2.1a) in transformed variables reads

$$\begin{aligned} \partial_{\hat{t}}(\widehat{hu_x}) + \nabla \cdot \left[(\widehat{u}_x, \widehat{\omega}) \otimes \widehat{hu_x} \right] + \tfrac{1}{2} g \nabla_{\hat{x}} \widehat{h}^2 \\ = -g\widehat{h} \nabla_{\hat{x}} \widehat{b} + \partial_{\hat{z}} \left(\tfrac{\mu}{\widehat{h}} \partial_{\hat{z}\widehat{u}_x} \right) \qquad \text{in } \widehat{\Omega} \times [0, T]. \end{aligned} \tag{2.6b}$$

The kinematic boundary conditions (2.1d), (2.1e) simplify to $\omega = 0$ on upper and lower boundaries of $\widehat{\Omega}$. Equation (2.2) reads in transformed coordinates

$$\partial_{\hat{t}} \widehat{h} + \nabla_{\hat{x}} \cdot (\textstyle\int_0^1 \widehat{hu_x} \, dz) = 0 \qquad \text{in } \widehat{\Omega_x} \times [0, T]. \tag{2.6c}$$

3 Discretization

In this paragraph, we present the LDG-discretization of the transformed system (2.6a)–(2.6c). We will close the system in terms of the transformed height function h and transformed horizontal momentum hu_x (we will omit $\hat{}$ from now on). Therefor, we need an alternative definition of ω which in contrast to (2.5) will not depend on the transformed vertical velocity component u_z any more.

Let $t \in [0,T]$ be a fixed time and $U_1 = (h, hu_x)$ be the vector of transformed height function and transformed horizontal momentum at time t. It is obvious that given U_1 we can calculate the average $\int_0^1 hu_x\,dz$. This also holds for the product $h\omega$. In fact, using (2.6a) and (2.6c) we can characterize $h\omega$ through the ODE

$$\partial_z(h\omega) = -\nabla_x \cdot \left(hu_x - \int_0^1 hu_x\,dz \right) \quad \text{in } \Omega, \tag{3.1a}$$

$$h\omega = 0 \quad \text{on } \partial\Omega|_{z=0} \text{ or } \partial\Omega|_{z=1}. \tag{3.1b}$$

Let $U_2 = (\int_0^1 hu_x\,dz, h\omega)$. Then, we rewrite the transformed system as

$$\partial_t U_1 + \nabla \cdot (F(U_1,U_2) + A(U_1, \nabla U_1)) = S_1(U_1), \tag{3.2a}$$

$$\partial_z U_2 = S_2(U_1), \tag{3.2b}$$

where (in case of two spatial dimensions)

$$F(U_1,U_2) := \begin{pmatrix} \int_0^1 hu_x\,dz & 0 \\ hu_x^2 + \frac{1}{2}gh^2 & hu_x\omega \end{pmatrix}, \qquad S_1(U_1) := \begin{pmatrix} 0 \\ -gh\partial_x b \end{pmatrix},$$

$$A(U_1, \nabla U_1) := \operatorname{diag}(0, \tfrac{\mu}{h}\partial_z u_x),$$

and

$$S_2(U_1) := \begin{pmatrix} 0 \\ -\nabla_x \cdot \left(hu_x - \int_0^1 hu_x\,dz \right) \end{pmatrix}.$$

Now, given U_1 the equation (3.2b) can be solved. This will be denoted by

$$U_2(\cdot,t) = \mathcal{L}_{data}[U_1](\cdot,t) \qquad \text{in } \Omega,$$

introducing some suitable spatial operator computing $\mathcal{L}_{data}[U_1]$. In our implementation this operator is evaluated using a Discontinuous Galerkin scheme as presented in [8].

We can now rewrite (3.2a)–(3.2b) as a first order system that is closed in terms of transformed water height h and transformed horizontal momentum hu_x:

$$U_2(\cdot,t) = \mathcal{L}_{data}[U_1](\cdot,t) \quad \text{in } \Omega, \tag{3.3a}$$
$$V_1(\cdot,t) = A(U_1, \nabla U_1)(\cdot,t) \quad \text{in } \Omega, \tag{3.3b}$$
$$V_2(\cdot,t) = \nabla \cdot V_1(\cdot,t) \quad \text{in } \Omega, \tag{3.3c}$$
$$V_3(\cdot,t) = -\nabla \cdot F(U_1,U_2)(\cdot,t) + S_1(U_1)(\cdot,t) \quad \text{in } \Omega, \tag{3.3d}$$
$$\partial_t U_1(\cdot,t) = V_3(\cdot,t) + V_2(\cdot,t) \quad \text{in } \Omega. \tag{3.3e}$$

The discretization of the equations (3.3b)–(3.3d) by the LDG-method is straight-forward as for any advection-diffusion equation. To increase the stability of the method, equation (3.3e) is integrated in time via a Semi-Implicit Runge-Kutta method as described in [2].

4 Implementation and numerical results

The discretization described above has been implemented using the free software packages DUNE [10] and DUNE-FEM [11]. The implementation consists of two parts: a generic prismatic grid package, and the LDG-solver.

The choice of prismatic grids seems quite natural regarding the geometry of the transformed fluid region (2.4). Figure 4.1 shows how the prismatic grid is constructed from a given two dimensional mesh and how partitions for parallel computations are chosen.

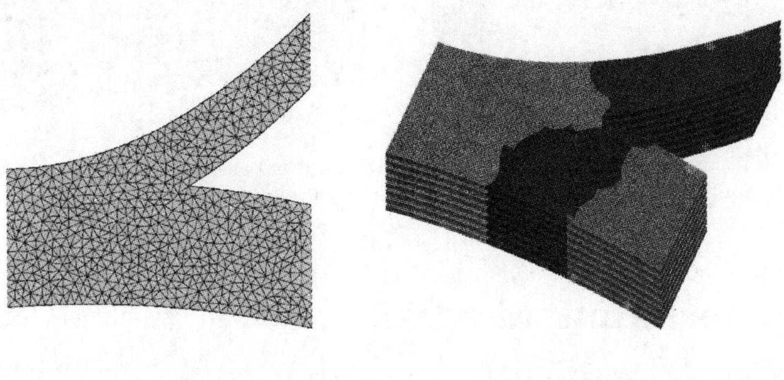

Figure 4.1 Two dimensional unstructured simplex grid (left) and resulting three dimensional prismatic grid with five partitions for parallel computations (right).

In Figures 4.2, 4.3 the results of an example computation are shown in original physical quantities. We point out that for the visualization a smoothing operator was applied to the discrete height funtion. The numerical solution was computed using piecewise linear basis funtions. Initial data is shown in the left picture in 4.2, the velocity field is zero. Finally, Figure 4.3 shows how the quality of the numerical solution increases with grid refinement.

Figure 4.2 3D representation of initial data (left) an solution to a latter time (right).

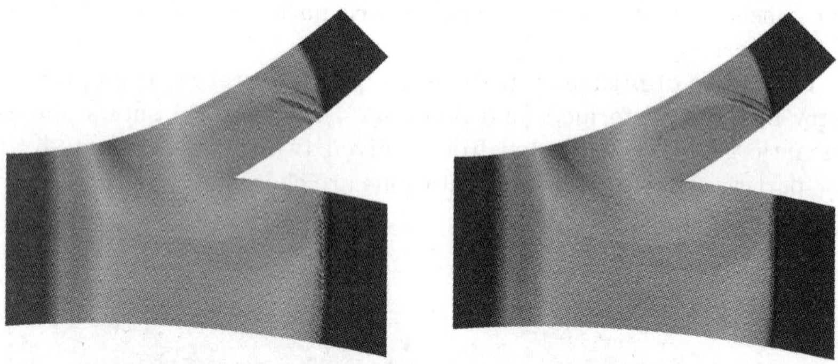

Figure 4.3 Numerical solution using piecewise linear basis functions on coarse grid (917760 elements) (left) and on finer grid (3671040 elements) (right).

5 Concluding remarks

In this paper, we presented a model for hydrostatic incompressible flows with a free surface. We showed the transformation in σ-coordinates and described a LDG discretization for the transformed system. From the numerical solution the original quantities can be computed. It is an open issue, however, whether this can be seen as an approximation to the hydrostatic incompressible Navier-Stokes equations.

References

[1] V. Aizinger and C. Dawson, The local discontinuous Galerkin method for three-dimensional shallow water flow. Comput. Methods Appl. Mech. Eng. 196 (2007), no. 4–6, 734–746.

[2] A. Burri, A. Dedner, D. Diehl, R. Klöfkorn and M. Ohlberger, A general object oriented framework for discretizing nonlinear evolution equations. Advances in High Performance Computing and Computational Sciences 93 (2006), 69–87.

[3] C. Dawson and V. Aizinger, A discontinuous Galerkin method for three-dimensional shallow water equations. J. Sci. Comput. 22–23 (2005), 245–267.

[4] S. Ferrari and F. Saleri, A new two-dimensional shallow water model including pressure effects and slow varying bottom topography. M2AN, Math. Model. Numer. Anal. 38 (2004), no. 2, 211–234.

[5] L. Fontana, E. Miglio, A. Quarteroni and F. Saleri, A finite element method for 3D hydrostatic water flows. Comput. Vis. Sci. 2 (1999), no. 2–3, 85–93.

[6] J.-F. Gerbeau and B. Perthame, Derivation of viscous Saint-Venant system for laminar shallow water; numerical validation. Discrete Contin. Dyn. Syst., Ser. B. 1 (2001), no. 1, 89–102.

[7] C. Gersbacher, Local Discontiunous Galerkin Verfahren zur Simulation flacher dreidimensionaler Strömungen mit freier Oberfläche. Diplomarbeit, Albert-Ludwigs-Universität Freiburg, 2008.

[8] C. Johnson, Numerical solution of partial differential equations by the nite element method. Cambridge etc.: Cambridge University Press (Orig. Studentlitteratur, Lund, Sweden), 1987.

[9] C. B. Vreugdenhil, Numerical methods for shallow-water ow. Water science and technology library, vol. 13, Kluwer Academic Publishers, Dordrecht, The Netherlands, 1994.

[10] DUNE — the Distributed and Unified Numerics Environment, `http://www.dune-project.org/`

[11] The DUNE-FEM module, `http://dune.mathematik.uni-freiburg.de/`

Analysis of Quasineutral Limits

Donatella Donatelli
Dipartimento di Matematica Pura ed Applicata
Università degli Studi dell'Aquila
67100 LAquila, Italy
E-mail: donatell@univaq.it

Pierangelo Marcati
Département de Mathématiques
Université de Metz Ile
du Saulcy, 57045 Metz cedex, France
E-mail: marcati@univaq.it

Abstract

In this paper we describe the main difficulties that arise in performing the quasineutral limit for the Navier Stokes Poisson system.

1 Introduction

In this paper we analyse the quasineutral limit ($\lambda \to 0$) of compressible Navier Stokes Poisson system,

$$\partial_s \rho + \operatorname{div}(\rho u) = 0, \tag{1.1}$$

$$\begin{aligned}\partial_s(\rho u) + \operatorname{div}(\rho u \otimes u) + \frac{1}{\gamma}\nabla(\rho)^\gamma &= \mu\Delta u \\ &+ (\nu + \mu)\nabla \operatorname{div} u + \rho\nabla V,\end{aligned} \tag{1.2}$$

$$\lambda^2 \Delta V = \rho - 1, \tag{1.3}$$

where $x \in \mathbb{R}^3$ and $s \geq 0$. The Navier Stokes Poisson system (1.1)–(1.3) is a simplified isentropic two-fluid model involving dissipation, which describes the dynamics of a plasma, where the compressible electron fluid interacts with its own electric field against a constant charged ion background. In fact in this contest $\rho(x,t)$ is the *negative charge density*, $m(x,t) = \rho(x,t)u(x,t)$ the *current density*, $u(x,t)$ the *velocity vector density*, $V(x,t)$ the *electrostatic potential*, μ, ν the *shear viscosity* and *bulk viscosity* respectively. The parameter λ is the so called *Debye length* (up to a constant factor). We recall that the Debye length is a characteristic physical parameter related to the phenomenon of the so called "Debye shielding". Any charged particle inside a plasma attracts other particles with opposite charge and repels those with the same charge, thereby creating a net cloud of opposite charges around itself. This cloud shields the particle's own charge from external view; it causes the particle's Coulomb field to fall off exponentially at large radii, rather than falling off as $1/r^2$. The physical meaning of the Debye length λ is the "screening" distance or the distance over which the usual Coulomb field $1/r$ is killed off exponentially by the polarization of the plasma. Such type of limit has been intensively studied. In the case of Euler Poisson system by Cordier and Grenier [3], Grenier [9], Cordier, Degond, Markowich, and Schmeiser [2], Loeper [13], Peng, Wang, and Yong [14]. Concerning the Navier Stokes Poisson, Wang [16] studied the quasineutral limit for smooth solutions with well-prepared initial data, Jiang and Wang in [10] studied the combined quasineutral and inviscid limit of the compressible Navier Stokes Poisson system for weak solutions with general initial data, Q. Ju, F. Li, and H. Li in [11] analyzed the quasineutral limit for the Navier Stokes Poisson system with heat conductivity. The main problem in such kind of analysis is the presence of high frequency waves, namely to the so called acoustic waves. Such waves which are harmless in practical applications complicate the mathematical analysis due to the lack of strong compactness. In order to understand this phenomena it is useful to look at the following example. Consider the linearized (around a constant state, $\rho = 1$) compressible Euler equation in 1-D, namely

$$\sigma_t + u_x = 0, \tag{1.4}$$

$$u_t + \frac{1}{\varepsilon^2}\sigma_x = V_x, \tag{1.5}$$

$$\lambda^2 V_{xx} = \sigma, \tag{1.6}$$

where $\varepsilon = 1/\sqrt{p'(1)}$ is the Mach number. By taking the Fourier transform with respect to the x-variable and by denoting with $\hat{\sigma}(\xi,t)$, $\hat{u}(\xi,t)$, the transforms of σ and u, we get that the system (1.4)–(1.6) reduces to a 2×2 system of the form

$$\begin{pmatrix} \hat{\sigma}_t \\ \hat{u}_t \end{pmatrix} + \begin{pmatrix} 0 & i\xi \\ \dfrac{i\xi}{\varepsilon^2} + \dfrac{i}{\lambda^2 \xi} & 0 \end{pmatrix} \begin{pmatrix} \hat{\sigma} \\ \hat{u} \end{pmatrix} = 0.$$

Its solutions are of the form

$$\begin{pmatrix} \hat{\sigma}(\xi,t) \\ \hat{u}(\xi,t) \end{pmatrix} = \begin{pmatrix} \hat{\sigma}_+(\xi) \\ \hat{u}_+(\xi) \end{pmatrix} e^{i\theta(\xi)t} + \begin{pmatrix} \hat{\sigma}_-(\xi) \\ \hat{u}_-(\xi) \end{pmatrix} e^{-i\theta(\xi)t},$$

where

$$\theta(\xi)=\left(\frac{\xi^2}{\varepsilon^2}+\frac{1}{\lambda^2}\right)^{1/2}.$$

Therefore the solution is given by the superposition of waves with frequency θ. As $\lambda\to 0$ or $\varepsilon\to 0$ we get that the solution has high frequency time oscillation. In [6] Donatelli and Marcati examine a double singular limit for the system (1.1)–(1.3) on the whole domain, where both the Mach number and the Debye length tend to zero. Here they overcome the trouble of the acoustic waves by using the dispersion of these waves at infinity yielding the strong convergence of u^ε in space. This short note is organized as follows. In Section 2 we recall the notations and basic tools we will need later on. In Section 3 we study the main steps of the work in [5]. In Section 4 we describe some future developments of this kind of problems.

2 Notations

For simplicity we give the basic notations that we are using later on. We will denote by $\mathcal{D}(\Omega\times\mathbb{R}_+)$ the space of test function $C_0^\infty(\Omega\times\mathbb{R}_+)$, by $\mathcal{D}'(\Omega\times\mathbb{R}_+)$ the space of Schwartz distributions and $\langle\cdot,\cdot\rangle$ the duality bracket between $\mathcal{D}'$ and $\mathcal{D}$. Moreover $W^{k,p}(\Omega)=(I-\Delta)^{-\frac{k}{2}}L^p(\Omega)$ and $H^k(\Omega)=W^{k,2}(\Omega)$ denote the nonhomogeneous Sobolev spaces for any $1\le p\le\infty$ and $k\in\mathbb{R}$. The notations $L_t^pL_x^q$ and $L_t^pW_x^{k,q}$ will abbreviate respectively the spaces $L^p([0,T];L^q(\Omega))$ and $L^p([0,T];W^{k,q}(\Omega))$. We denote by $L_2^p(\mathbb{R}^d)$ the Orlicz space defined as follows:

$$L_2^p(\mathbb{R}^d)=\{f\in L^1_{loc}(\mathbb{R}^d)\mid |f|\chi_{|f|\le\frac12}\in L^2(\mathbb{R}^d),\ |f|\chi_{|f|>\frac12}\in L^p(\mathbb{R}^d)\},\tag{2.1}$$

see [1], [12] for more details.

We define the Leray's projector P on the space of divergence-free vector fields and Q on the space of gradients vector fields by

$$Q=\nabla\Delta^{-1}\operatorname{div},\quad P=I-Q,\tag{2.2}$$

Let us recall the dispersive estimate we are going to use later on. Let us consider the following wave equation defined in the space $[0,T]\times\mathbb{R}^d$:

$$\begin{cases}(\partial_t^2-\Delta)\,w(t,x)=F(t,x),\\ w(0,\cdot)=f,\quad \partial_t w(0,\cdot)=g,\end{cases}$$

for some data f,g,F and time $0<T<\infty$. As is well known the wave equation belongs to the so called dispersive equations. In 1977 Strichartz

[15] realized that combining the dispersive properties of the wave equation with the restriction theorem of the Fourier transform on manifolds he could set up the following estimate:

$$\|w\|_{L^4_t L^4_x} + \|\partial_t w\|_{L^4_t W^{-1,4}_x} \lesssim \|f\|_{\dot{H}^{1/2}} + \|g\|_{\dot{H}^{-1/2}} + \|F\|_{L^{4/3}_t L^{4/3}_x}. \quad (2.3)$$

By using Duhamel's principle the previous estimate can be generalized in the following less standard one:

$$\|w\|_{L^4_{t,x}} + \|\partial_t w\|_{L^4_t W^{-1,4}_x} \lesssim \|f\|_{\dot{H}^{1/2}_x} + \|g\|_{\dot{H}^{1/2}_x} + \|F\|_{L^1_t L^2_x}. \quad (2.4)$$

3 Quasineutral and low Mach number limit

Here we consider the following Navier Stokes Poisson system:

$$\begin{cases} \partial_t \rho^\varepsilon + \operatorname{div}(\rho^\varepsilon u^\varepsilon) = 0, \\ \partial_t(\rho^\varepsilon u^\varepsilon) + \operatorname{div}(\rho^\varepsilon u^\varepsilon \otimes u^\varepsilon) + \dfrac{\nabla(\rho^\varepsilon)^\gamma}{\gamma\varepsilon^2} = \mu\Delta u^\varepsilon \\ \qquad +(\nu+\mu)\nabla \operatorname{div} u^\varepsilon + \dfrac{\rho^\varepsilon}{\varepsilon^2}\nabla V^\varepsilon, \\ \lambda^2 \Delta V^\varepsilon = \rho^\varepsilon - 1. \end{cases} \quad (3.1)$$

Our analysis is performed under the assumption that the previous small parameter ε (Mach number) is related to the Debye length λ (after suitable renormalization of the physical units) by the power law

$$\varepsilon^\beta = \lambda^2, \quad \text{where } \beta > 0. \quad (3.2)$$

To simplify our notation from now on we will denote by π^ε the renormalized pressure and by σ^ε the density fluctuation, namely

$$\pi^\varepsilon = \frac{(\rho^\varepsilon)^\gamma - 1 - \gamma(\rho^\varepsilon - 1)}{\varepsilon^2\gamma(\gamma-1)}, \qquad \sigma^\varepsilon = \frac{\rho^\varepsilon - 1}{\varepsilon}.$$

The system (3.1) is endowed with the following initial conditions:

$$\begin{cases} \rho^\varepsilon|_{t=0} = \rho_0^\varepsilon \geq 0, \ V^\varepsilon|_{t=0} = V_0^\varepsilon, \\ \rho^\varepsilon u^\varepsilon|_{t=0} = m_0^\varepsilon, \quad m_0^\varepsilon = 0 \text{ on } \{x \in \mathbb{R}^3 \mid \rho_0^\varepsilon(x) = 0\}, \\ \displaystyle\int_{\mathbb{R}^3} \left(\pi^\varepsilon|_{t=0} + \frac{|m_0^\varepsilon|^2}{2\rho_0^\varepsilon} + \varepsilon^{\beta-2}|V_0^\varepsilon|^2\right) dx \leq C_0, \\ \dfrac{m_0^\varepsilon}{\sqrt{\rho_0^\varepsilon}} \rightharpoonup u_0 \quad \text{weakly in } L^2(\mathbb{R}^3). \end{cases} \quad (3.3)$$

The existence of global weak solutions for fixed $\varepsilon > 0$ for the system (3.1), has been proved in the case of a bounded domain in [4] and in the case of whole domain in [7] and [8]. Now we are ready to state our main result.

Theorem 3.1. *Let $(\rho^\varepsilon, u^\varepsilon, V^\varepsilon)$ be a sequence of weak solutions in $\mathbb{R}^3$ of the system (3.1), assume that the initial data satisfy (3.3). Then*

(i) $\rho^\varepsilon \to 1$ *strongly in* $L^\infty([0,T]; L^k_2(\mathbb{R}^3))$.

(ii) *There exists* $u \in L^\infty([0,T]; L^2(\mathbb{R}^3)) \cap L^2([0,T]; \dot{H}^1(\mathbb{R}^3))$ *such that*

$$u^\varepsilon \rightharpoonup u \quad \text{weakly in } L^2([0,T]; \dot{H}^1(\mathbb{R}^3)).$$

(iii) *The gradient component* Qu^ε *of the vector field* u^ε *satisfies*

$$Qu^\varepsilon \to 0 \text{ strongly in } L^2([0,T]; L^p(\mathbb{R}^3)), \text{ for any } p \in [4,6),$$

provided $\beta < 1/2$.

(iv) *The divergence free component* Pu^ε *of the vector field* u^ε *satisfies*

$$Pu^\varepsilon \to Pu = u \quad \text{strongly in } L^2([0,T]; L^2_{loc}(\mathbb{R}^3)).$$

(v) $u = Pu$ *is a Leray weak solution to the incompressible Navier Stokes equation*

$$P(\partial_t u - \Delta u + (u\cdot\nabla)u) = 0 \quad \text{in } \mathcal{D}'([0,T]\times\mathbb{R}^3),$$

provided that

$$\begin{cases} \beta = \min\left\{\dfrac{1}{2}, -2 + \dfrac{4}{\gamma}\right\} & \text{if } \gamma < 2, \\ 0<\beta<\min\left\{\dfrac{1}{2}, \mu\left(\dfrac{3}{q} - \dfrac{1}{2}\right), \dfrac{1}{6} - \dfrac{2}{3}\mu\left(s_0 + \dfrac{7}{4} - \dfrac{3}{q}\right)\right\} & \text{if } \gamma \geq 2, \end{cases}$$

where $\mu > 0$, $s_0 \geq 3/2$, *and* $4 \leq q < 6$.

Proof. Here we will describe only the main steps of the proof, for details we refer to [6].

Step1. A priori estimate.

The following energy inequality holds for almost every $t \geq 0$,

$$\int_{\mathbb{R}^3}\left(\rho^\varepsilon\frac{|u^\varepsilon|^2}{2} + \pi^\varepsilon + \varepsilon^{\beta-2}|\nabla V^\varepsilon|^2\right)dx + \int_0^t\int_{\mathbb{R}^3}(\mu|\nabla u^\varepsilon|^2 + (\nu+\mu)|\operatorname{div} u^\varepsilon|^2)\,dxds \leq C_0. \tag{3.4}$$

As consequence we get the following bounds:

$$\sigma^\varepsilon \text{ is bounded in } L^\infty([0,T]; L^k_2(\mathbb{R}^3)),\ \ k = \min(\gamma, 2), \tag{3.5}$$

$$\nabla u^\varepsilon \text{ is bounded in } L^2_{t,x},\quad u^\varepsilon \text{ is bounded in } L^2_{t,x}\cap L^2_t L^6_x, \tag{3.6}$$

$$\sigma^\varepsilon u^\varepsilon \text{ is bounded in } L_t^2 H_x^{-1}. \tag{3.7}$$

Step 2. Density fluctuation wave equation.
In order to estimate the acoustic waves we recover the following nonhomogeneous wave equation structure for σ^ε:

$$\begin{aligned}\varepsilon^2 \partial_{tt}\sigma^\varepsilon - \Delta\sigma^\varepsilon = & - \varepsilon^2 \operatorname{div}(\mu\Delta u^\varepsilon + (\nu+\mu)\nabla \operatorname{div} u^\varepsilon) \\ & + \varepsilon^2 \operatorname{div}\operatorname{div}(\rho^\varepsilon u^\varepsilon \otimes u^\varepsilon) + \varepsilon^2(\gamma-1)\operatorname{div}\nabla\pi^\varepsilon \\ & - \varepsilon \operatorname{div}(\sigma^\varepsilon \nabla V^\varepsilon) - \operatorname{div}\nabla V^\varepsilon. \end{aligned} \tag{3.8}$$

By applying the Strichartz estimate (2.4) we are able to prove the following estimate:

$$\begin{aligned} & \varepsilon^{-\frac{1}{4}+\frac{\beta}{2}} \|\sigma^\varepsilon\|_{L_t^4 W_x^{-s_0-2,4}} + \varepsilon^{\frac{3}{4}+\frac{\beta}{2}} \|\partial_t \sigma^\varepsilon\|_{L_t^4 W_x^{-s_0-3,4}} \\ & \lesssim \varepsilon^{\frac{\beta}{2}} \|\sigma_0^\varepsilon\|_{H_x^{-1}} + \varepsilon^{\frac{\beta}{2}} \|m_0^\varepsilon\|_{H_x^{-1}} \\ & + \varepsilon^{1+\frac{\beta}{2}} T \|\operatorname{div}(\operatorname{div}(\sigma^\varepsilon u^\varepsilon \otimes u^\varepsilon) - (\gamma-1)\nabla\pi^\varepsilon)\|_{L_t^\infty H_x^{-s_0-2}} \\ & + \varepsilon^{1+\frac{\beta}{2}} \|\operatorname{div}\Delta u^\varepsilon + \nabla \operatorname{div} u^\varepsilon\|_{L_t^2 H_x^{-2}} \\ & + T \|\operatorname{div}\nabla V^\varepsilon\|_{L_t^\infty H_x^{-1}} + \varepsilon^{1+\frac{\beta}{2}} T \|\varepsilon^{\beta-2} \operatorname{div}(\sigma^\varepsilon V^\varepsilon)\|_{L_t^\infty H_x^{-s_0-1}}. \end{aligned} \tag{3.9}$$

Step 3. Strong convergence of Qu^ε and Pu^ε.
To prove the strong convergence of Qu^ε and Pu^ε one follows the same line of arguments of Section 5 of [6]. The main tool is that Qu^ε can be written in terms of the density fluctuation, namely $Q(\rho^\varepsilon u^\varepsilon) = \varepsilon\nabla\Delta^{-1}\partial_t\sigma^\varepsilon$ and by using the estimate (3.9) we are able to prove that

$$Qu^\varepsilon \to 0 \quad \text{strongly in } L_t^2 L_x^p \text{ for any } p \in [4,6) \text{ and } \beta < 1/2.$$

By proving equicontinuity in time properties for Pu^ε we get the following convergence result:

$$Pu^\varepsilon \to Pu \quad \text{strongly in } L_t^2 L_{loc,x}^2.$$

□

4 Future developments

In the analysis of the limit in Section 4 it was a crucial fact that the Mach number and Debye length are related through the relation (3.2). This power law allows us to recover the strong convergence for the density. If we simply consider the quasineutral limit for the system (1.1)–(1.3) without performing any kind of rescaling we can observe that the term $\rho\nabla V$ becomes very singular as $\lambda \to 0$ and if we consider ill prepared initial data. In order to understand this, at a formal level, it is enough to observe that by using the Poisson equation (1.3), $\rho\nabla V$ can be rewritten as

$$\rho\nabla V = \frac{1}{\lambda^2}\nabla\Delta^{-1}V.$$

One way to overcome this singular behaviour is to combine the dispersive nature of density fluctuation equation and its underlying oscillating properties. In this way we are able to recover the strong convergence of the velocity field. In particular we are able to prove the following result.

Theorem 4.1. *Let (ρ, u, V) be a sequence of weak solutions in $\mathbb{R}^3$ of the system (1.1)–(1.3), assume that the initial data satisfy*

$$\int_{\mathbb{R}^3}\left(\pi|_{t=0} + \frac{|m_0|^2}{2\rho_0} + \lambda^2|V_0|^2\right)dx \le C_0.$$

Then $\rho^\varepsilon \to 1$ strongly in $L_t^\infty L_{2,x}^k$, there exists $u \in L_t^\infty L_x^2 \cap L_t^2 H_x^1$ such that $u^\varepsilon \rightharpoonup u$ weakly in $L_t^2 H_x^1$. The gradient component Qu^ε converges strongly to 0 in $L_t^2 L_x^p$, for any $p \in [4, 6)$. The divergence free component Pu^ε of the vector field u^ε converges strongly to $Pu = u$ in $L_t^2 L_{loc}^2$. Moreover $u = Pu$ is a Leray weak solution to the incompressible Navier Stokes equation.

This theorem will be proved in a forth coming paper.

References

[1] R. A. Adams, Sobolev Spaces. Academic Press, New York, 1975.

[2] S. Cordier, P. Degond, P. Markowich and C. Schmeiser, Travelling wave analysis of an isothermal Euler-Poisson model. Ann. Fac. Sci. Toulouse Math. (6) 5 (1996), no. 4, 599–643.

[3] S. Cordier and E. Grenier, Quasineutral limit of an Euler-Poisson system arising from plasma physics. Comm. Partial Differential Equations. 25 (2000), no. 5–6, 1099–1113.

[4] D. Donatelli, Local and global existence for the coupled Navier-Stokes-Poisson problem. Quart. Appl. Math. 61 (2003), no. 2, 345–361.

[5] D. Donatelli and P. Marcati, A dispersive approach to the artificial compressibiltiy approximations of the navier stokes equations in 3d. J. Hyperbolic Differ. Equ. 3 (2006), no. 3, 575–588.

[6] D. Donatelli and P. Marcati, A quasineutral type limit for the Navier-Stokes-Poisson system with large data, Nonlinearity. 21 (2008), no. 1, 135–148.

[7] B. Ducomet, E. Feireisl, H. Petzeltová and I. Straškraba, Existence globale pour un fluide barotrope autogravitant. C. R. Acad. Sci. Paris Sér. I Math. 332 (2001), no. 7, 627–632.

[8] B. Ducomet, E. Feireisl, H. Petzeltová and I. Straškraba, Global in time weak solutions for compressible barotropic self-gravitating fluids. Discrete Contin. Dyn. Syst. 11 (2004), no. 1, 113–130.

[9] E. Grenier, Oscillations in quasineutral plasmas. Comm. Partial Differential Equations. 21(1996), no. 3–4, 363–394.

[10] S. Jiang and S. Wang, The convergence of the Navier-Stokes-Poisson system to the incompressible Euler equations. Comm. Partial Differential Equations. 31 (2006), no. 4–6, 571–591.

[11] Q. Ju, F. Li and H. Li, The quasineutral limit of compressible Navier-Stokes-Poisson system with heat conductivity and general initial data. J. Differential Equations. 247 (2009), no. 1, 203–224

[12] P. L. Lions, Mathematical Topics in Fluid Dynamics, Incompressible Models, Claredon Press, Oxford Science Pubblications, 1996.

[13] G. Loeper, Quasi-neutral limit of the Euler-Poisson and Euler-Monge-Ampère systems. Comm. Partial Differential Equations. 30 (2005), no. 7–9, 1141–1167.

[14] Y. J. Peng, Y. G. Wang and W. A. Yong, Quasi-neutral limit of the non-isentropic Euler-Poisson system. Proc. Roy. Soc. Edinburgh Sect. A. 136 (2006), no. 5, 1013–1026.

[15] R. S. Strichartz, Restrictions of Fourier transforms to quadratic surfaces and decay of solutions of wave equations. Duke Math. J. 44 (1977), no. 3, 705–714.

[16] S. Wang, Quasineutral limit of Euler-Poisson system with and without viscosity. Comm. Partial Differential Equations. 29 (2004), no. 3–4, 419–456.

On the Full Dissipative Property of the Vlasov-Poisson-Boltzmann System

Renjun Duan
Department of Mathematics, The Chinese University of Hong Kong
Shatin, Hong Kong, China
E-mail: rjduan@math.cuhk.edu.hk

Robert M. Strain
University of Pennsylvania, Department of Mathematics
209 South 33rd Street, Philadelphia, PA 19104-6395, USA
E-mail: strain@math.upenn.edu

Abstract

In this paper, we present a new approach of studying the full dissipative property of the Vlasov-Poisson-Boltzmann system over the whole space. The key part of this approach is to design the interactive functional to capture the dissipation of the system along the degenerate components. The developed approach is generally applicable to other relevant models arising from plasma physics both at the kinetic and fluid levels.

1 Introduction

In this paper we present a recent work [7] for the study of the optimal large-time behavior on the Vlasov-Poisson-Boltzmann system (VPB). The approach that we have developed in [7] is a combination of the time-decay property of the linearized nonhomogeneous system in terms of the Fourier analysis and the bootstrap to the nonlinear system. The key part of this approach is to design some interactive functionals to capture the dissipation of the system along the degenerate components. Moreover, the approach used in [7] has been applied to some models with the electromagnetic field in the context of plasma physics both at the kinetic and fluid levels, such as the Vlasov-Maxwell-Boltzmann system [8] and the Euler-Maxwell system with relaxation [3]. A new feature in the presence of the coupled Maxwell system which was found in [8] and [3] is that the full system admits the regularity-loss property. From the analysis of the Green's function in the case of the Euler-Maxwell system, this kind of regularity-loss property corresponds to the fact that the real part of eigenvalues along the electromagnetic component tends to zero as the frequency goes to infinity. It is believed that the ideas and methods that are established in [3, 7, 8] can also be applicable to many other relevant physical models.

In what follows we first formulate the problem considered in [7] and then give a review of the related references; the statement of the main results in [7] is left to the next subsection. The VPB system is a physical model describing the time evolution of dilute charged particles (e.g., electrons) under a given external magnetic field [20, 1]. For one-species of particles, it reads

$$\partial_t f + \xi \cdot \nabla_x f + \nabla_x \Phi \cdot \nabla_\xi f = Q(f, f), \tag{1.1}$$

$$\Delta_x \Phi = \int_{\mathbb{R}^3} f d\xi - \bar{\rho}(x), \tag{1.2}$$

with initial data

$$f(0, x, \xi) = f_0(x, \xi). \tag{1.3}$$

Here, the unknown $f = f(t, x, \xi)$ is a non-negative function standing for the number density of gas particles which have position $x = (x_1, x_2, x_3) \in \mathbb{R}^3$ and velocity $\xi = (\xi_1, \xi_2, \xi_3) \in \mathbb{R}^3$ at time $t > 0$. Q is the bilinear collision operator for the hard-sphere model defined by

$$Q(f, g) = \int_{\mathbb{R}^3 \times S^2} (f' g'_* - f g_*) |(\xi - \xi_*) \cdot \omega| d\omega d\xi_*,$$

$$f = f(t, x, \xi), \quad f' = f(t, x, \xi'), \quad g_* = g(t, x, \xi_*), \quad g'_* = g(t, x, \xi'_*),$$

$$\xi' = \xi - [(\xi - \xi_*) \cdot \omega]\omega, \quad \xi'_* = \xi_* + [(\xi - \xi_*) \cdot \omega]\omega, \quad \omega \in S^2.$$

The potential function $\Phi = \Phi(t, x)$ generating the self-consistent electric field in (1.1) is coupled with $f(t, x, \xi)$ through the Poisson equation (1.2). $\bar{\rho}(x)$ denotes the stationary background density satisfying $\bar{\rho}(x) \to \rho_\infty$ as $|x| \to \infty$ for a positive constant state $\rho_\infty > 0$.

The existence of stationary solutions to the system (1.1)–(1.2) and the nonlinear stability of solutions to the Cauchy problem (1.1)–(1.3) near the stationary state were obtained in [9] in terms of the energy method [10, 13, 14, 17, 18, 28, 29]. Notice that when $\bar{\rho}$ is a positive constant, the global existence was also studied in [14, 28, 29] in the energy space with time derivatives, but [9] used a solution space without initial layer. After [9], we obtained in [7] the optimal rate of convergence of solutions towards the stationary states. To consider this issue in a simple way, it is here supposed that

$$\bar{\rho}(x) \equiv \rho_\infty = 1, \quad x \in \mathbb{R}^3. \tag{1.4}$$

In this case, the VPB system (1.1)–(1.2) has a stationary solution (f_*, Φ_*) with

$$f_* = \mathbf{M} = (2\pi)^{-3/2} \exp\left(-|\xi|^2/2\right), \quad \Phi_* = 0,$$

where $\mathbf{M}$ is a normalized global Maxwellian.

The time rate of convergence to equilibrium is an important topic in the mathematical theory of the physical world. As pointed out in [26], there exist general structures in which the interaction between a conservative part and a degenerate dissipative part leads to convergence to equilibrium, where this property was called *hypocoercivity.* Our study in [7] indeed provides a concrete example of the hypocoercivity property for the nonlinear VPB system in the framework of perturbations. The key of the method to study hypocoercivity provided in [7] is to carefully capture the dissipative property for the perturbed macroscopic system of equations with the hyperbolic-parabolic structure, which is in the same spirit of the Kawashima's work [15].

There has been extensive investigations on the rate of convergence for the nonlinear Boltzmann equation or related spatially non-homogeneous kinetic equations with relaxations. We shall mention some of them; interested readers can refer to [7, 8, 22] and more references therein. In the context of perturbed solutions, the first result was given by Ukai [25], where the spectral analysis is used to obtain the exponential rates for the Boltzmann equation with hard potentials on torus. When there is an external force, a new approach of the energy-spectrum combination was developed by Duan-Ukai-Yang-Zhao [10]. The compensation function method of the Boltzmann equation was found by Kawashima [16]. This method was later applied by Glassey-Strauss [11] to the time decay of the linearized VPB system on torus, and the recent applications was made by Yang-Yu [27]. Here, we also mention Glassey-Strauss [12] for the study of the essential spectra of the solution operator of the VPB system. Moreover, Strain-Guo [24] developed a weighted energy method to get the exponential rate of convergence for the Boltzmann equation and Landau equation with soft potentials on the torus; the earlier work [23] was done by the same authors.

Another powerful tool is entropy method which has general applications in the existence theory for nonlinear equations. By using this method, Desvillettes-Villani [2] obtained first the almost exponential rate of convergence of solutions to the Boltzmann equation on torus with soft potentials for large initial data under the additional regularity conditions; see Villani [26] for extension and simplification of results. Recently, by finding some proper Lyapunov functional defined over the Hilbert space, Mouhot-Neumann [21] obtained the exponential rates of convergence for some kinetic models with general structures in the case of torus; see also [26] for the systematic study of this topic.

Besides those methods mentioned above for the study of rates of convergence, the method of Green's functions was also founded by Liu-Yu [19] to expose the pointwise large-time behavior of solutions to the Boltzmann equation in the full space $\mathbb{R}^3$.

2 Main results

As mentioned before, our work [7] is mainly concerned with the study of the large-time behavior for the Cauchy problem (1.1), (1.2) and (1.3) of the VPB system, especially the rates of convergence of solutions trending towards the global Maxwellian $\mathbf{M}$. The main result obtained in [7] is stated as follows.

Theorem 2.1. *Let $N \geq 4$ and $w(\xi) = (1+|\xi|^2)^{1/2}$. Assume that $f_0 \geq 0$ and $\|(f_0 - \mathbf{M})\mathbf{M}^{-1/2}\|_{H^N \cap L^2_w \cap Z_1}$ is sufficiently small. Let $f \geq 0$ be the solution to the Cauchy problem (1.1), (1.2) and (1.3) under the assumption (1.4). Then, f enjoys the estimate with algebraic rate of convergence:*

$$\|(f(t)-\mathbf{M})\mathbf{M}^{-1/2}\|_{H^N} \leq C\|(f_0-\mathbf{M})\mathbf{M}^{-1/2}\|_{H^N\cap Z_1}(1+t)^{-\frac{1}{4}}. \quad (2.1)$$

Furthermore, under the following additional conditions on f_0, f also enjoys some estimates with extra rates of convergence:

Case 1. If $\int_{\mathbb{R}^3}(f_0(x,\xi)-\mathbf{M})d\xi \equiv 0$ holds for any $x \in \mathbb{R}^3$, then one has

$$\|(f(t)-\mathbf{M})\mathbf{M}^{-1/2}\|_{H^N} \leq C\|(f_0-\mathbf{M})\mathbf{M}^{-1/2}\|_{H^N\cap Z_1}(1+t)^{-\frac{3}{4}}. \quad (2.2)$$

Case 2. Fix any $0 < \epsilon \leq 3/4$ and suppose that $\|(f_0 - \mathbf{M})\mathbf{M}^{-1/2}\|_{H^N_w \cap Z_1}$ is sufficiently small. Then one has

$$\begin{aligned}\|f(t)\mathbf{M}^{-1/2} - \sum_{j=0}^{4}\langle e_j, f(t)\rangle e_j \mathbf{M}^{1/2}\|_{H^N_w} + \sum_{j=1}^{4}\|\nabla_x \langle e_j, f(t)\rangle\|_{H^{N-1}_x} \\ + \|\langle e_0, f(t)-\mathbf{M}\rangle\|_{H^N_x} \\ \leq C\|(f_0-\mathbf{M})\mathbf{M}^{-1/2}\|_{H^N_w\cap Z_1}(1+t)^{-\frac{3}{4}+\epsilon}, \quad (2.3)\end{aligned}$$

where $\{e_j\}_{j=0}^4$ is the orthonormal set in $L^2(\mathbb{R}^3; \mathbf{M}d\xi)$ defined by

$$e_0 = 1, \quad e_j = \xi_j \ (1 \leq j \leq 3), \quad e_4 = \frac{|\xi|^2 - 3}{\sqrt{6}}.$$

Some remarks are given as follows. From the proof in [7], the condition in Case 1 in Theorem 2.1 can be weakened as

$$\iint_{\mathbb{R}^3\times\mathbb{R}^3}(f_0(x,\xi)-\mathbf{M})dxd\xi = 0.$$

Hence, the rate in (2.2) for the VPB system is the same as one for the Boltzmann equation provided that the zero-mass perturbation is postulated. On the other hand, from (2.1), for the general initial perturbation, the time-decay of the VPB system over $\mathbb{R}^3$ is slower than that of the Boltzmann equation. This is essentially caused by the self-induced potential force. Furthermore, if one decomposes f as the summation of three parts

$$f = \langle e_0, f\rangle e_0 \mathbf{M} + \sum_{j=1}^{4} \langle e_j, f\rangle e_j \mathbf{M} + \{f - \sum_{j=0}^{4} \langle e_j, f\rangle e_j \mathbf{M}\},$$

then by comparing (2.1) and (2.3), one can find out that the effect of the self-induced potential force on rates of convergence only occurs in the above second part which corresponds to the projections of f along the momentum and temperature components e_j, $1 \leq j \leq 4$.

3 Dissipative property of the linearized system

In this section we emphasize one of the key parts in the proof of Theorem 2.1. It is connected with the Fourier analysis of the following Cauchy problem of the linearized system with a nonhomogeneous source:

$$\begin{cases} \partial_t u + \xi \cdot \nabla_x u - \nabla_x \Phi \cdot \xi \sqrt{\mathbf{M}} = \mathbf{L} u + h, \\ \Delta_x \Phi = \displaystyle\int_{\mathbb{R}^n} \sqrt{\mathbf{M}} u d\xi, \quad t > 0, x \in \mathbb{R}^n, \xi \in \mathbb{R}^n, \\ u|_{t=0} = u_0, \quad x \in \mathbb{R}^n, \end{cases} \tag{3.1}$$

where $h = h(t, x, \xi)$ and $u_0 = u_0(x, \xi)$ are given, the spatial dimension $n \geq 2$ is arbitrary in order to determine how it enters into the time-decay rate at the level of linearization, and for simplicity we still use $\mathbf{M}$ to denote the normalized n-dimensional Maxwellian $\mathbf{M} = (2\pi)^{-n/2} e^{-|\xi|^2/2}$. Formally, the solution to the Cauchy problem (3.1) can be written as the mild form

$$u(t) = e^{t\mathbf{B}} u_0 + \int_0^t e^{(t-s)\mathbf{B}} h(s) ds,$$

where $e^{t\mathbf{B}}$ denotes the solution operator to the Cauchy problem of the linearized equation without source corresponding to (3.1) for $h \equiv 0$. A vital goal is to show that $e^{t\mathbf{B}}$ has the proposed algebraic decay properties as time tends to infinity. The idea for that purpose is to make energy estimates for pointwise time t and frequency variable k which corresponds to the spatial variable x. We decompose u as

$$\begin{cases} u(t,x,\xi) = \mathbf{P}u \oplus \{\mathbf{I}-\mathbf{P}\}u, \\ \mathbf{P}u = \{a^u + b^u\cdot\xi + c^u|\xi|^2\}\sqrt{\mathbf{M}}, \end{cases}$$

where $\mathbf{P}$ is the usual macroscopic projector, and a^u, b^u, c^u are the macro moment functions of u. Under the assumption of $\mathbf{P}h = 0$, one can obtain the macroscopic balance laws satisfied by a^u, b^u, c^u:

$$\begin{cases} \partial_t(a^u + nc^u) + \nabla_x\cdot b^u = 0, \\ \partial_t b_j^u + \partial_j(a^u + nc^u) + 2\partial_j c^u \\ \qquad + \sum_m \partial_m A_{jm}(\{\mathbf{I}-\mathbf{P}\}u) - \partial_j\Phi = 0, \\ \partial_t c^u + \frac{1}{n}\nabla_x\cdot b^u + \frac{1}{2n}\sum_j \partial_j B_j(\{\mathbf{I}-\mathbf{P}\}u) = 0, \\ \Delta_x\Phi = a^u + nc^u, \end{cases} \tag{3.2}$$

and also the high-order moment equations:

$$\begin{cases} \partial_t[A_{jj}(\{\mathbf{I}-\mathbf{P}\}u) + 2c^u] + 2\partial_j b_j^u = A_{jj}(R+h), \\ \partial_t A_{jm}(\{\mathbf{I}-\mathbf{P}\}u) + \partial_j b_m^u + \partial_m b_j^u = A_{jm}(R+h), \quad j\neq m, \\ \partial_t B_j(\{\mathbf{I}-\mathbf{P}\}u) + \partial_j c^u = B_j(R+h), \end{cases} \tag{3.3}$$

for $1\le j,m\le n$. Here, the velocity moment functions $A_{jm}(\cdot)$ and $B_j(\cdot)$ are given by

$$A_{jm}(u) = \langle(\xi_j\xi_m - 1)\sqrt{\mathbf{M}}, u\rangle, \quad B_j(u) = \frac{\langle[|\xi|^2 - (n+2)]\xi_j\sqrt{\mathbf{M}}, u\rangle}{2(n+2)},$$

where R is given by $R = -\xi\cdot\nabla_x\{\mathbf{I}-\mathbf{P}\}u + \mathbf{L}\{\mathbf{I}-\mathbf{P}\}u$. From the above system (3.2)–(3.3), we showed in [7] the following:

Theorem 3.1. *There exists an interactive time-frequency functional $\mathcal{E}_{\rm int}(\widehat{u}(t,k))$ whose modulus is bounded by the naturally existing Lyapunov functional of* (3.1), *that is,*

$$|\mathcal{E}_{\rm int}(\widehat{u}(t,k))| \le C\left(\|\widehat{u}(t,k)\|_{L^2_\xi}^2 + \frac{1}{|k|^2}|a^{\widehat{u}} + nc^{\widehat{u}}|^2\right),$$

such that

$$\begin{aligned} &\partial_t\mathfrak{R}\,\mathcal{E}_{\rm int}(\widehat{u}(t,k)) + \lambda\frac{|k|^2}{1+|k|^2}\left(|b^{\widehat{u}}|^2 + |c^{\widehat{u}}|^2\right) + \lambda|a^{\widehat{u}} + nc^{\widehat{u}}|^2 \\ &\quad \le C\left(\|\{\mathbf{I}-\mathbf{P}\}\widehat{u}\|_{L^2_\xi}^2 + \|\nu^{-1/2}\{\mathbf{I}-\mathbf{P}\}\widehat{h}\|_{L^2_\xi}^2\right) \end{aligned}$$

holds true for any $t\ge 0$ and $k\in\mathbb{R}^n$.

The proof of Theorem 3.1 is based on some observations from the previous work [4, 5, 6] together with the idea of carefully choosing a proper linear combination of the decomposed equations; this strategy was noted in [26]. Hence, the interactive time-frequency functional $\mathcal{E}_{\rm int}(\widehat{u}(t,k))$ indeed captures the dissipative property of all the degenerate components in the solution. The rest procedure to derive the time-decay of e^{tB} is based on energy estimates; refer to [7] for more details.

References

[1] C. Cercignani, R. Illner and M. Pulvirenti, The Mathematical Theory of Dilute Gases. Applied Mathematical Sciences, 106. Springer-Verlag, New York, 1994.

[2] L. Desvillettes and C. Villani, On the trend to global equilibrium for spatially inhomogeneous kinetic systems: The Boltzmann equation. Invent. Math. 159 (2) (2005), 245–316.

[3] R. J. Duan, Global smooth flows for the compressible Euler-Maxwell system: I. Relaxation case, arXiv:1006.3606v1 (2010).

[4] R. J. Duan, Hypocoercivity of linear degenerately dissipative kinetic equations, arXiv:0912.1733 (2009).

[5] R. J. Duan, On the Cauchy problem for the Boltzmann equation in the whole space: Global existence and uniform stability in $L^2_\xi(H^N_x)$. Journal of Differential Equations 244 (2008), 3204–3234.

[6] R. J. Duan, Stability of the Boltzmann equation with potential forces on torus. Physica D: Nonlinear Phenomena 238 (2009), 1808–1820.

[7] R. J. Duan and R. M. Strain, Optimal time decay of the Vlasov-Poisson-Boltzmann system in $\mathbb{R}^3$. Arch. Rat. Mech. Anal. doi:10.1007/s00205-010-0318-6 (2010).

[8] R. J. Duan and R. M. Strain, Optimal large-time behavior of the Vlasov-Maxwell-Boltzmann system in the whole space, arXiv:1006.3605 (2010).

[9] R. J. Duan and T. Yang, Stability of the one-species Vlasov-Poisson-Boltzmann system. SIAM Journal on Mathematical Analysis 41 (2010), no. 6, 2353–2387.

[10] R. J. Duan, S. Ukai, T. Yang and H. J. Zhao, Optimal decay estimates on the linearized Boltzmann equation with time-dependent forces and their applications. Communications in Mathematical Physics 277 (2008), no. 1, 189–236.

[11] R. Glassey and W. Strauss, Decay of the linearized Boltzmann-Vlasov system. Transport Theory Statist. Phys. 28 (1999), 135–156.

[12] R. Glassey and W. Strauss, Perturbation of essential spectra of evolution operators and the Vlasov-Poisson-Boltzmann system. Discrete Contin. Dynam. System. 5 (1999), 457–472.

[13] Y. Guo, The Boltzmann equation in the whole space. Indiana Univ. Math. J. 53 (2004), 1081–1094.

[14] Y. Guo, The Vlasov-Poisson-Boltzmann system near Maxwellians. Comm. Pure Appl. Math. 55 (2002), no. 9, 1104–1135.

[15] S. Kawashima, Systems of a hyperbolic-parabolic composite type, with applications to the equations of magnetohydrodynamics. Thesis, Kyoto University (1983).

[16] S. Kawashima, The Boltzmann equation and thirteen moments. Japan J. Appl. Math. 7 (1990), 301–320.

[17] T. P. Liu, T. Yang and S. H. Yu, Energy method for the Boltzmann equation. Physica D. 188 (2004), no. (3–4) 178–192.

[18] T. P. Liu and S. H. Yu, Boltzmann equation: micro-macro decompositions and positivity of shock profiles. Commun. Math. Phys. 246 (2004), no. 1 133–179.

[19] T. P. Liu and S. H. Yu, The Green's function and large-time behavior of solutions for the one-dimensional Boltzmann equation. Comm. Pure Appl. Math. 57 (2004), 1543–1608.

[20] P. A. Markowich, C. A. Ringhofer and C. Schmeiser, Semiconductor Equations. Springer-Verlag, Vienna, 1990.

[21] C. Mouhot and L. Neumann, Quantitative perturbative study of convergence to equilibrium for collisional kinetic models in the torus. Nonlinearity 19 (2006), no. 2, 969–998.

[22] R. M. Strain, The Vlasov-Maxwell-Boltzmann system in the whole space. Commun. Math. Phys. 268 (2006), no. 2, 543–567. (arXiv:math/0512002)

[23] R. M. Strain and Y. Guo, Almost exponential decay near Maxwellian. Comm. P.D.E. 31 (2006), 417–429.

[24] R.M. Strain and Y. Guo, Exponential decay for soft potentials near Maxwellian. Arch. Ration. Mech. Anal. 187 (2008), 287–339.

[25] S. Ukai, On the existence of global solutions of mixed problem for nonlinear Boltzmann equation. Proceedings of the Japan Academy. 50 (1974), 179–184.

[26] C. Villani, Hypocoercivity. Memoirs Amer. Math. Soc. 202 (2009). (arXiv: math/0609050v1)

[27] T. Yang and H. J. Yu, Optimal convergence rates of classical solutions for Vlasov-Poisson-Boltzmann system. Commun. Math. Phys. doi: 10.1007/s00220-010-1142-4 (2010).

[28] T. Yang, H. J. Yu and H. J. Zhao, Cauchy problem for the Vlasov-Poisson-Boltzmann system. Arch. Rational Mech. Anal. 182 (2006), 415–470.

[29] T. Yang and H. J. Zhao, Global existence of classical solutions to the Vlasov-Poisson-Boltzmann system. Commun. Math. Phys. 268 (2006), 569–605.

On the Space-Time Expansion Discontinuous Galerkin Method

P. Engel
Institut für Angewandte Analysis und Numerische Simulation
Universität Stuttgart, Pfaffenwaldring 57, 70569 Stuttgart, Germany
E-mail: patrick.engel@mathematik.uni-stuttgart.de

C. Rohde*
Institut für Angewandte Analysis und Numerische Simulation
Universität Stuttgart, Pfaffenwaldring 57, 70569 Stuttgart, Germany
E-mail: christian.rohde@mathematik.uni-stuttgart.de

Abstract

In this paper we review the Space-Time Expansion Discontinuous-Galerkin (STE-DG) method, introduced in [3, 6, 7]. The original approach achieved time-discretization using the Cauchy-Kovalevskaya technique. We present a more general expansion setting. For this framework new expansions are introduced. Moreover, it allows to derive an error estimate for all orders in space and time. The result applies for ansatz functions of arbitrary polynomial order $k \in \mathbb{N}$ provided the time step is sufficiently small. It gives a convergence rate of order $k + \frac{1}{2}$ in space and time.

1 Introduction

We consider in this paper a scalar conservation law in one space dimension. Precisely with $\Omega = (0, 1)$ (for simplicity) and $T > 0$ we search for a function $u : \Omega \times (0, T) \to \mathbb{R}$ such that

$$u_t + F(u)_x = 0 \qquad \text{in } \Omega \times (0,T), \tag{1.1a}$$

$$u(\cdot, 0) = u_0 \qquad \text{in } \Omega, \tag{1.1b}$$

$$u(0,t) = u(1,t) \qquad (t \in [0,T)) \tag{1.1c}$$

***Acknowledgment:** The authors thank the German Research Foundation (DFG) for financial support of the project within the Cluster of Excellence in Simulation Technology (EXC 310/1) at the University of Stuttgart.

holds. Here $F : \mathbb{R} \to \mathbb{R}$ is the, in general nonlinear, flux and $u_0 : \Omega \to \mathbb{R}$ the initial datum.

Discontinuous Galerkin schemes have been used successfully to resolve (1.1), see for example [1]. For the semi-discrete case (and for smooth solutions) a complete error analysis is available. More challenging is the fully-discrete case. Here — up to our knowledge — only the case of the Runge-Kutta time-discretization has been analysed [8, 9]. However, more other approaches for time-discretization have been suggested. We focus on the Space-Time Expansion method as introduced by [3, 6, 7]. In this paper we revise the method and present three different expansions (Cauchy-Kovalevskaya, characteristic and exact). The Cauchy-Kovalevskaya expansion is the original one from [3, 6, 7], which can get rather complex for nonlinear fluxes. Our new characteristic expansion seems to be a more easy to apply concept. The exact expansion uses a Godunov-type idea. We present a convergence result for the Cauchy-Kovalevskaya and characteristic expansion, which is valid for *all* orders in space *and* time (Section 4), unfortunately under a too restrictive time-step condition. Numerical experiments show that the new characteristic expansion gives results comparable to the original Cauchy-Kovalevskaya approach (Section 3.3).

2 Notations and function spaces

In the following section we will introduce basic notations we need to describe the method in Section 3. Let

$$\Gamma_h = \left\{0 = x_{\frac{1}{2}} < x_{\frac{3}{2}} < \cdots < x_{J+\frac{1}{2}} = 1\right\}$$

be a (not necessarily equidistant) grid and $J \in \mathbb{N}$ the number of mesh cells. Define for each $j \in \{1, \ldots, J\}$ the mesh cell I_j and the cell centre x_j by

$$I_j := (x_{j-\frac{1}{2}}, x_{j+\frac{1}{2}}), \qquad x_j := \frac{1}{2}(x_{j-\frac{1}{2}} + x_{j+\frac{1}{2}}).$$

Next define the grid parameter h as

$$h := \max_{1 \le j \le J} \left(x_{j+\frac{1}{2}} - x_{j-\frac{1}{2}}\right).$$

In the same way we introduce a time grid and for simplicity we choose a constant time-step $\tau > 0$ and define for $n \in \{0, \ldots, N\}$ the time levels t^n by $t^n = n\tau$. Choose $N \in \mathbb{N}$ and τ such that $t^N = N\tau = T$.
Let

$$C_\Gamma^k := \left\{v \colon [0,1] \to \mathbb{R} \,\middle|\, v \in C^k(I_j), \quad j = 1, 2, \ldots, J\right\}.$$

$V_\Gamma^k \subset C_\Gamma^k$ is the space of piecewise polynomials on Γ_h. All elements of these function spaces allow discontinuities at cell boundaries. So there is a need for the following notation: Define for all $v \in C_\Gamma^k$ and $x \in [0,1]$

$$v^+(x) := \lim_{y \downarrow x} v(y) \quad \text{and} \quad v^-(x) := \lim_{y \uparrow x} v(y).$$

We also need different norms and semi-norms on these function spaces. We use $\|\cdot\|_2$ to name the L^2-norm. By $\|\cdot\|_\infty$ we denote the maximum norm on Ω. Moreover we define for $v \in C_\Gamma^k$ the seminorm

$$|v|_{\Gamma_h} := \left(\sum_{1 \le j \le J} \left(v^+(x_{j+\frac{1}{2}})\right)^2 + \left(v^-(x_{j+\frac{1}{2}})\right)^2 \right)^{\frac{1}{2}}.$$

3 The numerical method

We introduce the STE-DG method in this section. Our presentation covers the original STE-DG method as it has been suggested in [3, 6, 7], but is more general: We present a number of new expansions.

3.1 Expansions: Definition and examples

The key element in the STE-DG method is the expansion. One can see it as a local approximation to the solution of (1.1a) on $[t^n, t^{n+1}) \times I_j$ only using the information $u_h(\cdot, t^n)|_{I_j}$, where $u_h(\cdot, t^n) \in V_h^k$ is the approximate solution (see Definition 3.6 below). In particular this implies that the exact solution generally does not provide a suitable example, as the exact weak solution depends on the values of $u_h(\cdot, t^n)$ in at least one of the neighbouring cells $I_{j\pm 1}$ (if the flux is non trivial). We use the following general framework.

Definition 3.1 (Expansion). We define the **space of time-dependent piecewise polynomial functions** by

$$\mathcal{V}_\Gamma^k := \left\{ v_h \in L^\infty([0,1] \times [0,\tau], \mathbb{R}) \,\middle|\, v_h(\cdot, t) \in V_h^k, 0 \le t \le \tau \right\}.$$

A function $E : V_\Gamma^k \to \mathcal{V}_\Gamma^k$ is called **expansion**, if the equation

$$E[v_h](\cdot, 0) = v_h$$

holds for all $v_h \in V_\Gamma^k$. For any $s \in [0,T]$ define further E_s as

$$E_s[v_h](x,t) := E[v_h](x, t-s)$$

for any $x \in \Omega$ and $t \in [s, s+\tau]$.

Remark 3.2. Note that we do not require the expansion to be a linear operator. In general the expansion is nonlinear. For a specific example we refer to Section 3.1.1.

Before we introduce the method, let us present some relevant choices for expansions.

3.1.1 Example 1: Cauchy-Kovalevskaya expansion (CK)

Lörcher et al. introduced the STE-DG method in [6] together with the CK expansion. To get this expansion one has to perform a Taylor expansion of the exact solution in space and time for each grid cell I_j. Then the time derivatives are replaced by space derivatives using the differential equation (1.1a). We need the following auxiliary notation.

Definition 3.3 (Derivational Operators)**.** Let $F \in C^k(\mathbb{R})$ in (1.1a). For any function $v \in C^k_\Gamma$ define the **derivational operators** in I_j, $1 \le j \le J$, associated with (1.1a) by

$$\tilde{\partial}_x v := v_x, \quad \tilde{\partial}_t v := -F(v)_x = -F'(v)v_x,$$
$$\tilde{\partial}_{xt} v := -(\tilde{\partial}_x F(v))_x = -(F'(v)v_x)_x = -F''(v)v_x^2 - F'(v)v_{xx}, \ldots$$

Note that the derivatives $v_x, v_{xx}, \ldots$ are well defined (only) for inner points of each cell. In the following we will not use this notation at cell boundaries of Γ_h.

Definition 3.4 (Cauchy-Kovalevskaya Expansion)**.** The **Cauchy-Kovalevskaya expansion** $E^{CK} : V^k_\Gamma \to \mathcal{V}^k_\Gamma$ on I_j, $1 \le j \le J$, is defined by

$$E^{CK}[v_h](y,s) := \sum_{l=0}^{k} \frac{1}{l!} \left((y - x_j)\tilde{\partial}_x + s\tilde{\partial}_t \right)^l v_h(x) \bigg|_{x=x_j}$$

for $y \in I_j$ and $s \in [0, \tau]$.

3.1.2 Example 2: Characteristic expansion (Ch)

Another way to get an expansion is to use characteristics. To illustrate this concept we define $k+1$ points in each cell I_j, $1 \le j \le J$, by

$$y_{j,l} := x_{j-\frac{1}{2}} + \frac{l}{k}\left(x_{j+\frac{1}{2}} - x_{j-\frac{1}{2}} \right), \quad (l = 0, \ldots, k).$$

For some given $v \in V^k_\Gamma$ and all $t \in [0, \tau]$ we define $p^t_j \in \mathbf{P}^k(I_j)$ as the polynomial that satisfies

$$p^t_j\left(y_{j,l} + tF'\left(v(y_{j,l})\right)\right) = v(y_{j,l}), \quad (l = 0, \ldots, k).$$

Note that p_j^t is well defined, if the set

$$\{y_{j,l} + tF'(v(y_{j,l}))\}_{l=0,\dots,k}$$

contains exactly $k+1$ elements. This is guaranteed, if

$$t \le \tau \le \frac{h}{(2k+1)\,\|F'(v)\|_\infty}$$

holds. Thus we need here already a CFL-like condition to ensure, that the expansion is well-defined. With this notation we define:

Definition 3.5 (Characteristic Expansion). The **characteristic expansion** $E^{Ch} : V_\Gamma^k \to \mathcal{V}_\Gamma^k$ is defined by

$$E^{Ch}[v](x,t) := p_j^t(x) \qquad \text{for } x \in I_j \text{ and } t \in [0,\tau].$$

For arbitrary $k \in \mathbb{N}$ we get on I_j, $1 \le j \le J$, the formulation

$$E^{Ch}[v_h](x,t) = \sum_{0\le l\le k} v_l \prod_{\substack{0\le m\le k\\ m\ne l}} \frac{x - y_{j,m} - tF'(v_m)}{y_{j,l} - y_{j,m} + t(F'(v_l) - F'(v_m))},$$

for all $x \in I_j$ and $0 \le t \le \tau$. Here $v_l = v_h(y_{j,l})$.

3.1.3 Example 3: Expansion with exact solution (Ex)

It is also possible to use not only the information of one cell, but also the neighboring cells. In a Godunov-like ansatz we may solve generalised Riemann problems for discontinuous initial data to get the unique entropy solution, which we can use as an expansion. Suppose $v_h \in V_h^k$. Let $v \in C\left([0,\tau), L^1(\Omega)\right)$ be the unique entropy solution of

$$\begin{aligned} v_t + F(v)_x &= 0 && \text{in } \Omega \times (0,\tau),\\ v(\cdot,0) &= v_h && \text{in } \Omega, \end{aligned}$$

assuming periodic boundary conditions. Then we are able to define the exact expansion by

$$E^{Ex}[v_h] := v.$$

Formally this is no expansion is the sense of Definition 3.1, because $E^{Ex}[v_h] \notin \mathcal{V}_\Gamma^k$. But one may weaken this assumption or just project v to $\mathcal{V}_\Gamma^k$.

3.2 Definition of the STE-DG method

In the following $H\colon \mathbb{R}^2 \to \mathbb{R}$ is a monotone numerical flux function [4]. With the definition of some expansion in Section 3.1 we are able to define the Space-Time Expansion DG method (STE-DG).

Definition 3.6 (Space-Time Expansion DG Method). Let $E : V_\Gamma^k \to \mathcal{V}_\Gamma^k$ be an expansion, $H : \mathbb{R}^2 \to \mathbb{R}$ a numerical flux function for F and Γ_h a given grid. A function $u_h : \Omega \times [0,T] \mapsto \mathbb{R}$ is called **solution of the STE-DG method of order $k \in \mathbb{N}_0$**, if the following properties hold:

1. For each $n \in \{0, \dots, N\}$ it holds, that
$$u_h^n := u_h(\cdot, t^n) \in V_\Gamma^k$$
and
$$u_h = E_{t^n}[u_h^n] \quad \text{on } \Omega \times (t^n, t^{n+1}).$$
2. For each $v_h \in V_\Gamma^k$ it holds, that
$$\int_\Omega u_h^0 \, v_h \, dx = \int_\Omega u_0 \, v_h \, dx.$$
3. For each $j \in \{1, \dots, J\}$, $n \in \{0, \dots, N\}$ and $v_h \in V_\Gamma^k$ it holds, that
$$\int_{I_j} (u_h^{n+1} - u_h^n) v_h \, dx = \mathcal{H}_j^n[u_h, v_h].$$

Here $\mathcal{H}_j^n[p, v_h]$ is defined for $p \in C^0([t^n, t^{n+1}], V_\Gamma^k)$ and $v_h \in V_\Gamma^k$ through

$$\int_{t^n}^{t^{n+1}} \left(\int_{I_j} F(p)\partial_x v_h \, dx - \left(H(p)v_h^-\right)_{j+\frac{1}{2}} + \left(H(p)v_h^+\right)_{j-\frac{1}{2}} \right) dt. \tag{3.1}$$

Remark 3.7. If the integral in (3.1) can not be calculated exactly, one has to use quadrature rules. For more details — for example, which order one should use — see [6]. We assume throughout the paper that the integral can be computed exactly.

3.3 Comparison between CK and Ch expansion

Let us mention that the Cauchy-Kovalevskaya and the characteristic expansion reduce to the standard finite volume method for $k = 0$. For $k = 1$ both methods are the same provided $F(u) = au$. We take as an example the Burgers-flux $F(u) = \frac{1}{2}u^2$ with initial condition $u_0(x) = \exp\left(-20(x - 1/2)^2\right)$ and measure the L^2-difference between the solutions using Cauchy-Kovalevskaya and characteristic expansion. From the following table one can clearly see that the two methods only differ in terms of order higher than the convergence order $k + 1$ (see Section 4).

	k	$J = 32$	$J = 64$	$J = 128$	$J = 256$
Error	1	$5.1 \cdot 10^{-5}$	$1.2 \cdot 10^{-5}$	$2.0 \cdot 10^{-6}$	$2.9 \cdot 10^{-7}$
EOC	1	2.1	2.6	2.7	
Error	2	$1.2 \cdot 10^{-4}$	$1.1 \cdot 10^{-5}$	$7.9 \cdot 10^{-7}$	$5.4 \cdot 10^{-8}$
EOC	2	3.4	3.8	3.9	

If we compare the two methods from an implementational point of view, we find that the characteristic expansion can be implemented straight forward by only knowing the first derivative of the flux F. On the other hand, one needs all derivatives of F up to order k to implement the Cauchy-Kovalevskaya method. We think this fewer need of regularity also shows up at shock waves, where the characteristic expansion does slightly better for $k \geq 2$. In the following table we compare the errors of both solutions for a shock wave solution with initial conditions $u_0(x) = 0$ for $x < \frac{1}{2}$ and $u_0(x) = 1$ for $x > \frac{1}{2}$.

	k	$J = 32$	$J = 64$	$J = 128$	$J = 256$
CK	1	$5.4 \cdot 10^{-2}$	$3.5 \cdot 10^{-2}$	$2.6 \cdot 10^{-2}$	$1.7 \cdot 10^{-2}$
Ch	1	$5.4 \cdot 10^{-2}$	$3.4 \cdot 10^{-2}$	$2.6 \cdot 10^{-2}$	$1.6 \cdot 10^{-2}$
CK	2	$3.9 \cdot 10^{-2}$	$3.3 \cdot 10^{-2}$	$2.1 \cdot 10^{-2}$	$1.7 \cdot 10^{-2}$
Ch	2	$3.7 \cdot 10^{-2}$	$3.2 \cdot 10^{-2}$	$1.7 \cdot 10^{-2}$	$1.5 \cdot 10^{-2}$

4 Theorem of convergence

Finally we formulate a theorem on convergence, which applies for the CK and Ch expansion.

Theorem 4.1 (Convergence). *Let $k \geq 1$. Let $u \in C^{k+2}(\Omega \times [0,T])$ be a classical solution of* (1.1a)+(1.1b)+(1.1c) *with $F \in C^{k+2}(\mathbb{R})$.*

Let $E : V_\Gamma^k \to \mathcal{V}_\Gamma^k$ be an expansion (either CK or Ch) of order k. Let $H : \mathbb{R}^2 \to \mathbb{R}$ be a monotone numerical flux function for F and Γ_h a given grid. For any $\gamma > 0$, there is a constant $C > 0$, such that the inequality

$$\max_{0 \leq n \leq N} \|u(\cdot, t^n) - u_h^n\|_2^2 \leq C h^{2k+1} \tag{4.1}$$

holds, provided we have h small enough and $\tau \leq \gamma h^2$.

Remark 4.2. The theorem is proven in [2] for a general class of expansion.

Remark 4.3. The convergence rate in (4.1) can be improved if we restrict ourselves to upwind numerical flux functions for F. Precisely we get as in [8] the rate

$$\max_{0 \leq n \leq N} \|u(\cdot, t^n) - u_h^n\|_2 \leq C h^{k+1}.$$

Remark 4.4. The STE-DG scheme is for the case $k = 0$ of piecewise constant ansatz functions nothing but a standard finite-volume scheme evolving cell averages. If the flux functions are, e.g., monotone in the sense of [4] extensive convergence results — including the case of discontinuous entropy solutions — are available. It is worth mentioning that estimate (4.1) gives the same L^2-rate that is already known for finite-volume methods provided the solution is smooth [5].

Remark 4.5. The time step restriction $\tau \leq \gamma h^2$ is not the one expected for first-order equations. The numerical experiments in [7] show that convergence of the STE-DG scheme can be expected provided just a classical CFL-condition of the form $\tau \leq \gamma h$ for some $\gamma > 0$ depending on F' controls the time step. The reason for the more restrictive condition is technical (see for instance also [8]). Let us point out [9], which avoids the problem and might also lead to the optimal condition in our case.

References

[1] B. Cockburn and C. W. Shu, TVB Runge-Kutta local projection discontinuous Galerkin finite element method for conservation laws. II. General framework. Math. Comp. 52 (1989), no. 186, S. 411–435.

[2] P. Engel and C. Rohde, Convergence of the Space-Time Expansion Discontinuous Galerkin Method for Scalar Conservation Laws. Preprint, http://www.simtech.uni-stuttgart.de/publikationen/preprints09.php, no. 2009-29.

[3] G. Gassner, F. Lörcher and C. D. Munz, A discontinuous Galerkin scheme based on a space-time expansion. II. Viscous flow equations in multi dimensions. J. Sci. Comput. 34 (2008), no. 3, 260–286.

[4] D. Kröner, Numerical schemes for conservation laws. Wiley-Teubner Series Advances in Numerical Mathematics, John Wiley & Sons Ltd., Chichester (1997).

[5] V. Jovanović and C. Rohde, Error estimates for finite volume approximations of classical solutions for nonlinear systems of hyperbolic balance laws. SIAM J. Numer. Anal. 43 (2006), no. 6, 2423–2449.

[6] F. Lörcher, G. Gassner and C. D. Munz, A discontinuous Galerkin scheme based on a space-time expansion. I. Inviscid compressible flow in one space dimension. J. Sci. Comput. 32 (2007), no. 2, 175–199.

[7] F. Lörcher, G. Gassner and C. D. Munz, An explicit discontinuous Galerkin scheme with local time-stepping for general unsteady diffusion equations. J. Comput. Phys. 227 (2008), no. 11, 5649–5670.

[8] Q. Zhang and C. W. Shu, Error estimates to smooth solutions of Runge-Kutta discontinuous Galerkin methods for scalar conservation laws. SIAM J. Numer. Anal. 42 (2004), no. 2, 641–666.

[9] Q. Zhang and C. W. Shu, Stability analysis and a priori error estimates to the third order explicit Runge-Kutta discontinuous Galerkin Mathod for scalar conservation laws. http://www.dam.brown.edu/scicomp/reports/2009-28/.

Energy Conservative and Stable Schemes for the Two-layer Shallow Water Equations

Ulrik Skre Fjordholm
ETH Zurich, Rämistrasse 101, 8092 Zurich, Switzerland
E-mail: ulrikf@sam.math.ethz.ch

Abstract

By extending the work in [4] we design an energy stable finite difference scheme for the two-layer shallow water equations.

1 Introduction

The two-layer shallow water equations model the flow of two fluids of different densities, superimposed on each other, under the influence of gravity. The main assumption in the derivation is that the horizontal length scales are much bigger than the vertical scales, and that one can therefore neglect variations in depth of density and velocity. Furthermore, it is assumed that no mixing occurs between the fluids. In one spatial dimension the equations have the form

$$
\begin{aligned}
(h_1)_t + (h_1u_1)_x = 0, \qquad & (h_2)_t + (h_2u_2)_x = 0, \\
(h_1u_1)_t + \left(\frac{1}{2}gh_1^2 + h_1u_1^2\right) &= -gh_1\left(b + rh_2\right)_x, \\
(h_2u_2)_t + \left(\frac{1}{2}gh_2^2 + h_2u_2^2\right) &= -gh_2\left(b + h_1\right)_x,
\end{aligned}
\tag{1.1}
$$

where subscripts 1 and 2 denote the lower and upper layers, respectively, $b = b(x)$ is the bottom topography, h is the layer height, u is layer velocity, $r := \frac{\rho_2}{\rho_1} < 1$ and ρ is the layer density, with $\rho_1 > \rho_2$. In this equation, mass (h_1 and h_2) and total momentum ($\rho_1h_1u_1 + \rho_2h_2u_2$) are the conserved variables. (1.1) is a *balance law*, which in general has the form

$$\mathbf{u}_t + \mathbf{f}(\mathbf{u}_x) = \mathbf{s}(\mathbf{u}, \mathbf{u}_x, x).$$

The system (1.1) is equipped with an entropy pair, the *energy* of the solution:

$$\eta = \frac{1}{2}\left(\rho_1\left(h_1u_1^2 + gh_1^2\right) + \rho_2\left(h_2u_2^2 + gh_2^2\right)\right) + g\rho_1h_1b + g\rho_2h_2(h_1+b),$$

$$q = \sum_{i=1}^{2}\rho_i\left(\frac{1}{2}h_iu_i^3 + gh_iu_i(h_i+b)\right) + g\rho_2h_1h_2(u_1+u_2). \tag{1.2}$$

The relevant entropy condition for (1.1) is therefore

$$\eta(\mathbf{u})_t + q(\mathbf{u})_x \le 0. \tag{1.3}$$

The two-layer shallow water equations entail several difficult problems:

- The right-hand side products (e.g., $h_1\partial_x h_2$) are undefined at discontinuities, as we get the product of a distribution (h_1) with a measure ($\partial_x h_2$), a *nonconservative product.* The theoretical framework for nonconservative products was set by DalMaso, LeFloch and Murat in [3], and numerical methods (so-called path-conservative schemes) have been developed to fit into this framework. However, as has been recently reported [1], such schemes may not converge to the correct weak solution.
- As (1.1) is a 4-by-4 system, a direct calculation of its eigenvalues can be hard. However, a first-order approximation in $u_2 - u_1$ was found in [2] to be

$$\lambda_{\text{int}}^{\pm} = U_c \pm \sqrt{g'\frac{h_1h_2}{h_1+h_2}\left(1 - \frac{(u_2-u_1)^2}{g'(h_1+h_2)}\right)},$$
$$\lambda_{\text{ext}}^{\pm} = U_m \pm \sqrt{g(h_1+h_2)}, \tag{1.4}$$

where $U_c := \frac{h_1u_2+h_2u_1}{h_1+h_2}$, $U_m := \frac{h_1u_1+h_2u_2}{h_1+h_2}$, $g' := g(1-r)$. From this we see that the system is only hyperbolic in the regime

$$\frac{(u_2-u_1)^2}{g'(h_1+h_2)} \le 1.$$

This loss of hyperbolicity can be linked with Kevin-Helmholtz instabilities — violent mixing that occurs when the relative difference in velocities between the two layers becomes too large.

From the above approximation of eigenvalues we also get a bound on the wave speeds of the system (1.1):

$$|\lambda| \le |U_m| + \sqrt{g(h_1+h_2)}.$$

This will be useful when determining CFL conditions and adding diffusion in numerical computations.

- It is easily seen that (1.1) has the steady state

$$u_1, u_2 \equiv 0, \quad b + h_1 \equiv \text{const}, \quad h_2 \equiv \text{const}. \tag{1.5}$$

This is the so-called *lake at rest* steady state. A major challenge in numerical schemes for shallow water models is the preservation of the lake at rest. Its importance is evident in the modelling of lakes and oceans, where the flows of interest are small perturbations of the lake at rest.

1.1 Numerical methods

We consider finite difference schemes to solve (1.1). Our spatial domain is partitioned into a uniform grid $\{x_j\}_j$ with $x_{j+1} - x_j \equiv \Delta x$, and we solve for the point values $u_j(t) \approx u(x_j, t)$. The general form of a finite difference scheme for (1.1) is then

$$\frac{d}{dt}\mathbf{u}_j + \frac{1}{\Delta x}\left(\mathbf{F}_{j+1/2} - \mathbf{F}_{j-1/2}\right) = \mathbf{S}_j, \tag{1.6}$$

where we suppress the t-dependence of u_j for notational convenience. Temporal integration is performed with the explicit Euler method for first-order schemes and strong stability preserving Runge Kutta method for second-order schemes.

Our main goal in this paper is to design a finite difference scheme that satisfies a discrete entropy inequality

$$\frac{d}{dt}\eta(\mathbf{u}_j) + \frac{1}{\Delta x}\left(Q_{j+1/2} - Q_{j-1/2}\right) \leq 0. \tag{1.7}$$

For some numerical entropy flux $Q_{j+1/2} = Q(\mathbf{u}_j, \mathbf{u}_{j+1})$ that is consistent with the entropy flux q. Such a scheme will be called *energy stable* after the work of Tadmor [6]. To this end, we first construct an *entropy conservative* scheme — one which satisfies

$$\frac{d}{dt}\eta(\mathbf{u}_j) + \frac{1}{\Delta x}\left(\widehat{Q}_{j+1/2} - \widehat{Q}_{j-1/2}\right) = 0, \tag{1.8}$$

and then add a numerical diffusion operator to obtain entropy stability. Define the *entropy variables* as $\mathbf{v}(\mathbf{u}) := \nabla\eta(\mathbf{u})$ and the *entropy potential* as $\psi(\mathbf{u}) := \mathbf{v}(\mathbf{u})^\top \mathbf{f}(\mathbf{u}) - q(\mathbf{u})$. For the two-layer shallow water equations, the entropy variables are

$$V = \begin{bmatrix} \rho_1\left(g(h_1 + rh_2 + b) - \dfrac{u_1^2}{2}\right) \\ \rho_2\left(g(h_1 + h_2 + b) - \dfrac{u_2^2}{2}\right) \\ \rho_1 u_1 \\ \rho_2 u_2 \end{bmatrix}. \tag{1.9}$$

It was shown in [6] that a scheme

$$\frac{d}{dt}\mathbf{u}_j + \frac{1}{\Delta x}\left(\mathbf{F}_{j+1/2} - \mathbf{F}_{j-1/2}\right) = 0 \tag{1.10}$$

is entropy conservative if its numerical flux $\mathbf{F}_{j+1/2}$ satisfies

$$[\![\mathbf{v}]\!]_{j+1/2}^{\top}\mathbf{F}_{j+1/2} = [\![\psi]\!]_{j+1/2},$$

where $[\![x]\!]_{j+1/2} := x_{j+1} - x_j$. Moreover, it was shown that if a numerical flux function $\mathbf{F}$ can be written as $\mathbf{F}_{j+1/2} = \mathbf{F}^*_{j+1/2} - D_{j+1/2}[\![\mathbf{v}]\!]_{j+1/2}$ for an entropy conservative flux $\mathbf{F}^*$ and a positive definite diffusion matrix $D_{j+1/2}$, then the resulting scheme is entropy *stable*. Using these results, the authors designed energy conservative and energy stable schemes for the shallow water equations in [4]. (The entropy used is the total energy of the solution, hence the word "energy".) A consistent discretization of the bottom topography source term in [5] led to energy conservative and energy stable, *well-balanced* methods for the shallow water equations with variable bottom topography.

2 The EEC scheme

It turns out that the generalization from the setting of [5] to the two-layer shallow water equations is more or less trivial. Without going into details we give the spatial discretization and refer to [4, 5] for further information:

$$\begin{aligned}
\mathbf{F}^*_{j+1/2} &= \begin{bmatrix} \overline{h_1}\,\overline{u_1} \\ \overline{h_2}\,\overline{u_2} \\ \frac{1}{2}g\overline{h_1^2} + \overline{h_1}\,\overline{u_1}^2 \\ \frac{1}{2}g\overline{h_2^2} + \overline{h_2}\,\overline{u_2}^2 \end{bmatrix}_{j+1/2}, \\
\mathbf{S}_j &= -\frac{g}{2\Delta x}\begin{bmatrix} 0 \\ 0 \\ \overline{(h_1)}_{j+1/2}[\![b + rh_2]\!]_{j+1/2} + \overline{(h_1)}_{j-1/2}[\![b + rh_2]\!]_{j-1/2} \\ \overline{(h_2)}_{j+1/2}[\![b + h_1]\!]_{j+1/2} + \overline{(h_2)}_{j-1/2}[\![b + h_1]\!]_{j-1/2} \end{bmatrix},
\end{aligned} \tag{2.1}$$

where we use the notation $\overline{x}_{j+1/2} := \frac{x_j + x_{j+1}}{2}$. We denote the scheme (1.6), (2.1) the *EEC* (Explicit Energy Conservative) scheme.

Theorem 2.1. *The EEC scheme is consistent, second-order accurate and energy conservative — solutions $\{\mathbf{u}_j\}_j$ satisfy the discrete entropy* *equality* (1.8) *with*

$$
\begin{aligned}
\widehat{Q}_{j+1/2} = \sum_{i=1}^{2} \rho_i \Bigg(& \frac{1}{2}\overline{(h_i)}_{j+1/2}\overline{(u_i)}_{j+1/2}(u_i)_j(u_i)_{j+1} \\
& + g\overline{(h_i)}_{j+1/2}\left(\frac{(u_i)_j(h_i+b)_{j+1}+(u_i)_{j+1}(h_i+b)_j}{2}\right)\Bigg) \\
& + \rho_2 g\Bigg(\overline{(h_1)}_{j+1/2}\left(\frac{(u_1)_j(h_2)_{j+1}+(u_1)_{j+1}(h_2)_j}{2}\right) \\
& + \overline{(h_2)}_{j+1/2}\left(\frac{(u_2)_j(h_1)_{j+1}+(u_2)_{j+1}(h_1)_j}{2}\right)\Bigg)
\end{aligned}
$$

(*compare to* (1.2)). *Furthermore, it is well-balanced — given initial data satisfying* (1.5), *the computed solution is constant in time.*

Proof. Consistency and order of accuracy is straight-forward to check. For entropy conservation, we take the inner product of (1.6) with $\mathbf{v}(\mathbf{u}_j)$ (see (1.9)) and obtain (1.8). Last, if the solution at some point of time satisfies the lake at rest conditions (1.5), then the two first components of the flux and source in (1.6) drop out immediately, and the third reduces to

$$
\begin{aligned}
&\overline{(h_1^2)}_{j+1/2} - \overline{(h_1^2)}_{j-1/2} + \overline{(h_1)}_{j+1/2}[\![b+rh_2]\!]_{j+1/2} + \overline{(h_1)}_{j-1/2}[\![b+rh_2]\!]_{j-1/2} \\
&= \frac{[\![h_1^2]\!]_{j+1/2}+[\![h_1^2]\!]_{j-1/2}}{2} + \overline{(h_1)}_{j+1/2}[\![b+rh_2]\!]_{j+1/2}+\overline{(h_1)}_{j-1/2}[\![b+rh_2]\!]_{j-1/2} \\
&= \overline{(h_1)}_{j+1/2}[\![b+h_1+rh_2]\!]_{j+1/2} + \overline{(h_1)}_{j-1/2}[\![b+h_1+rh_2]\!]_{j-1/2} \\
&= 0
\end{aligned}
$$

by (1.5). An analogous argument holds for the fourth component. Thus, we end up with $\frac{d}{dt}\mathbf{u}_j = 0$, whence $\mathbf{u}_j(t) \equiv \text{constant}$ for all j. □

2.1 Numerical experiments

We test the EEC scheme on a problem taken from [2]. The bottom topography is set to be

$$
b(x) = \begin{cases} (\cos(10\pi(x-0.5))+1)/5 & \text{if } \ |x-0.5| \leq 0.1 \\ 0 & \text{otherwise.} \end{cases}
$$

The initial data is the lake at rest data $u_1, u_2 \equiv 0$, $h_1 + b(x) \equiv 0.5$, $h_2 \equiv 0.1$. We discretize the domain $[0,1]$ into 100 grid cells and compute

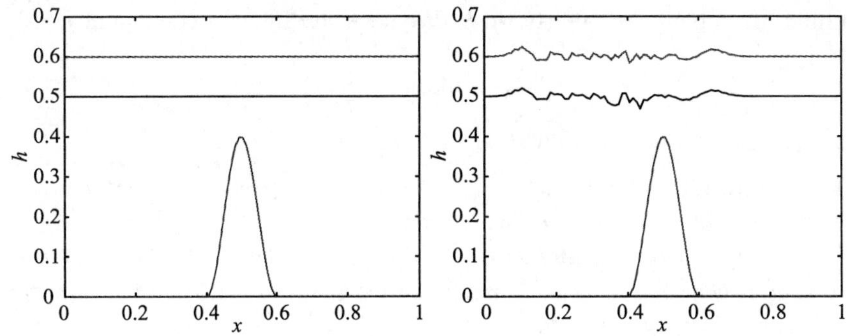

(a) Lake at rest

(b) Lake at rest with perturbation

Figure 2.1 Lake at rest computed with the EEC scheme.

up to $t = 1$. We set $g = 1$, $\rho_1 = 1$, and $\rho_2 = 0.9$ in this and the remaining experiments.

As can be seen in Figure 2.1(a), there is no unphysical generation of spurious waves, and hence the initial lake at rest data is preserved.

Figure 2.1(b) shows the same experiment at $t = 0.4$, but with a small initial perturbation of $+0.05$ in h_2 in the range $x \in [0.38, 0.42]$. The two resulting shock waves are resolved, but in-between there are unphysical oscillations that are due to a lack of numerical diffusion. To fix this we will add numerical diffusion to obtain an *energy stable* scheme.

3 The ERus scheme

To obtain energy decay near discontinuities we add a Rusanov-type numerical diffusion operator of the form $c_{j+1/2}[\![\mathbf{u}]\!]_{j+1/2}$ to the numerical flux, where $c_{j+1/2}$ is some approximation of the largest eigenvalues of the system. Assume for the moment that the bottom topography is flat, $b \equiv 0$. As the energy η is strictly convex, the mapping $\mathbf{u} \mapsto \mathbf{v}(\mathbf{u})$ is injective, and so the change-of-variables matrix $\partial_{\mathbf{v}}\mathbf{u}(\mathbf{v})$ is well-defined and positive definite. By the mean value theorem there is a state $\mathbf{v}_{j+1/2}$ such that

$$[\![\mathbf{u}]\!]_{j+1/2} = \partial_{\mathbf{v}}\mathbf{u}(\mathbf{v}_{j+1/2})[\![\mathbf{v}]\!]_{j+1/2}. \tag{3.1}$$

This is the form of our numerical diffusion operator, and we define the *ERus* (Energy stable Rusanov) flux to be

$$\mathbf{F}_{j+1/2} = \mathbf{F}^*_{j+1/2} - \frac{1}{2}c_{j+1/2}\partial_{\mathbf{v}}\mathbf{u}(\mathbf{v}_{j+1/2})[\![\mathbf{v}]\!]_{j+1/2}, \tag{3.2}$$

where $\mathbf{F}^*$ is the EEC flux (2.1). In the more general case $b \neq 0$ we still use the above expression, although the identity (3.1) no longer holds since $\mathbf{v}$, unlike $\mathbf{u}$, depends explicitly on b. We set the diffusion coefficient $c_{j+1/2}$ to be

$$c_{j+1/2} = \max(c_j, c_{j+1}),$$

where $c_j := |(U_m)_j| + \sqrt{g((h_1)_j + (h_2)_j)}$ (see (1.4)).

Theorem 3.1. *The ERus scheme is consistent, first-order accurate and energy stable — solutions* $\{\mathbf{u}_j\}_j$ *satisfy* (1.7) *with* $Q_{j+1/2} = \widehat{Q}_{j+1/2} - \frac{1}{2}\overline{\mathbf{v}}_{j+1/2}^\top D_{j+1/2}[\![\mathbf{v}]\!]_{j+1/2}$ *and* $D_{j+1/2} := \frac{1}{2}c_{j+1/2}\partial_{\mathbf{v}}\mathbf{u}(\mathbf{v}_{j+1/2})$. *Furthermore, it is well-balanced — given initial data satisfying* (1.5), *the computed solution is constant in time.*

Proof. First-order accuracy comes from the $\mathcal{O}(\Delta x)$ term $[\![\mathbf{v}]\!]_{j+1/2}$. The proof of the discrete entropy inequality follows [6, Theorem 5.2]: Multiply the ERus scheme (1.6) by $\mathbf{v}_j^\top$, use the entropy conservativity of the EEC flux $\mathbf{F}^*$ and rearrange the diffusion terms to obtain (1.7).

For well-balancedness, we note that when the solution satisfies (1.5), the entropy variables (1.9) are constant in space. Hence, the ERus flux reduces to the EEC flux, which we have already shown is well-balanced. □

3.1 Numerical experiments

We repeat the numerical experiments from the previous section with the ERus scheme. The lake at rest is preserved exactly; even after long-time simulations the relative error stays within 10^{-16}. Figure 3.1(a) shows the perturbed lake at rest experiment. Clearly, the unphysical oscillations of the EEC scheme are gone.

Figure 3.2 shows computed h_1 and u_1 on a problem from [2]. The results compare well to those of [2].

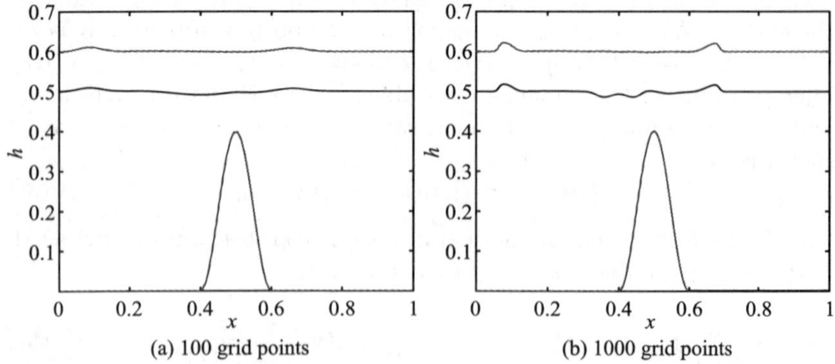

Figure 3.1 Perturbed lake at rest computed with the ERus scheme.

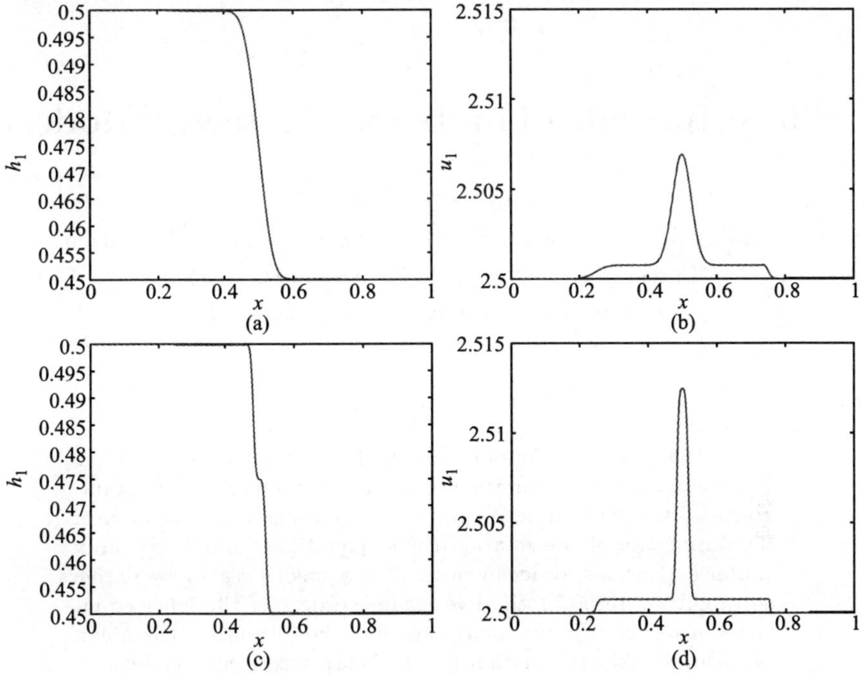

Figure 3.2 Problem from [2]. h_1 to the left, u_1 to the right. (a) and (b) with 500 grid points, (c) and (d) with 10000.

References

[1] R. Abgrall and S. Karni, A comment on the computation of non-conservative products. Journal of Computational Physics 229 (2010), 2759–2763.

[2] R. Abgrall and S. Karni, Two-layer shallow water system: A relaxation approach. SIAM Journal on Scientific Computing 31(3) (2009), 1603–1627.

[3] G. DalMaso, P. G. LeFloch and F. Murat, Definition and weak stability of nonconservative products. Journal de Mathematiques Pures et Appliquees 74(6) (1995), 483–548.

[4] U. S. Fjordholm, S. Mishra and E. Tadmor, Energy preserving and energy stable schemes for the shallow water equations. In: "Foundations of Computational Mathematics", Proc. FoCM held in Hong Kong 2008, volume 363 of London Math. Soc. Lecture Notes Series (2009), 93–139.

[5] U. S. Fjordholm, S. Mishra and E. Tadmor, Well-balanced, energy stable schemes for the shallow water equations with varying topography. Submitted 2010.

[6] E. Tadmor, The numerical viscosity of entropy stable schemes for systems of conservation laws. I. Mathematics of Computation 49 (1987), 91–103.

Sharp Interface Limits for Korteweg Models

Jan Giesselmann*
Institut für Angewandte Analysis und Numerische Simulation
Universität Stuttgart, 70569 Stuttgart, Germany
E-mail: giesselmann@ians.uni-stuttgart.de

Abstract

In this paper we consider the Korteweg regularization of the Euler equations. We perform a limit for the regularization parameter $\varepsilon \to 0$ by matching asymptotic expansions in the vicinity of discontinuities of the solution (inner equations) and in the bulk (outer expansions) order by order. For a special scaling we derive solvability criteria for the two leading orders of the inner equations in terms of jump conditions over the interface. These can be seen as a kinetic relation for the sharp interface model corresponding to the $\varepsilon = 0$ limit. Kinetic relations are a well-known concept, which is used to overcome the non-uniqueness of entropy solutions in the dynamics of phase boundaries.

1 Introduction

We consider the d-dimensional isothermal Euler equations

$$\begin{aligned} \rho_t + \nabla \cdot (\rho \mathbf{u}) &= 0, \\ (\rho \mathbf{u})_t + \nabla \cdot (\rho \mathbf{u} \otimes \mathbf{u}) + \nabla p(\rho) &= 0, \end{aligned} \quad \text{in } \mathbb{R}^d \times (0, \infty), \tag{1.1}$$

for density $\rho(x,t) > 0$ and velocity $\mathbf{u}(x,t) \in \mathbb{R}^d$. These equations are often used as a model for liquid vapor phase transitions, where one uses a non-monotone van-der-Waals pressure function $p(\rho)$. The phases correspond to the monotonely increasing branches of the pressure. For a non-monotone pressure function the Rankine-Hugoniot jump conditions and the entropy condition do not guarantee uniqueness of solutions. Also using Young-Laplace like equations taking into account surface tension instead of the Rankine-Hugoniot condition for the conservation of momentum does not yield uniqueness of solutions. To solve these problems

*The author would like to thank the German Research Foundation (DFG) for financial support of the project within the Cluster of Excellence in Simulation Technology (EXC 310/1) at the University of Stuttgart.

the notion of kinetic relation, which specifies one of the entropy solutions, was introduced in [1] and [14]. For example, imposing that there is no entropy dissipation at the interface guarantees local stability of solutions as was proven in [3]. For Riemann problems in one space dimension general kinetic relations were studied in [12].

Kinetic relations and hence uniqueness of solutions can also be obtained by examining limits of regularizations of the Euler equations. We will address the Navier-Stokes-Korteweg (NSK) regularization which means to add

$$\nabla \cdot \boldsymbol{\sigma}_{NS} + \nabla \cdot \boldsymbol{\sigma}_K \tag{1.2}$$

to the right hand side of $(1.1)_2$. Here

$$\boldsymbol{\sigma}_{NS} := \varepsilon^2 \lambda (\nabla \cdot \mathbf{u}) I + \varepsilon^2 \mu \left((\nabla \mathbf{u}) + (\nabla \mathbf{u})^T \right), \tag{1.3}$$

$$\boldsymbol{\sigma}_K := \varepsilon^2 \gamma \left(\left(\rho \Delta \rho + \frac{1}{2} |\nabla \rho|^2 \right) I - \frac{1}{2} \nabla \rho \otimes \nabla \rho \right) \tag{1.4}$$

are the Navier-Stokes and Korteweg tensor respectively, where λ, μ and γ are constants satisfying $\mu, \gamma > 0,\ \lambda + 2\mu \geq 0$ and ε is a small positive parameter, proportional to the width of the interfacial layer between both phases. We will call (1.1) with the regularization (1.2) the Navier-Stokes-Korteweg (NSK) equations.

In modern terminology such models were first introduced by Dunn and Serrin [9]. The existence of solutions for initial value problems was investigated in, e.g., [4] and [10]. Starting from the results on travelling wave solutions in [2], we study sharp interface limits of solutions of the (NSK) equations corresponding to $\varepsilon \to 0$.

Using matched asymptotic expansions we prove jump conditions including a kinetic relation at discontinuities of the solution. These contain classical Young-Laplace like terms for the surface tension, which are of the same order as new terms accounting for mass, momentum and energy attributed to the interfacial layer. Their main feature is the involvement of differential operators on the surface of discontinuity. The new contribution of this paper is the fact that we determine jump conditions up to first order in the dynamical case and that surface tension and entropy dissipation at the interface are of the same order. Usually sharp interface limits yield surface tension of order $\mathcal{O}(\varepsilon)$ and entropy dissipation of order $\mathcal{O}(1)$, this scaling has to be reversed for the stability result of [3]. In the case of equilibria our results agree with the results obtained in [8], [11] and [13]. Furthermore, these jump conditions imply a suitable sign for the entropy dissipation at the interface, whence they are in agreement with the second law of thermodynamics. Detailed proofs of the results stated here can be found in [7].

This paper is organized as follows: We state the general form of jump conditions in sharp interface models in Section 2. In Section 3 we introduce the coordinate transformation near the interface and the matching conditions. These imply the main result stated in Section 4. A sketch of its proof is given in Section 5.

2 General sharp interface models

For general isothermal sharp interface models without bulk viscosity the conservation laws of mass and momentum read in the bulk phases $\Omega^{\pm}$, where $\Omega^{+}(\Omega^{-})$ is the vapor (liquid) phase:

$$\rho_t + \nabla \cdot (\rho \mathbf{u}) = 0, \tag{2.1}$$

$$(\rho \mathbf{u})_t + \nabla \cdot (\rho \mathbf{u} \otimes \mathbf{u}) + \nabla p(\rho) = 0. \tag{2.2}$$

To allow for phase changes we have to use a non-monotone pressure law $p = p(\rho)$, which is related to a double-well free energy density function $W = W(\rho)$ by

$$p(\rho) = \rho g(\rho) - W(\rho), \quad \text{where } g(\rho) = W'(\rho). \tag{2.3}$$

Across the interface Γ between two adjacent phases $\Omega^{\pm}$ we have the corresponding jump conditions, constituting the conservation of mass and momentum across Γ. To state them we need the following notation for some quantity ψ having a jump at the interface. By $\psi^{\pm}$ we denote the limit at the interface from $\Omega^{\pm}$ and

$$[\![\psi]\!] := \psi^{+} - \psi^{-}, \qquad \{\psi\} := \frac{\psi^{+} + \psi^{-}}{2}.$$

Furthermore, each surface point is equipped with a tangential matrix $\mathbf{t}$ and a normal vector $\boldsymbol{\nu}$ and κ denotes the sum of the principal curvatures. The interfacial velocity $\mathbf{w}$ may be decomposed into normal- and tangential speed, i.e., $\mathbf{w} = w_t \mathbf{t} + w_\nu \boldsymbol{\nu}$. By $\rho_\Gamma, \boldsymbol{\sigma}_\Gamma, \gamma_\Gamma$ and W_Γ we denote the interfacial mass density, the surface stress vector, the surface tension and the interfacial free energy density respectively. Then the general forms of jump conditions derived in [6] read:

$$\begin{aligned} [\![\rho(u_\nu - w_\nu)]\!] &= -\frac{\partial \rho_\Gamma}{\partial t} - (\mathrm{div}_\Gamma(w_t) - \kappa w_\nu)\rho_\Gamma, \\ [\![\rho(u_\nu - w_\nu)(\mathbf{u} - \mathbf{w}) + p(\rho)\boldsymbol{\nu}]\!] &= -\rho_\Gamma \frac{\partial \mathbf{w}}{\partial t} + \mathrm{div}_\Gamma(\boldsymbol{\sigma}_\Gamma). \end{aligned} \tag{2.4}$$

Additionally div_Γ denotes the surface divergence. When Γ is one dimensional, it is defined by

$$\mathrm{div}_\Gamma(\psi) := \frac{1}{|\mathbf{t}|}\frac{\partial(|\mathbf{t}|\psi)}{\partial s}, \qquad \text{where } s \text{ parameterizes the interface.}$$

We are interested in the special case $\boldsymbol{\sigma}_\Gamma = \gamma_\Gamma \frac{\mathbf{t}}{|\mathbf{t}|^2}$. When additionally γ_Γ is constant $\mathrm{div}_\Gamma(\boldsymbol{\sigma}_\Gamma)$ becomes the Young-Laplace term $\gamma_\Gamma \kappa \boldsymbol{\nu}$.

Due to the second law of thermodynamics solutions of (2.1), (2.2) must satisfy the entropy inequality across the interface

$$\frac{\partial W_\Gamma}{\partial t} - (\gamma_\Gamma - W_\Gamma)(\mathrm{div}_\Gamma(w_t) - \kappa w_\nu) + \{j\} \left[\!\!\left[W'(\rho) + \frac{1}{2}|\mathbf{u} - \mathbf{w}|^2 \right]\!\!\right] + [\![j]\!] \left\{ W'(\rho) + \frac{1}{2}|\mathbf{u} - \mathbf{w}|^2 \right\} \leq 0. \quad (2.5)$$

Here $j := \rho(u_\nu - w_\nu)$ is the mass flux over the interface. Our aim is to recover (2.4) and (2.5) and to determine the exact form of $\rho_\Gamma, \boldsymbol{\sigma}_\Gamma$ and W_Γ in the sharp interface model arising as the limit of the (NSK) regularization, see (4.1), (4.2), and (4.4). Furthermore, we want to find an additional jump condition supplementing (2.4) to yield a kinetic relation, see (4.3). Finally we will show in Theorem 4.2 that these jump conditions are in agreement with the entropy inequality at the interface (2.5).

We assume the existence of outer expansions

$$\mathbf{u}_\varepsilon(t, x^1, x^2) = \sum_{i=0}^{\infty} \varepsilon^i \mathbf{u}_i(t, x^1, x^2), \quad \rho_\varepsilon(t, x^1, x^2) = \sum_{i=0}^{\infty} \varepsilon^i \rho_i(t, x^1, x^2).$$

Inserting them into the (NSK) equations and comparing coefficients of different powers of ε we obtain the outer equations:

$$\begin{aligned}
(\rho_0)_t + \nabla \cdot (\rho_0 \mathbf{u}_0) &= 0, \\
(\rho_0 \mathbf{u}_0)_t + \nabla \cdot (\rho_0 \mathbf{u}_0 \otimes \mathbf{u}_0) + \nabla p(\rho_0) &= 0, \\
(\rho_1)_t + \nabla \cdot (\rho_1 \mathbf{u}_0 + \rho_0 \mathbf{u}_1) &= 0, \\
(\rho_0 \mathbf{u}_1 + \rho_1 \mathbf{u}_0)_t + \nabla \cdot (\rho_1 \mathbf{u}_0 \otimes \mathbf{u}_0 + \rho_0 \mathbf{u}_1 \otimes \mathbf{u}_0 + \rho_0 \mathbf{u}_0 \otimes \mathbf{u}_1)& \\
+ \nabla(p'(\rho_0)\rho_1) &= 0.
\end{aligned}$$

3 Inner equations and matching

From now on we restrict ourselves to the two dimensional case. Let $\rho_\varepsilon, \mathbf{u}_\varepsilon$ be a local in time classical solution of the (NSK) equations. For $t \in [0, T)$ we define the interface and the bulk phases by

$$\begin{aligned}
\Omega_\varepsilon^-(t) &:= \{(x^1, x^2) \in \mathbb{R}^2 : \rho_\varepsilon(t, x^1, x^2) < \rho_*\}, \\
\Omega_\varepsilon^+(t) &:= \{(x^1, x^2) \in \mathbb{R}^2 : \rho_\varepsilon(t, x^1, x^2) > \rho_*\}, \\
\Gamma_\varepsilon(t) &:= \{(x^1, x^2) \in \mathbb{R}^2 : \rho_\varepsilon(t, x^1, x^2) = \rho_*\},
\end{aligned}$$

where ρ_* belongs to the decreasing part of p. Let $\mathbf{r}_\varepsilon(t)$ be a parameterization of Γ_ε, then points in a vicinity of the interface can be represented by

$$\begin{pmatrix} x \\ y \end{pmatrix}(t,s,z) = \mathbf{r}_\varepsilon(t,s) + \varepsilon z \boldsymbol{\nu}_\varepsilon(t,s), \tag{3.1}$$

where $\boldsymbol{\nu}_\varepsilon$ is the unit normal pointing into $\Omega_\varepsilon^-(t)$, thus, z denotes the distance from the interface in normal direction scaled by the small parameter ε.

Using this representation (3.1) we change variables from $(t, x^1, x^2) \leftrightarrow (\tau, s, z)$, where $t = \tau$, i.e., we just rename the time variable to stress in which coordinate system partial derivatives are calculated.

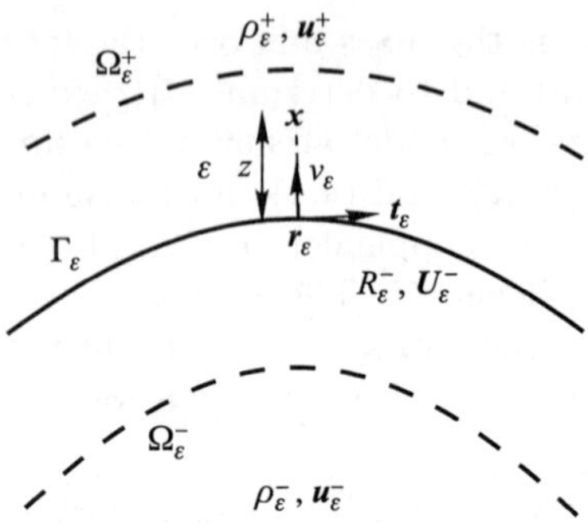

Figure 3.1 Boundary layer region.

When a scalar or a Cartesian component of a vector ψ is defined in inner and outer coordinates, i.e., $\psi(t, x^1, x^2) = \Psi(\tau, s, z)$, the partial derivatives transform as follows:

$$\begin{pmatrix} \dfrac{\partial \psi}{\partial x^1} \\ \dfrac{\partial \psi}{\partial x^2} \\ \dfrac{\partial \psi}{\partial t} \end{pmatrix} = \begin{pmatrix} (1+\varepsilon z \kappa_\varepsilon)\dfrac{1}{|\mathbf{t}_\varepsilon|^2} t_\varepsilon^1 & \varepsilon^{-1}\nu_\varepsilon^1 & 0 \\ (1+\varepsilon z \kappa_\varepsilon)\dfrac{1}{|\mathbf{t}_\varepsilon|^2} t_\varepsilon^2 & \varepsilon^{-1}\nu_\varepsilon^2 & 0 \\ -(1+\varepsilon z \kappa_\varepsilon)\left((w_t)_\varepsilon - \varepsilon z \dfrac{t_\varepsilon^i}{|\mathbf{t}_\varepsilon|^2}(\nu_\varepsilon^i)_\tau\right) & -\varepsilon^{-1}(w_\nu)_\varepsilon & 1 \end{pmatrix} \begin{pmatrix} \dfrac{\partial \Psi}{\partial s} \\ \dfrac{\partial \Psi}{\partial z} \\ \dfrac{\partial \Psi}{\partial \tau} \end{pmatrix},$$

where $\mathbf{t}_\varepsilon, \kappa_\varepsilon$ denote the tangent vector and the mean curvature respectively, $(w_t)_\varepsilon := w_\varepsilon^i \frac{t_\varepsilon^i}{|\mathbf{t}_\varepsilon|^2}$, $(w_\nu)_\varepsilon := w_\varepsilon^i \nu_\varepsilon^i$ and we neglect terms of order $\mathcal{O}(\varepsilon^2)$. We assume that there are expansions of $\mathbf{r}$:

$$\mathbf{r}_\varepsilon(\tau, s) = \sum_{i=0}^{\infty} \varepsilon^i \mathbf{r}_i(\tau, s), \tag{3.2}$$

which implies expansions of $\mathbf{t}_\varepsilon, \boldsymbol{\nu}_\varepsilon, \kappa_\varepsilon$, and of the density and velocity in inner coordinates:

$$\mathbf{U}_\varepsilon(\tau, s, z) = \sum_{i=0}^{\infty} \varepsilon^i \mathbf{U}_i(\tau, s, z), \quad R_\varepsilon(\tau, s, z) = \sum_{i=0}^{\infty} \varepsilon^i R_i(\tau, s, z). \tag{3.3}$$

Using (3.1) and (3.3) we get the inner equations by comparing coefficients of different powers of ε.

Inner- and outer quantities are matched by the usual procedure, see [5] for details. The upshot are the following matching conditions, which we assume to be attained exponentially fast:

$$\begin{aligned}
&\Psi_0(\tau,s,z) \overset{z\to\pm\infty}{\to} \psi_0^\pm(\tau,\mathbf{r}_0(\tau,s)),\\
&\Psi_1(\tau,s,z) \overset{z\to\pm\infty}{\to} \psi_1^\pm(\tau,\mathbf{r}_0(\tau,s)) + \left(\frac{\partial\psi_0}{\partial x^j}\right)^\pm(\tau,\mathbf{r}_0(\tau,s))(\nu_0^j(\tau,s)z + r_1^j(\tau,s)),\\
&\qquad\Psi_{0,z} \overset{z\to\pm\infty}{\to} 0,\\
&\qquad\Psi_{1,z} \overset{z\to\pm\infty}{\to} \left(\frac{\partial\psi_0}{\partial x^j}\right)^\pm(\tau,\mathbf{r}_0(\tau,s))\nu_0^j(\tau,s).
\end{aligned} \tag{3.4}$$

Here and in the sequel we sum over indices occurring twice. Defining

$$\psi_0^\pm(s,\tau) := \lim_{\pm z>0, z\to 0} \psi_0(\tau,\mathbf{r}_0(s,\tau) + z\boldsymbol{\nu}_0(s,\tau)), \tag{3.5}$$

we get the additional matching conditions:

$$\Psi_{0,\tau} \overset{z\to\pm\infty}{\to} (\psi_0^\pm)_\tau \quad \text{and} \quad \Psi_{0,s} \overset{z\to\pm\infty}{\to} (\psi_0^\pm)_s. \tag{3.6}$$

4 Main results

Theorem 4.1. *Let $\rho_0, \rho_1, \mathbf{u}_0, \mathbf{u}_1$ be solutions of the outer equations satisfying $\rho_0^-(u_{\nu 0}^- - w_{\nu 0}) \neq 0$. Let R_0, R_1, $\mathbf{U}_0, \mathbf{U}_1$ satisfy the inner equations and let them be linked by the matching conditions, then the following jump conditions at the phase boundary are fulfilled:*

$$[\![\rho_\varepsilon((u_\nu)_\varepsilon - (w_\nu)_\varepsilon)]\!] = -\frac{\partial\rho_\Gamma}{\partial\tau} - (\mathrm{div}_\Gamma(w_{t0}) - \kappa_0 w_{\nu 0})\rho_\Gamma, \tag{4.1}$$

$$[\![j_\varepsilon(\mathbf{u}_\varepsilon - \mathbf{w}_\varepsilon) + p(\rho_\varepsilon)\boldsymbol{\nu}_\varepsilon]\!] = -\rho_\Gamma\frac{\partial\mathbf{w_0}}{\partial\tau} + \mathrm{div}_\Gamma(\boldsymbol{\sigma}_\Gamma), \tag{4.2}$$

$$\begin{aligned}
\left[\!\!\left[\frac{1}{2}\left(\frac{j_\varepsilon}{\rho_\varepsilon}\right)^2 + g(\rho_\varepsilon)\right]\!\!\right] =& -\varepsilon\int_0^\infty \nu_0^j\left(U_0^j - (u_0^j)^+\right)_\tau dz\\
&-\varepsilon\int_{-\infty}^0 \nu_0^j\left(U_0^j - (u_0^j)^-\right)_\tau dz\\
&-\varepsilon(\lambda+2\mu)j_0\int_{-\infty}^\infty\left(\left(\frac{1}{R_0}\right)_z\right)^2 dz + \mathcal{O}(\varepsilon^2),
\end{aligned} \tag{4.3}$$

where

$$\rho_\Gamma := \varepsilon\int_0^\infty (R_0 - \rho_0^+)\,dz - \varepsilon\int_{-\infty}^0 (R_0 - \rho_0^-)\,dz + \mathcal{O}(\varepsilon^2),$$

$$\sigma_\Gamma^j := \varepsilon \int_0^\infty \left(\frac{t_0^j}{|\mathbf{t}_0|^2} \left(\frac{j_0^2}{R_0} - \frac{j_0^2}{\rho_0^+} + \gamma R_{0,z}^2 \right) \right) dz + \varepsilon \int_{-\infty}^0 \cdots dz + \mathcal{O}(\varepsilon^2).$$

Theorem 4.2. *Defining*

$$W_\Gamma := \varepsilon \int_0^\infty \left(W(R_0) - W(\rho_0^+) + \frac{1}{2}\frac{j_0^2}{R_0} - \frac{1}{2}\frac{j_0^2}{\rho_0^+} + \frac{\gamma}{2} R_{0,z}^2 \right) dz$$
$$+\varepsilon \int_{-\infty}^0 \left(W(R_0) - W(\rho_0^-) + \frac{1}{2}\frac{j_0^2}{R_0} - \frac{1}{2}\frac{j_0^2}{\rho_0^-} + \frac{\gamma}{2} R_{0,z}^2 \right) dz$$

and $\sigma_\Gamma^j =: \gamma_\Gamma \frac{t_0^j}{|\mathbf{t}_0|^2}$, *the jump conditions* (4.1)–(4.3) *imply*

$$\frac{\partial W_\Gamma}{\partial \tau} - (\gamma_\Gamma - W_\Gamma)\left(\operatorname{div}_\Gamma(w_{t0}) - \kappa_0 w_{\nu 0}\right)$$
$$+ [\![j_\varepsilon]\!] \left\{ W'(\rho_\varepsilon) + \frac{|\mathbf{u}_\varepsilon - \mathbf{w}_\varepsilon|^2}{2} \right\} + \{ j_\varepsilon \} \left[\!\!\left[W'(\rho_\varepsilon) + \frac{|\mathbf{u}_\varepsilon - \mathbf{w}_\varepsilon|^2}{2} \right]\!\!\right]$$
$$= -\varepsilon(\lambda + 2\mu) j_0^2 \int_{-\infty}^\infty \left(\left(\frac{1}{R_0} \right)_z \right)^2 dz + \mathcal{O}(\varepsilon^2) \leq 0, \tag{4.4}$$

thus, for ε *small enough, they are compatible with* (2.5) *and hence with the second law of thermodynamics.*

5 Sketch of the proof

We derive the jump conditions by determining solvability criteria for the inner equations in each order. The leading order of the inner equations is an ODE system for $R_0, \mathbf{U}_0$,

$$-w_{\nu 0} R_{0,z} + (U_{\nu 0} R_0)_z = 0, \tag{5.1}$$
$$-w_{\nu 0}(R_0 \mathbf{U}_0)_z + (R_0 U_{\nu 0} \mathbf{U}_0)_z + \boldsymbol{\nu}_0 p(R_0)_z = \boldsymbol{\nu}_0 \gamma R_0 R_{0,zzz}. \tag{5.2}$$

These are the equations satisfied by travelling wave solutions of the Euler equations regularized with the Korteweg term only. We see from (5.1) that $J_0 = R_0(U_{\nu 0} - w_{\nu 0})$ is constant and for $j_0 \neq 0$ we find $(U_{t0})_z = 0$. Thus, we can replace $\mathbf{U}_0$ in (5.2) and solve it for R_0. It was proven in [2], that there exist values $\rho_0^\pm$ satisfying

$$\begin{aligned} \frac{j_0^2}{\rho_0^-} + p(\rho_0^-) &= \frac{j_0^2}{\rho_0^+} + p(\rho_0^+), \\ \frac{1}{2}\left(\frac{j_0}{\rho_0^-}\right)^2 + g(\rho_0^-) &= \frac{1}{2}\left(\frac{j_0}{\rho_0^+}\right)^2 + g(\rho_0^+), \end{aligned} \tag{5.3}$$

provided j_0 is small enough, and that, when there are such values $\rho_0^\pm$, there exists a solution R_0 of (5.2) satisfying $R_0 \to \rho_0^\pm$ for $z \to \pm\infty$.

The next order of the inner equations forms a linear ODE system with boundary conditions given by (3.4) for R_1 and $\mathbf{U}_1$. By the Fredholm alternative theorem we get solvability conditions. They correspond to the $\mathcal{O}(\varepsilon)$ order of (4.1)–(4.3).

References

[1] R. Abeyaratne and J. K. Knowles, Kinetic relations and the propagation of phase boundaries in solids. Arch. Rational Mech. Anal. 114 (1991), no. 2, 119–154.

[2] S. Benzoni-Gavage, R. Danchin, S. Descombes and D. Jamet, Stability issues in the Euler-Korteweg model. Contemp. Math. 426 (2007), 103–127.

[3] S. Benzoni-Gavage and H. Freistühler, Effects of surface tension on the stability of dynamical liquid-vapour interfaces. Arch. Ration. Mech. Anal. 174 (2004), 111–150.

[4] D. Bresch, B. Desjardins and C. K. Lin, On some compressible fluid models: Korteweg, lubrication, and shallow water systems. Comm. Partial Differential Equations 28 (2003), 843–868.

[5] G. Caginalp and P. C. Fife, Dynamics of layered interfaces arising from phase boundaries. SIAM J. Appl. Math. 48 (1988), 506–518.

[6] W. Dreyer, On jump conditions at phase boundaries for ordered and disordered phases. WIAS preprint 869 (2003), http://www.wias-berlin.de/publications/wias-publ/run.jsp?template=abstract&type=Preprint&year=2003&number=869.

[7] W. Dreyer, J. Giesselmann, C. Kraus and C. Rohde, Asymptotic Analysis for Korteweg Models. In preparation.

[8] W. Dreyer and C. Kraus, The sharp interface limit of the van der Waals-Cahn-Hilliard phase model for fixed and time dependent domains. Proc. R. Soc. A, In press.

[9] J. E. Dunn and J. Serrin, On the thermomechanics of interstitial working. Arch. Rational Mech. Anal. 88 (1985), 95–133.

[10] H. Hattori and D. Li, The existence of global solutions to a fluid dynamic model for materials for Korteweg type, J. Partial Differential Equations 9 (1996), no. 4, 323–342.

[11] S. Luckhaus and L. Modica, The Gibbs-Thompson relation within the gradient theory of phase transitions. Arch. Rational Mech. Anal. 107 (1989), no. 1, 71–83.

[12] P. G. LeFloch and M. D. Thanh, Nonclassical Riemann solvers and kinetic relations. III. A nonconvex hyperbolic model for van der Waals fluids, Electron. J. Diff. Eqns. 72 (2000), 1–19.

[13] L. Modica, The gradient theory of phase transitions and the minimal interface criterion. Arch. Ration. Mech. Anal. 98 (1987), 123–142.

[14] L. Truskinovsky, Kinks versus shocks. In: Shock induced transitions and phase structures in general media, IMA Vol. Math. Appl. 52 (1993), 185–229.

On the Convergence of Semi-discrete High Resolution Schemes with Superbee Flux Limiter for Conservation Laws

Nan Jiang
Department of Mathematical Sciences, University of South Dakota
Vermillion SD 57069, USA
E-mail: njiang@usd.edu

Abstract

A class of high resolution schemes, using flux limiters for hyperbolic conservation laws, was introduced by Sweby [7] in 1980s. In the semi-discrete case, for the convex conservation laws with or without a source term, Yang and Jiang [2, 10] have shown the entropy consistence of the schemes of this class based on minmod limiter when the building block of the schemes is an arbitrary E-scheme, and based on Chakravarthy-Osher or van Leer's limiter when the building block of the schemes is the Godunov, the Engquist-Osher, or the Lax-Friedrichs. However, the convergence problems related to many other flux limiters, such as Roe's superbee, have been open. In this paper, we use the convergence criteria, established in [10], to show the convergence of the schemes with superbee limiter.

1 Introduction

In this paper, we show the entropy convergence of the semi-discrete high resolution schemes based on Roe's superbee flux limiter [6] for the initial value problems of hyperbolic conservation laws:

$$\begin{cases} w_t + f(w)_x = q(w), \\ w(x,0) = w_0(x), \end{cases} \tag{1.1}$$

where $f \in C^1(\mathbb{R})$ is convex, $q \in C^1(\mathbb{R})$, and $w_0 \in BV(\mathbb{R})$. Here BV stands for the subspace of L^1_{loc} consisting of functions with bounded total variation.

The semi-discrete high resolution schemes that approximate (1.1) have the form

$$\frac{du_j(t)}{dt} = -\frac{1}{h}(g_{j+\frac{1}{2}}(t) - g_{j-\frac{1}{2}}(t)) + q(u_j(t)), \tag{1.2}$$

where the numerical flux g is given by

$$g_{j+\frac{1}{2}}(t) = g(u_{j-p+1}(t), u_{j-p+2}(t), \cdots, u_j(t), \cdots, u_{j+p}(t)), \tag{1.3}$$

which is Lipschitz continuous with respect to its $2p$ arguments and is *consistent* with the conservation law in the sense that

$$g(u, u, \cdots, u) \equiv f(u). \tag{1.4}$$

A scheme (1.2)–(1.4) for the Cauchy problem (1.1) *converges* if, for every initial condition w_0 in BV and for each sequence of initial data $\{u_j^k(0), j \in \mathbb{Z}\}_{k=1}^{\infty}$ with uniformly bounded variations that converges in $L^1_{loc}(\mathbb{R})$ to w_0, the corresponding sequence of (extended) numerical solutions $\{u^k\}$ of the scheme converges in $L^1_{loc}(\mathbb{R} \times [0, T))$ to the unique entropy solution w of the problem (1.1) provided that the step sizes h_k of u^k vanish as $k \to \infty$.

The fundamental goal of the flux limiter method is to achieve high resolution solutions without the spurious oscillations that normally associated with the solutions computed by the classical second order schemes. Generally speaking, the construction of a scheme of this type involves hybridizing a high order scheme with an entropy satisfied lower order one. The methods of this form were proposed by van Leer and Harten-Zwas [9, 4] in the earlier time and later by Roe and Chakravarthy-Osher [6, 1]. In 1984, Sweby introduced a reasonably large class of flux limiter methods by means of adding a limited anti-diffusive flux to an entropy-satisfying first order scheme (whch is called the building block of the methods). Let $\psi(r)$ be a flux limiter in the Sweby's setting. His study has concluded that if ψ is Lipschitz continuous and its graph lies in the region D defined by

$$\begin{aligned} D := \{&(r, \psi_\Phi(r)) : \psi_\Phi(r) \\ &= \max(0, \min(\Phi r, 1), \min(r, \Phi)), 1 \leq \Phi \leq 2, r \in \mathbb{R}\}, \end{aligned} \tag{1.5}$$

then the flux limiter scheme is second order accurate away from discontinuity and spatial extrema of the solution, and is TVD (total variation diminishing). Therefore it is a high resolution scheme in the sense of Harten [3]. It is worth to mention that $\psi_2(r)$, the upper boundary of the region D, is exactly the Roe's superbee limiter:

$$\psi_{Roe}(r) = \max(0, \min(2r, 1), \min(r, 2)). \tag{1.6}$$

In [10], we have shown that the following separation property is a sufficient condition for the schemes (1.2)–(1.4) to be TVB (total variation bounded) schemes. This property was first derived by Tadmor [8] as a convenient TVD condition for the schemes (1.2)–(1.4) when $q(w) \equiv 0$.

Assumption 1.1. *The numerical fluxes* $g_{j+\frac{1}{2}}(t)$, $j = 0, \pm 1, \pm 2, \cdots$, *satisfy* $g_{j+\frac{1}{2}}(t) \geq f(u_j) \geq g_{j-\frac{1}{2}}(t)$ *if* $u_j(t) \geq u_{j\pm 1}(t)$, *and* $g_{j+\frac{1}{2}}(t) \leq f(u_j) \leq g_{j-\frac{1}{2}}(t)$ *if* $u_j(t) \leq u_{j\pm 1}(t)$.

We also have shown that the schemes satisfied the above assumption are extremum traceable. Therefore the WEI (wavewise entropy inequality) convergence criteria, established in [10], are applicable. Let $f[w; L, R]$ be the linear function interpolating of $f(w)$ at $w = L$ and $w = R$. For a convex conservation law, a pair of numbers $\{L, R\}$ is called a *rarefying pair* if $L < R$ and $f[w; L, R] > f(w)$ when $L < w < R$. A collection of data $\{v_j\}_{j=-p}^{n+p}$ is called a *rarefying collection with respect to the pair* $\{L, R\}$ if $L = v_0 \leq v_1 \leq \cdots \leq v_n = R$, and $L \leq v_{-1}$, and $R \geq v_{n+1}$. Let $\bar{g}_{j+1/2} := g(v_{j-p+1}, v_{j-p+2}, \cdots, v_{j+p})$, where g is the function given by (1.3). For the convenience of the application, we quote the powerful WEI convergence criterion as follows.

Theorem 1.2 (see Theorem 3.13 [10]). *A scheme of the form* (1.2)–(1.4) *satisfying Assumption* 1.1 *converges for convex conservation laws* (1.1) *if, for any rarefying pair* $\{L, R\}$, *there is a constant* $\delta > 0$ *such that the quadrature type inequality*

$$\sum_{j=0}^{n-1} (v_{j+1} - v_j)\bar{g}_{j+1/2} + \delta < \int_L^R f[w; L, R] dw \tag{1.7}$$

holds for all rarefying collections $\{v_j\}_{j=-p}^{n+p}$ *with respect to the pair* $\{L, R\}$.

The inequality (1.7) can be written as $\sum_{j=0}^{n-1} S_{j+\frac{1}{2}} < \bar{f}[L, R] - \delta$, by using the notations

$$\bar{f}[c, d] := \int_c^d \{f[w; c, d] - f(w)\} dw \tag{1.8}$$

for $c < d$, and

$$S_{j+\frac{1}{2}} := \int_{v_j}^{v_{j+1}} [\bar{g}_{j+\frac{1}{2}} - f(w)] dw. \tag{1.9}$$

In the following, to improve the readability, for any collection of data $\{u_j\}$ we denote $u_{j+\frac{1}{2}} = u_{j+1} - u_j$, $f_j = f(u_j)$, $f_{j+\frac{1}{2}} = f_{j+1} - f_j$, and $g^E_{j+\frac{1}{2}} = g^E(u_j, u_{j+1})$. We also set

$$f^+_{j+\frac{1}{2}} := f_{j+1} - g^E_{j+\frac{1}{2}}, \quad f^-_{j+\frac{1}{2}} := f_j - g^E_{j+\frac{1}{2}}; \tag{1.10}$$

$$r_j^+ := \frac{f_{j-\frac{1}{2}}^+}{f_{j+\frac{1}{2}}^+}, \quad r_j^- := \frac{f_{j+\frac{1}{2}}^-}{f_{j-\frac{1}{2}}^-}, \text{ and } f'_{j+\frac{1}{2}} := \frac{f_{j+\frac{1}{2}}}{u_{j+\frac{1}{2}}}. \tag{1.11}$$

Let ψ be a flux limiter in the framework of Sweby [7]. Using the notations (1.10)–(1.11), the *numerical flux with the limiter* ψ is given by

$$g_{j+\frac{1}{2}} = g_{j+\frac{1}{2}}^E + \frac{1}{2}\psi(r_j^+)f_{j+\frac{1}{2}}^+ + \frac{1}{2}\psi(r_{j+1}^-)f_{j+\frac{1}{2}}^-, \tag{1.12}$$

where $g^E(\cdot,\cdot)$ is the numerical flux of any E-scheme (see Osher [5]), i.e., it is Lipschitz continuous and, for all w between w_j and w_{j+1}, it satisfies

$$\operatorname{sgn}(w_{j+1} - w_j)(g^E(w_j, w_{j+1}) - f(w)) \leq 0. \tag{1.13}$$

Recall the flux of the Godunov scheme is given by

$$g^{God}(u_j, u_{j+1}) = \begin{cases} \min_{u_j \leq w \leq u_{j+1}} f(w) & \text{when} \quad u_j \leq u_{j+1}, \\ \max_{u_j \geq w \geq u_{j+1}} f(w) & \text{when} \quad u_j \geq u_{j+1}; \end{cases} \tag{1.14}$$

and, the flux of the Engquist-Osher scheme is given by

$$g^{EO}(u_j, u_{j+1}) = \int_0^{u_j} \max(f'(w), 0)dw + \int_0^{u_{j+1}} \min(f'(w), 0)dw + f(0). \tag{1.15}$$

Both $g^{God}(\cdot,\cdot)$ and $g^{EO}(\cdot,\cdot)$ are examples of E-fluxes.

In the next section, we will use the WEI convergence criterion, Theorem 1.2, to show the entropy consistency of the schemes (1.2)–(1.4) using Roe's superbee flux limiter when the building block of the schemes is the Godunov or the Engquist-Osher scheme.The results obtained here are valid for convex conservation laws with or without a source term.

2 The convergence result

Theorem 2.1. *The numerical solutions of the schemes* (1.2)–(1.4), *for the convex problems* (1.1), *converge provided that the numerical flux* $g_{j+\frac{1}{2}}$ *is defined by* (1.12), *where* $\psi = \psi_{Roe}$ *is the superbee flux limiter given by* (1.6), *and* $g^E(\cdot,\cdot)$ *is either the numerical flux of the Godunov or the Engquist-Osher scheme given by* (1.14) *and* (1.15) *respectively.*

Lemma 2.2. *For any rarefying collection* $\{v_j\}_{j=-p}^{n+p}$ *with respect to the rarefying pair* $\{L, R\}$, *the inequality*

$$\sum_{j=0}^{n-1} (S_{j+\frac{1}{2}} - \bar{f}[v_j, v_{j+1}]) \leq 0 \tag{2.1}$$

holds, where $\bar{f}[v_j, v_{j+1}]$ *and* $S_{j+\frac{1}{2}}$ *are given by* (1.8) *and* (1.9).

The proof of the Lemma will be given at the end of the section, where we keep the same notations $r_j^{\pm}$ and $f_{j+\frac{1}{2}}^{\pm}$ for $\{v_j\}$ instead of $\{u_j\}$.

Proof of Theorem 2.1. To show the convergence of the scheme, by Theorem 1.2, it suffices to show that for any rarefying pair $\{L, R\}$, there exists a $\delta > 0$ such that for any rarefying collection $\{v_j\}_{j=-p}^{n+p}$ with respect to the pair, the inequality (1.7) holds. That is the inequality

$$\sum_{j=0}^{n-1} S_{j+\frac{1}{2}} < \bar{f}[L, R] - \delta \tag{2.2}$$

holds. Now, by the convexity of f, the following inequality holds

$$\bar{f}[x, y] + \bar{f}[y, z] \leq \bar{f}[x, z] \quad \text{for} \quad x \leq y \leq z. \tag{2.3}$$

This and (2.1) yield

$$\sum_{j=0}^{n-1} S_{j+\frac{1}{2}} \leq \sum_{j=0}^{n-1} \bar{f}[v_j, v_{j+1}] \leq \bar{f}[L, v_i] + \bar{f}[v_i, R] \leq \bar{f}[L, R]. \tag{2.4}$$

Assume for each $\delta > 0$ there exists a rarefying collection $\{v_j\}_{j=-p}^{n+p}$, with respect to the pair $\{L, R\}$ for some positive integer n, violates (2.2). Then (2.4) implies that there exists a sequence of rarefying collections $\{v_j^\nu, j = -p, -p+1, \cdots, n^\nu + p\}_{\nu=1}^{\infty}$ with respect to $\{L, R\}$ such that

$$\lim_{\nu \to \infty} \sum_{j=0}^{n^\nu - 1} S_{j+\frac{1}{2}}^\nu = \lim_{\nu \to \infty} \sum_{j=0}^{n^\nu - 1} \bar{f}[v_j^\nu, v_{j+1}^\nu] = \bar{f}[L, R]. \tag{2.5}$$

This implies that there exists a sequence of integers $\{j^\nu\}$ satisfying $0 \leq j^\nu \leq n^\nu - 1$ such that

$$\lim_{\nu \to \infty} v_{j^\nu}^\nu = L \quad \text{and} \quad \lim_{\nu \to \infty} v_{j^\nu+1}^\nu = R. \tag{2.6}$$

For, otherwise, there would exist a constant $\rho > 0$, a subsequence of the rarefying collections, still denoted by $\{v_j^\nu, j = -p, -p+1, \cdots, n^\nu + p\}_{\nu=1}^{\infty}$, and a sequence of integers $\{i^\nu\}$ satisfying $1 \leq i^\nu \leq n^\nu - 1$ such that $L + \rho < v_{i^\nu}^\nu < R - \rho$. Then the convexity of f, the definition of rarefying pairs, and the inequality (2.4) would imply

$$\begin{aligned}
&\sum_{j=0}^{n^\nu - 1} \bar{f}[v_j^\nu, v_{j+1}^\nu] \leq \bar{f}[L, v_{i^\nu}^\nu] + \bar{f}[v_{i^\nu}^\nu, R] \\
&\leq H_\rho := \max(\bar{f}[L, L+\rho] + \bar{f}[L+\rho, R], \bar{f}[L, R-\rho] + \bar{f}[R-\rho, R]) \\
&< \bar{f}[L, R].
\end{aligned}$$

This would contradict (2.5) because H_ρ is independent of ν. Hence, we have confirmed the existence of a sequence $\{j^\nu\}$ satisfying (2.6). Now, combining (2.3), (2.4), (2.5) and (2.6), we obtain

$$\bar{f}[L,R] = \lim_{\nu\to\infty} \sum_{j=0}^{n^\nu-1} S^\nu_{j+\frac{1}{2}} = \lim_{\nu\to\infty} \left(\sum_{j=0}^{j^\nu-1} S^\nu_{j+\frac{1}{2}} + S^\nu_{j^\nu+\frac{1}{2}} + \sum_{j=j^\nu+1}^{n^\nu-1} S^\nu_{j+\frac{1}{2}} \right)$$
$$\le \lim_{\nu\to\infty} (\bar{f}[L, v^\nu_{j^\nu}] + S^\nu_{j^\nu+\frac{1}{2}} + \bar{f}[v^\nu_{j^\nu+1}, R]) = \lim_{\nu\to\infty} S^\nu_{j^\nu+\frac{1}{2}} \le \bar{f}[L,R].$$

This implies

$$\lim_{\nu\to\infty} S^\nu_{j^\nu+\frac{1}{2}} = \bar{f}[L,R]. \tag{2.7}$$

However, applying (2.6) and the definition of the rarefying collections with respect to the rarefying pair $\{L,R\}$ to (1.12), we get $\lim_{\nu\to\infty} \bar{g}^\nu_{j^\nu+\frac{1}{2}} = g^E(L,R)$, and hence, by (1.9) and (1.13), $\lim_{\nu\to\infty} S^\nu_{j^\nu+\frac{1}{2}} \le 0$. This contradicts (2.7) since $\bar{f}[L,R] > 0$ by the definition of the pair $\{L,R\}$. □

The proof of Lemma 2.2. We consider the g^{God} and the g^{EO} fluxes. Without loss of generality, let v_s be the sonic point such that $v_k \le v_s \le v_{k+1}$ for some integer k with $0 \le k \le n-1$. Then,

$$f^+_{j+\frac{1}{2}} = 0, \text{ if } 0 \le j \le k-1; \qquad f^+_{j+\frac{1}{2}} = f'_{j+\frac{1}{2}} v_{j+\frac{1}{2}}, \quad \text{if } n-1 \ge j \ge k+1;$$
$$f^-_{j+\frac{1}{2}} = 0, \text{ if } n-1 \ge j \ge k+1; \quad f^-_{j+\frac{1}{2}} = -f'_{j+\frac{1}{2}} v_{j+\frac{1}{2}}, \quad \text{if } 0 \le j \le k-1.$$

Next, for any E-flux, we have

$$\int_{v_j}^{v_{j+1}} (\, g^E_{j+\frac{1}{2}} - f[w; v_j, v_{j+1}]\,)\, dw = -\frac{f^-_{j+\frac{1}{2}} + f^+_{j+\frac{1}{2}}}{2} v_{j+\frac{1}{2}}.$$

In the following, we denote $v_{s-\frac{1}{2}} = v_s - v_k$ and $v_{s+\frac{1}{2}} = v_{k+1} - v_s$. We also set

$$f_{s-\frac{1}{2}} = f_s - f_k,\ f_{s+\frac{1}{2}} = f_{k+1} - f_s,\ f'_{s-\frac{1}{2}} = \frac{f_{s-\frac{1}{2}}}{v_{s-\frac{1}{2}}}, \text{ and } f'_{s+\frac{1}{2}} = \frac{f_{s+\frac{1}{2}}}{v_{s+\frac{1}{2}}}.$$

When $0 \le j \le k-1$, if $0 < r^-_{j+1} < 1$, we have $\psi(r^-_{j+1}) \le 1$ and

$$\sum_{j=0}^{k-1} (S_{j+\frac{1}{2}} - \bar{f}[v_j, v_{j+1}]) = \sum_{j=0}^{k-1} \frac{1}{2} f'_{j+\frac{1}{2}} (1 - \psi(r^-_{j+1})) v^2_{j+\frac{1}{2}} \le 0; \tag{2.8}$$

if $r^-_{j+1} \ge 1$, we have $\psi(r^-_{j+1}) \le r^-_{j+1}$ and, for $0 \le j \le k-2$,

$$\sum_{j=0}^{k-2} (S_{j+\frac{1}{2}} - \bar{f}[v_j, v_{j+1}]) \le \sum_{j=0}^{k-2} \left(g^E_{j+\frac{1}{2}} + \frac{1}{2} f^-_{j+\frac{3}{2}} - \frac{1}{2}(f_j + f_{j+1}) \right) v_{j+\frac{1}{2}}$$
$$\le \sum_{j=0}^{k-2} \frac{1}{4} f'_{j+\frac{3}{2}} (v_{j+\frac{1}{2}} - v_{j+\frac{3}{2}})^2 + \frac{1}{4} f'_{\frac{1}{2}} v^2_{\frac{1}{2}} - \frac{1}{4} f'_{k-\frac{1}{2}} v^2_{k-\frac{1}{2}} \le -\frac{1}{4} f'_{k-\frac{1}{2}} v^2_{k-\frac{1}{2}};$$

for $j = k-1$, we have $f^{+}_{k-\frac{1}{2}} = 0$ and

$$S_{k-\frac{1}{2}} - \bar{f}[v_{k-1}, v_k] \le \left(f_k + \frac{1}{2} f^{-}_{k+\frac{1}{2}} - \frac{1}{2}(f_{k-1} + f_k) \right) v_{k-\frac{1}{2}}$$
$$\le \frac{1}{4} f'_{s-\frac{1}{2}} (v_{k-\frac{1}{2}} - v_{s-\frac{1}{2}})^2 + \frac{1}{4} f'_{k-\frac{1}{2}} v^2_{k-\frac{1}{2}} - \frac{1}{4} f'_{s-\frac{1}{2}} v^2_{s-\frac{1}{2}}$$
$$\le \frac{1}{4} f'_{k-\frac{1}{2}} v^2_{k-\frac{1}{2}} - \frac{1}{4} f'_{s-\frac{1}{2}} v^2_{s-\frac{1}{2}};$$

thus

$$\sum_{j=0}^{k-1} (S_{j+\frac{1}{2}} - \bar{f}[v_j, v_{j+1}]) \le -\frac{1}{4} f'_{s-\frac{1}{2}} v^2_{s-\frac{1}{2}}. \tag{2.9}$$

When $j = k$, we have $r^{+}_{k} = 0$, $r^{-}_{k+1} = 0$ and

$$S_{k+\frac{1}{2}} - \bar{f}[v_k, v_{k+1}] = \left(g^{E}_{k+\frac{1}{2}} - \frac{1}{2}(f_k + f_{k+1}) \right) v_{k+\frac{1}{2}} \tag{2.10}$$
$$= (f_s - \frac{1}{2}(f_k + f_{k+1}))v_{k+\frac{1}{2}} = \frac{1}{2} f'_{s-\frac{1}{2}} v_{s-\frac{1}{2}} v_{k+\frac{1}{2}} - \frac{1}{2} f'_{s+\frac{1}{2}} v_{s+\frac{1}{2}} v_{k+\frac{1}{2}}.$$

When $k+1 \le j \le n-1$, if $0 < r^{+}_{j} < 1$, we have $\psi(r^{+}_{j}) \le 1$ and

$$\sum_{j=k+1}^{n-1} (S_{j+\frac{1}{2}} - \bar{f}[v_j, v_{j+1}]) \le 0; \tag{2.11}$$

if $r^{+}_{j} \ge 1$, we have $\psi(r^{+}_{j}) \le r^{+}_{j}$ and, for $k+2 \le j \le n-1$,

$$\sum_{j=k+2}^{n-1} (S_{j+\frac{1}{2}} - \bar{f}[v_j, v_{j+1}]) \le \frac{1}{4} f'_{k+\frac{3}{2}} v^2_{k+\frac{3}{2}};$$

for $j = k+1$, we have $f^{-}_{k+\frac{3}{2}} = 0$ and

$$S_{k+\frac{3}{2}} - \bar{f}[v_{k+1}, v_{k+2}] \le \frac{1}{2} f'_{s+\frac{1}{2}} v_{s+\frac{1}{2}} v_{k+\frac{3}{2}} - \frac{1}{2} f'_{k+\frac{3}{2}} v^2_{k+\frac{3}{2}}; \tag{2.12}$$

thus

$$\sum_{j=k+1}^{n-1} (S_{j+\frac{1}{2}} - \bar{f}[v_j, v_{j+1}]) \le \frac{1}{4} f'_{s+\frac{1}{2}} v^2_{s+\frac{1}{2}}. \tag{2.13}$$

Also, notice that we have the following two inequalities

$$\frac{1}{4} f'_{s+\frac{1}{2}} v^2_{s+\frac{1}{2}} - \frac{1}{4} f'_{s+\frac{1}{2}} v_{s+\frac{1}{2}} v_{k+\frac{1}{2}} = -\frac{1}{4} f'_{s+\frac{1}{2}} v_{s+\frac{1}{2}} v_{s-\frac{1}{2}} \le 0, \tag{2.14}$$

and

$$-\frac{1}{4}f'_{s-\frac{1}{2}}v^2_{s-\frac{1}{2}}+\frac{1}{4}f'_{s-\frac{1}{2}}v_{s-\frac{1}{2}}v_{k+\frac{1}{2}}=\frac{1}{4}f'_{s-\frac{1}{2}}v_{s-\frac{1}{2}}v_{s+\frac{1}{2}}\leq 0. \tag{2.15}$$

Finally, using (2.9), (2.10), (2.13) and (2.14)–(2.15), we have

$$\sum_{j=0}^{n-1}(S_{j+\frac{1}{2}}-\bar{f}[v_j,v_{j+1}])\leq\frac{1}{4}f'_{s-\frac{1}{2}}v_{s-\frac{1}{2}}v_{k+\frac{1}{2}}-\frac{1}{4}f'_{s+\frac{1}{2}}v_{s+\frac{1}{2}}v_{k+\frac{1}{2}}\leq 0$$

as desired. Other cases can be dealt similarly. The proof is completed. □

References

[1] S. Chakravarthy and S. Osher, High resolution applications of the Osher upwind scheme for the Euler equations. AIAA paper presented at 6th CFD conference 1983.

[2] N. Jiang and H. Yang, On convergence of semi-discrete high resolution schemes with van Leer's flux limiter for conservation laws. Methods and Applications of Analysis 12 (2005), No. 1, 89–102.

[3] A. Harten, J. M. Hyman and P. D. Lax, On finite-difference approximations and entropy conditions for shocks. Comm. Pure and Appl. Math. 29 (1976), 297–322.

[4] A. Harten and G. Zwas, Self-adjusting hybrid schemes for shock computations. J. Comput. Phys. 9 (1972), 568.

[5] S. Osher, Riemann solvers, the entropy condition and difference approximations. SIAM J. Numer. Anal. 21 (1984), 217–235.

[6] P. L. Roe, Some contributions to the modeling of discontinuous flows. Lect. Notes Appl. Math. 22 (1985), 163–193.

[7] P. K. Sweby, High resolution schemes using flux limiters for hyperbolic conservation laws. SIAM J. Numer. Anal. 21 (1984), 995–1011.

[8] E. Tadmor, Convenient total variation diminishing conditions for nonlinear difference schemes. SIAM J. Numer. Anal. 25 (1988), 1002–1014.

[9] B. van Leer, Towards the ultimate conservative difference scheme II. Monotonicity and conservation combined in a second order scheme. J. Comput. Phys. 14 (1974), 361–370.

[10] H. Yang and N. Jiang, On wavewise entropy inequalities for high-resolution schemes with source terms I: The Semi-discrete case. Methods and Applications of Analysis 10 (2003), No. 4, 487–512.

Asymptotic Limits of the Compressible Magnetohydrodynamic Equations*

Song Jiang
LCP, Institute of Applied Physics and Computational Mathematics
P.O. Box 8009, Beijing 100088, China
E-mail: jiang@iapcm.ac.cn

Qiangchang Ju
Institute of Applied Physics and Computational Mathematics
P.O. Box 8009-28, Beijing 100088, China
E-mail: qiangchang_ju@yahoo.com

Fucai Li
Department of Mathematics, Nanjing University
Nanjing 210093, China
E-mail: fli@nju.edu.cn

Abstract

Asymptotic limits of the weak solutions of the compressible magnetohydrodynamic equations are discussed. Several results are stated and the main ideas of proofs are given.

1 Introduction

Magnetohydrodynamics (MHD) studies the dynamics of compressible quasineutrally ionized fluids under the influence of electromagnetic fields. The compressible viscous MHD equations in the isentropic case take the form

$$\partial_t\tilde{\rho} + \mathrm{div}(\tilde{\rho}\tilde{\mathbf{u}}) = 0, \tag{1.1}$$

$$\partial_t(\tilde{\rho}\tilde{\mathbf{u}}) + \mathrm{div}(\tilde{\rho}\tilde{\mathbf{u}}\otimes\tilde{\mathbf{u}}) + \nabla\tilde{P} = (\mathrm{curl}\,\tilde{\mathbf{H}})\times\tilde{\mathbf{H}} + \tilde{\mu}\Delta\tilde{\mathbf{u}} + (\tilde{\mu}+\tilde{\lambda})\nabla(\mathrm{div}\tilde{\mathbf{u}}), \tag{1.2}$$

$$\partial_t\tilde{\mathbf{H}} - \mathrm{curl}\,(\tilde{\mathbf{u}}\times\tilde{\mathbf{H}}) = -\mathrm{curl}\,(\tilde{\nu}\,\mathrm{curl}\,\tilde{\mathbf{H}}), \quad \mathrm{div}\tilde{\mathbf{H}} = 0. \tag{1.3}$$

*The authors are partly supported by NSFC (No. 2005CB321700, No. 40890154, No. 10701011, No. 10971094).

Here $x \in \Omega$, Ω will be the whole space $\mathbb{R}^d$ or the torus $\mathbb{T}^d$, $d = 2$ or 3. The unknown $\tilde{\rho}$ denotes the density, $\tilde{\mathbf{u}} = (\tilde{u}_1, \dots, \tilde{u}_d) \in \mathbb{R}^d$ the velocity, and $\tilde{\mathbf{H}} = (\tilde{H}_1, \dots, \tilde{H}_d) \in \mathbb{R}^d$ the magnetic field, respectively. The constants $\tilde{\mu}$ and $\tilde{\lambda}$ are the shear and bulk viscosity coefficients of the flow, respectively, satisfying $\tilde{\mu} > 0$ and $2\tilde{\mu} + d\tilde{\lambda} > 0$; the constant $\tilde{\nu} > 0$ is the magnetic diffusivity acting as a magnetic diffusion coefficient of the magnetic field. $\tilde{P}(\tilde{\rho})$ is the pressure-density function and here we consider the case

$$\tilde{P}(\tilde{\rho}) = a\tilde{\rho}^{\gamma}, \tag{1.4}$$

where $a > 0$ and $\gamma > 1$ are constants.

From the physical point of view, one can formally derive the incompressible models from the compressible ones when the Mach number goes to zero and the density becomes almost constant. Based on this observation, Hu and Wang [4] proved the convergence of the weak solutions of the compressible MHD equations (1.1)–(1.3) to a weak solution of the viscous incompressible MHD equations. Our main purpose is to establish the asymptotic limits of the weak solutions of the compressible magnetohydrodynamic equations when the mach number goes to zero, and the shear viscosity coefficient and the magnetic diffusion coefficient go to some constants (including zero). Compared with the results of [4], here we require that the target equations exist a unique strong solution. In [5], We obtain the convergence towards the strong solution of the ideal incompressible MHD equations in the whole space by using the dispersion property of the wave equation for the general initial data. The periodic case was extensively studied in [6]. For this case, the acoustic waves will lose the dispersion property and will interact each other. Therefore, it is more complicated and difficult to prove the limit. First, we consider the well-prepared initial data for which the oscillations will never appear. We obtain that the weak solutions of the compressible MHD equations converge to the strong solution of the ideal incompressible MHD equations in the periodic domain if both the shear viscosity and the magnetic diffusion coefficients go to zero, as well as to the strong solution of the viscous incompressible MHD equations. Furthermore, we also give the rates of convergence which are not obtained in [4, 5]. Second, we consider the case of general initial data. Now we have to impose more regular conditions than L^2 on the initial data to control the oscillating parts. In addition, we have to assume that the Sobolev norm of the oscillating parts is comparable to the magnetic diffusion coefficient in order to deal with the general initial data. We obtain the convergence of the weak solutions of the compressible MHD equations to the strong solution of the incompressible MHD equations, as well as to the strong solution of the partial viscous incompressible MHD equations. The precise results and the main ideas of the proofs will be given in the next section.

2 The main results

In this section, we shall state our main results concerning the whole space and the periodic case, respectively. We begin with some formal analysis. Introduce the scalings

$$\tilde{\rho}(x,t) = \rho^\epsilon(x,\epsilon t), \quad \tilde{\mathbf{u}}(x,t) = \epsilon \mathbf{u}^\epsilon(x,\epsilon t), \quad \tilde{\mathbf{H}}(x,t) = \epsilon \mathbf{H}^\epsilon(x,\epsilon t)$$

and assume that the viscosity coefficients $\tilde{\mu}$, $\tilde{\xi}$, and $\tilde{\nu}$ are small constants and scaled like

$$\tilde{\mu} = \epsilon\mu^\epsilon, \quad \tilde{\lambda} = \epsilon\lambda^\epsilon, \quad \tilde{\nu} = \epsilon\nu^\epsilon, \tag{2.1}$$

where $\epsilon \in (0,1)$ is a small parameter and the normalized coefficients μ^ϵ, λ^ϵ, and ν^ϵ satisfy $\mu^\epsilon > 0$, $2\mu^\epsilon + d\lambda^\epsilon > 0$, and $\nu^\epsilon > 0$. With the preceding scalings, the pressure function (1.4), and the vector identities $\nabla(|\tilde{\mathbf{H}}|^2) = 2(\tilde{\mathbf{H}}\cdot\nabla)\tilde{\mathbf{H}} + 2\tilde{\mathbf{H}}\times \operatorname{curl}\tilde{\mathbf{H}}$, $\operatorname{curl}\operatorname{curl}\tilde{\mathbf{H}} = \nabla\operatorname{div}\tilde{\mathbf{H}} - \Delta\tilde{\mathbf{H}}$, and $\operatorname{curl}(\tilde{\mathbf{u}}\times\tilde{\mathbf{H}}) = \tilde{\mathbf{u}}(\operatorname{div}\tilde{\mathbf{H}}) - \tilde{\mathbf{H}}(\operatorname{div}\tilde{\mathbf{u}}) + (\tilde{\mathbf{H}}\cdot\nabla)\tilde{\mathbf{u}} - (\tilde{\mathbf{u}}\cdot\nabla)\tilde{\mathbf{H}}$, we can rewrite the compressible MHD equations (1.1)–(1.3) as

$$\partial_t\rho^\epsilon + \operatorname{div}(\rho^\epsilon\mathbf{u}^\epsilon) = 0, \tag{2.2}$$

$$\partial_t(\rho^\epsilon\mathbf{u}^\epsilon) + \operatorname{div}(\rho^\epsilon\mathbf{u}^\epsilon\otimes\mathbf{u}^\epsilon) + \frac{a\nabla(\rho^\epsilon)^\gamma}{\epsilon^2}$$

$$= (\mathbf{H}^\epsilon\cdot\nabla)\mathbf{H}^\epsilon - \frac{1}{2}\nabla(|\mathbf{H}^\epsilon|^2) + \mu^\epsilon\Delta\mathbf{u}^\epsilon + (\mu^\epsilon+\lambda^\epsilon)\nabla(\operatorname{div}\mathbf{u}^\epsilon), \tag{2.3}$$

$$\partial_t\mathbf{H}^\epsilon + (\operatorname{div}\mathbf{u}^\epsilon)\mathbf{H}^\epsilon + (\mathbf{u}^\epsilon\cdot\nabla)\mathbf{H}^\epsilon - (\mathbf{H}^\epsilon\cdot\nabla)\mathbf{u}^\epsilon = \nu^\epsilon\Delta\mathbf{H}^\epsilon, \quad \operatorname{div}\mathbf{H}^\epsilon = 0. \tag{2.4}$$

Moreover, by replacing ϵ by $\sqrt{a\gamma}\ \epsilon$, we can always assume $a = 1/\gamma$. Formally let $\epsilon \to 0$ in the equations (2.2)–(2.4), then we obtain from the momentum equation (2.3) that ρ^ϵ converges to some function $\bar{\rho}(t) \geq 0$. If we further assume that the initial datum ρ_0^ϵ is of order $1 + O(\epsilon)$, then we can expect that $\bar{\rho} = 1$. Thus, the continuity equation (2.2) gives $\operatorname{div}\mathbf{u} = 0$. Furthermore, using the assumption

$$\mu^\epsilon \to 0, \quad \nu^\epsilon \to 0 \quad \text{as} \quad \epsilon \to 0, \tag{2.5}$$

we obtain the following ideal incompressible MHD equations (suppose that the limits $\mathbf{u}^\epsilon \to \mathbf{u}$ and $\mathbf{H}^\epsilon \to \mathbf{H}$ exist)

$$\partial_t\mathbf{u} + (\mathbf{u}\cdot\nabla)\mathbf{u} - (\mathbf{H}\cdot\nabla)\mathbf{H} + \nabla p + \frac{1}{2}\nabla(|\mathbf{H}|^2) = 0, \tag{2.6}$$

$$\partial_t\mathbf{H} + (\mathbf{u}\cdot\nabla)\mathbf{H} - (\mathbf{H}\cdot\nabla)\mathbf{u} = 0, \tag{2.7}$$

$$\operatorname{div}\mathbf{u} = 0, \quad \operatorname{div}\mathbf{H} = 0. \tag{2.8}$$

If we assume that the shear and the bulk viscosity coefficients and the magnetic diffusivity coefficient satisfy

$$\mu^\epsilon \to \mu > 0, \quad \lambda^\epsilon \to \lambda, \quad \nu^\epsilon \to \nu > 0 \quad \text{as} \quad \epsilon \to 0, \tag{2.9}$$

then the compressible MHD equations (2.6)–(2.8) formally converge to the incompressible MHD equations (suppose that the limits $\mathbf{u}^\epsilon \to \mathbf{u}$ and $\mathbf{H}^\epsilon \to \mathbf{H}$ exist)

$$\partial_t \mathbf{u} + (\mathbf{u}\cdot\nabla)\mathbf{u} - \mu\Delta\mathbf{u} + \nabla p - (\mathbf{H}\cdot\nabla)\mathbf{H} + \frac{1}{2}\nabla(|\mathbf{H}|^2) = 0, \quad (2.10)$$

$$\partial_t \mathbf{H} + (\mathbf{u}\cdot\nabla)\mathbf{H} - (\mathbf{H}\cdot\nabla)\mathbf{u} - \nu\Delta\mathbf{H} = 0, \quad (2.11)$$

$$\operatorname{div}\mathbf{u} = 0, \quad \operatorname{div}\mathbf{H} = 0. \quad (2.12)$$

Some notations are used throughout the current paper. We denote the space $L_2^q(\Omega)$ by

$$L_2^q(\Omega) = \{f \in L_{loc}(\Omega) : f1_{\{|f|\geq 1/2\}} \in L^q, f1_{\{|f|\leq 1/2\}} \in L^2\}.$$

We use the letters C and C_T to denote various positive constants independent of ϵ, but C_T may depend on T. For convenience, we denote by $H^r \equiv H^r(\Omega)$ ($r \in \mathbb{R}$) the standard Sobolev space. For any vector field $\mathbf{v}$, we denote by $P\mathbf{v}$ and $Q\mathbf{v}$ the divergence-free part and the gradient part of $\mathbf{v}$, respectively. Namely, $Q\mathbf{v} = \nabla\Delta^{-1}(\operatorname{div}\mathbf{v})$ and $P\mathbf{v} = \mathbf{v} - Q\mathbf{v}$.

2.1 The whole space case

We prescribe the initial conditions to the compressible MHD equations (2.2)–(2.4) as

$$\rho^\epsilon|_{t=0} = \rho_0^\epsilon(x), \quad \rho^\epsilon\mathbf{u}^\epsilon|_{t=0} = \rho_0^\epsilon(x)\mathbf{u}_0^\epsilon(x) \equiv \mathbf{m}_0^\epsilon(x), \quad \mathbf{H}^\epsilon|_{t=0} = \mathbf{H}_0^\epsilon(x), \quad (2.13)$$

and assume that

$$\rho_0^\epsilon \geq 0, \ \rho_0^\epsilon - 1 \in L_2^\gamma, \ \rho_0^\epsilon|\mathbf{u}_0^\epsilon|^2 \in L^1, \ \mathbf{m}_0^\epsilon = 0 \text{ for a.e. } \rho_0^\epsilon = 0, \quad (2.14)$$

and

$$\mathbf{H}_0^\epsilon \in L^2, \ \operatorname{div}\mathbf{H}_0^\epsilon = 0. \quad (2.15)$$

Moreover, we assume that the initial data also satisfy the following uniform bound

$$\int_{\mathbb{R}^d}\Big[\frac{1}{2}\rho_0^\epsilon|\mathbf{u}_0^\epsilon|^2 + \frac{1}{2}|\mathbf{H}_0^\epsilon|^2 + \frac{a}{\epsilon^2(\gamma-1)}\big((\rho_0^\epsilon)^\gamma - 1 - \gamma(\rho_0^\epsilon - 1)\big)\Big]dx \leq C. \quad (2.16)$$

The initial energy inequality (2.16) implies that ρ_0^ϵ is of order $1 + O(\epsilon)$.

Under the above assumptions, we know that there are a sequence of global weak solutions $(\rho^\epsilon, \mathbf{u}^\epsilon, \mathbf{H}^\epsilon)$ of the compressible MHD equations (2.6)–(2.8) satisfying that $\rho^\epsilon - 1 \in L^\infty(0,\infty; L_2^\gamma) \cap C([0,\infty), L_2^r)$ for all $1 \leq r < \gamma$, $\rho^\epsilon|\mathbf{u}^\epsilon|^2 \in L^\infty(0,\infty; L^1)$, $\mathbf{H}^\epsilon \in L^\infty(0,\infty; L^2)$, $\mathbf{u}^\epsilon \in L^2(0,T; H^1)$, $\rho^\epsilon\mathbf{u}^\epsilon \in C([0,T], L_{weak}^{\frac{2\gamma}{\gamma+1}})$, and $\mathbf{H}^\epsilon \in L^2(0,T; H^1) \cap C([0,T],$

$L^{\frac{2\gamma}{\gamma+1}}_{weak})$ for all $T \in (0,\infty)$. Moreover, (2.6)–(2.8) hold in the sense of distributions, see [3] for example.

We impose the following conditions on the solution $(\rho^\epsilon, \mathbf{u}^\epsilon, \mathbf{H}^\epsilon)$ at infinity

$$\rho^\epsilon \to 1, \quad \mathbf{u}^\epsilon \to \mathbf{0}, \quad \mathbf{H}^\epsilon \to \mathbf{0} \quad \text{as} \quad |x| \to +\infty. \tag{2.17}$$

We write $\rho^\epsilon = 1 + \epsilon\varphi^\epsilon$ and denote

$$\Pi^\epsilon(x,t) = \frac{1}{\epsilon}\sqrt{\frac{2a}{\gamma-1}\big((\rho^\epsilon)^\gamma - 1 - \gamma(\rho^\epsilon - 1)\big)}.$$

We will use the above approximation $\Pi^\epsilon(x,t)$ instead of φ^ϵ, since we can not obtain any bound for φ^ϵ in $L^\infty(0,T;L^2)$ directly if $\gamma < 2$.

We have shown the following theorem.

Theorem 2.1 ([5]). *Let $s > d/2 + 2$. Assume that the shear viscosity μ^ϵ and the magnetic diffusion coefficient ν^ϵ satisfy*

$$\mu^\epsilon = \epsilon^\alpha, \quad \nu^\epsilon = \epsilon^\beta \tag{2.18}$$

for some constants $\alpha, \beta > 0$ satisfying $0 < \alpha + \beta < 2$. Moreover, we assume that $\sqrt{\rho_0^\epsilon}\mathbf{u}_0^\epsilon$ converges strongly in L^2 to some $\tilde{\mathbf{u}}_0$, $\mathbf{H}_0^\epsilon$ converges strongly in L^2 to some $\mathbf{H}_0$, and Π_0^ϵ converges strongly in L^2 to some φ_0. Let $(\mathbf{u}, \mathbf{H})$ be the smooth solution to the ideal incompressible MHD equations (2.6)–(2.8) defined on $[0, T^*)$ with $(\mathbf{u}, \mathbf{H})|_{t=0} = (\mathbf{u}_0, \mathbf{H}_0)$, $\mathbf{u}_0 = P\tilde{\mathbf{u}}_0$. Then, for any $0 < T < T^*$, the global weak solution $(\rho^\epsilon, \mathbf{u}^\epsilon, \mathbf{H}^\epsilon)$ of the compressible MHD equations (2.2)–(2.4) satisfies

1. *ρ^ϵ converges strongly to 1 in $C([0,T], L_2^\gamma(\mathbb{R}^d))$;*
2. *$\mathbf{H}^\epsilon$ converges strongly to $\mathbf{H}$ in $L^\infty(0,T;L^2(\mathbb{R}^d))$;*
3. *$P(\sqrt{\rho^\epsilon}\mathbf{u}^\epsilon)$ converges strongly to $\mathbf{u}$ in $L^\infty(0,T;L^2(\mathbb{R}^d))$;*
4. *$\sqrt{\rho^\epsilon}\mathbf{u}^\epsilon$ converges strongly to $\mathbf{u}$ in $L^r(0,T;L^2_{\text{loc}}(\mathbb{R}^d))$ for all $1 \le r < +\infty$.*

The proof of above results is based on the combination of the modulated energy method, motivated by Brenier [1], the Strichartz's estimate of linear wave equations [2] and the weak convergence method. Masmoudi [7] made use of such ideas to consider the incompressible, inviscid convergence of weak solutions of compressible Navier-Stokes equations to the strong solution of the incompressible Euler equations in the whole space and the torus. We shall follow and adapt the method used in [7] to prove our results. Besides the difficulties pointed out in [7], we have to overcome the additional difficulties caused by the strong coupling of the hydrodynamic motion and the magnetic fields. Very careful energy estimates are employed to estimate those nonlinear terms for large data. On the other hand, because the general initial data are considered in the present paper, the oscillations appear and propagate with the solution. We will use the Strichartz's estimate of linear wave equations to deal with such oscillations and their interactions with velocity and magnetic fields. Finally, the weak convergence method and refined energy analysis are employed to obtain the desired convergence results, see [5] for details.

2.2 The periodic case

Now we consider a sequence of solutions $(\rho^\epsilon, \mathbf{u}^\epsilon, \mathbf{H}^\epsilon)$ of the compressible MHD equations (2.6)–(2.8), satisfying the same conditions as in the whole space case (the functions are now periodic in space and all the integrations are performed over $\mathbb{T}^d$, (2.17) is not needed in this situation). First, we consider the well-prepared initial data.

Theorem 2.2 ([6]). *Let $s > d/2 + 2$, $\mu^\epsilon + \lambda^\epsilon > 0$, and $(\mathbf{u}, \mathbf{H})$ be the smooth solution to the ideal incompressible MHD equations (2.6)–(2.8) defined on $[0, T^*)$ with $(\mathbf{u}, \mathbf{H})|_{t=0} = (\mathbf{u}_0, \mathbf{H}_0)$. Assume that*

$$\int_{\mathbb{T}^d} |\rho_0^\epsilon - 1|^2 1_{(|\rho_0^\epsilon - 1| \le \delta)}\, dx + \int_{\mathbb{T}^d} |\rho_0^\epsilon - 1|^\gamma 1_{(|\rho_0^\epsilon - 1| > \delta)}\, dx \le C\epsilon^2, \tag{2.19}$$

$$||\sqrt{\rho_0^\epsilon}\mathbf{u}_0^\epsilon - \mathbf{u}_0||^2_{L^2(\mathbb{T}^d)} \le C\epsilon, \quad ||\mathbf{H}_0^\epsilon - \mathbf{H}_0||^2_{L^2(\mathbb{T}^d)} \le C\epsilon \tag{2.20}$$

for any $\delta \in (0,1)$. We assume that the shear viscosity μ^ϵ and the magnetic diffusion coefficient ν^ϵ satisfy

$$\mu^\epsilon = \epsilon^\alpha, \quad \nu^\epsilon = \epsilon^\beta \tag{2.21}$$

for some constants $\alpha, \beta > 0$ satisfying $0 < \alpha + \beta < 2$. Then, for any $0 < T < T^*$, the global weak solution $(\rho^\epsilon, \mathbf{u}^\epsilon, \mathbf{H}^\epsilon)$ of the compressible MHD equations (2.2)–(2.4) satisfies

$$\int_{\mathbb{T}^d} |\rho^\epsilon - 1|^2 1_{(|\rho^\epsilon - 1| \le \delta)}\, dx + \int_{\mathbb{T}^d} |\rho^\epsilon - 1|^\gamma 1_{(|\rho^\epsilon - 1| > \delta)}\, dx \le C_T\epsilon^2, \tag{2.22}$$

$$||\sqrt{\rho^\epsilon}\mathbf{u}^\epsilon - \mathbf{u}||^2_{L^2(\mathbb{T}^d)} \le C_T\epsilon^\sigma, \quad ||\mathbf{H}^\epsilon - \mathbf{H}||^2_{L^2(\mathbb{T}^d)} \le C_T\epsilon^\sigma \tag{2.23}$$

for any $t \in [0, T]$, where $\sigma = \min\{\alpha, \beta, 1 - (\alpha + \beta)/2\}$.

The proof of Theorem 2.2 is based on the combination of the modulated energy method, motivated by Brenier [1], the weak convergence method and the refined energy analysis.

Furthermore, we can use an idea similar to that described above to obtain the convergence of the compressible MHD equations (2.2)–(2.4) to the incompressible viscous MHD equations (2.10)–(2.12).

Theorem 2.3 ([6]). *Let $s > d/2 + 2$ and $\mu^\epsilon + \lambda^\epsilon > 0$. Assume that*

$$\int_{\mathbb{T}^d} |\rho_0^\epsilon - 1|^2 1_{(|\rho_0^\epsilon - 1| \le \delta)}\, dx + \int_{\mathbb{T}^d} |\rho_0^\epsilon - 1|^\gamma 1_{(|\rho_0^\epsilon - 1| > \delta)}\, dx \le C\epsilon^2, \quad (2.24)$$

$$||\sqrt{\rho_0^\epsilon}\mathbf{u}_0^\epsilon - \mathbf{u}_0||^2_{L^2(\mathbb{T}^d)} \le C\epsilon, \quad ||\mathbf{H}_0^\epsilon - \mathbf{H}_0||^2_{L^2(\mathbb{T}^d)} \le C\epsilon \quad (2.25)$$

for any $\delta \in (0,1)$ and for some $\mathbf{u}_0, \mathbf{H}_0 \in H^s(\mathbb{T}^d)$, satisfying $\mathrm{div}\mathbf{u}_0 = 0, \mathrm{div}\mathbf{H}_0 = 0$. *We also assume that the shear viscosity μ^ϵ and the magnetic diffusion coefficient ν^ϵ satisfy* (2.9). *Let* $(\mathbf{u}, \mathbf{H})$ *be the smooth solution to the incompressible MHD equations* (2.10)–(2.12) *with* $(\mathbf{u}, \mathbf{H})|_{t=0} = (\mathbf{u}_0, \mathbf{H}_0)$. *Then, for any $0 < T < T^{**}$ (T^{**} is the maximal time of existence for* (2.10)–(2.12)*), the global weak solution $(\rho^\epsilon, \mathbf{u}^\epsilon, \mathbf{H}^\epsilon)$ of the compressible MHD equations* (2.2)–(2.4) *satisfies that $\nabla \mathbf{u}^\epsilon$ and $\nabla \mathbf{H}^\epsilon$ converge strongly to $\nabla \mathbf{u}$ and $\nabla \mathbf{H}$ in $L^2(0, T; L^2(\mathbb{T}^d))$, respectively. Moreover, for any $t \in [0, T]$, we have*

$$\int_{\mathbb{T}^d} |\rho^\epsilon - 1|^2 1_{(|\rho^\epsilon - 1| \le \delta)}\, dx + \int_{\mathbb{T}^d} |\rho^\epsilon - 1|^\gamma 1_{(|\rho^\epsilon - 1| > \delta)}\, dx \le C_T\epsilon^2, \quad (2.26)$$

$$\begin{aligned} &||\sqrt{\rho^\epsilon}\mathbf{u}^\epsilon - \mathbf{u}||^2_{L^2(\mathbb{T}^d)} + ||\mathbf{H}^\epsilon - \mathbf{H}||^2_{L^2(\mathbb{T}^d)} \\ \le\, &C_T \frac{\epsilon}{\sqrt{\mu\nu}} + C_T\Big(\frac{|\mu^\epsilon - \mu|}{\sqrt{\mu}} + \frac{|\nu^\epsilon - \nu|}{\sqrt{\nu}}\Big). \end{aligned} \quad (2.27)$$

For the general initial data, we shall obtain the convergence of the compressible MHD equations (2.2)–(2.4) to the incompressible MHD equations (2.10)–(2.12) under the conditions that the oscillating parts of the initial data have higher regularity and the Sobolev norm of the oscillation parts is comparable to the magnetic diffusion coefficient. This implies that the influence of oscillations on the magnetic field can be balanced by the diffusive effect of the magnetic field, which is one of the new ingredients in our result.

Theorem 2.4 ([6]). *Let $s > 2 + d/2$ and $2\mu + d\lambda > 0$. We assume that $\sqrt{\rho_0^\epsilon}\mathbf{u}_0^\epsilon$ converges strongly in L^2 to some $\tilde{\mathbf{u}}_0$ satisfying $Q\tilde{\mathbf{u}}_0 \in H^{s-1}$, $\mathbf{H}_0^\epsilon$ converges strongly in L^2 to some $\mathbf{H}_0$ with $\int_{\mathbb{T}^d} \mathbf{H}_0(x)dx = 0$, $\Pi^\epsilon|_{t=0} = \Pi_0^\epsilon$ converges strongly in L^2 to some $\varphi_0 \in H^{s-1}$, and*

$$||\varphi_0||_{H^2} + ||Q\tilde{\mathbf{u}}_0||_{H^2} \le c_0\nu \quad (2.28)$$

for some constant $c_0 > 0$. Let $(\mathbf{u}, \mathbf{H})$ be the smooth solution to the incompressible MHD equations (2.10)–(2.12) *with $(\mathbf{u}, \mathbf{H})|_{t=0} = (\mathbf{u}_0, \mathbf{H}_0) \in H^s(\mathbb{T}^d)$ satisfying $\mathbf{u}_0 = P\tilde{\mathbf{u}}_0$ and* $\mathrm{div}\mathbf{H}_0 = 0$. *Then, for any $0 < T < T^{**}$ (T^{**} is the maximal time of existence for* (2.10)–(2.12)*), the global weak solution $(\rho^\epsilon, \mathbf{u}^\epsilon, \mathbf{H}^\epsilon)$ of the compressible MHD equations* (2.2)–(2.4) *satisfies*

1. ρ^ϵ *converges strongly to* 1 *in* $C([0,T], L_2^\gamma(\mathbb{T}^d))$;
2. $\nabla \mathbf{H}^\epsilon$ *converges strongly to* $\nabla \mathbf{H}$ *in* $L^2(0,T; L^2(\mathbb{T}^d))$;
3. $\mathbf{H}^\epsilon$ *converges strongly to* $\mathbf{H}$ *in* $L^\infty(0,T; L^2(\mathbb{T}^d))$;
4. $P(\sqrt{\rho^\epsilon}\mathbf{u}^\epsilon)$ *converges strongly to* $\mathbf{u}$ *in* $L^\infty(0,T; L^2(\mathbb{T}^d))$;
5. $\sqrt{\rho^\epsilon}\mathbf{u}^\epsilon$ *converges weakly to* $\mathbf{u}$ *in* $H^{-1}(0,T; L^2(\mathbb{T}^d))$.

By slightly modifying the proof of Theorem 2.4, we can obtain the convergence of compressible MHD equations to the partial viscous incompressible MHD equations when the shear viscosity goes to zero and the magnetic diffusion coefficient goes to a positive constant. The partial viscous incompressible MHD equations correspond to the case of turbulent flow with very high Reynolds number. For the details, see [6].

References

[1] Y. Brenier, Convergence of the Vlasov-Poisson system to the incompressible Euler equations. Comm. Partial Differential Equations 25 (2000), 737–754.

[2] B. Desjardins and E. Grenier, Low Mach number limit of viscous compressible flows in the whole space. R. Soc. Lond. Proc. Ser. A Math. Phys. Eng. Sci. 455 (1999), 2271–2279.

[3] X. P. Hu and D. H. Wang, Global existence and large-time behavior of solutions to the three-dimensional equations of compressible magnetohydrodynamic flows. Arch. Ration. Mech. Anal. 197 (2010), 203–238.

[4] X. Hu and D. H. Wang, Low Mach number limit of viscous compressible magnetohydrodynamic flows. SIAM J. Math. Anal. 41(2009), 1272–1294.

[5] S. Jiang, Q. C. Ju and F. C. Li, Incompressible limit of the compressible magnetohydrodynamic equations with vanishing viscosity coefficients. SIAM J. Math. Anal. 42 (2010), in press.

[6] S. Jiang, Q. C. Ju and F. C. Li, Incompressible limit of the compressible Magnetohydrodynamic equations with periodic boundary conditions. Comm. Math. Phys. 297 (2010), 371–400.

[7] N. Masmoudi, Incompressible, inviscid limit of the compressible Navier-Stokes system. Ann. Inst. H. Poincaré Anal. Non Linéaire 18 (2001), 199–224.

Waiting Time for a Non-Newtonian Polytropic Filtration Equation with Convection*

Chunhua Jin

School of Mathematical Sciences, Sourth China Normal University
Guangzhou, 510631, China
E-mail: jinchhua@126.com

Tong Yang

Department of Mathematics, City University of Hong Kong
Hong Kong, China

Jingxue Yin

School of Mathematical Sciences, Sourth China Normal University
Guangzhou, 510631, China

Abstract

In this paper, we discuss the waiting time phenomena for a class of non-Newtonian polytropic filtration equation with convection. According to the influence of the convection on the waiting time property, the convection is classified into three classes, that is, the strong convection, the mild convection and the weak convection. For different cases, the sufficient and necessary conditions on the initial data are given for the existence of waiting time respectively.

1 Introduction

This paper is concerned with a class of non-Newtonian polytropic filtration equation with convection

$$\frac{\partial u}{\partial t} = (|(u^m)_x|^{p-2}(u^m)_x)_x + \lambda(u^q)_x, \quad (x,t) \in Q, \tag{1.1}$$

*The authors are partly supported by NSFC (No. 10901070), partially supported by China Postdoctoral Science Foundation (No. 20080041110), and partially supported by Doctoral Program Foundation for New Teachers of Higher Education of China (No. 20090061120040).

with initial data given by

$$u(x,0) = u_0(x), \quad x \in \mathbb{R}. \tag{1.2}$$

Here, $Q = \mathbb{R} \times \mathbb{R}^+$, $p > 1$, $m(p-1) > 1$, $q > 0$, $\lambda \in \mathbb{R}$, and $u_0(x)$ is a non-negative and continuous function with compact support.

We aimed to discuss the waiting time phenomenon of solutions to the problem (1.1)–(1.2). For this purpose, set

$$\xi_-(t) = \inf\{x \in \mathbb{R}; u(x,t) > 0\}, \quad \xi_+(t) = \sup\{x \in \mathbb{R}; u(x,t) > 0\}.$$

Here, we assume that $u_0(x) > 0$ in the domain $(\xi_-(0), \xi_+(0))$. By the well known property of finite speed of propagation of perturbation in this kind of equations, cf. for example [1], both $\xi_-(t)$ and $\xi_+(t)$ are finite provided that the initial data have compact support under some general conditions on the parameter p, m, q as stated later. Here, $\xi_-(t)$ and $\xi_+(t)$ are called the interfaces of solutions. The appearance of the interfaces can be found in many physical situations, for example, the wetting fronts separating wet and dry regions of the porous medium in the soil-moisture infiltration; the leading edge of the fluid flow in a thin viscous film with the unknown variable representing the thickness of the film, cf. [2]. When the initial data are appropriately smooth near the boundary of the support, these interfaces may be stationary for certain time, called the waiting time. While if the support never expand for certain time, then we call it "forward" waiting time. (Note: It does not exclude the possibility that the support of the function shrinks.)

This waiting time phenomena has been investigated by some authors, for example, Knerr [3], Aronson-Caffarelli-Kamin [4], and Vazquez [5] considered the waiting time for the special case when $p = 2$, $\lambda = 0$ in the equation (1.1), namely the porous media equation

$$\frac{\partial u}{\partial t} = (u^m)_{xx}, \tag{1.3}$$

where $m > 1$. Some upper and lower bound estimates of the waiting time were also given in these works. Subsequently, Alvarez and Díaz [6] considered the waiting time property for the equations with convection, that is the special case of the equation (1.1) with $p = 2$, and some sufficient and necessary conditions for the existence of the "forward" waiting time are obtained. Besides this, there are also some papers concerning the waiting time phenomena for the higher order equation, see for example [7]. However, as far as we known, there are few papers concerning with the waiting time phenomena for the non-Newtonian fluid equation, even for the special case $m = 1$.

Due to the degeneracy and singularity, the problem (1.1)–(1.2) may not admit classical solutions, this lead to the following definition of weak solutions.

Definition 1.1. Let

$$E = \{u \in C(\overline{Q}_T); u^m \in W^{1,p}_{\text{loc}}(Q)\}.$$

A function $u \in E$ is called a local weak solution of the Cauchy problem for the above equation provided that for some $T > 0$ and any continuously differential function φ with $\varphi(x,t) = 0$ for large $|x|$ and $t = T$, it holds

$$\begin{aligned}&\iint_{Q_T} u\varphi_t dxdt + \int_{\mathbb{R}} u_0(x)\varphi(x,0)dx \\ = &\iint_{Q_T} |(u^m)_x|^{p-2}(u^m)_x\varphi_x dxdt + \iint_{Q_T} \lambda u^q \varphi_x dxdt.\end{aligned}$$

The following three comparison principles are the main tool to study the waiting time property of solutions.

Lemma 1.2. *Let u_1 and u_2 be two weak upper and lower solutions of the Cauchy problem with $\lambda = \lambda_1$ and $\lambda = \lambda_2$ respectively, where $\lambda_1 \geq \lambda_2$, $u_i(x,0) = u_{i0}(x) \in L^1(\mathbb{R})$ $(i = 1,2)$, and*

$$\int_{-\infty}^{x} u_{10}(s)ds \geq \int_{-\infty}^{x} u_{20}(s)ds \quad \textit{for any} \quad x \in \mathbb{R}.$$

Then for any $t > 0$, we have

$$\int_{-\infty}^{x} u_1(s,t)ds \geq \int_{-\infty}^{x} u_2(s,t)ds \quad \textit{for any} \quad x \in \mathbb{R}.$$

Lemma 1.3. *Let u_1 and u_2 be two weak upper and lower solutions of the Cauchy problem with $\lambda = \lambda_1$ and $\lambda = \lambda_2$ respectively, where $\lambda_1 \leq \lambda_2$, $u_i(x,0) = u_{i0}(x) \in L^1(\mathbb{R})$ $(i = 1,2)$, and*

$$\int_{x}^{+\infty} u_{10}(s)ds \geq \int_{x}^{+\infty} u_{20}(s)ds \quad \textit{for any} \quad x \in \mathbb{R}.$$

Then for any $t > 0$, we have

$$\int_{x}^{+\infty} u_1(s,t)ds \geq \int_{x}^{+\infty} u_2(s,t)ds \quad \textit{for any} \quad x \in \mathbb{R}.$$

Lemma 1.4. *It can also be verified that the above two lemmas remain true for the lateral boundary value problem in the domain $(-\infty,\delta] \times [0,T]$ or $[\delta,\infty) \times [0,T]$, if u_1, u_2 satisfy*

$$\int_{-\infty}^{\delta} u_1(s,t)ds \geq \int_{-\infty}^{\delta} u_2(s,t)ds \quad \textit{for any} \quad t \in (0,T]$$

or

$$\int_{\delta}^{+\infty} u_1(s,t)ds \geq \int_{\delta}^{+\infty} u_2(s,t)ds \quad \textit{for any} \quad t \in (0,T]$$

respectively, besides these conditions in the above two Lemmas.

It is easy to see that if u_1, u_2 satisfy

$$\int_{-\infty}^{x} u_1(s,t)ds \geq \int_{-\infty}^{x} u_2(s,t)ds, \quad \text{for any } x \in \mathbb{R},$$

then

$$\xi_{1-}(t) \leq \xi_{2-}(t);$$

similarly, if

$$\int_{x}^{+\infty} u_1(s,t)ds \geq \int_{x}^{+\infty} u_2(s,t)ds, \quad \text{for any } x \in \mathbb{R},$$

then

$$\xi_{1+}(t) \geq \xi_{2+}(t).$$

2 The main results

To discuss the influence of the convection on the waiting time property, we notice that the direction of the convection plays an important role. On the other hand, even though the case when $\lambda > 0$ is different from the case when $\lambda < 0$, they are just related by replacing x by $-x$. Thus, without loss of generality, it suffices to consider the case $\lambda \geq 0$. Notice that when $\lambda > 0$, the convection direction is in the opposite direction along the x-axis. We show that convection can be classified into three classes, the strong, mild and weak convections according to their influence on the waiting time. For this, there are two critical values

$$q_0 = 1, \quad q_c = \frac{(m+1)(p-1)}{p},$$

so that the three classes correspond to the following decomposition on the interval for q

$$(0, +\infty) = (0, q_0] \cup (q_0, q_c) \cup [q_c, +\infty).$$

The analysis in this paper shows that for the weak convection case when $q \in [q_c, +\infty)$, the waiting time property of solutions is completely analogous to the case without convection. That is,

Theorem 2.1. *Let $q \geq q_c$, and $u(x,t)$ be a weak solution of the Cauchy problem. Then there exists a waiting time $T > 0$ such that $\xi_-(t) = \xi_-(0)$ for all $t \in (0,T]$, if and only if*

$$\lim_{x \to \xi_-(0)^+} |x - \xi_-(0)|^{(m+1)(p-1)/(1-m(p-1))} \int_{-\infty}^{x} u_0(s)ds < \infty.$$

Theorem 2.2. *Let $q \geq q_c$, and $u(x,t)$ be a weak solution of the Cauchy problem. Then there exists a waiting time $t^* > 0$ such that $\xi_+(t) = \xi_+(0)$ for all $t \in [0,t^*]$, if and only if*

$$\lim_{x \to \xi_+(0)^-} |x - \xi_+(0)|^{(m+1)(p-1)/(1-m(p-1))} \int_{x}^{+\infty} u_0(s)ds < \infty.$$

For the mild convection when $q \in (q_0, q_c)$, the influence of the convection can not be ignored. And the behavior of the left and right interfaces are quite different because of the direction of the convection.

Theorem 2.3. *Let $q_0 < q < q_c$, and $u(x,t)$ be a weak solution of the Cauchy problem. Then there exists a waiting time $T > 0$ such that $\xi_-(t) = \xi_-(0)$ for all $t \in (0,T]$, if and only if*

$$\lim_{x \to \xi_-(0)^+} |x - \xi_-(0)|^{q/(1-q)} \int_{-\infty}^{x} u_0(s)ds < \infty.$$

Theorem 2.4. *Let $q_0 < q < q_c$, and $u(x,t)$ be a weak solution of the Cauchy problem. Denote*

$$\widehat{M}(q, u_0) \equiv \lim_{x \to \xi_+(0)^-} |x - \xi_+(0)|^{-r^*} \int_{x}^{+\infty} u_0(s)ds$$

with $r^* = \frac{(m+1)(p-1)-q}{m(p-1)-q}$. Then there exists a critical value $\widehat{M_c}(q)$ such that

(i) *if $\widehat{M}(q,u_0) < \widehat{M_c}(q)$, then there exists $t^* > 0$ such that $\xi_+(t) = \xi_+(0)$ for all $t \in ([0,t^*]$;*

(ii) *if $\widehat{M}(q,u_0) > \widehat{M_c}(q)$, then there exist $\widetilde{C}, \tilde{t} > 0$ such that*

$$\xi_+(t) \geq \xi_+(0) + \widetilde{C} t^{\frac{m(p-1)-q}{(m+1)(p-1)-qp}} \qquad \text{for any } \ t \in [0, \tilde{t}],$$

where $\widehat{M_c}(q) = \frac{1}{r^*}\left(\lambda\left(1 - \frac{q}{m(p-1)}\right)^{p-1}\right)^{\frac{1}{m(p-1)-q}}$.

We see that, for the right interface, the restriction on the initial data for the existence of waiting time is weaker and it depends on m, p, and q; but for the left interface, the convection dominates the diffusion, the restriction on the initial data depends only on the exponent index of the convection term.

Finally, for the strong convection case when $q \in (0, q_0]$:

Theorem 2.5. *Let $q \leq q_0$, and $u(x,t)$ be the solution of the Cauchy problem. Then the waiting time phenomenon will not happen for the left interface. Especially*

(i) *if $q < q_0$, $\xi_-(t)$ doesn't exist any more.*
(ii) *if $q = q_0$, $\xi_-(t) \leq \xi_-(0) - \lambda t$ for any $t > 0$.*

Theorem 2.6. *Let $q \leq q_0$, and $u(x,t)$ be the solution of the Cauchy problem. Let $\widehat{M}(q,u_0)$ and $\widehat{M}_c(q)$ be defined as in the above Theorem.*

(i) *If $\widehat{M}(q,u_0) < \widehat{M}_c(q)$, then there exist $\hat{t}, \hat{c} > 0$ such that*

$$\xi_+(t) \leq \xi_+(0) - \hat{c}t^{\frac{m(p-1)-q}{(m+1)(p-1)-pq}} \quad \text{for any } t \in (0, \hat{t}].$$

(ii) *If $\widehat{M}(q,u_0) > \widehat{M}_c(q)$, then there exist $\widetilde{c}, \tilde{t} > 0$ such that*

$$\xi_+(t) \geq \xi_+(0) + \widetilde{c}t^{\frac{m(p-1)-q}{(m+1)(p-1)-qp}} \quad \text{for any } t \in [0, \tilde{t}].$$

We note that the left interface exists only when $q = q_0$. On the other hand, even though the right interface always exists, it travels to either the left or the right in general, and the existence of waiting time appears only in some particular situations.

References

[1] J. N. Zhao, Lipschitz continuity of the free boundary of some nonlinear degenerate parabolic equations. Nonlin. Anal. TMA. 28 (1997), 1047–1062.

[2] B. H. Gilding, The occurrence of interfaces in nonlinear diffusion advection processes. Arch. Rational Mech. Anal. 100 (1988), 243–263.

[3] B. F. Knerr, The porous medium equation in one dimension. Trans. Amer. Math. Soc. 234 (1977), 381–415.

[4] D. G. Aronson, L. A. Caffarelli and S. Kamin, How an initial stationary interface begins to move in porous medium flow. SIAM. J. Math. Anal. 14 (1983), 639–658.

[5] J. L. Vazquez, The interfaces of one dimensional flows in porous media. Trans. Amer. Math. Soc. 285 (1984), 717–737.

[6] L. Alvarez and J. I. Díaz, Sufficient and necessary initial mass conditions for the existence of a waiting time in nonlinear-convection processes. J. Math. Anal. Appl. 155 (1991), 378–392.

[7] R. Dal Passo, L. Giacomelli and G. Grün, A waiting time phenomenon for thin film equations. Ann. Scuola Norm. Sup. Pisa Cl. Sci. 30 (2001), 437–463.

Stability of Rarefaction Waves to the 1D Compressible Navier-Stokes Equations with Density-dependent Viscosity*

Quansen Jiu
School of Mathematical Sciences, Capital Normal University
Beijing 100048, China
E-mail: qsjiumath@gmail.com

Yi Wang
Institute of Applied Mathematics, AMSS
& Hua Loo-Keng Key Laboratory of Mathematics
CAS, Beijing 100190, China
E-mail: wangyi@amss.ac.cn

Zhouping Xin
The Institute of Mathematical Sciences
Chinese University of Hong Kong
Shatin, N. T., Hong Kong, China
E-mail: zpxin@ims.cuhk.edu.hk

Abstract

In this paper, we study the asymptotic stability of rarefaction waves for the compressible isentropic Navier-Stokes equations with density-dependent viscosity. First, a weak solution around a rarefaction wave to the Cauchy problem is constructed by approximating the system and regularizing the initial values which may contain vacuum state. Then some global in time estimates on the weak solution are obtained. Based on these uniform estimates, the vacuum states are shown to vanish in finite time and the weak solution we constructed becomes a unique strong one. Consequently, the stability of the rarefaction wave is proven in a weak sense. The theory holds for large-amplitudes rarefaction waves and arbitrary initial perturbations.

*Q. S. Jiu is partially supported by the National Natural Sciences Foundation of China (No. 10871133) and Project of Beijing Education Committee. Y. Wang is partially supported by the National Natural Sciences Foundation of China (No. 10801128). Z. P. Xin is partially supported by Zheng Ge Ru Funds, Hong Kong RGC Earmarked Research Grant CUHK4042/08P and CUHK4040/06P, and a Focus Area Grant at The Chinese University of Hong Kong.

1 Introduction

In this paper, we consider the following compressible and isentropic Navier-Stokes equations with density-dependent viscosity

$$\begin{cases} \rho_t + (\rho u)_x = 0, & x \in \mathbf{R}, t > 0, \\ (\rho u)_t + (\rho u^2 + p(\rho))_x = (\mu(\rho) u_x)_x, \end{cases} \tag{1.1}$$

where $\rho(t,x) \geq 0$, $u(t,x)$ represent the density and the velocity of the gas, respectively. Assume that the pressure and viscosity function is given by

$$p(\rho) = A\rho^\gamma, \quad \mu(\rho) = B\rho^\alpha, \tag{1.2}$$

where $\gamma > 1$ denotes the adiabatic exponent, $\alpha > 0$ and $A, B > 0$ are the gas constants. Without loss of generality, we assume that $A = B = 1$.

We consider the Cauchy problem (1.1) with the initial values

$$(\rho, \rho u)(0,x) = (\rho_0, m_0)(x) \to (\rho_\pm, m_\pm) \ \text{ as } \ x \to \pm\infty, \tag{1.3}$$

where $\rho_\pm > 0$, $m_\pm$ are given constants. Thus we can define the far field velocity by $u_\pm = \frac{m_\pm}{\rho_\pm}$.

The large time asymptotic behavior of solutions to (1.1) is expected to be closely related to that of the corresponding Euler system

$$\begin{cases} \rho_t + (\rho u)_x = 0, \\ (\rho u)_t + (\rho u^2 + p(\rho))_x = 0. \end{cases} \tag{1.4}$$

The Euler system (1.4) is a strictly hyperbolic one for $\rho > 0$ whose characteristic fields are both genuinely nonlinear. The Jacobi matrix of the flux has two distinct eigenvalues $\lambda_1(\rho,u) = u - \sqrt{p'(\rho)}$, $\lambda_2(\rho,u) = u + \sqrt{p'(\rho)}$ with corresponding right eigenvectors $r_i(\rho,u) = (1, (-1)^i \frac{\sqrt{p'(\rho)}}{\rho})^t$, $(i = 1,2)$ such that

$$r_i(\rho,u) \cdot \nabla_{\rho,u}\lambda_i(\rho,u) = (-1)^i \frac{\rho p''(\rho) + 2p'(\rho)}{2\rho\sqrt{p'(\rho)}} \neq 0, \quad i = 1,2.$$

Define the i−Riemann invariant $(i = 1,2)$ by

$$\Sigma_i(\rho,u) = u + (-1)^{i+1} \int^\rho \frac{\sqrt{p'(s)}}{s} ds$$

such that $\nabla_{(\rho,u)}\Sigma_i(\rho,u) \cdot r_i(\rho,u) \equiv 0, \ \forall \rho > 0, u$.

There are two families of rarefaction waves to the Euler system (1.4). Here we only consider 2-rarefaction wave, which is characterized by the fact that 2-Riemann invariant $\Sigma_2(\rho, u)$ is constant in (x, t) and 2-characteristic speed, $\lambda_2(\rho, u)$ is increasing in x. Suppose that the end states $(\rho_\pm, u_\pm)$ of the initial values of (1.4) satisfy

$$\Sigma_2(\rho_-, u_-) = \Sigma_2(\rho_+, u_+), \qquad \lambda_2(\rho_+, u_+) > \lambda_2(\rho_-, u_-),$$

then the state (ρ_-, u_-) can be connected to the state (ρ_+, u_+) by a 2-rarefaction wave. The 2-rarefaction waves for (1.4) connecting (ρ_-, u_-) to (ρ_+, u_+) converge to each other time-asymptotically in supreme norm. For definiteness, we choose a particular 2-rarefaction wave $(\bar{\rho}, \bar{u})$ of (1.4) defined by (2.3) in the next section.

In the present paper, we are interested in the large time-asymptotic stability of the above rarefaction wave to the density-dependent Navier-Stokes equation (1.1) with large initial values which may contain vacuum states.

There have been large literature on the global existence and the large time behavior of solutions to the system (1.1) and even in the higher dimensional case when the viscosity $\mu(\rho)$ is constant. However, the possible appearance of the vacuum is one of the major difficulties to the global existence of the solutions to the compressible Navier-Stokes equations with constant viscosity. Hoff and Smoller [4] proved that weak solutions of the compressible Navier-Stokes equations in 1D case do not contain vacuum states in finite time if there is no vacuum initially. On the other hand, Xin [11] proved that there is no global smooth solution to the Cauchy problem to compressible Navier-Stokes equations with a nontrivial compactly supported initial density, which means that the solution will blow up in finite time in the presence of the vacuum states. Moreover, Hoff and Serre [3] proved the results of the failure of the continuous dependence of the weak solutions containing vacuum states on the initial values.

Thus, when the solution may contain vacuum states, it is natural to consider the compressible Navier-Stokes equations with density-dependent viscosity, as was derived from Chapman-Enskog expansions from the Boltzmann equation where the viscosity depends on the temperature and thus on the density for isentropic flows. Moreover, a one-dimensional viscous Saint-Venant system for the shallow water, derived rigorously from the incompressible Navier-Stokes equation with a moving free surface by Gerbeau-Perthame recently in [2], is expressed exactly as in (1.1)–(1.2) with $\alpha = 1$ and $\gamma = 2$.

There are many literatures on mathematical studies on (1.1)–(1.2). If the initial density is assumed to be connected to vacuum with discontinuities, Liu, Xin and Yang first obtained in [8] the local existence of weak solutions. The global existence of weak solutions was obtained later by many people. If the initial density connects to vacuum continuously, then new difficulty is encountered since no positive lower bound for the density is available. However, most of results concern with free boundary problems. Recently, initial-boundary-value problems for one-dimensional equations (1.1)–(1.2) with $\mu(\rho) = \rho^\alpha (\alpha > 1/2)$ was studied by Li, Li and Xin in [6] and the phenomena of vacuum vanishing and blow-up of solutions were found there. More recently, there are some results on Cauchy problem (1.1)–(1.3). The existence and uniqueness of global strong solutions to the compressible Navier-Stokes equations (1.1)–(1.3) were obtained by Mellet and Vasseur [10] where no vacuum is permitted in the initial density. However, the a priori estimates obtained in [10] depends on the time interval thus does not yield the time-asymptotic behavior of the solutions. The first result about the time-asymptotic behavior of the solutions to the Cauchy problem (1.1)–(1.3) is obtained by Jiu and Xin [5], where the global existence, large time-asymptotic behavior, the vanishing of the vacuum, and the blow-up phenomena of the weak solutions were considered in the case that $\rho_+ = \rho_-$ and $u_+ = u_- = 0$. It is well-known that the large time-asymptotic behavior of solutions to the system (1.1)–(1.3) with different far field states of the initial values is closely related to the corresponding Euler system (1.4). Rarefaction wave is one of the fundamental wave patterns to the Euler system (1.4). A natural question is how about the stability of rarefaction waves to the compressible Navier-Stokes equations (1.1)–(1.3) in the framework of weak solutions which may contain vacuum states.

In the case $\rho_\pm > 0$, that is, the rarefaction wave does not contain vacuum, we study in this paper the global existence, large time-asymptotic behavior, vanishing of the vacuum and the blow-up phenomena of weak solutions to the Cauchy problem (1.1)–(1.3). First we will construct a class of approximate solutions satisfying some uniform estimates and furthermore prove the global existence of weak solutions for Cauchy problem of (1.1)–(1.3). Moreover, the time-asymptotic behaviors of weak solutions are investigated. More precisely, it is proven that the density ρ tends to a rarefaction wave as $t \to \infty$. As a consequence, there exists a time $T_0 > 0$ such that when $t > T_0$, the vacuum states vanish and the global weak solution becomes a unique strong one. Moreover, the stability of the rarefaction wave is obtained in some weak sense. The key point in the proof is the uniform estimates in time and the approximations to the approximate solutions obtained by carefully combining the elementary energy estimates and the new entropy estimates. Note that the elementary energy estimates and the new entropy estimates uniformly in time are coupled each other due to the rarefaction waves. This is quite different from the previous works on the global existence and the time-asymptotic behavior of the solutions to the density dependent Navier-Stokes equations (1.1). Based on these uniform estimates, we successfully get the H_x^1 estimates to the density function like $\rho^{\alpha-\frac{1}{2}} - \bar{\rho}^{\alpha-\frac{1}{2}}$ and the dynamic of the vacuum states to the weak solutions.

2 Preliminaries and main results

Consider the solution to the following Cauchy problem for Burgers equation

$$\begin{cases} w_t + ww_x = 0, \quad t > 0, x \in \mathbf{R}, \\ w_0(x) := w(0,x) = \dfrac{w_+ + w_-}{2} + \dfrac{w_+ - w_-}{2} K_q \displaystyle\int_0^{\eta x} (1+y^2)^{-q}\, dy. \end{cases} \tag{2.1}$$

Here $q \geq 2$ is some fixed constant, and K_q is a constant such that $K_q \int_0^\infty (1+y^2)^{-q} dy = 1$, and η is a small positive constant to be determined later. It is easy to see that the solution to the above Burgers equation is given by

$$w(t,x) = w_0(x_0(t,x)), \qquad x = x_0(t,x) + w_0(x_0(t,x))t. \tag{2.2}$$

Define a 2-rarefaction wave $(\bar\rho, \bar u)(t,x)$ by

$$\begin{aligned} &\lambda_2(\rho_\pm, u_\pm) = w_\pm, \quad w_- < w_+, \\ &\lambda_2(\bar\rho(t,x), \bar u(t,x)) = w(1+t,x), \\ &\Sigma_2(\bar\rho(t,x), \bar u(t,x)) = \Sigma_2(\rho_\pm, u_\pm), \end{aligned} \tag{2.3}$$

thus $(\bar\rho, \bar u)(t,x)$ satisfies the system

$$\begin{cases} \bar\rho_t + (\bar\rho \bar u)_x = 0, \\ (\bar\rho \bar u)_t + (\bar\rho \bar u^2 + p(\bar\rho))_x = 0. \end{cases} \tag{2.4}$$

Lemma 2.1 ([9]). *The 2-rarefaction wave $(\bar\rho, \bar u)(t,x)$ satisfies*

i) $\bar\rho_x > 0, \quad \bar u_x > 0$;

ii) *For any p $(1 \leq p \leq \infty)$, there exists a constant C_{pq} such that*

$$\begin{aligned} &\|(\bar\rho_x, \bar u_x)(t,\cdot)\|_{L^p(\mathbf{R})} \leq C_{pq} \min\{\delta \eta^{1-\frac{1}{p}}, \delta^{\frac{1}{p}}(1+t)^{-1+\frac{1}{p}}\}, \\ &\|(\bar\rho_{xx}, \bar u_{xx})(t,\cdot)\|_{L^p(\mathbf{R})} \leq C_{pq} \min\{\delta \eta^{2-\frac{1}{p}}, \\ &\qquad \eta^{(1-\frac{1}{2q})(1-\frac{1}{p})} \delta^{-\frac{p-1}{2pq}} (1+t)^{-1-\frac{p-1}{2pq}} + \delta^{\frac{1}{p}} (1+t)^{-2+\frac{1}{p}}\}, \end{aligned}$$

where $\delta = |\rho_+ - \rho_-|$ is the strength of the rarefaction wave;

iii) $\lim\limits_{t\to\infty} \sup\limits_{x\in\mathbf{R}} \left|(\bar\rho, \bar u)(t,x) - (\rho^r, u^r)(\frac{x}{t})\right| = 0.$

Set

$$\Psi(\rho, \bar\rho) = \int_{\bar\rho}^{\rho} \frac{p(s) - p(\bar\rho)}{s^2} ds = \frac{1}{(\gamma-1)\rho}[\rho^\gamma - \bar\rho^\gamma - \gamma \bar\rho^{\gamma-1}(\rho - \bar\rho)]. \tag{2.5}$$

The initial data are assumed to satisfy:

$$\begin{cases} \rho_0 \geq 0; \quad m_0 = 0 \text{ a.e. on } \{x \in \mathbf{R} | \rho_0(x) = 0\}; \\ (\rho_0^{\alpha-\frac{1}{2}})_x \in L^2(\mathbf{R}), \quad \rho_0 \Psi(\rho_0, \bar{\rho}_0) \in L^1(\mathbf{R}); \\ \rho_0(\frac{m_0}{\rho_0} - \bar{u}_0)^2 \in L^1(\mathbf{R}), \quad \rho_0(\frac{m_0}{\rho_0} - \bar{u}_0)^3 \in L^1(\mathbf{R}), \end{cases} \tag{2.6}$$

where $\alpha > \frac{1}{2}$ and $(\bar{\rho}_0, \bar{u}_0) := (\bar{\rho}, \bar{u})(0, x)$ is the initial values of the 2-rarefaction wave $(\bar{\rho}, \bar{u})(t, x)$ constructed in section 2.1. Note that (2.6) implies that $\rho_0 \in C(\mathbf{R})$ which is the space of continuous functions.

Definition 2.1. A pair (ρ, u) is said to be a weak solution to (1.1)–(1.3) towards the rarefaction wave $(\bar{\rho}, \bar{u})(t, x)$ in (2.4), provided that

(1) $\rho \geq 0$ a.e., and

$$\rho \in L^\infty(0, T; L^\infty(\mathbf{R})) \cap C([0, \infty); W^{1,\infty}(\mathbf{R})^*),$$
$$(\rho^{\alpha-\frac{1}{2}})_x \in L^\infty(0, T; L^2(\mathbf{R})), \sqrt{\rho}(u - \bar{u}) \in L^\infty(0, T; L^2(\mathbf{R})),$$

where $W^{1,\infty}(\mathbf{R})^*$ is the dual space of $W^{1,\infty}(\mathbf{R})$;

(2) For any $t_2 \geq t_1 \geq 0$ and any $\zeta \in C_0^1(\mathbf{R} \times [t_1, t_2])$, the mass equation (1.1) holds in the following sense:

$$\int_{\mathbf{R}} (\rho - \bar{\rho})\zeta dx|_{t_1}^{t_2} = \int_{t_1}^{t_2} \int_{\mathbf{R}} [(\rho - \bar{\rho})\zeta_t + (\rho u - \bar{\rho}\bar{u}) \cdot \zeta_x] dx dt; \tag{2.7}$$

(3) For any $\psi \in C_0^\infty(\mathbf{R} \times [0, T))$, it holds that

$$\begin{aligned} &\int_{\mathbf{R}} (m_0 - \bar{\rho}_0 \bar{u}_0)\psi(0, \cdot) dx + \int_0^T \int_{\mathbf{R}} \{\sqrt{\rho}[\sqrt{\rho}(u - \bar{u})] + (\rho - \bar{\rho})\bar{u}\}\psi_t \\ &+ \{[\sqrt{\rho}(u - \bar{u})]^2 - 2\sqrt{\rho}\sqrt{\rho}(u - \bar{u})\bar{u} + (\rho - \bar{\rho})\bar{u}^2 + (\rho^\gamma - \bar{\rho}^\gamma)\}\psi_x dx dt \\ &+ < \rho^\alpha (u - \bar{u})_x, \psi_x > + \int_0^T \int_{\mathbf{R}} \rho^\alpha \bar{u}_x \psi_x dx dt = 0, \end{aligned} \tag{2.8}$$

where the diffusion term makes sense when written as

$$\begin{aligned} < \rho^\alpha (u - \bar{u})_x, \phi >= &- \int_0^T \int_R \rho^{\alpha - \frac{1}{2}} \sqrt{\rho}(u - \bar{u}) \phi_x dx dt \\ &- \frac{2\alpha}{2\alpha - 1} \int_0^T \int_R (\rho^{\alpha-\frac{1}{2}})_x \sqrt{\rho}(u - \bar{u}) \phi dx dt. \end{aligned} \tag{2.9}$$

Our main results read as

Theorem 2.1 (Existence of a weak solution). Let α and γ satisfy that $\frac{1}{2} < \alpha \leq \frac{\gamma+1}{2}$. Suppose that (2.6) holds. Then the Cauchy problem

(1.1)–(1.3) admits a global weak solution $(\rho(x,t), u(x,t))$ satisfying

$$\rho \geq 0, \quad \rho \in C(\mathbf{R} \times (0,T)), \quad \max_{(x,t)\in\mathbf{R}\times[0,T]} \rho \leq C, \tag{2.10}$$

$$\sup_{t\in[0,T]} \int_{\mathbf{R}} \Big(|\sqrt{\rho}(u-\bar{u})|^2 + (\rho^{\alpha-\frac{1}{2}})_x^2 + \frac{1}{\gamma-1}[\rho^\gamma - \bar{\rho}^\gamma - \gamma\bar{\rho}^{\gamma-1}(\rho-\bar{\rho})]\Big)dx$$

$$+\int_0^T \int_{\mathbf{R}} \big([(\rho^{\frac{\gamma+\alpha-1}{2}} - \bar{\rho}^{\frac{\gamma+\alpha-1}{2}})_x]^2 + \Lambda(x,t)^2\big)dxdt \leq C, \tag{2.11}$$

where C is an absolute constant depending on the initial data and $\Lambda(x,t) \in L^2(\mathbf{R}\times(0,T))$ satisfying

$$\int_0^T \int_{\mathbf{R}} \Lambda\varphi dxdxt = -\int_0^T \int_{\mathbf{R}} \rho^{\alpha-\frac{1}{2}} \sqrt{\rho}(u-\bar{u})\varphi_x dxdt$$

$$-\frac{2\alpha}{2\alpha-1}\int_0^T \int_{\mathbf{R}} (\rho^{\alpha-\frac{1}{2}})_x \sqrt{\rho}(u-\bar{u})\varphi dxdt. \tag{2.12}$$

Remark 1: Obviously, the case of shallow water equation, i.e., $\alpha = 1$, $\gamma = 2$, is included in our theorem.

Theorem 2.2. Suppose that $(\rho(x,t), u(x,t))$ is a weak solution of the Cauchy problem (1.1)–(1.3) satisfying (2.10) and (2.11). Then we have

$$\lim_{t\to\infty} \sup_{x\in\mathbf{R}} |\rho - \bar{\rho}| = 0. \tag{2.13}$$

Based on Theorem 2.2, it is easy to deduce that under the assumption that the approximate rarefaction wave satisfies $\inf_{x,t} \bar{\rho}(t,x) > 0$, there exists a time $T_0 > 0$ after which the density has a positive lower bound and the vacuum states vanish. Moreover, it will be shown that after the time $t = T_0$, the weak solution becomes a unique strong one. Precisely, we have

Theorem 2.3. Suppose that the assumptions of Theorem 2.1 hold. Let $(\rho(x,t), u(x,t))$ be a weak solution of the Cauchy problem (1.1)–(1.3) satisfying (2.10) and (2.11). Then for any $0 < \rho_1 < \inf_{t,x} \bar{\rho}(t,x)$, there exists a time T_0 such that

$$0 < \rho_1 \leq \rho(x,t) \leq C, \quad (x,t) \in \mathbf{R} \times [T_0, \infty), \tag{2.14}$$

where C is a constant same as in (2.10). Moreover, for $t \geq T_0$, the weak solution becomes a unique strong solution to (1.1)–(1.3), satisfying

$$\begin{aligned} &\rho - \bar{\rho} \in L^\infty(T_0, t; H^1(\mathbf{R})), \quad \rho_t \in L^\infty(T_0, t; L^2(\mathbf{R})), \\ &u - \bar{u} \in L^2(T_0, t; H^2(\mathbf{R})), \quad u_t \in L^2(T_0, t; L^2(\mathbf{R})), \\ &\sup_{x\in\mathbf{R}} |\rho - \bar{\rho}| + \|\rho - \bar{\rho}\|_{L^p(\mathbf{R})} + \|u - \bar{u}\|_{L^2(\mathbf{R})} \to 0 \end{aligned} \tag{2.15}$$

as $t \to \infty$, where $2 < p \leq \infty$.

Remark 2: It is interesting to note that there is no requirement on the sizes of the strength of the rarefaction wave and the perturbations. The class of initial perturbations given by (2.6) is quite large compared with those for the constant viscosity case [7, 9].

References

[1] D. Bresch, B. Desjardins and D. Gerard-Varet, On compressible Navier-Stokes equations with density dependent viscosities in bounded domains. J. Math. Pures Appl. 87 (2) (2007), 227–235.

[2] J. F. Gerbeau and B. Perthame, Derivation of viscous Saint-Venant system for laminar shallow water; numerical validation. Discrete Contin. Dyn. Syst. Ser. B 1 (2001), 89–102.

[3] D. Hoff and D. Serre, The failure of continuous dependence on initial data for the Navier-Stokes equations of compressible flow. SIAM J. Appl. Math. 51 (1991), 887–898.

[4] D. Hoff and J. Smoller, Non-formation of vacuum states for compressible Navier-Stokes equations. Comm. Math. Phys. 216 (2001), no. 2, 255–276.

[5] Q. S. Jiu and Z. P. Xin, The Cauchy problem for 1D compressible flows with density-dependent viscosity coefficients. Kinet. Relat. Models 1(2), (2008), 313–330.

[6] H. L. Li, J. Li and Z. P. Xin, Vanishing of vacuum states and blow-up phenomena of the compressible Navier-Stokes equations. Comm. Math. Phys. 281(2) (2008), 401–444.

[7] T. P. Liu and Z. P. Xin, Nonlinear stability of rarefaction waves for compressible Navier-Stokes equations. Comm. Math. Phys. 118 (1988), 451–465.

[8] T. P. Liu, Z. P. Xin and T. Yang, Vacuum states of compressible flow. Discrete Continuous Dynam. systems 4(1) (1998), 1–32.

[9] A. Matsumura and K. Nishihara, Global stability of the rarefaction wave of a one-dimensional model system for compressible viscous gas. Comm. Math. Phys. 144(2) (1992), 325–335.

[10] A. Mellet and A. Vasseur, Existence and uniqueness of global strong solutions for one-dimensional compressible Navier-Stokes equations. SIAM J. Math. Anal. 39 (2008), No. 4, 1344–1365.

[11] Z. P. Xin, Blow-up of smooth solution to the compressible Navier-Stokes equations with compact density. Comm. Pure Appl. Math. 51(1998), 229–240.

Ten-moment Two-fluid Plasma Model Agrees Well with PIC/Vlasov in GEM Problem*

E. A. Johnson
Department of Mathematics, University of Wisconsin
Madison WI 53717, USA
E-mail: ejohnson@math.wisc.edu

J. A. Rossmanith
Department of Mathematics, University of Wisconsin
Madison WI 53717, USA
E-mail: rossmani@math.wisc.edu

Abstract

We simulate magnetic reconnection in the GEM problem using a two-fluid model with 10 moments for the electron fluid as well as the proton fluid. We show that use of 10 moments for electrons gives good qualitative agreement with the electron pressure tensor components in published kinetic simulations.

1 Overview

This study is motivated by the following question: **What is the simplest fluid model that can accurately replicate kinetic (PIC and Vlasov) simulations of fast magnetic reconnection in two-species collisionless plasma?** The background for this question follows.

A plasma is a gas of charged particles interacting with an electromagnetic field. Simulations of plasma use a variety of models of plasma, which vary greatly in computational expense. For a given problem one seeks the computationally cheapest model that captures the phenomena of interest. We are specifically interested in the phenomenon of fast magnetic reconnection in "collisionless" (i.e., low-collision) plasma.

We discuss the following sequence of plasma models for two-species (e.g., electron/proton) plasmas: **kinetic** Vlasov/Boltzmann or PIC

*This work is supported in part by NSF Grant DMS-0711885.

(particle-in-cell), **ten-moment** two-fluid plasma, **five-moment** two-fluid plasma, and **MHD** (magnetohydrodynamics, a one-fluid model). Each model in this sequence can be regarded as a simplifying appoximation of its predecessor. The plasma community has used particle-in-cell (PIC) simulations as its standard first-principles plasma model. It is an approximation of the Vlasov/Boltzmann model, which we take as the "truth". The Vlasov model is the collisionless version of the Boltzmann model. Collisionless versions of each model are hyperbolic and conserve entropy for smooth solutions, whereas collisional versions are diffusive and produce entropy.

The Geospace Environmental Modeling (GEM) Magnetic Reconnection Challenge is a benchmark magnetic reconnection problem. It was formulated to test the ability of plasma models to resolve fast magnetic reconnection in collisionless plasma. The initial GEM studies showed that PIC simulations exhibit fast reconnection, whereas reconnection is much slower in MHD (with simple resistivity) [1].

2 Plasma models

We implemented five-moment and ten-moment two-fluid models and studied their ability to match published results of PIC and Vlasov simulations. Two-fluid models use gas-dynamics for each charged species. These fluids are coupled to Maxwell's equations by source terms. Hakim, Loverich, and Shumlak simulated the GEM problem with a five-moment two-fluid model [4, 6] and Hakim simulated the GEM problem with a two-fluid model using 10 moments for ions and 5 moments for electrons [5], but we are unaware of any previous studies that simulate the GEM problem with a ten-moment electron fluid.

Boltzmann/Vlasov model. The Boltzmann equation asserts conservation of particle number density $f_{\mathrm{s}}(\mathbf{x}, \tilde{\mathbf{v}}, t)$ in phase space:

$$\partial_t f_{\mathrm{s}} + \nabla_{\mathbf{x}} \cdot (\mathbf{v} f_{\mathrm{s}}) + \nabla_{\tilde{\mathbf{v}}} \cdot (\mathbf{a} f_{\mathrm{s}}) = C_{\mathrm{s}};$$

here $\tilde{\mathbf{v}} = \gamma \mathbf{v} \approx \mathbf{v}$ is (proper) velocity (where $\gamma = \sqrt{1 + (\tilde{v}/c)^2} \approx 1$ is the Lorentz factor), $\mathbf{a} = \frac{q_{\mathrm{s}}}{m_{\mathrm{s}}}(\mathbf{E} + \mathbf{v} \times \mathbf{B})$ is the acceleration due to the electric field $\mathbf{E}$ and the magnetic field $\mathbf{B}$, and C_{s} is a collision operator which operates on the function $(\tilde{\mathbf{v}}, \mathrm{p}) \mapsto f_{\mathrm{p}}(t, \mathbf{x}, \tilde{\mathbf{v}})$, where p ranges over all species. The Vlasov equation (collisionless Boltzmann equation) asserts that $C_{\mathrm{s}} = 0$. The relations $\mathbf{J} = \sum_{\mathrm{s}} q_{\mathrm{s}} \int_{\mathbf{v}} f_{\mathrm{s}} \mathbf{v}$ and $\sigma = \sum_{\mathrm{s}} q_{\mathrm{s}} \int_{\mathbf{v}} f_{\mathrm{s}}$ couple the Boltzmann equation to Maxwell's equations

$$\begin{aligned} \partial_t \mathbf{B} &= -\nabla \times \mathbf{E}, & \nabla \cdot \mathbf{B} &= 0, \\ \partial_t \mathbf{E} &= c^2 \nabla \times B - \mathbf{J}/\epsilon_0, & \nabla \cdot \mathbf{E} &= \sigma/\epsilon_0. \end{aligned}$$

Five-moment model. Generic physical equations for the gas-dynamic portion of the five-moment two-fluid model consist of conservation of mass and balance of momentum and energy for each species:

$$\partial_t \rho_s + \nabla\cdot(\rho_s \mathbf{u}_s) = 0,$$
$$\partial_t(\rho_s \mathbf{u}_s) + \nabla\cdot(\rho_s \mathbf{u}_s \mathbf{u}_s) + \nabla p_s = \frac{q_s}{m_s}\rho_s(\mathbf{E} + \mathbf{u}_s \times \mathbf{B}) + \mathbf{R}_s + \nabla\cdot\underline{\underline{\sigma}}_s,$$
$$\partial_t \mathcal{E}_s + \nabla\cdot(\mathbf{u}_s(\mathcal{E}_s + p_s)) + \nabla\cdot\mathbf{q}_s = \nabla\cdot(\underline{\underline{\sigma}}\cdot\mathbf{u}) + \mathbf{J}_s\cdot\mathbf{E} + \mathbf{u}_s\cdot\mathbf{R}_s + Q_s^f + Q_s^t,$$

where s is the species index (i for ions, e for electrons), ρ_s is mass density, $\mathbf{u}_s$ is fluid velocity, $\mathcal{E}_s$ is gas-dynamic energy, $q_s = \pm e$ is particle charge, m_s is particle mass, and the species current is $\mathbf{J}_s := (q_s/m_s)\rho_s \mathbf{u}_s$. For the pressure we assumed an isotropic monatomic gas: $(3/2)p_s = \left(\mathcal{E}_s - \rho_s u_s^2/2\right)$. A linear isotropic entropy-respecting viscous stress closure is $\underline{\underline{\sigma}}_s = 2\mu_s\left(\mathrm{Sym}\left(\nabla \mathbf{u}_s\right) - \nabla\cdot\mathbf{u}_s \mathbb{I}/3\right)$, where Sym denotes the symmetric part of its argument tensor, $\mathbb{I}$ is the identity tensor, and μ_s is the shear viscosity. In these five-moment simulations, however, we neglect all collisional effects. So we neglect viscosity ($\underline{\underline{\sigma}}_s = 0$), heat flux ($\mathbf{q}_s = 0$), resistive drag force ($\mathbf{R}_s = 0$), resistive heating ($Q_s^f = 0$), and interspecies thermal equilibration ($Q_s^t = 0$). To couple these equations to Maxwell's equations we use the relations

$$\mathbf{J} = \mathbf{J}_i + \mathbf{J}_e, \quad \sigma = \sigma_i + \sigma_e,$$

where $\sigma_s = (q_s/m_s)\rho_s$ is the charge density of each species.

Ten-moment model. Generic physical equations for the gas-dynamic portion of the ten-moment two-fluid model consist of conservation of mass and balance of momentum and energy tensor for each species:

$$\partial_t \rho_s + \nabla\cdot(\rho_s \mathbf{u}_s) = 0,$$
$$\partial_t(\rho_s \mathbf{u}_s) + \nabla\cdot(\rho_s \mathbf{u}_s \mathbf{u}_s + \mathbb{P}_s) = \frac{q_s}{m_s}\rho_s(\mathbf{E} + \mathbf{u}_s \times \mathbf{B}) + \mathbf{R}_s,$$
$$\partial_t \mathbb{E}_s + 3\nabla\cdot\mathrm{Sym}\left(\mathbf{u}_s \mathbb{E}_s\right) - 2\nabla\cdot(\rho_s \mathbf{u}_s \mathbf{u}_s \mathbf{u}_s) + \nabla\cdot\mathbb{q}_s$$
$$= \frac{q_s}{m_s} 2\,\mathrm{Sym}\left(\rho_s \mathbf{u}_s \mathbf{E} + \mathbb{E}_s \times \mathbf{B}\right) + \mathbb{R}_s + 2\,\mathrm{Sym}\left(\mathbf{u}_s \mathbf{R}_s\right) + \mathbb{Q}_s^f + \mathbb{Q}_s^t,$$

where $\mathbb{E}_s := \int_{\mathbf{v}} f_s \mathbf{v}\mathbf{v}$ is the energy tensor and $\mathbb{P}_s := \mathbb{E}_s - \rho_s \mathbf{u}_s \mathbf{u}_s$ is the pressure tensor. A linear isotropic entropy-respecting isotropization closure is $\mathbb{R}_s = \frac{1}{\tau_s}\left(\frac{1}{3}(\mathrm{tr}\,\mathbb{P}_s)\mathbb{I} - \mathbb{P}_s\right)$, where tr denotes the trace of its argument tensor and for the isotropization period we used $\tau_s = \tau_0\sqrt{\frac{\det \mathbb{P}_s}{\rho_s^5} m_s^3}$, which attempts to generalize the Braginskii closure; for the GEM problem this means that $\tau_i/\tau_e \approx (m_i/m_e)^{5/4}$. We neglect all other collisional terms: the heat flux tensors $\mathbb{q}_s$, the resistive drag forces $\mathbf{R}_s$, the frictional heating tensors $\mathbb{Q}_s^f$, and the temperature equilibration tensors $\mathbb{Q}_s^t$.

For small viscosity (or fast isotropization) the viscosity is related to the isotropization period by $\mu_s \approx p_s \tau_s$. We set $\tau_0 = 50$. The ten-moment model offers the advantage of *hyperbolic viscosity* — that is, viscosity can be implemented without numerically expensive diffusive terms.

3 GEM problem

The GEM problem is posed on a rectangular domain with periodic boundary conditions in the horizontal direction and with conducting wall boundary conditions for the upper and lower boundaries. The initial conditions are a Harris sheet equilibrium perturbed by "pinching" to form an X-point.

Nondimensionalization. The GEM problem nondimensionalizes time by the ion gyrofrequency $\Omega_i = \frac{eB_0}{m_i}$ and the space scale by the ion inertial length δ_i, which is the distance traveled in an ion gyroperiod $1/\Omega_i$ at the ion Alfvén speed $v_{A,i} := \frac{B_0}{\mu_0 m_i n_0}$ (where $\mu_0 := (c^2 \epsilon_0)^{-1}$ is magnetic permeability). Under this nondimensionalization the model equations above remain unchanged with the exception that ϵ_0 is replaced by $\epsilon := 1/c^2$ (where c is now the speed of light divided by the ion Alfvén speed).

Model parameters. The GEM problem specifies that the ion/electron mass ratio is $m_i/m_e = 25$ and the initial temperature ratio is $T_i/T_e = \sqrt{m_i/m_e} = 5$. The speed of light is not specified; we used the commonly used value of 20, chosen to be sufficiently high to exceed other wave speeds but computationally feasible.

Computational domain. The computational domain is the rectangular domain $[-L_x/2, L_x/2] \times [-L_y/2, L_y/2]$, where $L_x = 8\pi$ and $L_y = 4\pi$. The problem is symmetric under reflection across either the horizontal or vertical axis.

Boundary conditions. The domain is periodic in the x-axis. The boundaries perpendicular to the y-axis are thermally insulating conducting wall boundaries. A conducting wall boundary is a solid wall boundary (with slip boundary conditions in the case of ideal plasma) for the fluid variables, and the electric field at the boundary has no component parallel to the boundary. We also assume that magnetic field runs parallel to and so does not penetrate the boundary (this follows from Ohm's law of ideal MHD, but we assume it holds generally). So at the conducting wall boundaries

$$\begin{aligned} &\partial_y \rho_s = 0, && \partial_y B_x = B_y = \partial_y B_z = 0, \\ &\partial_y u_{sx} = u_{sy} = \partial_y u_{sz} = 0, && E_x = \partial_y E_y = E_z = 0. \end{aligned}$$

Initial conditions. The initial conditions are a perturbed Harris sheet equilibrium. The unperturbed equilibrium is given by

$$\mathbf{B}(y) = B_0 \tanh(y/\lambda)\mathbf{e}_x, \qquad p(y) = \frac{B_0^2}{2n_0}n(y),$$
$$n_i(y) = n_e(y) = n_0(1/5 + \mathrm{sech}^2(y/\lambda)), \qquad p_e(y) = \frac{T_e}{T_i + T_e}p(y),$$
$$\mathbf{E} = 0, \qquad p_i(y) = \frac{T_i}{T_i + T_e}p(y).$$

On top of this the magnetic field is perturbed by

$$\delta\mathbf{B} = -\mathbf{e}_z \times \nabla(\psi), \text{ where}$$
$$\psi(x, y) = \psi_0 \cos(2\pi x/L_x)\cos(\pi y/L_y).$$

In the GEM problem the initial condition constants are

$$\lambda = 0.5, \quad B_0 = 1, \quad n_0 = 1, \quad \psi_0 = B_0/10.$$

4 Method

To simulate the ten-moment and five-moment systems we implemented a Runge-Kutta Discontinuous Galerkin solver with third-order accuracy in space and time on a Cartesian mesh.

To suppress oscillations, after each time stage we limited the solution in the characteristic variables of the cell average using a modification of Krivodonova's method (beginning with the coefficients of the highest-order Legendre basis polynomials and descending to lower order if limiting occurs) [8].

To clean the magnetic field we used a correction potential ψ in Maxwell's equations, as suggested in [7]:

$$\partial_t\mathbf{B} + \nabla\times\mathbf{E} + \chi\nabla\psi = 0, \quad \partial_t\mathbf{E} - c^2\nabla\times\mathbf{B} = -\mathbf{J}/\epsilon_0, \quad \partial_t\psi + \chi c^2\nabla\cdot\mathbf{B} = 0.$$

These equations imply a wave equation that propagates the divergence constraint error $\nabla\cdot\mathbf{B}$ at the speed $c\chi$. We used $\chi = 1.05$.

Since the GEM problem is symmetric, we imposed symmetry and solved the equations on the quarter domain $[0, L_x/2] \times [0, L_y/2]$.

5 Results

We simulated the GEM problem with ten-moment and five-moment models and compared the results with the Vlasov simulations of [3] and the PIC simulations of [2]①. Their plots were made at a point

①We have to negate some quantities because we call the vertical axis y and the out-of-plane axis z, opposite to the convention of [2, 3].

in time when the flux through the positive x-axis approximately reaches 1 nondimensionalized unit②. This is shortly before the time when the reconnection rate peaks.

As measured against kinetic simulations the ten-moment model reconnects at about the correct rate and the five-moment model reconnects a bit too quickly, perhaps because in the five-moment model compression in the outflow direction automatically causes increased pressure in the perpendicular directions, opening up the outflow region and artificially increasing the rate of reconnection.

In contrast to the five-moment model the ten-moment model is capable of representing an anisotropic pressure tensor. Our plots of the ten-moment electron pressure tensor components look like somewhat smudged versions of the corresponding plots for the Vlasov model.

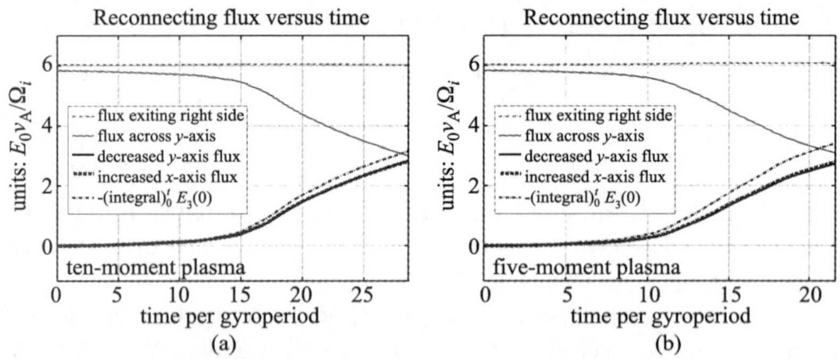

Figure 5.1 Ramp-up of reconnected flux for ten-moment and five-moment plasma. Fluxes are through the boundaries of the first quadrant of the domain. In each model the bottom three lines should coincide. Times at which the magnetic flux through the positive x-axis reached one nondimensionalized unit per model were 17.2 (our ten-moment), 13.0 (our five-moment), 17.7 (Vlasov [3]), 15.7 (PIC [2]), 17.6 (ten/five-moment [5]), 15.6 (five-moment [4]), and 15.3 (five-moment [6]).

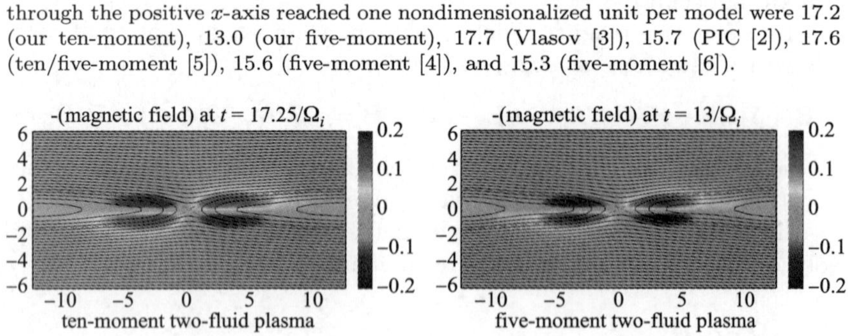

Figure 5.2 Magnetic field when one unit of flux passes through the positive x-axis. These results compare well with Plate 1 of [2] and the magnetic field shown in FIG. 2 of [3].

②The initial flux through the positive x-axis is $2\psi_0 = 0.2$ and the initial total flux through the positive y-axis is $\ln(\cosh(2\pi)) - (0.1(= \psi_0)) \approx 5.4900$. So the percent reconnection at this time is $0.8/5.49 \approx 14.6\%$.

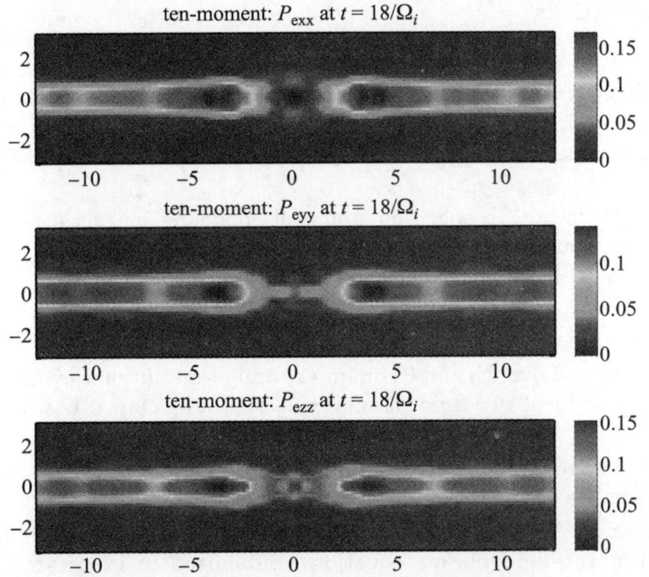

Figure 5.3 Diagonal components of the electron pressure tensor for ten-moment simulation at $\Omega_i t = 18$. These look like smudged versions of the plots in FIG. 5 of [3].

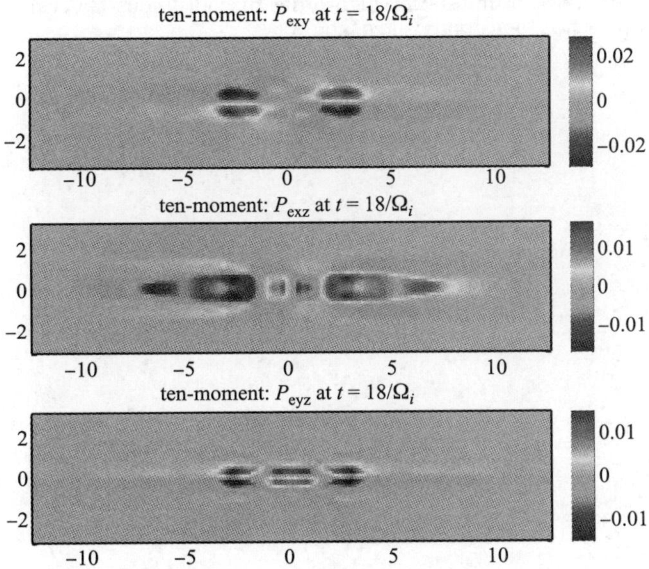

Figure 5.4 Off-diagonal components of the electron pressure tensor for ten-moment simulation at $\Omega_i t = 18$. These look like smudged versions of the plots in FIG. 6 of [3].

References

[1] J. Birn, J. F. Drake, M. A. Shay, B. N. Rogers, R. E. Denton, M. Hesse, M. Kuznetsova, Z. W. Ma, A. Bhattacharjee, A. Otto and P. L. Pritchett, Geospace environmental modeling (GEM) magnetic reconnection challenge. Journal of Geophysical Research – Space Physics. 106 (2001), 3715–3719.

[2] P. L. Pritchett, Geospace environment modeling magnetic reconnection challenge: Simulation with a full particle electromagnetic code. Journal of Geophysical Research. 106 (2001), no. A3, 3783–3798.

[3] H. Schmitz and R. Grauer, Kinetic Vlasov simulations of collisionless magnetic reconnection. Physics of Plasmas 13 (2006), 092309.

[4] A. Hakim, J. Loverich and U. Shumlak, A high-resolution wave propagation scheme for ideal two-fluid plasma equations. J. Comp. Phys. 219 (2006), 418–442.

[5] A. H. Hakim, Extended MHD modelling with the ten-moment equations. J. Fusion Energy. 27 (2008), 36–43.

[6] J. Loverich, A. Hakim and U. Shumlak, A discontinuous Galerkin method for ideal two-fluid plasma equations. Submitted to Physics of Plasmas, arXiv: 1003. 4542v1 (2010).

[7] A. Dedner, F. Kemm, D. Kröner, C. D. Munz, T. Schnitzer and M. Wesenberg, Hyperbolic divergence cleaning for the MHD equations. J. Comp. Phys. 175 (2002), 645–673.

[8] L. Krivodonova, Limiters for high-order discontinuous Galerkin methods. J. Comp. Phys. 226 (2007), 879–896.

Numerical Simulation of Nonclassical Shock Waves in Porous Media with a Heterogeneous Multiscale Method*

Frederike Kissling
Institut für Angewandte Analysis und Numerische Simulation, Univeristät Stuttgart, Pfaffenwaldring 57, D-70569 Stuttgart, Germany
E-mail: Frederike.Kissling@mathematik.uni-stuttgart.de

Christian Rohde
Institut für Angewandte Analysis und Numerische Simulation, Univeristät Stuttgart, Pfaffenwaldring 57, D-70569 Stuttgart, Germany
E-mail: Christian.Rohde@mathematik.uni-stuttgart.de

Abstract

We consider nonclassical shock waves appearing in two-phase flow in porous media. We are interested in solutions which are selected as sharp-interface limits for solutions of a rate-dependent regularization. We present a new multidimensional mass-conserving numerical method which belongs to the class of Heterogeneous Multiscale Methods in the sense of E&Engquist [4]. A key part of the approximation is a new special flux function for the multidimensional setting, which has been introduced in 1D in [1].

1 Introduction

We consider the infiltration of a wetting fluid (e.g., water) into an open set $\Omega \subseteq \mathbb{R}^2$ which is initially fully saturated by a nonwetting fluid (e.g., oil). We suppose that the interface between both fluids is described by a hypersurface $\Gamma = \Gamma(t)$, $t \in [0, T]$, $T > 0$. Neglecting gravitational forces and assuming the fluids to be immiscible and incompressible, the governing equations for this two-phase flow problem are given by mass balance laws. For the capillarity-free case these laws lead in the fractional flow formulation to an elliptic pressure equation and a hyperbolic

*The authors acknowledge support by the International Research Training Group NUPUS (GRK 1398) and the Cluster of Excellence SimTech (EXC 310/1).

saturation equation

$$\mathbf{v} = -\mathbf{K}\lambda\nabla p \quad \text{and} \quad \operatorname{div}(\mathbf{v}) = 0 \qquad \text{in } \Omega \times (0,T), \tag{1.1}$$

$$\phi S_t + \operatorname{div}(\mathbf{v} f^w(S)) = 0 \tag{1.2}$$

$$S(.,0) = S_0 \qquad \text{in } \Omega. \tag{1.3}$$

The total velocity $\mathbf{v} = \mathbf{v}(\mathbf{x},t) \in \mathbb{R}^2$ and the water saturation $S = S(\mathbf{x},t) \in [0,1]$ are the unknowns. $\mathbf{K}$ stands for the absolute permeability, ϕ is the porosity. We will use for all simulations $\mathbf{K} = \mathbf{Id}$ and $\phi = 1$. λ is the sum of both mobility functions, the wetting and non-wetting phase, λ^w and λ^n, respectively. The mobilities are given by $\lambda_i = k_{ri}/\mu_i$, $i = w, n$, for each phase, whereby k_{ri} is the relative permeability and μ_i the viscosity. The nonlinear fractional flow function is given by $f^w(S) = \frac{\lambda^w(S)}{\lambda^w(S)+\lambda^n(S)}$. Appropriate boundary conditions have to be added for Ω bounded. Principally we are interested in the hyperbolic saturation equation and turn our's attention to (1.2). (1.2) is known as the Buckley-Leverett (BL) equation.

The problem (1.1)–(1.3) can have multiple weak solutions, depending on the choice of mobilities. Of most interest is the nonconvex case. We use $k_{rw} = S^2$ and $k_{rn} = (1-S)^2$ for the relative permeabilities. There is a whole family of solutions involving non-Laxian so-called nonclassical transitional waves [10]. In contrast to the well understood case of classical Laxian waves nonclassical waves are not fully determined by the characteristic information. Let us note that these shock waves are indeed not mathematical artefacts, but have been experimentally observed as the leading front Γ separating both fluids [2]. Our main objective is to resolve this front and to compute the unique physically relevant solution which is selected as the sharp-interface limit of solutions of a nonstandard regularization [3, 6], see Figure 1.1 (a) for the planar case. To clarify why a special regularization has to be chosen consider the following situation. For the regularized equation capillary pressure effects have to be added

$$\phi S_t^\epsilon + \operatorname{div}(\mathbf{v} f^w(S^\epsilon)) = \operatorname{div}\left(\mathbf{K}\overline{\lambda}(S^\epsilon)\nabla p_c\left[S^\epsilon\right]\right), \tag{1.4}$$

where the water saturation $S^\epsilon = S^\epsilon(\mathbf{x},t) \in [0,1]$ is unknown, $\epsilon > 0$ is a regularization parameter and $\overline{\lambda}(S^\epsilon) = \lambda^n(S^\epsilon)f^w(S^\epsilon)$. The simplest choice for p_c is the static capillary pressure $p_c\left[S^\epsilon\right] = \epsilon S^\epsilon$. For this choice of p_c (1.4) is a (degenerate) parabolic equation and the solution S^ϵ of (1.4) converges against the classical Kruzkov solution of (1.2) as ϵ tends to 0. Figure 1.1 shows the numerical solution (view from above and cross section) of (1.1), (1.2) for Riemann-type initial data using a monotone Finite Volume scheme (Engquist-Osher), $\mu_w = \mu_n = 1$ and $\mathbf{v} = (0,1)^T$ constant. We observe a classical shock-rarefaction wave pattern. This is not the physically correct solution which consists of a leading nonclassical front followed by a wetting overshoot with some plateau saturation $\overline{S}$ (cf. Figure 1.1). In general nonclassical waves cannot be approximated by standard schemes. With this background it is reasonable that monotone standard schemes that rely on artificial viscosity, similar to the regularization (1.4) with the static capillary pressure fail. Let us give a short outline on the rest of the paper. After a short discussion of the specific regularized model in Sect. 2, in Sect. 3 we present the concept of the new HMM-algorithm in 2D and the new numerical flux function. In Sect. 4 we present numerical results.

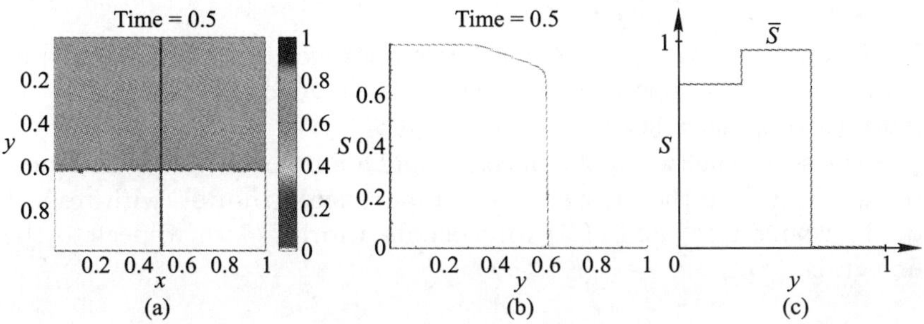

Figure 1.1 Numerical solution of (1.1), (1.2) with monotone FV scheme. Sketch of a nonclassical solution with saturation overshoot (c).

2 Extended model problem

We noted in Sect. 1 that (1.4) with the static capillary pressure p_c converges against the viscosity limit which cannot produce nonclassical fronts. Also experiments show that the static capillary pressure function do not give sufficiently accurate results but should be supplemented with rate-dependent terms. To model dynamic effects in the pressure difference between the two phases Hassanizadeh&Gray [5] suggest the choice

$$p_c\left[S^\epsilon\right] = \epsilon S^\epsilon - \epsilon^2 \tau S_t^\epsilon, \quad \tau > 0. \tag{2.1}$$

Inserting (2.1) into (1.4) we obtain a third-order evolution equation. We want to compute the singular limit $S := \lim_{\epsilon \to 0} S^\epsilon$ (which depends on τ). This is analytically reasonable due to the following theorem.

Theorem 1 [9]. Let $\{S^\epsilon\}_{\epsilon>0}$ be a family of smooth solutions to the Cauchy problem for

$$\phi S_t^\epsilon + \operatorname{div}\left(\mathbf{v} f^w(S^\epsilon)\right) = \epsilon \Delta S^\epsilon + \epsilon^2 \tau \Delta S_t^\epsilon$$

with $\mathbf{v} \in \mathbb{R}^2$. Then there is a function $S \in L^2(\mathbb{R}^2 \times (0,T))$ with

(i) $S^\epsilon \to S$ in $L^p_{loc}(\mathbb{R}^2 \times (0,T))$, $p \in [1,2)$,

(ii) S is a weak solution of (1.2), (1.3).

Since the internal layer of the front solution scales with the small parameter $\epsilon > 0$, it is not possible to solve (1.4), (2.1) directly and we suggest the following approach.

3 A heterogeneous multiscale method for the computation of nonclassical waves

We follow the heterogeneous multiscale approach of E&Engquist [4] as a guideline. This approach has already shown good results for the computation of nonclassical shock waves in 1D [7].

We use for the two models a coarse and a fine space-time grid respectively. We call in the following (1.2) **macroscale model** (with respect to the coarse grid) and (1.4) **microscale model** (with respect to the fine grid).

3.1 The heterogeneous multiscale algorithm

For the sake of simplicity choose $\Omega = \mathbb{R}^2$, $\mathbf{v}$ constant and let an unstructured conform triangulation $\mathcal{T}_h = \{\mathbf{T}_i\}_{i\in\mathcal{I}}$ of $\mathbb{R}^2$ be given. The approximation of the macroscale solution at discrete times $0 =: T^0 < T^1 < T^2 < \cdots$ is a function constant on each triangle $\mathbf{T}_i$. For the initialization on the macroscale we define $S_i^0 := \frac{1}{|\mathbf{T}_i|}\int_{\mathbf{T}_i} S_0(\mathbf{x})d\mathbf{x}$, $i \in \mathcal{I}$. The front $\Gamma(0)$ is chosen initially as part of $\partial\Omega$ (infiltration experiment). In the following the discrete front Γ^n at time T^n consists of connected edges e_{ij} of the grid (cf. Figure 3.1 (a), marked red). $\mathbf{n}_{ij}$ is the normal to the edge e_{ij} shared by triangles $\mathbf{T}_i$ and $\mathbf{T}_j$ pointing into $\mathbf{T}_j$. We only solve the expensive extended microscale model (1.4) with the capillary pressure (2.1) close to the front in normal direction $\mathbf{n}_{ij}$ for a microscale time step $t^n > 0$. On the macroscale we use a standard Finite Volume scheme.

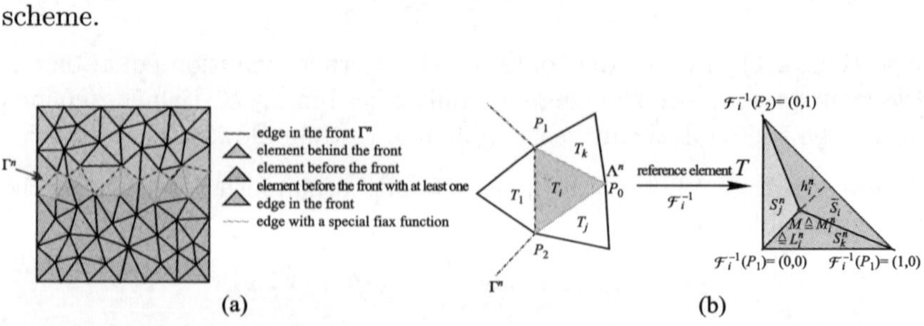

Figure 3.1 Macroscale computational domain Ω with front Γ^n (a). Construction of the numerical flux function (b).

Algorithm 1 below determines for all times T^n the macroscale approximation $S^n := \left(S_i^n\right)_{i\in\mathcal{I}}$ and the approximate front position Γ^n (see Figure 3.1 (a)) for the nonclassical shock wave.

Algorithm 1. Input: $n = 0$, ε, $\mathcal{T}_h$, $(S_i^0)_{i\in\mathcal{I}}$, Γ^0
WHILE $T^n < T$ {

1. **The reconstruction.** For all elements $\mathbf{T}_i, \mathbf{T}_j \in \mathcal{T}_h$ with joint edge in Γ^n we define
$$s_{ij}(x,0) := \begin{cases} S_i^n & \text{for} \quad x < 0, \\ S_j^n & \text{for} \quad x > 0. \end{cases}$$
2. **Solving the 1D microscale model.** For all elements $\mathbf{T}_i, \mathbf{T}_j \in \mathcal{T}_h$ with joint edge in Γ^n we solve the $\mathbf{n}_{ij}$-rotated 1D microscale model for Riemann initial data $s_{ij}(x,0)$ and denote the solution by $s_{ij}(x,t)$, $t \in (T^n, T^n + t^n)$.
3. **Data analysis.** For all elements $\mathbf{T}_i, \mathbf{T}_j \in \mathcal{T}_h$ with joint edge in Γ^n we determine approximately the plateau value $\overline{S} = \overline{S}_i$ in the solution s_{ij} (cf. Figure 1.1 for a sketch of a typical solution).
4. **Computing the macroscale time step.** We compute the time step ΔT^n such that $\Delta T^n \leqslant \mathrm{c} \min\limits_{\mathbf{T}_i\in\mathcal{T}_h} \left\{ \frac{|\mathbf{T}_i|}{\sum\limits_{\text{edges } e_{ij}} \max\{|e_{ij}|\mathbf{n}_{ij}\cdot\mathbf{v}f^{w\prime}(S_i^n),0\}} \right\}$ and $\Delta T^n \leqslant \mathrm{c} \min\limits_{\substack{\mathbf{T}_i\in\mathcal{T}_h \\ \partial\mathbf{T}_i\cap\Gamma^n\neq\emptyset}} \left\{ \frac{|\mathbf{T}_i|}{\sum\limits_{\text{edges } e_{ij}} \max\{|e_{ij}|\mathbf{n}_{ij}\cdot\mathbf{v}f^{w\prime}(\overline{S}_i),0\}} \right\}$ with a constant $\mathrm{c} \in (0,1)$ and set $T^{n+1} = T^n + \Delta T^n$.
5. **Upscaling process.** We get the new macroscale variables $\left\{S_i^{n+1}\right\}_{i\in\mathcal{I}}$ by evolving (1.2) with a Finite Volume scheme
$$S_i^{n+1} = S_i^n - \frac{\Delta T^n}{|\mathbf{T}_i|} \sum_{\text{edges } e_{ij}} g_{ij}\left(S_i^n, S_j^n, \overline{S}_i\right), \quad i \in \mathcal{I},$$
where g_{ij} is a special numerical flux function if $\mathbf{T}_i$ has at least one edge in Γ^n, otherwise an arbitrary monotone flux function (see (3.2) below).
6. **Moving the front $\Gamma^n \to \Gamma^{n+1}$.** With data from the construction of the numerical flux function we obtain the new location of the front (for more details, see [8]).
7. $n \to n + 1$ }

In step 5 a *special* numerical flux function g_{ij} is neceassary to obtain nonclassical shock waves, see Sect. 3.2 below. One constraint at the moving of the front in step 6 is that the edges have to be connected. It is not allowed that the front breaks open.

Remark. In general one uses an IMPES concept (implicit pressure - explicit saturation) to solve (1.1), (1.2) and solve the pressure equation and the saturation equation alternately. We only use $\mathbf{v}$ constant for the sake of simplicity.

3.2 Numerical flux function in 2D

In this part we want to highlight the main ideas of the new developed numerical flux function. The numerical flux function for 2D on an unstructured triangulation is motivated by the numerical flux function in 1D [1]. For detailed information we refer to [8].

To get a mass-conserving flux we have the following constraint

$$|\mathbf{T}_i|S_i^n = |\tilde{\mathbf{T}}_{ij}|\,\tilde{S}_{ij}^n + |\tilde{\mathbf{T}}_{ik}|\,\tilde{S}_{ik}^n + |\tilde{\mathbf{T}}_{il}|\,\tilde{S}_{il}^n. \tag{3.1}$$

Thereby the triangle $\mathbf{T}_i$ is divided in three subtriangles $\tilde{\mathbf{T}}_{ij}$, $\tilde{\mathbf{T}}_{ik}$, $\tilde{\mathbf{T}}_{il}$. $\tilde{S}_{ij}$, $\tilde{S}_{ik}$, $\tilde{S}_{il}$ are reconstructed data on the subtriangles. For the construction of the numerical flux function we have to distinguish how many edges of a triangle $\mathbf{T}_i$ belong to the front Γ^n. We only explain it at the example that one edge is in the front. Without loss of generality let the joint edge of $\mathbf{T}_i$ and $\mathbf{T}_l$ be the edge which belongs to the front Γ^n. The procedure and notations are illustrated in Figure 3.1 (b). The triangle $\mathbf{T}_i$ is mapped on a reference element $\mathbf{T}_0 = \{(x,y) \in \mathbb{R}^2 \mid x \in [0,1] \text{ and } x+y \leqslant 1\}$, i.e., $\mathcal{F}_i^{-1} : \mathbf{T}_i \to \mathbf{T}_0$. The sizes of the triangles $\tilde{\mathbf{T}}_{ij}$, $\tilde{\mathbf{T}}_{ik}$, $\tilde{\mathbf{T}}_{il}$ can be computed with the aid of the reference element $\mathbf{T}_0$, the constraint (3.1) and the condition that the intersection of the three subtriangles $M_i^n \in \mathbf{T}_0$ is located on the bisecting line of the angle at the origin. The underlying idea of the flux is that we update at the edges which do not belong to the front Γ^n with the adjacent saturation $\tilde{S}_{ij} = S_j^n$, $\tilde{S}_{ik} = S_k^n$, until the information of the plateau value $\tilde{S}_{il} = \overline{S}_i$ reaches the edges. Then we use $\overline{S}_i$ for the update. In the discrete case $\mathcal{F}_i(M_i^n)$ represents the placement of the front. The underlying idea is to compute when the point $\mathcal{F}_i(M_i^n)$ reaches the opposing point P_0. The numerical flux function is given for $\mathbf{v} \cdot \mathbf{n}_{ij} > 0$ by

$$\begin{aligned}\Delta T^n g_{ij}(S_i^n, S_j^n, \overline{S}_i) = {} & \min(\Delta T_{i+1/2}, \Delta T^n) f^w(S_j^n)\mathbf{v} \cdot \mathbf{n}_{ij} \\ & + \max(\Delta T^n - \Delta T_{i+1/2}, 0) f^w(\overline{S}_i)\mathbf{v} \cdot \mathbf{n}_{ij}. \end{aligned} \tag{3.2}$$

$\Delta T_{i+1/2}$ is the time that the point $\mathcal{F}_i(M_i^n)$ needs to reach P_0. This special constructed flux function is only possible if the intersection M_i^n is in the reference element $\mathbf{T}_0$. Otherwise a standard flux is used. With the flux (3.2) the Finite Volume like scheme in step 5 of Algorithm 1 conserves mass, i.e., $\sum_{i \in \mathcal{I}} S_i^n = \sum_{i \in \mathcal{I}} S_i^0$, $\forall n \in \mathbb{N}_0$.

4 Numerical results

In this section we test Algorithm 1 on the Buckley-Leverett problem and we examine Algorithm 1 on aspects of convergence and computational complexity. Concluding we apply the new developed method for an example where instabilities occur and so-called viscous fingers build out. For the examples in Sect. 4.1–4.3 we use the same viscosity $\mu_w = \mu_n = 1$ for both phases. To validate our numerical scheme and the numerical flux function (3.2), we have used for the simulations in Sect. 4.1–4.3 a constant velocity field $\mathbf{v}$. For all examples we use $\Omega = (0,1)^2$.

4.1 Buckley-Leverett problem

This simulation is the counterpart to the simulation in Figure 1.1, where we infiltrate water and get a planar front. There we have used a monotone Finite Volume scheme (Engquist-Osher) and obtained a monotone solution. Here we have used Algorithm 1 with the new developed flux function and we get a nonmonotone solution with a nonclassical shock wave (see Figure 4.1 (a)), that means also a saturation overshoot.

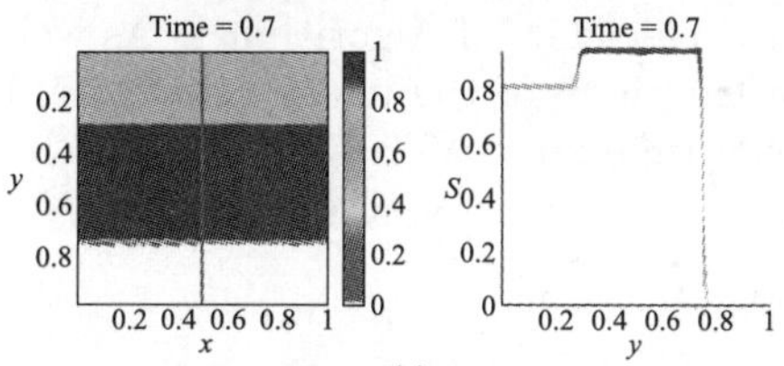

(a)

		CPU-Times
2D	BL with standard FV	38s
	BL with "kinetic relation" and new flux function	66s
	BL with HMM	24.7h
1D	Microscale problem over one edge	2.6s

(b)

Figure 4.1 Solution of (1.1), (1.2) with the HMM for Riemann initial data (a). Comparison of cpu-times for different schemes (b).

4.2 Grid convergence

The setting for the simulations in this example is the same like in the Sect. 4.1 before, but with different refined grids. In Figure 4.2 the

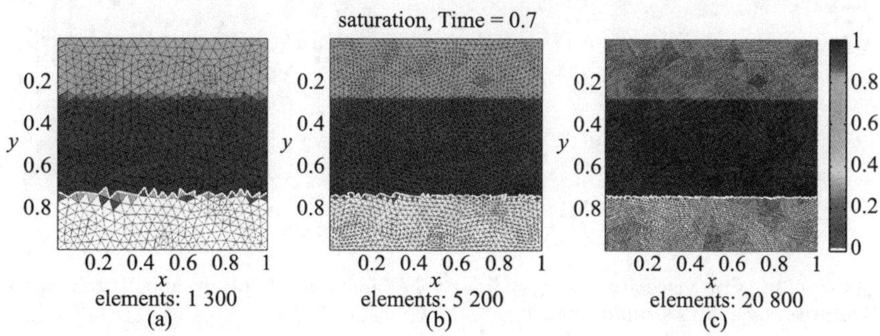

Figure 4.2 Simulation of the BL problem on different refined grids.

simulations are shown for a grid with 1 300, 5 200, and 20 800 elements respectively. It is visible that the shock wave is better resolved and we can observe grid convergence. We want to emphasize at this point that we have used an unstructured triangulation without preferred direction.

4.3 Computational complexity for the HMM

In this part we want to compare the complexity of the specific schemes to demonstrate the efficiency of the HMM. For this we have computed the cpu-time for the BL problem with a standard Finite Volume scheme and for the BL problem with the new developed numerical flux function. In this case we have given the plateau values. The cpu-times are listed in the table in Figure 4.1. The new flux function is only slightly slower. Furthermore we have computed the cpu-times by application of the HMM. Thereby the microscale model has been solved in each time step over each edge in the front. The time is in comparison to the other times very large, but the expenditure of time is in solving the microscale problem (1.4). However one can see immediately, that even for this simple example, it would be impossible to solve (1.4) with (2.1) in 2D on the whole domain. This demonstrates the efficiency of Algorithm 1. As well it has to be stressed that the time length t^n of the microscale domain is much smaller than the macroscale time step ΔT^n. This cannot be achieved for the direct approach.

4.4 Viscous fingering

In this example we use different viscosities for the wetting and nonwetting phase, $\mu_w = 1$ for the wetting phase and $\mu_n = 4$ for the nonwetting phase. Consequently we expect viscous fingering. Instabilities occur and fingers with saturation overshoots build out (cf. Figure 4.3).

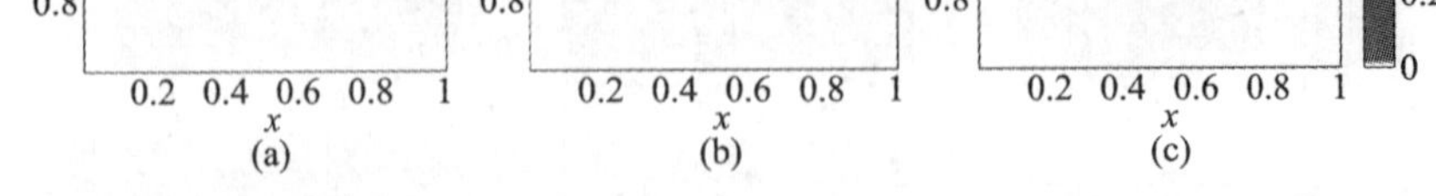

Figure 4.3 For viscosity ratio $\mu_w/\mu_n = 1/4$ instabilities occur and fingers with saturation overshoots build out.

References

[1] B. Boutin, C. Chalons, F. Lagoutière and P. G. LeFloch, Convergent and conservative schemes for nonclassical solutions based on Kinetic relations. Interfaces and Free Boundaries. 10 (2008), 399–421.

[2] D. DiCarlo, Experimental measurements of saturation overshoot on infiltration. Water Resour. Res. 40 (2004).

[3] C. J. van Duijn, L. A. Peletier and I. S. Pop, A new class of entropy solutions of the Buckley-Leverett equation. SIAM J. Math. Anal. 39 (2007), 507–536.

[4] W. E and B. Engquist, The heterogeneous multiscale methods. Domm. Math. Sci. No. 1 (2003), 87–132.

[5] S. M. Hassanizadeh and W. G. Gray, Thermodynamic basis of capillary pressure in porous media. Water Resour. Res. 29 (1993), 3389–3405.

[6] F. Kissling, P. G. LeFloch and C. Rohde, Kinetic decomposition for singular limits of non-local conservation laws. J. Diff. Eq. 247 (2009), 3338–3356.

[7] F. Kissling and C. Rohde, The computation of nonclassical shock waves with a heterogeneous multiscale method. Networks and Heterogeneous Media 5 (2010), no. 3, 661–674.

[8] F. Kissling and C. Rohde, The computation of nonclassical shock waves with a HMM: The multidimensional case. In preparation.

[9] F. Kissling and C. Rohde, Kinetic decomposition for singular limits in porous media. In preparation.

[10] P. G. LeFloch, Hyperbolic Systems of Conservation Laws. Lectures in Mathematics. Birkhäuser (2002).

Implicit Finite Difference Scheme for the Magnetic Induction Equation

Ujjwal Koley
CMA, University of Oslo
P.O. Box 1053, Oslo, Norway
E-mail: ujjwalk@cma.uio.no

Abstract

We describe high-order accurate and stable fully-discrete finite difference schemes for the initial-boundary value problem associated with the magnetic induction equations. We present various numerical experiments that demonstrate both the stability as well as high order of accuracy of the schemes.

1 Introduction

In this paper, we study the magnetic induction equation

$$\partial_t \mathbf{B} + \mathrm{curl}(\mathbf{B} \times \mathbf{u}) = -\mathbf{u}\mathrm{div}(\mathbf{B}), \tag{1.1}$$

where the unknown $\mathbf{B} = \mathbf{B}(\mathbf{x}, t) \in \mathbb{R}^3$ describes the magnetic field of a plasma in three space dimensions with coordinate $\mathbf{x} = (x, y, z)$. The above equation models the evolution of the magnetic field in the plasma which is moving with a prescribed velocity field $\mathbf{u}(\mathbf{x}, t)$. These equations arise in a wide variety of applications in plasma physics, astrophysics and electrical engineering. One important application is the equations of magneto-hydro dynamics (MHD). Observe that by taking divergence on both sides of (1.1) we get

$$(\mathrm{div}(\mathbf{B}))_t + \mathrm{div}\left(\mathbf{u}\mathrm{div}(\mathbf{B})\right) = 0. \tag{1.2}$$

Hence, if $\mathrm{div}(\mathbf{B}_0(\mathbf{x})) = 0$, also $\mathrm{div}(\mathbf{B}(\mathbf{x}, t)) = 0$ for $t > 0$.

There are many forms of induction equations available in literature (see [2, 3]). Here we are going to work with the following "conservative" symmetric form,

$$\partial_t \mathbf{B} + (\mathbf{u} \cdot \nabla)\, \mathbf{B} = M(D\mathbf{u})\mathbf{B}, \tag{1.3}$$

where the $D\mathbf{u}$ denotes the gradient of $\mathbf{u}$ and the matrix $M(D\mathbf{u})$ is given by

$$M(D\mathbf{u}) = \begin{pmatrix} -\partial_y u^2 - \partial_z u^3 & \partial_y u^1 & \partial_z u^1 \\ \partial_x u^2 & -\partial_x u^1 - \partial_z u^3 & +\partial_z u^2 \\ \partial_x u^3 & \partial_y u^3 & -\partial_x u^1 - \partial_y u^2 \end{pmatrix}.$$

We are aware of some other results in the literature related to induction equation; cf. [1, 2, 7, 4] and the references therein. But, boundary conditions were not considered either of the aforementioned papers. In [3], authors have described a high order accurate and stable finite difference schemes for the initial-boundary problem associated with the magnetic induction equations. The approach is based on a "semi-discrete" approximation where one discretizes the spatial variable, thereby reducing the equations to a system of ordinary differential equations. However, we stress that for numerical computations also this set of ordinary differential equations will have to be discretized in order to be solved. Thus in order to have a completely satisfactory numerical method, one seeks a fully discrete scheme that reduces the actual computation to a solution of a finite set of algebraic equations.

Our aim in this paper is to design stable and high-order accurate "fully-discrete" schemes for initial-boundary value problems corresponding to the magnetic induction equations by discretizing the non conservative symmetric form (1.3). The spatial derivatives are approximated by second and fourth-order SBP (Summation-By-Parts) operators. The boundary conditions are weakly imposed by using a SAT (Simultaneous Approximation Term) and backward Euler method used for temporal discretization. The SBP-SAT framework has been used to obtain stable and accurate high order schemes for a wide variety of hyperbolic problems.

The rest of this paper is organized as follows: In Section 2, we state the energy estimate for the initial-boundary value problem corresponding to (1.3). In Section 3, we present the fully-discrete SBP-SAT scheme and show stability. Numerical experiments are presented in Section 4 and conclusions are drawn in Section 5.

2 The continuous problem

For ease of notation, we shall restrict ourselves to two spatial dimensions in the remainder of this paper. Extending the results to three dimensions is straightforward.

In two dimensions, the non-conservative symmetric form (1.3) reads

$$\mathbf{B}_t + \Lambda_1 \mathbf{B}_x + \Lambda_2 \mathbf{B}_y - C\mathbf{B} = 0, \tag{2.1}$$

where

$$\Lambda_1 = \begin{pmatrix} u^1 & 0 \\ 0 & u^1 \end{pmatrix}, \quad \Lambda_2 = \begin{pmatrix} u^2 & 0 \\ 0 & u^2 \end{pmatrix}, \quad C = \begin{pmatrix} -\partial_y u^2 & \partial_y u^1 \\ \partial_x u^2 & -\partial_x u^1 \end{pmatrix}.$$

To begin with, we shall consider (2.1) in the domain $(x, y) \in \Omega = [0,1]^2$. We augment (2.1) with initial conditions,

$$\mathbf{B}(\mathbf{x}, 0) = \mathbf{B}_0(\mathbf{x}) \quad \mathbf{x} \in \Omega, \tag{2.2}$$

and Dirichlet boundary conditions,

$$\begin{aligned} &\mathbf{1}_{\{u^1(0,y,t)>0\}}\Big(\mathbf{B}(0,y,t) = \mathbf{g}(0,y,t)\Big), \\ &\mathbf{1}_{\{u^1(1,y,t)<0\}}\Big(\mathbf{B}(1,y,t) = \mathbf{g}(1,y,t)\Big), \\ &\mathbf{1}_{\{u^2(x,0,t)>0\}}\Big(\mathbf{B}(x,0,t) = \mathbf{g}(x,0,t)\Big), \\ &\mathbf{1}_{\{u^2(x,1,t)<0\}}\Big(\mathbf{B}(x,1,t) = \mathbf{g}(x,1,t)\Big), \end{aligned} \tag{2.3}$$

where $\mathbf{1}_A$ denotes the characteristic function of the set A. We have the following theorem. For a proof, see [3].

Theorem 2.1. *Assume that $\mathbf{B}_0 \in H^1(\Omega)$, that $\mathbf{g} \in H^1(\partial\Omega \times [0,T])$ for $T > 0$ and that u^1 and u^2 are in $H^2(\Omega \times [0,T])$. Then there exists a function $\mathbf{B} \in C([0,T], L^2(\Omega)) \cap L^\infty([0,T]; H^1(\Omega))$ which is the unique weak solution of* (2.1) *with the initial and boundary conditions* (2.2) *and* (2.3).

Furthermore, it satisfies the following stability estimate

$$\|\mathbf{B}(\cdot,t)\|^2_{H^1(\Omega)} \le e^{\alpha t}\left(\|\mathbf{B}_0\|^2_{H^1(\Omega)} + \|\mathbf{g}\|_{H^1(\partial\Omega\times(0,t))}\right), \tag{2.4}$$

where α is a positive constant.

3 Fully-discrete scheme

To simplify the treatment of the boundary terms we let the computational domain Ω be the unit square. It is straightforward to generalize our results to other domains by coordinate transformations (see [6]), and to three dimensions.

The SBP finite difference schemes for one-dimensional derivative approximations are as follows. Let $[0,1]$ be the domain discretized with $x_j = j\Delta x$, $j = 0, \ldots, N-1$. A scalar grid function is defined as $w = (w_0, \ldots, w_{N-1})$. To approximate $\partial_x w$ we use a summation-by-parts operator $D_x = P_x^{-1}Q_x$, where P_x is a diagonal positive $N \times N$ matrix, defining an inner product

$$(v,w)_{P_x} = v^T P_x w,$$

such that the associated norm $\|w\|_{P_x} = (w,w)_{P_x}^{1/2}$ is equivalent to the norm $\|w\| = (\Delta x \sum_k w_k^2)^{1/2}$. Furthermore, for D_x to be a summation-by-parts operator we require that

$$Q_x + Q_x^T = R_N - L_N,$$

where R_N and L_N are the $N \times N$ matrices: diag $(0,\ldots,1)$ and $(1,\ldots,0)$ respectively. Similarly, we can define a summation-by-parts operator $D_y = P_y^{-1}Q_y$ approximating ∂_y. Later we will also need the following Lemma, proven in [5].

Lemma 3.1. *Let u be a smooth grid function. Then*

$$\|D_x(u \circ w) - u \circ D_x w\|_{P_x} \leq C \|\partial_x u\|_{L^\infty([0,1])} \|w\|_{P_x}, \tag{3.1}$$

where $(u \circ v)_j = u_j v_j$.

Next, we move on to the two-dimensional case and discretize the unit square $[0,1]^2$ using NM uniformly distributed grid points $(x_i, y_j) = (i\Delta x, j\Delta y)$ for $i = 0,\ldots,N-1$ and $j = 0,\ldots,M-1$, such that $(N-1)\Delta x = (M-1)\Delta y = 1$. We order a scalar grid function $w(x_i,y_i) = w_{ij}$ as a column vector

$$w = \left(w_{0,0}, w_{0,1}, \ldots, w_{0,(M-1)}, w_{1,0}, \ldots, w_{(N-1),(M-1)}\right)^T.$$

Using Kronecker products, we can define 2D difference operators. Let I_n denote the $n \times n$ identity matrix, and define

$$\mathfrak{d}_x = D_x \otimes I_M, \quad \mathfrak{d}_y = I_N \otimes D_y.$$

For a smooth function $w(x,y)$, $(\mathfrak{d}_x w)_{i,j} \approx \partial_x w(x_i,y_j)$ and similarly $(\mathfrak{d}_y w)_{i,j} \approx \partial_y w(x_i,y_j)$.

Set $\mathbf{P} = P_x \otimes P_y$, define $(w,v)_{\mathbf{P}} = w^T \mathbf{P} v$ and the corresponding norm $\|w\|_{\mathbf{P}} = (w,w)_{\mathbf{P}}^{1/2}$. Also define $\mathcal{R} = R_N \otimes I_M$, $\mathcal{L} = L_N \otimes I_M$, $\mathcal{U} = I_N \otimes R_M$, and $\mathcal{D} = I_N \otimes L_M$.

For a vector valued grid function $\mathbf{V} = (V^1, V^2)$, we use the following notation

$$\mathfrak{d}_x \mathbf{V} = \begin{pmatrix} \mathfrak{d}_x V_1 \\ \mathfrak{d}_x V_2 \end{pmatrix},$$

and so on. In the same spirit, the $\mathbf{P}$ inner product of vector valued grid functions is defined by $(\mathbf{V},\mathbf{W})_{\mathbf{P}} = (V^1,W^1)_{\mathbf{P}} + (V^2,W^2)_{\mathbf{P}}$. We also introduce (a small) time step $\Delta t > 0$, and use the notation

$$D_t^+ p(t) = \frac{1}{\Delta t}\left(p(t+\Delta t) - p(t)\right),$$

for any function $p : [0,T] \to \mathbb{R}$. Write $t^n = n\Delta t$ for $n \in \mathbb{N}_0 = \mathbb{N} \cup \{0\}$. We will use the notation $V^1(x_i, y_j, t^n) = V^{1,n}_{ij}$ and so on.

The usefulness of summation by parts operators comes from this lemma, for a proof see [4].

Lemma 3.2. *For any grid functions v and w, we have*

$$\begin{aligned}(v, \partial_x w)_{\mathbf{P}} + (\partial_x v, w)_{\mathbf{P}} &= v^T \left[(\mathcal{R} - \mathcal{L})(I_N \otimes P_y)\right] w,\\ (v, \partial_y w)_{\mathbf{P}} + (\partial_y v, w)_{\mathbf{P}} &= v^T \left[(\mathcal{U} - \mathcal{D})(P_x \otimes I_M)\right] w.\end{aligned} \tag{3.2}$$

Before we define our numerical schemes, we collect some useful results in a lemma, for a proof see [4].

Lemma 3.3. *If u is a grid function, then*

$$\begin{aligned}(\mathbf{V}, u \circ \partial_x \mathbf{V})_{\mathbf{P}} &= \frac{1}{2}\mathbf{V}^T[(\mathcal{R} - \mathcal{L})(I_N \otimes P_y)]\,(u \circ \mathbf{V})\\ &\quad + \frac{1}{2}\left(u \circ \partial_x \mathbf{V} - \partial_x (u \circ \mathbf{V}), \mathbf{V}\right)_{\mathbf{P}},\\ (\mathbf{V}, u \circ \partial_y \mathbf{V})_{\mathbf{P}} &= \frac{1}{2}\mathbf{V}^T[(\mathcal{U} - \mathcal{D})(P_x \otimes I_M)]\,(u \circ \mathbf{V})\\ &\quad + \frac{1}{2}\left(u \circ \partial_y \mathbf{V} - \partial_y (u \circ \mathbf{V}), \mathbf{V}\right)_{\mathbf{P}}.\end{aligned} \tag{3.3}$$

The differential equation (2.1) will be discretized in an obvious manner. We incorporate the boundary conditions by penalizing boundary values away from the desired ones with a $\mathcal{O}(1/\Delta x)$ term. To this end set

$$\mathcal{B} = \left[\left(P_x^{-1} \otimes I_M\right)(\Sigma_{\mathcal{L}}\mathcal{L} + \Sigma_{\mathcal{R}}\mathcal{R}) + \left(I_N \otimes P_y^{-1}\right)(\Sigma_{\mathcal{D}}\mathcal{D} + \Sigma_{\mathcal{U}}\mathcal{U})\right],$$

where $\Sigma_{\mathcal{L}}$, $\Sigma_{\mathcal{R}}$, $\Sigma_{\mathcal{D}}$, and $\Sigma_{\mathcal{U}}$ are diagonal matrices, with components $(\sigma_{\mathcal{L}})_{jj}$ (and similarly for the other penalty matrices), to be specified later.

With this notation the scheme for the differential equation (2.1) reads

$$D_t^+\mathbf{V}^n + u^{1,n+1}\circ\partial_x\mathbf{V}^{n+1} + u^{2,n+1}\circ\partial_y\mathbf{V}^{n+1} - C^{n+1}\mathbf{V}^{n+1} = \mathcal{B}(\mathbf{V}^{n+1} - \mathbf{g}^{n+1}), \tag{3.4}$$

while $\mathbf{V}^0$ is given. Here C^n denotes the matrix

$$C^n = \begin{pmatrix} -\partial_y u^{2,n} & \partial_y u^{1,n} \\ \partial_x u^{2,n} & -\partial_x u^{1,n} \end{pmatrix}.$$

Theorem 3.1. *Let $\mathbf{V}$ be as solution to (3.4) with $\mathbf{g} = 0$. If the constants in $\mathcal{B}$ is chosen as*

$$\begin{aligned}(\sigma_{\mathcal{R}})_{N-1,j} &\le \frac{u^{1,-}(1,y_j)}{2}, & (\sigma_{\mathcal{L}})_{0,j} &\le -\frac{u^{1,+}(0,y_j)}{2},\\ (\sigma_{\mathcal{U}})_{i,M-1} &\le \frac{u^{2,-}(x_i,1)}{2}, & (\sigma_{\mathcal{D}})_{i,0} &\le -\frac{u^{2,+}(x_i,0)}{2},\end{aligned} \tag{3.5}$$

then

$$\|\mathbf{V}^n\|_{\mathbf{P}}^2 \le e^{KT}\left\|\mathbf{V}^0\right\|_{\mathbf{P}}^2, \tag{3.6}$$

where $u^{l,+} = (u^l \vee 0)$, $u^{l,-} = (u^l \wedge 0)$, *for* $l = 1,2$. K *is a constant chosen in such a way that* $(1-c\Delta t)^{-1} \le (1+K\Delta t)$ *for sufficiently small* Δt, *where* c *is a constant depending on* u^1, u^2, *and their derivative approximations, but not on* N *or* M.

Proof. Taking the $\mathbf{P}$ inner product of (3.4) and $\mathbf{V}^{n+1}$, we get

$$\begin{aligned}
&\frac{1}{2}\left\|\mathbf{V}^{n+1}\right\|_{\mathbf{P}}^2 - \frac{1}{2}\|\mathbf{V}^n\|_{\mathbf{P}}^2 + \frac{1}{2}\left\|\mathbf{V}^{n+1}-\mathbf{V}^n\right\|_{\mathbf{P}}^2 \\
&= -\Delta t\left(\mathbf{V}^{n+1}, u^{1,n+1}\circ \mathfrak{d}_x\mathbf{V}^{n+1}\right)_{\mathbf{P}} - \Delta t\left(\mathbf{V}^{n+1}, u^{2,n+1}\circ \mathfrak{d}_y\mathbf{V}^{n+1}\right)_{\mathbf{P}} \\
&\quad + \Delta t\left(\mathbf{V}^{n+1}, \mathcal{C}^{n+1}\mathbf{V}^{n+1}\right)_{\mathbf{P}} + \Delta t\left(\mathbf{V}^{n+1}, \mathcal{B}\mathbf{V}^{n+1}\right)_{\mathbf{P}}.
\end{aligned}$$

Using Lemma 3.3 we get

$$\begin{aligned}
&\frac{1}{2}\left\|\mathbf{V}^{n+1}\right\|_{\mathbf{P}}^2 - \frac{1}{2}\|\mathbf{V}^n\|_{\mathbf{P}}^2 + \frac{1}{2}\left\|\mathbf{V}^{n+1}-\mathbf{V}^n\right\|_{\mathbf{P}}^2 \\
&= -\Delta t\frac{1}{2}(\mathbf{V}^{n+1})^T\left[(\mathcal{R}-\mathcal{L})(I_N\otimes P_y)\right]\left(u^{1,n+1}\circ\mathbf{V}^{n+1}\right) \\
&\quad - \Delta t\frac{1}{2}(V^{n+1})^T\left[(\mathcal{U}-\mathcal{D})(P_x\otimes I_M)\right]\left(u^{2,n+1}\circ\mathbf{V}^{n+1}\right) \\
&\quad - \Delta t\frac{1}{2}\left(u^{1,n+1}\circ\mathfrak{d}_x\mathbf{V}^{n+1} - \mathfrak{d}_x(u^{1,n+1}\circ\mathbf{V}^{n+1}), \mathbf{V}^{n+1}\right)_{\mathbf{P}} \\
&\quad - \Delta t\frac{1}{2}\left(u^{2,n+1}\circ\mathfrak{d}_y\mathbf{V}^{n+1} - \mathfrak{d}_y(u^{2,n+1}\circ\mathbf{V}^{n+1}), \mathbf{V}^{n+1}\right)_{\mathbf{P}} \\
&\qquad + \Delta t\left(\mathbf{V}^{n+1}, \mathcal{C}\mathbf{V}^{n+1}\right)_{\mathbf{P}} + \Delta t\left(\mathbf{V}^{n+1}, \mathcal{B}\mathbf{V}^{n+1}\right)_{\mathbf{P}}.
\end{aligned}$$

Note that by (3.1),

$$\begin{aligned}
\left|\left(u^{1,n+1}\circ\mathfrak{d}_x\mathbf{V}^{n+1} - \mathfrak{d}_x(u^{1,n+1}\circ\mathbf{V}^{n+1}), \mathbf{V}^{n+1}\right)_{\mathbf{P}}\right| &\le c\left\|\mathbf{V}^{n+1}\right\|_{\mathbf{P}}^2, \\
\left|\left(u^{2,n+1}\circ\mathfrak{d}_y\mathbf{V}^{n+1} - \mathfrak{d}_y(u^{2,n+1}\circ\mathbf{V}^{n+1}), \mathbf{V}^{n+1}\right)_{\mathbf{P}}\right| &\le c\left\|\mathbf{V}^{n+1}\right\|_{\mathbf{P}}^2, \\
\left|\left(\mathbf{V}^{n+1}, \mathcal{C}\mathbf{V}^{n+1}\right)_{\mathbf{P}}\right| &\le c\left\|\mathbf{V}^{n+1}\right\|_{\mathbf{P}}^2,
\end{aligned} \tag{3.7}$$

for some constant c depending on the first derivatives of u^1 and u^2. Using the conditions (3.5) we arrive at

$$\left\|\mathbf{V}^{n+1}\right\|_{\mathbf{P}}^2 \le \|\mathbf{V}^n\|_{\mathbf{P}}^2 + c\Delta t\left\|\mathbf{V}^{n+1}\right\|_{\mathbf{P}}^2$$

Now we can use the fact that $(1-c\Delta t)^{-1} \le (1+K\Delta t)$ for sufficiently small Δt. Consequently this gives the required bound (3.6). □

4 Numerical experiment

In this experiment, we consider (2.1) with the divergence-free velocity field $\mathbf{u}(x,y) = (-y,x)^T$. The exact solution can be easily calculated by the method of characteristics and takes the form

$$\mathbf{B}(\mathbf{x},t) = R(t)\mathbf{B}_0(R(-t)\mathbf{x}), \tag{4.1}$$

where $R(t)$ is a rotation matrix with angle t and represents rotation of the initial data about the origin.

We consider the same test setup as in [7] and [2] by choosing the divergence free initial data,

$$\mathbf{B}_0(x,y) = 4\begin{pmatrix} -y \\ x-\frac{1}{2} \end{pmatrix} e^{-20\left((x-1/2)^2+y^2\right)}, \tag{4.2}$$

and the computational domain $[-1,1]\times[-1,1]$. Since the exact solution is known in this case, we can use this to specify the boundary data g.

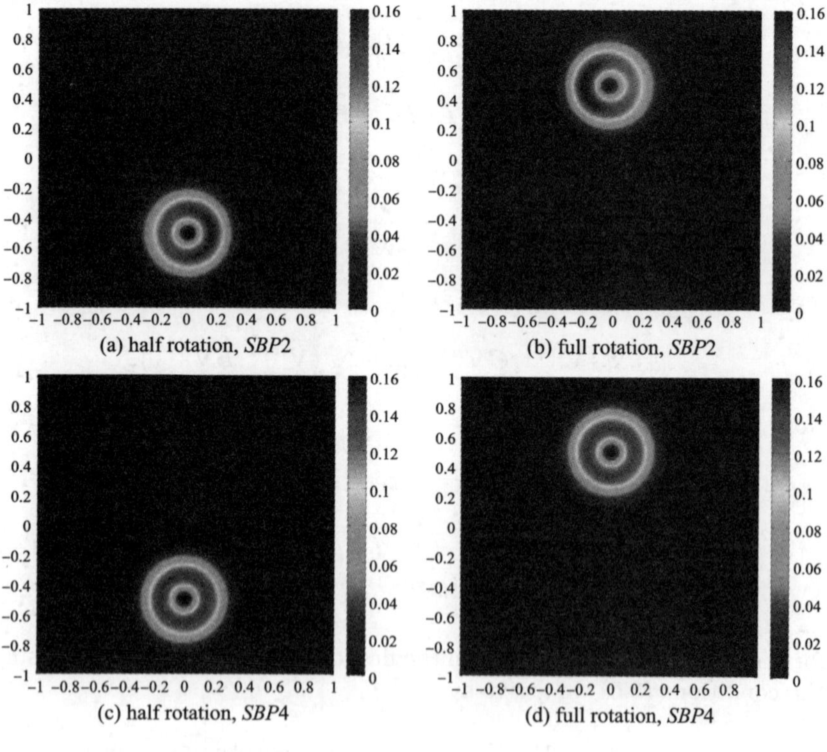

Figure 4.1 Numerical results for $|\mathbf{B}|$.

We run this test case with the second-order (first-order) accurate and the fourth-order (second-order) accurate SBP operators in the interior (boundary), denoted by $SBP2$ and $SBP4$ respectively, and present different sets of results. The theoretical orders of convergence are obtained.

Table 4.1 Relative percentage errors in l^2 for $|\mathbf{B}|$ at time $t = 2\pi$ and rates of convergence with $SBP2$ and $SBP4$ schemes.

Grid size	$SBP2$	rate	$SBP4$	rate
40×40	6.9×10^1		8.0×10^0	
80×80	2.1×10^1	1.7	5.0×10^{-1}	4.0
160×160	5.5×10^0	2.0	4.5×10^{-2}	3.5
320×320	1.3×10^0	2.0	5.1×10^{-3}	3.1
640×640	3.3×10^{-1}	2.0	6.4×10^{-4}	3.0

5 Conclusion

We have considered a fully-discrete scheme for the magnetic induction equations that arise as a submodel in the MHD equations of plasma physics. In future, our plan is to extend the semi-discrete scheme given in [4] to a semi-implicit fully-discrete scheme. We would like to show the stablity of that semi-implicit scheme in case of magnetic induction equations with resistivity.

References

[1] N. Besse and D. Kröner, Convergence of the locally divergence free discontinuous Galerkin methods for induction equations for the 2D-MHD system. M2AN Math. Model. Num. Anal. 39 (2005), no. 6, 1177–1202.

[2] F. Fuchs, K. H. Karlsen, S. Mishra and N. H. Risebro, Stable upwind schemes for the magnetic induction equation. Preprint, Submitted.

[3] U. Koley, S. Mishra, N. H. Risebro and M. Svärd, Higher order finite difference schemes for the magnetic induction equations. BIT Numer. Math. 49 (2009), 375–395.

[4] U. Koley, S. Mishra, N. H. Risebro and M. Svärd, Higher order finite difference schemes for the magnetic induction equations with resistivity. Preprint, Submitted.

[5] S. Mishra and M. Svärd, On stability of numerical schemes via frozen coefficients and magnetic induction equations. Preprint, Submitted.

[6] M. Svärd, On coordinate transformations for summation-by-parts operators. J. Sci. Comput. 20 (2004), 29–42.

[7] M. Torrilhon and M. Fey, Constraint-preserving upwind methods for multidimensional advection equations. SIAM. J. Num. Anal. 42 (2004), no.4, 1694–1728.

On Effects of Stochastic Regularization for the Pressureless Gas Dynamics*

Anastasia Korshunova
Mathematics and Mechanics Faculty, Moscow State University
Moscow 119992, Russia
E-mail: korshunova_aa@mail.ru

Olga Rozanova
Mathematics and Mechanics Faculty, Moscow State University
Moscow 119992, Russia
E-mail: rozanova@mech.math.msu.su

Abstract

It is well-known that the density component of solution to the pressureless gas dynamics system develops a delta-singularity when the velocity component loses its smoothness. We show that the stochastic perturbation along characteristics allows to introduce a specific pressure term which prevents the formation of the delta-singularity and has a regularizing effect in this sense. This pressure term is in some sense analogous to the pressure of barotropic monoatomic gas. This paper is an extension of [1].

1 Stochastic perturbation of the Burgers equation

Let us consider the Cauchy problem for the non-viscous Burgers equation:

$$\partial_t u + (u, \nabla)u = 0, \quad t > 0, \qquad u(x, 0) = u_0(x) \in C^1(\mathbb{R}^n) \cap C_b(\mathbb{R}^n). \tag{1.1}$$

Here $u(x,t) = (u_1, \ldots, u_n)(x,t)$ is a vector-function $\mathbb{R}^{n+1} \to \mathbb{R}^n$.

We associate with (1.1) the following system of stochastic differential equations:

$$dX_k(t) = U_k(t)dt + \sigma_1 d(W_k^1)_t, \quad dU_k(t) = \sigma_2 d(W_k^2)_t, \quad k = 1, \ldots, n, \tag{1.2}$$

*The authors are partly supported by DFG Award (436 RUS 113/823/0-1) and the Ministry of Education of the Russian Federation (2.1.1/1399).

$$X(0) = x, \quad U(0) = u, \quad t > 0,$$

where $X(t)$ and $U(t)$ are considered as random variables with given initial distributions, $(X(t), U(t))$ runs in the phase space $\mathbb{R}^n \times \mathbb{R}^n$, σ_1 and σ_2 are nonnegative constants such that $|\sigma| \neq 0$ ($\sigma = (\sigma_1, \sigma_2)$) and $(W^i)_t = (W_1^i, \ldots, W_n^i)_t$, $i = 1, 2$ are independent n-dimensional Brownian motions. In [1] we considered a particular case $\sigma_2 = 0$.

Let $P(t, x, u)$ be a probability density of joint distribution of random values (X, U), subject to initial data

$$P_0(x, u) = \delta(u - u_0(x))\rho_0(x) = \prod_{k=1}^{n} \delta(u_k - (u_0(x))_k)\rho_0(x), \tag{1.3}$$

where $\rho_0(x)$ is a bounded nonnegative function from $L^1_{loc}(\mathbb{R}^n)$.

Let us introduce functions

$$\rho_\sigma(t, x) = \int_{\mathbb{R}^n} P(t, x, u)\, du, \quad u_\sigma(t, x) = \frac{\int_{\mathbb{R}^n} uP(t, x, u)du}{\int_{\mathbb{R}^n} P(t, x, u)du}. \tag{1.4}$$

As will readily be observed $u_\sigma(0, x) = u_0(x)$. Certain properties of $u_\sigma(t, x)$ have been established in [2] and [3].

The density $P = P(t, x, u)$ obeys the Fokker-Planck equation

$$\frac{\partial P}{\partial t} = \left[-\sum_{k=1}^{n} u_k \frac{\partial}{\partial x_k} + \sum_{k=1}^{n} \frac{1}{2}\sigma_1^2 \frac{\partial^2}{\partial x_k^2} + \sum_{k=1}^{n} \frac{1}{2}\sigma_2^2 \frac{\partial^2}{\partial u_k^2} \right] P, \tag{1.5}$$

subject to initial data (1.3).

We apply the Fourier transform with respect to variables x and u in (1.5), (1.3) and obtain the following Cauchy problem for the Fourier transform $\tilde{P} = \tilde{P}(t, \lambda, \xi)$ of $P(t, x, u)$:

$$\frac{\partial \tilde{P}}{\partial t} = -\frac{1}{2}(\sigma_1^2|\lambda|^2 + \sigma_2^2|\xi|^2)\tilde{P} + \left(\lambda, \frac{\partial \tilde{P}}{\partial \xi}\right), \tag{1.6}$$

$$\tilde{P}(0, \lambda, \xi) = \int_{\mathbb{R}^n} e^{-i(\lambda, s)} e^{-i(\xi, u_0(s))} \rho_0(s)ds, \quad \lambda, \xi \in \mathbb{R}^n. \tag{1.7}$$

Equation (1.6) can easily be integrated and we obtain the solution:

$$\begin{aligned} &\tilde{P}(t, \lambda, \xi) \\ =&\tilde{P}(0, \lambda, \xi + \lambda t) \exp\left(-\frac{1}{2}\sigma_1^2|\lambda|^2 t - \frac{\sigma_2^2 t}{6}\left(|\lambda|^2 t^2 + 3t(\lambda, \xi) + 3|\xi|^2\right) \right). \end{aligned} \tag{1.8}$$

The inverse Fourier transform (in the distributional sense) allows to find the density $P(t,x,u)$:

$$P(t,x,u) = \frac{1}{(2\pi)^{2n}} \int_{\mathbb{R}^n} \int_{\mathbb{R}^n} e^{i(\lambda,x)} e^{i(\xi,u)} \tilde{P}(t,\lambda,\xi)\, d\lambda d\xi$$

$$= \frac{1}{(2\pi)^{2n}} \int_{\mathbb{R}^n} \rho_0(s) \int_{\mathbb{R}^n}\int_{\mathbb{R}^n} e^{-\frac{1}{2}\Lambda^2} e^{-\frac{1}{2}\Xi^2} e^{-\frac{|u_0(s)-u|^2}{2t\sigma_1^2} - \frac{6|(u_0(s)+u)\frac{t}{2}+s-x|^2}{12\sigma_1^2 t+\sigma_2^2 t^3}} d\lambda d\xi ds$$

$$= C \int_{\mathbb{R}^n} \rho_0(s)\, e^{-\frac{|u_0(s)-u|^2}{2t\sigma_1^2} - \frac{6|(u_0(s)+u)\frac{t}{2}+s-x|^2}{12\sigma_1^2 t+\sigma_2^2 t^3}} ds, \qquad (1.9)$$

where

$$\Lambda = \Lambda(t,\lambda,s) = \frac{\sqrt{12\sigma_1^2 t+\sigma_2^2 t^3}}{2\sqrt{3}}\lambda + \frac{2i\sqrt{3}}{\sqrt{12\sigma_1^2 t+\sigma_2^2 t^3}} \left((u_0(s)+u)\frac{t}{2}+s-x\right),$$

$$\Xi = \Xi(t,\xi,s) = \sigma_2\sqrt{t}\xi + \frac{1}{\sigma_2\sqrt{t}}\left(\frac{\sigma_2^2}{2}\lambda t^2 + iu_0(s) - u\right),$$

$$C = \left(\frac{\sqrt{3}}{\pi\sigma_2 t\sqrt{12\sigma_1^2+\sigma_2^2 t^2}}\right)^n.$$

From (1.9) we have

$$\rho_\sigma(t,x) = \left(\frac{\sqrt{3}}{(\sqrt{2\pi t(3\sigma_1^2+\sigma_2^2 t^2)}}\right)^n \int_{\mathbb{R}^n} \rho_0(s) e^{-\frac{3|u_0(s)t+s-x|^2}{2t(3\sigma_1^2+\sigma_2^2 t^2)}} ds, \qquad (1.10)$$

$$u_\sigma(t,x) = \frac{1}{2(3\sigma_1^2+\sigma_2^2 t^2)} \frac{\int_{\mathbb{R}^n} F(t,x,s)\rho_0(s) e^{-\frac{3|u_0(s)t+s-x|^2}{2t(3\sigma_1^2+\sigma_2^2 t^2)}} ds}{\int_{\mathbb{R}^n} \rho_0(s) e^{-\frac{3|u_0(s)t+s-x|^2}{2t(3\sigma_1^2+\sigma_2^2 t^2)}} ds}, \qquad (1.11)$$

where $F(t,x,s) = 6\sigma_1^2 u_0(s) - \sigma_2^2 t(tu_0(s) + 3(s-x))$.

2 Properties of the "perturbed velocity"

Proposition 2.1. *Let u_0 and $\rho_0 > 0$ be functions of class $C^1(\mathbb{R}^n) \cap C_b(\mathbb{R}^n)$. If $t_*(u_0) > 0$ is a moment of time such that the solution to the Cauchy problem* (1.1) *with the initial condition u_0 keeps this smoothness for $0 < t < t_*(u_0) \le +\infty$, then $\hat{u}_\sigma(t,x)$ tends to a solution of problem* (1.1) *as $|\sigma| \to 0$ for any fixed $(t,x) \in \mathbb{R}^{n+1}$, $0 < t < t_*(u_0)$.*

We do dwell on the proof of this proposition very shortly, it is quite analogous to [1]. Let us denote by $J(u_0(x)))$ the Jacobian matrix of the map $x \longmapsto u_0(x)$. If $J(u_0(x))$ has at least one eigenvalue which is negative for a certain point $x \in \mathbb{R}^n$, then the classical solution to (1.1) fails to exist beyond a positive time $t_*(u_0)$. Otherwise, $t_*(u_0) = \infty$. The matrix $C(t,x) = (I + tJ(u_0(x)))$, where I is the identity matrix, fails to be invertible for $t = t_*(u_0)$.

From (1.11) and (1.10) we have $u_\sigma(t,x)\rho_\sigma(t,x) = (u^I_\sigma(t,x) - u^{II}_\sigma(t,x))$ $\rho_\sigma(t,x)$, where $u^I_\sigma(t,x) = K\int_{\mathbb{R}^n} u_0(s)\rho_0(s)e^{-\frac{3|u_0(s)t+s-x|^2}{2t(3\sigma_1^2+\sigma_2^2t^2)}}ds$ and $K = \frac{\sqrt{3}}{\sqrt{2\pi t(3\sigma_1^2+\sigma_2^2t^2)}}$. Using the weak convergence of measures and the fact that ρ_0 and u_0 are continuous and bounded, we get for $u^I_\sigma(t,x)$:

$$\lim_{|\sigma|\to 0} u^I_\sigma(t,x) = \frac{\int_{\mathbb{R}^n} u_0(s)\rho_0(s)\delta_{p(t,x,s)}ds}{\int_{\mathbb{R}^n} \rho_0(s)\delta_{p(t,x,s)}ds},$$

with $p(t,x,s) = u_0(s)t + s - x$, where δ_y is the Dirac measure as $y \in \mathbb{R}^n$. We can then on the basis on the invertibility of $C(t,x)$ use locally the implicit function theorem and find $s = s_{t,x}(p)$. Therefore, $\lim\limits_{|\sigma|\to 0} u^I_\sigma(t,x) = u_0(s_{t,x}(0))$. Further, it can be readily proved that $\lim\limits_{|\sigma|\to 0} u^{II}_\sigma(t,x) = 0$. Let us introduce a new notation $s_0(t,x) \equiv s_{t,x}(0)$. Then the following vectorial equation holds:

$$u_0(s_0(t,x))t + s_0(t,x) - x = 0. \tag{2.1}$$

The function $u(t,x) = u_0(s_0(t,x))$ satisfies the Burgers equation. To check this fact it is enough to differentiate (2.1) with respect to t and x_j and then find $s_{0,t}$ and $s_{0,x}$. The more detailed proof one can find in [1], Proposition 1. □

It is important to note that $s_0(t,x)$ is unique for all t for which the solution to the Burgers equation $u(t,x)$ is smooth.

Remark 2.2. Proposition 2.1 can naturally be extended to the class of functions ρ_0 such that there exists a sequence $\rho_0^\varepsilon \in C^1(\mathbb{R}^n) \cap C_b(\mathbb{R}^n)$ converging to ρ_0 as $\varepsilon \to 0$ almost everywhere. In this case $u_\sigma(t,x)$ tends to a solution of problem (1.1) as $\sigma \to 0$ almost everywhere on $(t,x) \in \mathbb{R}^{n+1}$, $0 < t < t_*(u_0)$.

Theorem 2.3. *The scalar function $\rho_\sigma(t,x)$ and the vector-function u_σ, defined in* (1.4), *solve the following system:*

$$\frac{\partial \rho_\sigma}{\partial t} + div_x(\rho_\sigma u_\sigma) = \frac{1}{2}\sigma_1^2 \sum_{k=1}^n \frac{\partial^2 \rho_\sigma}{\partial x_k^2}, \tag{2.2}$$

$$\frac{\partial(\rho_\sigma u_{\sigma,i})}{\partial t} + div_x(\rho_\sigma u_{\sigma,i}\, u_\sigma) = \frac{1}{2}\sigma_1^2 \sum_{k=1}^n \frac{\partial^2(\rho_\sigma u_{\sigma,i})}{\partial x_k^2} - I_\sigma, \tag{2.3}$$

where $i = 1,..,n$, $t \geq 0$, *and* $I_\sigma = \int_{\mathbb{R}^n}(u_i - u_{\sigma,i})((u - u_\sigma), \nabla_x P(t,x,u))du$.

Proof. Integrating (1.5) with respect to u we get:

$$\frac{\partial \rho_\sigma}{\partial t} + div_x(\rho_\sigma u_\sigma) = \frac{1}{2}\sigma_1^2 \sum_{k=1}^{n} \frac{\partial^2 \rho_\sigma}{\partial x_k^2} + \frac{1}{2}\sigma_2^2 \sum_{k=1}^{n} \int_{\mathbb{R}^n} \frac{\partial^2 P(t,x,u)}{\partial u_k^2} du.$$

The last integral is equal to zero, because $P(t, x, u)$ and all its derivatives tend to zero as $|u| \to \infty$. Thus, we obtain (2.2).

To prove (2.3) we note that the definitions of $u_\sigma(t, x)$ and $\rho_\sigma(t, x)$ imply

$$\begin{aligned}\frac{\partial(\rho_\sigma u_{\sigma,i})}{\partial t} &= \frac{\partial}{\partial t}\int_{\mathbb{R}^n} u_i P\, du = \int_{\mathbb{R}^n} u_i P_t\, du = -\int_{\mathbb{R}^n} u_i (u, \nabla_x P) du \\ &+ \frac{1}{2}\sigma_1^2 \sum_{k=1}^{n} \int_{\mathbb{R}^n} u_i \frac{\partial^2 P}{\partial x_k^2} du + \frac{1}{2}\sigma_2^2 \sum_{k=1}^{n} \int_{\mathbb{R}^n} u_i \frac{\partial^2 P}{\partial u_k^2} du, \qquad (2.4)\end{aligned}$$

where $P_t \equiv \frac{\partial}{\partial t} P$. Early we noted that the last integral is equal to zero.

Further, for $i = 1, \dots, n$ we have

$$\begin{aligned}&div_x(\rho\, u_{\sigma,i}\, u_\sigma) = u_{\sigma,i} div_x(\rho_\sigma u_\sigma) + u_\sigma \partial_x(\rho_\sigma u_{\sigma,i}) - u_\sigma u_{\sigma,i} \partial_x \rho_\sigma \\ &= \int_{\mathbb{R}^n} u_{\sigma,i}(u, \nabla_x P)\, du + \int_{\mathbb{R}^n} u_i (u_\sigma, \nabla_x P)\, du - \int_{\mathbb{R}^n} u_{\sigma,i}(u_\sigma, \nabla_x P)\, du,\end{aligned} \qquad (2.5)$$

where $i, k = 1, \dots, n$. Equation (2.3) follows immediately from (2.4) and (2.5). □

Let us set $\rho(t, x) = \lim\limits_{|\sigma|\to 0} \rho_\sigma(t, x)$ and $\bar{u}(t, x) = \lim\limits_{|\sigma|\to 0} u_\sigma(t, x)$.

Theorem 2.4. *Assume that $(\rho(t, x), \bar{u}(t, x))$, the limits of (ρ_σ, u_σ) as $|\sigma| \to 0$, are C^1 — smooth bounded functions for $(t, x) \in \Omega := [0, t_*(u_0)) \times \mathbb{R}^n$, $t_*(u_0) \le \infty$. Then they solve in Ω the pressureless gas dynamics system*

$$\partial_t \rho + div_x(\rho \bar{u}) = 0, \quad \partial_t(\rho \bar{u}) + \nabla_x(\rho \bar{u} \otimes \bar{u}) = 0. \qquad (2.6)$$

Proof. As follows from Proposition 2.1, the function $\bar{u}(t, x)$ is a C^1 — solution of the non-viscous Burgers equation. Further, (2.2) is a linear parabolic equation with respect to ρ_σ, hence the limit as $|\sigma| \to 0$ reduces it to the first equation in (2.6) (continuity equation). The second equation in (2.6) is a corollary of the non-viscous Burgers equation and continuity equation for smooth solutions. □

Remark 2.5. Proposition 2.4 implies that the integral term on the right-hand side of (2.3) vanishes as $|\sigma| \to 0$ in the case of smooth limit functions ρ and $\bar{u}$.

3 Regularization of the pressureless gas dynamics

After the moment of shock formation in the solution to the Burgers equation the limit of (2.2)–(2.3) as $|\sigma| \to 0$ is not equivalent to the pressureless gas dynamics system (2.6), the integral term in (2.3) does not vanish and can be interpreted as a gradient of special (spurious) pressure. It is well known that for the solutions the pressureless gas dynamic system after the moment of singularity formation the question arises: how to define a generalized solution. The answer depends on the prescribed type of interaction between particles. For the sticky particles model, where two particles move together after interaction, one need to introduce a strongly singular solution with a delta-singularity in the density component. If we assume that particles do not interact, we get the generalized solution in the sense of free particles, whose singularities are only shocks. In this sense we can call the limit of system (2.2)–(2.3) as $|\sigma| \to 0$ a regularization of the pressureless gas dynamics equations. There exists an algorithm how to pass from the model of free particles (FP) to the model of sticky particles (see [1] for details in the case $\sigma_2 = 0$, the general case is analogous).

The generalized solution in the sense of free particles for discontinuous initial data can be introduced in the following way (see [1]):

Definition 3.1. We call the couple of functions $(\rho_{FP}(t,x), u_{FP}(t,x))$ the generalized solution to the Cauchy problem for (2.6) in the sense of free particles (FP-generalized solution) subject to initial data $(\rho_0(x), u_0(x)) \in \mathbb{L}^2_{loc}(\mathbb{R}^n) \cap \mathbb{L}_\infty(\mathbb{R}^n)$, if for almost all $(t,x) \in \mathbb{R}_+ \times \mathbb{R}^n$

$$\rho_{FP}(t,x) = \lim_{\varepsilon\to 0}(\lim_{\sigma\to 0} \rho^\varepsilon_\sigma(t,x)), \quad u_{FP}(t,x) = \lim_{\varepsilon\to 0}(\lim_{\sigma\to 0} u^\varepsilon_\sigma(t,x)),$$

where $(\rho^\varepsilon_\sigma(t,x), u^\varepsilon_\sigma(t,x))$ correspond to initial data $\rho^\varepsilon_0 = \eta_\varepsilon * \rho_0, u^\varepsilon_0 = \eta_\varepsilon * u_0$, and $\eta_\varepsilon(x)$ is the standard averaging kernel.

It can be proved [1] that the FP-solution $(f_{FP}(t,x), u_{FP}(t,x))$ to the Cauchy problem for (2.6) with initial data $(f_0(x), u_0(x)) \in \mathbb{L}^2_{loc}(\mathbb{R}^n) \cap \mathbb{L}_\infty(\mathbb{R}^n)$ does not depend of the choice of the monotonic approximation $(f^\varepsilon_0, u^\varepsilon_0)$.

It is very important that the double limit procedure used in the definition of the FP-solution plays role only for the case of a central rarefaction wave, where it helps to find a unique stable solution in the vacuum domain (we refer to [1] again). For the compression wave the FP-solution can be computed directly by formulae (1.10), (1.11) using only one limit pass $|\sigma| \to 0$. Thus, after the moment of the shock formation $t_*(u_0)$ in the Burgers equation the solution to the respective conservation law

$$\partial_t\rho + div_x(\rho\bar u) = 0, \quad \partial_t(\rho\bar u) + \nabla_x(\rho\bar u\otimes\bar u) + \nabla p = 0, \quad \nabla p = I := \lim_{|\sigma|\to 0} I_\sigma \tag{3.1}$$

can be found from (1.10), (1.11) as $|\sigma| \to 0$, I_σ is defined in the statement of Theorem 2.3. This term does not vanish and therefore the solution to (3.1) is the FP-solution to (2.6), thus (3.1) can be considered as a regularization of (2.6). Thus in this case the pressure term arises naturally. This term prevents the formation of delta-shocks, however of course it does not prevent the formation of ordinary shocks, in finite time.

Let us illustrate this fact on example of the Riemann problem in 1D (see [1] for the case $\sigma_2 = 0$, the general case is analogous):

$$\rho_0(x) = \rho_1 + \rho_2\theta(x), \quad u_0(x) = u_1 + u_2\theta(x), \tag{3.2}$$

where ρ_i, u_i, $i = 1, 2$ are constants. We compute the FP-solution according to Definition 3.1. For the case $u_2 > 0$ the term $I = 0$ and

$$\rho_{FP}(t,x) = \begin{cases} \rho_1, & x < u_1 t, \\ 0, & u_1 t < x < (u_1 + u_2)t, \\ \rho_1 + \rho_2, & x > (u_1 + u_2)t. \end{cases}$$

$$u_{FP}(t,x) = \begin{cases} u_1, & x < u_1 t, \\ \dfrac{x}{t}, & u_1 t \le x \le (u_1 + u_2)t, \\ u_1 + u_2, & x > (u_1 + u_2)t. \end{cases}$$

For $u_2 < 0$ we have

$$I = -\frac{\rho_1(\rho_1+\rho_2)u_2^2}{(2\rho_1+\rho_2)}(\delta(x-(u_1+u_2)t) - \delta(x-u_1 t)),$$

$$\rho_{FP}(t,x) = \rho_1 + (\rho_1+\rho_2)\theta(x-u_1 t) - \rho_1\theta(x-(u_1+u_2)t),$$

$$u_{FP}(t,x) = u_1 + \frac{\rho_1+\rho_2}{2\rho_1+\rho_2}u_2\theta(x-u_1 t) - \frac{\rho_1}{2\rho_1+\rho_2}u_2\theta(x-(u_1+u_2)t),$$

$$p(t,x) = \frac{\rho_1(\rho_1+\rho_2)}{2\rho_1+\rho_2}u_2^2(\theta(x-(u_1+u_2)t) - \theta(x-u_1 t)).$$

Thus, at the points where the velocity is formally multivalued, the original conservation law no longer yields and there arises an additional pressure term.

4 The 1D model for polytropic gas

In this section we study an analogy between solutions to (3.1) after the moment of the shock formation $t_*(u_0)$ in the relative Burgers equation and the system of polytropic gas. We consider the 1D case, nevertheless the hypothesis is that the analogy holds in any dimensions. Let us *assume* that the gas that obeys system (3.1) is polytropic, i.e., the state equation is $p = A\rho^\gamma$, with adiabatic exponent γ and a positive constant A. Therefore we consider the first and second equations in (3.1) with an equation for pressure:

$$\partial_t p + u\partial_x p + \gamma p \partial_x u = 0.$$

Proposition 4.1. *Let us consider the Riemann problem* (3.2) *with* $u_2 < 0$. *Assumed that the state equation is polytropic, the adiabatic exponent* $\gamma \to 3$ *as the ratio* $\frac{\rho_1}{\rho_2} \to 0$.

Proof. Now we use the result of the previous section. We have three Hugoniot conditions, that follow from the mass and momentum conservation (3.1) and the conservation of total energy $E = \frac{1}{2}\rho u^2 + \frac{p}{\gamma-1}$. We denote by f_- and f_+ the values of $f(x)$ before and after shock, respectively. For the shock moving with the speed $D = u_1 + u_2$ we have $\rho_+ = 2\rho_1 + \rho_2$, $\rho_- = \rho_1$, $u_+ = u_1 + \frac{\rho_1+\rho_2}{2\rho_1+\rho_2}u_2$, $u_- = u_1$, $p_+ = \frac{\rho_1(\rho_1+\rho_2)}{2\rho_1+\rho_2}u_2^2$, $p_- = 0$. The third Hugoniot condition gives $D = \frac{[Eu+pu]}{[E]}$ and $\gamma = \frac{3\rho_1+\rho_2}{\rho_1+\rho_2}$. This implies the proposition. The consideration of the second shock gives the same result. □

The adiabatic exponent $\gamma = 1 + \frac{2}{n}$, where n is the space dimension, corresponds to the monoatomic gas.

References

[1] S. Albeverio, A. Korshunova and O. Rozanova, Probabilistic model associated with the pressureless gas dynamics. Submitted, arXiv:0908.2084.

[2] S. Albeverio and O. Rozanova, The non-viscous Burgers equation associated with random positions in coordinate space: a threshold for blow up behaviour. Math. Meth. Mod. Appl. Sci. 19 (2009), 1–19.

[3] S. Albeverio and O. Rozanova, Suppression of unbounded gradients in a SDE associated with the Burgers equation. Proc. Amer. Math. Soc. 138 (2010), 241–251.

Jump Conditions across Phase Boundaries for the Navier-Stokes Korteweg System

Dietmar Kroener
Abteilung für Angewandte Mathematik, University of Freiburg
Hermann-Herder-Straße 10, D–79104 Freiburg i. Br., Germany
dietmar@mathematik.uni-freiburg.de

Abstract

In this contribution we will study the behaviour of the velocity and the pressure across phase boundaries in liquid-vapour flows for the Navier-Stokes-Korteweg model.

1 Introduction

In this paper we will consider a mathematical model for liquid-vapour flows including phase transition which was proposed by Korteweg already in 1901 [9] and which is known as the Navier-Stokes-Korteweg model. It is an extension of the compressible Navier-Stokes equation and given by the following system:

$$\begin{aligned} \partial_t \rho + \nabla \cdot (\rho v) &= 0, \\ \partial_t(\rho v) + \nabla \cdot (\rho v v^t + p(\rho) I) &= \mu \Delta v + \gamma \varepsilon^2 \rho \nabla \Delta \rho \end{aligned} \tag{1.1}$$

in $\Omega_T := \Omega \times [0, T]$. This is a one fluid model where $\rho, v, p(\rho)$, and μ denote the density, velocity, pressure, and the viscosity of the fluid/vapour respectively. Compared to the original compressible Navier-Stokes equation the system (1.1) contains the additional term $\gamma \varepsilon^2 \rho \nabla \Delta \rho$ which is supposed to model capillarity effects close to phase transitions. The pressure $p(\rho)$ as a function of the density ρ is defined as

$$p(\rho) = \rho^2 \psi'(\rho), \tag{1.2}$$

where ψ is a smooth function of ρ such that $\rho\psi(\rho)$ is the total free energy density and of the form as presented in Figure 1.1, 1.2.

The values α_1 and α_2 in Figure 1.1, 1.2 are defined by the extrema of p and β_1 and β_2 are the Maxwell points at which the slope of the tangent line to $\rho\psi(\rho)$ is equal to the corresponding difference quotient (see Figure 1.1). The conservation of energy is neglected in (1.1). Different phases

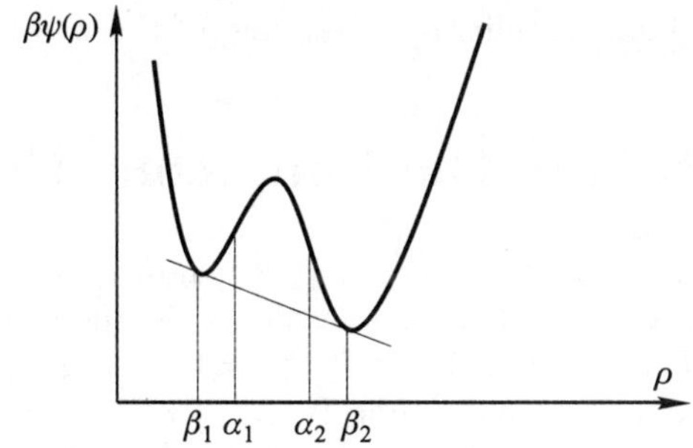

Figure 1.1 Graph of the total free energy.

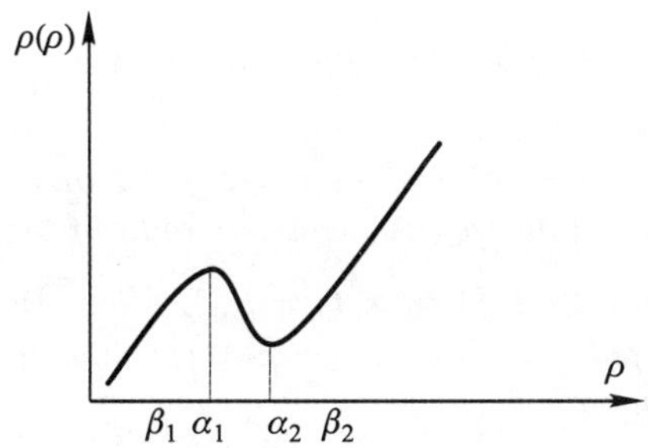

Figure 1.2 Graph of the pressure.

of the fluid are defined by the size of ρ. If $\rho \leq \alpha_1$ the system is in the vapour and if $\rho \geq \alpha_2$ in the liquid phase. The equation (1.2) is known as the van der Waals equation of state. For a rigorous derivation of the system (1.1) we refer to [1, 4].

Existence and uniqueness results for Navier-Stokes-Korteweg models have been shown in [2], [3], [10]. Diehl [4] has studied numerical experiments for the Navier-Stokes-Korteweg model in 2D and 3D. The limit $\varepsilon \to 0$ in the static case has been considered in [5], [12] and in combination with the low Mach number limit in [8]. It turned out that there will be no pressure jumps across the interface in the limit $\varepsilon \to 0$ for static solutions of (1.1). If the limit $\varepsilon \to 0$ is coupled with the zero Mach number limit in a suitable way, the classical jump condition, i.e., the Young-Laplace law, can be derived formally [8]. If in (1.1) the parameters μ and ε tend to zero simultaneously then the limit will depend in general on the relation between μ and ε. This has been investigated for model problems in [6].

In this paper we will assume that the solution $\rho_\varepsilon, v_\varepsilon$ of (1.1) converge to some limit ρ_0, v_0 and in Section 2 we will investigate the behavior of ρ_0, v_0 across the phase boundary. In Section 3 we will do the same for a different scaling, i.e., phasefield like scaling, of the pressure in (1.1).

2 Derivation of the jump conditions

First let us fix some notations: $\tilde{W}(\rho) :=$ free energy density (double well) $= \rho\psi(\rho)$ (see Figure 1.1). Then there exist β_1, β_2, and a affine linear function l such that $l(\beta_i) = \tilde{W}(\beta_i)$ and $l'(\beta_i) = \tilde{W}'(\beta_i), i = 1, 2$ (see Figure 1.1). Let $l(\rho) =: d_0\rho + d_1$ and define $W(\rho) := \rho\psi(\rho) - l(\rho)$. Throughout this paper we will make the following assumptions.

Assumptions 2.1. *For a sequence* $\varepsilon \to 0$ *we assume that*

(1) ρ_ε *and* v_ε *are smooth solutions of* (1.1) *in* $\Omega_T := \Omega \times]0, T[$.

(2) $\rho_\varepsilon(\cdot, t) \to \rho_0(\cdot, t)$ *for all* $t \in [0, T]$ *a.e. in* Ω, *and in* $L^1(\Omega), |\rho_\varepsilon(x, t)|$ *for all* $(x, t) \in \Omega_T$ *and* c *does not depend on* ε. *Furthermore, we assume that the limit* $\rho_0(\cdot, t)$ *consists only of the two values* β_1, β_2.

(3) *Define* $E_0(t) := \{x \in \Omega | \rho_0(x, t) = \beta_1\}$. *Then we have* $\rho_0(\cdot, t) = \beta_1$ *in* $E_0(t)$ *and* $\rho_0(\cdot, t) = \beta_2$ *in* $\Omega - E_0(t)$; *the boundary* $\partial E_0(t) \cap \Omega$ *is sufficiently smooth.*

(4) $v_\varepsilon \to v_0$ *a.e. in* Ω_T *and* v_0 *is sufficiently smooth in* E_0 *and* $\Omega \backslash E_0$.

(5) $\lim_\varepsilon \frac{1}{\varepsilon} \int_\Omega \left(\varepsilon^2 |\nabla \rho_\varepsilon|^2 + W(\rho_\varepsilon)\right) dx$ *exists uniformly in* t.

The condition (5) is at least satisfied in the stationary case (see [11], (3)). In this section we are going to prove the following theorem.

Theorem 2.2. *Let us assume the conditions in Assumption* 2.1 *and let* $\gamma(x, t)$ *be the velocity of the interface in the point* (x, t). *Then the limits* v_0, ρ_0 *satisfy in* $E_0(t)$ *and in* $\Omega \backslash E_0(t)$ *respectively the following equations, which are similar to the incompressible Navier-Stokes equations.*

$$\rho_0 \partial_t v_0 + \nabla \cdot \left(\rho_0 v_0 v_0^t + p(\rho_0) I)\right) = \mu \Delta v_0 \text{ in } E_0(t) \text{ and in } \Omega \backslash E_0(t),$$
$$div\ v_0 = 0 \text{ in } E_0(t) \text{ and in } \Omega \backslash E_0(t). \quad (2.1)$$

On the interface Γ *they satisfy*

$$n_+ [(\gamma(x, t) - v_0)\rho_0] = 0, \quad (2.2)$$
$$[\rho_0 v_0]\gamma - [\rho_0 v_0 v_0^t + p(\rho_0)]n_+ = -\mu n_+ [\nabla v_0], \quad (2.3)$$
$$[v_0] = 0. \quad (2.4)$$

Here [ ] *denotes the jump across the interface* Γ *and* n_+ *the outer normal vector on* Γ *with respect to* $E_0(t)$.

Remark 2.3. a) *In particular we see that there will be no pressure jump in the static case* $v_0 = 0$.
b) *Due to Assumption* 2.1 *the pressure in each phase turns out to be constant. The situation can be different for the scaling in Section* 3 *or/and if the zero Mach number limit is taken into account and combined in a suitable way with the limit* $\varepsilon \to 0$ (*see* [8]).

Now we will present some ideas for the proof of Theorem 2.2. In order to control the terms in the momentum equation in (1.1) if $\varepsilon \to 0$ we multiply (1.1) by a smooth testfunction ψ, $\psi \in C_0^\infty(\Omega_T, \mathbb{R}^n), \Omega \subset \mathbb{R}^n$ open, integrate over Ω_T and integrate by parts. Let us use the notation $v_\varepsilon = v, \rho_\varepsilon = \rho$. We obtain

$$\begin{aligned} &-\int_0^T \int_\Omega \rho_\varepsilon v_\varepsilon \partial_t \psi + \left(\rho v_\varepsilon v_\varepsilon^t + p(\rho_\varepsilon) I\right) \nabla \psi dx dt \\ &= \int_0^T \int_\Omega \left(\mu v_\varepsilon \Delta \psi + \gamma \varepsilon^2 \rho_\varepsilon \nabla \Delta \rho_\varepsilon \psi\right) dx dt. \end{aligned} \tag{2.5}$$

The terms in the left hand side in (2.5) converge to

$$-\int_0^T \int_\Omega \rho_0 v_0 \partial_t \psi + \left(\rho_0 v_0 v_0^t + p(\rho_0) I\right) \nabla \psi dx dt \tag{2.6}$$

and the first term on the right hand side in (2.5) to $\int_0^T \int_\Omega \mu v_0 \Delta \psi dx dt$. It is much more complicated to control the last term in (2.5):

$$R := \int_0^T \int_\Omega \gamma \varepsilon^2 \rho_\varepsilon \nabla \Delta \rho_\varepsilon \psi dx dt. \tag{2.7}$$

This term can be written as

$$\begin{aligned} R := \varepsilon^2 \int_0^T \int_\Omega &\partial_k \rho_\varepsilon \partial_j \rho_\varepsilon \partial_k \psi_j - (\partial_k \rho_\varepsilon)^2 \partial_j \psi_j + \frac{3}{2}\left((\partial_k \rho_\varepsilon)^2 + \frac{1}{\varepsilon^2} W(\rho_\varepsilon)\right) \partial_j \psi_j \\ &- \frac{3}{2\varepsilon^2} W(\rho_\varepsilon) \partial_j \psi_j - \frac{1}{2} \rho_\varepsilon^2 \partial_j \partial_k^2 \psi_j dx dt. \end{aligned} \tag{2.8}$$

Due to Assumption 2.1 and the extension of the results in [11] Lemma 2 to the timedependent case, all the terms in (2.8) can be controlled and we obtain that $R_\varepsilon \to 0$ for $\varepsilon \to 0$. In particular we use the following lemma in order to control the first two terms on the right-hand side in (2.8).

Lemma 2.4. *Let* $\Gamma_t := \partial E_0(t) \cap \Omega$, n_x *the outer unit normal vector on* Γ_t, $c_0 := \int_{\beta_1}^{\beta_2} W^{\frac{1}{2}}(s) ds$ *and* κ_t *the mean curvature of* Γ_t. *If the interface*

Γ_t is smooth in supp$(\psi(\cdot,t))$ then we have

$$\lim_{\varepsilon\to 0+} \varepsilon \int_\Omega \partial_k\rho_\varepsilon(\cdot,t)\partial_j\rho_\varepsilon(\cdot,t)\partial_k\psi_j(\cdot,t) - (\partial_k\rho_\varepsilon(\cdot,t))^2\partial_j\psi_j(\cdot,t)\ dx$$

$$= c_o \int_{\Gamma_t} \kappa_t\psi(\cdot,t)n_x dH_{n-1} \tag{2.9}$$

for all $t \in]0,T]$.

This implies that in the limit $\varepsilon \to 0$ we obtain

$$\int_0^T \int_\Omega (\rho_0\partial_t\psi + (\rho_0 v_0)\nabla\psi)\, dxdt = 0, \tag{2.10}$$

$$-\int_0^T \int_\Omega \rho_0 v_0 \partial_t\psi + \left(\rho_0 v_0 v_0^t + p(\rho_0)I)\right) \nabla\psi dxdt$$

$$= \int_0^T \int_\Omega \mu v_0 \Delta\psi dxdt. \tag{2.11}$$

Now we choose test functions ψ with compact support in $\{(x,t)|x \in E_0(t),\, t\in[0,T]\}$ and in $\{(x,t)|x \in \Omega\setminus E_0(t),\, t\in[0,T]\}$ respectively in (2.10), (2.11) in order to get the conditions as stated in (2.1). For proving the jump conditions (2.2), (2.3), and (2.4) we choose test functions with compact support in a small neighborhood (in space and time) of the interface, use this test function in (2.10), (2.11) and apply the Reynolds transport theorem.

3 Phasefield like scaling

In the static case (1.1) reduces to $\nabla\cdot(p(\rho)I) = \gamma\varepsilon^2\rho\nabla\Delta\rho$. Using $\tilde{W}(\rho) = \rho\psi(\rho)$ and $p(\rho) = \rho^2\psi'(\rho)$ (see (1.2)) we see that this equation is equivalent to $\tilde{W}'(\rho) = \gamma\varepsilon^2\Delta\rho + \lambda_\varepsilon$ which is just the Euler-Lagrange equation with the Lagrange multiplier λ_ε of the following minimization problem:

Minimize

$$\tilde{J}_\varepsilon(\rho) := \int_\Omega \left(W(\rho) + \gamma\varepsilon^2\frac{|\nabla\rho|^2}{2}\right) dx + \tilde{M} \tag{3.1}$$

under the constraint $\int_\Omega \rho\ dx = M_0$ for a given constant M_0 (conservation of mass), where W and l are defined above and $\tilde{M} = \int_\Omega l(\rho)dx$.

Now instead of (3.1) we can consider the functional with a different scaling

$$I_\varepsilon(\rho) := \frac{1}{\varepsilon}\int_\Omega W(\rho) + \frac{\varepsilon^2}{2}|\nabla\rho|^2 dx + \tilde{M} \to \text{Minimum}. \tag{3.2}$$

Due to the Assumption 2.1 this functional (i.e., the energy) is still bounded uniformly in ε. It turns out that the functional in (3.2) satisfies $I_\varepsilon(\rho) = \int_\Omega \rho\psi_\varepsilon(\rho) + \frac{\varepsilon}{2}|\nabla\rho|^2 dx$, where $\psi_\varepsilon(\rho)$ is defined by $\rho\psi_\varepsilon(\rho) = \frac{1}{\varepsilon}\left(\rho\psi(\rho) - l(\rho)\right) + l(\rho)$ and the corresponding pressure p_ε is given by (compare (1.2)) $p_\varepsilon(\rho) := \rho^2\psi_\varepsilon'(\rho) = \frac{1}{\varepsilon}p(\rho) + \frac{1-\varepsilon}{\varepsilon}d_1$.

The Euler-Lagrange equation of (3.2) is $\frac{d}{d\rho}\left(\rho\psi_\varepsilon(\rho)\right) - \varepsilon\Delta\rho = \tilde{\lambda}_\varepsilon$ where $\tilde{\lambda}_\varepsilon$ is the Lagrange multiplier with respect to the constraint $\int_\Omega \rho = M_0$. In order to see the relation to the static version of the Navier-Stokes-Korteweg equation we use

$$\nabla p_\varepsilon(\rho) = 2\rho\nabla\rho\psi_\varepsilon'(\rho) + \rho^2\psi_\varepsilon''(\rho)\nabla\rho \tag{3.3}$$

and the Euler-Lagrange equation to obtain $\nabla p_\varepsilon(\rho) = \varepsilon\rho\nabla\Delta\rho$ and for the dynamical case

$$\begin{aligned} \partial_t\rho + \nabla\cdot(\rho v) &= 0, \\ \partial_t(\rho v) + \nabla\cdot(\rho v v^t + p_\varepsilon(\rho)I) &= \mu\Delta v + \gamma\varepsilon\rho\nabla\Delta\rho. \end{aligned} \tag{3.4}$$

In addition to Assumption 2.1 we need the following:

Assumptions 3.1. $p_\varepsilon(\rho_\varepsilon(x,t)) \to p(x,t)$ *for* $(x,t) \in \Omega_T$, $p(\cdot,t) \in C^1(E_t)$ *and* $p(\cdot,t) \in C^1(\Omega\backslash E_t)$.

Now we can state the following theorem.

Theorem 3.2. *Let* ρ, v *be a solution of* (3.4), *which satisfy the conditions* (2), ..., (5) *in Assumption* 2.1 *and Assumption* 3.1. *Then the limits* v_0, ρ_0 *satisfy in* $E_0(t)$ *and in* $\Omega\backslash E_0(t)$ *respectively the following equations, which are (similar to) the incompressible Navier-Stokes equations.*

$$\begin{aligned} \rho_0\partial_t v_0 + \nabla\cdot\left(\rho_0 v_0 v_0^t + pI\right) &= \mu\Delta v_0 \text{ in } E_0(t) \text{ and in } \Omega\backslash E_0(t), \\ \operatorname{div} v_0 &= 0 \text{ in } E_0(t) \text{ and in } \Omega\backslash E_0(t). \end{aligned} \tag{3.5}$$

On the interface Γ *they satisfy*

$$n_+\left[(\gamma(x,t) - v_0)\rho_0\right] = 0 \quad \text{on} \quad \Gamma, \tag{3.6}$$

$$[\rho_0 v_0]\gamma n_+ - [\rho_0 v_0 v_0^t + p(\rho_0)]n_+ = \mu n_+[\nabla v_0] + \frac{5}{2}c_o\kappa_t n_+ \text{on } \Gamma, \tag{3.7}$$

$$[v_0] = 0 \quad \text{on} \quad \Gamma. \tag{3.8}$$

Remark 3.3. *In particular we see that we recover the classical Young-Laplace law in the static case* $v_0 = 0$, *i.e., that the pressure jump is proportional to the mean curvature.*

For the proof of Theorem 3.2 we proceed in a similar way as for Theorem 2.2. Now due to the different scaling in (3.2) we have to control the term

$$\frac{1}{\varepsilon}R = \int_0^T \int_\Omega \gamma\varepsilon\rho_\varepsilon\nabla\Delta\rho_\varepsilon\psi \, dxdt$$

instead of (2.7). An extension of Lemma 2.4, [7], [11] implies in this case

$$\lim_{\varepsilon\to 0}\frac{1}{\varepsilon}R = \lim_{\varepsilon\to 0}\int_0^T\int_\Omega \gamma\varepsilon\rho_\varepsilon\nabla\Delta\rho_\varepsilon\psi\ dxdt$$
$$= \int_0^T c_o\int_{\Gamma_t}\kappa_t\psi(\cdot,t)n_x dH_{n-1}dt + \frac{3}{2}c_0\int_0^T\int_{\Gamma_t}\partial_j\psi_j(x)dH_{n-1}.$$

Using Assumption 3.1 we obtain for $\varepsilon\to 0$ in the weak form of the momentum equation in (3.4) (for the conservation of mass we argue in a similar way)

$$-\int_0^T\int_\Omega \rho_0 v_0\partial_t\psi + \left(\rho v_0 v_0^t + p(x,t)I)\right)\nabla\psi dxdt - \int_0^T\int_\Omega \mu v_0\Delta\psi dxdt$$
$$= \int_0^T c_o\int_{\Gamma_t}\kappa_t\psi(\cdot,t)n_x dH_{n-1}dt + \frac{3}{4}c_0\int_0^T\int_{\Gamma_t}\partial_j\psi_j(x)dH_{n-1}. \quad (3.9)$$

As before we have to use test functions ψ which have compact support in E_0, $\Omega_T\backslash E_0$ and in a neighborhood of the interface Γ respectively to obtain the conditions (3.5), (3.6), (3.7), and (3.8) in Theorem 3.2. In particular we obtain on the interface Γ

$$\int_0^T\int_\Gamma \psi\left([\gamma(\rho_0 v_0)]\,n_+ - \left[\rho_0 v_0 v_0^t + pI\right]n_+ + \mu[\nabla v_0]n_+\right)$$
$$-\int_0^T\int_\Gamma \mu\nabla\psi[v_0]n_+$$
$$= \int_0^T c_o\int_{\Gamma_t}\kappa_t\psi(\cdot,t)n_+ dH_{n-1}dt + \frac{3}{2}c_0\int_0^T\int_{\Gamma_t}\partial_j\psi_j(x)dH_{n-1}. \quad (3.10)$$

Now we choose a smooth test function $\phi:\Gamma\to\mathbb{R}^n$ and continue it to a function ψ such that $\psi\in C_0^\infty(\Omega)$ and $n_x\nabla\psi = 0$ on Γ. Due to the Gauss theorem on surfaces we obtain

$$\int_{\Gamma_t}\partial_j\psi_j(x)d\sigma = \int_{\Gamma_t}\nabla_{\Gamma_t}\cdot\psi d\sigma = \int_{\Gamma_t}\nabla_{\Gamma_t}\cdot n_x\psi n_x d\sigma = \int_{\Gamma_t}\kappa_t\psi n_x d\sigma.$$

Using this in (3.10), we obtain the jump condition (3.7). Similarly (3.6), (3.8) can be proved.

References

[1] L. K. Antanovskii, Microscale theory of surface tension. Phys. Rev. E54(1996), 6285–6290.

[2] S. Benzoni-Gavage, R. Danchin and S. Descombes, On the well-posedness for the Euler-Korteweg model in several space dimensions. Indiana Univ. Math. J. 56 (2007), no. 4, 1499–1579.

[3] D. Bresch, B. Desjardin and C. K. Lin, On some compressible fluid models: Korteweg, lubrication, and shallow water systems. (English) Commun. Partial Differ. Equations. 28(2003), no. 3–4, 843–868.

[4] D. Diehl, Higher order schemes for simulation of compressible liquid-vapour flows with phase change. PhD thesis, Freiburg, Abteilung für Angewandte Mathematik, 2007.

[5] W. Dreyer and C. Kraus, The sharp interface limit of the van der Waals-Cahn-Hilliard phase model for fixed and time dependent domains. Preprint 2007, WIAS.

[6] B. T. Hayes and P. G. LeFloch, Nonclassical shocks and kinetic relations: Strictly hyperbolic systems. (English) SIAM J. Math. Anal. 31 (2000), no. 5, 941–991.

[7] J. E. Hutchinson and Y. Tonegawa, Convergence of phase interfaces in the van der Waals-Cahn-Hilliard theory. Calc. Var. Partial Differ. Equ. 10 (2000), no. 1, 49–84.

[8] K. Hermsdörfer, C. Kraus and D. Kröner, Pressure jump conditions for limits of the Navier-Stokes-Korteweg model. Preprint 2008, Dep. of Appl. Math., University of Freiburg.

[9] D. J. Korteweg, Sur la forme que prennent les quations du mouvement des fluids si l'on tient compte des forces capillaires causes par les variations de densite [on the form the equations of motions of fluids assume if account is taken of the capillary forces caused by density variations], Archives Nerlandaises des Sciences Exactes et naturelles, Series II, volume 6 (1901), 1–24.

[10] M. Kotschote, Strong well-posedness for a Korteweg-type model for the dynamics of a compressible non-isothermal fluid. Preprint Leipzig 2006.

[11] S. Luckhaus, L. Modica, The Gibbs-Thompson relation within the gradient theory. Arch. Rat. Mech. Anal. 107 (1989), 7183.

[12] L. Modica, The gradient theory of phase transitions and the minimal interface criterion. Arch. Rat. Mech. Anal. 98 (1987), 123–142.

Numerical Treatment of a Non-classical Transport Equation Modelling Radiative Transfer in Atmospheric Clouds

Kai Krycki

Center for Computational Engeneering Sciences
RWTH Aachen University, Schinkelstrasse 2, 52062 Aachen, Germany
E-mail: krycki@mathcces.rwth-aachen.de

Martin Frank

Center for Computational Engeneering Sciences
RWTH Aachen University, Schinkelstrasse 2, 52062 Aachen, Germany
E-mail: frank@mathcces.rwth-aachen.de

Abstract

We present a numerical treatment of a non-classical transport equation which models radiative transfer in atmospheric clouds. The medium is assumed to be correlated which results in a non-exponential attenuation law. A diffusive scaling is introduced and we use a HLL scheme to solve the associated M1 moment model numerically. The P1-model is used for a diffusion preconditioning of a source iteration-like iterative process.

1 Introduction

In this paper we present a first numerical treatment of the non-classical transport equation

$$\partial_s \phi(s,x,v) + v \nabla_x \phi(s,x,v) + \Sigma_t(s)\phi(s,x,v) = 0, \tag{1.1a}$$

$$\phi(0,x,v) = c \int_0^\infty \int \sigma(v,v')\Sigma_t(s')\phi(s',x,v')dv'ds' + Q(x,v), \tag{1.1b}$$

where $s \in (0,\infty)$, $x \in \mathbb{R}^N$, $v \in S^{N-1}$ the unit sphere in $\mathbb{R}^N$. Larsen [5] suggested this equation to model the radiative transfer in atmospheric clouds. Here, $\phi(s,x,v)$ is the angular particle flux at point x into direction v. What makes this equation "non-classical" is that the cross section Σ_t depends explicitly on the additional variable s, which is the path length travelled by the particle since its previous interaction. The

variable s can also be seen as the time elapsed since the previous interaction, hence (1.1) can be interpreted as an initial value problem.
The only possible interaction between a photon and the background medium is the collision with a scattering center, a water droplet in this case. The probability that a particle that has already travelled a distance s hits a water droplet is given by $\Sigma_t(s)$. Then, by a given constant probability $0 < c < 1$ the photon with velocity v scatters and changes its direction to v' due to a probability given by the scattering kernel $\sigma(v, v')$. With a probability of $1 - c$ the particle experiences absorption. After the scattering the particle travels another distance until the next scattering event occurs. Equation (1.1b) states that all photons which are scattered or injected into the system have zero distance to the last scattering event, while equation (1.1a) models the transport of the particles. A probability density p for the distance between collisions can be defined in terms of the cross section Σ_t and vice versa:

$$p(s) = \Sigma_t(s) \exp\left(\int_0^\infty \Sigma_t(\tau) d\tau \right) \quad \text{or} \quad \Sigma_t(s) = \frac{p(s)}{1 - \int_0^s p(\tau) d\tau}. \qquad (1.2)$$

Current observations suggest that p decays algebraically, which results in non-exponential decay. It has been suggested that this is rooted in a correlation of the locations of the scattering centres. For Σ_t constant we have classical transport with an exponential attenuation law.

This model was first suggested by Larsen in [5], existence and uniqueness of a solution were proven in [3]. What we present here is an initial numerical treatment for one space dimension in a diffusive regime, using a moment model ansatz and a HLL discretization scheme. The solution is computed by successive iteration from a carefully chosen initial value. The slow convergence of the iteration forces us to introduce a diffusion preconditioning. Numerical results are presented in the final section.

2 Scaling and the diffusion limit

The details of the scaling and an analysis of the diffusion asymptotics can be found in [3], we only give a sketch of it here. The first and second moment of this probability are defined as

$$\langle s \rangle = \int_0^\infty s p(s) ds, \quad \langle s^2 \rangle = \int_0^\infty s^2 p(s) ds.$$

An atmospheric cloud is optically thick and dominated by scattering, therefore we introduce the following scaling. Let $0 < \varepsilon \ll 1$, then we scale the parameters Q, Σ_t, and c as follows:

$$Q \to \varepsilon Q, \quad \Sigma_t(s) \to \frac{1}{\varepsilon} \Sigma_t(s/\varepsilon), \quad c \to 1 - \kappa\varepsilon^2,$$

where $\kappa > 0$. By additionally scaling the pseudo time-scale by

$$\phi_\varepsilon(s,x,v) = \phi(s\varepsilon, x, v),$$

the new scaled unknown satisfies

$$\partial_s \phi_\varepsilon(s,x,v) + \varepsilon v \nabla_x \phi_\varepsilon(s,x,v) + \Sigma_t(s)\phi_\varepsilon(s,x,v) = 0;$$
$$\phi_\varepsilon(0,x,v) = (1-\kappa\varepsilon^2)\int_0^\infty \int \sigma(v,v')\Sigma_t(s')\phi_\varepsilon(s',x,v')dv'ds' + \varepsilon Q(x,v).$$

This equation can be further simplified. By setting

$$\Phi_\varepsilon = \varepsilon\langle s\rangle \exp\left(\int_0^s \Sigma_t(\tau)d\tau\right)\phi_\varepsilon(s,x,v)$$

we make use of (1.2), and the equation now reads

$$\partial_s \Phi_\varepsilon + \varepsilon v \nabla_x \Phi_\varepsilon = 0; \tag{2.1a}$$
$$\Phi_\varepsilon(0,x,v) = (1-\kappa\varepsilon^2)\int_0^\infty \int \sigma(v,v')p(s)\Phi_\varepsilon(s',x,v')ds'dv'. \tag{2.1b}$$

Applying a formal Hilbert expansion in the parameter ε by

$$\Phi_\varepsilon = \Phi_0 + \varepsilon\Phi_1 + \varepsilon^2\Phi_2 + \cdots$$

and letting $\varepsilon \to 0$, one derives the following diffusion approximation (see again [3] or [5] for the details):

$$-\frac{1}{3}\left(\frac{\langle s^2\rangle}{2} + \frac{\bar{\mu}_0}{1-\bar{\mu}_0}\right)\nabla_x^2\Phi_0 + \kappa\Phi_0 = \langle s\rangle Q(x), \tag{2.2}$$

with $\Phi_0 = \Phi_0(x)$ and $\bar{\mu}_0$ being the mean scattering cosine. We will always consider an isotropic medium with constant probability for the outgoing angle and in this case we simply have $\bar{\mu}_0 = 0$. All combined, Φ_0 is an approximation of the classical radiation flux up to first order due to

$$\int_0^\infty \int \phi(s,x,v)dvds = \Phi_0(x) + O(\varepsilon).$$

Note that this diffusion limit is not valid anymore as soon as

$$p(s) \geq \frac{C}{s^3},$$

because in this case $\langle s^2\rangle = \infty$. We will study this case in the numerics section.

3 Moment models

Introducing the first three angular moments of function ϕ given by

$$\phi^{(0)}(s,x) = \int \phi(s,x,v)dv, \quad \phi^{(1)}(s,x) = \int v\phi(s,x,v)dv,$$
$$\phi^{(2)}(s,x) = \int (v \otimes v)\phi(s,x,v)dv,$$

and by multiplying equation (2.1) by v and integrating over v, we obtain the following coupled system of linear transport equations:

$$\partial_s \phi^{(0)} + \varepsilon \nabla_x \phi^{(1)} = 0, \quad \partial_s \phi^{(1)} + \varepsilon \nabla_x \phi^{(2)} = 0;$$
$$\phi^{(0)}(0,x) = \varepsilon^2 \langle s \rangle Q(x) + (1-\kappa\varepsilon^2) \int_0^\infty p(s)\phi^{(0)}(s,x)ds,$$
$$\phi^{(1)}(0,x) = 0.$$

Note that this system of equations does not depend on v anymore, hence it is a major simplification of the initial problem. In this situation of three unknowns and only two equations it is usual to use a moment closure of the form $\phi^{(2)} = \phi^{(2)}(\phi^{(0)}, \phi^{(1)})$. The linear P1-model is based on the idea that $\phi^{(2)}$ is polynomial in $\phi^{(0)}$ and $\phi^{(1)}$ and assumes that

$$\phi^{(2)} \approx \chi(\phi^{(0)}, \phi^{(1)}) = \frac{1}{3}\phi^{(0)}.$$

Another moment closure is the M1-model based on the minimum entropy principle (see [2]). It leads to a coupled system of nonlinear partial differential equations by setting $f = \frac{\phi^{(1)}}{\phi^{(0)}}$ and

$$\phi^{(2)} \approx \chi(\phi^{(0)}, \phi^{(1)}) = \frac{3+4f^2}{5+2\sqrt{4-3f^2}}\phi^{(0)}.$$

Hence, the system becomes

$$\partial_s \phi^{(0)} + \varepsilon \nabla_x \phi^{(1)} = 0, \quad \partial_s \phi^{(1)} + \varepsilon \nabla_x \chi(\phi^{(0)}, \phi^{(1)}) = 0;$$
$$\phi^{(0)}(0,x) = \varepsilon Q(x) + (1-\kappa\varepsilon^2) \int_0^\infty p(s)\phi^{(0)}(s,x)ds;$$
$$\phi^{(1)}(0,x) = 0;$$

with χ being linear in the P1- and nonlinear in the M1-model.

4 Numerical method

To simplify the notation we set $E(s,x) = \phi^{(0)}(s,x)$ and $F(s,x) = \phi^{(1)}(s,x)$. Furthermore, we choose a uniform time-space grid (s_i, x_n)

with $s_{i+1} = s_i + \Delta s$ and $x_{i+1} = x_i + \Delta x$. Then we use the abbreviations $E_i^n = E(s_n, x_i)$ and $\chi_i^n = \chi(E_i^n, F_i^n)$. Using HLL as a guideline (see [4]), one derives the following discretization scheme for both the P1- and the M1-model:

$$\frac{E_i^{n+1} - E_i^n}{\Delta s} + \varepsilon \frac{F_{i+1}^n - F_{i-1}^n}{2\Delta x} - \varepsilon^2 \frac{E_{i+1}^n - 2E_i^n + E_{i-1}^n}{2\Delta x} = 0, \tag{4.1a}$$

$$\frac{F_i^{n+1} - F_i^n}{\Delta s} + \varepsilon \frac{\chi_{i+1}^n - \chi_{i-1}^n}{2\Delta x} - \varepsilon^2 \frac{F_{i+1}^n - 2F_i^n + F_{i-1}^n}{2\Delta x} = 0, \tag{4.1b}$$

$$E_i^0 = (1 - \kappa\varepsilon^2) \sum_n \lambda_n p_n E_i^n \Delta s + \varepsilon^2 \langle s \rangle Q(x), \quad F_i^0 = 0, \tag{4.1c}$$

for some quadrature formula given by the weights λ_n. There are two obvious ways to solve this system:

(a) Directly, leading to a large matrix.

(b) Iteratively, requires the choice of an initial setting for $E(s, x)$.

Option (a) results in a large linear system, and our experiments have shown that a very large space interval is needed such that the solution tends to 0 at the boundaries as one would expect. Instead of an explicit Euler an implicit method is needed because the condition number of the matrix becomes very large even for moderately small ε. Thinking of extensions to the 2D and 3D case, option (a) seems to be too expensive for simulations.

The iteration process works as follows: starting with an initial setting for $E(s, x)$ we can calculate $E(0, x)$ using (4.1c). Then we calculate new values for E and F "forward in time" applying (4.1a) and (4.1b). Now we can insert these values into (4.1c) again and iterate this process. In [3] it is proven that this iteration with initial value $E(s, x) = 0$ is monotone and bounded, hence convergent. Main problem with option (b) is the choice of an initial setting for $E(s, x)$. Choosing $E(s, x) = 0$ as initial value leads to very slow convergence and even wrong results. In order to find a suitable initial value we employ the following idea due to [1]. We set

$$\Pi(x) = \int_0^\infty p(s) E(s, x) ds,$$

and we calculate formally, using the P1-model:

$$\begin{aligned} \Pi'(x) &= \int_0^\infty p(s) \partial_x E(s, x) ds = -\frac{3}{\varepsilon} \int_0^\infty p(s) \partial_s F(s, x) ds \\ &= -\frac{3}{\varepsilon} \left[\partial_s p(0) E(0, x)\right]_0^\infty + \frac{3}{\varepsilon} \int_0^\infty \partial_s p(s) F(s, x) ds \\ &= \frac{3}{\varepsilon} \int_0^\infty \partial_s p(s) F(s, x) ds. \end{aligned}$$

Now the second derivative of Π can be computed as

$$\begin{aligned}\Pi''(x) = &\frac{3}{\varepsilon}\int_0^\infty \partial_s p(s)\partial_x F(s,x)ds = -\frac{3}{\varepsilon^2}\int_0^\infty \partial_s p(s)\partial_s E(s,x)ds \\ = &-3\partial_s p(0)\langle s\rangle Q(x) + 3\kappa\partial_s p(0)\Pi(x) \\ &+\left(-\frac{3}{\varepsilon^2}\partial_s p(0)\Pi(x) + \frac{3}{\varepsilon^2}\int_0^\infty \partial_s^2 p(s)E(s,x)ds\right),\end{aligned}$$

and, if $\Sigma_t(s) \equiv \Sigma$, the last term becomes 0 and this equation reads

$$\Pi''(x) = 3\kappa\Sigma^2\Pi(x) - 3\Sigma^2\langle s\rangle Q(x). \tag{4.2}$$

Since in this case $\langle s\rangle = 1/\Sigma$ and $\langle s^2\rangle = 2/\Sigma^2$, equation (4.2) even equals the diffusion approximation (2.2). For nonconstant Σ_t the last term does not disappear, but we neglect it and take the numerical solution of

$$\Pi''(x) = -3\kappa\partial_s p(0)\Pi(x) + 3\partial_s p(0)\langle s\rangle Q(x)$$

as an initial value for $\int_0^\infty p(s)E(s,x)ds$ in the iteration of the M1-model. The numerical results confirm this approach, it can be thought of as a diffusion preconditioning.

5 Numerical results

For our numerical experiments we always chose the following setting. The source term Q is the step function

$$Q(x) = \begin{cases} 1, & x \in [-1,1], \\ 0, & \text{otherwise.}\end{cases}$$

It models a radiation source at the center of the space interval. The space interval is $[-30, 30]$ and the $s-$ or (pseudo-)time interval $[0, 10]$, as step size we chose $\Delta x = 0.1$ and $\Delta s = 0.05$. Furthermore, we enforce a non-exponential attenuation law by setting

$$p(s) = \frac{\alpha-1}{(s+1)^\alpha}, \quad \text{hence } \Sigma_t(s) = \frac{\alpha-1}{s+1},$$

with $\alpha > 1$ due to the relationship (1.2). Choosing $\Sigma_t(s) \equiv \Sigma > 0$ constant results in an exponential attenuation law $p(s) = \Sigma\exp(-s\Sigma)$. We always chose periodic boundary conditions for all equations involved. The further parameters were chosen as $\kappa = 1 - c$ and $c = 0.9$.

Within this setting we have made the following observations:

(1) Trying to solve the system directly requires a very large space interval such that the solution becomes 0 at the boundaries. Also for small ε the matrix becomes ill-conditioned and that has to be fixed by an implicit discretization. This is rather inefficient and too expensive in terms of computational time and resources.

(2) The iterative process converges if we choose a suitable initial value (see Section 4), but the convergence is still very slow. The contraction rate seems to tend towards $1 - \varepsilon^2(1 - c)$.

(3) For values of α in a range between 3 and 1, the second moment of the probability density p does not exist, hence the diffusion coefficient of the limit equation formally tends to "infinity" and in that case we observe that the solution tends to 0.

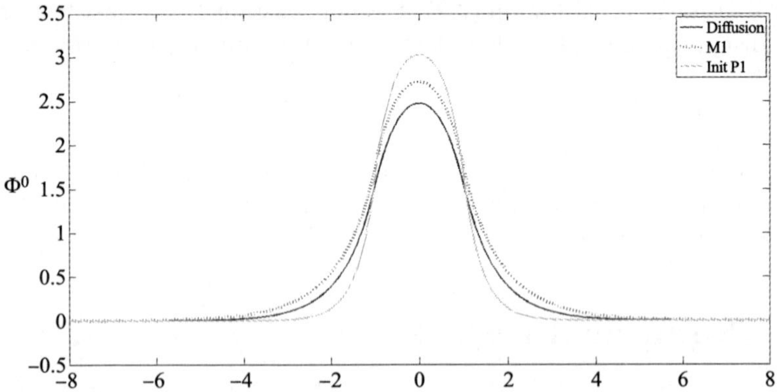

Figure 5.1 This is the particle density Φ_0 for $\alpha = 5$ and $\varepsilon = 0.5$. The maximum of the curve lies between the initial value and the diffusion limit.

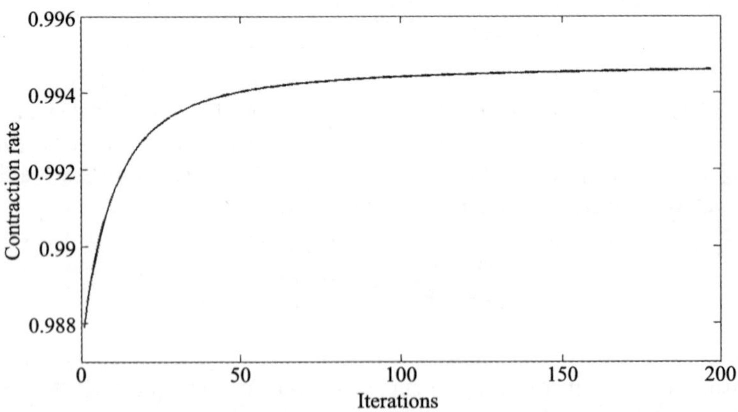

Figure 5.2 Contraction rate for $\alpha = 5$ and $\varepsilon = 0.3$. The value where it stabilizes is of order $1 - \varepsilon^2(1 - c)$.

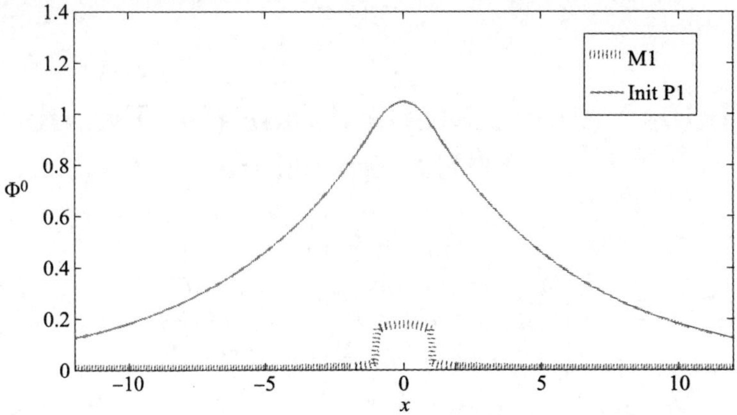

Figure 5.3 This is the particle density Φ_0 for $\alpha = 1.1$ and $\varepsilon = 0.5$ after 200 iterations, suggesting that this function tends to 0.

References

[1] C. Berthon and R. Turpault, private communication.

[2] B. Dubroca and J. L. Feugeas, Étude théorique et numérique d'une hiérarchie de modèles aux moments pour le transfert radiatif. C. R. Acad. Paris 1999, 915–920.

[3] M. Frank and T. Goudon, On a generalized Boltzmann equation for non-classical particle transport. Kinet. Relat. Models 3. (2010), 395–407.

[4] A. Harten, P. Lax and B. van Leer, On upstream differencing and Godunov-type schemes for hyperbolic conservation laws. SIAM Review. 25(1983) no. 1, 35–61.

[5] E. W. Larsen, A generalized Boltzmann equation for "non-classical" particle transport. Joint Int. Meeting on Math. Comp. Supercomp. In: Nucl. App., Monterey, California, on CD-ROM, Am. Nuc. Soc., LaGrange Park, IL 2007.

Finite Volume Methods for the Two-fluid MHD Equations

Harish Kumar
Seminar for Applied Mathematics
ETH Zurich, CH-8092, Switzerland
E-mail: harish@math.ethz.ch

Abstract

Two fluid (TF) ideal magnetohydrodynamics (MHD) equations are a generalized form of the ideal MHD equations in which the electrons and ions are considered as separate species. The design of efficient numerical schemes for these equations is complicated on account of the non-linearities and the presence of stiff source terms, particularly for realistic charge to mass ratios. We design novel finite volume schemes based on an implicit-explicit (IMEX) time stepping routine. The special structure of the two-fluid MHD equations enable us to split the source terms carefully in order to ensure that only local (in each cell) equations need to be solved at each time step. Furthermore, these equations are solved exactly. Benchmark numerical experiments are presented to illustrate the efficiency of new approach.

1 Introduction

Flows in plasmas are frequently modeled by the equations of ideal magnetohydrodynamics (MHD). These equations combine the Euler equations for compressible flow together with Maxwell's equations for magnetic fields. These equations assume *quasi-neutrality*, i.e., the difference in number density of ions and electrons is ignored.

However, in many applications like fast magnetic reconnection, the assumption of quasi-neutrality is violated. In such cases, one resorts to extended MHD models for the plasmas. A popular example models the flow of two different species, one electron and one ion, separately. This result in the so-called two fluid MHD (TF-MHD) equations (see [1, 2]). In non-dimensional form, these equations are

$$\frac{\partial \rho_i}{\partial t} + \nabla \cdot (\rho_i \vec{v}_i) = 0, \tag{1.1a}$$

$$\frac{\partial(\rho_i \vec{v}_i)}{\partial t} + \nabla \cdot \left(\rho_i \vec{v}_i \vec{v}_i^\top + p_i \mathbf{I}\right) = \frac{1}{\hat{r}_g} \rho_i (\vec{E} + \vec{v}_i \times \vec{B}), \tag{1.1b}$$

$$\frac{\partial E_i}{\partial t} + \nabla \cdot ((E_i + p_i)\vec{v}_i) = \frac{1}{\hat{r}_g} \rho_i (\vec{E} \cdot \vec{v}_i), \tag{1.1c}$$

$$\frac{\partial \rho_e}{\partial t} + \nabla \cdot (\rho_e \vec{v}_e) = 0, \tag{1.1d}$$

$$\frac{\partial(\rho_e \vec{v}_e)}{\partial t} + \nabla \cdot \left(\rho_e \vec{v}_e \vec{v}_e^\top + p_e \mathbf{I}\right) = -\frac{\lambda_m}{\hat{r}_g} \rho_e (\vec{E} + \vec{v}_e \times \vec{B}), \tag{1.1e}$$

$$\frac{\partial E_e}{\partial t} + \nabla \cdot ((E_e + p_e)\vec{v}_e) = -\frac{\lambda_m}{\hat{r}_g} \rho_e (\vec{E} \cdot \vec{v}_e), \tag{1.1f}$$

$$\frac{\partial \vec{B}}{\partial t} + \nabla \times \vec{E} + \kappa \nabla \psi = 0, \tag{1.1g}$$

$$\frac{\partial \vec{E}}{\partial t} - \hat{c}^2 \nabla \times \vec{B} + \xi \hat{c}^2 \nabla \phi = -\frac{1}{\hat{\lambda}_d^2 \hat{r}_g} (r_i \rho_i \vec{v}_i + r_e \rho_e \vec{v}_e), \tag{1.1h}$$

$$\frac{\partial \phi}{\partial t} + \xi \nabla \cdot \vec{E} = \frac{\xi}{\hat{\lambda}_d^2 \hat{r}_g} (r_i \rho_i + r_e \rho_e), \tag{1.1i}$$

$$\frac{\partial \psi}{\partial t} + \kappa \hat{c}^2 \nabla \cdot \vec{B} = 0. \tag{1.1j}$$

Here subscript $\{i, e\}$ refers to the ion and electron species respectively, $\rho_{\{i,e\}}$ are densities, $\vec{v}_{\{i,e\}} = (v^x_{\{i,e\}}, v^y_{\{i,e\}}, v^z_{\{i,e\}})$ are velocities, $E_{\{i,e\}}$ are the energies, $p_{\{i,e\}}$ are the pressures, $\vec{B} = (B^x, B^y, B^z)$ is magnetic field, $\vec{E} = (E^x, E^y, E^z)$ is electric field, ϕ, ψ are the potentials and ξ, κ are the speeds for Maxwell's equations. Also $r_i = \frac{q_i}{m_i}$ and $r_e = \frac{q_e}{m_e}$ are charge to mass ratios of ions and electrons respectively.

The Eqns. (1.1a)–(1.1c) represent the conservation of mass, momentum and energy for the ions. The source term in (1.1b) represents the Lorentz force acting on ions due to the electric and magnetic fields. Similarly Eqns. (1.1d)–(1.1f) represent conservation of these quantities for electrons. Eqns. (1.1g)–(1.1j) are the perfectly hyperbolic Maxwell's (PHM) equations (see [4]). These equations satisfy the divergence free condition approximately, and are consistent with the hyperbolic structure of the fluid equations. In addition we assume that both ions and electrons satisfy ideal gas law with gas constant $\gamma = \frac{5}{3}$.

Many physically significant parameters appear in the non dimensionalized form of the TF equations. Here, $\hat{r}_g = \frac{m_i v_i^T}{q_i B_0 x_0}$ is the normalized ion Larmor radius, $\lambda_m = m_i/m_e$ is the ion-electron mass ratio, $\hat{c} = c/v_i^T$ is the normalized speed of light and $\hat{\lambda}_d = \lambda_d/r_g$ is the ion Debye length normalized with Larmor radius. Also v_i^T is the reference thermal velocity of ions, B_0 is the reference magnetic field and x_0 is the reference length. Ion mass m_i is assumed to be 1. Note that when Larmor radius

$\hat{r}_g \to 0$, the TF model approaches the MHD limit. Similarly for $\hat{r}_g \to \infty$ (1.1) reduce to simple flow (Euler) equations for ions and electrons. TF equations model the intermediate physics between these two limits.

For simplicity, we will focus on the TF equations in one space dimension for the rest of this paper. Eqns. (1.1) can be written as a system of balance laws,

$$\mathbf{U}_t + \mathbf{F}(\mathbf{U})_x = \mathbf{S}(\mathbf{U}). \tag{1.2}$$

Here the vector of unknowns $\mathbf{U}$, the flux vector $\mathbf{F}$, and the source vector $\mathbf{S}$ can be read from (1.1). It is well known that solutions of (1.2) consist of discontinuities in the form of shocks and contact discontinuities. Numerical algorithms have to take into account the formation of these complex waves and their interactions. Finite volume methods (FVM) have been quite successful in approximating balance laws of the form (1.2) (see [3]). In particular, one has to device suitable numerical flux functions in order to approximate these waves.

However, the most challenging issue in designing efficient algorithms for (1.1) is the stiffness of the source terms. As an example, consider a generic situation involving a mass ratio of 1832.6, non-dimensonal Debye length of 0.01, and Larmor radius of 0.005. Assuming that all the other quantities are of $O(1)$, the strength of source term is 3.6652×10^9. Thus, explicit time stepping will be extremely expensive computationally due to small time steps.

The TF equations have received some attention in recent years. In [1], the authors used a second order operator splitting approach and a fourth order Runge-Kutta (RK4) method to discretize source terms in time. This approach is easy to implement but computationally expensive. In [2], the authors treat source terms implicitly and flux terms explicitly. The resulting equations are solved using Newton iterations. This method might be diffusive and may require many iterations for each time step.

In this paper, we propose a novel implicit-explicit(IMEX) scheme to discretize (1.1) based on the following ingredients:

1. The numerical fluxes are designed by exploiting the split structure of the flux in (1.1) and are integrated in time explicitly.

2. The source terms are treated implicitly. We observe and exploit the special structure of the source terms in order to device an implicit scheme that requires solving cell-wise linear system of equations. The local mass matrices are inverted symbolically and are evaluated at each time step.

3. High order spatial accuracy is obtained by non-oscillatory piecewise linear reconstruction in each cell. High order time integration is performed using SSP-RK methods (see [5]).

Rest of the paper is organized as follows: In Section 2, we describe the proposed IMEX algorithm. In Section 3, we present the numerical results for two numerical test examples.

2 Finite volume methods

Consider a computational domain $(0, L)$ (for some $L > 0$), discretized uniformly(for simplicity), with mesh size Δx in to cells $[x_{i-\frac{1}{2}}, x_{i+\frac{1}{2}}]$. We aim to approximate the cell averages $\mathbf{U}_i^n \approx \frac{1}{\Delta x}\int_{x_{i-\frac{1}{2}}}^{x_{i+\frac{1}{2}}} \mathbf{U}(x, t^n)dx$ at time level t^n. Denote the time-step Δt^n at time level t^n, then a first order IMEX scheme for (1.1) is written as

$$\frac{\mathbf{U}_i^{n+1} - \mathbf{U}_i^n}{\Delta t} = -\frac{1}{\Delta x}\left(\mathbf{F}_{i+\frac{1}{2}}^n - \mathbf{F}_{i-\frac{1}{2}}^n\right) + \mathbf{S}(\mathbf{U}_i^{n+1}). \tag{2.1}$$

2.1 Numerical flux function

Denote $\mathbf{U} = \{\mathbf{U}^f, \mathbf{U}^m\}$, and $\mathbf{F}(\mathbf{U}) = \{\mathbf{F}^f(\mathbf{U}), \mathbf{F}^m(\mathbf{U})\}$ with

$$\begin{aligned}
\mathbf{U}^f &= \{\rho_i, \rho_i\vec{v}_i, E_i, \rho_e, \rho_e\vec{v}_e, E_e\}^\top,\\
\mathbf{F}^f &= \{\rho_i v_i^x, \rho_i v_i^{x2} + p_i, \rho_i v_i^x v_i^y, \rho_i v_i^x v_i^z, (E_i + p_i)v_i^x,\\
&\qquad \rho_e v_e^x, \rho_e v_e^{x2} + p_e, \rho_e v_e^x v_e^y, \rho_e v_e^x v_e^z, (E_e + p_e)v_e^x\}^\top,\\
\mathbf{U}^m &= \{\vec{B}, \vec{E}, \phi, \psi\}^\top,\\
\mathbf{F}^m &= \{\kappa\psi, -E^y, E^x, \xi\hat{c}^2\phi, \hat{c}^2 B^y, -\hat{c}^2 B^x, \xi E^x, \kappa\hat{c}^2 B^x\}^\top.
\end{aligned}$$

We exploit the split structure of the flux in (1.1) to design a four-wave HLL type approximate Riemann solver. Consider the Euler HLL speeds,

$$\begin{aligned}
b_l^f &= \min_j \min_\alpha(\lambda_j^\alpha(\mathbf{U}_i), \lambda_j^\alpha(\mathbf{U}_a), 0),\\
b_r^f &= \max_j \max_\alpha(\lambda_j^\alpha(\mathbf{U}_{i+1}), \lambda_j^\alpha(\mathbf{U}_a), 0),
\end{aligned}$$

with $\lambda_j^\alpha \in \{v_\alpha^x - c_\alpha, v_\alpha^x, v_\alpha^x + c_\alpha\}$, $\alpha \in \{i, e\}$, $c_\alpha = \sqrt{\gamma p_\alpha/\rho_\alpha}$ and the Maxwell HLL speeds $b_l^m = -\hat{c}$, $b_r^m = \hat{c}$. Here $\mathbf{U}_a$ represent the simple average of states $\mathbf{U}_i$ and $\mathbf{U}_{i+1}$. Then the four wave HLL type numerical flux for (1.1) is

$$\mathbf{F}_{i+\frac{1}{2}}^{n,\beta} = \frac{b_r^\beta \mathbf{F}^\beta(\mathbf{U}_i^{n,\beta}) - b_l^\beta \mathbf{F}^\beta(\mathbf{U}_{i+1}^{n,\beta})}{b_r^\beta - b_l^\beta} + \frac{b_r^\beta b_l^\beta}{b_r^\beta - b_l^\beta}(\mathbf{U}_{i+1}^{n,\beta} - \mathbf{U}_i^{n,\beta}), \tag{2.2}$$

with $\beta \in \{f, m\}$.

2.2 Processing of source terms

Denote $\mathbf{U} = \{\mathbf{V}_1, \mathbf{V}_2, \mathbf{V}_3\}$ with

$$\begin{aligned}
\mathbf{V}_1 &= \{\rho_i, \rho_e, B^x, B^y, B^z, \psi\}^\top, \\
\mathbf{V}_2 &= \{\rho_i v_i^x, \rho_i v_i^y, \rho_i v_i^z, \rho_e v_e^x, \rho_e v_e^y, \rho_e v_e^z, E^x, E^y, E^z\}^\top, \\
\mathbf{V}_3 &= \{E_i, E_e, \phi\}^\top.
\end{aligned}$$

We observe that (2.1) can be rewritten in the following 3 blocks,

$$\mathbf{V}_{1,i}^{n+1} = \mathbf{G}_1(\mathbf{U}_{i-1}^n, \mathbf{U}_i^n, \mathbf{U}_{i+1}^n), \tag{2.3a}$$

$$\mathbf{V}_{2,i}^{n+1} = \mathbf{G}_2(\mathbf{U}_{i-1}^n, \mathbf{U}_i^n, \mathbf{U}_{i+1}^n) + \mathbb{A}(\mathbf{V}_{1,i}^n)\mathbf{V}_{2,i}^{n+1}, \tag{2.3b}$$

$$\mathbf{V}_{3,i}^{n+1} = \mathbf{G}_3(\mathbf{U}_{i-1}^n, \mathbf{U}_i^n, \mathbf{U}_{i+1}^n) + \mathbf{H}(\mathbf{V}_{1,i}^{n+1}, \mathbf{V}_{2,i}^{n+1}). \tag{2.3c}$$

Here $\mathbf{G}_1, \mathbf{G}_2$ and $\mathbf{G}_3$, are the explicit flux updates using (2.1) of variables $\mathbf{V}_1, \mathbf{V}_2$ and $\mathbf{V}_3$ respectively. The Eqns. (2.3) are then solved in sequential manner:

I) Equation (2.3a) is updated explicitly as it involves the evaluation of terms at the previous time level.

II) Note that $\mathbb{A}(\mathbf{V}_{1,i}^{n+1})$ in Eqn. (2.3b) is

$$\begin{bmatrix}
0 & \frac{B^{z,n+1}}{\hat r_g} & -\frac{B^{y,n+1}}{\hat r_g} & 0 & 0 & 0 & \frac{\rho_i^{n+1}}{\hat r_g} & 0 & 0 \\
-\frac{B^{z,n+1}}{\hat r_g} & 0 & \frac{B^{x,n+1}}{\hat r_g} & 0 & 0 & 0 & 0 & \frac{\rho_i^{n+1}}{\hat r_g} & 0 \\
\frac{B^{y,n+1}}{\hat r_g} & -\frac{B^{x,n+1}}{\hat r_g} & 0 & 0 & 0 & 0 & 0 & 0 & \frac{\rho_i^{n+1}}{\hat r_g} \\
0 & 0 & 0 & 0 & \frac{B^{z,n+1}}{\hat r_{1g}} & -\frac{B^{y,n+1}}{\hat r_{1g}} & \frac{\rho_e^{n+1}}{\hat r_{1g}} & 0 & 0 \\
0 & 0 & 0 & -\frac{B^{z,n+1}}{\hat r_{1g}} & 0 & \frac{B^{x,n+1}}{\hat r_{1g}} & 0 & \frac{\rho_e^{n+1}}{\hat r_{1g}} & 0 \\
0 & 0 & 0 & \frac{B^{y,n+1}}{\hat r_{1g}} & -\frac{B^{x,n+1}}{\hat r_{1g}} & 0 & 0 & 0 & \frac{\rho_e^{n+1}}{\hat r_{1g}} \\
\frac{-r_i}{K} & 0 & 0 & \frac{-r_e}{K} & 0 & 0 & 0 & 0 & 0 \\
0 & \frac{-r_i}{K} & 0 & 0 & \frac{-r_e}{K} & 0 & 0 & 0 & 0 \\
0 & 0 & \frac{-r_i}{K} & 0 & 0 & \frac{-r_e}{K} & 0 & 0 & 0
\end{bmatrix} \tag{2.4}$$

All the quantities in the matrix are already computed in step I. So, we can rewrite Eqn. (2.3b) as

$$\mathbf{V}_{2,i}^{n+1} = \left(\mathbf{I} - (\Delta t)\mathbb{A}(\mathbf{V}_{1,i}^{n+1})\right)^{(-1)} \left(\mathbf{G}_2(\mathbf{U}_{i-1}^n, \mathbf{U}_i^n, \mathbf{U}_{i+1}^n)\right). \tag{2.5}$$

The term $\left(\mathbf{I} - (\Delta t)\mathbb{A}(\mathbf{V}_{1,i}^{n+1})\right)^{(-1)} \left(\mathbf{G}_2(\mathbf{U}_{i-1}^n, \mathbf{U}_i^n, \mathbf{U}_{i+1}^n)\right)$ is evaluated symbolically. The Eqn. (2.3c) is now updated for $\mathbf{V}_3^{n+1}$ by evaluating $\mathbf{H}(\mathbf{V}_{1,i}^{n+1}, \mathbf{V}_{2,i}^{n+1})$.

III) High-order discretization: We use standard piecewise linear second order reconstruction using the minmod limiter (see [3]). High order (second and third order) time integration is achieved by using SSP-RK methods (see [5]), where each intermediate update is computed using the IMEX algorithm described above.

3 Numerical results

In this section, we consider two benchmark test cases for illustrating the efficiency of the IMEX scheme (2.1) in approximating Eqns. (1.1). The first example is a generalization of the MHD Brio-Wu shock tube problem (see [1, 2]). The second example is a simulation of the soliton propagation in plasma flows (see [1]). For the comparison we use fully explicit schemes, i.e., $\mathbf{S}(\mathbf{U}_i^{n+1})$ is replaced by $\mathbf{S}(\mathbf{U}_i^n)$ in (2.1).

3.1 Generalized shock tube Riemann problem

The initial conditions for the Riemann problem are

$$\mathbf{U}_{\text{left}} = \begin{cases} \rho_i = 1.0 \\ p_i = 5 \times 10^{-5} \\ \rho_e = 1.0 \; m_e/m_i \\ p_e = 5 \times 10^{-5} \\ B^x = 0.75 \\ B^y = 1.0 \\ \vec{v}_i = \vec{v}_e = \vec{E} = 0 \\ \phi = \psi = B^z = 0 \end{cases} \qquad \mathbf{U}_{\text{right}} = \begin{cases} \rho_i = 0.125 \\ p_i = 5 \times 10^{-6} \\ \rho_e = 0.125 \; m_e/m_i \\ p_e = 5 \times 10^{-6} \\ B^x = 0.75 \\ B^y = -1.0 \\ \vec{v}_i = \vec{v}_e = \vec{E} = 0 \\ \phi = \psi = B^z = 0 \end{cases} \tag{3.1}$$

on a domain $(0, 1)$ with $\mathbf{U} = \mathbf{U}_{\text{left}}$ for $x < 0.5$ and $\mathbf{U} = \mathbf{U}_{\text{right}}$ for $x > 0.5$. The electron-ion mass ratio is taken to be $m_i/m_e = 1832.6$. The problem is nondimensionalized using $p_0 = 10^{-4}$. Non-dimensional Debye length is taken to be 0.01. We compute solutions using two non-dimensional Larmor radii of 50 and 0.005. Neumann boundary conditions are used.

Results are presented in Figure 3.1 using second order explicit scheme (o2exp) and second order IMEX scheme (o2imp) on a mesh with 20 000 cells. As expected, solutions with high Larmor radius of 50 (see Figure 3.1(a)) are close to the Euler limit. In this case source terms are not stiff. Hence the time step is dictated by the convective CFL number, based on the maximum wave speed. The simulation time for o2exp is 16 073.35 seconds and for o2imp it is marginally higher at 17 482.41 seconds. For the lower Larmor radius of 0.005 solutions are close to the MHD solution. Also the o2imp solution has resolved all the dispersion

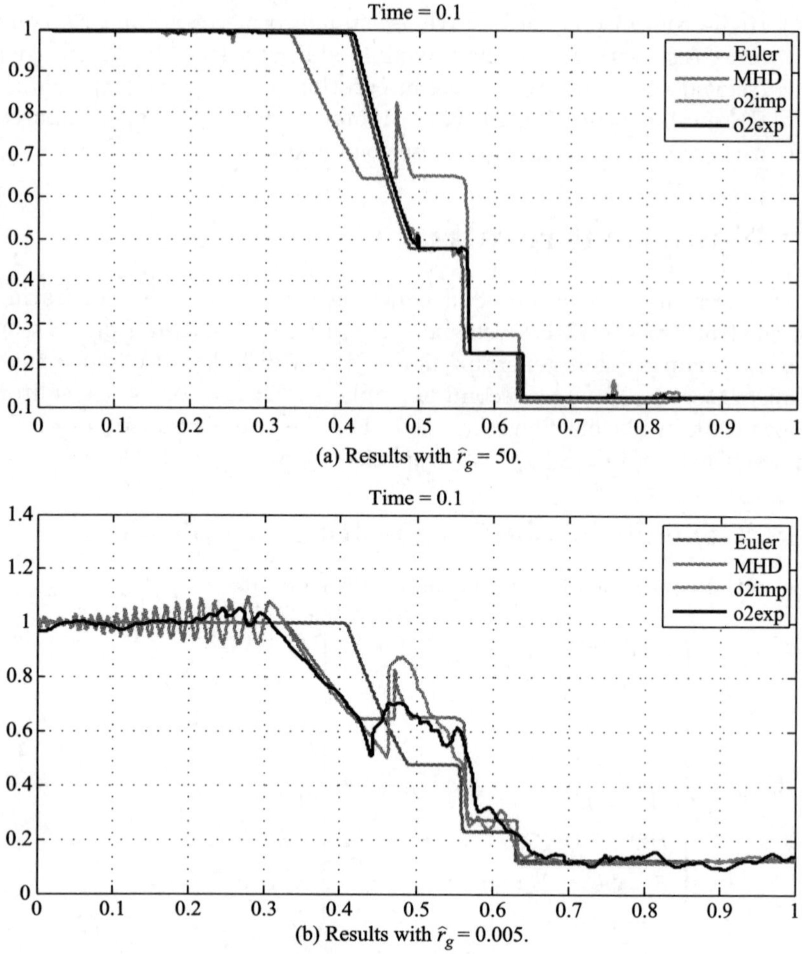

Figure 3.1 Ion densities of Riemann problem (3.1) for o2exp and o2imp schemes with 20000 cells. Euler and MHD solutions are also plotted.

effects whereas o2exp solution appears to be under resolved. In this case the source terms are stiff, hence the time step is dictated by the source terms. The simulation time for o2exp is 27 377.6 sec and for o2imp it is 13 613 sec. So the IMEX scheme is almost twice as fast compared to the fully explicit scheme.

3.2 Soliton propagation

Initially the plasma is assumed to be stationary with ion density,

$$\rho_i = (1.0 + \exp(-25.0|x - L/3.0|)) \tag{3.2}$$

and mass ratio $m_i/m_e = 25$ on domain $D = (0, L)$ with $L = 12.0$. Electron pressure is $p_e = 5.0\rho_i$ with ion-electron pressure ratio of 1/100. Normalized Debye length is taken to be 1.0 and nomalized Larmor radius is 0.01. Periodic boundary conditions were used. Numerical results are presented in Figure 3.2 using second (o2exp), third (o3exp), fourth

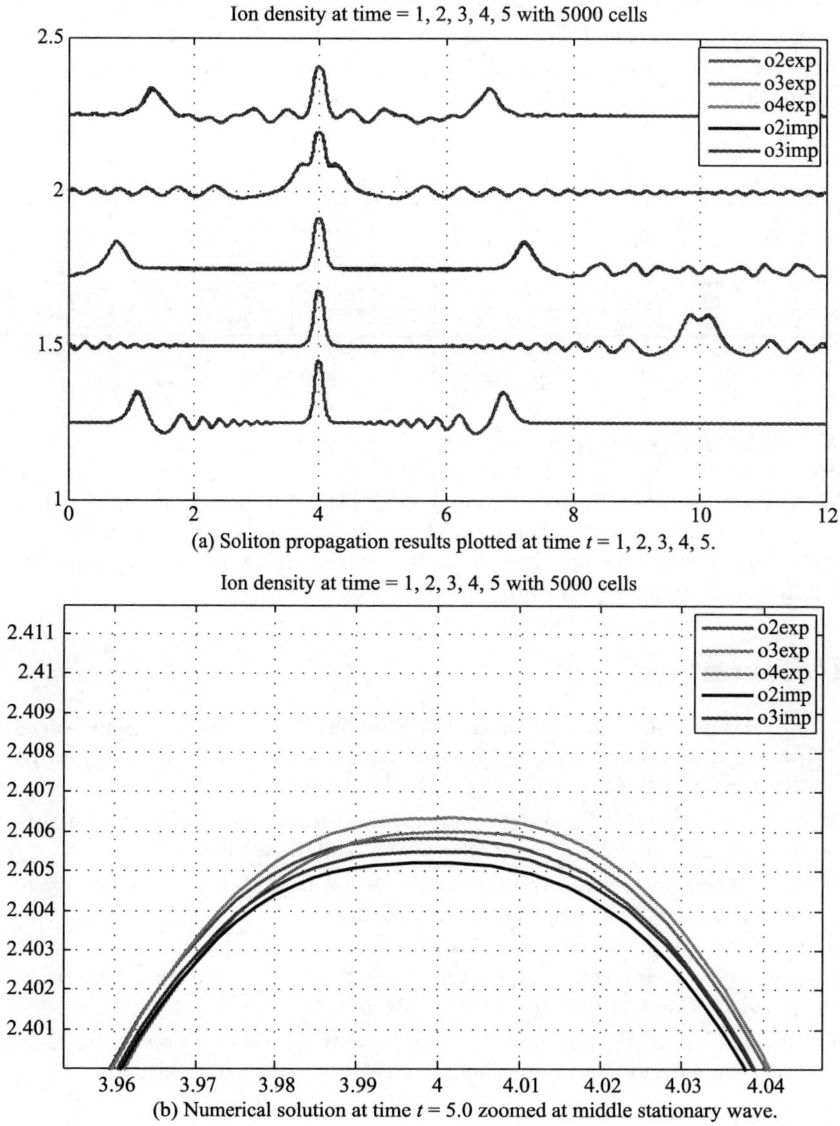

Figure 3.2 Ion densities for explicit schemes o2exp, o3exp, o4exp and implicit schemes o2imp, o3imp using 5000 cells.

(o4exp) order explicit time stepping and second (o2imp), third (o3imp) order IMEX schemes. In Figure 3.2(a) we observe that all schemes perform comparably. However, when we zoom in the middle wave (see Figure 3.2(b)), we observe that second (o2imp), third (o3imp) order IMEX schemes are slightly more diffusive compared to explicit schemes. Table 1 presents the simulation times for different schemes. In the case of low resolution (None of cells less than 1000), time step is dictated by the source terms, so the IMEX schemes are faster. However, as we compute for finer resolutions the CFL condition dictates the time step, so simulation times of both schemes are comparable.

Table 1 Simulation times for various schemes at different resolution.

Cells	o2exp	o3exp	o4exp	o2imp	o3imp
500	10.65	15.96	21.82	3.15	5.32
1000	21.25	31.52	44.09	12.36	18.58
2000	46.12	69.68	95.29	49.71	75.01
4000	185.05	277.53	396.77	200.46	299.41

4 Conclusion

We have presented finite volume based IMEX schemes for TF Eqns. (1.1). These schemes are then benchmarked using two numerical examples. These schemes are shown to be computationally faster than the fully explicit schemes on under resolved meshes.

References

[1] A. Hakim, J. Loverich and U. Shumlak, A high-resolution wave-propagation scheme for ideal two-fluid plasma equations. J. Comp. phys. 219 (2006), 418–442.

[2] U. Shumlak and J. Loverich, Approximate Riemann solvers for the two-fluid plasma model. J. Comp. Phys. 187 (2003), 620–638.

[3] R. J. LeVeque, Finite Volume Methods for Hyperbolic Problems. Cambridge University Press, Cambridge, 2002.

[4] C. D. Munz, P. Omnes, R. Schneider, E. Sonnendru ker and U. Voβ, Divergence correction techniques for Maxwell solvers based on a hyperbolic model. J. Comp. Phys. 161 (2000), 484–511.

[5] S. Gottlieb, C. W. Shu and E. Tadmor, Strong stability-preserving high-order time discretization methods. SIAM Rev. 43 (2001), 89–112.

Nonlinear Stability of Large Amplitude Traveling Waves to Hyperbolic-parabolic System Modeling Chemotaxis

Tong Li
Department of Mathematics, University of Iowa
Iowa City, IA 52242, USA
E-mail: tli@math.uiowa.edu

Zhi-An Wang
Department of Applied Mathematics
The Hong Kong Polytechnic University, Hong Kong, China
E-mail: mawza@polyu.edu.hk

Abstract

Traveling wave (band) behavior driven by chemotaxis was observed experimentally by Adler [1, 2] and was modeled by Keller and Segel [13]. For a quasilinear hyperbolic-parabolic system that arises as a non-diffusive limit of the Keller-Segel model with nonlinear kinetics, we establish the existence and nonlinear stability of traveling wave solutions with large amplitudes.

1 Introduction

Chemotaxis describes a process in which cells change their states of movement reacting to the presence of a chemical substance, approaching chemically favorable environment and avoiding unfavorable ones. It is a fundamental cellular process which plays an essential role in embryonic development, immune response, progression of diseases, tissue homeostasis, wound healing, as well as in finding food, repellent action and forming the multicellular body of protozoa [3].

The mathematical modeling of chemotaxis dates to Keller and Segel [11] from the macroscopic perspective and to Patlak [21] based on the microscopic(statistical) description. The original model in [11] comprised four partial differential equations and was proposed to describe the aggregation of cellular slime molds *Dictyostelium discoideum.* A simplified formulation, also most extensively studied in the literature [8], is the following two parabolic equations:

$$\begin{aligned} u_t &= \nabla \cdot (D\nabla u - \chi u \nabla \Phi(c)), \\ \tau c_t &= d\Delta c + g(u,c), \end{aligned} \tag{1.1}$$

where $u(t,x)$, $c(t,x)$ denote the cell density and the chemical concentration, respectively. $D > 0$ is the diffusivity of cells and $d \geq 0$ is the diffusion rate of the chemical substance. $\tau \geq 0$ is a relaxation time scale such that $1/\tau$ is the rate towards equilibrium states. The function $\Phi(c)$ is called the chemotactic potential function describing the mechanism of signal detection. The function $g(u,c)$ describes the chemical kinetics. The constant χ, often referred to as chemosensitivity, is a measure of the strength of chemical signals, and $\chi > 0 (< 0)$ corresponds to attractive (repulsive) chemotaxis.

There are two major limiting cases of the Keller-Segel(KS) model (1.1). The first one is when the chemical substance c relaxes so fast that it reaches its equilibrium instantaneously, i.e., $\tau = 0$. This gives rise to the parabolic-elliptic system

$$\begin{aligned} u_t &= \nabla \cdot (D\nabla u - \chi u \nabla \Phi(c)), \\ 0 &= d\Delta c + g(u,c). \end{aligned} \tag{1.2}$$

There were vast results to the model (1.2) on the global existence, blow up behavior and the stationary solutions. The most relevant results are summarized in a review article [8] and a textbook [22].

The other limiting case is when the diffusion of the chemical substance is so small that it is negligible, i.e., $d = 0$. Then the model becomes a partial differential equation coupled with an ordinary differential equation

$$\begin{aligned} u_t &= \nabla \cdot (D\nabla u - \chi u \nabla \Phi(c)), \\ c_t &= g(u,c), \end{aligned} \tag{1.3}$$

where we have assumed that $\tau = 1$ without loss of generality. The model (1.3) is also called the parabolic-degenerate chemotaxis model [4, 5]. This model had already been used to study the chemotactic traveling bands of bacteria in [13] whereby $g(u,c) = -kuc^m$ ($m \geq 1$) and has received increasing attention recently (see [24]). The above system (1.3) is used to model haptotaxis where cells move towards immobilized, substratum-bound such as laminin and fibronectin. It also has essential applications in the modeling of cancer invasion [3] as well as the gliding movement of myxobacteria [19] towards slime trails which is a non-diffusive chemical substance.

In this paper, we consider the one-dimensional case where $D > 0$, $d = 0$, $\chi > 0$, and $\tau > 0$. Assume that $\Phi(c) = \log c$ and $g(u,c) = -f(u)c$

where f is a smooth function satisfying $f(u) \geq 0,\ f'(u) > 0, f''(u) \geq -\beta$ for some $\beta > 0$. We derive a system of quasilinear hyperbolic-parabolic system from the KS model (1.1)

$$\begin{cases} u_t - (uv)_x = Du_{xx}, \\ v_t - f(u)_x = 0 \end{cases} \tag{1.4}$$

through a change of variable

$$v = -c^{-1}c_x = -(\ln c)_x \tag{1.5}$$

which was first introduced in [25] for a chemotaxis model proposed in [14] describing the chemotactic movement for non-diffusible chemicals, and under the scalings

$$\tilde{t} = \chi t, \quad \tilde{x} = \sqrt{\chi}x, \quad \tilde{v} = \sqrt{\chi}v \tag{1.6}$$

where the tildes are dropped.

This paper is concerned with the existence and the asymptotic nonlinear stability of traveling wave solutions to the above system of conservation laws (1.4). The experimental study of traveling wave (band) behavior driven by chemotaxis dates to Adler's works [1, 2]. The mathematical modeling and analysis of the traveling waves was first presented by Keller and Segel [13] and a flurry of works then followed, see [9] and references therein and [26].

Now we define the traveling wave ansatz

$$(u, v)(x, t) = (U, V)(z), \quad z = x - st$$

where s denotes the wave speed and z is the traveling wave variable. Substituting the above ansatz into (1.4), one obtains the traveling wave equations

$$\begin{cases} -sU_z - (UV)_z = DU_{zz}, \\ -sV_z - f(U)_z = 0. \end{cases} \tag{1.7}$$

Consider the boundary conditions

$$(U, V)(z) \ \rightarrow \ (u_\pm, v_\pm) \text{ as } z \rightarrow \pm\infty \tag{1.8}$$

where $u_\pm \geq 0$ and $v_\pm \leq 0$ for biological relevance. The wave speed s is determined by

$$\begin{cases} -s(u_+ - u_-) - (u_+v_+ - u_-v_-) = 0, \\ -s(v_+ - v_-) - (f(u_+) - f(u_-)) = 0. \end{cases} \tag{1.9}$$

The entropy condition for the shock of second characteristic family is

$$\lambda_2(u_+, v_+) < s < \lambda_2(u_-, v_-) \tag{1.10}$$

where $\lambda_2(u,v)$ is the characteristic speed of the second characteristic family of iniviscid (1.4).

Consider the initial data

$$(u,v)(x,0) = (u_0,v_0)(x) \to \begin{cases} (u_-,v_-) & \text{as } x \to -\infty \\ (u_+,v_+) & \text{as } x \to +\infty \end{cases} \tag{1.11}$$

which is a small initial perturbation of a traveling wave (U,V)

$$\int_{-\infty}^{+\infty} \begin{pmatrix} u_0(x) - U(x) \\ v_0(x) - V(x) \end{pmatrix} dx = x_0 \begin{pmatrix} u_+ - u_- \\ v_+ - v_- \end{pmatrix} + \eta \mathbf{r_1}(u_-,v_-) \tag{1.12}$$

where $\mathbf{r_1}(u_-,v_-)$ is the first eigenvector evaluated at (u_-,v_-), x_0 and η are constants which can be determined uniquely. We assume that $\eta = 0$ as in [7, 10].

Theorem 1.1. *Let* (1.9) *and* (1.10) *hold. If β is small, then there exists a monotone shock profile $(U,V)(x-st)$ to system* (1.7) *with boundary conditions* (1.8), *which is unique up to a translation and satisfies $U_z < 0$ and $V_z > 0$. If $u_+ > 0$, then there exists a constant $\varepsilon_0 > 0$ such that if $\|u_0 - U\|_1 + \|v_0 - V\|_1 + \|(\phi_0,\psi_0)\| \le \varepsilon_0$, the Cauchy problem* (1.4) *and* (1.11) *has a unique global solution $(u,v)(x,t)$ satisfying $u(x,t) \ge \delta_0 > 0$ for some $\delta_0 > 0$ for all $x \in R$, $t \ge 0$, and*

$$(u-U, v-V) \in (C([0,\infty);H^1) \cap L^2([0,\infty);H^1))^2.$$

Furthermore, the solution (u,v) has the following asymptotic behavior

$$\sup_{x \in R} |(u,v)(x,t) - (U,V)(x-st)| \to 0 \quad \text{as} \quad t \to +\infty. \tag{1.13}$$

Remark 1.2. It is worthwhile to remark that the above nonlinear stability results hold true regardless of the strengths of the waves. That is, the wave amplitude $|u_+ - u_-| + |v_+ - v_-|$ can be large, in contrast to the previous results related to the nonlinear stability of traveling waves to conservation laws, where various smallness conditions on wave strengths were imposed (e.g., see Ref. [7, 15, 18, 20, 23]).

2 Reformulation of the problem

We use the phase plane analysis to prove the existence of traveling wave solutions of (1.4) as stated in Theorem 1.1, see [16, 17]. The proof of the stability results in Theorem 1.1 is based on iterative L^2 energy estimates. The method of energy estimates for the nonlinear stability of small-amplitude viscous shock profiles of conservation laws was first introduced independently by Matsumura and Nishihara in [20] and by Goodman in

[7]. It has been further developed over the years, see [10, 15, 18, 23]. We establish the stability estimates without the smallness assumption on the wave strengths, see [16, 17].

In view of (1.12) in which $\eta = 0$, we decompose the solution of (1.4) and (1.11) in the form

$$(u, v)(x, t) = (U, V)(x - st) + (\phi_x, \psi_x)(x, t) = (U, V)(z) + (\bar{\phi_z}, \bar{\psi_z})(z, t) \tag{2.1}$$

with $(\bar{\phi}, \bar{\psi})$ in some functional space which will be defined below. For simplicity of notation, we will omit the bars in $(\bar{\phi}, \bar{\psi})$.

Substituting (2.1) into (1.4), using (1.7) and integrating the resulting equations with respect to z, we obtain the equations for (ϕ, ψ)

$$\begin{cases} \phi_t = D\phi_{zz} + (s + V)\phi_z + U\psi_z + \phi_z\psi_z, \\ \psi_t = s\psi_z + f'(U)\phi_z + F(U, \phi_z) \end{cases} \tag{2.2}$$

where $F(U, \phi_z) = f(U + \phi_z) - f(U) - f'(U)\phi_z$. The initial data is given by

$$(\phi, \psi)(z, 0) = (\phi_0, \psi_0)(z), \; z \in R \tag{2.3}$$

where (ϕ_0, ψ_0) is defined as

$$(\phi_0, \psi_0)(x) = \int_{-\infty}^{x} (u_0 - U, v_0 - V)(y)dy. \tag{2.4}$$

We seek solutions of the reformulated problem (2.2)–(2.3) in the following solution space:

$$\begin{aligned} X(0, T) = \{(\phi(z, t), \psi(z, t)) : \phi, \psi \in C([0, T); H^2) \cap C^1((0, T); H^1), \\ \phi_z \in L^2((0, T); H^2), \psi_z \in L^2((0, T); H^1)\} \end{aligned}$$

with $0 \leq T \leq +\infty$.

Let

$$N(t) = \sup_{0 \leq \tau \leq t} \{\|\phi(\cdot, \tau)\|_2 + \|\psi(\cdot, \tau)\|_2\}. \tag{2.5}$$

By the Sobolev embedding theorem, we have, for $t \geq 0$,

$$\sup_{z \in \mathbb{R}} \{|\phi|, |\phi_z|, |\psi|, |\psi_z|\} \leq N(t). \tag{2.6}$$

Theorem 1.1 is proved based on the following theorem.

Theorem 2.1. *Let $(\phi_0, \psi_0) \in (H^2(R))^2$ and $\beta > 0$ small. If $u_+ > 0$, then there exists a constant $\delta_1 > 0$ such that if*

$$N(0) \leq \delta_1, \tag{2.7}$$

the Cauchy problem (2.2)–(2.3) has a unique global solution $(\phi, \psi) \in X(0, +\infty)$ *which satisfies*

$$\begin{aligned}&\|(\phi(\cdot,t),\psi(\cdot,t))\|_2^2+\int_0^t\int|U_z(z)||(\phi(z,\tau),\psi(z,\tau))|^2dzd\tau\\&+D\int_0^t\|\phi_z(\cdot,\tau)\|_2^2d\tau+\int_0^t\|\psi_z(\cdot,\tau)\|_1^2d\tau\le CN^2(0)\end{aligned}\tag{2.8}$$

for all $t \in [0, +\infty)$, *where* $C > 0$ *is a constant. Moreover the following asymptotic behavior holds*

$$\sup_{z\in\mathbb{R}}|(\phi_z,\psi_z)(z,t)|\to 0 \quad \text{as} \quad t\to+\infty.\tag{2.9}$$

The global existence of (ϕ, ψ) announced in Theorem 2.1 follows from the local existence theorem and the *a priori* estimates.

Proposition 2.2 (Local existence). *For any* $\delta_2 > 0$, *there exists a positive constant* T *depending on* δ_2 *such that if* $(\phi_0, \psi_0) \in H^2$ *with* $N(0) \le \delta_2/2$, *then the problem (2.2)–(2.3) has a unique solution* $(\phi, \psi) \in X(0, T)$ *satisfying*

$$N(t) < 2N(0)\tag{2.10}$$

for any $0 \le t \le T$.

Lemma 2.3. *Let* $(\phi_0, \psi_0) \in (H^2(R))^2$ *and* (ϕ, ψ) *be a solution of (2.2)–(2.3). If* $u_+ > 0$ *and* $\beta > 0$ *is small, then there exists a constant* $C > 0$, *such that*

$$\begin{aligned}&\|\phi(\cdot,t)\|_2^2+\|\psi(\cdot,t)\|_2^2+\int_0^t\int|U_z(z)||(\phi(z,\tau),\psi(z,\tau))|^2dzd\tau\\&+D\int_0^t\|\phi_z(\cdot,\tau)\|_2^2d\tau+\int_0^t\|\psi_z(\cdot,\tau)\|_1^2d\tau\\&\le C\Big(\|\phi_0\|_2^2+\|\psi_0\|_2^2+\int_0^t\int(|\phi|+|\phi_z|+|\phi_{zz}|)|\phi_z\psi_z|dzd\tau\Big)\\&+CN(t)\Big(\int_0^t\|\phi_z(\cdot,\tau)\|_2^2d\tau+\int_0^t\|\psi_z(\cdot,\tau)\|_1^2d\tau\Big).\end{aligned}\tag{2.11}$$

3 Conclusions

This paper established the existence and nonlinear stability of traveling wave solutions for a system of hyperbolic-parabolic equations which was derived from the well known Keller-Segel model through a transformation. We proved the existence of traveling wave solutions and showed the asymptotic nonlinear stability of traveling wave solutions without the smallness assumption on the wave strength via the method of energy estimates.

This paper is focused on the transformed system (1.4) instead of the original Keller-Segel model. Our ultimate goal is to translate the results of system (1.4) back to the original Keller-Segel model (1.1) using the transformation (1.5).

References

[1] J. Adler, Chemotaxis in bacteria. Science, 153 (1966), 708–716.

[2] J. Adler, Chemoreceptors in bacteria. Science, 166 (1969), 1588–1597.

[3] M. A. J. Chaplain and G. Lolas, Mathematical modelling of cancer invasion of tissue: dynamic heterogeneity. Net. Hetero. Med. 1 (2006), 399–439.

[4] L. Corrias, B. Perthame and H. Zaag, A chemotaxis model motivated by angiogenesis. C. R. Acad. Sci. paris, Ser. I. 336 (2003), 141–146.

[5] L. Corrias, B. Perthame and H. Zaag, Global solutions of some chemotaxis and angiogenesis systems in high space dimensions. Milan J. Math. 72 (2004), 1–28.

[6] S. S. Willard and P. N. Devreotes, Signalling pathways mediating chemotaxis in the social amoeba, Dictyostelium discoideum. Euro. J. Cell. Biol. 85 (2006), 897–904.

[7] J. Goodman, Nonlinear Asymptotic Stability of Viscous Shock Profiles for Conservation Laws. Arch. Rat. Mech.Anal. 95 (1986), 325–344.

[8] D. Horstmann, From 1970 until present: the Keller-Segel model in chemotaxis and its consequences: I. Jahresber. Deutsch. Math.-Verein. 105 (2003), 103–165.

[9] D. Horstmann, From 1970 until present: the Keller-Segel model in Chemotaxis and its consequences: II, Jahresber. Deutsch. Math.-Verein. 106 (2004), 51–69.

[10] S. Kawashima and A. Matsumura, Stability of shock profiles in viscoelasticity with non-convex constitutive relations. Comm. Pure Appl. Math. 47 (1994), 1547–1569.

[11] E. F. Keller and L. A. Segel, Initiation of slime mold aggregation viewed as an instability. J. Theor. Biol. 26 (1970), 399–415.

[12] E. F. Keller and L. A. Segel, Model for Chemotaxis. J. Theor. Biol. 30 (1971), 225–234.

[13] E. F. Keller and L. A. Segel, Traveling bands of chemotactic bacteria: a theoretical analysis. J. Theor. Biol. 30 (1971), 235–248.

[14] H. A. Levine and B. D. Sleeman, A system of reaction diffusion equations arising in the theory of reinforced random walks. SIAM J. Appl. Math. 57 (1997), 683–730.

[15] T. Li, Stability of Traveling Waves in Quasi-Linear Hyperbolic Systems with Relaxation and Diffusion. SIAM J. Math. Anal. 40 (2008), 1058–1075.

[16] T. Li and Z. A. Wang, Nonlinear stability of traveling waves to a hyperbolic-parabolic system modeling chemotaxis. SIAM J. Appl. Math. 70 (2009), 1522–1541.

[17] T. Li and Z. A. Wang, Nonlinear stability of large amplitude viscous shock waves of a generalized hyperbolic-parabolic system arising in chemotaxis. Math. Model. Methods. Appl. Sci. accepted, 2010.

[18] T. P. Liu and Y. Zeng, Time-asymptotic behavior of wave propagation around a viscous shock profile. Comm. Math. Phy. 290 (2009), 23–82.

[19] H. Othmer and A. Stevens, Aggregation, blowup and collapse: The ABC's of taxis in reinforced random walks. SIAM J. Appl. Math. 57 (1997), 1044–1081.

[20] A. Matsumura and K. Nishihara, On the stability of travelling wave solutions of a one-dimensional model system for compressible viscous gas. Japan J. Appl. Math. 2 (1985), 17–25.

[21] C. S. Patlak, Random Walk with Persistence and External Bias. Bull. Math. Biophys. 15 (1953), 311–338.

[22] B. Perthame, Transport Equations in Biology. Birkhäuser Verlag, Basel, 2007.

[23] A. Szepessy and Z. P. Xin, Nonlinear stability of viscous shock waves, Arch. Rat. Mech. Anal. 122 (1993), 53–103.

[24] A. Stevens and J. J. L. Velazquez, Partial differential equations and non-diffusive structures. Nonlinearity, 21 (2008), 283–289.

[25] Z. A. Wang and T. Hillen, Shock formation in a chemotaxis model. Math. Meth. Appl. Sci. 31 (2008), 45–70.

[26] M. Zhang and C. J. Zhu, Global existence of solutions to a hyperbolic-parabolic system. Proc. Amer. Math. Soc. 135 (2006), 1017–1027.

Quasi-neutral Limits of the Two-fluid Euler-Poisson Systems*

Yong Li
College of Applied Sciences, Beijing University of Technology
Beijing 100124, China
E-mail: yli@bjut.edu.cn

Song Jiang
Institute of Applied Physics and Computational Mathematics
P.O.Box 8009-28, Beijing 100088, China
E-mail: jiang@iapcm.ac.cn

Qiangchang Ju
Institute of Applied Physics and Computational Mathematics
P.O.Box 8009-28, Beijing 100088, China
E-mail: qiangchang_ju@yahoo.com

Hailiang Li
Department of Mathematics, Capital Normal University
Beijing 100048, China
E-mail: hailiang_li@yahoo.com

Abstract

Quasi-neutral limits of the multi-dimensional two-fluid isentropic and non-isentropic Euler-Poisson systems are rigorously justified. For well-prepared initial data, the convergence of the two-fluid Euler-Poisson systems to the compressible Euler equations (isentropic or non-isentropic) is proved as the Debye length goes to zero.

*The authors is partly supported by NSFC (Grant No. 10901011, No. 40890154, No. 10701011, No. 10431060), the National Basic Research Program (Grant No. 2005CB321700), the Beijing Nova program, the NCET support of the Ministry of Education of China, the Huo Ying Dong Foundation (Grant No. 111033), the support of Institute of Mathematics and Interdisciplinary Science at CNU, BNSF (Grant No. 1102009), the Foundation for Talents of Beijing (No. 20081D0501500171) and Funds of Beijing University of Technology.

1 Introduction

We consider an unmagnetized plasma consisting of electrons with density n_e and charge $q_e = -1$, and a single species of ions with density n_i and charge $q_i = 1$. We denote by u_e (respectively, u_i) the scaled mean velocity of the electrons (respectively, ions), and by $\phi = \phi(x,t)$ the scaled electric potential. Then the model describing the motion of the plasma is described by the following two-fluid isentropic compressible Euler-Poisson system:

$$\begin{cases} \partial_t n_\alpha + \text{div}\,(n_\alpha u_\alpha) = 0, \\ \partial_t(n_\alpha u_\alpha) + \text{div}\,(n_\alpha u_\alpha \otimes u_\alpha) + \nabla p_\alpha(n_\alpha) = -q_\alpha n_\alpha \nabla\phi - \dfrac{n_\alpha u_\alpha}{\tau_\alpha}, \\ -\lambda^2 \Delta\phi = n_i - n_e, \end{cases} \tag{1.1}$$

for $\alpha = e, i$ and $x \in \mathbb{T}^d \subset \mathbb{R}^d$ $(d = 2,3)$, the d-dimensional torus, where $\tau_\alpha > 0$ are the scaled relaxation time constants and $\lambda > 0$ is the Debye length. The pressure functions $p_\alpha(n_\alpha), \alpha = e, i$, are usually of the form

$$p_\alpha(n_\alpha) = a_\alpha^2 n_\alpha^{\gamma_\alpha}, \quad n_\alpha \geq 0,$$

where $\gamma_\alpha \geq 1$ and $a_\alpha > 0$ are constants.

The re-scaled two-fluid non-isentropic Euler-Poisson system is given as follows:

$$\begin{cases} \partial_t n_\alpha + \text{div}(n_\alpha u_\alpha) = 0, \\ \partial_t(n_\alpha u_\alpha) + \text{div}(n_\alpha u_\alpha \otimes u_\alpha) + \nabla p_\alpha = -q_\alpha n_\alpha \nabla\phi, \\ \partial_t w_\alpha + \text{div}((w_\alpha + p_\alpha)u_\alpha) = -q_\alpha n_\alpha u_\alpha \cdot \nabla\phi, \\ -\lambda^2 \Delta\phi = n_i - n_e, \end{cases} \tag{1.2}$$

for $\alpha = e, i$ and $x \in \mathbb{T}^d$. The re-scaled equations of state are written as

$$w_i = \frac{p_i}{\gamma_i - 1} + \frac{n_i|u_i|^2}{2}, \quad w_e = \frac{p_e}{\gamma_e - 1} + \frac{n_e|u_e|^2}{2},$$

with $p_i = n_i\theta_i$, $p_e = n_e\theta_e$, respectively. θ_i and θ_e are temperature for ions and electrons, respectively. For simplicity, we do not include the damping term in the model (1.2).

Usually, in plasma and semiconductor physics, one is concerned with large scale structures with respect to the Debye length (typically $\lambda \approx 10^{-4}$). For such scales, the plasma is electrically neutral, i.e., there is no charge separation or electric field. Therefore, the small Debye length asymptotic limit $\lambda \to 0$ (quasi-neutral limit) is of particular interest in the description of dynamics of solutions to the compressible Euler-Poisson systems.

The quasi-neutral limit of various plasma models has attracted considerable attention in recent years. For the unipolar (i.e., one-fluid) Euler-Poisson system, the quasi-neutral limit of the steady problem was studied by Peng [7] for a potential flow in multi-dimension without the formation of boundary layers, and Peng-Wang [8] considered the case with boundary layers. For the transient model, Wang [13] studied the zero Debye length limit for smooth solutions to the isentropic Euler-Poisson system with doping profile being constant and well-prepared initial data by an energy method, and Peng-Wang [9] also dealt with the same limit with well-prepared initial data and a general doping profile for isentropic flow by the complete asymptotic expansion and energy methods. Peng et al. [10] extended the similar limit to non-isentropic flows with well-prepared initial data. Recently, Li et al. [6] studied the quasi-neutral limit for the non-isentropic Euler-Poisson system with general initial data. All of the limit systems for the unipolar model are the isentropic or non-isentropic incompressible Euler equations.

There are very limited works on the quasi-neutral analysis for the bipolar model. Slemrod-Sternberg [12] proved the quasi-neutral limit for the steady bipolar Euler-Poisson system with well-prepared boundary data as the ion velocity reaches the ion-sound speed in a one-dimensional bounded domain. When the electron density is described by the Maxwell-Boltzmann relation with the electrostatic potential, Cordier-Grenier [1] studied the limit for local smooth solutions of a one-dimensional and isothermal model for plasmas by using the pseudodifferential energy estimates. From the numerical view point, Crispel et al. [2] proposed an asymptotic preserving scheme for the bipolar isentropic Euler-Poisson model in the quasi-neutral limit.

Different from the unipolar case, the formal quasi-neutral limit of the bipolar Euler-Poisson system is a model of compressible type [3, 11]. The rigorous justification of the quasi-neutral limit is an open challenging problem [3, 11]. Indeed, it is shown for the unipolar Euler-Poisson system that the velocity of the solution is approximated, for small Debye length $\lambda > 0$, by the solution of the incompressible Euler equations plus a term of fast singular oscillating gradient vector fields which are not determined by a singular Klein-Gordon equation due to the effect of electric field [6]. It is rather complicated, however, to study the quasi-neutral limit for the bipolar Euler-Poisson system. The constraint of the Poisson equation gives rise to the quasi-neutrality of the densities for two carriers, namely, $n_i = n_e$ in the quasi-neutral limit. Thus, one can expect that the limiting system is the compressible Euler equations [3].

The purpose of our papers [4, 5] are to investigate the quasi-neutral limits for the two-fluid isentropic (or non-isentropic) Euler-Poisson system in multi-dimension. Using an asymptotic expansion, we formally derive the compressible Euler equations for the well-prepared initial data.

Thus this formal limits are very different from the unipolar case for which the limits are the incompressible type Euler equations. Furthermore, under the compatibility conditions which guarantee no initial layer to be formed, we obtain the existence of solutions to the multi-dimensional two-fluid Euler-Poisson system and prove the convergence when the Debye length λ goes to zero in the time interval where the smooth solution to the compressible Euler equations exits. To get the uniform error estimates, the weighted norm estimates are employed due to the presence of electric field. This is the first rigorous proof for the quasi-neutral limits of the systems. It is still remained open to prove rigorously the limits for the general initial data.

In our papers we shall assume that $\tau_e = \tau_i$. Without loss of generality, we assume that $\tau_\alpha = 1$ for $\alpha = e, i$. We use the superscript λ to indicate that the unknowns are dependent on λ.

2 Quasi-neutral limit of isentropic case

We impose the initial conditions for the isentropic system (1.1) as

$$n_\alpha^\lambda \mid_{t=0} = n_\alpha^\lambda(x), \quad u_\alpha^\lambda \mid_{t=0} = u_\alpha^\lambda(x), \quad \alpha = e, i. \tag{2.1}$$

We denote by $(n_\alpha^\lambda, u_\alpha^\lambda, \phi^\lambda)$ a solution to the system (1.1) with initial data (2.1). We will study its formal expansions when $\lambda \to 0$. Thus, we make the following ansatz:

$$(n_\alpha^\lambda, u_\alpha^\lambda, \phi^\lambda)(t,x) = \sum_{j\geq 0} \lambda^{2j}(n_\alpha^j, u_\alpha^j, \phi^j), \quad \alpha = e, i. \tag{2.2}$$

Plugging the expansions (2.2) into the system (1.1), we obtain the leading profiles $(n_\alpha^0, u_\alpha^0, \phi^0)$ satisfy the following system:

$$\begin{cases} \partial_t n_\alpha^0 + \operatorname{div}(n_\alpha^0 u_\alpha^0) = 0, \\ \partial_t u_\alpha^0 + (u_\alpha^0 \cdot \nabla) u_\alpha^0 + \nabla h_\alpha(n_\alpha^0) + u_\alpha^0 = -q_\alpha \nabla \phi^0, \\ n_e^0 = n_i^0 \end{cases} \tag{2.3}$$

for $\alpha = e, i$. Denote by $n^0 = n_e^0 = n_i^0$. Then (2.3) is written into

$$\begin{cases} \partial_t n^0 + \operatorname{div}(n^0 u_\alpha^0) = 0, \\ \partial_t u_\alpha^0 + (u_\alpha^0 \cdot \nabla) u_\alpha^0 + \nabla h_\alpha(n^0) + u_\alpha^0 = -q_\alpha \nabla \phi^0, \end{cases} \tag{2.4}$$

with the initial conditions

$$(n^0, u_\alpha^0)(0,x) = (n_0, u_{\alpha,0})(x), \quad \alpha = e, i. \tag{2.5}$$

Here we have assumed $n_{e,0}(x) = n_{i,0}(x)$ which is denoted by $n_0(x)$. The following result is proved in [11].

Lemma 2.1. *Assume that $n^0, u_e^0, u_i^0, \phi^0$ are smooth solutions of the Cauchy problems (2.4)-(2.5) with $n^0 > 0$ in $[0,T] \times \mathbb{T}^d$ for some $T > 0$. If $u_{e,0} = u_{i,0} = u_0$, then in the class $m(2\phi^0 - h_e(n^0) + h_i(n^0)) = 0$, we have $u_e^0 = u_i^0 = u^0$ and $2\phi^0 = h_e(n^0) - h_i(n^0)$ for all $(t,x) \in [0,T] \times \mathbb{T}^d$. Thus, (n^0, u^0) is a unique smooth solution of the damped compressible Euler equations*

$$\begin{cases} \partial_t n^0 + \operatorname{div}(n^0 u^0) = 0, \\ \partial_t u^0 + (u^0 \cdot \nabla) u^0 + \nabla h_+(n^0) + u^0 = 0, \end{cases} \tag{2.6}$$

with the initial data

$$n^0(x,0) = n_0(x), \quad u^0(x,0) = u_0(x), \tag{2.7}$$

where $h_\pm(n^0) = \frac{1}{2}(h_e(n^0) \pm h_i(n^0))$.

For $j \geq 1$, we find that

$$-\Delta \phi^{j-1} = n_i^j - n_e^j, \tag{2.8}$$

and (n_α^j, u_α^j) satisfy the following linear system for $\alpha = e, i$:

$$\begin{cases} \partial_t n_\alpha^j + \operatorname{div}(n^0 u_\alpha^j + n_\alpha^j u^0) = -\displaystyle\sum_{k=1}^{j-1} \operatorname{div}(n_\alpha^k u_\alpha^{j-k}), \\ \partial_t u_\alpha^j + (u^0 \cdot \nabla) u_\alpha^j + (u_\alpha^j \cdot \nabla) u^0 + \nabla(h'_\alpha(n^0) n_\alpha^j) + u_\alpha^j + q_\alpha \nabla \phi^j \\ \qquad = -\displaystyle\sum_{k=1}^{j-1} (u_\alpha^k \cdot \nabla) u_\alpha^{j-k} - \nabla(h_\alpha^{j-1}((n_\alpha^k)_{k \leq j-1})), \end{cases} \tag{2.9}$$

with initial conditions

$$(n_\alpha^j, u_\alpha^j)(0,x) = (n_{\alpha,j}, u_{\alpha,j})(x), \tag{2.10}$$

which satisfy the following compatibility condition:

$$-\Delta \phi_{j-1} = n_{i,j} - n_{e,j}. \tag{2.11}$$

Next we have to prove the existence of (n_α^j, u_α^j) to the problem for $\alpha = e, i$ and $j \geq 1$. We get the following theorem in [5].

Theorem 2.2. *Let T^* be the time length of the solution exists and let $s > 1 + d/2$. For any $j \geq 1$, assume that*

$$(n_{\alpha,k}, u_{\alpha,k}) \in H^{s+3(j-k)}(\mathbb{T}^d), \qquad 0 \leq k \leq j,$$

and the compatibility condition (2.11) holds for any $0 \leq k \leq j$. Then there exists a unique solution such that

$$\begin{aligned} (n_\alpha^j, u_\alpha^j) &\in C([0,T^*], H^s(\mathbb{T}^d)) \cap C^1([0,T^*], H^{s-1}(\mathbb{T}^d)), \\ \phi^j &\in C([0,T^*], H^{s+1}(\mathbb{T}^d)). \end{aligned}$$

For any fixed $m \in \mathbb{N}$, we denote

$$n_{\alpha,m}^{\lambda} = \sum_{j=0}^{m} \lambda^{2j} n_{\alpha}^{j}(t,x), \quad u_{\alpha,m}^{\lambda} = \sum_{j=0}^{m} \lambda^{2j} u_{\alpha}^{j}(t,x), \quad \phi_{m}^{\lambda} = \sum_{j=0}^{m} \lambda^{2j} \phi^{j}(t,x).$$

Then we can rigorously justify the formal asymptotic expansion of solutions $(n_\alpha^\lambda, u_\alpha^\lambda, \phi^\lambda)$ to the Cauchy problem (1.1) and (2.1). In [5], we obtain the existence of solutions $(n_\alpha^\lambda, u_\alpha^\lambda, \phi^\lambda)$ to the multi-dimensional bipolar Euler-Poisson system (1.1) and prove the convergence of $(n_\alpha^\lambda, u_\alpha^\lambda, \phi^\lambda)$ when the Debye length λ goes to zero in the time interval where the smooth solutions to the compressible Euler equations exist.

Theorem 2.3. *Let $s > 1 + d/2$. Suppose that for all sufficiently small λ, $(n_\alpha^\lambda, u_\alpha^\lambda)(x) \in H^s(\mathbb{T}^d)$ and*

$$\|(n_\alpha^\lambda - n_{\alpha,m}^\lambda, u_\alpha^\lambda - u_{\alpha,m}^\lambda)(0,\cdot)\|_{H^s(\mathbb{T}^d)} \leq C\lambda^{2m}$$

with $m > s$. Then there is a $\lambda_0 > 0$ such that the problem (1.1), (2.1) admits a unique classical solution $(n_\alpha^\lambda, u_\alpha^\lambda)$ in $C([0,T^*], H^s(\mathbb{T}^d))$ for any $\lambda \leq \lambda_0$, and there is a constant $K > 0$ independent of λ, such that

$$\|(n_\alpha^\lambda - n_{\alpha,m}^\lambda, u_\alpha^\lambda - u_{\alpha,m}^\lambda)(t)\|_{H^s(\mathbb{T}^d)} + \|(\nabla\phi^\lambda - \nabla\phi_m^\lambda)(t)\|_{H^s(\mathbb{T}^d)} \leq K\lambda^{2m-2s}$$

for any $t \in [0, T^*]$.

3 Quasi-neutral limit of non-isentropic case

We consider the initial data for the non-isentropic Euler-Poisson system,

$$n_\alpha^\lambda |_{t=0} = \bar{n}_\alpha^\lambda(x), \quad u_\alpha^\lambda |_{t=0} = \bar{u}_\alpha^\lambda(x), \quad \theta_\alpha^\lambda |_{t=0} = \bar{\theta}_\alpha^\lambda(x) \tag{3.1}$$

for $\alpha = i, e$, and we make the following ansatz:

$$(n_\alpha^\lambda, u_\alpha^\lambda, \theta_\alpha^\lambda, \phi^\lambda)(t,x) = \sum_{j\geq 0} \lambda^{2j}(n_\alpha^j, u_\alpha^j, \theta_\alpha^j, \phi^j). \tag{3.2}$$

Plugging the above expansions (3.2) into the system (1.2), then we obtain the leading profiles $(n_\alpha^0, u_\alpha^0, \theta_\alpha^0, \phi^0)$ satisfy the following system:

$$\begin{cases} \partial_t n_\alpha^0 + \mathrm{div}(n_\alpha^0 u_\alpha^0) = 0, \\ \partial_t u_\alpha^0 + (u_\alpha^0 \cdot \nabla) u_\alpha^0 + \dfrac{1}{n_\alpha^0}\nabla(n_\alpha^0 \theta_\alpha^0) = -q_\alpha \nabla\phi^0, \\ \partial_t \theta_\alpha^0 + u_\alpha^0 \cdot \nabla\theta_\alpha^0 + (\gamma - 1)\theta_\alpha^0 \mathrm{div} u_\alpha^0 = 0, \\ n_e^0 = n_i^0 \end{cases} \tag{3.3}$$

with initial data

$$(n_\alpha^0, u_\alpha^0, \theta_\alpha^0)(0,x) = (\bar{n}_{\alpha,0}, \bar{u}_{\alpha,0}, \bar{\theta}_{\alpha,0})(x). \tag{3.4}$$

Remark 3.1. *Here we assume $\gamma_i = \gamma_e = \gamma$. In fact, if the difference of the two ratios of specific heats is small, for instance, $\gamma_i = \gamma + \lambda^2 O(1)$ and $\gamma_e = \gamma + \lambda^2 O(1)$, we can similarly deal it.*

Denoted $n^0 = n_e^0 = n_i^0$, assumed $\bar{n}_{e,0}(x) = \bar{n}_{i,0}(x) = \bar{n}_0(x)$. Then we have the following existence for the leading profile (n^0, u^0, θ^0) in [4].

Lemma 3.2. *Let $T^* > 0$, and assume that $(n^0, u_i^0, u_e^0, \theta_i^0, \theta_e^0, \phi^0)$ are smooth solutions of the initial value problem (3.3) and (3.4) with $n^0 > 0$ in $[0, T^*] \times \mathbb{T}^d$. If $\bar{u}_{e,0} = \bar{u}_{i,0} = \bar{u}_0$ and $\bar{\theta}_{e,0} = \bar{\theta}_{i,0} = \bar{\theta}_0$, then we have $u_e^0 = u_i^0 = u^0$, $\theta_e^0 = \theta_i^0 = \theta^0$, and $\nabla \phi^0 = 0$ for all $(t, x) \in [0, T^*] \times \mathbb{T}^d$. Moreover, (n^0, u^0, θ^0) is the unique smooth solution of the compressible non-isentropic Euler equations*

$$\begin{cases} \partial_t n^0 + \operatorname{div}(n^0 u^0) = 0, \\ \partial_t u^0 + (u^0 \cdot \nabla) u^0 + \dfrac{1}{n^0} \nabla (n^0 \theta^0) = 0, \\ \partial_t \theta^0 + u^0 \cdot \nabla \theta^0 + (\gamma - 1) \theta^0 \operatorname{div} u^0 = 0 \end{cases}$$

with initial data $n^0(x, 0) = \bar{n}_0(x)$, $u^0(x, 0) = \bar{u}_0(x)$, $\theta^0(x, 0) = \bar{\theta}_0(x)$.

For $j \geq 1$, we prove the following existence of $(n_\alpha^j, u_\alpha^j, \theta_\alpha^j, \phi^j)$ in [4].

Theorem 3.3. *Let $T^* > 0$ and $s > 1 + d/2$. For any $j \in \mathbb{N}$, assume*

$$(\bar{n}_{\alpha,k}, \bar{u}_{\alpha,k}, \bar{\theta}_{\alpha,k}) \in H^{s+3(j-k)}(\mathbb{T}^d), \quad 0 \leq k \leq j.$$

and the compatibility conditions hold for $0 \leq k \leq j$. Then there exists a unique solution, such that

$$(n_\alpha^j, u_\alpha^j, \theta_\alpha^j) \in C([0, T^*], H^s(\mathbb{T}^d)) \cap C^1([0, T^*], H^{s-1}(\mathbb{T}^d)),$$
$$\phi^j \in C([0, T^*], H^s(\mathbb{T}^d)).$$

Then we can obtain the following results for the quasi-neutral limit of two-fluid non-isentropic Euler-Poisson system in [4].

Theorem 3.4. *Let $s > 1 + d/2$. Suppose that $(\bar{n}_\alpha^\lambda, \bar{u}_\alpha^\lambda, \bar{\theta}_\alpha^\lambda) \in H^s(\mathbb{T}^d)$ for all λ sufficiently small and*

$$\|(\bar{n}_\alpha^\lambda - n_{\alpha,m}^\lambda, \bar{u}_\alpha^\lambda - u_{\alpha,m}^\lambda, \bar{\theta}_\alpha^\lambda - \theta_{\alpha,m}^\lambda)(0, \cdot)\|_{H^s(\mathbb{T}^d)} \leq C \lambda^{2m}$$

with $m > s$. Then there exists $\lambda_0 \in (0, \lambda_1]$, such that the problem admits a unique classical solution $(n_\alpha^\lambda, u_\alpha^\lambda, \theta_\alpha^\lambda)$ in $C([0, T^*], H^s(\mathbb{T}^d))$ for any $\lambda \leq \lambda_0$, which satisfies

$$\|(n_\alpha^\lambda - n_{\alpha,m}^\lambda, u_\alpha^\lambda - u_{\alpha,m}^\lambda, \theta_\alpha^\lambda - \theta_{\alpha,m}^\lambda)(t, \cdot)\|_{H^s(\mathbb{T}^d)} \leq K \lambda^{2m-2s}$$

for $t \in [0, T^*]$, where $K > 0$ is a constant independent of λ.

Remark 3.5. *The electric potential ϕ is unique up to a constant, one can immediately obtain the following estimate:*

$$\|(\nabla \phi^\lambda - \nabla \phi_m^\lambda)(t, \cdot)\|_{H^{s-1}(\mathbb{T}^d)} \leq K \lambda^{2m-2s+2}.$$

References

[1] S. Cordier and E. Grenier, Quasineutral limit of an Euler-Poisson system arising from plasma physics. Commun. Partial Differential Equations. 23 (2000), 1099–1113.

[2] P. Crispel, P. Degond and M. Vignal, An asymptotic preserving scheme for the two-fluid Euler-Poisson model in the quasineutral limit. J. Comput. Phys. 223 (2007), 208–234.

[3] I. Gasser and P. Marcati, A quasi-neutral limit in a hydrodynamic model for charged fluids, Monatsh. Math. 138 (2003), 189–208.

[4] S. Jiang, Q. C. Ju, H. L. Li and Y. Li, Quasi-neutral limit of the full bipolar Euler-Poisson system. Accepted by Science in China Series A: Mathematics.

[5] Q. C. Ju, H. L. Li, Y. Li and S. Jiang, Quasi-neutral limit of the two-fluid Euler-Poisson system. Commun. Pure Appl. Anal. 9 (2010), no.5, 1577–1590.

[6] H. L. Li, C. K. Lin and N. Masmoudi, Asymptotic of the compressible non-isentropic Euler-Poisson system for small debye length. Matimyàs Mat. 26 (2003), 71–79.

[7] Y. J. Peng, Some asymptotic analysis in steady-state Euler-Poisson equations for potential flow. Asymptotic Anal. 158 (2003), 75–92.

[8] Y. J. Peng and Y. G. Wang, Boundary layers and quasi-neutral limit in steady state Euler-Poisson equations for potential flows. Nonlinearity, 17 (2004), 835–849.

[9] Y. J. Peng and Y. G. Wang, Convergence of compressible Euler-Poisson equations to incompressible type Euler equations. Asymptotic Anal. 41 (2005), 141–160.

[10] Y. J. Peng, Y. G. Wang and W. A. Yong, Quasi-neutral limit of the non-isentropic euler-Poisson system. Proc. Roy. Soc. Edinburgh Sect. A, 136 (2006), 1013–1026.

[11] Y. Peng and S. Wang, Asymptotic expansions in two-fluid compressible Euler-Maxwell equations with small parameters. Discrete and Continuous Dynamical Systems - A 23 (2009), 415–433.

[12] M. Slemrod and N. Sternberg, Quasi-neutral limit for Euler-Poisson system. J. Nonlinear Sci. 11 (2001), 193–209.

[13] S. Wang, Quasineutral limit of Euler-Poisson system with and without viscosity. Commun. Partial Differential Equations. 29 (2004), 419–456.

Dynamically Adaptive Simulation of Solute Transport Driven by Shallow Flows

Qiuhua Liang
School of Civil Engineering and Geosciences, Newcastle University
Newcastle upon Tyne NE1 7RU, England, UK
E-mail: Qiuhua.Liang@ncl.ac.uk

Abstract

This paper presents a new adaptive grid based numerical model for simulating solute transport driven by shallow flows. The model solves a 4×4 matrix form of the hyperbolic conservation laws consisting of the 2D shallow water and advection-diffusion equations using a finite volume Godunov-type scheme. The governing equations and numerical scheme are chosen so that they can provide well-balanced solutions and maintain non-negative water depth for simulations involving wetting and drying over complex topographies. A simple adaptive but structured Cartesian grid system is proposed to improve further the performance of the model. Compared with the uniform grid counterpart, the new adaptive grid based model is able to produce results with similar accuracy but at a much lower computational cost.

1 Introduction

Solute transport is a common process taking place in most of the shallow water bodies like rivers, lakes, and estuaries. It may affect the water quality and have great impacts on the local environment and ecosystem. Therefore understanding the shallow flow driven solute transport processes is an important topic in hydraulic and environmental engineering.

Nowadays, numerical modelling has become a common tool in predicting and understanding shallow flows and the relevant processes, including solute transport. For simulations involving complex flow and solute dynamics, e.g., discontinuous waves and contact interfaces, a numerical scheme that solves simultaneously the shallow water and diffusion-advection equations in an integrated form is more appropriate [1, 2]. Several such numerical schemes have been developed and reported in literature (e.g., [2–6]). Among these, the well-balanced finite volume Godunov-type scheme presented recently by Liang [6] is incorporated with a non-negative reconstruction technique and so is more suitable for simulations involving wetting and drying over complex domain topographies. The model is developed on uniform Cartesian grids and its performance may be improved further by incorporating a properly designed adaptive grid system, which is the purpose of this work.

2 Mathematical model

Mathematically, the shallow flow hydrodynamics can be described by the shallow water equations while the movement of passive solute concentration may be governed by the depth-averaged advection-diffusion equation. The two sets of equations can be written in a 4 × 4 matrix form as

$$\frac{\partial \mathbf{q}}{\partial t} + \frac{\partial \mathbf{f}}{\partial x} + \frac{\partial \mathbf{g}}{\partial y} = \frac{\partial \mathbf{f}_d}{\partial x} + \frac{\partial \mathbf{g}_d}{\partial y} + \mathbf{s}, \tag{2.1}$$

where t is the time, x and y are the Cartesian coordinates, $\mathbf{q}$ represents the vector of flow variables, $\mathbf{f}$ and $\mathbf{g}$ are flux vectors in the x and y-directions, $\mathbf{f}_d$ and $\mathbf{g}_d$ contain the diffusion fluxes and s denotes the vector of source terms. The vectors may be given as follows

$$\begin{aligned}
\mathbf{q} &= [\eta \ \ q_x \ \ q_y \ \ q_c]^T, \ \mathbf{f} = \left[q_x \ \ uq_x + \frac{g}{2}(\eta^2 - 2\eta z_b) \ \ uq_y \ \ uq_c\right]^T, \\
\mathbf{g} &= \left[q_y \ \ vq_x \ \ vq_y + \frac{g}{2}(\eta^2 - 2\eta z_b) \ \ vq_c\right]^T, \\
\mathbf{f}_d &= \left[0 \ \ 0 \ \ 0 \ \ hD_{xx}\frac{\partial c}{\partial x} + hD_{xy}\frac{\partial c}{\partial y}\right]^T, \\
\mathbf{g}_d &= \left[0 \ \ 0 \ \ 0 \ \ hD_{yx}\frac{\partial c}{\partial x} + hD_{yy}\frac{\partial c}{\partial y}\right]^T, \\
\mathbf{s} &= \left[0 \ \ -g\eta\frac{\partial z_b}{\partial x} - \frac{\tau_{bx}}{\rho} \ \ -g\eta\frac{\partial z_b}{\partial y} - \frac{\tau_{by}}{\rho} \ \ s_c\right]^T,
\end{aligned} \tag{2.2}$$

where η is the water level, u and v are the x and y components of the depth-averaged velocity, c is the depth-averaged solute concentration, $h = \eta - z_b$ gives the total water depth with z_b denoting the bottom elevation, q_x $(= uh)$ and q_y $(= vh)$ are the unit-width discharge in the two Cartesian directions, q_c $(= ch)$ is the solute volume per unit area, g represents the acceleration due to gravity, D_{xx}, D_{xy}, D_{yx}, and D_{yy} are the diffusion coefficients in different directions, $-\partial z_b/\partial x$ and $-\partial z_b/\partial y$ are the bed slopes, τ_{bx} and τ_{by} give the bed friction stresses and are estimated using the following empirical formulae:

$$\tau_{bx} = \rho C_f u\sqrt{u^2 + v^2} \quad \text{and} \quad \tau_{by} = \rho C_f v\sqrt{u^2 + v^2}, \tag{2.3}$$

in which ρ is the density of water and $C_f = gn^2/h^{1/3}$ is the bed roughness coefficient with n being the Manning coefficient. The shallow water equations (2.1) and (2.2) directly employ the water level η as a flow variable and therefore ensure well-balanced property of a numerical scheme, i.e., preserve the solution of lake at rest [7], for wet-bed applications. As reported in [8,9], the formulation also facilitates a non-negative Godunov-type scheme for dry-bed applications.

3 Grid generation

To enable solution adaption, a simulation may start from a non-uniform grid recently reported by Liang [10], which is generated according to the following four steps:

1) Fit the computational domain into a rectangle (or square);
2) Discretize the rectangular domain by a coarse uniform grid;
3) Check in turn each cell on the background grid and allocate specific subdivision levels;
4) Regularize the mesh to ensure that any cell does not have a neighbour that is more than two times bigger or smaller.

4 Numerical scheme

The uniform grid based non-negative finite volume Godunov-type scheme introduced in [6] is implemented on an aforementioned structured but non-uniform grid. As shown in Figure 4.1, to calculate the x-direction face values in ic on a uniform grid template, flow information is required at ic, e, and w. Obviously the flow variables at ic and e are directly available and those at w are easily obtained using linear interpolation. During flow calculation, attention must be also given to maintain the conservative property of a finite volume scheme. In the same figure, a typical example is to calculate $\mathbf{F}_E$, which should be calculated as $\mathbf{F}_E = (\mathbf{F}_{E1} + \mathbf{F}_{E2})/2$ to ensure mass and momentum conservation.

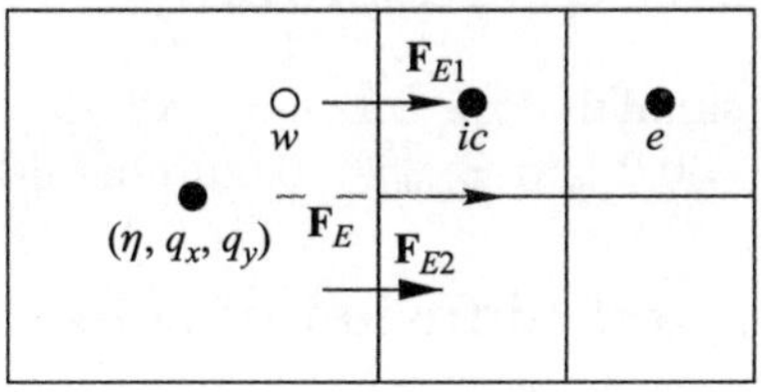

Figure 4.1 Flow calculation on the adaptive grid system.

With the above simple non-uniform grid based flow calculation, solution adaptation only involves variation of grid configuration, i.e., addition and removal of grid cells according to error indicators. Herein the averaged gradients of water level and solution concentration are used as error indicators:

$$\begin{aligned} G_{\eta_{ic}} &= \sqrt{(\partial\eta/\partial x)_{ic}^2 + (\partial\eta/\partial y)_{ic}^2} \\ \text{and}\quad G_{c_{ic}} &= \sqrt{(\partial c/\partial x)_{ic}^2 + (\partial c/\partial y)_{ic}^2}\ . \end{aligned} \tag{4.1}$$

Both of them are then nondimensionalized as

$$\Theta_{\eta_{ic}} = G_{\eta_{ic}}/P_q(\mathbf{G}_\eta) \quad \text{and} \quad \Theta_{c_{ic}} = G_{c_{ic}}/P_q(\mathbf{G}_c)\ , \tag{4.2}$$

where $\mathbf{G}_\eta$ and $\mathbf{G}_c$ are the vectors storing corresponding grid-based gradients in the entire computational domain. $P_q(\mathbf{X})$ returns the qth percentile of $\mathbf{X}$ with $q = 1 - s_a$. s_a indicates the sensitivity to the error indicator. Finally the error indicator is defined as

$$\Theta = \max(\ \Theta_{\eta_{ic}}\ , \Theta_{c_{ic}}\ )\ . \tag{4.3}$$

Apart from s_a, another critical value Θ_{coar} should also be prescribed for grid coarsening. During a simulation, grid adaptation is undertaken at every time step as:

1) If a background cell is not maximum subdivided and has at least one child with $\Theta_{ic} > 1$, it will be marked for refinement and its subdivision level will become $lev + 1$;
2) When a subdivided background cell has all its children with $\Theta_{ic} > \Theta_{\text{coar}}$, it will be marked for coarsening and its subdivision level will be changed to $lev - 1$.

Independently, cells are automatically refined to capture moving wet-dry fronts. Flow variables in a newly created cell are obtained by linear interpolation.

5 Results and discussions

All of the tests are simulated on a Lenovo X61 laptop. $g = 9.81$ m/s^2 and CFL =0.75. $s_a = 0.2$ and $\Theta_{\text{coar}} = 0.6$ are used for grid adaption.

5.1 Advection and diffusion of solute

In an 800 m $\times$ 200 m frictionless domain with a horizontal bed, a simple uniform flow ($h = 1$ m, $u = 1$ m/s, $v = 0$) is given and remains steady throughout the simulation. In such a simple flow field, advection and diffusion of a solute cloud may be described analytically as:

$$c(x, y) = \frac{c_0/h}{4\pi t\sqrt{D_{xx}D_{yy}}} \exp\left(-\frac{(x - x_0 - ut)^2}{4D_{xx}t} - \frac{(y - y_0)^2}{4D_{yy}t}\right). \quad (5.1)$$

Herein, $C_0 = 233.06, D_{xx} = 1.02$ m^2/s, $Dyy = 0.094$ m^2/s.$(x_0, y_0) = (0, 100)$ is the location of peak concentration at $t = 0$. Simulation starts at $t = 60$ s and the initial conditions for starting the simulation are calculated by (5.1). The computational domain is first decomposed by a coarse background grid of 100 × 25 cells and maximum subdivision level is set to 3 so that the minimum cell is 1 m × 1 m. Transmissive boundary conditions are imposed at the four lateral walls.

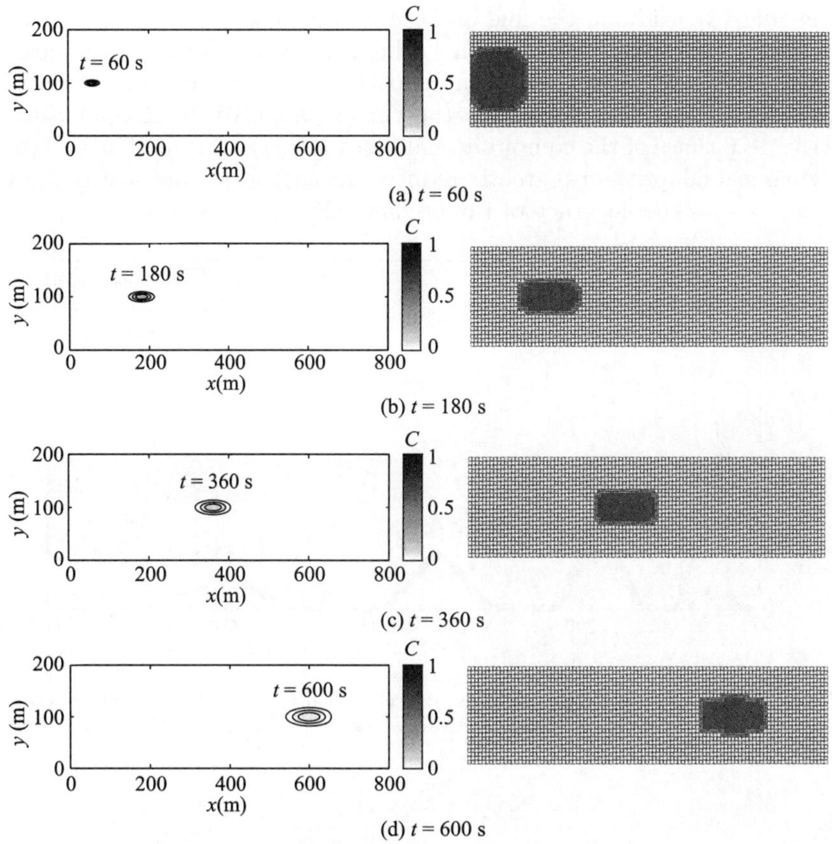

Figure 5.1 Advection and diffusion: numerical predictions at different output times.

Simulation is performed for 600 s. Figure 5.1 presents the numerical predictions at different output times with Figure 5.1(a) actually being the initial conditions and the original grid when the simulation starts. As shown by the concentration contours, the solute disperses while being transported to downstream by the uniform flow, which is as expected. From Figure 5.1, it is obvious that the grid evolves according to the movement of the solute and generates high-resolution mesh in those areas with higher solute concentration while coarse background mesh is used in other zones where the concentration is near zero. This grid adaptation strategy optimizes the computational resources such that high-resolution simulation is only used when necessary. Figure 5.2 demonstrates the numerically predicted concentration profiles output at different times, compared with the analytical solution and numerical results obtained on the fine mesh with 800×200 cells on which the grid size is the same as the finest cell on the adapted grid. The numerical predictions obtained on the adaptive grid and the fine uniform grid is identical and both match perfectly the analytical solution. The 157.8 s CPU time is required by the adaptive grid simulation. On the fine uniform grid, however, 2625.2 s is needed. Therefore, the current adaptive mesh simulation saves 16.6 times of the computational cost for this test case. The current dynamical adaptive grid greatly reduces the computational cost but not compromises the accuracy of the prediction.

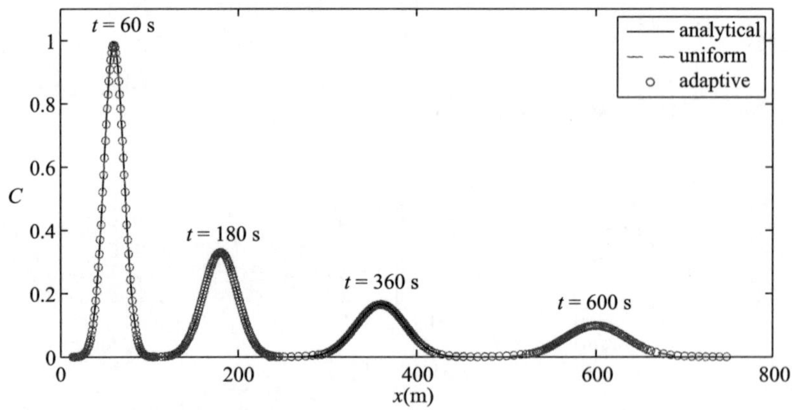

Figure 5.2 Advection and diffusion: concentration profiles alone the x-direction central line, compared with the analytical solution and predictions obtained on corresponding uniform grid.

5.2 Dam-Break induced solute advection

This test case involves discontinuous dam-break flow over a non-uniform topography and therefore represents a more realistic flow in the real world. This case has been simulated by Liang [6] to validate the current finite-volume Godunov-type scheme implemented on uniform grids and is reconsidered here to demonstrate the benefit of using the current adaptive mesh simulation. The details of the case can be found in [6]. For the adaptive mesh simulation, the computational domain is approximated by a coarse background grid of 50 × 20 cells and is allowed to be refined up to a maximum subdivision level of 2 so that the uniform grid of similar resolution contains 200 × 80 cells. At the beginning, the highest-level mesh is generated to represent the dam so that the discontinuous water surface can be more accurately described. Simulation is conducted for 30 s.

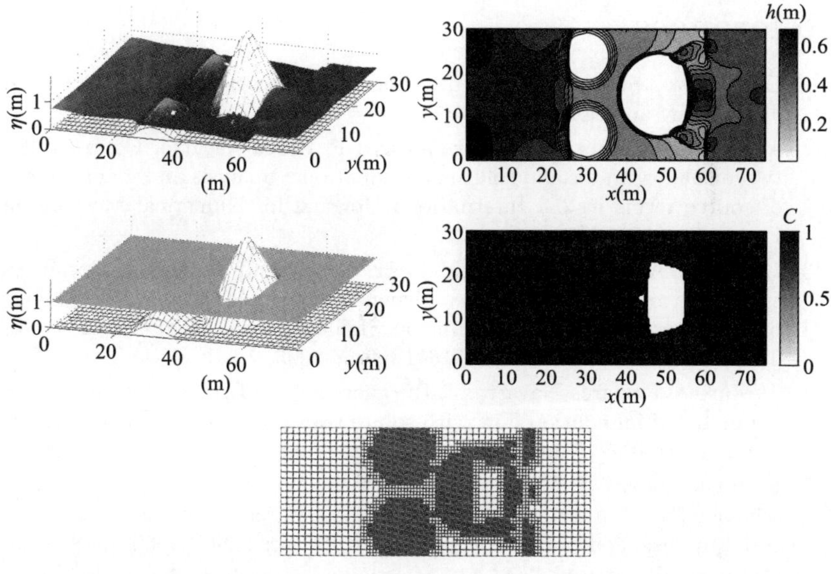

Figure 5.3 Dam break over three humps: numerical predictions at $t = 30$ s.

Figure 5.3 present the numerical predictions at $t = 30$ s. The numerically predicted wave patterns are identical to those obtained on uniform grids [6] and compare well with those produced by other numerical schemes (refer to the references in [6]). In this case, the fine uniform grid has 16 000 cells and requires 277.3 s CPU time to finish the simulation while the adaptive mesh simulation consumes only 100.1 s of CPU time. Therefore the adaptive computation is about 3 times more efficient and still represents a significant improvement in computational efficiency.

6 Conclusions

This paper presents a new adaptive grid based Godunov-type scheme for solving the integrated shallow water and advection-diffusion equations.

The adaptive mesh simulation starts from a non-uniform grid with local structured property. Initial grid generation and grid adaptation are both straightforward to achieve. The solution scheme is identical to that on a uniform Cartesian grid after employing a linear interpolation scheme. Due to the use of the depth-positivity-preserving Godunov-type scheme and dynamical adaptive grid, the model is suitable for simulations involving discontinuous flows and contact wet-dry front or solute interface and significantly improves computational efficiency compared with its uniform grid based counterpart.

References

[1] E. F. Toro, Shock-capturing Methods for Free-surface Shallow Flows. John Wiley and Sons, 2001.

[2] J. Murillo, J. Burguete, P. Brufau and P. Carca-Navarro, Coupling between shallow water and solute flow equations: analysis and management of source terms in 2D. International Journal for Numerical Methods in Fluids 49 (2005), 267–299.

[3] Q. Liang, A. G. L. Borthwick, P. H. Taylor and J. Huang, Godunov-type quadtree model of species dispersion in shallow flows. Proceedings of the International Symposium on Shallow Flows, Delft University of Technology, The Netherlands, 16–18 June 2003.

[4] J. Murillo, P. Carca-Navarro, J. Burguete and P. Brufau, A conservative 2D model of inundation flow with solute transport over dry bed. International Journal for Numerical Methods in Fluids 52 (2006), 1059–1092.

[5] F. Benkhaldoun, I. Elmahi and M. Sead, Well-balanced finite volume schemes for pollutant transport by shallow water equations on unstructured meshes. Journal of Computational Physics 226(1) (2007), 180–203.

[6] Q. Liang, A well-balanced and non-negative numerical scheme for solving the integrated shallow water and solute transport equations. Communications in Computational Physics 7(5) (2010), 1049–1075.

[7] J. M. Greenberg and A.-Y. LeRoux, A well-balanced scheme for the numerical processing of source terms in hyperbolic equations. SIAM Journal on Numerical Analysis 33 (1996), 1–16.

[8] Q. Liang and A. G. L. Borthwick, Adaptive quadtree simulation of shallow flows with wet-dry fronts over complex topography. Computers and Fluids 38(2) (2009), 221–234.

[9] Q. Liang, Flood simulation using a well-balanced shallow flow model. Journal of Hydraulic Engineering 136(9) (2010), 669–675.

[10] Q. Liang, A structured but non-uniform cartesian grid based model for the shallow water equations. International Journal for Numerical Methods in Fluids(2010), DOI: 10.1002/fld.2266.

Asymptotic Stability of Rarefaction Waves in Radiative Hydrodynamics*

Chunjin Lin
Department of Mathematics, College of Sciences, Hohai University
Nanjing 210098, Jiangsu, China
E-mail: cjlin@hhu.edu.cn

Abstract

In this work we study the asymptotic stability of rarefaction waves for solutions to a one-dimensional radiative hydrodynamic system which couples hyperbolic-elliptic equations. We assume that the initial data tend to constant states at $x = \pm\infty$, respectively, and the corresponding Riemann problem for the compressible Euler equations admits a continuous rarefaction wave solution with small strength. If the initial perturbation is small, the solution is proved to tend to the rarefaction wave as $t \to +\infty$. The proof is based on the L^2-energy method and elliptic estimates.

1 Introduction and main results

In this work, we consider a model describing the interaction of an inviscid compressible fluid with photons through energy exchanges. The fluid is described by its specific volume v, its velocity u, and its specific internal energy e. The evolution of the fluid is governed by the following system of equations, written in Lagrangian coordinates:

$$\begin{cases} v_t - u_x = 0\,, \\ u_t + p_x = 0\,, \\ (e + u^2/2)_t + (p\,u + q)_x = 0\,, \end{cases} \tag{1.1}$$

where p is the pressure of the fluid, and q denotes the radiative heat flux which satisfies the following elliptic equation, still written in Lagrangian coordinates:

$$-\left(\frac{q_x}{v}\right)_x + vq + \left(\theta^4\right)_x = 0, \tag{1.2}$$

*This proceeding is the subject of the paper to be published in Comm. Math. Science. The author is partly supported by the Fundamental Research Funds for the Central Universities (B1020275), NSF of Hohai University (No.2008432811), and NSFC (No. 10926162, No. 10871059).

where $\theta > 0$ is the absolute temperature of the fluid. We assume that the pressure p and the specific internal energy e are related to the specific volume v and the temperature θ by

$$p(v,\theta) = \frac{R\theta}{v}, \quad e = \frac{R}{\gamma - 1}\theta := c_v\theta, \tag{1.3}$$

where R is the perfect gas constant, $\gamma > 1$ is the adiabatic constant, and $c_v = R/(\gamma - 1)$ is the specific heat at constant volume.

System (1.1)–(1.2), describing the case that the fluid is not in the thermodynamical equilibrium with the radiation, can be formally derived by asymptotic arguments, starting from a more complete system coupling fluid dynamics to a kinetic equation describing the evolution of the specific intensity of radiation, see the Appendix of [9] and the references therein. We also refer to [12] for the physical background. We mention that a simplified version of system (1.1)–(1.2) has been introduced and studied in [5, 6, 16] and references therein. This simple model, consisting in a Burgers' type equation coupled to a linear diffusion equation, can be viewed as a prototype for discussing of the coupled system (1.1)–(1.2). In this work we try to extend the results in [6, 16] to the more physical model (1.1)–(1.2).

We introduce some known results to system (1.1)–(1.2). Kawashima, Nikkuni, and Nishibata in [4] showed the global in time existence of solutions in the neighborhood of a constant state in Eulerian coordinates. Such results can be regarded as the stability of the equilibrium states. The situation when the data leads to a shock has been investigated recently in [9], where the existence of a shock profile is established and in [10] the stability of the shock profile is analyzed. For a general hyperbolic system coupled with a linear elliptic equation, the existence of shock profile was given in [7], and in [13] the stability of shock profile.

In this work we are interested in the stability of rarefaction waves. Such problem has been well studied for viscous regularization of hyperbolic systems and the relaxation approximation of conservation laws. For example, the stability of the rarefaction wave for the compressible Navier-Stokes equations has been studied in [3, 14]. See also in [1, 11, 15]. Inspired from [14], we wish to deal with the full system of hydrodynamic equations coupled with a nonlinear elliptic equation (1.1)–(1.2).

Before stating our main result, we introduce some variations and an approximation of system (1.1)–(1.2). We focus on the asymptotic stability of rarefaction waves. To this end, it is convenient to work with the equation for the specific entropy s. The second law of thermodynamics assets that

$$\theta \mathrm{d}s = \mathrm{d}e + p\mathrm{d}v. \tag{1.4}$$

Thanks to (1.3), the expression of the specific entropy s can be obtained:

$$A \exp \frac{s}{c_v} = pv^{\gamma}$$

with a constant A. We choose (v, s) as independent variables and view the pressure p, the internal energy e and the temperature θ as functions of (v, s), $p = \tilde{p}(v, s)$, $e = \tilde{e}(v, s)$, $\theta = \tilde{\theta}(v, s)$. Note that assumption (1.3) yields many algebraic simplifications, but we believe that our results still hold for a general pressure law satisfying the usual requirements of thermodynamics. Thus for smooth solutions, system (1.1) is equivalent to

$$\begin{cases} v_t - u_x = 0, \\ u_t + \tilde{p}(v, s)_x = 0, \\ s_t + \dfrac{q_x}{\tilde{\theta}(v, s)} = 0. \end{cases} \tag{1.5}$$

In what follows we study the coupled system (1.2),(1.5) which is completed by imposing the initial data

$$(v,\ u,\ s)(0, x) = (v_0, u_0, s_0)(x) \to (v_\pm,\ u_\pm,\ s_\pm), \text{ as } x \to \pm\infty, \tag{1.6}$$

where $v_\pm > 0$, $u_\pm$, and $s_\pm$ are constants which can be connected by a rarefaction wave. Since we focus on the stability of rarefaction waves, throughout this paper, we assume that $s_+ = s_- = \bar{s}$.

An approximation of the coupled system (1.2),(1.5) can be obtained by neglecting the derivatives q_x and q_{xx}, i.e,

$$\begin{cases} v_t - u_x = 0, \\ u_t + \tilde{p}(v, s)_x = 0, \\ s_t = 0, \end{cases} \tag{1.7}$$

which is the usual compressible Euler equations in Lagrangian coordinates. We can still associate to the solution of (1.7) the following quantity:

$$q = -\left(\tilde{\theta}^4(v, s)\right)_x / v. \tag{1.8}$$

There are two families of rarefaction waves for (1.7) which are solutions of the compressible Euler equations (1.7) with the Riemann data

$$(v,\ u,\ s)(0, x) = (v_0^R,\ u_0^R, s_0^R)(x) = (v_\pm,\ u_\pm,\ \bar{s}), \text{ for } x \gtrless 0. \tag{1.9}$$

For illustration, we only consider the 1-rarefaction wave. In this case, we assume that the asymptotic data satisfies $v_- < v_+$. The rarefaction wave $(V^R, U^R, S^R, \Theta^R)(t,x)$ is characterized by

$$\begin{cases} \lambda_1(V^R(t,x),\bar{s}) = -\sqrt{-\tilde{p}_v(V^R(t,x),\bar{s})} = w^R(t,x), \\ S^R(t,x) = \bar{s}, \\ U^R(t,x) = u_\pm - \displaystyle\int_{v_\pm}^{V^R(t,x)} \lambda_1(z,s)dz, \\ \Theta^R(t,x) = \tilde{\theta}(V^R(t,x),\bar{s}), \\ \lambda_{1x}(V^R(t,x),S^R(t,x)) > 0, \end{cases} \tag{1.10}$$

where $\lambda_1(v,s)$ is the first characteristic speed of the compressible Euler system (1.7) and $w^R(t,x)$ is the unique solution to the following Riemann equation:

$$\begin{cases} w_t^R + \left(\dfrac{(w^R)^2}{2}\right)_x = 0, \\ w^R(0,x) = w_0^R(x) = w_\pm := \lambda_1(v_\pm,\bar{s}), \text{ for } x \gtrless 0. \end{cases} \tag{1.11}$$

Since $w_- < w_+$, system (1.11) admits a unique continuous solution.

Associated with the rarefaction wave $(V^R, U^R, S^R, \Theta^R)(t,x)$, the radiative flux Q^R is defined by the equality (1.8) with $\tilde{\theta}(v,s)$ replaced by $\Theta^R = \tilde{\theta}(V^R, S^R)$. The case for the 3-rarefaction wave can be discussed similarly. For more details about the rarefaction wave of compressible Euler equations, we refer to [17].

Under the above preparation, our asymptotic stability result on rarefaction wave $(V^R, U^R, S^R)(t,x)$ states in the following:

Theorem 1.1. *Let $(V^R, U^R, S^R)(t,x)$ be the 1-rarefaction wave solution to Riemann problem of the compressible Euler system (1.7) and (1.9). Assume that the initial data $(v_0, u_0, s_0)(x)$ of the Cauchy problem (1.2),(1.5), and (1.6) satisfies*

$$(v_0(x) - v_\pm, u_0(x) - u_\pm) \in L^2(\mathbb{R}_\pm)$$
$$(\partial_x v_0(x), \partial_x u_0(x)) \in H^1(\mathbb{R}).$$

Let $(V,U,\Theta)(t,x)$ denote the smooth approximation of the rarefaction wave constructed in Section 2. It satisfies the classical Euler system (1.7). We suppose also that

$$\|(v_0(x) - V(0,x), u_0(x) - U(0,x), s_0(x) - \bar{s})\|_{H^2(\mathbb{R})}$$

is sufficiently small. Then the Cauchy problem (1.5)-(1.6) admits a unique global smooth solution $(v, u, s, q)(t,x)$ satisfying

$$\lim_{t\to+\infty} \sup_{x\in\mathbb{R}} \left\{ \left| \left(v(t,x) - V^R(t,x),\ u(t,x) - U^R(t,x),\ s(t,x) - \bar{s}\right) \right| \right\} = 0.$$

Remark 1.2. Theorem 1.1 states the stability of weak rarefaction wave. In fact $V(0,x)$ defined in Section 2, satisfies

$$\|V(0,x) - v_\pm\|_{L^2(\mathbb{R}_\pm)} = O(1)\frac{|v_+ - v_-|}{\varepsilon},$$

where $\varepsilon > 0$ is a parameter intended to be small. Thus the assumption about the initial perturbation suggests $|v_+ - v_-| = o(\varepsilon)$. In the same way, we need $|u_+ - u_-| = o(\varepsilon)$.

Remark 1.3. The stability result for the full system of hydrodynamic equations coupled with a nonlinear elliptic equation (1.1)-(1.2), obtained in Theorem 1.1, is similar to that in [6] for the simple model. Hence we extend the known result for the simple model to the more physical system (1.1)–(1.2).

In the next section, following [3, 14], we define a smooth approximation of the rarefaction wave solutions. Then we reformulate system (1.1)–(1.2) by considering the perturbation from the smooth approximation of the rarefaction wave and state the stability result for the reformulated system. That stability result can be proven by the standard energy method combined with elliptic estimates. Notice that these elliptic estimates are crucial and make the analysis different from the case of compressible Navier-Stokes [3, 14]. We will use H^l $(l > 0)$ to denote the usual Sobolev space with the norm $\|\cdot\|_l$. Let $\|\cdot\|$ denote the usual L^2 norm, and $|\cdot|_p$ denote the L^p norm. Note that $\|\cdot\|_0 = \|\cdot\| = |\cdot|_2$.

2 Smooth approximation and the reformulation problem

Following [3, 14], to prove the stability result, we make use of a smooth approximation of the 1-rarefaction wave $(V^R,\ U^R,\ S^R)(t,x)$. Given a small but fixed constant $\varepsilon > 0$, let $w(x,t)$ be the unique global smooth solution to the Cauchy problem:

$$\begin{cases} w_t + ww_x = 0, \\ w(0,x) = w_0(x) := \dfrac{w_- + w_+}{2} + \dfrac{w_+ - w_-}{2}\tanh(\varepsilon x), \end{cases} \tag{2.1}$$

where $w_\pm = \lambda_1(v_\pm, \bar{s})$. Then, $V(t,x)$, $U(t,x)$, $S(t,x)$, and $\Theta(t,x)$ are defined by (1.10) with $w^R(t,x)$ replaced by its smooth approximation $w(t,x)$. Similarly we can define Q, the smooth approximation of radiative flux, by (1.8) with replaced $\tilde{\theta}(v,s)$ by Θ. Thus we get the smooth approximation (V,U,S,Θ,Q) which verifies the approximate system (1.7)–(1.8).

In fact $w(t,x)$ which is well-defined and fulfilled some regularity proprieties, can be viewed as a good approximation of $w^R(t,x)$, the continuous solution of the Riemann problem (1.11), see [11]. Consequently (V,U,S,Θ) is globally well-posed and smooth, hence it is the smooth approximation of the 1-rarefaction wave (V^R,U^R,S^R,Θ^R), and also verifies some regularity properties, see [8] and references cited therein.

Under the above preparations, let us state the reformation problem. To this end we set

$$(\varphi,\psi,\phi,\xi,z)(t,x) = (v-V, u-U, \theta-\Theta, s-\bar{s}, q-Q)(t,x),$$

which represents the perturbation of the rarefaction wave and solves

$$\begin{cases} \varphi_t - \psi_x = 0, \\ \psi_t + [p(v,\theta)-p(V,\Theta)]_x = 0, \\ c_v\phi_t + [p(v,\theta)-p(V,\Theta)]U_x + p(v,\theta)\psi_x + z_x = \left(\dfrac{(\Theta^4)_x}{V}\right)_x, \\ vz - \left(\dfrac{z_x}{v}\right)_x + (\theta^4-\Theta^4)_x = \dfrac{(\Theta^4)_x}{V}\varphi - \left\{\left(\dfrac{(\Theta^4)_x}{V}\right)_x \dfrac{1}{v}\right\}_x, \end{cases} \tag{2.2}$$

with the initial data

$$\begin{aligned} (\varphi,\psi,\phi)(0,x) &= (\varphi_0,\psi_0,\phi_0)(x) \\ &:= (v_0(x)-V(0,x), u_0-U(0,x), \theta_0(x)-\Theta(0,x))(x). \end{aligned} \tag{2.3}$$

In what follows we choose φ, ψ, ϕ, and z as independent variables. For any $T>0$, we define the functional space:

$$X(0,T) = \left\{ (\varphi,\psi,\phi,z)(t,x) \;\middle|\; \begin{array}{c} (\varphi,\psi,\phi) \in L^\infty(0,T;H^2(\mathbb{R})) \\ (\varphi_x,\psi_x,\ \phi_x) \in L^2(0,T;H^1(\mathbb{R})) \\ z \in L^\infty(0,T;H^3(\mathbb{R})) \bigcap L^2(0,T;H^3(\mathbb{R})) \end{array} \right\}.$$

The global existence to (2.2)-(2.3) follows by combining a local result and the a priori estimate. The local existence of solutions of such a hyperbolic-elliptic system is proved by the standard iteration method, we omit the detail and refer to [4]. Hence, to complete the proof of Theorem 1.1, the crucial step is to show the a priori estimate. To this end, we set

$$N(t)^2 = \sup_{\tau\in[0,t]} \left(\|(\varphi,\ \psi,\ \phi)(\tau)\|_2^2 + \|z(\tau)\|_3^2\right).$$

Proposition 2.1. *Let $T>0$ and let $(\varphi,\ \psi,\ \phi,\ z) \in X(0,T)$ be the unique solution to system (2.2)–(2.3). Then there exists a constant $\hat{\varepsilon}$ in-*

dependent of T such that if $N(T) \leq \hat{\varepsilon}, \varepsilon \leq \hat{\varepsilon}$, then the following estimate

$$\begin{aligned}&\|(\varphi,\ \psi,\ \phi)(t)\|_2^2 + \|z(t)\|_3^2 + \int_0^t \|V_t^{1/2}(\tau)(\varphi,\phi)(\tau)\|^2 d\tau \\ &+ \int_0^t \left(\|(\varphi_x,\ \psi_x,\ \phi_x)(\tau)\|_1^2 + \|z(\tau)\|_3^2\right) d\tau \\ &\leq C(\|(\varphi_0,\ \psi_0,\ \phi_0)(t)\|_2^2 + \varepsilon^{\frac{1}{4}}),\end{aligned}$$

holds for any $t \in [0, T]$, where the constant C does not depend on T.

It follows from Proposition 2.1 and the standard continuation argument (cf. [2]) that we get the proof of Theorem 1.1. The proof of Proposition 2.1 is based on the energy method combined with the elliptic estimates. More precisely, the $L^\infty(H^2)$ estimates of the fluid perturbation variables (φ, ψ, ϕ) can be obtained by the symmetrization of quasi-linear form of (2.2). Notice that linearizing the equations around the smooth approximation of the rarefaction wave is not valid. As pointed in [10], this is because the dissipation introduced by the elliptic equation is not strong enough to control the linearization errors. But the standard symmetrization of the quasi-linear form of the perturbation system (2.2) works well. And then following [2], we can also get the $L^2(H^1)$ estimate of the derivatives of the fluid perturbation variables (φ, ψ). Eventually the elliptic inequalities can be derived by using the elliptic equation. Detailed proofs of the above results may be found in [8].

Acknowledgement

This work is a continuation of the PhD thesis of the author. The author would like to thank his supervisors Thierry Goudon and Jean-François Coulombel for their encouragement and guidance.

References

[1] S. Jiang, G. Ni and W. Sun, Vanishing viscosity limit to rarefaction waves for the Navier-Stokes equations of one-dimensional compressible heat-conducting fluids. SIAM J. Math. Anal. 38 (2006), no. 2, 368–284 (electronic).

[2] S. Kawashima and A. Matsumura, Asymptotic stability of traveling wave solutions of systems for one-dimensional gas motion. Comm. Math. Phys. 101(1985), no. 1, 97–127.

[3] S. Kawashima, A. Matsumura and K. Nishihara, Asymptotic behavior of solutions for the equations of a viscous heat-conductive gas. Proc. Japan Acad. Ser. A Math. Sci. 62(1986), no. 7, 249–252.

[4] S. Kawashima, Y. Nikkuni and S. Nishibata, Large-time behavior of solutions to hyperbolic-elliptic coupled systems. Arch. Ration. Mech. Anal. 170(2003), no. 4, 297–329.

[5] S. Kawashima and S. Nishibata, Shock waves for a model system of the radiating gas. SIAM J. Math. Anal. 30(1999), no. 1, 95–117 (electronic).

[6] S. Kawashima and Y. Tanaka, Stability of rarefaction waves for a model system of a radiating gas. Kyushu J. Math. 58(2004), no. 2, 211–250.

[7] C. Lattanzio, C. Mascia and D. Serre, Shock waves for radiative hyperbolic-elliptic systems. Indiana Univ. Math. J. 56(2007), no. 5, 2601–2640.

[8] C. Lin, Asymptotic stability of rarefaction waves in radiative hydrodynamics. To appear in Commun. Math. Sci., Preprint(2010).

[9] C. Lin, J. F. Coulombel and T. Goudon, Shock profiles for non-equilibrium radiating gases. Phys. D 218(2006), no. 1, 83–94.

[10] C. Lin, J. F. Coulombel and T. Goudon, Asymptotic stability of shock profiles in radiative hydrodynamics. C. R. Math. Acad. Sci. Paris 345(2007), no. 11, 625–628.

[11] A. Matsumura and K. Nishihara, Asymptotics toward the rarefaction waves of the solutions of a one-dimensional model system for compressible viscous gas. Japan J. Appl. Math. 3(1986), no. 1, 1–13.

[12] D. Mihalas and B. W. Mihalas, Foundations of Radiation Hydrodynamics. Oxford University Press, New York 1984.

[13] T. Nguyen, R. Plaza and K. Zumbrun, Stability of radiative shock profiles for hyperbolic-elliptic coupled systems. Phys. D 239(2010), no. 8, 428–453.

[14] K. Nishihara, T. Yang and H. Zhao, Nonlinear stability of strong rarefaction waves for compressible Navier-Stokes equations. SIAM J. Math. Anal. 35(2004), no. 6, 1561–1597.

[15] K. Nishihara, H. Zhao and Y. Zhao, Global stability of strong rarefaction waves of the Jin-Xin relaxation model for the p-system. Comm. Partial Differential Equations 29(2004), no. 9–10, 1607–1634.

[16] L. Ruan and J. Zhang, Asymptotic stability of rarefaction wave for hyperbolic-elliptic coupled system in radiating gas. Acta Math. Sci. Ser. B Engl. Ed. 27(2007), no. 2, 347–360.

[17] J. Smoller, Shock Waves and Reaction-diffusion Equations. Springer-Verlag, New York, 1994.

A Symplecticity-preserving Gas-kinetic Scheme for Hydrodynamic Equations under External Forcing Field*

Jun Luo, Kun Xu
School of Mathematical Sciences, Fudan University
Shanghai 200433, China
E-mail: xxx@fudan.edu.cn

Abstract

In this paper, based on the BGK equation, Liouville's theorem, and symplecticity preserving property of a Hamiltonian flow, a well-balanced kinetic scheme for the compressible Navier-Stokes equations under external forcing field is developed. In order to construct such a scheme, the physical process of particles transporting through a potential barrier at a cell interface will be modeled with penetration and reflection according to incident particle velocity, and the symplectic property of phase space volume. For the simulation of a gravitational gas system, the current scheme has both a well-balanced property for keeping hydrostatic solution and the shock capturing capability for unsteady flow. In this paper, we will present a symplecticity preserving BGK scheme with the inclusion of particle collision in the gas evolution stage.

1 Introduction

Generally, flows equations with source terms in one-dimensional case can be written as

$$U_t + F(U)_x = S, \tag{1.1}$$

where U is the vector of conservative flow variables with corresponding fluxes $F(U)$ and S is the source term. For a gas flow under an external time-independent gravitational field, there is a special hydrostatic solution which satisfies

$$F(U)_x = S. \tag{1.2}$$

*The authors are partly supported by Hong Kong Research Grant Council 621709, the National Natural Science Foundation of China (Project No. 10928205), and National Key Basic Research Program (2009CB724101).

The scheme which could keep such a solution is called a well-balanced method. In order to develop an accurate flow solver for a slowly evolving gravitational gas dynamic system, such as the galaxy evolution problem in astrophysics, the numerical scheme should not only have the shock-capturing ability, but also be a well-balanced one.

There have been many attempts to construct well-balanced gas dynamic codes that preserve the hydrostatic solution [1, 6]. The schemes in [1, 6] are designed based on the condition Eq. (1.2), such as to re-balance the quantities which are not in hydrostatic balance after the original updating. However, for a transition flow, the direct use of Eq. (1.2) in the design of the scheme is problematic, because in general case Eq. (1.2) cannot be enforced in a physical evolution process at all. So, our aim of this paper is to design a general scheme with correct particle transport physics across a potential barrier, which becomes automatically a well-balanced one under the hydrostatic situation.

In the past years, gas-kinetic BGK scheme has been developed successfully for compressible Euler and Navier-Stokes equations [4]. In an early approach, a gas kinetic scheme (GKS) for gravitational system [3] has been constructed with the inclusion of particle acceleration in the transport process. The above scheme much improved the solution in comparison with operator splitting method. Unfortunately, the assumption of a continuous distribution of the gravitational potential makes the scheme in [3] not be a precise well-balanced method. In a previous paper [5], a well-balanced kinetic flux vector splitting scheme for gravitational Euler equations has been developed. However, in the above approach, only a few simple moments of a gas distribution function are needed, and these simple moments can be intuitively guessed rather than derived from a solid physical and mathematical foundation. In order to extend the above scheme to high-order and solve the gravitational Navier-Stokes equations, the gas-kinetic BGK model with the inclusion of particle collision has to be solved. In developing such a scheme, much more high-order moments have to be evaluated. It can be hardly obtained through institutive guessing. Therefore, we need to use the symplectic property of a Hamiltonian system and Liouville's theorem to evaluate high-order moment of a gas distribution function on both sides of a cell interface which is associated with a potential jump. Furthermore, to develop a scheme based on a collision model is much more challenge than that for the collision-less case. Around a potential jump, a multiple equilibrium states have to be constructed on both sides of a jump. In the construction of such an equilibrium state for the BGK model, the second law of thermodynamics has to be explicitly satisfied.

2 The 1D BGK scheme without external forcing field

The BGK equation without external forcing field is

$$f_t + u f_x = \frac{g - f}{\tau}, \tag{2.1}$$

where f is the gas distribution function and g is the equilibrium state approached by f. The particle collision time τ is related to the viscosity and heat conduction coefficients, i.e., $\tau = \mu/p$ where μ is the dynamic viscosity coefficient and p is the pressure. The relation between mass ρ, momentum ρU, and energy ρE densities with the distribution function f is

$$W_\alpha = \int \psi_\alpha f d\Xi, \quad \alpha = 1, 2, 3, \tag{2.2}$$

where $W = (\rho, \rho U, \rho E)^T$, $\psi = (\psi_1, \psi_2, \psi_3)^T = (1, u, \frac{1}{2}(u^2 + \xi^2))^T$, $d\Xi = dud\xi_1 d\xi_2 \dots d\xi_K$ is the volume element in the phase space. For 1D flow, the total number of degrees of freedom K in ξ is equal to $(3-\gamma)/(\gamma-1)$, and γ is the ratio of specific heat. The integral solution of (2.1) is

$$f(x, t, u, \xi) = \frac{1}{\tau} \int_0^t g(x', t', u, \xi) e^{-(t-t')/\tau} dt' + e^{-t/\tau} f_0(x - ut, u, \xi), \tag{2.3}$$

where $x' = x - u(t - t')$ is the particle trajectory. The solution f in (2.3) solely depends on the modeling of f_0 and g.

By using the MUSCL-type limiter, a discontinuous reconstruction for the macroscopic flow variables can be obtained around a cell interface. Based on the Chapmann-Enskog expansion to the Navier-Stokes order, the initial gas distribution function f_0 in (2.3) on both sides of a cell interface can be constructed as

$$\begin{aligned} f_0^l(x, u, \xi) &= g_0^l(1 + a^l(x - x_{j+1/2}) - \tau(a^l u + A^l)), \quad x \le x_{j+1/2}, \\ f_0^r(x, u, \xi) &= g_0^r(1 + a^r(x - x_{j+1/2}) - \tau(a^r u + A^r)), \, x > x_{j+1/2}. \end{aligned} \tag{2.4}$$

Here $g_0^{l,r}$ are the corresponding Maxwellians at different sides of the interface. The equilibrium distribution functions around the cell interface $x_{j+1/2}$ can be modeled as

$$\begin{aligned} g^l(x, t, u, \xi) &= g_{j+1/2}^l(1 + \overline{a}^l(x - x_{j+1/2}) + \overline{A}^l t), \quad x \le x_{j+1/2}, \\ g^r(x, t, u, \xi) &= g_{j+1/2}^r(1 + \overline{a}^r(x - x_{j+1/2}) + \overline{A}^r t), \, x > x_{j+1/2}. \end{aligned} \tag{2.5}$$

In the case without external forcing term, $g_{j+1/2}^l$ and $g_{j+1/2}^r$ in the above equation are the same distribution functions, i.e., $g_{j+1/2}^l = g_{j+1/2}^r$, which are obtained by

$$\int g^{l,r}_{j+1/2}\psi d\Xi = \int_{u>0} f^l_0(x_{j+1/2},u,\xi)\psi d\Xi + \int_{u<0} f^r_0(x_{j+1/2},u,\xi)\psi d\Xi. \tag{2.6}$$

Therefore, at the cell interface the final distribution function can be fully determined using the integral solution

$$\begin{aligned} f^l(x_{j+1/2},t,u,\xi) &= \tfrac{1}{\tau}\int_0^t g^l(x',t',u,\xi)e^{-(t-t')/\tau}dt' \\ &+e^{-t/\tau}f^l_0(x_{j+1/2}-ut), \quad u\geq 0, \end{aligned} \tag{2.7}$$

$$\begin{aligned} f^r(x_{j+1/2},t,u,\xi) &= \tfrac{1}{\tau}\int_0^t g^r(x',t',u,\xi)e^{-(t-t')/\tau}dt' \\ &+e^{-t/\tau}f^r_0(x_{j+1/2}-ut), \quad u< 0, \end{aligned} \tag{2.8}$$

which can be used to evaluate the fluxes

$$F^{l,r}_{j+1/2}(t) = \int_{u>0} uf^l(x_{j+1/2},t,u,\xi)\psi d\Xi + \int_{u<0} uf^r(x_{j+1/2},t,u,\xi)\psi d\Xi, \tag{2.9}$$

and update the cell averaged conservative variables

$$W^{n+1}_j = W^n_j + \frac{1}{\Delta x}\int_{t_n}^{t_{n+1}}\left[F^r_{j-1/2}(t) - F^l_{j+1/2}(t)\right]dt. \tag{2.10}$$

In summary, without external forcing term, all particles can transport freely across the interface. Therefore, the equilibrium states and fluxes at the interface have unique values $g^l_{j+1/2} = g^r_{j+1/2}$ and $F^l_{j+1/2}(t) = F^r_{j+1/2}(t)$.

3 The symplectic preserving BGK scheme

If there is a gravitational field, the potential $\phi(x)$ is modeled as a piecewise constant function, i.e., in jth cell $\phi(x)=\phi_j$ with a potential jump at the cell interface $x_{j+1/2}$. The potential jump gives a critical speed $U_c = \sqrt{|\phi_j - \phi_{j+1}|}$, which provides a threshold for the particle movement. Because of the potential jump, not all the particles colliding with the cell interface will pass through it. Without losing generality, we only discuss the case for $\phi_j < \phi_{j+1}$. From the discussion of the previous subsection, we refer to the distribution function without external forcing field at the cell interface as original distribution function. It is

$$f(x_{j+1/2},t,u,\xi) = \begin{cases} f_j(x_{j+1/2},t,u,\xi), & u\geq 0, \\ f_{j+1}(x_{j+1/2},t,u,\xi), & u<0. \end{cases} \tag{3.1}$$

The particle collision with the potential jump will change the original distribution functions to $f^l_{j+1/2}(t,u,\xi)$ and $f^r_{j+1/2}(t,u,\xi)$ at the left and right hand sides of the cell interface respectively,

$$f^l_{j+1/2}(t,u,\xi) = \begin{cases} f_j(x_{j+1/2},t,u,\xi), & u > 0, \\ \tilde{f}_j(x_{j+1/2},t,u,\xi), & 0 \geq u > -U_c, \\ \overline{f}_{j+1}(x_{j+1/2},t,u,\xi), & u \leq -U_c, \end{cases} \tag{3.2}$$

and

$$f^r_{j+1/2}(t,u,\xi) = \begin{cases} \overline{f}_j(x_{j+1/2},t,u,\xi), & u \geq 0, \\ f_{j+1}(x_{j+1/2},t,u,\xi), & u < 0. \end{cases} \tag{3.3}$$

In (3.2), $\tilde{f}_j$ is the reflected particle in the jth cell, whose original distribution function is f_j with positive velocity being less than U_c. $\overline{f}_{j+1}$ is the particles in the jth cell coming from the distribution f_{j+1} of the $(j+1)$th cell with negative velocity. Here $\overline{f}_j$ is the particles in the $(j+1)$th cell coming from jth cell with original velocity being higher than U_c. So, the effect of the potential jump modifies the distribution function, but the moments of the modified and original distribution functions are related through the following physical principles.

With the original particle distribution function $f(u)$, after impacted by the potential jump, the particle velocity u is changed to u' and the distribution function becomes $\overline{f}(u')$. Since the energy conservation for a particle movement under a conservative potential field, the Hamiltonian function H of the particle keeps a constant

$$H(t) = \frac{1}{2}u(t)^2 + \phi(x(t)) = constant \rightarrow \frac{1}{2}u^2 + \phi = \frac{1}{2}(u')^2 + \phi', \tag{3.4}$$

from which the relation between u and u' can be obtained. Also, Liouville's theorem states that the probability density of particle in phase space keeps a constant along their movement trajectory,

$$\overline{f}(u') = f(u). \tag{3.5}$$

Most importantly, the symplectic preserving property of the Hamiltonian phase flow shows

$$\int\int_{D'} dx'du' = \int\int_{D} dxdu, \tag{3.6}$$

where D' and D are the phase volume on the trajectory of the Hamiltonian phase flow. Based on $D = (u_1,u_2) \times (ut_1,ut_2)$ and $D' = (u'_1,u'_2) \times (u't_1,u't_2)$, Eq. (3.6) goes to

$$\int_{u'_1}^{u'_2} u'du' = \int_{u_1}^{u_2} udu. \tag{3.7}$$

Then from (3.5), (3.7), and (3.4),

$$n\text{th-order } u \text{ moment} = \int_{u_1'}^{u_2'} \overline{f}(u')(u')^n du' = \int_{u_1(u_1')}^{u_2(u_2')} f(u)(u'(u))^{n-1} u du. \quad (3.8)$$

So, using $f_j = f_0^l$ and $f_{j+1} = f_0^r$, we can get the conservative variables at different sides of an interface

$$W_{j+1/2}^l = \int f_{j+1/2}^l \psi d\Xi, \quad \text{and} \quad W_{j+1/2}^r = \int f_{j+1/2}^r \psi d\Xi.$$

From which, two Maxwellians $g_{j+1/2}^l$ and $g_{j+1/2}^r$ in the equilibrium states (2.5) can be fully determined. Then, following the method used in the development of BGK-NS scheme [4], the final gas distribution at the left and right hand sides of a cell interface, i.e., f^l and f^r in (2.8), can be obtained. If choosing $f_j = f^l$ and $f_{j+1} = f^r$, we can get the corresponding fluxes at different sides of the cell interface,

$$F_{j+1/2}^l = \int u f_{j+1/2}^l \psi d\Xi, \quad \text{and} \quad F_{j+1/2}^r = \int u f_{j+1/2}^r \psi d\Xi.$$

Note that with the potential jump, in general we have $g_{j+1/2}^l \neq g_{j+1/2}^r$ and $F_{j+1/2}^l \neq F_{j+1/2}^r$. Finally, we can use (2.10) to update the cell average conservative variables.

4 Numerical examples

Perturbation of the 1D isothermal equilibrium solution: We consider an ideal gas with $\gamma = 1.4$ on an initial isothermal hydrostatic state, $\rho_0(x) = p_0(x) = e^{-x}$, and $u_0(x) = 0$, for $x \in [0, 1]$. Initially, the pressure is perturbed by $p(x, t = 0) = p_0(x) + \eta e^{\alpha(x-x_0)^2}$, where $\alpha = 100$, $x_0 = 0.5$, and η is the amplitude of the perturbation. The computation is conducted with 100 grid points in the whole domain and stops at time $t = 0.25$. The potential jump at each cell interface is $\Delta\phi = -G\Delta x = 0.01$. Figure 4.1 shows the result for the case $\eta = 0.001$.

Rayleigh-Taylor instability: Consider an isothermal equilibrium idea gas ($\gamma = 1.4$) in a 2D polar coordinate,

$$\rho_0(r) = e^{-\alpha(r+r_0)}, p_0(r) = \frac{1.5}{\alpha} e^{-\alpha(r+r_0)}, u_0 = 0,$$

where

$$\begin{cases} \alpha = 2.68, & r_0 = 0.258, \quad r \leq r_1, \\ \alpha = 5.53, & r_0 = -0.308, \quad r > r_1, \end{cases}$$

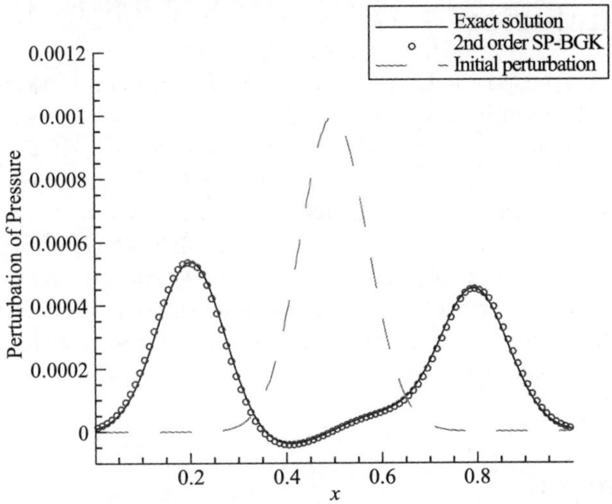

Figure 4.1 Perturbation of pressure of isothermal equilibrium solution for $\eta = 0.001$.

$$\begin{cases} r_1 = 0.6(1 + 0.02\cos(20\theta)), & \text{for density,} \\ r_1 = 0.62324965, & \text{for pressure.} \end{cases}$$

The potential satisfies $-\nabla\phi(r) = 1.5$. The results at times $t = 0, 0.8, 1.4$, and 2.0 are shown in Figure 4.2.

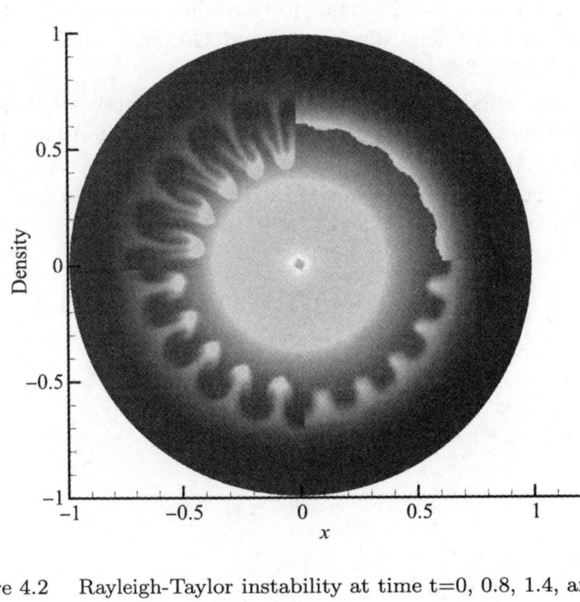

Figure 4.2 Rayleigh-Taylor instability at time t=0, 0.8, 1.4, and 2.0.

5 Conclusion

In this paper, based on Liouville's theorem and symplectic preserving property of a Hamiltonian flow, a kinetic scheme has been developed for the gas dynamic equations under external forcing field, where the potential is approximated as a step function at the cell interface. The physical mechanism of particle transport across the potential barrier has been explicitly used in the construction of gas distribution function. The method presented in this paper is a well-balanced and shock-capturing scheme for the Navier-Stokes equations under external forcing field. The design of the scheme is based on principles of physics, instead of directly using the well-balanced condition as a guideline in the construction of such a scheme.

References

[1] R. J. LeVeque and D. S. Bale, Wave propagation methods for conservation laws with source terms. In: Proc. 7th International Conference on Hyperbolic Problems, Zurich, February (1998).

[2] A. Slyz and K. H. Prendergast, Time-independent gravitational fields in the BGK scheme for hydrodynamics. Astron. Astrophys. Suppl. Ser. 139 (1999), 199–217.

[3] C. L. Tian, K. Xu, K. L. Chan and L. C. Deng, A three-dimensional multidimensional gas kinetic scheme for the Navier-Stokes equations under gravitational fields. J. Comput. Phys. 226 (2007), 2003–2027.

[4] K. Xu, A gas-kinetic BGK scheme for the Navier-Stokes equations, and its connection with artificial dissipation and Godunov method. J. Comput. Phys. 171 (2001), 289–335.

[5] K. Xu, J. Luo and S. Z. Chen, A well-balanced kinetic scheme for gas dynamic equations under gravitational field. To appear in Advance in Applied Mathematics and Mechanics, 2010.

[6] M. Zingale, et. al., Mapping initial hydrostatic models in Godunov codes. Astro. Phys. J. Supple 143 (2002), 539–565.

Shock Wave Admissibility and Shock-induced Phase Transitions in a van der Waals Fluid*

Andrea Mentrelli, Tommaso Ruggeri
Department of Mathematics & CIRAM, University of Bologna
Bologna 40123, Italy
E-mail: mentrelli@ciram.unibo.it, ruggeri@ciram.unibo.it

Masaru Sugiyama
Graduate School of Engineering, Nagoya Institute of Technology
Nagoya 466-8555, Japan
E-mail: sugiyama@nitech.ac.jp

Nanrong Zhao
College of Chemistry, Sichuan University
Chengdu 610064, China
E-mail: zhaonanr@scu.edu.cn

Abstract

Shock waves in a van der Waals fluid are studied in the framework of the theory and numerics of hyperbolic systems. The focus is on the influence of the unperturbed state of the shock, taken in the gas phase, on the admissibility of negative (rarefaction) shocks and shock-induced phase transitions.

1 Introduction

We analyze the propagation of shock waves in a van der Waals fluid with the aim of investigating the role played by the unperturbed state of the shock — taken in the gas phase — on the admissibility of negative (rarefaction) shocks [1, 2, 3] and on shock-induced phase transitions

*The authors are partly supported by: the National Natural Science Foundation of China (NSFC), No. 20973119 (N. Z.); GNFM/INdAM "Young Researchers Project 2009"(A. M.); Japan Society for the Promotion of Science (JSPS), No. PE07046 (A. M.) and No. 20560054 (M. S.); MIUR/PRIN Project "Nonlinear Propagation and Stability in Thermodynamical Processes of Continuous Media"(T. R.).

[4, 5]. We claim that it is possible to identify some sets of states, being each set made of unperturbed states that lead to shock waves with similar features from the point of view of the above-mentioned phenomena. In particular, there exist sets of unperturbed states such that: (i) all and only the positive shocks are admissible and no phase transition is allowed, just like in an ideal gas; (ii) all the negative (positive) shocks are inadmissible (admissible) and the positive (negative) ones may be admissible or inadmissible, depending on the strength of the shock; (iii) both positive and negative shocks may be admissible or inadmissible, depending on the strength of the shock. Moreover, for each of the last three scenarios, gas/liquid ($G \to L$) and gas/coexistence ($G \to G/L$) shock-induced phase transitions may occur.

The analysis presented here relies on the application of the well-established theory of hyperbolic systems as well as on numerical calculations.

2 The van der Waals model

Van der Waals fluids are described by the following equations of state:

$$e = c_v T - a\rho, \quad p = RT\rho/(1 - b\rho) - a\rho^2, \tag{2.1}$$

where e is the specific internal energy, c_v is the specific heat at fixed volume, T is the absolute temperature, ρ is the mass density, p is the pressure, R is the specific gas constant, a and b are material-dependent parameters. Assuming c_v to be constant, we may define the dimensionless quantity $\delta \equiv R/c_v$, such that $e = RT/\delta - a\rho$.

It is convenient to introduce the so-called *reduced variables* $\hat{\rho}$, $\hat{p}$ and $\hat{T}$ defined as follows: $\hat{\rho} = \rho/\rho_{cr}$, $\hat{p} = p/p_{cr}$, $\hat{T} = T/T_{cr}$, where $\rho_{cr} = 1/3b$, $p_{cr} = a/(27b^2)$, $T_{cr} = 8a/(27Rb)$ are the *critical values* of density, pressure and temperature, respectively. In terms of reduced variables, the caloric and thermal equations of state take the form ($\hat{e} = (\rho_{cr}/p_{cr})e$):

$$\hat{e} = 8\hat{T}/(3\delta) - 3\hat{\rho}, \quad \hat{p} = 8\hat{T}\hat{\rho}/(3 - \hat{\rho}) - 3\hat{\rho}^2.$$

In order to model phase transitions, the equations of state written above must be suitably modified, and the so-called *coexistence region* (where a mixture of gas and liquid phases exists), bounded by the coexistence curve $\mathcal{C}$, is thus introduced. It suffices here to mention that for any temperature $\hat{T} < 1$ there are a *vaporization point* and a *liquefaction point*, with densities $\hat{\rho}_G \equiv \hat{\rho}_G(\hat{T})$ and $\hat{\rho}_L \equiv \hat{\rho}_L(\hat{T})$, respectively, belonging to the same isotherm, between which the fluid may undergo a gas/liquid phase transition at constant pressure, $\hat{p}_{coe} \equiv \hat{p}_{coe}(\hat{T})$. The equations of state (2.1) are thus replaced, in the coexistence region, by:

$$p_{coe} = R\frac{T\rho_G}{1-\rho_G} - a\rho_G^2, \quad e_{coe} = \frac{R}{\delta}T - a\left(\rho_G + \rho_L - \frac{\rho_G\rho_L}{\rho}\right). \tag{2.2}$$

In the following, we will conveniently assume that the gas (G) and the liquid (L) regions are bounded, in the $\hat{\rho}\hat{p}$ plane, by the coexistence curve and by the isotherm $\hat{T} = 1$, respectively below and above the critical point. When not in gas or liquid regions, we assume the fluid to be in a coexistence state (G/L), thus neglecting metastable states.

3 Rankine-Hugoniot conditions and shock admissibility in a van der Waals fluid

The Euler equations, i.e., the conservation laws of mass, momentum and energy for a perfect fluid read, in their one-dimensional form, as

$$\partial_t \mathbf{u} + \partial_x \mathbf{F}(\mathbf{u}) = 0, \quad \mathbf{u} \equiv (\rho, \rho v, E)^T, \quad \mathbf{F} \equiv \left(\rho v, \rho v^2 + p, (E+p)\, v\right)^T,$$

where v is the velocity and E is the total energy ($E = \rho e + \rho v^2/2$).

Denoting as *unperturbed state*, $\mathbf{u}_0$, the state before the shock and as *perturbed state*, $\mathbf{u}_1$, the state after the shock, it is well-known that a shock wave must satisfy the Rankine-Hugoniot (RH) conditions [6]:

$$-s[\![\mathbf{u}]\!] + [\![\mathbf{F}]\!] = 0, \tag{3.1}$$

where $[\![\varphi(\mathbf{u})]\!] = \varphi_1 - \varphi_0$ represents the discontinuity of a quantity φ across the shock front, being $\varphi_1 \equiv \varphi(\mathbf{u}_1)$ and $\varphi_0 \equiv \varphi(\mathbf{u}_0)$ the quantity φ evaluated, respectively, in the perturbed and unperturbed states.

The RH conditions admit a one-parameter family of solutions: letting ξ be the parameter, the state $\mathbf{u}_1$ that can be connected to a given state $\mathbf{u}_0$ through a shock, and the velocity s of the shock, may be written as $\mathbf{u}_1 \equiv \mathbf{u}_1(\mathbf{u}_0, \xi)$ and $s \equiv s(\mathbf{u}_0, \xi)$. Given $\mathbf{u}_0$, the set of the states $\mathbf{u}_1$ satisfying Eq. (3.1) is called the *Hugoniot locus* for the point $\mathbf{u}_0$ and is denoted as $\mathcal{H}(\mathbf{u}_0)$. In the literature, the shock velocity, s, is often replaced by the Mach number of the unperturbed state, $M_0 = (s - v_0)/c_0$, where c_0 is the sound velocity in the unperturbed state [7].

Without loss of generality, due to the Galilean invariance, we assume $v_0 = 0$ and we consider only the shock wave propagating in the positive x-direction. Moreover, we focus on the case in which the unperturbed state is in the gas phase ($\mathbf{u}_0 \in G$). In order to explicitly write down the solutions of the RH conditions, we distinguish two cases:

(i) *The perturbed state,* $\mathbf{u}_1$, *is in the gas or liquid phase ($G \rightarrow G$ and $G \rightarrow L$ solutions of the Rankine-Hugoniot conditions).*

In this case, the equations of state to be considered are those given in Eq. (2.1). Given an unperturbed state, $\mathbf{u}_0 \equiv (\rho_0, 0, \rho_0 e_0)^T$, in the gas region, and taking as parameter the density of the perturbed state $\mathbf{u}_1 \equiv \left(\rho_1, \rho_1 v_1, \rho_1 e_1 + \rho_1 v_1^2/2\right)^T$, i.e., letting $\xi \equiv \rho_1$, the solution of the RH conditions, in terms of dimensionless quantities, is the following:

$$\hat{v}_1 = \hat{c}_0 M_0 (\hat{\rho}_1 - \hat{\rho}_0)/\hat{\rho}_1, \quad \hat{p}_1 = \hat{p}_0 + \hat{c}_0^2 M_0^2 \hat{\rho}_0 \left(\hat{\rho}_1 - \hat{\rho}_0\right)/\hat{\rho}_1, \tag{3.2}$$

where the unperturbed Mach number, M_0, is given by

$$M_0 = \frac{1}{\hat{c}_0}\sqrt{6\hat{\rho}_1 \frac{\hat{p}_0\,(1+\delta) + \hat{\rho}_0 \hat{\rho}_1\,(\hat{\rho}_1 + \hat{\rho}_0 + 3\delta - 3)}{\hat{\rho}_0\,(2\hat{\rho}_0\,(3-\hat{\rho}_1) + 3\delta\,(\hat{\rho}_0 - \hat{\rho}_1))}}.$$

(ii) *the perturbed state, $\mathbf{u}_1$, is in the gas/liquid coexistence phase ($G \to G/L$ solution of the Rankine-Hugoniot conditions).*

In this case, the equations of state (2.2) must be considered for the perturbed state $\mathbf{u}_1$. The perturbed fluid velocity and pressure are still given by Eq. (3.2), but the Mach number must be numerically calculated once ρ_G and ρ_L, appearing in Eq. (2.2), are known.

It is well-known that not every solution of the RH conditions corresponds to a physically meaningful shock wave. Thus, we need a criterion to select which of the states $\mathbf{u}_1 \in \mathcal{H}(\mathbf{u}_0)$ are perturbed states that, together with $\mathbf{u}_0$ form *admissible* shocks. When locally linearly degenerate waves, i.e., waves such that[①] $\nabla\lambda \cdot \mathbf{r} = 0$ for some $\mathbf{u}$, are involved — as in the van der Waals case — the selection rule is given by the *Liu condition* [8], stating that a shock wave is admissible if its velocity is not smaller than the velocity of any other shock with the same unperturbed state $\mathbf{u}_0$ and with perturbed state $\mathbf{u}_*$ lying on $\mathcal{H}(\mathbf{u}_0)$ between $\mathbf{u}_0$ and $\mathbf{u}_1$. It is proven [9] that in this case both positive and negative (i.e., compressive and rarefaction) shocks may be admissible, depending on the unperturbed state, and the locus of states $\mathbf{u}$ such that $\nabla\lambda \cdot \mathbf{r} = 0$ is crucial in the study of the admissibility. For a van der Waals fluid, it is found that this locus (*local exceptionality curve*, $\mathcal{L}_\delta$) is given by:

$$\hat{p} = \hat{\rho}^2 \left(2\,(3-\hat{\rho})^2 / \left(2 + 3\delta + \delta^2\right) - 3\right).$$

We note that the curve $\mathcal{L}_\delta$ is meaningful when $0 < \delta < \delta_C \simeq 0.06$ since, as $\delta \to \delta_C$, the region enclosed between the $\mathcal{L}_\delta$ and $\mathcal{C}$ curves vanishes.

The investigation of weak shock admissibility in a van der Waals fluid leads to results may still be summarized as follows: (i) The locus of the states such that $\nabla\lambda \cdot \mathbf{r} = 0$ separates the regions where all the positive and all the negative weak shocks are admissible; (ii) Both positive

①Here, λ is the kth characteristic velocity and $\mathbf{r}$ the corresponding eigenvector, i.e., $\lambda \equiv \lambda^{(k)}$, $\mathbf{r} \equiv \mathbf{r}^{(k)}$, and $\nabla \equiv \partial/\partial\mathbf{u}$.

and negative weak shocks are admissible (inadmissible) when the unperturbed state is taken on the $\mathcal{L}_\delta$ curve and $\lambda \equiv \lambda(\xi)$ has a maximum (minimum) on this curve.

For non-weak shocks a satisfactory classification requires a closer look at the features of the unperturbed state (see [11]); moreover, in the case of a van der Waals fluid, in contrast to the case of an ideal gas, the possibility of shock-induced phase transitions is to be considered.

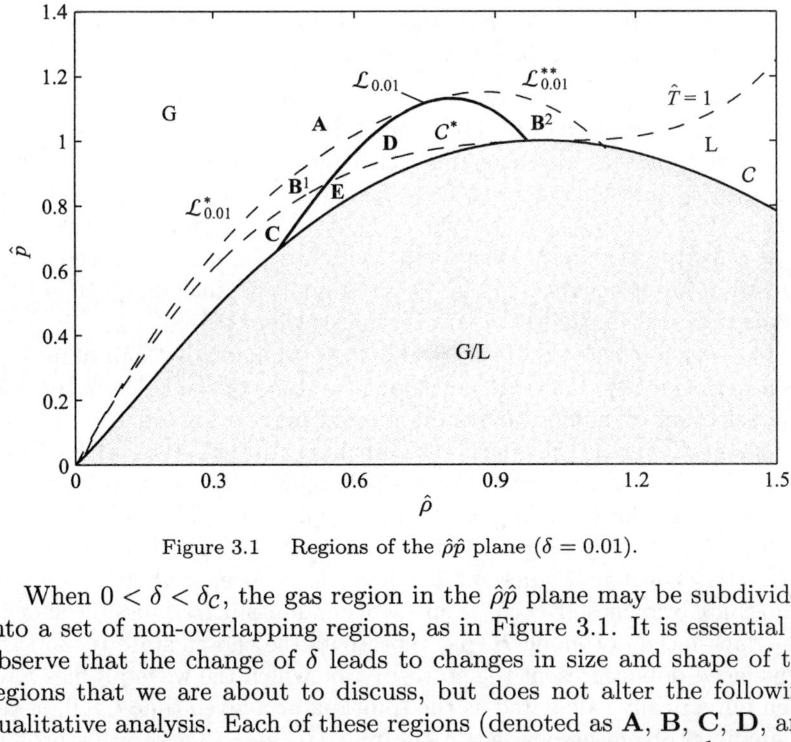

Figure 3.1 Regions of the $\hat{\rho}\hat{p}$ plane ($\delta = 0.01$).

When $0 < \delta < \delta_C$, the gas region in the $\hat{\rho}\hat{p}$ plane may be subdivided into a set of non-overlapping regions, as in Figure 3.1. It is essential to observe that the change of δ leads to changes in size and shape of the regions that we are about to discuss, but does not alter the following qualitative analysis. Each of these regions (denoted as $\mathbf{A}$, $\mathbf{B}$, $\mathbf{C}$, $\mathbf{D}$, and $\mathbf{E}$, assuming the region $\mathbf{B}$ as split up into the two subregions $\mathbf{B}^1$ and $\mathbf{B}^2$) contains states that, if taken as unperturbed states, lead to Hugoniot loci with similar features from the point of view of the shock admissibility:

– *region* $\mathbf{A} \cup \mathbf{B} \cup \mathbf{C}$. This is the region where $(\nabla\lambda \cdot \mathbf{r})_0 > 0$. When $\hat{\mathbf{u}}_0$ belongs to this region, positive (negative) weak shocks are admissible (inadmissible). A deeper analysis shows that when $\hat{\mathbf{u}}_0 \in \mathbf{A} \cup \mathbf{B}$, $\mathcal{H}(\hat{\mathbf{u}}_0)$ does not go through the coexistence region nor the liquid region, so no shock-induced phase transition is possible. More in detail, when $\hat{\mathbf{u}}_0 \in \mathbf{A}$, all positive (negative) shocks are admissible (inadmissible). In this case, the shock admissibility is *qualitatively* the same as in the ideal gas case. When $\hat{\mathbf{u}}_0 \in \mathbf{B}$, $\mathcal{H}(\hat{\mathbf{u}}_0)$ crosses the local exceptionality curve. As a result, not all the positive shocks are admissible (but all the negative ones are inadmissible) for $\hat{\mathbf{u}}_0 \in \mathbf{B}^1$, and not all the negative shocks are admissible (but all the positive ones are admissible) for $\hat{\mathbf{u}}_0 \in \mathbf{B}^2$. Finally, when $\hat{\mathbf{u}}_0 \in \mathbf{C}$, $\mathcal{H}(\hat{\mathbf{u}}_0)$ crosses both the coexistence and local exceptionality curves, so phase transitions may be allowed and inadmissible positive shocks are encountered.

– *region* $\mathbf{D} \cup \mathbf{E}$. This is the region where $(\nabla\lambda \cdot \mathbf{r})_0 < 0$. When $\hat{\mathbf{u}}_0$ belongs to this region, negative (positive) weak shocks are admissible (inadmissible). More in detail, when $\hat{\mathbf{u}}_0 \in \mathbf{D}$, $\mathcal{H}(\hat{\mathbf{u}}_0)$ does not cross the coexistence region nor the liquid region, so no phase transition is allowed; when $\hat{\mathbf{u}}_0 \in \mathbf{E}$, $\mathcal{H}(\hat{\mathbf{u}}_0)$ goes through the coexistence and liquid region, so phase transitions may be allowed.

When $\delta \geqslant \delta_{\mathcal{C}} \simeq 0.06$, the local exceptionality curve, $\mathcal{L}_\delta$, is not relevant, as already discussed. The regions $\mathbf{B}$, $\mathbf{D}$, and $\mathbf{E}$ vanish as $\delta \to \delta_{\mathcal{C}}$ and it is seen that this case may still be described as the case with $0 < \delta < \delta_{\mathcal{C}}$ just letting $\mathbf{B} \equiv \mathbf{D} \equiv \mathbf{E} \equiv \emptyset$ for $\delta_{\mathcal{C}} \leq \delta \leq \delta_{\mathcal{P}} \simeq 0.16$ or letting $\mathbf{B} \equiv \mathbf{C} \equiv \mathbf{D} \equiv \mathbf{E} \equiv \emptyset$ for $\delta_{\mathcal{P}} \leq \delta \leq 2/3$. That is, in the latter case, only the region $\mathbf{A}$ exists and therefore the shock-induced phase transition never occurs. A more detailed analysis, including also all the proofs that are omitted here, are given elsewhere [10, 11].

In order to check the theoretical results, numerical calculations have been performed with a code developed by some of the authors [12, 13]. The selection of numerical results offered here is limited to the cases of regions $\mathbf{C}$ and $\mathbf{E}$. For each case, an unperturbed state $\hat{\mathbf{u}}_0$ has been selected and several perturbed states belonging to $\mathcal{H}(\hat{\mathbf{u}}_0)$ have been chosen, with the aim of picking out some peculiarities of the shock admissibility. Figure 3.2 concerns the case of $\hat{\mathbf{u}}_0 \in \mathbf{C}$. The Hugoniot locus $\mathcal{H}(\hat{\mathbf{u}}_0)$ is shown in Figure 3.2(a); here, as well as in Figure 3.3, the thin/thick branches of $\mathcal{H}(\hat{\mathbf{u}}_0)$ represent inadmissible/admissible shocks. The dark-faced dot along $\mathcal{H}(\hat{\mathbf{u}}_0)$ represents the chosen state $\hat{\mathbf{u}}_0$ and the light-faced dots represent the states $\hat{\mathbf{u}}_1$ for which the wave profiles have been numerically calculated as the solution, at a given time $t > 0$, of the Riemann problem associated to the pair $(\hat{\mathbf{u}}_0, \hat{\mathbf{u}}_1)$. These wave profiles are shown in Figure 3.2(c–h). The behavior of the unperturbed Mach number, M_0, and of the dimensionless characteristic velocity, $\lambda^* = \lambda/c_0$, as functions of the parameter $\xi \equiv \hat{\rho}$ are shown in Figure 3.2(b)②. As it is seen, the Hugoniot locus crosses the coexistence region; in this case both $G \to L$ and $G \to G/L$ transitions may be allowed. An example of a $G \to L$ transition is given in Figure 3.2(h), while a stable shock with no phase transition is shown in Figure 3.2(d). The wave profile shown in Figure 3.2(c) represents a wave profile associated to an inadmissible shock, while in Figure 3.2(e–g) *shock-splitting* phenomena associated to inadmissible compressive shocks are shown. It is worth mentioning that

②Since $v_0 = 0$, the unperturbed Mach number is proportional to the shock velocity: $M_0 = s/c_0$.

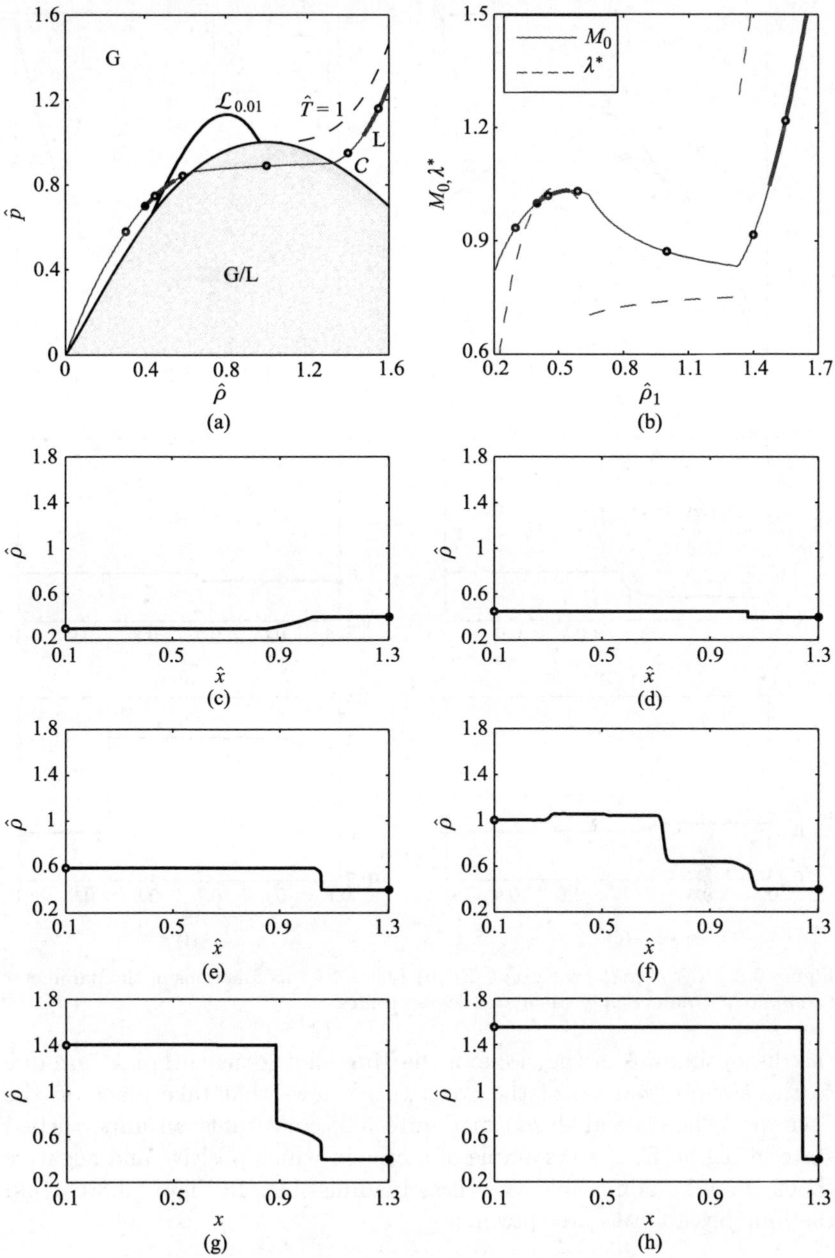

Figure 3.2 (a) $\mathcal{H}(\hat{\mathbf{u}}_0)$, with $\hat{\mathbf{u}}_0 \in \mathbf{C}$; (b) M_0 and λ^* as functions of the parameter $\xi \equiv \hat{\rho}_1$; (c–h): numerically calculated wave profiles.

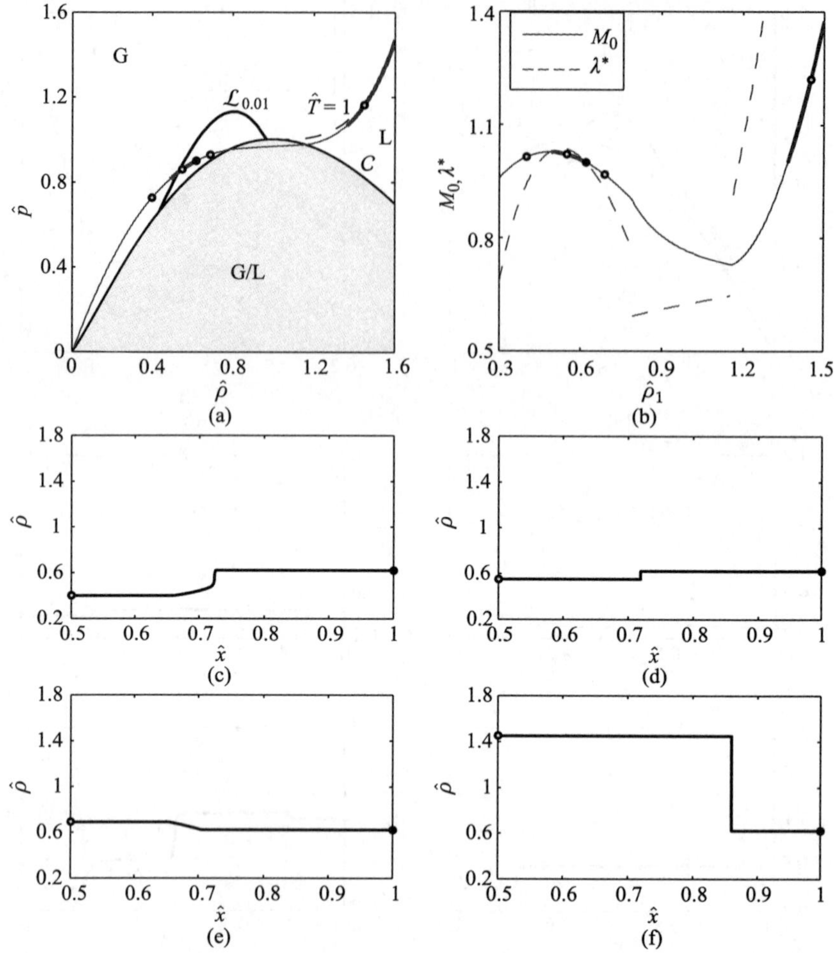

Figure 3.3 (a) $\mathcal{H}(\hat{\mathbf{u}}_0)$, with $\mathbf{u}_0 \in \mathbf{E}$; (b) M_0 and λ^* as functions of the parameter $\xi \equiv \hat{\rho}_1$; (c–f): numerically calculated wave profiles.

the discontinuities of the slope of the Hugoniot locus and of λ^* are due to the abrupt changes of the constitutive laws that take place on the $\mathcal{C}$ curve. The case analyzed in Figure 3.3, concerning an unperturbed state in region **E**, is an example of a case in which positive and negative shocks may be admissible as well as inadmissible. In Figure 3.3(c–f) all the four possibilities are shown.

4 Conclusions

We have analyzed, by means of the well-established theory of hyperbolic systems and by numerical calculations, the role played by the unperturbed state of a shock on the admissibility of negative shocks and on shock induced phase transitions in a van der Waals fluid. The gas region has been subdivided into several regions, each of which contains states that, when taken as unperturbed states, lead to shock waves with similar features as far as the above-mentioned phenomena are concerned.

References

[1] P. A. Thompson and K. C. Lambrakis, Negative shock waves. J. Fluid Mech. 60 (1973), 187–208.

[2] A. A. Borisov, A. L. Borisov, S. S. Kutateladze and V. E. Nakoyakov, Rarefaction shock wave near the thermodynamic critical point. J. Fluid Mech. 126 (1983), 59–73.

[3] C. Zamfirescu, A. Guardone and P. Colonna, Admissibility region for rarefaction shocks in dense gases. J. Fluid Mech. 599 (2008), 363–381.

[4] M. Slemrod, Admissibility criteria for propagating phase boundaries in van der Waals fluid. Arch. Rational Mech. Anal. 81 (1983), 301–315.

[5] G. E. A. Meier, Liquefaction shock waves. In: Shock Wave Science and Technology 1 (2007), Springer Verlag, 231–267.

[6] P. D. Lax, Hyperbolic Systems of Conservation Laws and the Mathematical Theory of Shock Waves, CBMS-NSF, Regional Conference Series in Applied Mathematics 11 (1973).

[7] S. Taniguchi, A. Mentrelli, T. Ruggeri, M. Sugiyama and N. Zhao, Prediction and simulation of compressive shocks with lower perturbed density for increasing shock strength in real gases. Phys. Rev. E 82 (2010), 036324.

[8] T. P. Liu, The entropy condition and the admissibility of shocks. J. Math. Anal. Appl. 53 (1976), 78–88.

[9] T. Ruggeri, A. Muracchini and L. Seccia, Continuum approach to phonon gas ans shape changes of second sound via shock waves theory. Il Nuovo Cimento 16 (1994), 15–44.

[10] S. Taniguchi, A. Mentrelli, N. Zhao, T. Ruggeri and M. Sugiyama, Shock-induced phase transition in systems of hard spheres with internal degrees of freedom. Phys. Rev. E 81 (2010), 066307.

[11] N. Zhao, A. Mentrelli, T. Ruggeri and M. Sugiyama, Admissible shock waves and shock-induced phase transitions in a van der Waals fluid. In preparation.

[12] A. Mentrelli and T. Ruggeri, Asymptotic behavior of riemann and riemann with structure problems for a 2×2 hyperbolic dissipative system. Suppl. Rend. Circ. Mat. Palermo II/78 (2006), 201–226.

[13] A. Mentrelli, T. Ruggeri, M. Sugiyama and N. Zhao, Interaction between a shock and an acceleration wave in a perfect gas for increasing shock strength. Wave Motion 45 (2008), 498–517.

L^1 Stability for Scalar Balance Laws; Application to Pedestrian Traffic

Magali Mercier
Université de Lyon, Université Lyon 1, Institut Camille Jordan
43 blvd. du 11 Novembre 1918, 69622 Villeurbanne Cedex, France
E-mail: mercier@math.univ-lyon1.fr

Abstract

We present here a stability result for the solutions of scalar balance laws. The estimates we obtained are then used to study the continuity equation with a non-local flow, which appears for example in a new model of pedestrian traffic.

1 Introduction

We consider the Cauchy problem for scalar balance laws of the form $\partial_t u + \mathrm{Div} f(t,x,u) = F(t,x,u)$, which often appear in physics. Thanks to Kružkov's theorem [8, Theorem 1 & 5] we know that this kind of equation admits a unique weak entropy solution and we can describe the dependence on the initial condition of the solution.

In the first part, we describe the dependence of the solution with respect to the flow f and the source F. Some cases were already studied: for example Lucier [9] or Bouchut & Perthame [2] have considered the case in which the flow depends only on u and in which there is no source. We treat here the general case, which includes the preceding results. These results come from a collaboration with R. Colombo and M. Rosini and are more precisely described in [5].

The second part is devoted to the study of the continuity equation with a non-local flow. Using estimates of the first part, we show not only that this model admits a unique weak entropy solution, but also that the linearized equation admits a weak entropy solution. Furthermore, the non-linear local semi-group obtained by solving the initial value problem is Gâteaux-differentiable with respect to the initial condition and the Gâteaux-derivative is the solution of the linearized equation. This fact allows us to characterize the minima or maxima of a given cost functional depending on the initial condition. This is of interest in pedestrian traffic if for example we want to minimize the time of exit out of a room, avoiding high density in the crowd. These results come from a collaboration with R. Colombo and M. Herty and are presented in [4].

2 $\mathbf{L}^1$ stability for scalar balance laws

Below, for a vector valued function $f = f(x,u)$ with $u = u(x)$, $\mathrm{Div} f$ stands for the total divergence whereas $\mathrm{div}\, f$ and ∇f denote the partial divergence and gradient with respect to the x variable. Moreover, ∂_u and ∂_t are the usual partial derivatives with respect to the variables u and t. Hence, if $u \in \mathscr{C}^1(\mathbb{R}^N, \mathbb{R})$, we have $\mathrm{Div} f = \mathrm{div}\, f + \partial_u f \cdot \nabla u$.

We study here Cauchy problems for scalar balance laws:

$$\begin{cases} \partial_t u + \mathrm{Div} f(t,x,u) = F(t,x,u) & (t,x) \in \mathbb{R}_+^* \times \mathbb{R}^N, \\ u(0,x) = u_0(x) \in \mathbf{L}^1 \cap \mathbf{L}^\infty & x \in \mathbb{R}^N, \end{cases} \tag{2.1}$$

where $f \in \mathscr{C}^2([0,T] \times \mathbb{R}^N \times \mathbb{R}; \mathbb{R}^N)$ is the flow and $F \in \mathscr{C}^1([0,T] \times \mathbb{R}^N \times \mathbb{R}; \mathbb{R})$ is the source. The properties of this kind of equation have already been intensively investigated, see for example [7, 8]. Here we want to describe the dependence of the solutions with respect to flow and source.

2.1 Previous results

Let us first recall the Kružkov Theorem [8, Theorem 5]:

Theorem 2.1 (Kružkov). *Let $T > 0$. For any $A \geqslant 0$, we denote $\Omega_A = [0,T] \times \mathbb{R}^N \times [-A, A]$. Under the conditions $f \in \mathscr{C}^2$, $F \in \mathscr{C}^1$ and:*

$$(\mathbf{K}) \forall A > 0 : \partial_u f, \quad \partial_u (F - \mathrm{div}\, f), \quad F - \mathrm{div}\, f \text{ are bounded on } \Omega_A$$

there exists a unique weak entropy solution $u \in \mathbf{L}^\infty([0,T]; \mathbf{L}^1(\mathbb{R}^N; \mathbb{R}))$ of (2.1) right-continuous in time.

Let $v_0 \in (\mathbf{L}^1 \cap \mathbf{L}^\infty)(\mathbb{R}^N; \mathbb{R})$. Let u be the solution associated to the initial condition u_0 and v be the solution associated to the initial condition v_0. Let M be such that $M \geqslant \sup(\|u\|_{\mathbf{L}^\infty([0,T] \times \mathbb{R}^N; \mathbb{R})}, \|v\|_{\mathbf{L}^\infty([0,T] \times \mathbb{R}^N; \mathbb{R})})$. Then, for all $t \in [0,T]$, with $\gamma = \|\partial_u F\|_{\mathbf{L}^\infty(\Omega_M)}$, we have

$$\|(u-v)(t)\|_{\mathbf{L}^1} \leqslant e^{\gamma t} \|u_0 - v_0\|_{\mathbf{L}^1}.$$

We also know some other results concerning the dependence of the solution with respect to flow and source. The following was first proved by Lucier [9], and later improved by Bouchut & Perthame [2]. Their results are about the homogeneous conservation laws: the flow depends only on u and there is no source. More precisely, if $f, g : \mathbb{R} \to \mathbb{R}^N$

are globally lipschitz, then for all $u_0, v_0 \in \mathbf{L}^1 \cap \mathbf{L}^1 ff8\infty(\mathbb{R}^N;\mathbb{R})$ initial conditions for

$$\partial_t u + \mathrm{Div} f(u) = 0, \quad \partial_t v + \mathrm{Div} g(v) = 0,$$

with furthermore $v_0 \in \mathbf{BV}(\mathbb{R}^N;\mathbb{R})$, we have for all $t \geqslant 0$,

$$\|(u-v)(t)\|_{\mathbf{L}^1} \leqslant \|u_0 - v_0\|_{\mathbf{L}^1} + C\,t\ \mathrm{TV}\,(v_0)\ \mathbf{Lip}\,(f-g).$$

A flow depending also on x was considered by Chen & Karlsen [3], in the special case $f(x,u) = \lambda(x) l(u)$. There, under appropriate hypotheses, with $f(t,x,u) = \lambda(x)\,l(u)$, $g(t,x,v) = \mu(x)\,m(v)$, and without source ($F = G = 0$), they obtained the estimate

$$\begin{aligned}\|(u-v)(t)\|_{\mathbf{L}^1} \leqslant \|u_0 - v_0\|_{\mathbf{L}^1} &+ C_1\,t\,(\|\lambda-\mu\|_{\mathbf{L}^\infty} + \|\lambda-\mu\|_{\mathbf{W}^{1,1}} \\ &+ \|l-m\|_{\mathbf{L}^\infty} + \|l-m\|_{\mathbf{W}^{1,\infty}})\end{aligned}$$

where $C_1 = C\sup_{[0,T]}(\mathrm{TV}\,(u(t)), \mathrm{TV}\,(v(t)))$. However, this general settings contains the Cauchy problem: $\partial_t u + \partial_x(\cos x) = 0$ with $u_0 = 0$, is $u(t,x) = t\sin x$ for which $\mathrm{TV}\,(u_0) = 0$ and $\mathrm{TV}\,(u(t)) = +\infty$ for any $t > 0$. Hence, the coefficient C_1 is also $+\infty$. This fact motivated us for searching first an estimate on the total variation in the case the flow and source depend on the three variables t, x, and u.

2.2 Estimate on the total variation

Let us recall here the definition of total variation.

Definition 2.2. For $u \in \mathbf{L}^1_{loc}(\mathbb{R}^N;\mathbb{R})$ we denote the total variation of u:

$$\mathrm{TV}\,(u) = \sup\left\{\int_{\mathbb{R}^N} u \,\mathrm{div}\,\Psi; \quad \Psi \in \mathscr{C}^1_c(\mathbb{R}^N;\mathbb{R}^N), \quad \|\Psi\|_{\mathbf{L}^\infty} \leqslant 1\right\}.$$

Then we introduce the space of function with bounded variation $\mathbf{BV}(\mathbb{R}^N;\mathbb{R}) = \{u \in \mathbf{L}^1_{loc};\mathrm{TV}\,(u) < \infty\}$.

When f and F depend only on u we already know that $u_0 \in \mathbf{L}^\infty \cap \mathbf{BV}$ implies that for all $t \geqslant 0$, $u(t) \in \mathbf{L}^\infty \cap \mathbf{BV}$, with the same notation as in Theorem 2.1, we have $\mathrm{TV}\,(u(t)) \leqslant \mathrm{TV}\,(u_0)e^{\gamma t}$, where $\gamma = \|\partial_u F\|_{\mathbf{L}^\infty(\Omega_M)}$.

Now we give a more general estimate on the total variation. Let $W_N = \int_0^{\pi/2}(\cos\theta)^N \mathrm{d}\theta$, $\Omega = \Omega_\infty$ and

$$(\mathbf{H1}): \begin{cases} f \in \mathscr{C}^2(\Omega;\mathbb{R}^N), & F \in \mathscr{C}^1(\Omega;\mathbb{R}), \\ \nabla\partial_u f \in \mathbf{L}^\infty(\Omega;\mathbb{R}^{N\times N}), & \partial_t \,\mathrm{div}\, f \in \mathbf{L}^\infty(\Omega;\mathbb{R}), \\ \partial_t\partial_u f \in \mathbf{L}^\infty(\Omega;\mathbb{R}^N), & \partial_t F \in \mathbf{L}^\infty(\Omega;\mathbb{R}). \\ \displaystyle\int_{\mathbb{R}_+^*}\int_{\mathbb{R}^N} \|\nabla(F - \mathrm{div}\, f)(t,x,\cdot)\|_{\mathbf{L}^\infty(\mathbb{R};\mathbb{R}^N)}\,dxdt < +\infty \end{cases}$$

Theorem 2.3 (see Theorem 2.5 in [5]). *Assume* (f,F) *satisfies* **(K)** *and* **(H1)**. *Let* $u_0 \in (\mathbf{L}^\infty \cap \mathbf{BV})(\mathbb{R}^N;\mathbb{R})$, $M = \|u\|_{\mathbf{L}^\infty([0,T]\times\mathbb{R}^N)}$, $\kappa_0 = (2N+1)\|\nabla\partial_u f\|_{\mathbf{L}^\infty(\Omega_M)} + \|\partial_u F\|_{\mathbf{L}^\infty(\Omega_M)}$. *Then for all* $t \in [0,T]$, $u(t) \in (\mathbf{L}^\infty \cap \mathbf{BV})(\mathbb{R}^N;\mathbb{R})$ *and*

$$\begin{aligned}&\mathrm{TV}\,(u(t)) \leqslant \mathrm{TV}\,(u_0)e^{\kappa_0 t}\\ &+ NW_N \int_0^t e^{\kappa_0(t-\tau)} \int_{\mathbb{R}^N} \|\nabla(F - \operatorname{div} f)(\tau,x,\cdot)\|_{\mathbf{L}^\infty([-M,M])}\,dx d\tau.\end{aligned}$$

Remark 2.4. In some cases, we recover known estimates. When f depends only on u and $F = 0$, we have a result similar to the one that was already known: $\mathrm{TV}\,(u(t)) \leqslant \mathrm{TV}\,(u_0)$.

When f, F do not depend on u, we have in fact an ODE $\partial_t u = (F - \operatorname{div} f)(t,x)$. The solution is consequently $u(t,x) = u_0(x) + \int_0^t (F - \operatorname{div} f)(\tau,x)d\tau$. Meanwhile, the bound above reduces to $\mathrm{TV}\,(u(t)) \leqslant \mathrm{TV}\,(u_0) + NW_N \int_0^t \int_{\mathbb{R}^N} |(F - \operatorname{div} f)(\tau,x)|\,d\tau$ which is essentially what we expected.

Remark 2.5. The set of hypotheses **(H1)** is in fact very strong. We expect it can be relaxed to

$$\textbf{(H1*)}: \begin{cases} f \in \mathscr{C}^0(\Omega;\mathbb{R}^N), \quad F \in \mathscr{C}^0(\Omega;\mathbb{R}), \text{ and } \forall A > 0,\\ \nabla\partial_u f \in \mathbf{L}^\infty(\Omega_A;\mathbb{R}^{N\times N}), \quad \partial_u F \in \mathbf{L}^\infty(\Omega_A;\mathbb{R}),\\ \displaystyle\int_0^T \int_{\mathbb{R}^N} \|\nabla(F - \operatorname{div} f)(t,x,\cdot)\|_{\mathbf{L}^\infty([-A,A];\mathbb{R}^N)} < \infty, \end{cases} \tag{2.2}$$

which is useful for example in [4]. This is a work in progress [11].

2.3 L^1 stability of the solution

Now, we can study the dependence of the solution with respect to flow and source. Let us introduce the set of hypotheses:

$$\textbf{(H2)}: \begin{cases} f \in \mathscr{C}^1(\Omega;\mathbb{R}^N), \qquad F \in \mathscr{C}^0(\Omega;\mathbb{R}),\\ \partial_u F \in \mathbf{L}^\infty(\Omega;\mathbb{R}), \quad \partial_u f \in \mathbf{L}^\infty(\Omega;\mathbb{R}^N),\\ \displaystyle\int_{\mathbb{R}_+^*} \int_{\mathbb{R}^N} \|(F - \operatorname{div} f)(t,x,\cdot)\|_{\mathbf{L}^\infty(\mathbb{R};\mathbb{R})}\,dx dt < +\infty. \end{cases}$$

Theorem 2.6 (see Theorem 2.6 in [5]). *Assume* (f,F) *and* (g,G) *satisfy* **(K)**, (f,F) *satisfies* **(H1)** *and* $(f-g, F-G)$ *satisfies* **(H2)**. *Let* $u_0, v_0 \in (\mathbf{L}^1 \cap \mathbf{L}^\infty \cap \mathbf{BV})(\mathbb{R}^N;\mathbb{R})$. *We denote*

$$\kappa = 2N\|\nabla\partial_u f\|_{\mathbf{L}^\infty(\Omega_M)} + \|\partial_u F\|_{\mathbf{L}^\infty(\Omega_M)} + \|\partial_u(F-G)\|_{\mathbf{L}^\infty(\Omega_M)}.$$

Let u and v be the solutions associated to (f, F) and (g, G) respectively and with initial conditions u_0 and v_0. Then for all $t \in [0, T]$:

$$\begin{aligned}
\|(u-v)(t)\|_{\mathbf{L}^1} &\leqslant e^{\kappa t}\|u_0 - v_0\|_{\mathbf{L}^1} + \frac{e^{\kappa_0 t} - e^{\kappa t}}{\kappa_0 - \kappa} \mathrm{TV}(u_0) \|\partial_u (f-g)\|_{\mathbf{L}^\infty} \\
&+ \int_0^t \frac{e^{\kappa_0 (t-\tau)} - e^{\kappa(t-\tau)}}{\kappa_0 - \kappa} \int_{\mathbb{R}^N} \|\nabla(F - \operatorname{div} f)(\tau, x, \cdot)\|_{\mathbf{L}^\infty([-M,M])} \, dx d\tau \\
&\times N W_N \|\partial_u (f-g)\|_{\mathbf{L}^\infty} \\
&+ \int_0^t e^{\kappa(t-\tau)} \int_{\mathbb{R}^N} \|((F-G) - \operatorname{div}(f-g))(\tau, x, \cdot)\|_{\mathbf{L}^\infty([-M,M])} \, dx d\tau.
\end{aligned}$$

Remark 2.7. As in Remark 2.4, we recover known estimates in some particular cases

1. f and g depend only on u, $F = G = 0$,
2. f, g, F, and G do not depend on u.

Remark 2.8. As in Remark 2.5, we think the set of hypotheses **(H2)** can be slightly weakened (see [11]) into

$$(\mathbf{H2}*): \begin{cases} \forall A > 0, \quad \partial_u F \in \mathbf{L}^\infty(\Omega_A; \mathbb{R}^N), \\ \displaystyle\int_{\mathbb{R}_+^*} \int_{\mathbb{R}^N} \|(F - \operatorname{div} f)(t, x, \cdot)\|_{\mathbf{L}^\infty([-A,A];\mathbb{R})} \, dx dt < +\infty. \end{cases}$$

Furthermore, κ can be replaced by $\kappa^* = \|\partial_u G\|_{\mathbf{L}^\infty(\Omega_U;\mathbb{R})}$ where $U = \sup(\|u\|_{\mathbf{L}^\infty}, \|v\|_{\mathbf{L}^\infty})$.

3 The continuity equation with a non-local flow

This section is a short version of [4]. We study the continuity equation:

$$\partial_t u + \operatorname{Div}(u\, V(x, u(t))) = 0, \qquad u(0, \cdot) = u_0 \in \mathbf{L}^1 \cap \mathbf{L}^\infty \cap \mathbf{BV}, \quad (3.1)$$

where $V : \mathbb{R}^N \times \mathbf{L}^1(\mathbb{R}^N; \mathbb{R}) \to \mathscr{C}^2(\mathbb{R}^N; \mathbb{R})$ is a non-local averaging functional, for example, if $\varphi : \mathbb{R} \to \mathbb{R}$ is a regular function:

1. $V(u) = \varphi\left(\int_{\mathbb{R}} u \, \mathrm{d}x\right)$ for a supply-chain [1],
2. $V(x, u) = \varphi(\eta *_x u) w(x)$ for pedestrian traffic. This model follows several other macroscopic models [6, 10, 12].

Our goals are: First, prove existence and uniqueness of a weak entropy solution. Second, find the extrema of a cost functional depending on the initial condition. The second point leads us to differentiate the semi-group in the Gâteaux sense with respect to initial conditions.

3.1 Existence and uniqueness of a solution

Let us introduce the following sets of hypotheses:
(V1) There exists $C \in \mathbf{L}^\infty_{loc}(\mathbb{R}_+;\mathbb{R}_+)$ such that for all $u \in \mathbf{L}^1(\mathbb{R}^N;\mathbb{R})$

$$V(u) \in \mathbf{L}^\infty, \qquad \|\nabla V(u)\|_{\mathbf{L}^\infty} \leqslant C(\|u\|_{\mathbf{L}^\infty}),$$
$$\|\nabla V(u)\|_{\mathbf{L}^1} \leqslant C(\|u\|_{\mathbf{L}^\infty}), \qquad \left\|\nabla^2 V(u)\right\|_{\mathbf{L}^1} \leqslant C(\|u\|_{\mathbf{L}^\infty}),$$

and or all $u_1, u_2 \in \mathbf{L}^1(\mathbb{R}^N;\mathbb{R})$

$$\|V(u_1) - V(u_2)\|_{\mathbf{L}^\infty} \leqslant C(\|u_1\|_{\mathbf{L}^\infty})\,\|u_1 - u_2\|_{\mathbf{L}^1}\,,$$
$$\|\nabla(V(u_1) - V(u_2))\|_{\mathbf{L}^1} \leqslant C(\|u_1\|_{\mathbf{L}^\infty})\,\|u_1 - u_2\|_{\mathbf{L}^1}\,.$$

(V2) There exists a positive function $C \in \mathbf{L}^\infty_{loc}(\mathbb{R}_+;\mathbb{R}_+)$ such that $\left\|\nabla^2 V(u)\right\|_{\mathbf{L}^\infty} \leqslant C(\|u\|_{\mathbf{L}^\infty})$ and $\left\|\nabla^3 V(u)\right\|_{\mathbf{L}^\infty} \leqslant C(\|u\|_{\mathbf{L}^\infty})$.

Theorem 3.1 (see Theorem 2.2 in [4]). *Let $u_0 \in \mathbf{L}^\infty \cap \mathbf{L}^1 \cap \mathbf{BV}$. If V satisfies* **(V1)**, *then there exists a time $T_{ex} > 0$ and a unique entropy solution $u \in \mathscr{C}^0([0, T_{ex}[; \mathbf{L}^1 \cap \mathbf{L}^\infty \cap \mathbf{BV})$ to (3.1) and we denote $S_t u_0 = u(t,\cdot)$. Besides, we have*

$$T_{ex} \geqslant \sup\left\{\sum_n \frac{\ln(\alpha_{n+1}/\alpha_n)}{C(\alpha_{n+1})};(\alpha_n)_n \textit{ strict. increasing}, \alpha_0 = \|u_0\|_{\mathbf{L}^\infty}\right\},$$

where the function C is the one appearing in **(V1)**.

If furthermore, V satisfies **(V2)** *then*

$$u_0 \in \mathbf{W}^{2,1} \cap \mathbf{L}^\infty \Rightarrow \forall t \in [0, T_{ex}[, \quad u(t) \in \mathbf{W}^{2,1}.$$

Idea of the proof: We introduce the space $X_\alpha = \mathbf{L}^1 \cap \mathbf{BV}(\mathbb{R}^N;[0,\alpha])$ and the application $\mathcal{Q}$ that associates to $w \in \mathcal{X}_\beta = \mathscr{C}^0([0,T[, X_\beta)$ the solution $u \in \mathcal{X}_\beta$ of the Cauchy problem

$$\partial_t u + \mathrm{Div}(uV(w)) = 0, \quad u(0,\cdot) = u_0 \in X_\alpha.$$

For $w_1, w_2 \in \mathcal{X}_\beta$, we obtain, thanks to the estimate of Theorem 2.6:

$$\|\mathcal{Q}(w_1) - \mathcal{Q}(w_2)\|_{\mathbf{L}^\infty([0,T[,\mathbf{L}^1)} \leqslant f(T)\,\|w_1 - w_2\|_{\mathbf{L}^\infty([0,T[,\mathbf{L}^1)}\,,$$

where f is increasing, $f(0) = 0$ and $f \to \infty$ when $T \to \infty$. Then we apply the Banach Fixed Point Theorem.

3.2 Gâteaux derivative of the semi-group

Let us recall the standard (local) situation: the semi-group generated by a conservation law is in general lipschitz continuous and *not* differentiable. Here, the non-local property gives us more regularity and we are able to differentiate the semi-group in the Gâteaux sense. Let us first recall the definition of Gâteaux differentiability.

Definition 3.2. The application $S : \mathbf{L}^1(\mathbb{R}^N;\mathbb{R}) \to \mathbf{L}^1(\mathbb{R}^N;\mathbb{R})$ is said to be $\mathbf{L}^1$ *Gâteaux differentiable in* $u_0 \in \mathbf{L}^1$ *in the direction* $r_0 \in \mathbf{L}^1$ if there exists a linear continuous application $DS(u_0) : \mathbf{L}^1 \to \mathbf{L}^1$ such that

$$\left\| \frac{S(u_0 + hr_0) - S(u_0)}{h} - DS(u_0)(r_0) \right\|_{\mathbf{L}^1} \to 0 \quad \text{when } h \to 0.$$

Formally, we expect the Gâteaux derivative of the semi-group to be the solution of the linearized problem:

$$\partial_t r + \mathrm{Div}(rV(u) + uDV(u)(r)) = 0, \quad r(0,\cdot) = r_0.$$

In order to give sense to this equation, we have first to require the differentiability of V. Let us introduce stronger hypotheses:
(V3) $V : \mathbf{L}^1 \to \mathscr{C}^2$ is differentiable and there exists $C \in \mathbf{L}^\infty_{loc}$ such that for all $u, r \in \mathbf{L}^1$,

$$\begin{aligned}\|V(u+r) - V(u) - DV(u)(r)\|_{\mathbf{W}^{2,\infty}} &\leqslant C\left(\|u\|_{\mathbf{L}^\infty} + \|u+r\|_{\mathbf{L}^\infty}\right)\|r\|^2_{\mathbf{L}^1},\\ \|DV(u)(r)\|_{\mathbf{W}^{2,\infty}} &\leqslant C(\|u\|_{\mathbf{L}^\infty})\,\|r\|_{\mathbf{L}^1}.\end{aligned}$$

(V4) There exists $C \in \mathbf{L}^\infty_{loc}(\mathbb{R}_+;\mathbb{R}_+)$ such that for all $u, \tilde{u}, r \in \mathbf{L}^1$,

$$\begin{aligned}\|\mathrm{div}\,(V(\tilde{u}) - V(u) - DV(u)(\tilde{u}-u))\|_{\mathbf{L}^1} &\leqslant C(\|\tilde{u}\|_{\mathbf{L}^\infty} + \|u\|_{\mathbf{L}^\infty})\,\|\tilde{u}-u\|^2_{\mathbf{L}^1}\\ \|\mathrm{div}\,(DV(u)(r))\|_{\mathbf{L}^1} &\leqslant C(\|u\|_{\mathbf{L}^\infty})\,\|r\|_{\mathbf{L}^1}.\end{aligned}$$

We show that the linearized problem admits a unique entropy solution:

Theorem 3.3 (see Proposition 2.8 in [4]). *Assume that* V *satisfies* **(V1)**, **(V3)**. *Let* $u \in \mathscr{C}^0([0,T_{ex}[;\mathbf{W}^{1,\infty} \cap \mathbf{W}^{1,1})$, $r_0 \in (\mathbf{L}^1 \cap \mathbf{L}^\infty)(\mathbb{R}^N;\mathbb{R})$. *Then the linearized Cauchy problem*

$$\partial_t r + \mathrm{Div}(rV(u) + uDV(u)(r)) = 0, \qquad \text{with } r(0,x) = r_0 \tag{3.2}$$

admits a unique entropy solution $r \in \mathscr{C}^0([0,T_{ex}[;\mathbf{L}^1(\mathbb{R}^N;\mathbb{R}))$ *and we denote* $\Sigma^u_t r_0 = r(t,\cdot)$.

If furthermore V *satisfies* **(V2),** *and* $r_0 \in \mathbf{W}^{1,1}$, *then* $\forall t \in [0,T_{ex}[$, $r(t) \in \mathbf{W}^{1,1}$.

Now, we can prove that the solution of the linearized equation is really the derivative of the semi-group.

Theorem 3.4 (see Theorem 2.10 in [4]). *Assume that* V *satisfies* **(V1)**, **(V2)**, **(V3)**, *and* **(V4)**. *Let* $u_0 \in \mathbf{W}^{1,\infty} \cap \mathbf{W}^{2,1}$, $r_0 \in \mathbf{W}^{1,1} \cap \mathbf{L}^\infty$ *and let* T_{ex} *be the time of existence for the initial problem given by Theorem* 3.1.

Then, for all $t \in [0,T_{ex}[$ *the local semi-group of the pedestrian traffic problem is* $\mathbf{L}^1$ *Gâteaux differentiable in the direction* r_0 *and*

$$DS_t(u_0)(r_0) = \Sigma^{S_t u_0}_t r_0.$$

Idea of the proof: Let u and u_h be the solutions of the Cauchy problem $\partial_t u + \mathrm{Div}(uV(u)) = 0$ with initial conditions $u_0, u_0 + hr_0$. Let r be the solution of the linearized equation (3.2), with $r(0) = r_0$. We define then $z_h = u + hr$ that satisfies $z_h(0) = u_0 + hr_0$ and

$$\partial_t z_h + \mathrm{Div}\left(z_h(V(u) + hDV(u)(r))\right) = h^2 \mathrm{Div}(rDV(u)(r)).$$

Next, we use Theorem 2.6 to compare u_h and z_h. We obtain

$$\begin{aligned} \|u_h - z_h\|_{\mathbf{L}^\infty([0,T[,\mathbf{L}^1)} \leqslant & F(T)\left[\|u_h - u\|^2_{\mathbf{L}^\infty(\mathbf{L}^1)} + \|u_h - z_h\|_{\mathbf{L}^\infty(\mathbf{L}^1)}\right] \\ & + h^2 C(\beta) T e^{C(\beta)T} \|r\|_{\mathbf{L}^\infty(\mathbf{W}^{1,1})} \|r\|_{\mathbf{L}^\infty(\mathbf{L}^1)}, \end{aligned}$$

where F is increasing and $F(0) = 0$. After a good choice of T, we can divide by h, make $h \to 0$, and conclude.

References

[1] D. Armbruster, P. Degond and C. Ringhofer, A model for the dynamics of large queuing networks and supply chains. SIAM J. Appl. Math. 66 (2006), no. 3, 896–920.

[2] F. Bouchut and B. Perthame, Kružkov's estimates for scalar conservation laws revisited. Trans. Amer. Math. Soc. 350 (1998), no. 7, 2847–2870.

[3] G. Q. Chen and K. H. Karlsen, Quasilinear anisotropic degenerate parabolic equations with time-space dependent diffusion coefficients. Commun. Pure Appl. Anal., 4 (2005), no. 2, 241–266.

[4] R. M. Colombo, M. Herty and M. Mercier, Control of the continuity equation with a non-local flow. Esaim-Control Opt. Calc. Var., 2010.

[5] R. M. Colombo, M. Mercier and M. D. Rosini, Stability and total variation estimates on general scalar balance laws. Comm. Math. Sciences 7 (2009), no. 1, 37–65.

[6] R. M. Colombo and M. D. Rosini, Pedestrian flows and non-classical shocks. Math. Methods Appl. Sci. 28 (2005), no. 13, 1553–1567.

[7] C. M. Dafermos, Hyperbolic Conservation Laws in Continuum Physics. Springer-Verlag, 2005.

[8] S. N. Kružkov, First order quasilinear equations with several independent variables. Mat. Sb. (N.S.) 81 (1970), no. 123, 228–255.

[9] B. J. Lucier, A moving mesh numerical method for hyperbolic conservation laws. Math. Comp. 46 (1986), no. 173, 59–69.

[10] B. Maury, A. Roudneff-Chupin and F. Santambrogio, A macroscopic crowd motion model of the gradient-flow type. M3AS, 2009.

[11] M. Mercier, Improved stability estimates on general scalar balance laws. Preprint, 2010.

[12] B. Piccoli and A. Tosin, Pedestrian flows in bounded domains with obstacles. Contin. Mech. Thermodyn. 21 (2009), no. 2, 85–107.

Solver for the Generalized Riemann Problem for Balance Laws with Stiff Source Terms: The Scalar Case

Gino Montecinos
Laboratory of Applied Mathematics
University of Trento, Italy
E-mail: montecin@ing.unitn.it

Eleuterio Toro
Laboratory of Applied Mathematics
University of Trento, Italy
E-mail: toroe@ing.unitn.it

Abstract

We propose a new method for solving approximately the generalized Riemann problem (GRP) for a scalar balance law with stiff source term. The solution at the origin $x = 0$ is a function of time and an approximation is constructed in terms of an implicit Taylor series expansion in time. Then, this approximate solution is used locally to construct explicit, one-step high-order schemes of the ADER type to solve the full initial-boundary value problem. Some properties of the solver are proved and some numerical examples are given.

1 Introduction

We are concerned with the approximate solution of the generalized Riemann problem (GRP) for a scalar balance law with stiff source term. The GRP is a Cauchy problem that generalizes the classical Riemann problem in two ways: (i) the equation has a source term (stiff in the present case) and (ii) the initial condition is piecewise smooth, typically polynomials of arbitrary degree. The solution at the origin $x = 0$ is a function of time. In this paper we propose a solution method to find an approximate solution. The method is inspired by the work of Scott [8], who was concerned with numerical methods for stiff ordinary differential equations. The approximate semi-analytical solution obtained is then used locally to construct explicit, one-step numerical methods of the ADER type [2] to solve the general initial-boundary value problem for the equations (scalar here) of interest. The local generalized Riemann problems have polynomial initial data that comes from a nonlinear reconstruction procedure of the WENO type. Some properties of the solver are proved for special cases and some numerical examples are given for the test problem proposed by LeVeque and Yee [7], comparing the new scheme with state-of-the-art solvers and the exac solution.

2 ADER schemes and the generalized Riemann problem

We first recall the finite volume method and then study the generalized Riemann problem.

2.1 Finite volume framework

We are interested in solving the general initial-boundary value problem (IBVP)

$$\left.\begin{array}{l} PDE: \partial_t q(x,t) + \partial_x f(q(x,t)) = s(q(x,t)), \quad x \in [a,b], \quad t > 0, \\ IC: \quad q(x,0) = q_0(x), \\ BCs: \ q(a,t) = q_L(t), \quad q(b,t) = q_R(t), \end{array}\right\} \tag{2.1}$$

with prescribed flux $f(q(x,t))$, source $s(q(x,t))$, initial condition $q_0(x)$, and boundary conditions $q_L(t)$ and $q_R(t)$. The finite volume version of the ADER method results from integrating the PDE in (2.1) in space and time in the control volume $[x_{i-\frac{1}{2}}, x_{i+\frac{1}{2}}] \times [t^n, t^{n+1}]$ to obtain

$$q_i^{n+1} = q_i^n - \frac{\Delta t}{\Delta x}[f_{i+\frac{1}{2}} - f_{i-\frac{1}{2}}] + \Delta t s_i, \tag{2.2}$$

where q_i^n is the spatial-integral average

$$q_i^n = \frac{1}{\Delta x}\int_{x_{i-\frac{1}{2}}}^{x_{i+\frac{1}{2}}} q(x,t^n)dx, \tag{2.3}$$

$f_{i+\frac{1}{2}}$ is the time-integral average

$$f_{i+\frac{1}{2}} = \frac{1}{\Delta t}\int_{t^n}^{t^{n+1}} f(q(x_{i+\frac{1}{2}},t))dt \tag{2.4}$$

and s_i is the space-time integral average

$$s_i = \frac{1}{\Delta t}\frac{1}{\Delta x}\int_{t^n}^{t^{n+1}}\int_{x_{i-\frac{1}{2}}}^{x_{i+\frac{1}{2}}} s(q(x,t))dxdt, \tag{2.5}$$

with $\Delta x = x_{i+\frac{1}{2}} - x_{i-\frac{1}{2}}$ and $\Delta t = t^{n+1} - t^n$. Formula (2.2) is exact if definitions (2.3)-(2.5) are used. The finite volume ADER scheme begins by interpreting (2.2) as a numerical formula to update approximations to cell averages q_i^n within cell $I_i = [x_{i-\frac{1}{2}}, x_{i+\frac{1}{2}}]$, with cell or volume centre $x_i = \frac{1}{2}(x_{i-\frac{1}{2}} + x_{i+\frac{1}{2}})$ and cell interfaces $x_{i-\frac{1}{2}}$ and $x_{i+\frac{1}{2}}$, mesh length Δx and time step Δt. The numerical flux is computed from (2.4) to high accuracy, in which the integrand is found by solving a local generalized Riemann problem at the interface. A similar procedure is followed to evaluate the numerical source from (2.5) to high accuracy. For an introduction to ADER methods see chapters 19 and 20 of [3].

2.2 The generalized Riemann problem

Consider a scalar generalized Riemann problem

$$\begin{aligned} &PDE: \partial_t q(x,t) + \partial_x f(q(x,t)) = s(q(x,t)), \\ &IC: \quad q(x,0) = \begin{cases} q_L(x) & if\ x < 0, \\ q_R(x) & if\ x > 0. \end{cases} \end{aligned} \tag{2.6}$$

The functions $q_L(x)$ and $q_R(x)$ are smooth functions, polynomials of arbitrary degree for example. The aim is to find an approximation $q_{LR}(t) := q(0,t)$ to the solution at $x = 0$, as function of time. First we express $q_{LR}(t)$ as an implicit time Taylor series expansion. This step is a generalization of Scott's approach [8] for ordinary differential equations. So we write

$$\begin{aligned} q(0,0_+) &= lim_{t_- \to t} q(0, t - t_-) \\ &= lim_{t_- \to t} \left[q(0,t) - t_- \partial_t q(0,t) + \sum_{k=1}^{K} \frac{(-t_-)^k}{k!} \partial_t^{(k)} q(0,t) \right]. \end{aligned}$$

So we obtain

$$q(0,t) = q(0,0_+) - \sum_{k=1}^{K} \frac{(-t)^k}{k!} \partial_t^{(k)} q(0,t). \tag{2.7}$$

This is the implicit Taylor expansion about $q(0,0_+)$. To compute the term $\partial_t^{(k)} q(0,t)$ we express them in terms of space derivatives via the Cauchy-Kowalewski procedure, yielding

$$q_{LR}(t) = q(0,0_+) - \sum_{k}^{K} \frac{(-t)^k}{k!} G^k(q_{LR}(t), \partial_x q(0,t), \ldots, \partial_x^{(k)} q(0,t)).$$

Arguments of the functional G^l are time dependent and are computed via a forward Taylor series expansion

$$\partial_x^{(l)} q(0,t) = \partial_x^{(l)} q(0,0_+) + \sum_{j=1}^{K-l} \frac{t^j}{j!} H^j(\partial_x^{(l)} q(0,0_+),\ldots,\partial_x^{(l+j)} q(0,0_+)) \,.$$

We denote by H^l the functional resulting from the Cauchy-Kowalewski procedure for the homogeneous problem given by (2.6) with $s(q) \equiv 0$. The leading term $q(0,0_+) = \partial_x^{(0)} q(0,0_+)$ and the spatial derivatives $\partial_x^{(l)} q(0,0_+)$ are determinated as the self-similar solutions at the interface of the classical Riemann problem

$$\begin{aligned} &PDE: \partial_t(\partial_x^{(l)} q(x,t)) + f'(q(0,0_+))\partial_x(\partial_x^{(l)} q(x,t)) = 0, \\ &IC: \quad \partial_x^{(l)} q(x,0) = \begin{cases} q_L^{(l)}(0) & if\ x < 0, \\ q_R^{(l)}(0) & if\ x > 0. \end{cases} \end{aligned} \tag{2.8}$$

Then $q_{LR}(t)$ is found from the solution to the nonlinear equation

$$\mathbf{J}(q_{LR}(t)) = 0, \tag{2.9}$$

where

$$\begin{aligned} \mathbf{J}(q_{LR}(t)) &= q(0,0_+) - q_{LR}(t) \\ &\quad - \sum_{k}^{K} \frac{(-t)^k}{k!} G^k(q_{LR}(t), \partial_x q(0,t), \ldots, \partial_x^{(k)} q(0,t)). \end{aligned}$$

3 Properties of the MT Solver

Proposition 3.1. *For the homogeneous linear advection equation the GRP solution at $x = 0$ given by the present method coincides with that obtained from the Toro-Titarev solver* [4].

Proof. We denote the Toro-Titarev (TT) solution and the present solution (MT) respectively by

$$Q_{LR,TT}(t) = q(0,0_+) + \sum_{k=1}^{N} \frac{t^t}{k!} \partial_t^{(k)} q(0,0_+), \tag{3.1}$$

and

$$Q_{LR,MT}(t) = q(0,0_+) - \sum_{k=1}^{N} \frac{(-t)^k}{k!} \partial_t^{(k)} q(0,t). \tag{3.2}$$

Then we will prove that the following expression is zero

$$Q_{LR,TT}(t) - Q_{LR,MT}(t) = \sum_{k=1}^{K} \frac{t^k}{k!}(1+(-1)^k)\partial_t^{(k)} q(0,0_+) + \sum_{j=1}^{K}\sum_{r=1}^{K-j} \frac{(-1)^j t^{j+r}}{j!r!}\partial_t^{(j+r)} q(0,0_+).$$

The result follows by first maniputating the right-hand side to obtain expressions of the type $\sum_{k=0}^{n} \frac{(-b)^k a^{n-k}}{k!(n-k)!}$. These are zero when $a = b = 1$, and the result is obtained. □

Proposition 3.2. *For the special linear case of no advection, that is $q'(t) = \beta q(t)$, with $t \in [0,\infty)$ and $\beta < 0$, the numerical solution obtained with ADER with the MT solver satisfies*

$$q_i^n = \begin{cases} q_i^0 & if\ n = 0, \\ 0 & if\ n > 0, \end{cases} \tag{3.3}$$

for all i, as $|\beta|$ tends to ∞.

Proof. The one-step numerical scheme has the form

$$q_i^{n+1} = q_i^n + \sum_j \sum_{p=-k}^{k} a_{i+p}^{n,i} q_{i+p}^n \omega_j \frac{r}{1 - r_j + \cdots + \frac{(-r_j)^k}{k!}}, \tag{3.4}$$

where $r_j = \eta_j r$ and ω_j and η_j are the weights and points of the quadrature. We will assume that $q_i^0 \geq 0$, $\forall i$. We have written the leading term as $q(0,0_+) = \sum_{p=-k}^{k} a_{i+p}^{n,i} q_{i+p}^n$, in which $a_{i+p}^{n,i}$ do not depend on r and the result from the reconstruction procedure. Simple manipulations allows us to show that $q_i^n \geq q_i^{n+1} \geq 0$. Then the sequence $\{q_i^n\}_n$, with fixed i, has a convergence subsequence whose limit is zero. From (3.4) we can prove that $\{q_i^n\}_n$ is a Cauchy sequence and therefore its limit is zero. □

4 Numerical implementation in ADER

For assessing the performance of the present MT solver we implement it into the finite volume ADER scheme. We solve numerically the test problem proposed in [7], using 550 cells and CFL number 0.4. The problem is

$$\begin{aligned} &PDE: \partial_t q + \lambda \partial_x q = \beta q(q-1)(q-\tfrac{1}{2}), \quad x \in [0,1], t > 0, \\ &IC: \qquad q(x,0) = q_0(x) = \begin{cases} 1 & if\ x \leq 0.3, \\ 0 & if\ x > 0.3, \end{cases} \end{aligned} \tag{4.1}$$

where $\beta \leq 0$ is given. The test has exact solution $q(x,t) = q_0(x-t)$. Most numerical schemes are unable to capture the discontinuity with the correct speed. However the present ADER-MT method captures the shock wave correctly, as does the ADER-DET method [6], see Figure 4.1. For assessing the convergence rates (empirically) we use the model linear advection-reaction problem with $\lambda = 1$, CFL number 0.1 and Gaussian initial condition $h_0(0) = \exp(-50x^2)$, with periodic boundary conditions; the domain is $[-1.4, 1.4]$. In Table 1 we show CPU times of the present solver relative to the CPU time of the ADER-DET. Also shown there are errors and convergence rates for $t = 0.25$, $\beta = -1$, and $t = 0.01$, $\beta = -10$. Note that the analytical solution $q(x,t) = \exp(-50(x-\lambda t)^2 + \beta t)$ tends to zero very fast; therefore it is necessary to choose a short simulation time for capturing the behaviour of the numerical solution.

Table 1 Convergence rates for the ADER-MT scheme applied to the linear advection-reaction equation. Top part shows results for $t = 0.25$ and $\beta = -1$. Bottom part shows results for $t = 0.01$ and $\beta = -10$.

$\mathcal{O}_6$	L_∞	p_∞	L_2	p_2	CPU
10 cells	1.2656e-01	-	5.5770e-02	-	0.09
20 cells	2.7077e-02	2.22	7.6751e-03	2.86	0.21
40 cells	2.7298e-03	3.31	7.6664e-04	3.32	0.41
80 cells	4.6258e-05	5.88	1.0158e-05	6.24	0.81
160 cels	6.6972e-07	6.11	1.2714e-07	6.03	0.80
$\mathcal{O}_6$	L_∞	p_∞	L_2	p_2	CPU
10 cells	8.6735e-02	-	3.4837e-02	-	0.14
20 cells	2.8779e-02	1.59	9.1281e-03	1.93	0.21
40 cells	6.7070e-04	5.42	1.4574e-04	5.97	0.36
80 cells	1.9741e-05	5.09	4.2756e-06	5.09	0.42
160 cells	3.0632e-07	6.01	5.4264e-08	6.30	0.71

5 Concluding remarks

We have presented a new solver for the GRP for a scalar non-linear equation with stiff source term. Then we have incorporated the solver in the ADER method for solving the corresponding IBVP. For simplified cases we have proved some useful properties of the solver. We have also demonstrated that our numerical method performs very satisfactorily when solving the exceedingly demanding test problem suggested by Leveque and Yee, for which most methods fail. Our solver reconciles stiffness and high order of accuracy and is comparable to the state-of-the-art ADER-DET scheme [6], but more efficient in CPU time. Future research will involve the application of the proposed methodology to non-linear hyperbolic systems with stiff source terms.

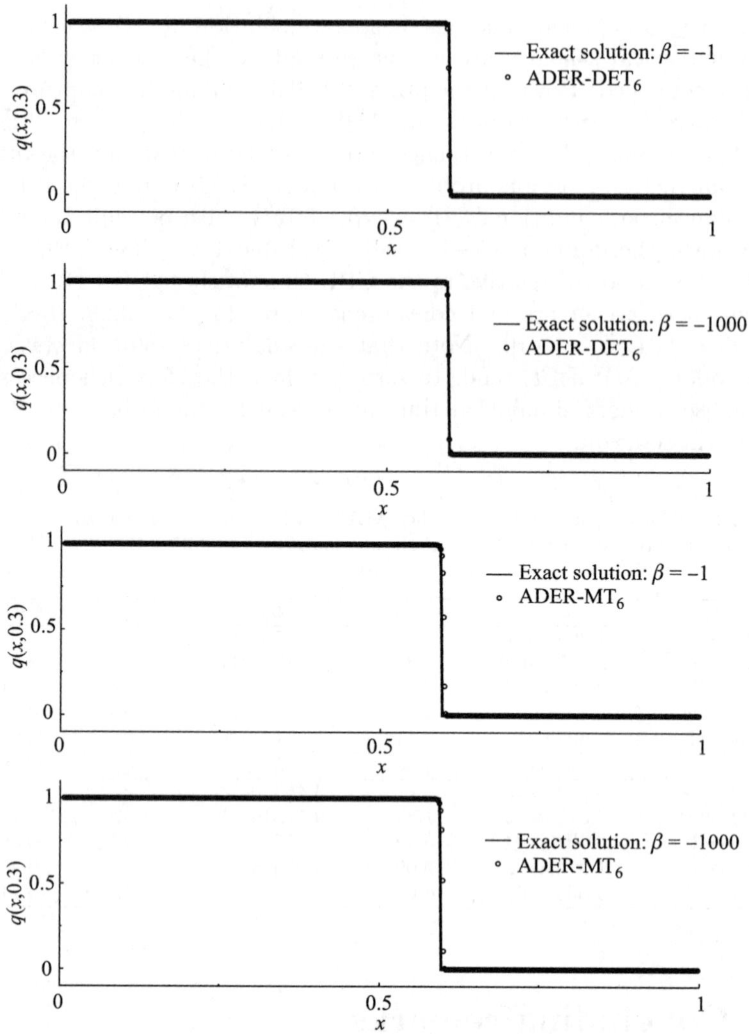

Figure 4.1 LeVeque-Yee test. Exact (line) and numerical solutions (symbols) by the present ADER-MT scheme and ADER-DET, both of order six in space and time.

References

[1] S. K. Godunov, A finite difference method for the computation of discontinuous solutions of the equations of fluids dynamics. Math. Sbornik 47 (1959), 357–393.

[2] E. F. Toro, R. C. Millington and L. A. M. Nejad, Towards very high-order Godunov schemes. In: Godunov Methods: Theory and Applications. E. F. Toro ed., Kluwer Academic/Plenum Publishers. Conference in Honour of

S. K. Godunov, Vol. 1, 897–902. New York, Boston, and London, 2001.

[3] E. F. Toro, Riemann Solvers and Numerical Methods for Fluid Dynamics. A Practical Introduction. Third edition, Springer-Verlag, Berlin Heidelberg, 2009.

[4] E. F. Toro and V. A. Titarev, Solution of the generalised Riemann problem for advection-reaction equations. Proceedings of the Royal Society of London. Series A 458 (2002), 271–281.

[5] E. F. Toro and V. A. Titarev, Derivative Riemann solvers for systems of conservation laws and ADER methods. Journal of Computational Physics 212 (2006), 150–165.

[6] M. Dumbser, C. Enaux and E. F. Toro, Finite volume schemes of very high order of accuracy for stiff hyperbolic balance laws. Journal of Computational Physics 227 (2008), 3971–4001.

[7] R. J. LeVeque and H. C. Yee, A study of numerical methods for hyperbolic conservation laws with stiff source terms. Journal of Computational Physics 86 (1990), No. 1, 187–210.

[8] J. R. Scott, Solving ODE initial value problems with implicit Taylor series methods. Glenn Research Center. Cleveland, Ohio, 2002, NASA/ TM2000–209400.

Related Problems for TV-estimates for Conservation Laws on Surfaces*

Thomas Mueller
Abteilung für Angewandte Mathematik, University of Freiburg
Hermann-Herder-Straße 10, D79104 Freiburg i. Br., Germany
E-mail: mueller@mathematik.uni-freiburg.de

Dietmar Kroener
Abteilung für Angewandte Mathematik, University of Freiburg
Hermann-Herder-Straße 10, D79104 Freiburg i. Br., Germany
E-mail: dietmar@mathematik.uni-freiburg.de

Abstract

The study of nonlinear conservation laws on surfaces is motivated by many applications such as geophysical flows, general relativity, fractions in porous media, sound waves on surfaces and transport processes on cells. Since huge difficulties in the theoretical analysis of systems of nonlinear conservation laws arise, even in the Euclidean space, we focus on nonlinear scalar conservation laws posed on closed Riemannian manifolds. The fluxes of the conservation laws are modelled by vector fields on the manifold which depend both on the unknown function as a parameter and on the space variable, since the vector fields have to be in the tangent space of the manifold at every point.

In order to prove error estimates for finite volume schemes, one requires estimates of the total variation of the solution. In this contribution we present a proof (that is different from the one in [4]) for an estimate of the total variation for the two dimensional case. Therefore we consider the viscous approximation of the conservation law. Due to the space-dependence of the Laplace-Beltrami operator, new second-order derivatives in the unkown occur. The key idea to control these terms is to reuse the differential equation in a sophisticated way. The result is an estimate of the total variation of the entropy solution of the conservation law. We construct an example which shows that we cannot expect

*The authors are partly supported by the Deutsche Forschungsgemeinschaft via SFB/Transregio 71 "Geometric Partial Differential Equations".

a TVD property on a manifold. This is solely a consequence of the fact that the flux-vector depends on the spatial coordinate.

In addition we implemented the finite volume scheme proposed by Amorim et al. [1] for the case of a sphere. On the basis of Ben-Artzi et al. [3] we constructed a class of test problems which are equivalent to one-dimensional conservation laws in order to calculate EOCs.

1 Introduction

There are numerous applications of conservation laws on surfaces. Consider for example the flow of a fluid in the curved fraction of a porous medium. Other applications are the global air and water flows that can be described by the shallow water equations on the rotating 2-sphere. There are further examples in astrophysics where curved spaces occur.

All these phenomena are modelled by systems of conservation laws on manifolds. Since the theoretical analysis of systems even in the Euclidean case is extremely involved, we focus in this paper on scalar conservation laws posed on a Riemannian manifold as a model problem for systems.

Estimates of the total variation (TV) of entropy solutions and of their viscous approximation play a central role for conservation laws, since they form not only an important part in the existence proof, but are also used in several proofs for error estimates for finite volume schemes.

In this paper we present a new proof for a TV-estimate for scalar conservation laws on two-dimensional manifolds which is obtained through a uniform TV-estimate for a regularized equation. Our proof is different from Ben-Artzi's and LeFloch's one [4]. This estimate for the regularized equation provides existence of an entropy solution and a TV-estimate for it.

The paper is organized as follows. In Section 2 we give a short overview of Riemannian geometry as it will be necessary later on. Section 3 contains the main theorem that provides existence of entropy solutions on manifolds and the TV-estimate. In Section 4 we present some results, including EOCs, of the numerical experiments we made.

2 Basics of Riemannian geometry

Throughout this paper (M, g) denotes a compact smooth d-dimensional Riemannian manifold without boundary. This means in particular that M is locally homeomorphic to an open subset of the $\mathbb{R}^d$ and can be covered by such homeomorphisms. Thus, locally we can choose coordinates $(x^j)_{1\leq j\leq d}$ and do local calculations in these coordinates. Here, g denotes the Riemannian metric which is a smooth $(2,0)$-tensor field such that at each point $x \in M$

$$g(x) : T_xM \times T_xM \to \mathbb{R}$$

is a scalar product, and hence defines a norm $\|\cdot\|_g$, on the tangent space T_xM at x. Local coordinates induce the partial derivatives $\frac{\partial}{\partial x^j}\big|_x$ which form a basis of the tangent space T_xM. Given such a basis, the Riemannian metric g is represented by a symmetric, positive definite matrix $(g_{ij})_{1\le i,j\le d}$.

The inverse of the matrix $(g_{ij})_{1\le i,j\le d}$ is denoted by $\left(g^{ij}\right)_{1\le i,j\le d}$. This enables us to introduce some differential operators. Recall that the gradient operator ∇_g is defined by

$$\nabla_g f = g^{ij}\frac{\partial f}{\partial x^i}\frac{\partial}{\partial x^j}$$

for $f \in C^\infty(M)$. As in the whole paper we use implicit summation here. The divergence operator div_g is defined by

$$\operatorname{div}_g \mathbf{X} = \frac{1}{\sqrt{|g|}}\frac{\partial}{\partial x^i}\left(X^i\sqrt{|g|}\right),$$

where $\mathbf{X}$ is a vector field on M, $|g|$ denotes the determinant of $(g_{ij})_{1\le i,j\le d}$ and X^i are the components of $\mathbf{X}$ in the basis $\frac{\partial}{\partial x^j}$. Finally, we define the Laplace-Beltrami operator Δ_g which locally fullfills

$$\Delta_g u := \operatorname{div}_g\left(\nabla_g u\right) = g^{ij}\frac{\partial^2 u}{\partial x^i \partial x^j} - g^{ij}\Gamma^k_{ij}\frac{\partial u}{\partial x^k}, \tag{2.1}$$

where $u \in C^\infty(M)$ and the Christoffel symbols Γ^k_{ij} are given by

$$\Gamma^k_{ij} := \frac{1}{2}g^{mk}\left(\frac{\partial}{\partial x^i}g_{jm} + \frac{\partial}{\partial x^j}g_{mi} - \frac{\partial}{\partial x^m}g_{ij}\right).$$

Now we can introduce the total variation of a function $u \in L^1(M)$ as

$$\operatorname{TV}(u) := \sup_{\mathbf{X}} \int_M u \operatorname{div}_g(\mathbf{X})dv_g,$$

where the supremum is taken over all smooth vector fields $\mathbf{X}$ on M satisfying $\sup_M \|\mathbf{X}\|_g \le 1$.

3 TV-estimate and existence of entropy solutions

We consider the following initial value problem. Find $u : M \times \mathbb{R}^+ \to \mathbb{R}$ with

$$\partial_t u + \operatorname{div}_g \mathbf{f}(u, \cdot) = 0 \qquad \text{in } M \times \mathbb{R}^+, \tag{3.1}$$

$$u(\cdot, 0) = u_0 \qquad \text{on } M, \tag{3.2}$$

where $\mathbf{f}(\bar{u}, \cdot)$ for every $\bar{u} \in \mathbb{R}$ is a smooth, divergence-free vector field tangential to M and $u_0 : M \to \mathbb{R}$ are initial values.

As in the Euclidean case we cannot expect strong solutions of (3.1) and (3.2) that exist globally in time. Therefore we introduce entropy solutions.

Definition 3.1. A pair $(U, \mathbf{F})$ is called convex entropy pair if $U : \mathbb{R} \to \mathbb{R}$ is a convex, Lipschitz-continuous function and $\mathbf{F} = \mathbf{F}(u, x)$ is a vector field, such that for a.e. $\bar{u} \in \mathbb{R}$ and $x \in M$

$$\partial_u \mathbf{F}(\bar{u}, x) = U'(\bar{u}) \partial_u \mathbf{f}(\bar{u}, x).$$

A function $u \in L^\infty(M \times \mathbb{R}^+)$ is called entropy solution of (3.1) and (3.2) if the entropy inequalities

$$\int_M \int_{\mathbb{R}^+} U(u)\partial_t \varphi + g(\mathbf{F}(u, \cdot), \nabla_g \varphi) dt dv_g + \int_M U(u_0)\varphi(\cdot, 0) dv_g \geq 0 \tag{3.3}$$

hold for all convex entropy pairs $(U, \mathbf{F})$ and nonnegative test functions $\varphi \in C_0^\infty(M \times [0, \infty))$.

The central result of this work is the following theorem.

Theorem 3.2 (Existence and TV-Estimate). *If M is two-dimensional, then for initial values $u_0 \in L^\infty(M) \cap \mathrm{BV}(M)$ there is a unique entropy solution $u \in L^\infty(M \times \mathbb{R}^+)$ of (3.1) and (3.2) which satisfies*

$$\mathrm{TV}(u(t)) \leq [C_1 t + C_2] \exp(C_3 t) \qquad \textit{for a.e. } t \in \mathbb{R}^+ \tag{3.4}$$

for positive constants C_1, C_2, and C_3, which only depend on u_0, $\mathbf{f}$ and (M, g).

Remark 3.3. There is a similar theorem stated by Ben-Artzi and LeFloch [4], but since unresolved problems in their proof occured we developed a different proof which we present now.

Proof of Theorem 3.2. We consider a regularization of equation (3.1),

$$\partial_t u^\varepsilon + \operatorname{div}_g \mathbf{f}(u^\varepsilon,\cdot) = \varepsilon\Delta_g u^\varepsilon \qquad \text{on } M\times\mathbb{R}^+. \tag{3.5}$$

For the regularized initial value problem (3.5) and (3.2) existence, uniqueness and L^∞-stability were established by Ben-Artzi and LeFloch [4]. We want to consider the vanishing viscosity limit, i.e., we want to let $\varepsilon \to 0$ in (3.5). To establish convergence for the sequence $(u^\varepsilon)_{\varepsilon>0}$ we need to provide in ε uniform estimates of the time and space derivatives of u^ε in order to apply a compactness argument. Let us turn to the estimates now.

Denoting $z^\varepsilon := \partial_t u^\varepsilon$ and differentiating (3.5) w.r.t. t leads to

$$\partial_t z^\varepsilon + \operatorname{div}_g \partial_u \mathbf{f}(u^\varepsilon,\cdot) z^\varepsilon = \varepsilon\Delta_g z^\varepsilon \qquad \text{on } M\times\mathbb{R}^+. \tag{3.6}$$

Multiplying (3.6) by a monotone regularization $s_\delta(z^\varepsilon)$ of the sign-function for which the derivative can be estimated by $\frac{C}{\delta}$ and integrating over M leads to

$$\underbrace{\int_M s_\delta(z^\varepsilon)\partial_t z^\varepsilon}_{=:I_1} + \underbrace{\int_M s_\delta(z^\varepsilon) \operatorname{div}_g\left(\partial_u \mathbf{f}(u^\varepsilon,\cdot)z^\varepsilon\right)}_{=:I_2} = \underbrace{\varepsilon\int_M s_\delta(z^\varepsilon)\Delta_g z^\varepsilon}_{=:I_3} \quad \text{in } \mathbb{R}^+.$$

With $r(y) := \int_0^y s_\delta(\xi)d\xi$ and partial integration we see that

$$\begin{aligned}
I_1 &= \int_M s_\delta(z^\varepsilon)\partial_t z^\varepsilon = \int_M \partial_t\left(r_\delta(z^\varepsilon)\right) \to \frac{d}{dt}\int_M |z^\varepsilon| \text{ for } \delta\to 0,\\
|I_2| &= \left|\int_M s_\delta(z^\varepsilon)\operatorname{div}_g \partial_u \mathbf{f}(u^\varepsilon,\cdot)z^\varepsilon\right| = \left|\int_M s'_\delta(z^\varepsilon) g(\nabla_g z^\varepsilon, \partial_u \mathbf{f}(u^\varepsilon,\cdot))z^\varepsilon\right|\\
&\le \sup|g(\nabla_g z^\varepsilon, \partial_u \mathbf{f})| \int_{\{0<|z^\varepsilon|\le\delta\}} \underbrace{|z^\varepsilon|}_{\le\delta}\,\underbrace{|s'_\delta(z^\varepsilon)|}_{\le\frac{C}{\delta}} \to 0 \text{ for } \delta\to 0,\\
I_3 &= \varepsilon\int_M s_\delta(z^\varepsilon)\Delta_g z^\varepsilon = -\varepsilon\int_M g\left(\nabla_g z^\varepsilon, \nabla_g z^\varepsilon\right) s'_\delta(z^\varepsilon) \le 0.
\end{aligned}$$

For $\delta \to 0$ this implies

$$\frac{d}{dt}\|\partial_t u^\varepsilon\|_{L^1} = \frac{d}{dt}\int_M |z^\varepsilon| \le 0.$$

Now we turn to the more involved estimate of the space derivatives which leads to the restriction to two-dimensional manifolds. We prove the estimate

$$\operatorname{TV}(u^\varepsilon(t)) \le [C_1 t + C_2]\exp(C_3 t). \tag{3.7}$$

Therefore, we again consider (3.5). Since M is compact, there exists a finite atlas and a subordinate partition of unity. From now on we consider a fixed coordinate ball $\Omega \subseteq \mathbb{R}^2$, i.e., Ω is homeomorphic to an open subset of M. Denoting $u_k^\varepsilon := \partial_k u^\varepsilon$ we differentiate (3.5) on $\Omega \times \mathbb{R}^+$ w.r.t. the kth coordinate x^k, multiply it by a cut-off function ψ (given by the partition of unity) and by $s_\delta(u_k^\varepsilon)$ and integrate over Ω:

$$\underbrace{\int_\Omega \psi \partial_t u_k^\varepsilon s_\delta(u_k^\varepsilon)}_{=:I_{1,k}} + \underbrace{\int_\Omega \psi \partial_k \left(\operatorname{div}_g \mathbf{f}(u^\varepsilon, \cdot)\right) s_\delta(u_k^\varepsilon)}_{=:I_{2,k}} = \underbrace{\int_\Omega \psi \varepsilon \partial_k \left(\Delta_g u^\varepsilon\right) s_\delta(u_k^\varepsilon)}_{=:I_{3,k}}.$$

The term $I_{1,k}$ can be dealt with analogously to I_1. After some calculation one can see that $I_{2,k}$ can be estimated by some constants plus terms that depend linearly on $|u_k^\varepsilon|$. To see this, one has to use the L^∞-stability of u^ε. Now we turn to the more problematic term $I_{3,k}$. By (2.1) and the product rule we can write

$$I_{3,k} =: T_{1,k} + \underbrace{\varepsilon \int_\Omega \psi \partial_k g^{ij} u_{ij}^\varepsilon s_\delta(u_k^\varepsilon)}_{=:T_{2,k}},$$

where, after partial integration, $T_{1,k}$ can be estimated similarly to I_1. To get an expression that is equivalent to the norm of the gradient of u^ε we sum over k from 1 to the dimension d. Now we need the restriction to two-dimensional manifolds, i.e., $d = 2$, in order to deal with the term $\sum_{k=1}^2 T_{2,k}$. Then we have

$$\sum_{k=1}^2 T_{2,k} = \sum_{k=1}^2 \epsilon \int_\Omega \psi \partial_k g^{ij} u_{ij}^\varepsilon s_\delta(u_k^\varepsilon) = \underbrace{\epsilon \int_\Omega \psi \partial_1 g^{22} u_{22}^\varepsilon s_\delta(u_1^\varepsilon)}_{A} + B,$$

where, after partial integration, the terms in B can be estimated similarly to I_1, respectively analogously to A. Note that in the term A the derivative of u^ε that occurs in the argument of s_δ is w.r.t. x^1 while both derivatives of the second u^ε are w.r.t. x^2. Thus, A cannot be dealt with by the technique we used for I_1. But, if we solve the original differential equation (3.5) for u_{22}^ε and replace it in A by the resulting term, we can apply this technique. Thus, all terms can be estimated so that we can apply a Gronwall's lemma, which proves (3.7). Existence of an entropy solution of (3.1) and (3.2) now follows from a compactness argument that provides a convergent subsequence of u^ε. The TV-Estimate (3.4) is obtained by letting $\varepsilon \to 0$ in (3.7). For uniqueness we refer to [4]. □

Remark 3.4. For scalar conservation laws posed in the Euclidean space one has a TVD property for the entropy solution, if the flux function does not depend explicitly on space. If the flux function depends explicitly on space we do not have this property in general (see Example 3.5). Since the flux function on manifolds is tangential at every point, it depends on space in general. Therefore, a TVD property on manifolds cannot be expected.

Example 3.5. The function $u(x_1, x_2, t) := u_0(x_1 - \sin(x_2)t, x_2)$ is an entropy solution of the initial value problem (3.1) and (3.2) for $M = \mathbb{R}^2$ and

$$\begin{aligned} \mathbf{f}(\bar{u}, x_1, x_2) &:= \bar{u}\,(\sin(x_2), 0)^T\,, \\ u_0(x_1, x_2) &:= \Psi(x_1)\Psi(x_2), \end{aligned}$$

where

$$\Psi(z) := \mathbb{1}_{\{|z-\pi|<\pi/2\}} \cos^2(z).$$

One can compute that the total variation increases in time.

4 Numerics

Finite volume schemes for conservation laws on Riemannian manifolds can be motivated as in the Euclidean case by averaging the conservation law over an element $K \subset M$ and applying the divergence theorem for manifolds. Using the averaged values as approximations and introducing numercal flux function one gets a finite volume scheme that formally looks like one in the Euclidean case. For more details see Amorim et al. [1].

We implemented this scheme for several test problems on the 2-sphere using DUNE [2]. This software package enabled us to do parallel computations on 16 CPUs for grids from 632 elements to 3.028.862 elements (see Table 1) in an adequate time.

We will present the results of the numerical experiments for the following two test problems. We define $\mathbf{V}(x, y, z) := (-y, x, 0)^T$.

Test Problem 1. Find a solution u of the initial value problem

$$\begin{aligned} \partial_t u + \operatorname{div}_g (2\pi u \mathbf{V}) &= 0 && \text{on } \mathbb{S}^2 \times \mathbb{R}^+, \\ u(x, y, z, 0) &= 4 \ _{\{x \geq 0,8\}}(x, y, z) && \text{for } (x, y, z) \in \mathbb{S}^2. \end{aligned}$$

Test Problem 2. Find a solution u of the initial value problem

$$\begin{aligned} \partial_t u + \operatorname{div}_g \left(\pi u^2 \mathbf{V}\right) &= 0 && \text{on } \mathbb{S}^2 \times \mathbb{R}^+, \\ u(x(\varphi, \theta), 0) &= \sin(\varphi) \sin^2(3\theta) \mathbb{1}_{\{|\theta - \pi/2| < \pi/6\}}(\theta) && \text{for } x(\varphi, \theta) \in \mathbb{S}^2, \end{aligned}$$

where the initial values are given in polar coordinates.

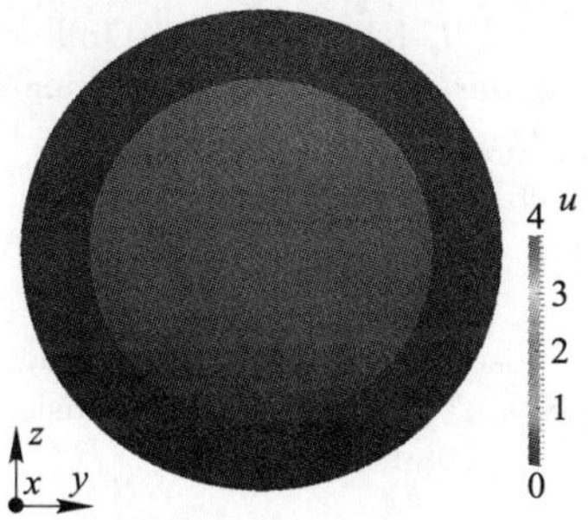

Figure 4.1 Initial values for test problem 1 (front view).

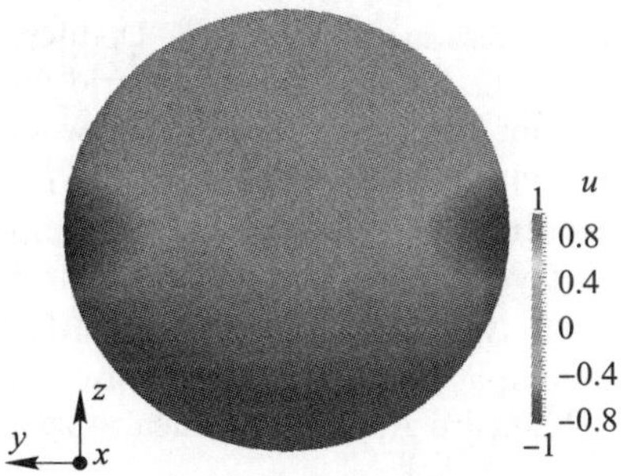

Figure 4.2 Initial values for test problem 2 (rear view).

The experimental order of convergence (EOC) for both problems can be seen in Table 1. It shows that convergence is obtained at a rate that one would expect for similar test problems posed in the Euclidean case.

Table 1 L^1-error and EOCs

grid elements	$\bar{h}$	Test Problem 1		Test Problem 2	
		$\|u-u^h\|_{L^1}$	EOC	$\|u-u^h\|_{L^1}$	EOC
632	0.2161	5.4033	—	0.3513	—
2.628	0.1061	4.2605	0.33	0.2136	0.70
11.164	0.0515	3.2533	0.37	0.1081	0.94
45.102	0.0256	2.1792	0.57	0.0544	0.98
187.682	0.0125	1.6105	0.42	0.0272	0.97
747.416	0.0063	1.1624	0.47	0.0141	0.95
3.028.862	0.0031	0.8357	0.47	0.0070	0.99

5 Conclusions

In this paper, we have proven a TV-estimate for entropy solutions of conservation laws posed on Riemannian manifolds. This estimate implies existence of a solution. Furthermore, we implemented a finite volume scheme for the case of a 2-sphere and tested it by computing the experimental order of convergence.

References

[1] P. Amorim, M. Ben-Artzi and P. G. LeFloch, Hyperbolic conser vation laws on manifolds: total variation estimates and the finite volume method. Methods Appl. Anal. 12 (2005), no. 3, 291–323.

[2] P. Bastian, M. Blatt, A. Dedner, C. Engwer, R. Kloefkorn, M. Ohlberger and O. Sander, A generic grid interface for parallel and adaptive scientific computing. I: Abstract framework. Computing 82 (2008), no. 2–3, 103–119.

[3] M. Ben-Artzi, J. Falcovitz and P. G. LeFloch, Hyperbolic conservation laws on the sphere. A geometry-compatible finite volume scheme. J. Comput. Phys. 228 (2009), no. 16, 5650–5668.

[4] M. Ben-Artzi and P. G. LeFloch, Well-posedness theory for geometry-compatible hyperbolic conservation laws on manifolds. Ann. Inst. H. Poincaré Anal. Non Linéaire 24 (2007), no. 6, 989–1008.

Properties of Path-conservative Schemes for Hyperbolic Systems of Balance Laws*

María Luz Muñoz-Ruiz
Dept. Matemática Aplicada, Universidad de Málaga
29071 Málaga, Spain
E-mail: munoz@anamat.cie.uma.es

Carlos Parés Madroñal
Dept. Análisis Matemático, Universidad de Málaga
29071 Málaga, Spain
E-mail: pares@anamat.cie.uma.es

Abstract

This paper is concerned with the numerical approximation of hyperbolic systems of balance laws in one space dimension. In order to make easier the obtaining of well-balanced numerical schemes, these systems can be considered as a particular case of hyperbolic nonconservative systems. For these more general problems, the chosen concept of weak solutions is that introduced by Dal Maso, LeFloch, and Murat [6], which is based on the prescription of a family of paths in the phases space. This prescription is also useful to construct the so called path-conservative schemes introduced in [8]. In [4] it was shown that, for general nonconservative systems, a rather strong convergence assumption is needed to prove a convergence result. Nevertheless, the numerical results obtained there for systems of balance laws suggested that, for that particular case, weaker hypotheses are enough to prove such a result, as we establish here. In addition, we study the relationship between the well-balanced properties of path-conservative schemes and the prescribed family of paths in this particular case of systems of balance laws. Finally, we discuss the well-balanced properties of some families of path-conservative schemes for these systems.

*The authors are partly supported by the Spanish Government Research Project MTM 2009-11923.

1 Introduction

This work deals with the numerical approximation of Cauchy problems for one-dimensional systems of conservation laws with source terms or balance laws

$$U_t + F(U)_x = S(U)\,\sigma_x, \tag{1.1}$$

where $U(x,t) \in \Omega$, with Ω being an open convex set of $\mathbf{R}^N$, $\sigma(x)$ is a known function from $\mathbf{R}$ to $\mathbf{R}$, F is a regular function from Ω to $\mathbf{R}^N$, and S is also a function from Ω to $\mathbf{R}^N$.

It is well known that standard methods that solve correctly systems of conservation laws can fail in solving (1.1), specially when approaching equilibria or near to equilibria solutions. In the context of shallow water equations, Bermúdez and Vázquez-Cendón introduced in [1] the condition called *C-property*: a scheme is said to satisfy this condition if it solves correctly the steady-state solutions corresponding to water at rest. This idea of constructing numerical schemes that preserve some equilibria, which are called in general *well-balanced* schemes, has been studied by many authors. The design of numerical schemes with good properties for problems of the form (1.1) is a very active front of research, as PDE systems of this nature arise in many flow models (see [3]).

Here, we will study systems (1.1) as a particular case of the more general systems

$$W_t + A(W)W_x = 0, \tag{1.2}$$

where the unknown $W(x,t)$ takes values on an open convex set and A is a smooth matrix function. This approach will allow us the discretization of the system as a whole, making thus easier to balance correctly the numerical flux and the source terms.

The presence of the nonconservative products $A(W)W_x$ in (1.2) makes difficult both the definition of weak solutions and the dessign of appropriate numerical schemes. We use the theory introduced by Dal Maso, LeFloch, and Murat in [6] in order to define the concept of weak solutions. This theory is based on the prescription of a family of paths in the phases space. The introduction of a family of paths not only gives a way to properly define weak solutions of nonconservative systems, it is also useful in the theoretical framework introduced in [8] for the design of finite difference methods for systems of the form (1.2). This framework is based on the concept of path-conservative numerical scheme, which is a generalization of the usual concept of conservative method for conservation laws.

In [4] it was shown that, for general nonconservative systems a rather strong convergence assumption is needed to prove a Lax-Wendroff type convergence theorem. In this work we prove that, for systems of balance laws (1.1) with smooth enough σ, a Lax-Wendroff type result can be proved for path-conservative numerical schemes assuming the same hypotheses of convergence of the classical theorem for conservative problems, as the numerical results shown in [4] seemed to suggest. Together with the hypothesis concerning the convergence of the approximations, some minimal assumptions both on the chosen family of paths and on the numerical scheme are required, so that if σ is constant the usual notion of weak solution and a conservative numerical scheme are recovered.

In general, the assumptions on the family of paths required to prove the Lax-Wendroff theorem are not enough to ensure the well-balanced property. In this work we also present some general results showing that, in order to obtain well-balanced schemes, some further assumptions have to be imposed to the family of paths, concerning the relationship between the paths and the integral curves of the linearly degenerate field of the system associated to the null eigenvalue. Finally, the well-balanced properties of some particular families of path-conservative numerical schemes are discussed.

2 Weak solutions

The trivially satisfied $\sigma_t = 0$ is added to system (1.1) in order to rewrite it in the nonconservative form form (1.2) with

$$W = \begin{bmatrix} U \\ \sigma \end{bmatrix}, \quad A(W) = \begin{bmatrix} J(U) & -S(U) \\ 0 & 0 \end{bmatrix}, \quad J(U) = \frac{\partial F}{\partial U}(U). \tag{2.1}$$

Clearly, the eigenvalues of $A(W)$ will be the N eigenvalues of $J(U)$ plus the eigenvalue 0. We will suppose that the ones corresponding to $J(U)$ are distinct and non-vanishing real numbers, in order to obtain a strictly hyperbolic system. We will also suppose that the only linearly degenerate characteristic field is the $(N+1)$th one.

For discontinuous σ, the term $S(U)\sigma_x$ does not make sense in the distributions sense. Nevertheless, we are interested in the study of discontinuous σ for two reasons. The first one is that piecewise constant approximations of σ will be considered in the numerical schemes considered here and therefore, numerical schemes based on exact or approximate Riemann solvers have to deal with the lack of sense in the distributional framework of the source term. The second one is that the well-balanced property of the numerical schemes is strongly related to their ability to handle with the contact discontinuities added to the system when σ is allowed to be discontinuous.

In order to define weak solutions of (1.2) for discontinuous σ, according to the theory developed in [6], a family of paths

$$\widetilde{\Phi}(s; W_L, W_R) = \begin{bmatrix} \Phi(s; W_L, W_R) \\ \Phi_{N+1}(s; W_L, W_R) \end{bmatrix},$$

where

$$\Phi(s; W_L, W_R) = \begin{bmatrix} \Phi_1(s; W_L, W_R) \\ \vdots \\ \Phi_N(s; W_L, W_R) \end{bmatrix},$$

$W_L = [U_L, \sigma_L]^T$ and $W_R = [U_R, \sigma_R]^T$, has to be chosen.

A family of paths is a locally Lipschitz map

$$\widetilde{\Phi}\colon [0,1] \times (\Omega \times \mathbb{R}) \times (\Omega \times \mathbb{R}) \to (\Omega \times \mathbb{R})$$

satisfying some natural properties as

$$\widetilde{\Phi}(0; W_L, W_R) = W_L, \quad \widetilde{\Phi}(1; W_L, W_R) = W_R.$$

The family of paths $\widetilde{\Phi}$ allows one to give a sense to the nonconservative product $A(W)W_x$ as a bounded measure for $W \in (L^\infty(\mathbb{R} \times \mathbb{R}^+) \cap BV(\mathbb{R}\times\mathbb{R}^+))^N$. A notion of entropy weak solution, related to the chosen paths, can now be given. Across a discontinuity, a weak solution must satisfy a generalized Rankine-Hugoniot condition that can be written as

$$\begin{cases} \xi(U^+ - U^-) = F(U^+) - F(U^-) - S_{\widetilde{\Phi}}(W^-, W^+), \\ \xi(\sigma^+ - \sigma^-) = 0, \end{cases} \tag{2.2}$$

where ξ is the speed of propagation of the discontinuity, W^- and W^+ are the left and right limits of the solution at the discontinuity, and

$$S_{\widetilde{\Phi}}(W^-, W^+) = \int_0^1 S(\Phi(s; W^-, W^+) \frac{\partial \Phi_{N+1}}{\partial s}(s; W^-, W^+))\, ds. \tag{2.3}$$

According to (2.2), the discontinuities appearing in weak solutions have to be either stationary or develop in regions where σ is continuous.

Any arbitrary choice of the family of paths leads to a consistent mathematical meaning of weak solutions. Nevertheless, following [7], it is natural to impose some requirements on that family, such as

$$\Phi_{N+1}(s; W_L, W_R) = \bar{\sigma} \quad \forall s \in [0,1], \tag{2.4}$$

if W_L and W_R are such that $\sigma_L = \sigma_R = \bar{\sigma}$. As a consequence of those requirements, the solutions of Riemann problems are found to be the same for every family of paths satisfying them, as the families of simple curves are independent of the family of paths. In addition, a family of paths satisfying the required conditions can be constructed at least for close enough states. In the sequel, we will assume that the family of paths chosen to define weak solutions satisfies (2.4) and the other requirements established in [7]. In addition, for continuous σ, every family of paths satisfying (2.4) defines the same concept of weak solution.

3 Numerical schemes

We consider numerical schemes that are path-conservative in the sense introduced in [8]. These schemes can be written for the particular case of systems (1.1) in the form

$$U_i^{n+1} = U_i^n - \frac{\Delta t}{\Delta x}\big(D_{i-1/2}^{n,+} + D_{i+1/2}^{n,-}\big), \tag{3.1}$$

where $D_{i+1/2}^{n,\pm} = D^{\pm}\big(W_{i-q}^n, \ldots, W_{i+p}^n\big)$, with D^- and D^+ being two continuous functions satisfying

$$D^{\pm}(W, \ldots, W) = 0 \quad \forall W \in \Omega \times \mathbb{R}, \tag{3.2}$$

together with

$$\begin{aligned} D^-(W_{-q}, \ldots, W_p) &+ D^+(W_{-q}, \ldots, W_p) \\ &= F(U_1) - F(U_0) - S_{\widetilde{\Phi}}(W_0, W_1), \end{aligned} \tag{3.3}$$

for $W_i \in \Omega \times \mathbb{R}$, $i = -q, \ldots, p$.

The path-conservative schemes constructed using the family of paths used to define weak solutions can be characterized by

$$\begin{aligned} D^-\big(W_{-q}, \ldots, W_p\big) &+ D^+\big(W_{-q}, \ldots, W_p\big) \\ &= F(U_1) - F(U_{1/2}^+) + F(U_{1/2}^-) - F(U_0), \end{aligned} \tag{3.4}$$

where $W_{1/2}^- = [U_{1/2}^-, \sigma_0]^T$ and $W_{1/2}^+ = [U_{1/2}^+, \sigma_1]^T$ are, respectively, the left and right limits at $x = 0$ of the solution of the Riemann problem associated to W_0 and W_1.

Although it should be desirable to use the same family of paths to define weak solutions and to construct the numerical scheme, the construction of $\widetilde{\Phi}$-conservative schemes is not always possible or reasonable from the computational point of view, as the election of $\widetilde{\Phi}$ is restricted by the mathematical conditions mentioned at the end of the previous section, and a path-conservative scheme for an arbitrary family of paths $\widetilde{\Psi}$ has to be considered.

In fact, at least for $q = 0$ and $p = 1$, the $\widetilde{\Psi}$-conservative schemes which reduce to a conservative scheme in regions where σ is constant can be characterized and can be written in the form

$$U_i^{n+1} = U_i^n - \frac{\Delta t}{\Delta x}\big(G_{i+1/2}^n - G_{i-1/2}^n\big) + \frac{\Delta t}{\Delta x}\big(H_{i-1/2}^{n,+} + H_{i+1/2}^{n,-}\big)\,, \tag{3.5}$$

where $G_{i+1/2}^n = G\big(U_i^n, U_{i+1}^n\big)$ and $H_{i+1/2}^{n,\pm} = H^{\pm}\big(W_i^n, W_{i+1}^n\big)$, with G, H^-, and H^+ being Lipschitz continuous functions such that

$$G(U, U) = F(U)\,, \tag{3.6}$$

$$H^{\pm}(W, W) = 0\,, \tag{3.7}$$

$$H^{\pm}(\overline{W}_0, \overline{W}_1) = 0\,, \tag{3.8}$$

and

$$H^{-}(W_0, W_1) + H^{+}(W_0, W_1) = S_{\tilde{\Psi}}(W_0, W_1)\,. \tag{3.9}$$

The notation $\overline{W}$ has been used for W with $\sigma = \bar{\sigma}$, for a fixed value $\bar{\sigma}$.

The convergence of path-conservative schemes for general nonconservative systems (1.2) has been recently studied in [4]. In fact, it has been proven that, if the approximations generated by a path-conservative scheme converge in the uniform sense of graphs (see [6] for more details on this notion of convergence), then the limit is a weak solution of the nonconservative system. Unfortunately, this convergence is rather strong and usually fails in applications (although it holds for random choice methods). It can also be proved that, if the approximations just converge almost everywhere (which is a more realistic assumption from the practical point of view), a convergence error term appears in the limiting system. In the particular case of systems of balance laws this lack of convergence was not appreciated in the numerical experiments developed in [4] if a correct choice of the family of paths is done.

In fact, we can prove that, for continuous σ, any family of paths satisfying condition (2.4) provides a numerical scheme that converges to the weak solutions of the problem. Nevertheless, further assumptions on the paths have to be imposed if we also want the scheme to be well-balanced.

In particular, a numerical scheme of the form (3.5) is well-balanced for an integral curve of the linearly degenerate field if and only if, for every W_0, W_1 in that curve:

$$H^{-}(W_0, W_1) = G(U_0, U_1) - F(U_0)$$

and

$$H^{+}(W_0, W_1) = F(U_1) - G(U_0, U_1).$$

It can also be proved that a necessary condition to obtain a scheme that is well-balanced for a particular integral curve of the linearly degenerate field is:

(C) The path connecting two states belonging to the integral curve is a parametrization of the arc of the curve connecting the states.

Although this condition is not sufficient to obtain a well-balanced scheme, in most cases it ensures the well-balanced property of a numerical scheme, as we will see in the following section.

4 Examples

The extension of the classical Godunov method to systems of balance laws presented in [7] is path-conservative for the family of paths used to define weak solutions and is exactly well-balanced for any integral curve of the linearly degenerate field.

A Roe scheme which is path-conservative for a family of paths $\widetilde{\Psi}$ can be written in the form (3.5) with the choices

$$G^n_{i+1/2} = \frac{1}{2}(F(U^n_i) + F(U^n_{i+1})) - \frac{1}{2}\left|J^n_{i+1/2}\right|(U^n_{i+1} - U^n_i),$$
$$H^{n,\pm}_{i+1/2} = \mathcal{P}^{n,\pm}_{i+1/2} \cdot S^n_{i+1/2}(\sigma_{i+1} - \sigma_i),$$

where $J^n_{i+1/2}$ is a Roe matrix for the homogeneous problem, i.e., a matrix satisfying

$$J^n_{i+1/2}(U^n_{i+1} - U^n_i) = F(U^n_{i+1}) - F(U^n_i),$$

$S^n_{i+1/2}$ is a vector satisfying

$$S^n_{i+1/2}(\sigma_{i+1} - \sigma_i) = S_{\widetilde{\Phi}}(W^n_i, W^n_{i+1}),$$

and

$$\mathcal{P}^{n,\pm}_{i+1/2} = \frac{1}{2}\left(Id \pm \left|J^n_{i+1/2}\right|(J^n_{i+1/2})^{-1}\right).$$

Here, Id represents the identity matrix and

$$|J^n_{i+1/2}| = K^n_{i+1/2}\left|L^n_{i+1/2}\right|(K^n_{i+1/2})^{-1},$$

where $\left|L^n_{i+1/2}\right|$ is the diagonal matrix whose coefficients are the absolute value of the eigenvalues of $J^n_{i+1/2}$ and $K^n_{i+1/2}$ a $N \times N$ matrix whose columns are associated eigenvectors.

A Roe scheme is well-balanced for an integral curve of the linearly degenerate field if the family of paths satisfies the condition (C). If the family of straight segments is chosen, the numerical scheme is still exactly well-balanced for the integral curves which are straight lines in the phases space. Moreover, all the smooth stationary solutions are solved up to second order (see [9]).

A path-conservative extension of the classical Lax-Friedrichs scheme based on a Roe linearization is given by (3.5) with the choices

$$G^n_{i+1/2} = \frac{1}{2}\left(F(U^n_i) + F(U^n_{i+1}) - \frac{\Delta x}{\Delta t}(U^n_{i+1} - U^n_i)\right),$$
$$H^{n,\pm}_{i+1/2} = \frac{1}{2}\left(Id \pm \frac{\Delta x}{\Delta t}(J^n_{i+1/2})^{-1}\right) S^n_{i+1/2}(\sigma_{i+1} - \sigma_i),$$

where $J^n_{i+1/2}$ and $S^n_{i+1/2}$ are defined as above. This scheme is also well-balanced for an integral curve of the linearly degenerate field if (C) is satisfied.

In [5] the Hydrostatic Reconstruction technique, introduced for the shallow water system in [2], has also been generalized to obtain a numerical scheme which is exactly well-balanced for a set of integral curves of the linearly degenerate field.

References

[1] A. Bermúdez and M. E. Vázquez, Upwind methods for hyperbolic conservation laws with source terms. Computers and Fluids 23 (1994), 1049–1071.

[2] E. Audusse, F. Bouchut, M. O. Bristeau, R. Klein and B. Perthame, A fast and stable well-balanced scheme with hydrostatic reconstruction for shallow water flows. SIAM J. Sci. Comp. 25 (2004), 2050–2065.

[3] F. Bouchut, Nonlinear Stability of Finite Volume Methods for Hyperbolic Conservation Laws and Well-Balanced Schemes for Sources. Birkhäuser, 2004.

[4] M. J. Castro, P. G. LeFloch, M. L. Muñoz-Ruiz and C. Parés, Why many theories of shock waves are necessary: Convergence error in formally path-consistent schemes. J. Comput. Phys. 227 (2008), 8107–8129.

[5] M. J. Castro, A. Pardo and C. Parés, Well-balanced numerical schemes based on a generalized hydrostatic reconstruction technique. Math. Mod. Meth. App. Sci. 17 (2007), 2055–2113.

[6] G. Dal Maso, P. G. LeFloch and F. Murat, Definition and weak stability of nonconservative products. J. Math. Pures Appl. 74 (1995), 483–548.

[7] M. L. Muñoz-Ruiz and C. Parés, Godunov method for nonconservative hyperbolic systems. Math. Model. Numer. Anal. 41 (2007), 169–185.

[8] C. Parés, Numerical methods for nonconservative hyperbolic systems: A theoretical framework. SIAM J. Numer. Anal. 44 (2006), 300–321.

[9] C. Parés and M. J. Castro, On the well-balanced property of Roe's method for nonconservative hyperbolic systems. Applications to shallow water systems. Math. Model. Numer. Anal. 38 (2004), 821–852.

A Fast Sweeping Method for Computing the Effective Hamiltonian

Martin Nolte

Section of Applied Mathematics, University of Freiburg
Hermann-Herder-Straße 10, 79104 Freiburg, Germany
E-mail: nolte@mathematik.uni-freiburg.de

Abstract

Fast sweeping methods have proven to be very efficient solvers of Dirichlet problems for stationary Hamilton-Jacobi equations. We propose a fast sweeping method to approximate a periodic solution of the cell problem for the effective Hamiltonian and, hence, its value. An outline of the convergence proof as well as numerical evidence for the efficiency of the method are given.

1 Introduction

The last two decades have seen a rising interest in homogenization. For example, in [5] the limit $\varepsilon \to 0$ of (a generalization of) the advection-diffusion-reaction equation

$$\partial_t T_t^\varepsilon + V\Big(\frac{x}{\sqrt{\varepsilon}}\Big) \cdot \nabla T^\varepsilon - \varepsilon \Delta T^\varepsilon - \frac{T(1-T)}{\varepsilon} = 0$$

with non-zero, compactly supported initial data were studied. It turned out that the limiting function is $\{0,1\}$-valued. The evolution of a level set function u describing the front separating these two states is governed by the equation

$$\partial_t u = F(\nabla u), \qquad F(p) := \inf_{\lambda>0} \lambda\Big[\bar{\mathrm{H}}\Big(\frac{p}{\lambda}\Big) - f'(0)\Big].$$

Here, $\bar{\mathrm{H}}$ is the effective Hamiltonian of

$$\mathrm{H}(x,p) = -\kappa|p|^2 + V(x) \cdot p,$$

which is defined as follows:

Definition 1.1. Let $\mathrm{H} = \mathrm{H}(x, p)$ be a 1-periodic Hamiltonian. Its effective Hamiltonian $\bar{\mathrm{H}}$ maps each $p \in \mathbb{R}^n$ to the unique constant $\bar{\mathrm{H}}(p)$ such that

$$\mathrm{H}(x, p + \nabla u) = \bar{\mathrm{H}}(p) \tag{1.1}$$

has a 1-periodic (viscosity) solution u.

Solving (1.1) numerically turns out to be difficult for the following reasons: there are two unknowns (u and $\bar{\mathrm{H}}(p)$), u is in general not unique (not even up to a constant), and the problem has no boundary values.

Several techniques for approximating the effective Hamiltonian have been suggested. It is possible to base numerical schemes on the inf-sup-formula

$$\bar{\mathrm{H}}(p) = \inf_{u \in C^\infty} \sup_{x \in \mathbb{R}^n} \mathrm{H}(x, p + \nabla u), \tag{1.2}$$

where the infimum is taken over all smooth, 1-periodic functions on $\mathbb{R}^n$. Directly discretizing this formula leads to a classical minimax problem (see [3]). In [2], a highly nonlinear, elliptic problem was derived from (1.2).

Another approach is solving the evolution problem

$$\partial_t u + \mathrm{H}(x, p + \nabla u) = 0$$

for arbitrary initial data. In the large time limit, $\partial_t u$ converges to $-\bar{\mathrm{H}}(p)$. Derivation of this equation with respect to x yields a system of hyperbolic conservation laws. This approach was, e.g., applied in [4].

Finally, we mention the possibility of regularizing (1.1) as

$$\lambda u_\lambda + \mathrm{H}(x, p + \nabla u_\lambda) = 0 \tag{1.3}$$

for some small $\lambda > 0$.

In this paper we suggest to use a fast sweeping method to solve (1.3) numerically. Fast sweeping schemes are known to be very efficient in solving Dirichlet problems for stationary Hamilton-Jacobi equations (see, e.g., [7]). We will show that they can also be applied to aquire periodic solutions of (1.3). Numerical evidence shows that the indepence of the number of iterations from the grid size is retained.

The paper is organized as follows: In Section 2.1, we outline how solutions of (1.3) approximate the effective Hamiltonian. Then, we briefly describe the fast sweeping scheme in Section 2.2 and sketch the convergence proof for the method in the periodic context in Section 2.3. We then validate this scheme numerically in Section 3.1 and demonstrate in Section 3.2 its practical performance on the homogenization problem described above. The presentation ends with some concluding remarks in Section 4.

2 A Fast sweeping scheme

2.1 Construction of a solution

To obtain a numerical approximation of the effective Hamiltonian, we consider the regularized cell problem (1.3). For $\lambda > 0$, this equation has a unique 1-periodic solution u_λ. The solution u_λ can be proved to be uniformly Lipschitz continuous in λ and, hence, by the Arzelà-Ascoli theorem, there is a subsequence (which we also denote by u_λ) such that

$$u_\lambda - \min u_\lambda \xrightarrow{\lambda \to 0} \bar{u}.$$

Since $-\lambda u_\lambda$ thus converges to some constant, it is easy to see that $\bar{u}$ is a 1-periodic solution of (1.1). Moreover, the following convergence rate can be proved (see, e.g., [6]):

$$\|\bar{\mathrm{H}}(p) + \lambda u_\lambda\|_\infty = \mathcal{O}(\lambda).$$

2.2 A brief description of the scheme

Fast sweeping schemes (see, e.g., [7]) have proven to be very efficient methods for solving Dirichlet problems for stationary Hamilton-Jacobi equations. In this section we briefly describe a simple fast sweeping scheme for obtaining periodic solutions of (1.3). The key difference to standard fast sweeping methods is the periodic boundary conditions.

For the sake of presentation, we will restrict ourselves to a 2D Cartesian grid and consider only Hamiltonians of the form

$$\mathrm{H}(x,p) = \frac{|p|^2}{2} + E(x).$$

Let us introduce a 2D Cartesian grid for the unit square with N cells in each direction and denote the vertices by $x_{i,j} = (ih, jh)$ with $h = \frac{1}{N}$. Using a Godunov-type Hamiltonian, we arrive at the discrete equation

$$\lambda u_{i,j} + \frac{1}{2}\Big(\frac{u_{i,j} - u_{i\pm,j}}{h} + p_1\Big)_+^2 + \frac{1}{2}\Big(\frac{u_{i,j} - u_{i,j\pm}}{h} + p_2\Big)_+^2 + E(x_{i,j}) = 0, \tag{2.1}$$

where $u_{i\pm,j} := \min\{u_{i+1,j}, u_{i-1,j}\}$ and $u_{i,j\pm} := \min\{u_{i,j+1}, u_{i,j-1}\}$. To satisfy the periodicity condition, we have to set $u_{i+N\mathbb{Z},j+N\mathbb{Z}} = u_{i,j}$.

It can be proved (see, e.g., [6]) that (2.1) is uniquely solvable and that

$$\lambda\|u_\lambda - u_{\lambda,h}\|_\infty = \mathcal{O}(h^{\frac{1}{2}}).$$

A central building block of the fast sweeping method is the local solvability of (2.1), i.e., given $u_{i-1,j}$, $u_{i+1,j}$, $u_{i,j-1}$, $u_{i,j+1}$, this equation can be solved for $u_{i,j}$. The reason is that the left hand side of (2.1) is strictly increasing in $u_{i,j}$.

The fast sweeping method then works as follows: Let u^n be a supersolution of (2.1) and denote by $s_{i,j}^{n+1}$ the result of the local solver applied to $w_{i\pm1,j}$, $w_{i,j\pm1}$ with

$$w_{k,l} = \begin{cases} u_{k,l}^{n+1}, & \text{if already computed,} \\ u_{k,l}^{n}, & \text{otherwise.} \end{cases} \tag{2.2}$$

We then set $u_{i,j}^{n+1} := \min\{u_{i,j}^n, s_{i,j}^{n+1}\}$. This computation is done in alternating sweeping directions

1. $i = 0, \ldots, N\,; j = 0, \ldots, N$,
2. $i = N, \ldots, 0\,; j = 0, \ldots, N$,
3. $i = 0, \ldots, N\,; j = N, \ldots, 0$,
4. $i = N, \ldots, 0\,; j = N, \ldots, 0$.

These sweeping directions are cycled periodically.

2.3 Convergence

In this section we will outline the convergence of the iterates u^n to the unique 1-periodic solution of the discrete problem (2.1).

The sequence $(u_{i,j}^n)_n$ is decreasing and bounded from below and, hence, has a limit $u_{i,j}^* \in \mathbb{R}$. Obviously, u^* is a fixpoint of the iteration.

Moreover, each iterate u^n is a discrete supersolution, i.e.,

$$\lambda u_{i,j} + \frac{1}{2}\Big(\frac{u_{i,j} - u_{i\pm,j}}{h} + p_1\Big)_+^2 + \frac{1}{2}\Big(\frac{u_{i,j} - u_{i,j\pm}}{h} + p_2\Big)_+^2 + E(x_{i,j}) \geq 0. \tag{2.3}$$

To see this, assume that we want to compute $u_{i,j}^{n+1}$ and that w defined in (2.2) is a discrete supersolution. We want to show that

$$\tilde{w}_{k,l} := \delta_{k,i}\delta_{l,j}u_{i,j}^{n+1} + (1 - \delta_{k,i}\delta_{l,j})w_{k,l}$$

is a supersolution, too. Indeed, (2.3) holds in (i,j) because $\tilde{w}_{i,j}$ is either the old value or the local solution. For any other (k,l), $\tilde{w}$ also satisfies (2.3), because its left hand side is decreasing in $u_{i\pm,j}$ and $u_{i,j\pm}$. By induction we thus see that u^{n+1} is a discrete supersolution.

If $u_{i,j}^{n+1} = u_{i,j}^n$ then $s_{i,j}^{n+1} \geq u_{i,j}^n$ and u^{n+1} is a local subsolution. Hence, a fixpoint of the iteration is a subsolution of (2.1).

Putting these results together, we have proved that u^* is a subsolution and a supersolution of (2.1) and, thus, a solution.

3 Numerical results

3.1 Validation of the scheme

In this section we will test the fast sweeping scheme described in Section 2.2 on a Hamiltonian for which the effective Hamiltonian is known analytically. This will also give insight into the convergence properties of the scheme.

In the one-dimensional case, the effective Hamiltonian can be computed analytically (see, e.g., [1]):

Lemma 3.1. *Let* $\mathrm{H}(x,p) = \frac{1}{2}|p|^2 + E(x)$ *be a continuous,* 1*-periodic,* 1*-dimensional Hamiltonian and define*

$$\mu(\alpha) := \int_0^1 \sqrt{2(\alpha - E(x))}\, dx.$$

Then, denoting $\mu^* := \mu(\max E)$*, the effective Hamiltonian is given by the following formula:*

$$\bar{\mathrm{H}}(p) = \begin{cases} \max E, & \text{if } |p| \le \mu^*, \\ \mu^{-1}(|p|), & \text{if } |p| \ge \mu^*. \end{cases}$$

Moreover, for a Hamiltonian of the form $\mathrm{H}(x,p) = \sum_{i=1}^n \mathrm{H}_i(x_i, p_i)$, the effective Hamiltonian is given by $\bar{\mathrm{H}}(p) = \sum_{i=1}^n \bar{\mathrm{H}}_i(p_i)$.

In order to validate the numerical scheme, we choose the Hamiltonian

$$\mathrm{H}_i(x,p) = \mathrm{H}_{1\mathrm{D}}(x,p) := \frac{|p|^2}{2} + \sin 2\pi x.$$

The effective Hamiltonian for $\mathrm{H}_{1\mathrm{D}}$ for $p \in [0.5, 2]$ is plotted in Figure 3.1.

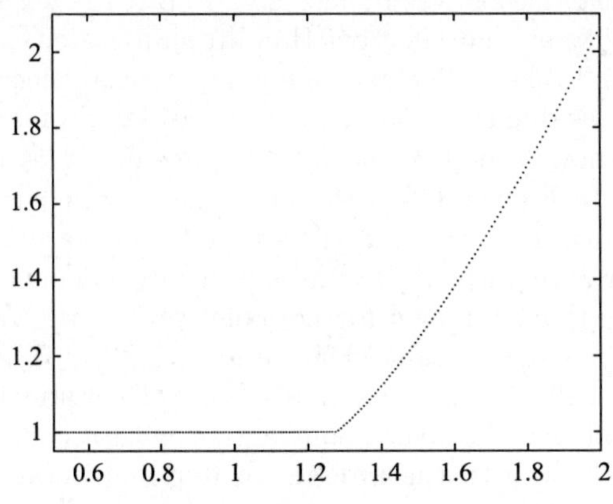

Figure 3.1 $\bar{\mathrm{H}}_{1\mathrm{D}}$ for $p \in [0.5, 2]$.

Table 1 Number of iterations and L^∞-error for the approximation of the effective Hamiltonian of $\mathrm{H}(x,p) = \frac{|p|^2}{2} + \sin 2\pi x_1 + \sin 2\pi x_2$ for different values of λ and h in three different points $p \in \mathbb{R}^2$.

(a) Iterations for $p = (0.8, 0.8)$

Iterations	$h = \frac{1}{32}$	$h = \frac{1}{128}$	$h = \frac{1}{512}$
$\lambda = 0.16$	17	17	17
$\lambda = 0.04$	17	17	17
$\lambda = 0.01$	17	17	17

(d) Error for $p = (0.8, 0.8)$

Error	$h = \frac{1}{32}$	$h = \frac{1}{128}$	$h = \frac{1}{512}$
$\lambda = 0.16$	0.05318	0.04869	0.04722
$\lambda = 0.04$	0.01341	0.01224	0.01195
$\lambda = 0.01$	0.00336	0.00306	0.00299

(b) Iterations for $p = (1.27, 1.27)$

Iterations	$h = \frac{1}{32}$	$h = \frac{1}{128}$	$h = \frac{1}{512}$
$\lambda = 0.16$	201	17	17
$\lambda = 0.04$	17	17	17
$\lambda = 0.01$	17	17	17

(e) Error for $p = (1.27, 1.27)$

Error	$h = \frac{1}{32}$	$h = \frac{1}{128}$	$h = \frac{1}{512}$
$\lambda = 0.16$	0.00721	0.00280	0.00145
$\lambda = 0.04$	0.00174	0.00061	0.00028
$\lambda = 0.01$	0.00042	0.00013	0.00006

(c) Iterations for $p = (1.6, 1.6)$

Iterations	$h = \frac{1}{32}$	$h = \frac{1}{128}$	$h = \frac{1}{512}$
$\lambda = 0.16$	1105	1097	1057
$\lambda = 0.04$	4369	4357	4185
$\lambda = 0.01$	17437	17397	16701

(f) Error for $p = (1.6, 1.6)$

Error	$h = \frac{1}{32}$	$h = \frac{1}{128}$	$h = \frac{1}{512}$
$\lambda = 0.16$	0.000783	0.000241	0.000106
$\lambda = 0.04$	0.000185	0.000049	0.000015
$\lambda = 0.01$	0.000046	0.000012	0.000003

In the interval $[-\frac{4}{\pi}, \frac{4}{\pi}]$, the effective Hamiltonian is constantly 1, while outside of this interval it is strictly convex. In $\pm\frac{4}{\pi}$ is not differentiable.

As a test, we approximate the effective Hamiltonian in the flat region at $p = (0.8, 0.8)$, near the kink at $p = (1.27, 1.27)$, and in the strictly convex region at $p = (1.6, 1.6)$. The number of iterations required such that the L^∞-distance between successive iterates reaches 10^{-8} as well as the error between the effective Hamiltonian and its approximation are displayed in Table 1. The extremely fast convergence within the flat region seems to be due to the alignment of the Hamiltonian and the grid.

Figure 3.2 shows a plot of the L^∞-distance of successive iterates for $p = (1.6, 1.6)$. In Figure 3.2(a) the grid size is varied while keeping λ constant. The distances between successive iterates are almost parallel, i.e., the number of required iterations is basically independent of the grid size, a property that is known for Dirichlet problems. Varying λ while keeping the grid size constant yields Figure 3.2(b). The dashed lines correspond to $c_1 e^{-c_2 \lambda x}$ for constants c_1, c_2. Heuristically, this shows that the method behaves like a contraction with a contraction rate of $1 - c_2\lambda$. In other words, the number of required iterations is roughly proportional to λ^{-1}.

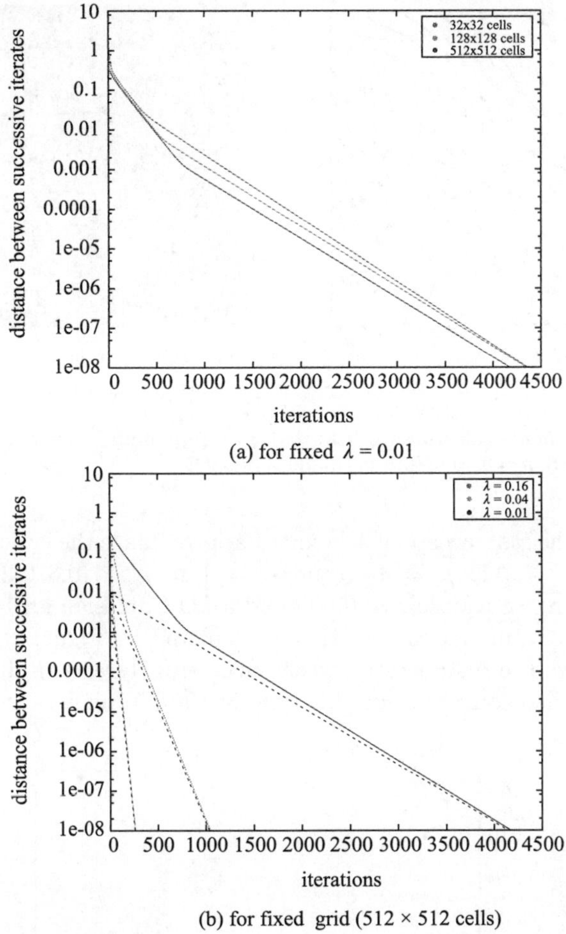

Figure 3.2 Distances between successive iterates for $p = (1.6, 1.6)$.

3.2 Speed of a flame front

In this section we return to our initial example and consider the Hamiltonian

$$\mathrm{H}(x,p) := \kappa|p|^2 + \big(\bar{v} + \mu v(x)\big) \cdot p.$$

As in [4], we choose the small scale velocity field $v = (-\partial_2\psi, \partial_1\psi)$ to be given by the stream lines of

$$\psi(x) = \sin(2\pi x_1)\sin(2\pi x_2) + \delta\cos(2\pi x_1)\cos(2\pi x_2).$$

For $\delta = 0.5$, a plot of this velocity field is shown in Figure 3.3(a).

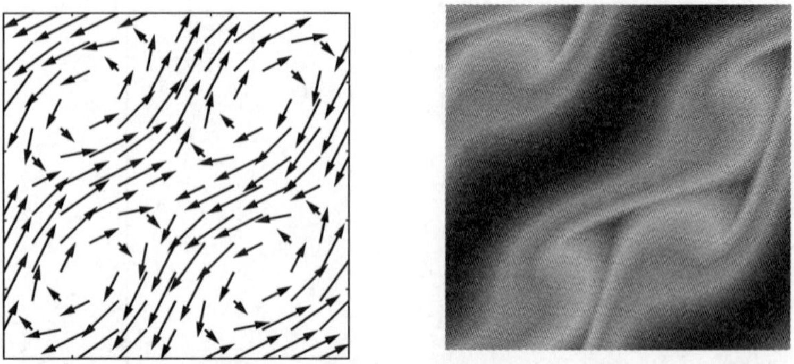

(a) small scale velocity field v

(b) numerical solution u_λ of (1.3) on a 2D Cartesian grid with 65536 cells

Figure 3.3 Small scale volecity field and a corresponding numerical solution for $\delta = 0.5$, $\mu = 1.6$, $\kappa = 1$, $p = (0.3, 0.3)$, and $\lambda = 0.005$.

We use the fast sweeping scheme to approximate the effective Hamiltonian for $\delta = 0.5$, $\mu = 1.6$, and $\kappa = 1$ in $p = (0.3, 0.3)$. For the discretization, we choose $\lambda = 0.005$ and a 2D Cartesian grid with 256×256 cells. The numerical solution is depicted in Figure 3.3(b). Figure 3.4 shows the distances between successive iterates, indicating that the scheme also converges for this type of Hamiltonian.

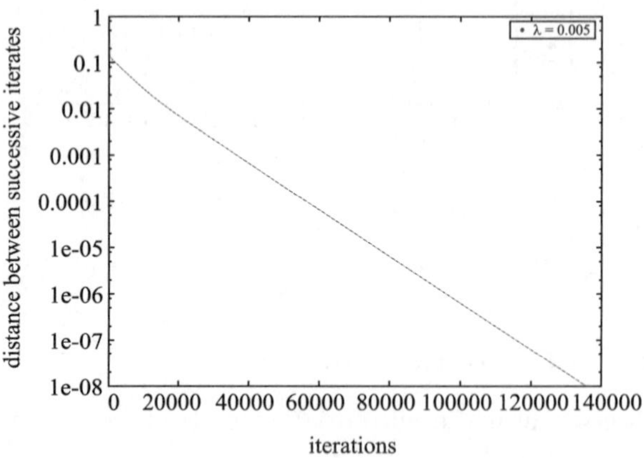

Figure 3.4 Distance between successive iterates in $p = (0.3, 0.3)$.

4 Concluding remarks

In this paper, we have discussed an approximation of the effective Hamiltonian through a fast sweeping method. In particular, the convergence of the fast sweeping method in the presence of periodic boundary conditions has been proved. Apart from numerical validation of the scheme, we gave numerical evidence for the independence of the number of required iterations from the grid size. Finally, we showed that the method is well capable of handling relevant applications.

References

[1] M. C. Concordel, Periodic homogenization of Hamilton-Jacobi equations: Additive eigenvalues and variational formula. Indiana Univ. Math. J. 45 (1996), no. 4, 1095–1117.

[2] L. C. Evans, Some new PDE methods for weak KAM theory. Calc. Var. Partial Differ. Equ. 17 (2003), no. 2, 159–177.

[3] D. Gomes and A. M. Oberman, Computing the effective Hamiltonian using a variational approach. SIAM J. Crontrol Optim. 43 (2004), no. 3, 792–812.

[4] B. Khouider and A. Bourlioux, Computing the effective Hamiltonian in the Majda-Souganidis model of turbulent premixed flames. SIAM J. Numer. Anal. 40 (2002), no. 4, 1330–1353.

[5] A. J. Majda and P. E. Souganidis, Large scale front dynamics for turbulent reaction-diffusion equations with separated velocity scales. Nonlinearity 7 (1994), no. 1, 1–30.

[6] M. Nolte, Efficient numerical approximations of the effective Hamiltonian. Ph.D. thesis, Albert-Ludwigs-Universität Freiburg, To appear.

[7] H. Zhao, A fast sweeping method for eikonal equations. Math. Comput. 74 (2005), no. 250, 603–627.

A Spectral Difference Method with Dubiner Polynomials on Triangulations*

Martina Pawlak, Thomas Sonar
Institut Computational Mathematics
Technische Universität Braunschweig, 38106 Braunschweig, Germany
E-mail: m.pawlak@tu-bs.de, t.sonar@tu-bs.de

Abstract

This current work focusses on developing a Spectral Difference Method using Dubiner expansion instead of Lagrange interpolation. The Dubiner basis is introduced and modifications occurring in the update scheme are presented. Furthermore, attention is paid to possible distributions for the flux points on a standard triangle. First numerical tests for the linear advection equation show convergence and the need for stabilization for high orders.

1 Introduction

The Spectral Difference Method (SDM) has been recently introduced by Z. J. Wang et. al. [3, 5] as an alternative to the Discontinuous Galerkin Method with lower computational costs for hyperbolic conservation laws. The main part is the flux reconstruction at certain flux points using Lagrange interpolation polynomials. Riemann solvers are used to couple cells with each other, and the solution is updated by a classical time stepping scheme. In this work, we choose Dubiner instead of Lagrange polynomials for the flux reconstruction and show the modifications in the update scheme, especially different possible distributions for the flux points on a standard triangle. Our aim is to compare different basis polynomials respective their properties in the SDM and, later on, to stabilize this method using spectral filtering based on the chosen expansion basis [6]. In this work, first numerical results for the linear case with Dubiner expansion are presented.

This paper is organized as follows: In the first part, the classical SDM is introduced and some aspects of the conservation property are inspected. The second part deals with the extension of this method to

*The authors are partly supported by DFG.

Dubiner polynomials, where first the Dubiner basis is introduced and then modifications in the formulation are presented. Finally, numerical tests for the linear advection equation are carried out.

2 The spectral difference method

The SDM is a numerical method for general hyperbolic conservation laws $u_t + \nabla \cdot F[u] = 0$. Both single equations and systems are covered. In this work, we will restrict ourselves to two space dimensions, that means our flux vector $F[u] = (F_1, F_2)^T$ is a two-dimensional vector. Furthermore, we will only deal with triangulations on a certain domain, although the spectral difference method can be used with cartesian or mixed grids, also. Our aim is to solve general conservation laws or systems of conservation laws (which may be nonlinear) using a simple formulation and therefore being computationally efficient. Additionally, we want to achieve a high order method. A method is said to be of order n if it is exact for conservative variables $u \in \mathbb{P}_{n-1} = \{x^l y^m; 0 \leq l + m \leq n - 1\}$.

2.1 Basic formulation

The basic idea of this method is to discretize the right-hand side of the underlaying conservation law

$$u_t = -\nabla \cdot F[u] \tag{2.1}$$

at certain solution points $\vec{x}_j$ in each time step t. Then, one can use an arbitrary explicit time-stepping scheme (like Runge-Kutta) to solve the resulting ODE (in t) at each $\vec{x}_j$. Since we use the derivative of the flux F to update our solution u, we need to be exact for $F \in \mathbb{P}_n$ to achieve a method of order n.

Let's have a look at the right-hand side $-\nabla \cdot F[u]$. To approximate the flux F we reconstruct it in each cell V_i of our triangulation $\mathcal{T}$: We choose N_F flux points $\vec{x}_k$ (that means points where we compute the flux value) and compute the corresponding Lagrange polynomials L_k such that we can expand each component of the true flux $F[u] = (F_1, F_2)^T$ as

$$\begin{pmatrix} F_1(\vec{x}) \\ F_2(\vec{x}) \end{pmatrix} = \sum_{k=1}^{N_F} \begin{pmatrix} F_{k,1} \\ F_{k,2} \end{pmatrix} L_k(\vec{x}), \tag{2.2}$$

where $F_{k,i} = F_i(\vec{x}_k)$ for $i = 1, 2$ are the values of flux F at flux points $\vec{x}_k$.

Inserting (2.2) into (2.1) yields, due to linearity,

$$u_t(\vec{x}) = -\sum_{k=1}^{N_F} \begin{pmatrix} F_{k,1} \\ F_{k,2} \end{pmatrix} \cdot \nabla L_k(\vec{x}). \tag{2.3}$$

To avoid computing the partial derivatives of L_k at each point $\vec{x}_j$ in each cell, we introduce universal coefficients by the coordinate transform ψ (compare Figure 2.1) which maps an arbitrary triangle to the standard triangle $T := \{(\xi,\eta); 0 \le \xi, \eta \le 1, \xi + \eta \le 1\}$. Then, we can rewrite the gradient of L_k as $\nabla L_k = \mathcal{J}\psi \nabla_{\xi,\eta} L_k$, where $\mathcal{J}\psi$ is the Jacobian of ψ. Thus, we get the universal update scheme

$$u_t(\vec{x}) = -\sum_{k=1}^{N_F} \begin{pmatrix} F_{k,1} \\ F_{k,2} \end{pmatrix} \cdot \mathcal{J}\psi \nabla_{\xi,\eta} L_k(\xi,\eta), \tag{2.4}$$

where for each cell only the coefficient matrix $\mathcal{J}\psi$ has to be stored.

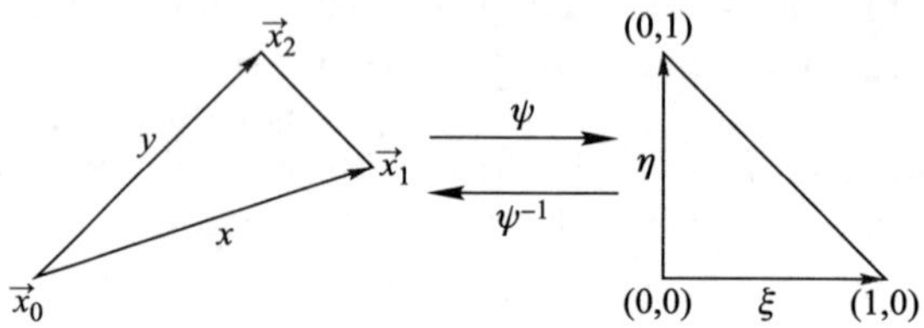

Figure 2.1 Transformation to standard triangle.

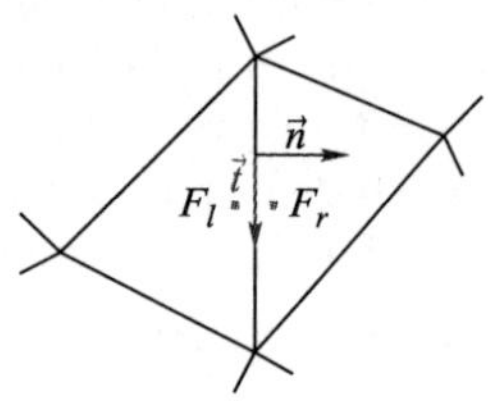

Figure 2.2 Flux at cell interfaces.

2.2 Integral conservation

We now look at the conservation property of this method. The Gaussian integral law $\int_{V_i} \frac{\partial u}{\partial t} \mathrm{d}V = -\int_{\partial V_i} F \cdot \vec{n}_i \mathrm{d}S$ should hold numerically both locally in each cell V_i and globally in our domain. One can show that the SDM is indeed conservative within a cell [4]. To couple neighbouring cells with each other, we have to place at least some of the flux points at each edge of a cell. Then, Riemann solver can be used to compute a common normal component F_n and, finally, the numerical flux function F^{num} with $F_n = F_l^{num} \cdot \vec{n} = F_r^{num} \cdot \vec{n}$ (Figure 2.2). Placing sufficient flux points at quadrature points of degree n for the surface integral leads to global conservation. In spite of most other classical methods, the whole flux vector is needed here such that we have to find a second condition to determine F^{num}. We claim F^{num} to keep the tangential component of the original flux value, that means $F_l \cdot \vec{t} = F_l^{num} \cdot \vec{t}$, and hence can compute the numerical flux value at each edge point.

3 Extended spectral difference method

We will now extend the SDM by assuming a more general approach for the flux reconstruction. In doing so we want to keep the simple formulation of the SDM while avoiding the computation of the Lagrange basis. We have two main aims: On the one hand, we want to compare different basis expansions in order to find indicators for discontinuities occurring in the solution in the context of nonlinear conservation laws. On the other hand, high order spectral filter based on the chosen basis (as developed in [6]) shall be implemented to stabilize the method.

3.1 Dubiner expansion

The Dubiner Polynomials [2] (also called Proriol or Koornwinder Polynomials) are an orthogonal set of polynomials on the triangle $\mathbb{T}^2 := \{(r,s) \in \mathbb{R}^2 | r, s \geq -1, r+s \leq 0\}$. They result as a kind of "warped product" from the transformation shown in Figure 3.1. Since the Jacobi Polynomials $P_n^{\alpha,\beta}$ are known as an orthogonal and stable basis on $[-1,1]$, we consider the product on $[-1,1]^2$ and apply the transform $\psi : [-1,1]^2 \to \mathbb{T}^2$. This yields the Dubiner Polynomials

$$\phi_k(r,s) := g_{\ell,m}(r,s) := P_\ell^{0,0}\left(\frac{2(1+r)}{(1-s)} - 1\right)(1-s)^\ell P_m^{2\ell+1,0}(s). \quad (3.1)$$

They are an orthogonal basis of $\mathbb{P}_N = \text{span}\{r^\ell s^m | 0 \leq \ell + m \leq N\}$ with respect to the scalar product

$$\langle \phi_i, \phi_j \rangle = \int_{\mathbb{T}^2} \phi_i(r,s)\phi_j(r,s) \mathrm{d}(r,s). \quad (3.2)$$

We avoid the singularity of our transformation in $(-1,1)$ by computing the values of ϕ_k at this point directly using the coefficients of the Jacobi Polynomials instead of (3.1).

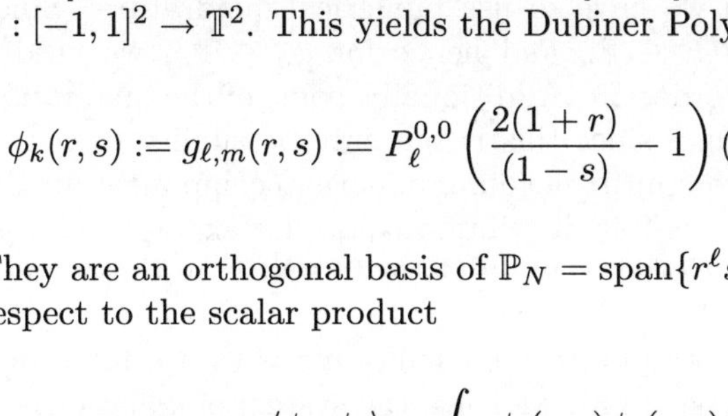

Figure 3.1 Dubiner transformation.

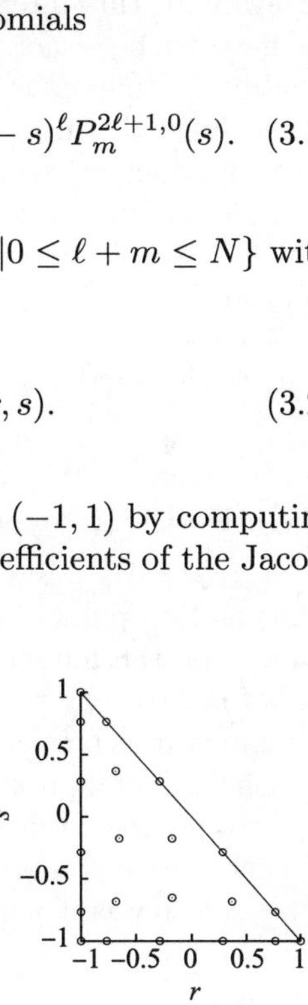

Figure 3.2 Interpolation points.

3.2 Modifications within the formulation

We now assume the flux approximation

$$\begin{pmatrix} F_1(\vec{x}) \\ F_2(\vec{x}) \end{pmatrix} = \sum_{k \in K} \begin{pmatrix} \hat{F}_{k,1}(\vec{x}) \\ \hat{F}_{k,2}(\vec{x}) \end{pmatrix} \phi_k(\vec{x}) \tag{3.3}$$

instead of (2.2) where K is a set of indices needed for the desired approximation order. The following modifications occur in our formulation: Since the standard element changes to $\mathbb{T}^2$, we get a factor 2 in our update scheme. The derivation with respect to r, s instead of ξ, η does not change the scheme. The main difference can be found in (3.3): How do we compute, for $i = 1, 2$, the coefficients $\hat{F}_{k,i}$?

One ansatz could be projection, which yields (using (3.2))

$$\hat{F}_{k,i} = \frac{1}{\|\phi_k\|^2} \langle F_i, \phi_k \rangle = \frac{1}{\|\phi_k\|^2} \int_{\mathbb{T}^2} F_i(r,s) \phi_k(r,s) \mathrm{d}(r,s). \tag{3.4}$$

To evaluate this integral we have to use numerical quadrature. Since we have to be exact for $F \in \mathbb{P}_n$ and hence for $\phi_k \in \mathbb{P}_n$, we need a quadrature rule on $\mathbb{T}^2$ of order $2n$. Additionally, some of the quadrature points must lie on the edges since this is the only possibility to couple cells with each other by computing a common normal component for the numerical flux function. Possible distributions are, for example, Gauss-Lobatto product rules or closed Newton-Cotes formulae on the triangle as proposed in [7].

Another approach to compute these coefficients is to use interpolation. Here, starting from eq. (3.3), we solve the system of equations

$$\mathbf{F_i} = \mathbf{V} \cdot \hat{\mathbf{F}}_\mathbf{i} \tag{3.5}$$

where $\mathbf{F} = (F_i(\vec{x}_j))_j$, $\hat{\mathbf{F}}_\mathbf{i} = \left(\hat{F}_{k,i}\right)_k$ for each component $i = 1, 2$, and $\mathbf{V} = (\phi_k(\vec{x}_j))_{k,j}$ is the Vandermonde matrix. Now we have to find interpolation points $\vec{x}_j$ which lead to "good" numerical behaviour of V_i. Especially, the number of flux points has to be equal to the number of basis functions ϕ_k to get a square matrix. The common disadvantange of famous interpolation points like Fekete points or electrostatic points is their high computation cost. Therefore, we choose a two-dimensional Lobatto points extension proposed in [1] arised from a geometrical point of view. These points are pretty easy to implement and have a very good Lebesgue constant which indicates good interpolation properties.

4 Numerical results

In this section, we will present first numerical results with the SDM using Dubiner expansion.

The first test case is the linear advection equation $u_t + u_x + u_y = 0$ with two different initial conditions $u_0^1(x,y) = \sin(\pi(x+y))$ and $u_0^2(x,y) = \exp((x-0.2)^2 + (y-0.3)^2)$. Orders 2 to 5 and grids containing 238 to 5798 triangles are considered. The time step has been chosen small enough such that the error is dominated by the spacial error. All figures show the SDM using Dubiner polynomials and interpolation points (compare Figure 3.1).

The sine wave as initial condition yields very good results even for 5th order methods as can be seen in Figure 4.2. The corresponding L_∞-error can be found in Figure 4.6 and shows convergence both in grid and order refinement. Similar results can be achieved with SDM using quadrature points, where the method slows due to more points which have to be used in this case.

Problems arise now in the second test case: While the 3rd order method still produces good results (Figure 4.4), we get some oscillations with 4th and 5th order method (Figure 4.5) due to the instability of the SDM.

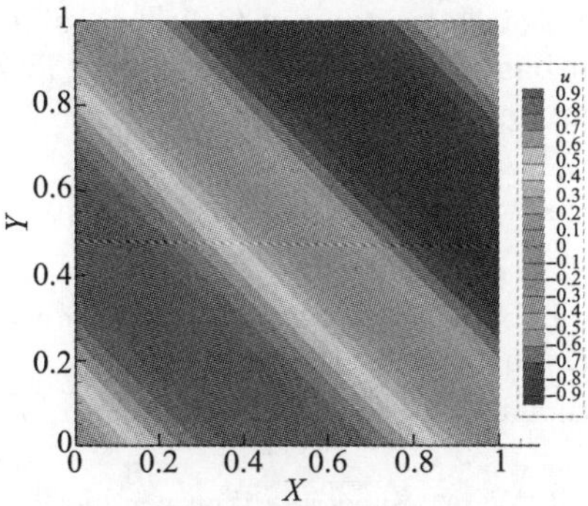

Figure 4.1 Test case 1, time 0.

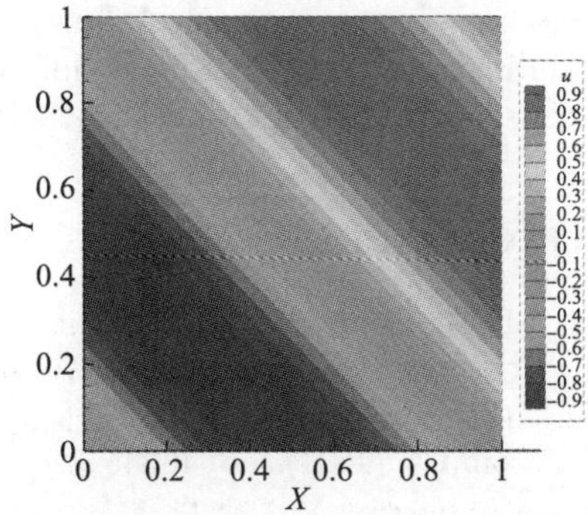

Figure 4.2 Test case 1, time 0.5.

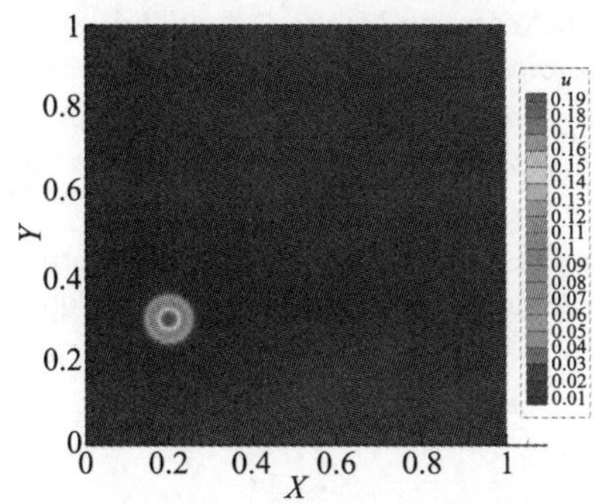

Figure 4.3 Test case 2, time 0.

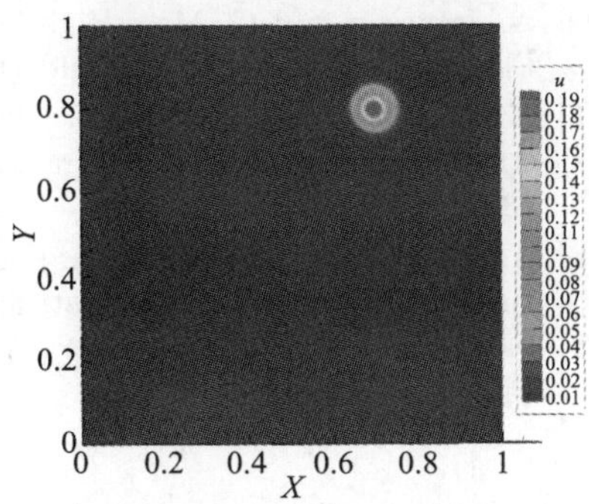

Figure 4.4 Test case 2, time 0.5.

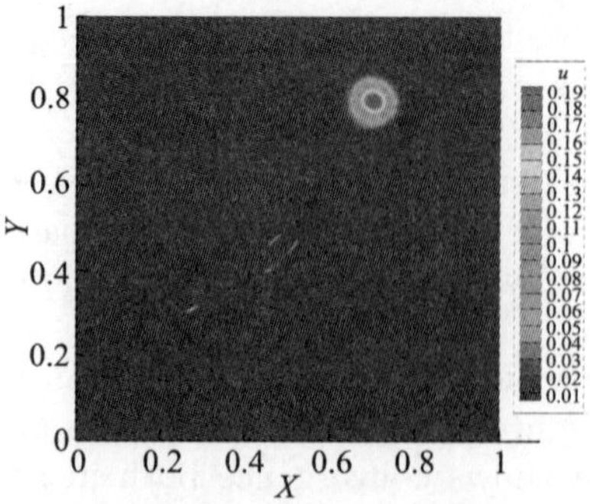

Figure 4.5 Test case 2, time 0.5.

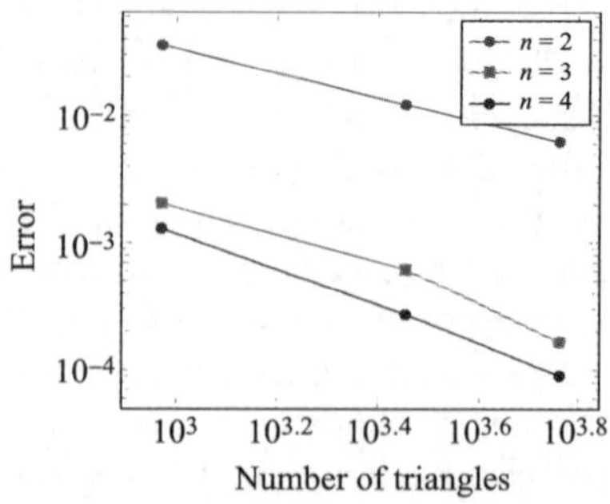

Figure 4.6 L_∞-error SDM Dubiner.

5 Conclusions

We have introduced the classical SDM and our extension to Dubiner polynomials. Modifications occurring in the formulation and possible interpolation points have been presented. Furthermore, first numerical tests have been carried out. Further research is required to find possible stabilizations and to compare different basis polynomials used in the flux reconstruction with respect to their numerical properties.

References

[1] M. G. Blyth and C. Pozrikidis, A Lobatto interpolation grid over the triangle. IMA Journ. Appl. Math. (2006), no. 71, 153–169.

[2] M. Dubiner, Spectral methods on triangles and other domains. Journ. Sc. Comp. 6 (1991), no. 4, 345–390.

[3] A. Jameson, Y. Liu, G. May and Z. J. Wang, Spectral difference method for unstructured grids II: Extension to the Euler Equations. Journ. Sc. Comp. 32 (2007), no. 1.

[4] A. Jameson and G. May, High-Order-Accurate Methods for High-Speed Flow. 17th AIAA Comp. Fluid Dynamics Conf. (2005).

[5] Y. Liu, M. Vinokur and Z. J. Wang, Spectral difference method for unstructured grids I: Basic formulation. Journ. Comp. Physics 216 (2006), 780–801.

[6] A. Meister, S. Ortleb and Th. Sonar, On spectral filtering for discontinuous Galerkin methods on unstructured triangular grids. Preprint (2009).

[7] P. Silvester, Symmetric quadrature formulae for simplexes. Math. Comp. 24 (1970), no. 109, 95–100.

Adaptive Grids and the Entropy Error Indicator*

G. Puppo
Politecnico di Torino
Corso Duca degli Abruzzi 24, 10129 Torino, Italy
E-mail: gabriella.puppo@polito.it

M. Semplice
Università dell'Insubria
Via Valleggio 11, 22010 Como, Italy
E-mail: matteo.semplice@uninsubria.it

Abstract

In this work we explore the coupling of high order non-oscillatory schemes with grid adaptivity. In particular, we focus on the shortcomings due to the low resolution of local extrema due to numerical diffusion induced by standard finite difference schemes. We show that avoiding extrema-clipping and excessive coarsening on extrema is essential to capture features of the solution that might otherwise be lost.

1 Introduction

This work is concerned with the solution of hyperbolic systems of conservation laws with numerical schemes based on adaptive grids. The numerical integration of hyperbolic systems is particularly suited to the coupling with adaptive grids, because the solution can exhibit localized structures with a marked unsteady behavior. Thus the coupling of a numerical solver with an adaptive grid can yield highly resolved solutions without the need to use extremely fine meshes to catch travelling discontinuities.

Adaptive algorithms are based on a numerical scheme (here a second order finite volume scheme) coupled with an error indicator which gives information on the local quality of the numerical solution to decide whether the grid should be locally refined, coarsened or is already

*The authors are partly supported by PRIN07 N. 200774A7LH-002.

adequate. The purpose of this work is to show that standard algorithms may give poor results when coupled with adaptive mesh refinement. In particular, here we concentrate on the influence of slope limiters on the behavior of the grid.

In this paper, as an example, we study the resolution of the adaptive algorithm proposed in [2], which is driven by an error indicator based on the evaluation of the local numerical entropy production.

We thus consider a finite volume scheme for the hyperbolic system $u_t + f_x(u) = 0$, which we assume to be endowed with an entropy pair (η, ψ), where η is the (convex) entropy and ψ is the corresponding entropy flux, satisfying the compatibility condition $(\nabla^T \eta) f'(u) = \nabla^T \psi$. Let $U(t,x)$ be the numerical solution and denote with $\overline{U}_j^n$ its cell average on the interval $I_j = (x_j - \frac{1}{2}h_j, x_j + \frac{1}{2}h_j)$ at time t^n. We write the scheme as

$$\overline{U}_j^{n+1} = \overline{U}_j^n - \lambda \left(F_{j+1/2} - F_{j-1/2}\right), \tag{1.1}$$

where $F_{j+1/2}$ is a suitable numerical flux consistent with the exact flux $f(u)$ and $\lambda = \Delta t/h$ is the mesh ratio. Following [3], we define the numerical entropy production

$$S_j^n = \frac{1}{\Delta t}\left(\overline{\eta(U^{n+1})}_j - \overline{\eta(U^n)}_j + \lambda[\Psi_{j+1/2} - \Psi_{j-1/2}]\right), \tag{1.2}$$

where $\Psi_{j+1/2}$ is a numerical entropy flux. In [2] it proven that, if $\Psi_{j+1/2}$ is consistent with the exact entropy flux $\psi(u)$, then S_j^n converges to zero with the same rate of the local truncation error in regions of smoothness. On the other hand, if a shock occurs in the control volume $V_j^n = I_j \times (t^n, t^{n+1})$, then the local entropy production is of order $S_j^n = O(\frac{1}{\Delta t})$. For these reasons, the quantity S_j^n can be considered as an a posteriori error indicator for the numerical solution obtained in (1.1). In fact, it contains information on the size of the local truncation error and it signals the presence of singularities.

2 Adaptive algorithm

In this work and in [2], the numerical solution is computed with a second order finite volume scheme, based on a piecewise linear reconstruction in space, and a second order Runge-Kutta method in time. To prevent spurious oscillations, the linear reconstruction in [2] is based on MinMod limited slopes. The adaptive scheme (see [2]) is based on a threshold S_{ref} and at each time step:

1. Computes the numerical solution $\overline{U}_j^{n+1}$ and the corresponding entropy production S_j^n on each cell of the current non-uniform grid.

2. Marks cells for refinement if $|S_j^n| > S_{\text{ref}}$. ($S_{\text{ref}} = 0.005$ in this paper.)
3. Refines all marked cells, and recomputes locally the solution and the numerical entropy production on the new grid.
4. Repeat steps 2, 3 until no cell is marked for refinement.
5. Close the current step coarsening the grid if neighboring cells are characterized by an entropy production $|S_j^n| < \frac{1}{4}S_{\text{ref}}$.

This strategy yields good results on Riemann problems test cases (see [2]), but performs more poorly on more complex initial data as in the test problem described in [5], where an acoustic wave interacts with a strong shock. This is due to the fact that the entropy dissipation on local extrema on smooth problems is very small: it is $O(h^4)$ for a second order scheme, compared to $O(h^2)$ on smooth regions, but away from extrema (see [2]). Moreover, TVD limiters for second order schemes introduce clipping on local extrema, thus deteriorating to first order accuracy. These two effects combine in the coarsening step, and the algorithm soon looses resolution on local extrema due to excessive coarsening.

Another source of error is formed when a wave moves into a constant state. The algorithm maintains a very coarse grid in the constant region, and important details in the wave evolution may be lost before the adaptive algorithm has the opportunity of refining the grid enough.

The problems illustrated in this paper with the adaptive algorithm proposed in [2] may be found in other locally adaptive schemes, which have difficulty in recognizing weak waves and local extrema as troublesome points.

Table 1 summarizes the schemes compared in this paper, which explore different ideas to improve the resolution of the original one.

Table 1 Summary of the schemes under comparison.

Label	Description
Reference	fixed grid, MinMod limiter
MM	with MinMod limiter as in [2]
S&E	Entropy indicator plus extrema detection (2.1)
ENO	ENO2 limiter (2.3)
CT	Cada-Torrilhon limiter (2.4)
AdLim	Adaptive limiter as in (2.5)

Detecting local extrema. Since entropy production is $O(h^4)$ on extrema, the scheme S&E uses as an indicator for refinement/derefinement the quantity

$$\mathcal{I}_j^n = |S_j^n| + E_j^n, \tag{2.1}$$

where S_j^n is as before the numerical entropy production and the quantity E_j^n detects extrema on the density component ρ of the solution:

$$E_j^n = \begin{cases} 1 & \text{if } \rho^n \text{ or } \rho^{n+1} \text{ have an extrema at } j, \\ 0 & \text{otherwise,} \end{cases}$$

where the extrema in ρ are computed marking those cells for which $(\rho_{j+1} - \rho_j)(\rho_j - \rho_{j-1}) < 0$. The additional cost of this correction is negligible.

TVB limiters. The piecewise linear reconstruction used in semidiscrete schemes computes two extrapolated data at each cell edge, based on a slope corrected by a limiter function ϕ, to which the numerical fluxes will be applied. One defines slopes and regularity indicators as

$$\delta_{j+1/2} = \frac{\overline{u}_{j+1} - \overline{u}_j}{x_{j+1} - x_j}, \qquad \theta_j = \frac{\delta_{j-1/2}}{\delta_{j+1/2}},$$

and reconstruct the values at the cell edges as

$$u^-_{j+1/2} = \overline{u}_j + \frac{h_j}{2}\delta_{j+1/2}\phi(\theta_j), \qquad u^+_{j-1/2} = \overline{u}_j - \frac{h_j}{2}\delta_{j-1/2}\phi(1/\theta_j). \tag{2.2}$$

Note that, if $\phi(1/\theta) = \phi(\theta)/\theta$, one can simplify the second equation to $u^+_{j-1/2} = \overline{u}_j - \delta_{j+1/2}\phi(\theta_j)h_j/2$ and avoid an explicit evaluation of $\phi(1/\theta_j)$; moreover many limiters can be applied without computing the division in θ_j but simply comparing the slopes $\delta_{j\pm1/2}$.

For TVD limiters such as MinMod, $\phi_{\text{MM}}(\theta) = 0$ for $\theta \leq 0$ and therefore the reconstruction becomes first order accurate at local extrema (clipping phenomenon). To prevent this phenomenon, and to decrease numerical diffusion, several non TVD limiters were proposed. The simplest limiter which does not produce the clipping phenomenon is the ENO2 limiter, see, e.g., [4], which simply selects the minimum slope:

$$\phi_{\text{ENO2}}(\theta) = \begin{cases} 1 & |\theta| > 1, \\ \theta & \text{otherwise.} \end{cases} \tag{2.3}$$

Another TVB limiter which does not reduce to first order on critical points is, for instance, the one introduced in [1]:

$$\phi_{\text{CT}}(\theta) = \max\left[0, \min\left[\frac{2+\theta}{3}, \max\left[-\frac{\theta}{2}, \min\left[2\theta, \frac{2+\theta}{3}, \frac{3}{2}\right]\right]\right]\right]. \tag{2.4}$$

This limiter gives rise to the most expensive reconstruction procedure considered in this paper, since it does not satisfy $\phi(1/\theta) = \phi(\theta)/\theta$ and it requires the explicit evaluation of the ratio θ_j.

Adaptive use of limiters. A more radical strategy is to use a TVD limiter only across strong shocks and strong contact waves, where spurious oscillations would have large amplitudes, and use unlimited slopes otherwise. Since contacts are characterized by $O(1)$ entropy production, this strategy will be simply to turn off the limiter if $|S_j^n| < 1.0$. This amounts to considering the limiter function

$$\phi_{\mathrm{AdLim}}(\theta) = \begin{cases} \phi_{\mathrm{MM}}(\theta) & |S_j| > 1.0, \\ \dfrac{1+\theta}{2} & |S_j| \leq 1.0. \end{cases} \tag{2.5}$$

3 Results

We illustrate the differences between the techniques described on two benchmark problems: in Lax' shock tube test many numerical schemes tend to produce spurious oscillations, while in the Shu-Osher shock-acoustic interaction [5] the main difficulty is to represent the rich smooth structure just behind the strong schock, without generating spurious oscillations on the small shocks emerging from the main structure.

Table 2 Number of cells at final time and average number of cells (in parenthesis), mesh ratio λ, and CPU time (in seconds).

	Lax, $t = 0.15$			Shu-Osher, $t = 0.20$		
Scheme	N cells	λ	CPU	N cells	λ	CPU
Ref	4096	0.2	228.390	8192	0.2	1156
MM	189 (170)	0.2	8.65	774 (677)	0.2	47
CT	128 (131)	0.2	12.37	654 (677)	0.2	128
ENO	185 (170)	0.2	8.59	748 (664)	0.1	89
S&E	207 (195)	0.2	10.30	806 (732)	0.2	50
AdLim	233 (200)	0.2	10.55	781 (715)	0.1	98

The reference solution is obtained with a fixed grid of 4096 cells and 8192 cells respectively. All adaptive solutions have been generated with the algorithm described in [2] in dt-mode. They all started with 256 cells on [0, 1] and were allowed to use cell sizes between 1/64 and 1/2048. The performance of the algorithms is summarized in Table 2.

Lax shock tube. In Figure 3.1, we show two details of the density profile at time $t = 0.15$: the rarefaction wave on the left and the plateaux between the shock and the contact in the right panel.

All schemes correctly detect the position of the contact and the shock wave. The entropy production (bottom part of the figure) shows that the error indicator correctly signals the presence of the two discontinuities with sharp peaks. The solution obtained switching on-off limiters

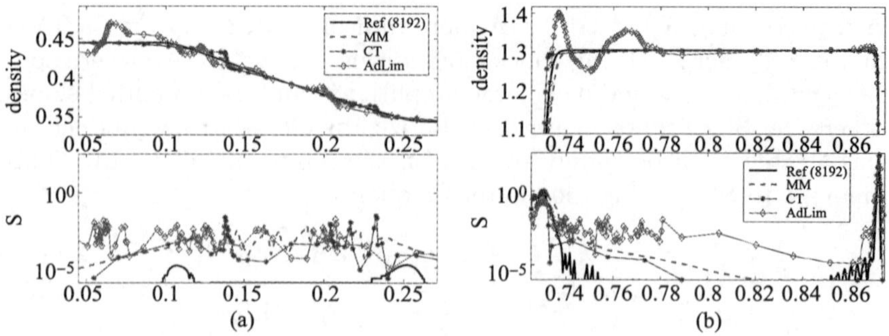

Figure 3.1 Comparison of the methods on the Lax problem (density, numerical entropy production)

(AdLim) shows that this criterion is too crude and causes spurious oscillations. The CT solution has a better resolution of the contact wave, while ENO and S&P (not shown) are almost identical to the one obtained with MM.

The resolution of the rarefaction wave is more troublesome, because the rarefaction advances into a constant state, where the mesh is very coarse: the mesh should be refined fast enough to resolve the corner of the rarefaction correctly. Again, the AdLim strategy performs very poorly, and the CT scheme produces a step in the middle of the rarefaction. Again, ENO and S&E (not shown) are very similar to MM. On the other hand, the rarefaction wave is well resolved by the uniform grid (see also the very low values of the entropy production, which is plotted on the lower part of the figure). Obviously, the cost of the solution with the uniform grid is much higher, since it is using a much larger number of grid points. (Note that for $S_{\text{ref}} = 0.001$, the rarefaction wave is correctly resolved by all schemes and allowing adaptive schemes to use cells of width $1/8192$ as in the reference solution would improve dramatically the resolution of all schemes.)

Figure 3.2 shows the complete density component of the solution for the shock-acoustic wave interaction (a) together with the cell levels employed. Note that in the area behind the big shock, cells are kept to the maximum refinement level, while some coarsening is allowed in the smooth area between the small shocks on the left. On the right part of the figure (b), we show the number of cells used by the different adaptive algorithms as a function of time. It is clear that all algorithms use approximately the same number of cells, and this number increases with time, as the solution evolves and more structures emerge.

In Figure 3.3 we enlarge two areas of the solution to highlight the differences among the schemes. We observe that the MM method performs quite poorly on this problem, because of the clipping of extrema by the

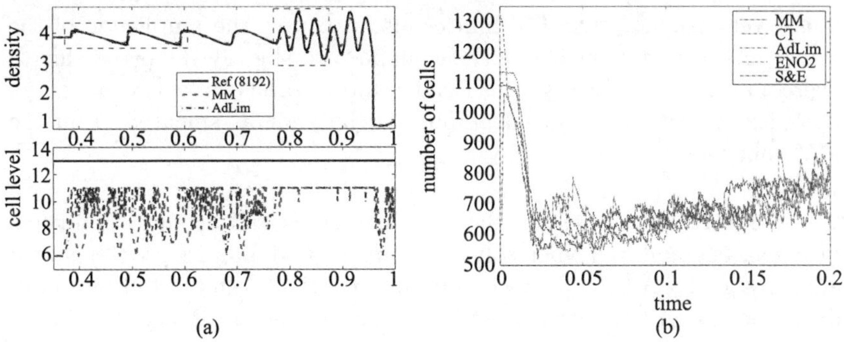

Figure 3.2 Shu-Osher problem: (a) density and cell levels, (b) number of cells versus time. (The dashed rectangles indicate zoomed areas in Figure 3.3.)

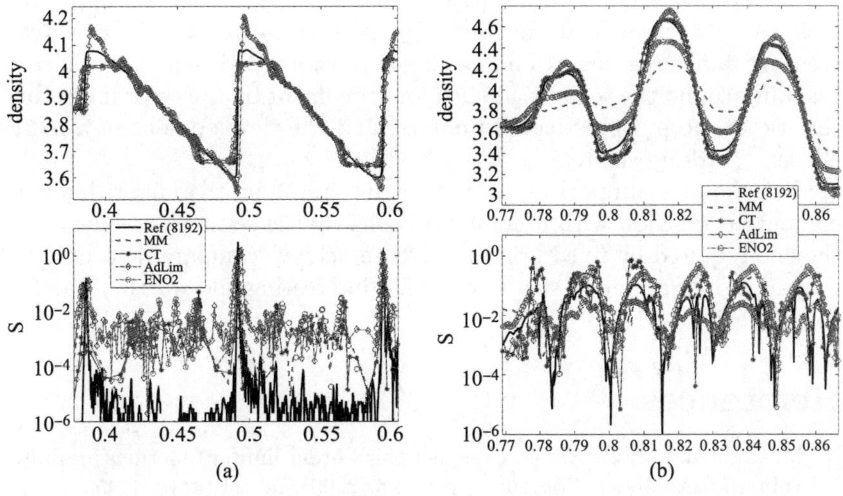

Figure 3.3 Comparison of the methods on the Shu-Osher problem (density, numerical entropy production).

MinMod limiter: the acoustic wave is smeared due to grid coarsening before entering into the shock and the waves do not have the correct amplitude after the shock (b).

The results do not improve if coarsening is prevented on local extrema and the S&E solution (not shown) almost coincide with the MM one. It is apparent that the decay to first order accuracy is the main source of error in this case. Resolution in fact improves dramatically behind the main shock, if the limiter is applied only locally, as in the AdLim strategy. The price however is that small spurious oscillations appear on the weak shocks seen in the left panel of the figure (a). Moreover, the mesh ratio λ had to be lowered, which indicates that the method is not very robust. The CT limiter also captures the amplitude of the waves in the turbulence, but their shapes are slightly distorted due to overcompression. Finally, the ENO limiter stands in between, giving turbulence waves of reduced amplitude, but not as smeared as in the MM solution.

Conclusions. The results shown in this work call attention on some problems encountered when solving complex gas dynamics problems with adaptive grids. Although the quality of numerical results would improve dramatically with a more careful selection of the thresholds, the initial number of cells, and the maximum levels of grid refinement, we are interested in developing a robust algorithm, avoiding an excessive dependence on parameters tailored on single problems.

The results we find do not indicate a clear direction for the improvement of adaptive algorithms, but suggest some future lines of research: creating buffer regions extending the cells where grid refinement is carried out around the regions labelled for refinement by the error indicator, the use of more sophisticated limiters, and the development of a more efficient marking strategy.

It is however important to notice that the adaptive algorithm can compute a solution with a number of cells that is so much smaller than the one required by fixed grid methods to achieve comparable accuracy: this makes the adaptive strategy is quite interesting and worth of careful consideration and investigation.

References

[1] M. Čada and M. Torrilhon, Compact third-order limiter functions for finite volume methods. J. Comput. Phys. 228 (2009), no. 11, 4118–4145.

[2] P. Gabriella and S. Matteo, Numerical entropy and adaptivity for finite volume schemes. Submitted to CICP.

[3] G. Puppo, Numerical entropy production for central schemes. SIAM J. Sci. Comput. 25 (2003/04), no. 4, 1382–1415.

[4] C. W. Shu, High order ENO and WENO schemes for computational fluid dynamics. High-order methods for computational physics (1999), Springer, 439–582.

[5] C. W. Shu and S. Osher, Efficient implementation of essentially nonoscillatory shock-capturing schemes. J. Comput. Phys. 77 (1988), no. 2, 439–471.

Large-time Behavior of Solutions to the Inflow Problem of Full Compressible Navier-Stokes Equations*

Xiaohong Qin

Depart. of Math, Nanjing University of Science and Technology
Nanjing 210094, China
E-mail: xqin@amss.ac.cn

Yi Wang

Institute of Applied Mathematics, AMSS, CAS
Beijing 100190, China
E-mail: wangyi@amss.ac.cn

Abstract

Large-time behavior of the solutions to the inflow problem of full compressible Navier-Stokes equations is considered on the half line $\mathbb{R}_+ = (0, \infty)$. First, we give the existence (or non-existence) of the boundary layer solution to the inflow problem when the right end state (ρ_+, u_+, θ_+) belongs to the subsonic, transonic, and supersonic regions, respectively. Then some wave structures involving the the boundary layer solution (subsonic and transonic cases), the viscous contact waves and the rarefaction waves to the inflow problem are described and the asymptotic stability of these wave patterns are proved under some smallness conditions. The proofs are given by the elementary energy method based on underlying wave structures.

1 Introduction

In this paper, we consider an initial-boundary-value problem (IBVP) for full compressible Navier-Stokes equations in *Eulerian* coordinates $(t, \bar{x})$

*This research is supported by NSFC (grant No. 10801128 and No. 10901083), and NUST Research Funding No. 2010GJPY025 and No. 2010ZYTS063.

on the half line $\mathbb{R}_+ = (0, +\infty)$

$$\begin{cases} \rho_t + (\rho u)_{\bar{x}} = 0, \\ (\rho u)_t + \left(\rho u^2 + p\right)_{\bar{x}} = (\mu u_{\bar{x}})_{\bar{x}}, \\ \left[\rho\left(e + \frac{1}{2}u^2\right)\right]_t + \left[\rho u\left(e + \frac{1}{2}u^2\right) + pu\right]_{\bar{x}} = (\kappa\theta_{\bar{x}} + \mu u u_{\bar{x}})_{\bar{x}}, \end{cases} \tag{1.1}$$

where $\rho > 0$, u, $\theta > 0$, $p > 0$, and $e > 0$ represent the mass density, the velocity, the absolute temperature, the pressure, and the specific internal energy of the gas, respectively. $\mu > 0$ is the coefficient of viscosity, $\kappa > 0$ the coefficient of heat conduction. Assume that both μ and κ are positive constants. Let $v = \frac{1}{\rho}(> 0)$ and s denote the specific volume and the entropy of the gas, respectively. For the ideal polytropic gas,

$$p = Rv^{-1}\theta = Av^{-\gamma}\exp\left(\frac{\gamma - 1}{R}s\right), \quad e(v, \theta) = \frac{R}{\gamma - 1}\theta, \tag{1.2}$$

where $\gamma > 1$ is the adiabatic exponent, A and R are positive constants.

We consider the IBVP of (1.1) with the initial data

$$\begin{aligned} &(\rho_0, u_0, \theta_0)(\bar{x}) = (\rho, u, \theta)(0, \bar{x}) \to (\rho_+, u_+, \theta_+) \quad \text{as} \quad \bar{x} \to +\infty, \\ &\inf_{\bar{x}\in\mathbb{R}_+} (\rho_0, \theta_0)(\bar{x}) > 0, \end{aligned} \tag{1.3}$$

where $\rho_+ > 0$, u_+ and $\theta_+ > 0$ are given constants. As for the boundary condition, we consider the following case which is pointed out in [6]:

$$u(t, \bar{x})|_{\bar{x}=0} = u_- > 0, \quad \rho(t, \bar{x})|_{\bar{x}=0} = \rho_-, \quad \theta(t, \bar{x})|_{\bar{x}=0} = \theta_-. \tag{1.4}$$

Here $\rho_- > 0$, u_- and $\theta_- > 0$ are also constants and we assume that the initial values (1.3) and the boundary conditions (1.4) satisfy the compatibility condition at the origin. The IBVP (1.1), (1.3), and (1.4) is called *inflow problem.*

Recently, the IBVP of (1.1) attracts increasing interest because it has more physical meanings and produces some new mathematical difficulties due to the boundary effect. Not only basic wave patterns but also a new wave, which is called boundary layer solution (BL-solution for brevity) [6], may appear in the IBVP. Matsumura [6] proposes a criterion on the question when the BL-solution forms to the isentropic Navier-Stokes equations. The argument in [6] for the isentropic case can also be applied to the full case (1.1), for example, see [2].

In this paper, firstly we give the existence (or non-existence) of the BL-solution when the right end state (ρ_+, u_+, θ_+) belongs to the subsonic, transonic, and supersonic regions, respectively. Secondly, we are interested in the stability of wave patterns to the inflow problem when (ρ_-, u_-, θ_-) belongs to the regions. In [8] we show the stability of the single contact wave and the superposition of the BL-solution (subsonic case), viscous contact wave and 3-rarefaction wave. But due to the fact that the first characteristic speed on the boundary is coincident with the speed of the moving boundary in the transonic BL-solution case, a part of the nonlinear waves in the first characteristic field may appear, which is quite different from the the regime that (ρ_-, u_-, θ_-) belongs to the subsonic region [8], where the waves in the first characteristic field must be absent. Here we just assume that the 1-rarefaction wave appear in the first characteristic field. Correspondingly, some new mathematical difficulties occur due to the degeneracy of the transonic BL-solution and its interactions with other wave patterns in the superposition wave. In particularly, the transonic BL-solution is attached with 1-rarefaction wave for all time, so the interaction of these two waves should be carefully treated in the stability analysis.

Under the *Lagrangian* coordinate (t, x)

$$t = t, \qquad x = \int_{(0,0)}^{(t,\bar{x})} \rho(\tau, y)\, dy - (\rho u)(\tau, y)\, d\tau,$$

the inflow problem of (1.1)–(1.4) can be transformed into a moving boundary problem. In order to fix the moving boundary $x = \sigma t$, we introduce a new variable $\xi = x - \sigma t$. Then we have the half-space problem

$$\begin{cases} v_t - \sigma v_\xi - u_\xi = 0, \\ u_t - \sigma u_\xi + p_\xi = \mu \left(\frac{u_\xi}{v}\right)_\xi, \qquad\qquad t > 0, \xi \in \mathbb{R}_+, \\ \left(\frac{R}{\gamma-1}\theta + \frac{1}{2}u^2\right)_t - \sigma \left(\frac{R}{\gamma-1}\theta + \frac{1}{2}u^2\right)_\xi + (pu)_\xi = \kappa \left(\frac{\theta_\xi}{v}\right)_\xi + \mu \left(\frac{u u_\xi}{v}\right)_\xi, \\ (v_0, u_0, \theta_0)(\xi) = (v, u, \theta)(0, \xi) \to (v_+, u_+, \theta_+) \text{ as } \xi \to +\infty, \\ (v, u, \theta)(t, \xi = 0) = (v_-, u_-, \theta_-), \quad u_- > 0, \end{cases} \tag{1.5}$$

where $\sigma := -\frac{u_-}{v_-} < 0$.

Given the right end state (v_+, u_+, θ_+), we can define the following wave curves in the phase space (v, u, θ) with $v > 0$ and $\theta > 0$.

- Boundary layer curve:

$$BL(v_+, u_+, \theta_+) := \left\{ (v, u, \theta) \middle| \ \frac{u}{v} = \frac{u_+}{v_+} \right\},$$

where (v_+, u_+, θ_+) satisfies Theorem 3.1 below.

- Contact wave curve:

$$CD(v_+, u_+, \theta_+) := \{(v, u, \theta) | \, u = u_+, p = p_+, v \neq v_+\},$$

- i–Rarefaction wave curve $(i = 1, 3)$:

$$R_i(v_+, u_+, \theta_+) := \left\{ (v, u, \theta) \middle| \quad \lambda_i < \lambda_{i+}, \ s(v, \theta) = s_+, \right.$$

$$u = u_+ - \int_{v_+}^{v} \lambda_i(\eta, s_+)\, d\eta \Big\},$$

where $s_+ = s(v_+, \theta_+)$ and $\lambda_i = \lambda_i(v, s)$ is the ith characteristic speed of the hyperbolic part of (1.1).

We divide the phase space $\{(v, u, \theta) | \, v > 0, \theta > 0\}$ into three parts

$$\begin{cases} \Omega_{sub} := \{(v, u, \theta) \mid M < 1\}, \\ \Gamma_{trans} := \{(v, u, \theta) \mid M = 1\}, \\ \Omega_{super} := \{(v, u, \theta) \mid M > 1\}, \end{cases}$$

which are named subsonic, transonic, and supersonic regions, respectively. Then our main stability result is, roughly speaking, as follows:

(1) If the left state $(v_-, u_-, \theta_-) \in \Omega_{sub}^+$ and it satisfies with right state (v_+, u_+, θ_+) that $(v_-, u_-, \theta_-) \in BL\text{-}CD\text{-}R_3(v_+, u_+, \theta_+)$, then there exist the unique two middle states $(v_*, u_*, \theta_*) \in \Omega_{sub}^+$ and (v^*, u^*, θ^*), such that $(v_-, u_-, \theta_-) \in BL(v_*, u_*, \theta_*)$, $(v_*, u_*, \theta_*) \in CD(v^*, u^*, \theta^*)$ and $(v^*, u^*, \theta^*) \in R_3(v_+, u_+, \theta_+)$ and the superposition of the BL-solution, the viscous contact wave, and the rarefaction wave is asymptotically stable provided that the strengths of BL-solution and the viscous contact wave are weak but the rarefaction wave is not necessarily weak.

(2) If the left state $(v_-, u_-, \theta_-) \in \Omega_{sub}^+$ and it satisfies with right state (v_+, u_+, θ_+) that $(v_-, u_-, \theta_-) \in BL\text{-}R_1\text{-}CD\text{-}R_3(v_+, u_+, \theta_+)$, then there exist the unique three states (v_*, u_*, θ_*), (v_m, u_m, θ_m), and (v^*, u^*, θ^*), such that $(v_-, u_-, \theta_-) \in BL(v_*, u_*, \theta_*)$, $(v_*, u_*, \theta_*) \in R_1(v_m, u_m, \theta_m)$, $(v_m, u_m, \theta_m) \in CD(v^*, u^*, \theta^*)$, and $(v^*, u^*, \theta^*) \in R_3(v_+, u_+, \theta_+)$, then the superposition of the four wave patterns: the transonic BL-solution, 1-rarefaction wave, 2-viscous contact wave and 3-rarefaction wave is time-asymptotically stable provided that the wave strength $\delta = |(v_+ - v_-, u_+ - u_-, \theta_+ - \theta_-)|$ is suitably small.

This paper is organized as follows. Section 2 is devoted to some preliminaries on BL-solution, viscous contact wave, and viscous rarefaction waves. In Section 3, we state the main results.

2 Preliminaries

First of all, we introduce the BL-solution. When $(v_-, u_-, \theta_-) \in \Omega_{sub}^+ \cup \Gamma_{trans}^+$, we have

$$\lambda_1(v_-, \theta_-) = -\frac{\sqrt{R\gamma\theta_-}}{v_-} \le -\frac{u_-}{v_-} = \sigma < 0, \tag{2.1}$$

hence a stationary solution $(V^b, U^b, \Theta^b)(\xi)$ to (1.5) is expected

$$\begin{cases} -\sigma V^b_\xi - U^b_\xi = 0, \\ -\sigma U^b_\xi + P^b_\xi = \mu\Big(\dfrac{U^b_\xi}{V^b}\Big)_\xi, \\ -\sigma\left(\dfrac{R}{\gamma-1}\Theta^b + \dfrac{1}{2}\left(U^b\right)^2\right)_\xi + \left(P^bU^b\right)_\xi = \kappa\Big(\dfrac{\Theta^b_\xi}{V^b}\Big)_\xi + \mu\Big(\dfrac{U^bU^b_\xi}{V^b}\Big)_\xi, \\ (V^b, U^b, \Theta^b)(0) = (v_-, u_-, \theta_-), \quad (V^b, U^b, \Theta^b)(+\infty) = (v_+, u_+, \theta_+), \end{cases} \tag{2.2}$$

where $P^b := p(V^b, \Theta^b) = \frac{R\Theta^b}{V^b}$. The stationary solution is named boundary layer solution.

Next, from [3] the viscous version of the above contact discontinuity, called viscous contact wave $(V^d, U^d, \Theta^d)(t, x)$, can be defined by

$$\begin{cases} V^d(t,x) = \dfrac{R\Theta^{\text{sim}}(t,x)}{p_+}, \\ U^d(t,x) = u_+ + \dfrac{(\gamma-1)\kappa\Theta^{\text{sim}}_x(t,x)}{\gamma\Theta^{\text{sim}}(t,x)}, \\ \Theta^d(t,x) = \Theta^{\text{sim}}\Big(\dfrac{x}{\sqrt{1+t}}\Big) + R\Big(\mu - \dfrac{(\gamma-1)\kappa}{R\gamma}\Big)\Theta^{\text{sim}}_t, \end{cases} \tag{2.3}$$

where $\Theta^{\text{sim}}\left(\frac{x}{\sqrt{1+t}}\right)$ is the unique self-similar solution to the following nonlinear diffusion equation

$$\begin{cases} \Theta_t = \dfrac{(\gamma-1)\kappa p_+}{R^2\gamma}\left(\dfrac{\Theta_x}{\Theta}\right)_x, \\ \Theta(t, \pm\infty) = \theta_\pm. \end{cases} \tag{2.4}$$

Then the viscous contact wave $(V^d, U^d, \Theta^d)(t, x)$ satisfies

$$\begin{cases} V^d_t - \sigma V^d_\xi - U^d_\xi = 0, \\ U^d_t - \sigma U^d_\xi + P^d_\xi = \mu\Big(\dfrac{U^d_\xi}{V^d}\Big)_\xi, \quad t > 0, \xi \in \mathbb{R}_+, \\ \dfrac{R}{\gamma-1}(\Theta^d_t - \sigma\Theta^d_\xi) + P^dU^d_\xi = \kappa\left(\dfrac{\Theta^d_\xi}{V^d}\right)_\xi + \mu\dfrac{(U^d_\xi)^2}{V^d} + H^d, \end{cases} \tag{2.5}$$

where $P^d := p(V^d, \Theta^d)$ and

$$H^d = O(1)\delta_d(1+t)^{-2}\exp\left(-\frac{C_d(\xi+\sigma t)^2}{1+t}\right).$$

At last, consider the following Burgers equation

$$\begin{cases} w_t + ww_x = 0, \quad t > 0, x \in \mathbb{R}, \\ w_0(x) := w(0,x) = \begin{cases} w_-, & x < 0, \\ w_- + C_q(w_+ - w_-)\displaystyle\int_0^x y^q e^{-y}\,dy, & x \geq 0. \end{cases} \end{cases} \tag{2.6}$$

Here $q \geq 14$ is a constant to be determined, and C_q is a constant such that $C_q \int_0^{+\infty} y^q e^{-y} dy = 1$. If $w_- < w_+$, then the solution to the above Burgers equation can be expressed by

$$w(t,x) = w_0(x_0(t,x)), \qquad x = x_0(t,x) + w_0(x_0(t,x))t. \tag{2.7}$$

Now the i–rarefaction wave $(V^{r_i}, U^{r_i}, \Theta^{r_i})(t,x)$ $(i = 1,3)$ to the inflow problem (1) can be defined by

$$\begin{cases} \lambda_i(V^{r_i}, \Theta^{r_i})(t,x) = w(1+t, x+\sigma), \\ s(V^{r_i}, \Theta^{r_i})(t,x) = s_+ = s(v_+, \theta_+), \\ U^{r_i}(t,x) = u_+ - \displaystyle\int_{v_+}^{V^{r_i}(t,x)} \lambda_i(\eta, s_+) d\eta. \end{cases} \tag{2.8}$$

Then the i–rarefaction wave $(V^{r_i}, U^{r_i}, \Theta^{r_i})(t,x)$, $(i = 1,3)$ defined in (2.8) satisfies the system

$$\begin{cases} V^{r_i}_t - \sigma V^{r_i}_\xi - U^{r_i}_\xi = 0, \\ U^{r_i}_t - \sigma U^{r_i}_\xi + P^{r_i}_\xi = 0, \\ \left[\dfrac{R}{\gamma-1}\Theta^{r_i} + \dfrac{1}{2}(U^{r_i})^2\right]_t - \sigma\left[\dfrac{R}{\gamma-1}\Theta^{r_i} + \dfrac{1}{2}(U^{r_i})^2\right]_\xi \\ \qquad +(P^{r_i}U^{r_i})_\xi = 0, \\ (V^{r_i}, U^{r_i}, \Theta^{r_i})(t, \xi = 0) = (v_-, u_-, \theta_-), \\ (V^{r_i}, U^{r_i}, \Theta^{r_i})(t, \xi) \to (v_+, u_+, \theta_+) \quad \text{as} \quad \xi \to +\infty, \end{cases} \tag{2.9}$$

where $P^{r_i} := p(V^{r_i}, \Theta^{r_i})$.

3 Main results

Now we state our main results.

Theorem 3.1 (Existence of BL-solution) [8]. *Suppose that $v_\pm > 0$, $u_- > 0$, $\theta_\pm > 0$ and let $\delta_b := |(u_+ - u_-, \theta_+ - \theta_-)|$. If $u_+ \leq 0$, then there is no solution to (2.2). If $u_+ > 0$, then there exists a suitably small constant $\delta_0 > 0$ such that if $0 < \delta_b \leq \delta_0$, then the existence and non-existence of solutions to (2.2) is divided into three cases according to the location of (u_+, θ_+):*

Case I : $(u_+, \theta_+) \in \Omega^+_{sup}$. *Then there is no solution to (2.2).*

Case II : $(u_+, \theta_+) \in \Gamma^+_{trans}$. *Then (u_+, θ_+) is a saddle-knot point to (2.2). Precisely, there exists a unique trajectory Σ tangent to the straight line $\mu u_+(u - u_+) - \kappa(\gamma - 1)(\theta - \theta_+) = 0$ at the point (u_+, θ_+). For each $(u_-, \theta_-) \in \Sigma(u_+, \theta_+)$, there exists a unique solution (U^b, Θ^b) satisfying*

$$U^b_\xi > 0, \quad \Theta^b_\xi > 0,$$

and

$$\left|\frac{d^n}{d\xi^n}(U^b - u_+, \Theta^b - \theta_+)\right| = O(1)\frac{\delta_b^{n+1}}{(1+\delta_b\xi)^{n+1}}. \tag{3.1}$$

Case III : $(u_+, \theta_+) \in \Omega^+_{sub}$. *Then* (u_+, θ_+) *is a saddle point to* (2.2). *Precisely, there exists a center-stable manifold* $\mathcal{M}$ *tangent to the line*

$$(1 + a_2c_2u_+)(u - u_+) - a_2(\theta - \theta_+) = 0$$

on the opposite directions at the point (u_+, θ_+). *Here* c_2 *is one of the solutions to the equation*

$$y^2 + \left(\frac{M_+^2\gamma - 1}{M_+^2R\gamma} - \frac{\mu}{\kappa(\gamma-1)}\right)y - \frac{\mu}{M_+^2R\gamma\kappa} = 0$$

and $a_2 = -\frac{R}{\mu(\lambda_J^1 - \lambda_J^2)}$ *with* $\lambda_J^1 > 0,\ \ \lambda_J^2 < 0$ *are the two eigenvalues of the linearized matrix of* (2.2). *Only when* $(u_-, \theta_-) \in \mathcal{M}(u_+, \theta_+)$, *does there exist a unique solution* $(U^b, \Theta^b) \subset \mathcal{M}(u_+, \theta_+)$ *satisfying*

$$\left|\frac{d^n}{d\xi^n}(U^b - u_+, \Theta^b - \theta_+)\right| = O(1)\delta_b e^{-c\xi}, \quad n = 0, 1, 2, \ldots. \tag{3.2}$$

Remark 3.1. This Lemma is the first one for the classification of the BL-solution to the inflow problem (1.1)–(1.4). The stability of the single BL-solution in the subsonic case (Case III) is proved in [2].

Theorem 3.2 (Stability of superposition of three waves) [8]. *Assume that* $(v_-, u_-, \theta_-) \in BL\text{-}CD\text{-}R_3(v_+, u_+, \theta_+)$. *If the strengths of BL-solution and the viscous contact wave are weak and there exists a small positive constant* δ_0 *such that the initial values satisfy*

$$\|(v_0 - V_0, u_0 - U_0, \theta_0 - \Theta_0)\|_{H^1} \le \delta_0, \tag{3.3}$$

the inflow problem (1.1)–(1.4) *has a unique global solution* $(v, u, \theta)(t, \xi)$ *satisfying*

$$\begin{cases} (v - V, u - U, \theta - \Theta)(t, \xi) \in C\big([0, \infty); H^1(\mathbb{R}^+)\big), \\ (v - V)_\xi(t, \xi) \in L^2\big(0, \infty; L^2(\mathbb{R}^+)\big), \\ (u - U, \theta - \Theta)_\xi(t, \xi) \in L^2\big(0, \infty; H^1(\mathbb{R}^+)\big). \end{cases} \tag{3.4}$$

Furthermore,

$$\lim_{t\to\infty} \sup_{\xi\in\mathbb{R}_+} |(v - V, u - U, \theta - \Theta)(t, \xi)| = 0, \tag{3.5}$$

where the superposition wave $(V, U, \Theta)(t, \xi)$ *is*

$$\begin{pmatrix} V \\ U \\ \Theta \end{pmatrix}(t,\xi) = \begin{pmatrix} V^b + V^d + V^{r_3} \\ U^b + U^d + U^{r_3} \\ \Theta^b + \Theta^d + \Theta^{r_3} \end{pmatrix}(t,\xi) - \begin{pmatrix} v_* + v^* \\ u_* + u^* \\ \theta_* + \theta^* \end{pmatrix}.$$

Theorem 3.3 (Stability of superposition of four waves). *Assume that* $(v_-, u_-, \theta_-) \in BL\text{-}R_1\text{-}CD\text{-}R_3(v_+, u_+, \theta_+)$ *and satisfies Case* III *of Theorem* 3.1. *Then there exists a small positive constant* δ_0 *such that if the initial values and the wave strength* $\delta = |(v_+ - v_-, u_+ - u_-, \theta_+ - \theta_-)|$ *satisfy*

$$\delta + \|(v_0 - V_0, u_0 - U_0, \theta_0 - \Theta_0)\|_{H^1} \le \delta_0, \tag{3.6}$$

the inflow problem (1.5) *has a unique global solution* $(v, u, \theta)(t, \xi)$ *satisfying*

$$\begin{cases} (v - V, u - U, \theta - \Theta)(t, \xi) \in C\big([0, \infty); H^1(\mathbb{R}^+)\big), \\ (v - V)_\xi(t, \xi) \in L^2\big(0, \infty; L^2(\mathbb{R}^+)\big), \\ (u - U, \theta - \Theta)_\xi(t, \xi) \in L^2\big(0, \infty; H^1(\mathbb{R}^+)\big). \end{cases} \tag{3.7}$$

Furthermore,

$$\lim_{t \to \infty} \sup_{\xi \in \mathbb{R}_+} |(v - V, u - U, \theta - \Theta)(t, \xi)| = 0, \tag{3.8}$$

where the superposition wave $(V, U, \Theta)(t, \xi)$ *is*

$$\begin{pmatrix} V \\ U \\ \Theta \end{pmatrix}(t,\xi) = \begin{pmatrix} V^b + V^{r_1} + V^d + V^{r_3} \\ U^b + U^{r_1} + U^d + U^{r_3} \\ \Theta^b + \Theta^{r_1} + \Theta^d + \Theta^{r_3} \end{pmatrix}(t,\xi) - \begin{pmatrix} v_* + v_m + v^* \\ u_* + u_m + u^* \\ \theta_* + \theta_m + \theta^* \end{pmatrix}.$$

Remark 3.2. In Theorem 3.3, $\delta = |(v_+ - v_-, u_+ - u_-, \theta_+ - \theta_-)|$ is suitably small which is equivalent to a fact that the amplitudes of the four waves are all suitably small.

References

[1] F. Huang, J. Li and A. Matsumura, Asymptotic stability of combination of viscous contact wave with rarefaction waves for 1-D compressible Navier-Stokes system. To appear in Arch. Ration. Mech. Anal.

[2] F. Huang, J. Li and X. Shi, Asymptotic behavior of solutions to the full compressible Navier-Stokes equations in the half space. Commu. Math. Sci. 8 (2010), 639–654.

[3] F. Huang, Z. Xin and T. Yang, Contact discontinuity with general perturbations for gas motions. Adv. Math. 219 (2008), 1246–1297.

[4] S. Kawashima, A. Matsumura and K. Nishihara, Asymptotic behavior of solutions for the equations of a viscous heat-conductive gas. Proc. Japan Acad. 62A (1986), 249–252.

[5] T. Liu and Z. Xin, Nonlinear stability of rarefaction waves for compressible Navier-Stokes equations. Comm. Math. Phys. 118 (1988), 451–465.

[6] A. Matsumura, Inflow and outflow problems in the half space for a one-dimensional isentropic model system of compressible viscous gas. Methods Appl. Anal. 8 (2001), 645–666.

[7] K. Nishihara, T. Yang and H. Zhao, Nonlinear stability of strong rarefaction waves for compressible Navier-Stokes equations. SIAM J. Math. Anal. 35 (2004), 1561–1597.

[8] X. Qin and Y. Wang, Stability of wave patterns to the inflow problem of full compressible Navier-Stokes equations. SIAM J. Math. Anal. 41 (2009), 2057–2087.

Mechanism of Singularity Formation for Linearly Degenerate Quasilinear Hyperbolic Systems*

Peng Qu
School of Mathematical Sciences, Fudan University
Shanghai 200433, China
E-mail: qupeng_sygs@163.com

Abstract

For the Cauchy problem of 1-D first order quasilinear hyperbolic linearly degenerate systems, a new mechanism of singularity formation is given to show that all the $W^{1,p}$ $(1 < p \leq +\infty)$ norms of the C^1 solution should blow up simultaneously. It gives a way to verify the property of ODE singularity by directly using the energy method in the framework of C^1 solution.

1 Introduction

We discuss the formation of singularity of classical solutions to the following Cauchy problem for 1-D quasilinear strictly hyperbolic systems:

$$\begin{cases} \dfrac{\partial u}{\partial t} + A(u)\dfrac{\partial u}{\partial x} = F(u), \quad x \in \mathbb{R}, t \geq 0, & (1.1) \\ t = 0: \ u = u_0(x), \quad x \in \mathbb{R}, & (1.2) \end{cases}$$

where $u = (u_1, \ldots, u_n)^T$ is an unknown vector function of (t, x), the initial data $u_0(x)$ is continuously differentiable with bounded C^1 norm

$$\|u_0\|_{C^1} = \|u_0\|_{L^\infty} + \|u_0'\|_{L^\infty} < +\infty. \tag{1.3}$$

$A(u)$ is a C^2 $n \times n$ matrix function with n distinct real eigenvalues

$$\lambda_1(u) < \lambda_2(u) < \cdots < \lambda_n(u), \quad \forall u \in \mathbb{R}^n, \tag{1.4}$$

which implies that the left eigenvectors $l_i(u) = (l_{i1}(u), \ldots, l_{in}(u))(i = 1, \ldots, n)$ and the right eigenvectors $r_i(u) = (r_{1i}(u), \ldots, r_{ni}(u))^T(i =$

*The author is partially supported by Fudan University Creative Student Cultivation Program in Key Disciplinary Areas No. EHH1411208.

$1, \ldots, n)$ form two bases of $\mathbb{R}^n$ respectively, and $\lambda_i(u), l_i(u), r_i(u)$ $(i = 1, \ldots, n)$ have the same regularity as $A(u)$. Without loss of generality, we assume

$$l_i(u)r_j(u) \equiv \delta_{ij}, \quad \forall u \in \mathbb{R}^n, \forall 1 \leq i, j \leq n, \tag{1.5}$$

where δ_{ij} stands for the Kronecker's symbol.

The local existence of C^1 solution to the Cauchy problem is well-known (See [10]). If all the characteristics are linearly degenerate in the sense of P.D. Lax:

$$\nabla\lambda_i(u)r_i(u) \equiv 0, \quad \forall u \in \mathbb{R}^n, \forall 1 \leq i \leq n, \tag{1.6}$$

then Cauchy problem (1.1)–(1.2) with small and decaying initial data admits a unique global C^1 solution $u = u(t, x)$ for all $t \in \mathbb{R}$ (See [6], [11], [14]). However, for general C^1 initial data, the C^1 solution may blow up in a finite time (See [4] and [8] for examples). A. Majda has proposed the following conjecture (originally, for systems of conservation laws) in [12].

Conjecture. Under hypotheses (1.3)–(1.6), if the C^1 solution $u = u(t, x)$ to Cauchy problem (1.1)–(1.2) blows up in a finite time $T_* < +\infty$, then the C^0 norm of u must blow up at T_*, i.e., $\sup_{0 \leq t < T_*} \|u(t, \cdot)\|_{C^0} = +\infty$.

This is a different kind of singularity from the shock formation as for the Burger's equation

$$\frac{\partial u}{\partial t} + u\frac{\partial u}{\partial x} = 0$$

and is called to be the ODE singularity in [1] or the semilinear behavior in [2] and [3], since the solution to the Riccati ordinary differential equation and the classical solution to semilinear hyperbolic system blow up in such a way. In this paper we will use the name "ODE singularity".

To verify this conjecture is still an open problem, but some results around it have been made in recent years, the author recommend the recent work [8], [9], and [13] for the characteristic method, [2] and [3] for the energy method and [7] for review in this subject. All these known results are for specific systems, while in this paper, for general systems, we will show a weaker result than the ODE singularity, that if the singularity occurs, then all the $W^{1,p}$ $(1 < p \leq +\infty)$ norms of the C^1 solution should blow up simultaneously.

Theorem 1. Under hypotheses (1.3)–(1.6) and suppose that, without loss of generality, the initial data u_0 has a compact support, then the C^1 solution $u = u(t, x)$ to Cauchy problem (1.1)–(1.2) blows up in a finite time $t = T_* < +\infty$, if and only if all the $W^{1,p}(1 < p \leq +\infty)$ norms of $u = u(t, x)$ blow up at the time T_*:

$$\sup_{0 \leq t < T_*} \|u(t, \cdot)\|_{W^{1,p}} = \sup_{0 \leq t < T_*} (\|u(t, \cdot)\|_{L^p} + \|\partial_x u(t, \cdot)\|_{L^p}) = +\infty.$$

We will prove the following equivalent theorem in Section 2.

Theorem 2. Under the hypotheses of Theorem 1, for any fixed $T_0 > 0$, if for any given $T\,(0 < T < T_0)$, Cauchy problem (1.1)–(1.2) admits a unique C^1 solution $u = u(t,x)$ on $0 \le t \le T$, which satisfies the uniform a priori estimate

$$\|u(t,\cdot)\|_{C^0} = \|u(t,\cdot)\|_{L^\infty} \le C_0, \quad \forall t \in [0,T], \tag{1.7}$$

and there exists real numbers $p\,(1 < p \le +\infty)$ and $C_p > 0$, such that

$$\|\partial_x u(t,\cdot)\|_{L^p} \le C_p, \quad \forall t \in [0,T], \tag{1.8}$$

then we have the following uniform a priori estimate

$$\|\partial_x u(t,\cdot)\|_{C^0} \le C_1, \quad \forall t \in [0,T], \tag{1.9}$$

where $C_0 = C_0(T_0), C_p = C_p(T_0)$, and $C_1 = C_1(T_0)$ denote positive numbers independent of T, but possibly depending on T_0.

With the help of Theorem 2, in order to get the property of ODE singularity for the C^1 solution to Cauchy problem (1.1)–(1.2), we can directly use the energy method in the framework of C^1 solution so that there is no need of applying the higher order Sobolev norm estimation and embedding theory as in [2] and [3]. In Section 3, we will make some generalization, and then in Section 4, as an application of this method, we will discuss the generalized Kerr–Debye model.

2 Sketch of the proof of Theorem 2

In this section, we will give the sketch of the proof for Theorem 2 by means of a method coming from [6], [11], and especially [14]. Obviously, we need only to consider the case $1 < p < +\infty$.

Firstly, setting

$$w_i = l_i(u)\frac{\partial u}{\partial x} \quad (i = 1,\ldots,n), \tag{2.1}$$

for $i = 1,\ldots,n$ we have the wave decomposition formulas (cf. [5] and [6])

$$\frac{\partial w_i}{\partial t} + \frac{\partial}{\partial x}(\lambda_i(u)w_i) = \sum_{j,k=1}^{n} \Gamma_{ijk}(u)w_j w_k + \sum_{j=1}^{n} H_{ij}(u)w_j \tag{2.2}$$

and

$$\frac{\partial w_i}{\partial t} + \lambda_i(u)\frac{\partial w_i}{\partial x} = \sum_{j,k=1}^{n} \gamma_{ijk}(u)w_j w_k + \sum_{j=1}^{n} H_{ij}(u)w_j, \tag{2.3}$$

where

$$\Gamma_{ijk}(u) = \frac{1}{2}(\lambda_j(u) - \lambda_k(u))l_i(u)(\nabla r_k(u)r_j(u) - \nabla r_j(u)r_k(u)), \quad (2.4)$$

$$H_{ij}(u) = l_i(u)(\nabla F(u)r_j(u) - \nabla r_j(u)F(u)) \quad (2.5)$$

and

$$\gamma_{ijk}(u) = \Gamma_{ijk}(u) - \nabla\lambda_k(u)r_j(u)\delta_{ik}. \quad (2.6)$$

By the linear degeneracy (1.6), we have

$$\gamma_{ijj}(u) \equiv \Gamma_{ijj}(u) \equiv 0, \quad \forall 1 \le i, j \le n. \quad (2.7)$$

Noting (1.7), there exist positive constants $\delta_1 = \delta_1(T_0) > 0$ and $M = M(T_0) > 0$ depending only on T_0, such that

$$|\lambda_i(u)| \le M, \quad \forall x \in \mathbb{R}, \forall t \in [0,T], \forall 1 \le i \le n,$$

$$|\lambda_i(u) - \lambda_j(u)| > \delta_1, \quad \forall x \in \mathbb{R}, \forall t \in [0,T], \forall 1 \le i \ne j \le n$$

and all the coefficients are bounded.

By (1.8), we can use the Hölder inequality to obtain

$$\int_{x_1-L}^{x_1+L} \left|\frac{\partial u}{\partial x}(t,x)\right| \mathrm{d}x \le \left\|\frac{\partial u}{\partial x}\right\|_{L^p[x_1-L,x_1+L]} \cdot \|1\|_{L^{p'}[x_1-L,x_1+L]} \le C_p(2L)^{\frac{1}{p'}},$$

where $L > 0$ is a small positive number to be chosen later and p' is the constant satisfying

$$\frac{1}{p} + \frac{1}{p'} = 1.$$

Secondly, by localizing Lemma 3.2 of [14] in a bounded domain, we can estimate the interaction terms

$$Q(t_1; t_0, x_0) = \sup_{i \ne j} \int_{t_0}^{t_1} \int_{x_0-L+M(t-t_0)}^{x_0+L-M(t-t_0)} |w_i(t,x)|\,|w_j(t,x)|\mathrm{d}x\mathrm{d}t, \quad (2.8)$$

where $0 \le t_0 \le \bar{t}_0$, $t_0 \le t_1 \le \bar{t}_1, x_0 \in \mathbb{R}$ and $\bar{t}_0 = \max\{T - L/M, 0\}, \bar{t}_1 = \min\{t_0 + L/M, T\}$, and get

$$Q(t_1; t_0, x_0) \le \frac{1}{\delta_1}\left(\epsilon(L) + CQ(t_1; t_0, x_0)\right)^2,$$

here and hereafter, $\epsilon(L)$ denotes a small positive number, depending on L with $\lim_{L\to 0}\epsilon(L) = 0$ and C is a positive constant.

Now, choosing $L > 0$ small enough, solving this algebraic inequality and noting that $Q(t_1; t_0, x_0)$ is continuous with respect to t_1 and $Q(t_0; t_0, x_0) = 0$, finally we get

$$Q(t_1; t_0, x_0) \le \epsilon(L), \quad \forall t_0 \in [0, \bar{t}_0], \forall t_1 \in [t_0, \bar{t}_1], \forall x_0 \in \mathbb{R}. \quad (2.9)$$

Thirdly, using the corresponding exterior differential form of (2.2), namely,

$$\mathrm{d}(|w_i|(\mathrm{d}x-\lambda_i(u)\mathrm{d}t)) = -\mathrm{sgn}(w_i)\Big(\sum_{a,b=1,a\neq b}^{n}\Gamma_{iab}w_aw_b+\sum_{a=1}^{n}H_{ia}w_a\Big)\mathrm{d}x\wedge\mathrm{d}t$$

and the Stokes formula on the small "triangular" domain surrounded by characteristic curves $\mathfrak{C}_i$, $\mathfrak{C}_j$ and $t=t_0$, we obtain

$$\sup_{i\neq j}\int_{\mathfrak{C}_j}|w_i|\mathrm{d}t \leq \epsilon(L), \tag{2.10}$$

where $\mathfrak{C}_j$ is the jth characteristic curve:

$$\frac{\mathrm{d}x_j(t)}{\mathrm{d}t} = \lambda_j(u(t,x_j(t))), \quad t_0\leq t\leq \bar{t}_1.$$

Finally, multiplying $\mathrm{sgn}(w_i)$ to the both sides of (2.3), integrating it along the ith characteristic curve, we can easily get

$$\sup_{t_0\leq t\leq t_1}\sup_{x\in\mathbb{R}}\max_{1\leq i\leq n}|w_i(t,x)| \leq 2\sup_{x\in\mathbb{R}}\max_{1\leq i\leq n}|w_i(t_0,x)|,$$

where $t_0=\max\{t_1-L/M,0\}$. Noting

$$\frac{\partial u}{\partial x}=\sum_{k=1}^{n}w_kr_k,$$

we reach the desired result (1.9). □

3 Generalization

In this section, We will generalize the previous results to systems without strict hyperbolicity but having some weaker properties. In this case, we do not require (1.4), but we still assume the hyperbolicity of the system, i.e., the coefficient matrix $A(u)$ possesses n real eigenvalues $\lambda_i(u)\,(i=1,\ldots,n)$, and the related left eigenvectors $l_i(u)\,(i=1,\ldots,n)$ and right eigenvectors $r_i(u)\,(i=1,\ldots,n)$ suitably chosen form two bases of $\mathbb{R}^n$ respectively. Moreover, we assume that the eigenvalues and eigenvectors satisfy (1.5) and have the same regularity as the coefficient matrix $A(u)$.

Theorem 3. Under the hypotheses of Theorem 2 (except (1.4) and (1.6)), suppose furthermore that

$$\Gamma_{ijk}(u)\equiv\gamma_{ijk}(u)\equiv 0, \quad \forall u\in\mathbb{R}^n, \forall 1\leq i\leq n, \forall(j,k)\in\mathcal{A}, \tag{3.1}$$

then the conclusion of Theorem 2 is still valid. Here,

$$\mathcal{A} = \{(i,j) \in \mathbb{Z} \times \mathbb{Z} : 1 \le i,j \le n, \exists \alpha \in \mathbb{R}^n \text{ s.t. } \lambda_i(\alpha) = \lambda_j(\alpha)\} \quad (3.2)$$

is the set of index pairs, whose corresponding eigenvalues violate the strict hyperbolicity.

Proof. We need only to replace the definition of $Q(t_1; t_0, x_0)$ by

$$\widetilde{Q}(t_1; t_0, x_0) = \sup_{(i,j)\notin\mathcal{A}} \int_{t_0}^{t_1} \int_{x_0-L+M(t-t_0)}^{x_0+L-M(t-t_0)} |w_i(t,x)|\,|w_j(t,x)| \mathrm{d}x\mathrm{d}t$$

and repeat the whole procedure in Section 2.

Remark 1. Taking $i = j = k$ in (3.1) and noting (2.6), condition (3.1) implies the linear degeneracy (1.6).

Remark 2. Condition (3.1) can be regarded as a generalization of the strict hyperbolicity, because it means the eigenvalues that have the same quantity at some points never interact each other in the wave decomposition formulas.

By Theorem 3, we can easily deduce the generalization of Theorem 1 as follows.

Theorem 4. Under the hypotheses of Theorem 3, the C^1 solution $u = u(t,x)$ to Cauchy problem (1.1)–(1.2) blows up in a finite time $t = T_* < +\infty$, if and only if $u = u(t,x)$ satisfies

$$\sup_{0 \le t < T_*} \|u(t,\cdot)\|_{W^{1,p}} = +\infty, \quad \forall 1 < p \le +\infty.$$

Remark 3. Noting that the strict hyperbolicity of the system is equivalent to

$$(i,j) \in \mathcal{A} \iff i = j,$$

it is easy to see that all the results mentioned in the previous sections can be regarded as special cases of the results in this section.

Remark 4. Since the condition of eigenvalues with constant multiplicity (See [9] for definition) can be expressed as

$$\lambda_i(u) \equiv \lambda_j(u), \quad \forall u \in \mathbb{R}^n, \forall (i,j) \in \mathcal{A}, \quad (3.3)$$

all the results mentioned above are valid for this kind of system.

At the end of this section, we point out that it is easy to give some examples for the Burgers' equation that, if the system is not linearly degenerate or not strictly hyperbolic, then the previous results fail generically.

4 Application

The previous results provide us a way to directly use the energy method of estimating the H^1 norm of the solution in the framework of C^1 solution to verify the ODE singularity. As an application, we consider the generalization of the Kerr–Debye model (cf. [3])

$$\begin{cases} A_0(\chi)\partial_t v + A_1 \partial_x v = \phi(u), \\ \partial_t \chi = \psi(u), \end{cases} \quad x \in \mathbb{R}, t \geq 0 \tag{4.1}$$

with the initial data

$$t = 0 : \ u = u_0(x) = \begin{pmatrix} v_0(x) \\ \chi_0(x) \end{pmatrix}, \quad x \in \mathbb{R}, \tag{4.2}$$

where $u = \begin{pmatrix} v \\ \chi \end{pmatrix} \in \mathbb{R}^{n-r} \times \mathbb{R}^r$ is the unknown vector function, $A_0(\chi) \in C^2$ is a symmetric and positive definite matrix, A_1 is a reversible constant symmetric matrix, $\Phi(u) = \begin{pmatrix} \phi(u) \\ \psi(u) \end{pmatrix} \in C^2$, and $u_0(x)$ is a compactly supported C^1 vector function.

It is easy to verify that system (4.1) satisfies the generalized strict hyperbolicity condition (3.1), So we can apply Theorem 3 to show the property of ODE singularity for Cauchy problem (4.1)–(4.2). For this purpose, we choose $p = 2$ and we know that for H^2 initial data, the boundedness of the H^2 norm of the solution has been shown in [3]. Now we need only to get the estimate of H^1 norm of the solution for C^1 initial data, which can be accomplished by repeating the corresponding part of [3].

Remark 5. The result here is slightly different from that given in [3]. We do not need the estimate on $\dot{H}^2$ norm and we only need $u_0 \in C^1$ instead of H^2, but we require that the eigenvalues and eigenvectors can be chosen to have the C^2 smoothness.

Acknowledgements. The author would like to thank Prof. Ta-Tsien Li, Prof. Yi Zhou, and Prof. Zhen Lei for their valuable suggestions and guidance.

References

[1] S. Alinhac, Blowup for Nonlinear Hyperbolic Equations. Birkhäuser, 1995.

[2] G. Carbou and B. Hanouzet, Semilinear behaviour for a quasilinear hyperbolic system: the Kerr Debye model (English summary). Comptes Rendus Mathematique 343 (2006), 243–247.

[3] G. Carbou, B. Hanouzet and R. Natalini, Semilinear behavior for totally linearly degenerate hyperbolic systems with relaxation. Journal of Differential Equations 246 (2009), 291–319.

[4] A. Jeffrey, Breakdown of the solution to a completely exceptional system of hyperbolic equations. Journal of Mathematical Analysis and Applications 45 (1974), 375–381.

[5] F. John, Formation of singularities in one-dimensional nonlinear wave propagation. Communications on Pure and Applied Mathematics 27 (1974), 377–405.

[6] T. T. Li, Global Classical Solutions for Quasilinear Hyperbolic Systems. Masson/John Wiley, 1994.

[7] T. T. Li, Mechanism of the formation of singularities for quasilinear hyperbolic systems. International Journal of Numerical Analysis and Modeling 6 (2009), 520–530.

[8] T. T. Li and F. G. Liu, Singularity caused by eigenvectors for quasilinear hyperbolic systems. Communications in Partial Differential Equations 28 (2003), 477–503.

[9] T. T. Li, Y. J. Peng, Y. F. Yang and Y. Zhou, Mechanism of the formation of singularities for quasilinear hyperbolic systems with linearly degenerate characteristic fields. Mathematical Methods in the Applied Sciences 31 (2008), 193–227.

[10] T. T. Li and W. C. Yu, Boundary Value Problems for Quasilinear Hyperbolic Systems. Duke University Mathematics Series V, 1985.

[11] T. T. Li, Y. Zhou and D. X. Kong, Global classical solutions for general quasilinear hyperbolic systems with decay initial data. Nonlinear Analysis, Theory, Methods & Applications 28 (1997), 1299–1332.

[12] A. Majda, Compressible Fluid Flow and Systems of Conservation Laws in Several Space Variables. Springer-Verlag, 1984.

[13] Y. F. Yang, Mechanism of the formation of singularities for diagonal systems with linearly degenerate characteristic fields. Communications on Pure and Applied Analysis 9 (2009), 757–768.

[14] Y. Zhou, Global classical solutions to quasilinear hyperbolic systems with weak linear degeneracy. Chinese Annals of Mathematics 25B (2004), 37–56.

Extension of Harten's Artificial Compression Method and of an Entropy-based Detector of Discontinuity to Unstructured Triangular Grids, using Arminjon-Viallon's Staggered Central Scheme

Olivier Rouch
Département de Mathématique et de Statistique
Université de Montréal, Montréal, Canada
E-mail: rouch@dms.umontreal.ca, olivier.rouch@hotmail.ca

Paul Arminjon
Centre de Recherche en Mathématiques
Université de Montréal, Montréal, Canada
E-mail: arminjon@crm.umontreal.ca

Abstract

We herein treat of hyperbolic conservation laws, endowed with a convex entropy function. The discretization is based on triangles, meshing the computational domain. The conservation law is numerically solved using Arminjon-Viallon's staggered central scheme [1], which relies on two dual meshes, featuring barycentric cells and diamond-shaped cells. We are then interested in the characterization of regions of the domain where shocks or contact discontinuities are expected. For this purpose, we use a detector of discontinuities (DoD) based on the entropy production rate [2] and two complementary meshes [3, 4, 5], one having a space resolution twice less than the other. Finally, in the obtained region, we use an artificial compression method (ACM), directly inspired from Harten's work [6, 7]. Numerical results show that the detection of both the shock and the contact discontinuities are sharp enough to make them noticeably benefit from the subsequent artificial compression, without damaging the solution in its continuous regions.

1 Introduction

The detection of discontinuities, whether they are shocks or contacts, is a tricky problem in the numerical treatment of hyperbolic equations. Actually, a wrong determination of shock position can lead to substantially altered solutions, even non-physical profiles and propagation speeds in detonation problems, for example [8]. The capture of contact discontinuities is even more difficult, as they get diffused along with the time iterations [9]; they can actually get so degraded, that one cannot make the difference with a rarefaction (see Section 5). Furthermore, these difficulties, as bad as they already are for one-dimensional problems, get even worse for the two-dimensional case we present.

The method used in this work is based on three independent modules: the main computation scheme (MCS), which numerically solves the hyperbolic problem, advancing in time, the detector of discontinuities (DoD), which collects data from the numerical solution and returns sub-regions of the computational domain that potentially feature discontinuities, and the artificial compression method (ACM), which corrects the profile of the solution in these sub-regions, in a stationary manner. Therefore, it follows the same paradigm as what was earlier adopted for one-dimensional [3] and two-dimensional cartesian-grid [4] problems. This construction principle makes it possible and rather easy to combine each module with any other method. For example, the ACM and DoD presented here could be associated with one's favorite MCS ... or if one wants to build a different DoD ... or use a compressive limiter in only the sub-regions returned by the DoD ... these are just a sample of all the possible combinations.

In the first part of this paper, we give a short reminder on the MCS we use, namely the Arminjon-Viallon staggered central scheme. The second section will show the reader how we construct our DoD, based on the entropy production rate, using two complementary superimposed meshes with different space resolutions. Then, we shall turn to the artificial compression method, which has to be extended from the one-dimensional case to the two-dimensional barycentric and diamond-shaped cells, used by our MCS. Finally, numerical results will be shown.

2 The Arminjon-Viallon MCS

We consider hyperbolic conservation laws of the form:

$$\partial_t u \ + \ \nabla \cdot F(u) \ = \ 0 \ , \text{ with } \ F(u) = (f(u), g(u))^T. \qquad (2.1)$$

Having performed a standard triangulation of the computational domain, we then build the corresponding barycentric cells $\{C_i\}$ and diamond-

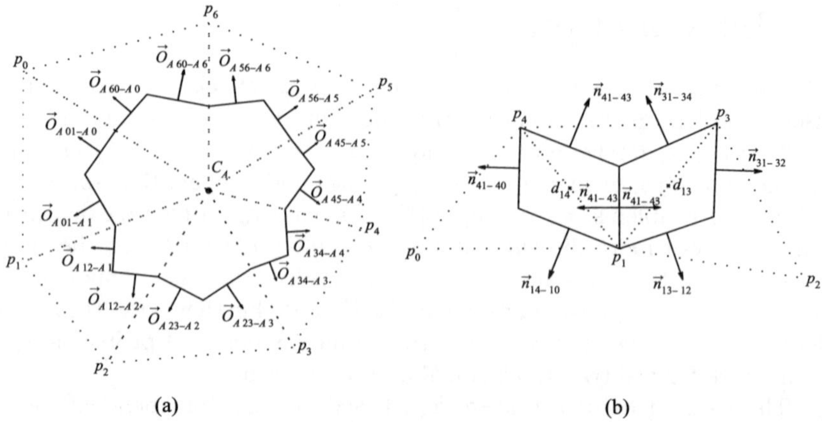

Figure 2.1 Barycentric ((a), red) and diamond-shaped ((b), green) cells, built on triangles (dotted).

shaped cells $\{L_{ij}\}$, as exposed in Figure 2.1. The initial condition is supposed to be available at the center of each C_i. Advancing the solution from u_i^0 at $t = 0$ on the $\{C_i\}$'s, to u_{ij}^1 at $t = t^1$ on the $\{L_{ij}\}$'s, is performed in two steps, a predictor and a corrector. They follow directly from the integration of (2.1) over the control volume $L_{ij} \times [t^0, t^1]$, and the application of Green's theorem:

$$u(x, y, t^{1/2}) = u_i^0 - (\Delta t/2)\, (f'(u), g'(u))^T\Big|_{(x,y,t^0)} \cdot \hat{\nabla} u_i^0, \tag{2.2}$$

$$\text{with } u(x, y, t^0) \approx u_i^0 + (x - x_i, y - y_i)^T \cdot \hat{\nabla} u_i^0,$$

$$A(L_{ij}) u_{ij}^1 = \iint_{L_{ij} \cap C_i} u(x, y, t^0)\big|_{C_i}\, dA + \iint_{L_{ij} \cap C_j} u(x, y, t^0)\big|_{C_j}\, dA$$
$$- \Delta t \int_{\partial L_{ij}} (f(u), g(u))^T\Big|_{(x,y,t^{1/2})} \cdot \vec{n}\, ds, \tag{2.3}$$

where (2.3) uses, for the last part of its R.H.S., the midpoint rule on each side of L_{ij}. The limitation $\hat{\nabla}$ of the gradient of u is performed using, for example, the least square method. Naturally, integrating again (2.1), but over $C_i \times [t^1, t^2]$ advances the solution one more step forward in time, staggering it back on barycentric cells. We refer the reader to [1] for further details.

This scheme has the advantage of neither requiring the computation of eigenvalues and eigenvectors of f and g, nor solving the associated two-dimensional Riemann problems, thus saving a lot of computational time. But it also has the inconvenient of being more diffusive than its upwind counterparts, which is a particularly big handicap in the neighborhood of discontinuities. Therefore, the benefits of being able to balance the diffusivity of the Arminjon-Viallon's scheme, by improving its capture of discontinuities, is easily measurable.

3 The entropy-based DoD

The equation 2.1 can be endowed with an entropy function ϕ and a vectorial entropy flux ψ satisfying:

$$\partial_t\phi + \nabla \cdot \psi(\phi) \leq 0 \text{ , with } (\nabla_u \psi)^T = (\nabla_u \phi(u))^T \; J(F(u)). \qquad (3.1)$$

The determination of the entropy production rate Φ_{ij}^{n+1} over a diamond-shaped computational cell L_{ij} follows the same main ideas as in the one-dimensional case [2]. We add an extra equation to our system by introducing a quantity $\partial_t \Phi$ on the R.H.S. of (3.1), transforming it into an equality. Then we still use the same MCS to solve the new system, *as if it was homogeneous.* This will give, starting from the $\{\phi_i^0 = \phi(u_i^0)\}$'s, the updated quantities $\{\phi_{ij}^1\}$. But since we also have an analytic formula for ϕ, depending on u, we can compute the "real" $\phi(u_{ij}^1)$. We finally define the entropy production rate Φ_{ij}^1 as:

$$\Phi_{ij}^1 = \left(\phi(u_{ij}^1) - \phi_{ij}^1\right)/\Delta t. \qquad (3.2)$$

For the next step, we keep the values $\{\phi(u_{ij}^1)\}$ as initial data, and use the same approach to determine $\{\Phi_i^2\}$ on the barycentric cells. It can be shown [5] that, just as in the one-dimensional case [2], the quantities $\{\Phi_{i|ij}^n\}$ behave as follows:

$\Phi_{i|ij}^n = \mathcal{O}(1/\Delta t)$ close to shocks,
$\Phi_{i|ij}^n = \mathcal{O}(1)$ close to contact discontinuities,
$\Phi_{i|ij}^n = \mathcal{O}(\Delta t^r)$ in smooth regions (r — order of the MCS).

Using these properties, one could try to detect discontinuities by setting a threshold, e.g., around Δt, in order to separate them from smooth regions. However, this would imply the definition of a different threshold for each new problem, each new initial condition. Moreover, this threshold would become very difficult to set, when discontinuities of various strengths appear in the numerical solution. Therefore, we rely in this paper on the ratio between two entropy production rates $\tilde{\Phi}_{i|ij}^n$ and $\Phi_{i|ij}^n$, respectively computed using a rough triangulation of space resolution $\tilde{h}$ and a refined grid of space resolution h, with $\tilde{h} \approx 2h$, as shown in Figure 3.1.

Notice that the rough mesh counts about three times less points, and that computations on both meshes can be performed independently, in parallel, although they share a lot of data. Moreover, as soon as

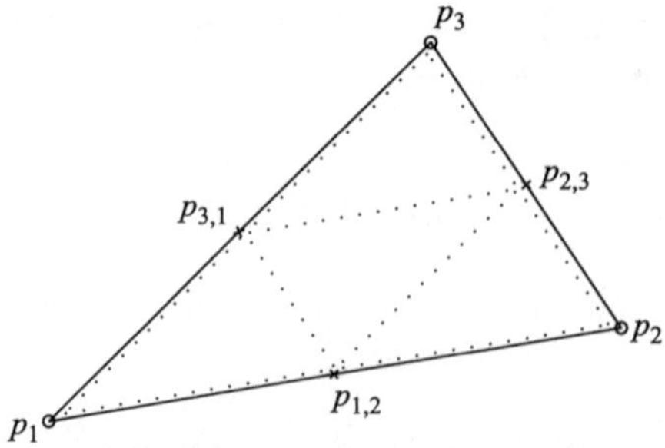

Figure 3.1 Rough mesh (full) and refined mesh (dotted).

computations on the rough mesh are finished, the available processor can begin computing the $\phi(u^n_{i|ij})$'s, which are responsible for the most important part of the computational overhead associated to this DoD. In fact, this kind of algorithm fully benefits from the multi-core technology of modern processors.

In order to compare the quantities obtained on the two complementary meshes, we use a local $\|\cdot\|_\infty$, which is actually extracting the maximum absolute value in a close neighborhood of the cell considered (two-cell away on the fine mesh, one-cell away on the rough mesh). We then get the following table of ratios:

Table 1 Characterizing ratios for smooth regions, contacts, and shocks.

u	Smooth	Contact	Shock
$\left\|\tilde{\Phi}^n_{i\|ij}\right\|_\infty / \left\|\Phi^n_{i\|ij}\right\|_\infty$	$\approx \left(\frac{1}{2}\right)^r$	≈ 1	≈ 2

The characterization of the regions holding a discontinuity obtained this way is sharp enough to permit the use of a highly compressive correction method, like the one based on Harten's one-dimensional ACM presented hereafter, that would otherwise damage continuous regions of the numerical solution.

4 An extension of Harten's ACM

Let us first remember that Harten's artificial compression method (ACM) [7] is conducted in the vicinity of a shock to reduce its viscous profile [10], or close to a contact discontinuity to provide it with such a viscous profile [6]. Indeed, the width of the viscous profile of a shock between two states u_L and u_R is given by:

$$W(u_-, u_+) = \Delta t \int_{u_-}^{u_+} \frac{\beta(u,\lambda)}{(s(u_L,u) - S)\,(u - u_L)} du, \qquad (4.1)$$

where u_- and u_+ are two successive states along the path from u_L to u_R; $\beta(u, \lambda)$ is related to the viscous version of the one-dimensional reduction of equation 2.1; S is the propagation speed of the discontinuity; and $s(u,v)(u-v) = f(u) - f(v)$. A close inspection of the denominator of (4.1) leads to the idea that, by increasing its value, one can reduce $W(u_-, u_+)$. Therefore, we introduce a compressive flux $K(u)$, which will modify the value of f only in discontinuous regions of u, so that the solution does not get disturbed elsewhere. The resolution of the augmented conservation law $\partial_t u + \nabla \cdot (f(u) + K(u)) = 0$ is then split into two independent problems: $\partial_t u + \nabla \cdot f(u) = 0$ the original one, and $\partial_t u + \nabla \cdot K(u) = 0$ the compressive and stationary one. Verifying that K is compressive holds in three points:

$K = \mathcal{O}(1)$ and $\text{sign}(K) = \text{sign}(u_R - u_L)$ in discontinuous regions,

$\partial_x K = \mathcal{O}(\Delta x)$ between discontinuous and smooth regions,

$K = 0$ in smooth regions of u.

Solving the compression equation then requires a numerical method that has a best possible resolution of stationary shocks, as one does not want ACM to introduce ... further diffusion! Harten uses the classic upwind method.

Starting from this one-dimensional version of the ACM, we want to build its two-dimensional extension, based on the classic unstaggered donor-cell upwind (DCU) method [11], since it has a good resolution of stationary shocks, and apply it either on the diamond-shaped cells or on the barycentric cells, using a compression flux $\vec{K} = (K^x, K^y)$, yet to define. Let $(u_B, u_F, \vec{S}(x,y), \vec{n}(x,y))$ characterize a two-dimensional discontinuity, i.e., a curve, at a fixed time $t = t^n$, separating two states u_B and u_F, moving following the speed vector $\vec{S}(x,y)$, and having a normal $\vec{n}(x,y)$. Along a particular normal to the discontinuity, the same idea of a viscous profile can be used, and to reduce it, the same kind of compressive flux can be defined, which will have to obey:

$\|\vec{K}\| = \mathcal{O}(1)$ and $\vec{K} \cdot \vec{n} > 0$ in discontinuous regions,

$\nabla \cdot \vec{K} = \mathcal{O}(h)$ between discontinuous and smooth regions,

$\|\vec{K}\| = 0$ in smooth regions of u.

We use $\vec{K} = \nabla^{lim} u$. It verifies the three conditions ahead, provided we keep it to zero where the DoD has not found a potentially discontinuous region. Remark that all this does not hold in the case of interacting discontinuities, but we are confident that two better captured discontinuities will interact in a more physical way. A criterion to differentiate regions where multiple discontinuities interact is presently studied.

Then, we use the DCU method to solve the compression equation:

$$\partial_t u + \nabla \cdot \vec{K}(u) = 0 \text{ , with } \vec{K}(u) = \nabla^{lim} u. \tag{4.2}$$

One time step over a diamond-shaped cell then amounts to:

$$u_{ij}^K = u_{ij} - \frac{\Delta t^K}{2} \sum_{L_{mn}} \begin{pmatrix} K_{mn}^x - \left|K_{mn}^x - K_{ij}^x\right| \; s_{ij,mn} \\ K_{mn}^y - \left|K_{mn}^y - K_{ij}^y\right| \; s_{ij,mn} \end{pmatrix} \cdot \left(\frac{\vec{n_{ij,mn}}}{\|\vec{n_{ij,mn}}\|^2} \right), \tag{4.3}$$

where $L_{mn} \in \mathcal{N}(ij)$ browses the set of the neighbors of L_{ij}; the length of the interface between L_{ij} and L_{mn} is $\|\vec{n_{ij,mn}}\|$; and $s_{ij,mn} = sign(u_{mn} - u_{ij})$. The quantity $\Delta t^K = h\, \lambda_K$ is determined as follows:

$$\lambda_K \max_{u_{ij} \neq u_{mn},\; L_{mn} \in \mathcal{N}(ij)} \frac{|K_{mn}^x - K_{ij}^x| + |K_{mn}^y - K_{ij}^y|}{|u_{mn} - u_{ij}|} \leq 1. \tag{4.4}$$

The same principle applies for barycentric cells and gives similar formulas, from u_i to u_i^K.

5 Numerical results

We present the very well known test case of the shocktube, using Euler's equations of an ideal gas. The initial conditions are $u_L = ()^T$ and $u_R = ()^T$. Notice that the final times are not exactly the same for both computations (although they are very close), indeed, since Δt is dynamically computed at each time step, the solutions substantially differ, depending on whether ACM is used or not. Because the mesh is unstructured, the problem becomes genuinely two-dimensional, which is different from what happens with a cartesian grid, which is aligned with the discontinuities. The borders of the domain are not compressed, so the status of the DoD on the border is irrelevant.

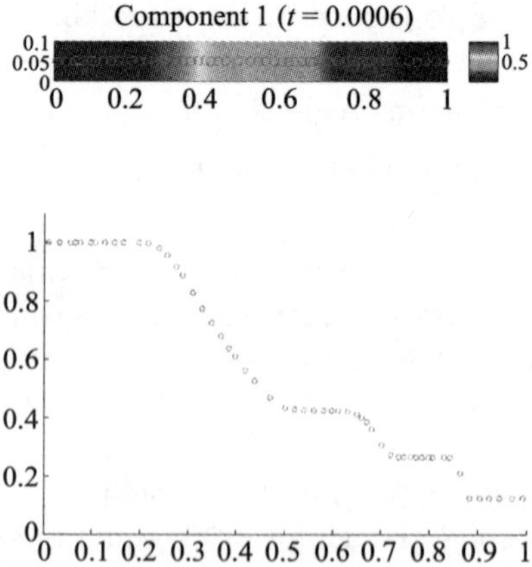

Figure 5.1 Without ACM.

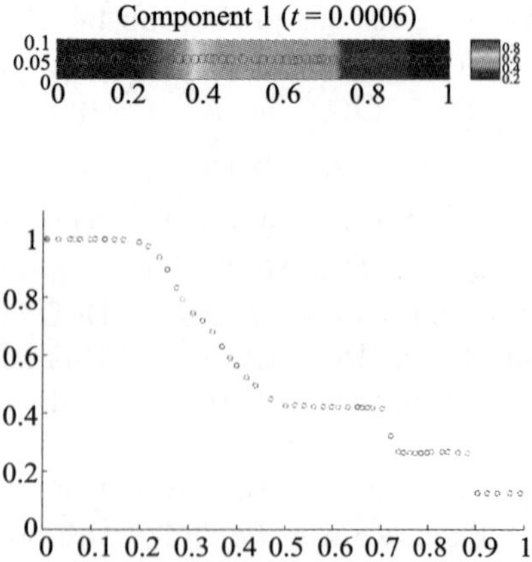

Figure 5.2 With DoD/ACM.

From Figures 5.1 and 5.2, one can appreciate the difference, especially in the capture of the contact discontinuity. The rarefaction is slightly perturbed by the ACM, but not degenerated as what happens when the DoD is not used. Figure 5.3 shows the rather good reliability of the DoD, that correctly identifies compressible regions, and gives wrong positives only in the constant states between waves.

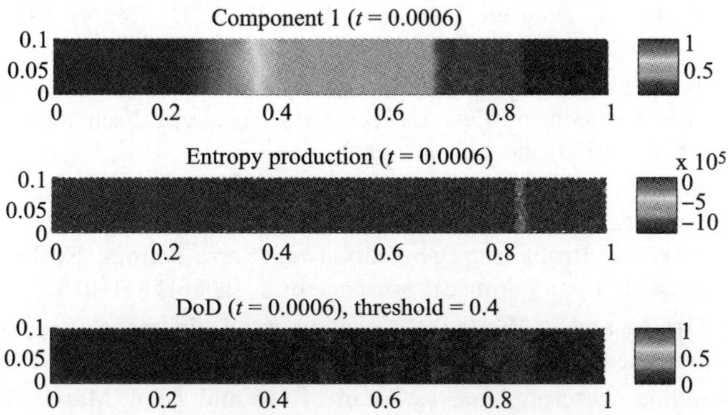

Figure 5.3 Density (top), Φ (middle), and DoD (bottom).

6 Conclusions

In this paper, we presented numerical results for the Arminjon-Viallon scheme, where the effect of the DoD and ACM is particularly interesting. However, the reader will notice that these two last tools are adaptable to virtually any eulerian MCS.

References

[1] P. Arminjon, M. C. Viallon and A. Madrane, A finite volume extension of the Lax-Friedrichs and Nessyahu-Tadmor schemes for conservation laws on unstructured grids. Int. J. of Comp. Fluid Dyn. 9 (1997), 1–22.

[2] G. Puppo, Numerical entropy production on shocks and smooth transitions. J. of Sci. Comput. 17 (2002), 263–271.

[3] O. Rouch and P. Arminjon, Using the entropy production rate to enhance artificial compression. Proc. of the 12th International Conference on Hyperbolic Problems. Tadmor-Liu-Tzavaras eds., University of Maryland (USA), Amer. Math. Soc. (AMS) pub. (2008), 899–909.

[4] O. Rouch, Enhancing the capture of two-dimensional, shock induced detonation fronts using Harten's artificial compression method on under-resolved Cartesian grids. Proc. of the 2009 ICNAAM Internat. Conf.

on Numer. Anal. and Appl. Math., Simos-Psihoyios-Tsitouras editors, Rethymno (Crete,Greece), Amer. Instit. of Phys. (AIP) pub. 2 (2009), 1112–1116.

[5] O. Rouch, Méthodes numériques pour la capture et la résolution des discontinuités dans les solutions des systèmes hyperboliques. Université de Montréal, PhD thesis in correction.

[6] A. Harten, The method of artificial compression: 1. Shocks and Contact Discontinuities, Courant Institute, AEC R. & D. Report C00-3077-50 (1974), New York U.

[7] A. Harten, The artificial compression method for computation of shocks and contact discontinuities: III. self-adjusting hybrid schemes. Math. of Comput. 32 (1978), no.142, 363–389.

[8] O. Rouch, M. O. Saint-Hilaire and P. Arminjon, Toward an improved capture of stiff detonation waves, Proc. of the 11th International Conference on Hyperbolic Problems, Benzoni-Gavage-Serre editors, École Normale Supérieure de Lyon (France), Springer pub. (2006), 911–918.

[9] S. A. Orszag and L. W. Jayne, Local errors of difference approximations to hyperbolic equations. J. of Comput. Phys. 14 (1974), 361–370.

[10] G. Jennings, Discrete shocks. Comm. Pure and Appl. Math. 27 (1974), 25–37.

[11] R. J. LeVeque, Finite Volume Methods for Hyperbolic Problems. Cambridge University Press, 2002.

Layered Structures: Instability of the Shock Waves and Elecrtodynamical Instability

Roman Semenko
Novosibirsk State University
Novosibirsk, 630090, Russia
E-mail: rsem86@mail.ru

Abstract

We are concerned with a hydrodynamical model of layered structures at the presence of the electric current. We formulate a linearized stability problem for shock waves and prove its ill-posedness that means instability of shock waves for the given model of layered structures. Also we prove the electrodynamical instability of the layered structures at the presence of the alternating electric current.

1 Introduction

It is known that the increase of oil reservoir performance is of great importance for modern power engineering. The difficulty in the resolution of this problem is that in chinked zones of exploiting collectors there appear water-oil layered structures which block transport structures of collectors and, therefore, eject significant petroliferous areas from modes of water replacement. The restoration of collector's permeability is only possible under the destruction of layered water-oil structures. It should be noted that it is expedient to provide the power action on a layered structure by using methods of electrometry. The point is that water-oil layered structures are anisotropic dielectrics weakly conducting the electric current. The goal of this work is studying two possible methods of destruction of this layered structures: shock waves and electrodynamical instability at the presence of the small-amplitude alternating current. In this work the hydrodynamical model of layered structures at the presence of the electric current is concerned. We formulate a linearized stability problem for shock waves and prove its ill-posedness that means instability of shock waves for the given model of layered structures. Also we formulate a linearized stability problem for layered structures and prove that solutions of this problem infinitely grow, which means destruction of layered structures at the presence of the small-amplitude alternating electric current.

2 Ill-posedness of the stability problem for shock waves

Let's consider the system of electrohydrodynamical equations (in reversible approximation) for **water-oil layered structures**. The basis of this way is **continuous approximation**. In detail this approximation is described in monograph [1] and based on the validity of **the conservation laws of mass, impulse, energy, and entropy** and **Maxwell equations**. Considering a mixture of water and oil as an **one-velocity layered continuum**, a mathematical model of such layered structures was proposed in [2, 3, 4]:

$$\rho_t + div(\rho \boldsymbol{v}) = 0, \tag{2.1a}$$

$$(\rho w)_t + div(\rho w \boldsymbol{v}) = 0, \tag{2.1b}$$

$$(\rho v_i)_t + \left\{ p + \frac{|\boldsymbol{E}|^2}{8\pi} \right\}_{x_i} + \sum_{k=1}^{3} (\rho v_i v_k - \Sigma_{ik})_{x_k} = 0, i = 1, 2, 3, \tag{2.1c}$$

$$(\rho s)_t + div(\rho s \boldsymbol{v}) = \frac{R}{T}, \tag{2.1d}$$

$$\left\{ \rho(\tilde{U} + \frac{|\boldsymbol{v}|^2}{2}) \right\}_t + div\boldsymbol{Q} = (\boldsymbol{j}, \boldsymbol{E}), \tag{2.1e}$$

$$rot\boldsymbol{E} = 0, \tag{2.1f}$$

$$div\boldsymbol{D} = 4\pi\rho_e, \tag{2.1g}$$

$$\frac{\partial \rho_e}{\partial t} + div\boldsymbol{j} = 0, \tag{2.1h}$$

$$\boldsymbol{j} = \rho_e(\boldsymbol{v} + \boldsymbol{E}\,\Omega). \tag{2.1i}$$

Here
ρ is the density of the layered continuum,
$\boldsymbol{v} = (v_1, v_2, v_3)$ is the medium velocity vector,
p is the pressure,
Σ_{ik} is the stress tensor,
s is the entropy,
R is the dissipative function,
T is the temperature,
ε_0 is the mass internal energy,
$|\boldsymbol{v}|^2 = (\boldsymbol{v}, \boldsymbol{v})$,

$\boldsymbol{E} = (E_1, E_2, E_3)$ is the electric force,
$\boldsymbol{D} = (D_1, D_2, D_3)$ is the vector of electric induction,
$\boldsymbol{j} = (j_1, j_2, j_3)$ is the vector of current density,
ρ_e is the volume charge density,
$w(\boldsymbol{x}, t) = \theta$, $\boldsymbol{x} = (x_1, x_2, x_3)$, where θ is the parameter identifying layers.

By standard arguments (see [1, 5, 6, 7]) we get the following Rankine-Hugoniot conditions for equations (2.1a) – (2.1i):

$$[\rho(v_n - \mathcal{D}_n)] = [\mathrm{J}] = 0, \tag{2.2a}$$

$$\mathrm{J}[w] = 0, \tag{2.2b}$$

$$\mathrm{J}[v_n] + [\Pi] = 0, \tag{2.2c}$$

$$\mathrm{J}[v_\tau] - [(\boldsymbol{C}, \boldsymbol{\tau})] = 0, \tag{2.2d}$$

$$\mathrm{J}\left[\tilde{U} + \frac{|\boldsymbol{v}|^2}{2}\right] + [G] = 0, \tag{2.2e}$$

$$[j_n] = \frac{\partial \omega}{\partial t}, \tag{2.2f}$$

$$[D_n] = -4\pi\omega, \tag{2.2g}$$

$$[E_\tau] = 0. \tag{2.2h}$$

Here
$\tilde{f} = (f(t, y, z) - x) = (f(t, x_2, x_3) - x_1)$ is a surface of strong discontinuity,
$\boldsymbol{n} = (n_1, n_2, n_3) = \frac{1}{\sqrt{1 + f_{x_2}^2 + f_{x_3}^2}}(-1, f_{x_2}, f_{x_3})$ is the unit normal to the discontinuity surface,
$[j] = (j - j_\infty)$ denotes the jump of j on the discontinuity surface, with j_∞ the value of j from the left of the discontinuity (i.e., when $(f(t, x_2, x_3) - x_1) \to +0$),
$v_n = (\boldsymbol{v}, \boldsymbol{n})$, $v_\tau = (\boldsymbol{v}, \boldsymbol{\tau})$,
$\mathcal{D}_n = -\frac{f_t}{\sqrt{1 + f_{x_2}^2 + f_{x_3}^2}}$ is the normal component of the discontinuity speed,
$\boldsymbol{\tau} = \begin{cases} (f_{x_2}, 1, 0) \\ (f_{x_3}, 0, 1) \end{cases}$ is the vectors orthogonal to the normal $\boldsymbol{n}$ and tangent to the discontinuity surface,

$$\Pi = p + \frac{|\boldsymbol{E}|^2}{8\pi} - (\boldsymbol{C}, \boldsymbol{n}) = p + \frac{|\boldsymbol{E}|^2}{8\pi} + b_n\left(f_n - \frac{\partial \Phi}{\partial n}\right) + \Phi(\boldsymbol{n}\mathcal{B}, \boldsymbol{n}) - \frac{E_n D_n}{4\pi},$$

$$\boldsymbol{C} = ((\boldsymbol{\Sigma}_1, \boldsymbol{n}), (\boldsymbol{\Sigma}_2, \boldsymbol{n}), (\boldsymbol{\Sigma}_3, \boldsymbol{n})),\ \boldsymbol{\Sigma}_i = (\Sigma_{i1}, \Sigma_{i2}, \Sigma_{i3}),\ i = 1, 2, 3,$$

$$\boldsymbol{b} = (b_1, b_2, b_3) = \nabla w,\ \boldsymbol{d} = \frac{\boldsymbol{b}}{|\boldsymbol{b}|},$$

$$\Phi = \frac{N}{q^2}\triangle w,\ \boldsymbol{f} = \frac{B}{2q^2}(\frac{|\boldsymbol{b}|^2}{q^2} - 1) + \frac{\sigma q}{b_3}(d_1, d_2, \frac{-d_1^2 - d_2^2}{d_3}),$$

$$\mathcal{B} = (\frac{\partial^2 w}{\partial x_i \partial x_k}) = \mathcal{B}^*,\ i, k = 1, 2, 3,\ b_n = (\boldsymbol{b}, \boldsymbol{n}),\ f_n = (\boldsymbol{f}, \boldsymbol{n}),$$

$$E_n = (\boldsymbol{E}, \boldsymbol{n}),\ D_n = (\boldsymbol{D}, \boldsymbol{n}) = (\boldsymbol{E}\tilde{\Psi}, \boldsymbol{n}),\ \frac{\partial \Phi}{\partial n} = (\boldsymbol{n}, \nabla\Phi),$$

$$-(\boldsymbol{C}, \boldsymbol{\tau}) = b_\tau\left(f_n - \frac{\partial\Phi}{\partial n}\right) + \Phi(\boldsymbol{n}\mathcal{B}, \boldsymbol{\tau}) - \frac{E_\tau D_n}{4\pi},\ b_\tau = (\boldsymbol{b}, \boldsymbol{\tau}),\ E_\tau = (\boldsymbol{E}, \boldsymbol{\tau}),$$

$$G = \tilde{p} v_n + \Phi\left(\boldsymbol{b}, \frac{\partial \boldsymbol{v}}{\partial n}\right) + \Phi(\boldsymbol{n}\mathcal{B}, \boldsymbol{v}),$$

$$v_b = (\boldsymbol{v}, \boldsymbol{b}),\ \frac{\partial \boldsymbol{v}}{\partial n} = (\boldsymbol{n}, \nabla)\boldsymbol{v},$$

$$j_n = (\boldsymbol{j}, \boldsymbol{n}) = \rho_e((\boldsymbol{v} + \boldsymbol{E}\Omega), \boldsymbol{n}) = \rho_e v_n + \rho_e(\boldsymbol{E}\Omega, \boldsymbol{n}),$$

$\tilde{p} = p - \frac{1}{2}(\boldsymbol{E}, \boldsymbol{E}\Psi)$, B, N — elastic modules, Ψ is the matrix of dielectric sensivity, $\tilde{\Psi} = (I_3 + 4\pi\Psi)$.

While getting the jump conditions for system (2.1a)–(2.1i) we assumed the existence of a **surface charge** $\omega = \omega(t, \boldsymbol{x}')$(where$\boldsymbol{x}' = (x_2, x_3)$) and the absence of a **surface current** on the discontinuity surface (see [5]).

We are interested in strong discontinuities of the type of **shock wave** (see [8]), i.e., when

$$[\rho] \neq 0, \mathrm{J} \neq 0.$$

We consider a stationary flow of a layered dielectric with a surface of shock wave. Such a flow is described, for example, by a piecewise constant solution

$$\begin{aligned}
&\boldsymbol{v} = (\hat{v}_1, \hat{v}_2, 0),\ \rho = \hat{\rho},\ s = \hat{s},\\
&w = \hat{q}z,\ \hat{q} = q(\hat{V}, \hat{s}),\ \hat{V} = \frac{1}{\hat{\rho}},\\
&\hat{\boldsymbol{E}} = (\hat{E}_1, \hat{E}_2, \hat{E}_3),\ \rho_e = 0 \quad \text{with } x_1 > 0,\\
&\boldsymbol{v} = (\hat{v}_{1\infty}, \hat{v}_{2\infty}, 0),\ \rho = \hat{\rho}_\infty,\ s = \hat{s}_\infty,\\
&w = \hat{q}_\infty z,\ \hat{q}_\infty = q(\hat{V}_\infty, \hat{s}_\infty),\ \hat{V}_\infty = \frac{1}{\hat{\rho}_\infty},\\
&\hat{\boldsymbol{E}} = (\hat{E}_{1\infty}, \hat{E}_{2\infty}, \hat{E}_{3\infty}),\ \rho_{e\infty} = 0 \quad \text{with } x_1 < 0.
\end{aligned} \tag{2.3}$$

This solution should satisfy the conditions (2.2a)–(2.2h).

Now we'll linearise our equations (2.1a)–(2.1i) and conditions (2.2a)–(2.2h) on this basic solution. Following [9, 10, 11], the question on the stability of shock waves for the electrohydrodynamical model of a layered dielectric is reduced to the study of the **well-posedness** of the linearized stability problem. The proof of ill-posedness follows from the construction of **Hadamard-type ill-posedness examples** (see [12, 13]). From the physical point of view the presence of such examples means the instability of shock waves in layered structures.

As usual (see [14]), in the case of linear problems, we construct Hadamard examples by seeking for **particular exponential solutions**. In this work we restrict ourself to the case of the **version of the basic problem**. For the 2D case the version of the basic problem looks as follows. One seeks solutions of the linear dimentionless system

$$\begin{gathered}(E_{1\infty})_z - (E_{3\infty})_x = 0,\\ (E_{1\infty}\tilde{\Psi}_{11\infty})_x + (E_{3\infty}\tilde{\Psi}_{33\infty})_z = 0,\ t > 0,\ (x,z) \in \mathbb{R}^2_-;\end{gathered} \tag{2.4}$$

$$\begin{gathered}Lw + v_3 = 0,\\ Ls = \rho_e,\\ Lv_1 + \frac{\partial \tilde{p}}{\partial x} = \hat{A}_1\rho_e,\\ Lv_3 + \hat{\beta}\frac{\partial \tilde{p}}{\partial z} = \hat{N}\Delta^2 w - \hat{B}\hat{\lambda}^2\frac{\partial^2 w}{\partial z^2} - \hat{\sigma}\frac{\partial^2 w}{\partial x^2} + \hat{\Sigma}\frac{\partial s}{\partial z},\\ L\tilde{p} + \frac{\partial v_1}{\partial x} + \hat{\beta}\frac{\partial v_3}{\partial z} = R_1\rho_e,\\ (E_1)_z - (E_3)_x = 0,\\ (E_1\tilde{\Psi}_{11})_x + (E_3\tilde{\Psi}_{33})_z = 4\pi\rho_e,\\ L\rho_e + \hat{\Lambda}_1\frac{\partial \rho_e}{\partial x} = 0,\ t > 0,\ (x,z) \in \mathbb{R}^2_+,\end{gathered} \tag{2.5}$$

satisfying the boundary conditions

$$\begin{gathered}w = 0,\\ f_t = \frac{\mu}{\hat{\mu}}\tilde{p} - \frac{\mu_E}{\hat{\mu}}E_1 + \frac{\mu_{E\infty}}{\hat{\mu}}E_{1\infty} = 0,\\ v_1 + dM_1\tilde{p} + d_E M_1 E_1 - d_{E\infty}M_1E_{1\infty} = 0,\\ (M_1^2 - \hat{\sigma})b_1 + \hat{N}\Delta b_1 - M_1\Theta E_3 + M_1\bar{\mu}f_z = 0,\\ s = \nu\tilde{p} + \nu_E E_1 - \nu_{E\infty}E_{1\infty},\\ \omega_t = -\nu_\omega\rho_e,\\ \omega = \frac{1}{4\pi\hat{\chi}}(\tilde{\Psi}_{11}E_1 - \bar{A}\tilde{\Psi}_{11\infty}E_{1\infty}),\\ E_{3\infty} = \frac{E_3}{\bar{A}} + W_E f_z\end{gathered} \tag{2.6}$$

for $t > 0$, $x = 0$, $z \in \mathbb{R}_1$.

Here

$$L = \frac{\partial}{\partial t} + M_1 \frac{\partial}{\partial x_1} + M_2 \frac{\partial}{\partial x_2}, \; M_i = \frac{\hat{v}_i}{\hat{c}_s}, \; i = 1, 2,$$

$$L_\infty = K_\infty \frac{\partial}{\partial t} + M_{1\infty} \frac{\partial}{\partial x_1} + M_{2\infty} \frac{\partial}{\partial x_2}, \; M_{i\infty} = \frac{\hat{v}_{i\infty}}{\hat{c}_{s\infty}}, \; i = 1, 2, \; K_\infty = \frac{\hat{c}_s}{\hat{c}_{s\infty}},$$

$$\mathbb{R}^2_- = \{\boldsymbol{x} \in \mathbb{R}^2, \; x_1 < 0\}, \quad \mathbb{R}^2_+ = \{\boldsymbol{x} \in \mathbb{R}^2, \; x_1 > 0\},$$

$\hat{c}_s$, $\hat{c}_{s\infty}$ — sound speeds, $\hat{A}_1$, $\hat{\beta}$, $\hat{N}$, $\hat{B}$, $\hat{\lambda}$, $\hat{\sigma}$, $\hat{\Sigma}$, R_1, $\hat{\Lambda}_1$, μ_E, $\mu_{E\infty}$, $\hat{\mu}$, $\bar{\mu}$, d, d_E, $d_{E\infty}$, Θ, ν, ν_E, $\nu_{E\infty}$, ν_ω, $\bar{\chi}$ — some constants (see [3]).

In [3] it is proved that this problem has non-trivial solutions (2.4) in the form

$$\begin{pmatrix} E_{1\infty} \\ E_{3\infty} \end{pmatrix} (t, \, x, \, z) = \begin{pmatrix} \mathcal{E}_{1\infty} \\ \mathcal{E}_{3\infty} \end{pmatrix} exp\{n(\tau t + r x + i z)\}, \; x < 0, \tag{2.7}$$

$$\begin{pmatrix} w \\ s \\ v_1 \\ v_3 \\ \tilde{p} \\ E_1 \\ E_3 \\ \rho_e \end{pmatrix} (t, \, x, \, z) = \begin{pmatrix} W_0 \\ S \\ V_1 \\ V_3 \\ \tilde{\mathcal{P}} \\ \mathcal{E}_1 \\ \mathcal{E}_3 \\ R_e \end{pmatrix} exp\{n(\tau t + R x + i z)\}, \; x > 0, \tag{2.8}$$

where $\mathcal{E}_{1\infty}$, $\mathcal{E}_{3\infty}$, W_0, S, V_1, V_3, $\tilde{\mathcal{P}}$, $\mathcal{E}_1$, $\mathcal{E}_3$, R_e, τ, r, R are constants, with

$$Re\,\tau > 0, \; Re\,r > 0, \; Re\,R < 0, \; n \in \mathbb{Z}. \tag{2.9}$$

The small perturbations of the shock front and the surface charge are sought in the form

$$\begin{aligned} f(t, z) &= F\, exp\{n(\tau t + i z)\}, \\ \omega(t, z) &= \Omega\, exp\{n(\tau t + i z)\}. \end{aligned} \tag{2.10}$$

Here F and Ω are constants. That means shock waves are unstable in the given model of layered structures.

3 Electrodynamical instability of the layered structures

Let us assume now that we have an alternating electric current in the layered structures. We consider this basic solution of

$$\begin{gathered} \rho = \hat{\rho}, \; s = \hat{s}, \; v_1 = 0, \; v_2 = 0, \; v_3 = 0, \\ w = \hat{w} = \hat{q} z, \\ \boldsymbol{E} = \hat{E} = (0, 0, \hat{E}_0 \sin(\omega t)), \; \rho_e = 0. \end{gathered} \tag{3.1}$$

As in the previous case, we linearise system on this solution. We can reduce this system to

$$\hat{\rho}_e' + \hat{\Lambda}\sin(\omega t) i\omega_3 \hat{\rho}_e = 0, \text{ or}$$
$$\hat{\rho}_e = \hat{\rho}_e(t, \boldsymbol{\omega}) = \hat{\rho}_{e0}(\boldsymbol{\omega}) exp\left\{\frac{\cos(\omega t) - 1}{\omega} i\omega_3 \hat{\Lambda}\right\}, \tag{3.2a}$$

$$\begin{aligned} &\hat{v}_3'' + (\hat{\sigma}|\boldsymbol{\omega}_\perp|^2 + \hat{B}\hat{\lambda}^2\omega_3^2 + \hat{N}|\boldsymbol{\omega}|^4)\hat{v}_3 + i\omega_3\hat{\beta}\hat{\tilde{p}}' \\ &= \hat{\Sigma}_1 \sin^2(\omega t) i\omega_3 \hat{\rho}_e + \hat{\rho}_e \hat{A}_3 \omega \cos(\omega t), \end{aligned} \tag{3.2b}$$

$$\hat{\tilde{p}}'' + |\boldsymbol{\omega}_\perp|^2 \hat{\tilde{p}} + \hat{\beta} i\omega_3 \hat{v}_3' = -R_1 \hat{\Lambda} \sin^3(\omega t) i\omega_3 \hat{\rho}_e + R_1 \hat{\rho}_e \omega \sin(2\omega t), \tag{3.2c}$$

where

$$|\boldsymbol{\omega}_\perp|^2 = \omega_1^2 + \omega_2^2,\ |\boldsymbol{\omega}|^2 = \omega_1^2 + \omega_2^2 + \omega_3^2,\ \boldsymbol{\omega} = (\omega_1, \omega_2, \omega_3),\ \boldsymbol{\omega}_\perp = (\omega_1, \omega_2),$$

$$\hat{\Sigma}_1 = \hat{\Sigma} - \frac{\hat{s}\hat{T}}{\hat{c}_s^2}.$$

Here we use a **Fourier transform**:

$$\hat{\rho}_e(t, \boldsymbol{\omega}) = \int_{\mathbb{R}^3} e^{-2\pi i(\boldsymbol{x}, \boldsymbol{\xi})} \rho_e(t, \boldsymbol{x}) d\boldsymbol{x},\ \boldsymbol{\omega} = 2\pi\boldsymbol{\xi},$$

$$\hat{v}_3(t, \boldsymbol{\omega}) = \int_{\mathbb{R}^3} e^{-2\pi i(\boldsymbol{x}, \boldsymbol{\xi})} v_3(t, \boldsymbol{x}) d\boldsymbol{x},$$

$$\hat{\tilde{p}}(t, \boldsymbol{\omega}) = \int_{\mathbb{R}^3} e^{-2\pi i(\boldsymbol{x}, \boldsymbol{\xi})} \tilde{p}(t, \boldsymbol{x}) d\boldsymbol{x}.$$

Let's consider the Cauchy problem for the system (3.2a) – (3.2c) with the initial data:

$$\begin{aligned} &\hat{v}_3\Big|_{t=0} = \hat{v}_{30}(\boldsymbol{\omega}), \quad \frac{\partial \hat{v}_3}{\partial t}\Big|_{t=0} = \hat{v}_{31}(\boldsymbol{\omega}), \\ &\hat{\tilde{p}}\Big|_{t=0} = \hat{\tilde{p}}_0(\boldsymbol{\omega}), \quad \frac{\partial \hat{\tilde{p}}}{\partial t}\Big|_{t=0} = \hat{\tilde{p}}_1(\boldsymbol{\omega}), \\ &\hat{\rho}_e\Big|_{t=0} = \hat{\rho}_{e0}(\boldsymbol{\omega}). \end{aligned} \tag{3.3}$$

It is proved in [4] that
if $|\boldsymbol{\omega}_\perp|^2 = \omega^2$, or $|\boldsymbol{\omega}_\perp|^2 = 4\omega^2$, then $|\hat{\tilde{p}}|$ infinitely grows as $t \to \infty$;
if $|\boldsymbol{\omega}_\perp|^2 + \hat{B}\hat{\lambda}^2\omega_3^2 + \hat{N}|\boldsymbol{\omega}|^4 = \omega^2$, then $|\hat{v}_3|$ infinitely grows as $t \to \infty$.

So, within some conditions we can find a infinitely growing solutions. That means the destruction of the layered structures.

References

[1] A. M. Blokhin and V. N. Dorovsky, Problems of the mathematical simulation in theory of the multi-velocity continuum. RAS, Sib. dep, United Institute of geology, geophysics and mineralogy, Institute of mathematics, Novosibirsk, 1994 (Russian).

[2] V. N. Dorovsky and S. V. Dorovsky, A hydrodynamic model of water-oil layered systems containing gas. Math. Comput. Mod. 35 (2002), 751–757.

[3] A. M. Blokhin and R. E. Semenko, On the stability of shock waves in layered structures at the presence of the electric current. Quart. Appl. Math. 67 (2009), 441–475.

[4] A. M. Blokhin and R. E. Semenko, Stability of the layered structures at the presence of the electric current. Vestnik NGU. 9 (2009), 24–34 (Russian).

[5] V. V. Gogosov and V. A. Polansky, Electrohydrodynamics: problem and applications, fundamental equations, discontinuous solutions. Fluid mechanics (Results of science and technics) 10 (1976), 5–85 (Russian).

[6] L. D. Landau and E. M. Lifshits, Theoretical Physics. Vol. VI, Hydrodynamics, Moscow, Nauka, 1986 (Russian).

[7] L. N. Sedov, Mechanics of Continua. Vol. 1, Moscow, Nauka, 1970 (Russian).

[8] L. V. Ovsyannikov, Lectures on the Foundations of the Gas Dynamics. Moscow, Nauka, 1981 (Russian).

[9] A. M. Blokhin and S. V. Dorovsky, Shock waves stability in layered structures. Comp. and Math. with Appl. 47 (2004), 427–440.

[10] A. M. Blokhin, S. V. Dorovsky and E. V. Ovechkin, Stability of shock waves in layered structures. Part 2. Comp. and Math. with Appl. 47 (2004), 1379–1387.

[11] A. M. Blokhin, I. Z. Merazhov and Yu. L. Trakhinin, Investigation on stability of electrohydrodynamic shock waves. Le Matematiche LII (1997), Fasc. I, 87–114.

[12] A. M. Blokhin, Energy integrals and their applications to gas dynamics problems. Nauka, Sibirsk. Otdel., Novosibirsk, 1986 (Russian).

[13] A. M. Blokhin, The elements of the theory of hyperbolic systems and equations, Novosibirsk. Gos. Univ., Novosibirsk, 1995 (Russian).

[14] S. K. Godunov, Equations of Mathematical Physics. Nauka, Moscow, 1979 (Russian).

Global Classical Discontinuous Solutions to the Generalized Riemann Problem for Linearly Degenerate Hyperbolic Conservation Laws under Small BV Perturbations of the Initial Data*

Zhi-Qiang Shao
Department of Mathematics, Fuzhou University
Fuzhou 350002, China
E-mail: zqshao_fzu@yahoo.com.cn

Abstract

In this paper, we prove the existence and uniqueness of global piecewise C^1 solution containing only n contact discontinuities to a class of the generalized Riemann problem, which can be regarded as a small BV perturbation of the corresponding Riemann problem, for general $n \times n$ linearly degenerate quasilinear hyperbolic system of conservation laws; moreover, this solution has a global structure similar to the one of the self-similar solution $u = U(\frac{x}{t})$ to the corresponding Riemann problem. Our result indicates that this kind of Riemann solution $u = U(\frac{x}{t})$ mentioned above for general $n \times n$ quasilinear hyperbolic systems of conservation laws possesses a global nonlinear structure stability under a small BV perturbation of the Riemann initial data. Applications include the one-dimensional Born-Infeld system arising in the string theory and high energy physics.

1 Introduction and main result

Consider the following quasilinear hyperbolic system of conservation laws:

$$\partial_t u + \partial_x f(u) = 0, \quad x \in \mathbf{R}, \ t > 0, \tag{1.1}$$

*The author is partly supported by NSFC (No. 70371025), the Scientific Research Foundation of the Ministry of Education of China (No. 02JA790014), the Natural Science Foundation of Fujian Province of China (No. 2009J01006), and the Science and Technology Developmental Foundation of Fuzhou University (No. 2004-XQ-16).

where $u = (u_1, \dots, u_n)^T$ is the unknown vector-valued function of (t, x), $f : \mathbf{R}^n \to \mathbf{R}^n$ is a given C^3 vector function of u.

It is assumed that system (1.1) is strictly hyperbolic, i.e., for any given u on the domain under consideration, the Jacobian $A(u) = \nabla f(u)$ has n real distinct eigenvalues

$$\lambda_1(u) < \lambda_2(u) < \cdots < \lambda_n(u). \tag{1.2}$$

Let $l_i(u) = (l_{i1}(u), \dots, l_{in}(u))$ (resp. $r_i(u) = (r_{i1}(u), \dots, r_{in}(u))^T$) be a left (resp. right) eigenvector corresponding to $\lambda_i(u) (i = 1, \dots, n)$:

$$l_i(u)A(u) = \lambda_i(u)l_i(u) \qquad (resp.\ A(u)r_i(u) = \lambda_i(u)r_i(u)). \tag{1.3}$$

We have

$$det|l_{ij}(u)| \neq 0 \qquad (equivalently,\ det|r_{ij}(u)| \neq 0). \tag{1.4}$$

Without loss of generality, we may assume that on the domain under consideration

$$l_i(u)r_j(u) \equiv \delta_{ij} \quad (i, j = 1, \dots, n) \tag{1.5}$$

and

$$r_i^T(u)r_i(u) \equiv 1 \quad (i = 1, \dots, n), \tag{1.6}$$

where δ_{ij} stands for the Kronecker's symbol.

Clearly, all $\lambda_i(u)$, $l_{ij}(u)$, and $r_{ij}(u) (i, j = 1, \dots, n)$ have the same regularity as $A(u)$, i.e., C^2 regularity.

We also assume that on the domain under consideration, each characteristic field is linearly degenerate in the sense of Lax (cf. [1]):

$$\nabla \lambda_i(u)r_i(u) \equiv 0. \tag{1.7}$$

We are concerned with the existence and uniqueness of global piecewise C^1 solutions to the generalized Riemann problem for system (1.1), which is a Cauchy problem with a piecewise C^1 initial data of the form:

$$t = 0 : u = \begin{cases} u_0^-(x), & x \leq 0, \\ u_0^+(x), & x \geq 0, \end{cases} \tag{1.8}$$

where $u_0^-(x)$ and $u_0^+(x)$ are C^1 vector functions defined for $x \leq 0$ and $x \geq 0$ respectively with

$$u_0^-(0) \neq u_0^+(0). \tag{1.9}$$

Problem (1.1) and (1.8) may be regarded as a perturbation of the corresponding Riemann problem (1.1) and

$$t = 0 : u = \begin{cases} \widehat{u}_-, & x \leq 0, \\ \widehat{u}_+, & x \geq 0, \end{cases} \tag{1.10}$$

in which

$$\widehat{u}_{\pm} = u_0^{\pm}(0). \tag{1.11}$$

Let

$$\theta = |\widehat{u}_- - \widehat{u}_+|. \tag{1.12}$$

When $\theta > 0$ is suitably small, by Lax [1], the Riemann problem (1.1) and (1.10) admits a unique self-similar solution composed of $n+1$ constant states $\widehat{u}^{(0)} = \widehat{u}_-, \widehat{u}^{(1)}, \ldots, \widehat{u}^{(n-1)}, \widehat{u}^{(n)} = \widehat{u}_+$ separated by contact discontinuities, i.e.,

$$u = U(\frac{x}{t}) = \begin{cases} \widehat{u}^{(0)}, & x \leq \widehat{\lambda}_1 t, \\ \widehat{u}^{(j)}, & \widehat{\lambda}_j t \leq x \leq \widehat{\lambda}_{j+1} t \quad (j = 1, \ldots, n-1), \\ \widehat{u}^{(n)}, & x \geq \widehat{\lambda}_n t, \end{cases} \tag{1.13}$$

where $x = \widehat{\lambda}_j t$ stands for the jth contact discontinuity ($j = 1, \ldots, n$).

For the self-similar solution of the Riemann problem of general quasilinear hyperbolic systems of conservation laws, the local nonlinear structure stability has been proved by Li and Yu [2] for one-dimensional case, and by Majda [3] for multidimensional case. If the system (1.1) is strictly hyperbolic and linearly degenerate, Li and Kong [4] prove the existence and uniqueness of global piecewise C^1 solution containing only n contact discontinuities with small amplitude to the generalized Riemann problem for general linearly degenerate quasilinear hyperbolic systems of conservation laws with small decay initial data; moreover, this solution has a global structure similar to the self-similar solution of the corresponding Riemann problem. The result shows that in this case the self-similar solution with small amplitude $u = U(\frac{x}{t})$ possesses a global nonlinear structural stability under perturbation (1.8) satisfying (1.10). Precisely speaking, they obtained the following well-known result.

Theorem 1.1. Suppose that system (1.1) is strictly hyperbolic and all characteristic fields are linearly degenerate. Suppose furthermore that $u_0^-(x)$ and $u_0^+(x)$ are all C^1 vector functions on $x \leq 0$ and on $x \geq 0$ respectively, satisfying that there exists a constant $\mu > 0$ such that

$$\begin{aligned} \theta \overset{\triangle}{=} & \sup_{x \leq 0} \{(1+|x|)^{1+\mu}(|u_0^-(x)| + |u_0^{-'}(x)|)\} \\ & + \sup_{x \geq 0} \{(1+|x|)^{1+\mu}(|u_0^+(x)| + |u_0^{+'}(x)|)\} < \infty. \end{aligned} \tag{1.14}$$

Suppose finally that $f(u)$ is a C^3 vector function and (1.9) holds. Then there exists $\theta_0 > 0$ so small that for any given $\theta \in (0, \theta_0]$, the generalized Riemann problem (1.1) and (1.8) admits a unique global piecewise C^1 solution $u = u(t, x)$ containing only n contact discontinuities with small amplitude $x = x_i(t)$ $(i = 1, \ldots, n)$ on $t \geq 0$. This solution possesses a global structure similar to that of the self-similar solution $u = U(\frac{x}{t})$ of the corresponding Riemann problem (1.1) and (1.10). Precisely speaking,

$$u = u(t,x) = \begin{cases} u^{(0)}(t,x), & x \leq x_1(t), \\ u^{(i)}(t,x), & x_i(t) \leq x \leq x_{i+1}(t) \quad (i = 1, \ldots, n-1), \\ u^{(n)}(t,x), & x \geq x_n(t), \end{cases} \tag{1.15}$$

where $u^{(i)}(t,x)$ $(i = 0, 1, \ldots, n)$ are all C^1 solutions to system (1.1) on each corresponding domains respectively and for $i = 1, \ldots, n$, $u^{(i-1)}(t,x)$ and $u^{(i)}(t,x)$ are connected to each other by the ith contact discontinuity $x = x_i(t)$. Moreover,

$$x_i'(0) = \widehat{\lambda}_i \quad (i = 1, \ldots, n) \tag{1.16}$$

and

$$u^{(i)}(0,0) = \widehat{u}^{(i)} \quad (i = 0, 1, \ldots, n). \tag{1.17}$$

Remark 1.1. Recently, under the assumption that the self-similar solution of the Riemann problem for the system (1.1) only contains non-degenerate shocks and contact discontinuities but no centered rarefaction waves and other weak discontinuities, Kong [5] proved the existence and uniqueness of global piecewise C^1 solution with small amplitude to the generalized Riemann problem for general quasilinear hyperbolic systems of conservation laws with small decay initial data; moreover, this solution has a global structure similar to the self-similar solution of the corresponding Riemann problem. The result shows that this kind of Lax's Riemann solution with small amplitude possesses a global nonlinear structural stability.

However, it is well known that the BV space is a suitable framework for one-dimensional Cauchy problem for the hyperbolic systems of conservation laws (see Bressan [6], Glimm [7]), the result in Bressan [8] suggests that one may achieve global smoothness even if the C^1 norm of the initial data is large. So the following question arises naturally: Can we obtain the existence and uniqueness of global piecewise C^1 solution containing only contact discontinuities to a class of the generalized Riemann problem, which can be regarded as a small BV perturbation of the corresponding Riemann problem, for system (1.1) with the following piecewise C^1 initial data:

$$t = 0 : u = \begin{cases} \widehat{u}_- + u_-(x), & x \leq 0, \\ \widehat{u}_+ + u_+(x), & x \geq 0, \end{cases} \tag{1.18}$$

where $u_\pm(x) \in C^1$ with both bounded C^1 norm and small bounded variation, such that

$$||u_-(x)||_{C^1}, \quad ||u_+(x)||_{C^1} \leq M, \tag{1.19}$$

for some $M > 0$ bounded but possibly large, and also such that

$$\int_0^{+\infty} |u'_+(x)|dx, \int_{-\infty}^0 |u'_-(x)|dx \leq \varepsilon, \tag{1.20}$$

for some $\varepsilon > 0$ sufficiently small? Here, it is important to mention that the global existence of weak solutions to a strictly hyperbolic system of conservation laws in one space dimension when the initial data is a small BV perturbation of a solvable Riemann problem has been proved by Schochet [9], unfortunately his method is not useful to show that the solutions are still contact discontinuities. An analogous result on stability of contact discontinuities under perturbations of small bounded variation is stated by Corli and Sable-Tougeron [10]. In this paper we exploit to some extent the ideas of Bressan [8], and we will develop the method of using continuous Glimm's functional to solve this problem globally and to provide a new concise proof of the above mentioned problem of the stability of contact discontinuities. The basic idea we will use here is to combine the techniques employed by Li and Kong [4], especially both the decomposition of waves and the global behavior of waves on the contact discontinuity curves, with the method of using continuous Glimm's functional. However, we must modify Glimm's functional in order to take care of the presence of contact discontinuities. This makes our new analysis more complicated than those for the C^1 solutions of the Cauchy problem for linearly degenerate quasilinear hyperbolic systems in Bressan [8], Zhou [11], Dai and Kong [12].

Our main results can be summarized as follows:

Theorem 1.2. Suppose that system (1.1) is strictly hyperbolic and all characteristic fields are linearly degenerate. Suppose furthermore that $u_-(x)$ and $u_+(x)$ are all C^1 vector functions on $x \leq 0$ and on $x \geq 0$ respectively with both bounded C^1 norm and

$$u_-(0) = u_+(0) = 0. \tag{1.21}$$

Suppose finally that

$$\theta = |\widehat{u}_+ - \widehat{u}_-| = |u_0^+(0) - u_0^-(0)| > 0 \tag{1.22}$$

is suitably small. Then for any constant $M > 0$, there exists a positive constant ε so small that if (1.19) holds together with

$$\int_0^{+\infty} |u'_+(x)|dx, \int_{-\infty}^0 |u'_-(x)|dx \leq \varepsilon, \tag{1.23}$$

then the generalized Riemann problem (1.1) and (1.18) admits a unique global piecewise C^1 solution $u = u(t,x)$ only containing n contact discontinuities $x = x_i(t)$ $(x_i(0) = 0, i = 1,\ldots,n)$ defined for all $x \in \mathbf{R}$ and all $t \geq 0$, which has a global structure similar to the one of the self-similar solution $u = U(\frac{x}{t})$ of the corresponding Riemann problem (1.1) and (1.10). Precisely speaking,

$$u = u(t,x) = \begin{cases} u^{(0)}(t,x), & x \leq x_1(t), \\ u^{(i)}(t,x), & x_i(t) \leq x \leq x_{i+1}(t) \quad (i = 1,\ldots,n-1), \\ u^{(n)}(t,x), & x \geq x_n(t), \end{cases} \tag{1.24}$$

where $u^{(i)}(t,x)$ $(i = 0,1,\ldots,n)$ are all C^1 solutions to system (1.1) on each corresponding domains respectively and for $i = 1,\ldots,n$, $u^{(i-1)}(t,x)$ and $u^{(i)}(t,x)$ are connected to each other by the ith contact discontinuity $x = x_i(t)$.

Remark 1.2. Our result indicates that the self-similar solution $u = U(\frac{x}{t})$ of the Riemann problem (1.1) and (1.10) has a global nonlinear structure stability under a small BV perturbation of the Riemann initial data.

Remark 1.3. Suppose that (1.1) is a non-strictly hyperbolic system with characteristics with constant multiplicity, say, on the domain under consideration,

$$\lambda_1(u) \equiv \cdots \equiv \lambda_p(u) < \lambda_{p+1}(u) < \cdots < \lambda_n(u) \quad (1 \leq p \leq n). \tag{1.25}$$

Then the conclusion of Theorem 1.2 still holds (cf. [12]).

Some of the known results related to these topics are listed below. G. Q. Chen et al. investigated the asymptotic stability of Riemann waves for hyperbolic conservation laws. L. Hsiao and S. Q. Tang investigated the construction and qualitative behavior of the solution of the perturbated Riemann problem for the system of one-dimensional isentropic flow with damping. Z. Xin, F. Huang et al. proved the nonlinear stability of contact discontinuities in systems of conservation laws. Smoller et al. investigated the instability of rarefaction shocks in systems of conservation laws. For the overcompressive shock waves, T. P. Liu proved the nonlinear stability and instability. Bressan and LeFloch investigated the structural stability and regularity of entropy solutions to hyperbolic systems of conservation laws. Lions et al. proved the existence and stability of entropy solutions for the hyperbolic systems of isentropic gas dynamics in Eulerian and Lagrangian coordinates. Recently, L^1 stability for systems of hyperbolic conservation laws was investigated by A. Bressan, T. P. Liu and T. Yang. Liu and Xin proved the nonlinear stability of discrete shocks for systems of conservation laws. Dafermos studied the entropy and the stability of classical solutions of hyperbolic systems of conservation laws. For a relaxation system in several space dimensions, Luo and Xin proved the nonlinear stability of shock fronts. Liu and Xin investigated the nonlinear stability of rarefaction waves for compressible Navier-Stokes equations. L. Hsiao and R. H. Pan investigated the nonlinear stability of rarefaction waves for a rate-type viscoelastic system. Moreover, the nonlinear stability of an undercompressive shock for complex Burgers equation was studied by Liu and Zumbrun. For the viscous conservation laws, the theory of nonlinear stability of shock waves was established by T. P. Liu and Z. Xin.

Finally, we remark that the starting motivation of our study comes from two concrete applications. They are both the one-dimensional Born-Infeld system (cf. Li et al. [13], Brenier [14]) and the system of the motion of relativistic closed strings in the Minkowski space-time R^{1+n}(cf. Kong et al. [15]), for which our work is of importance. Here, we skip the details and refer the reader to the above mentioned papers [13–15].

2 Unsolved problem

In this paper, we have proved the existence and uniqueness of global piecewise C^1 solution containing only n contact discontinuities to the generalized Riemann problem for general $n \times n$ linearly degenerate quasilinear hyperbolic system of conservation laws under a small BV perturbation of the Riemann initial data. However, we left open the question of the existence and uniqueness of global piecewise C^1 solution to the generalized Riemann problem for general $n \times n$ quasilinear hyperbolic system of conservation laws whose some of the characteristic fields are genuinely nonlinear under a small BV perturbation of the Riemann initial data. This problem is worthy of our studying in the future.

References

[1] P. D. Lax, Hyperbolic systems of conservation laws II. Comm. Pure Appl. Math. 10 (1957), 537–556.

[2] T. T. Li and W. C. Yu, Boundary Value Problems for Quasilinear Hyperbolic Systems, Duke University Mathematics Series V. Duke University, Durham, 1985.

[3] A. Majda, The stability of multidimensional shock fronts. Mem. Amer. Math. Soc. 41 (275) (1983), 1–95.

[4] T. T. Li and D. X. Kong, Global classical discontinuous solutions to a class of generalized Riemann problem for general quasilinear hyperbolic

systems of conservations laws. Comm. Partial Differential Equations 24 (1999), 801–820.

[5] D. X. Kong, Global structure stability of Riemann solutions of quasilinear hyperbolic systems of conservation laws: shocks and contact discontinuities. J. Differential Equations 188 (2003), 242–271.

[6] A. Bressan, Hyperbolic Systems of Conservation Laws: The one-dimensional Cauchy Problem. Oxford University Press, Oxford, 2000.

[7] J. Glimm, Solutions in the large for nonlinear hyperbolic systems of equations. Comm. Pure Appl. Math. 18 (1965), 697–715.

[8] A. Bressan, Contractive metrics for nonlinear hyperbolic systems. Indiana Univ. Math. J. 37 (1988), 409–421.

[9] S. Schochet, Sufficient condition for local existence via Glimm's scheme for large BV data. J. Differential Equations 89 (1991), 317–354.

[10] A. Corli and M. Sable-Tougeron, Stability of contact discontinuities under perturbations of bounded variation. Rend. Sem. Mat. Univ. Padova 97 (1997), 35–60.

[11] Y. Zhou, Global classical solutions to quasilinear hyperbolic systems with weak linear degeneracy. Chinese Ann. Math. 25B (2004), 37–56.

[12] W. R. Dai and D. X. Kong, Global existence and asymptotic behavior of classical solutions of quasilinear hyperbolic systems with linearly degenerate characteristic fields. J. Differential Equations 235 (2007), 127–165.

[13] T. T. Li, Y. J. Peng and J. Ruiz, Entropy solutions for linearly degenerate hyperbolic systems of rich type. J. Math. Pures Appl. 91 (2009), 553–568.

[14] Y. Brenier, Hydrodynamic structure of the augmented Born-Infeld equations. Arch. Rational Mech. Anal. 172 (2004), 65–91.

[15] D. X. Kong, Q. Zhang and Q. Zhou, The dynamics of relativistic strings moving in the Minkowski space R^{1+n}. Comm. Math. Phys. 269 (2007), 153–174.

One Class of Systems of Conservation Laws Admitting Delta-shocks*

V. M. Shelkovich
Department of Mathematics
St. Petersburg State Architecture and Civil Engineering University
190005, St. Petersburg, Russia
E-mail: shelkv@vs1567.spb.edu

Abstract

We study a new class of systems of conservation laws admitting δ-shocks. We introduce integral identities which give the definition of δ-shocks for these systems and derive the corresponding Rankine-Hugoniot conditions. It is proved that the "area" transport processes between the moving singular one-dimensional δ-shock wave front and the region outside the front are going on. Balance relations describing these processes are derived.

1 Introduction

As is well known, even in the case of smooth initial data, in general, there exist no smooth and global in time solutions of this system. This fact leads to the necessity of introducing a notion of L^∞-*generalized solution* (weak solution) of the Cauchy problem in the sense of the integral identities. Moreover, there are "nonclassical" situations where, in contrast to Lax's and Glimm's classical results, the Cauchy problem for a system of conservation laws *either does not possess a weak* L^∞-*solution or possesses it for some particular initial data.* In order to solve the Cauchy problem in these "nonclassical" situations, it is necessary to seek solutions in the form of δ-*shocks.* Roughly speaking, a δ-*shock* is a solution whose components *contain Dirac delta functions.* Problems related to δ-shocks have been intensively studied in the last 12 years (see [3], [7], [9], [13], [14], [16] and the references therein). Recently, in [12], a concept of $\delta^{(n)}$-shock wave type solutions was introduced, $n = 1, 2, \ldots$. It is a *new type of singular solution* of a system of

*The author was supported in part by DFG Project 436 RUS 113/895 and by Analytical Departmental Special Program "The development of scientific potential of the Higher School", Project 2.1.1/1399.

conservation laws such that its components contain delta functions and their derivatives up to nth order.

Systems of conservation laws admitting such singular solutions have a physical context. In numerous papers δ-shocks were studied for the zero-pressure gas dynamics (see the above references). This system was used to describe the *formation of large-scale structures of the universe*; for modeling the *formation and evolution of traffic jams*; for modeling media which can be considered as *having no pressure* (for example, dusty gases, two-phase flows with solid particles or droplets) [10]. δ-Shocks arise in the model of non-classical shallow water flows [4], in model of granular gases [6], in the system of *nonlinear chromatography* [11].

In this paper we study a new type of systems of conservation laws

$$(u_j)_t + \big(u_j\, f_j(\mu_1 u_1 + \cdots + \mu_n u_n)\big)_x = 0, \quad x \in \mathbb{R}, \quad t \geq 0, \tag{1.1}$$

admitting δ-shock wave type solution, where $f_j(\cdot)$ is a smooth function, μ_j is a constant, $j = 1, 2, \ldots, n$. Systems (1.1) constitute a new class admitting δ-shocks. This class includes some *Temple type system* (see [1], [15]). In particular, the system of *nonlinear chromatography* for $f_j(w) = 1 + \frac{a_j}{1+w}$, where a_j is Henry's constant, $\mu_j = 1$, $j = 1, 2, \ldots, n$ (see [1], [11]); the system for *Isotachophoresis* [2].

In Section 2, we introduce Definition 2.1 of δ-shocks for system (1.1), derive the corresponding Rankine-Hugoniot conditions (2.2) and give some motivation for our definition. In Section 3, we prove that the *"area" transport process is going on* between the region $\Omega_t^- \cup \Omega_t^+ = \{x \in \mathbb{R} : x \neq \phi(t)\}$ and the moving singular point $x = \phi(t)$ (one-dimensional δ-shock wave front). Moreover, the total "area" $S_j(t) + \dot{e}_j(t)$ is independent of time. In the case of nonlinear chromatography the balance relations (3.3) have a physical sense.

2 δ-Shock type solutions

Definition of δ-shocks. Suppose that $\Gamma = \{\gamma_i : i \in I\}$ is a set of curves in the first quadrant $\{(x,t) : x \geq 0,\ t \in [0,\infty)\} \in \mathbb{R}^2$ containing smooth arcs $\gamma_i = \{(x,t) : S_i(x,t) = 0\}$ of the class C^1, $(S_i)_x \neq 0$, $i \in I$, and I is a finite set. Let I_0 be a subset of I such that the arcs γ_k for $k \in I_0$ start from points of the x-axis and $\Gamma_0 = \{x_k^0 : k \in I_0\}$ be the set of initial points of the arcs γ_k, $k \in I_0$. It is clear that $-G = \frac{S_t}{|S_x|}$ is the *velocity of the moving point* $\Gamma_t = \{x \in \mathbb{R} : S(x,t) = 0\}$ in the direction $\nu = \frac{S_x}{|S_x|}$.

For system (1.1), we will use the *δ-shock type initial data*

$$u^0 = (u_1^0, \ldots, u_n^0), \quad \text{where} \quad u_j^0(x) = \widehat{u}_j^0(x) + e_j^0 \delta(\Gamma_0), \tag{2.1}$$

and $\widehat{u}_j^0 \in L^\infty(\mathbb{R};\mathbb{R})$; $e_j^0\delta(\Gamma_0) \stackrel{def}{=} \sum_{k\in I_0} e_{j;k}^0 \delta(x - x_k^0)$, $e_{j;k}^0$ is a constant; $k \in I_0$; $j = 1, 2, \ldots, n$.

Motivation of the definition of δ-shocks. As noted in the Evans' book [5], "the great difficulty in this subject is discovering a proper notion of weak solution for the initial problem". In order to introduce classical integral identities which give a definition of L^∞-generalized solution to systems of conservation laws, we use some *inexact reasoning* (see [5]). According to the above cited papers, δ-shock wave type solutions for system (1.1) are distributions of the form

$$u_j(x,t) = \widehat{u}_j(x,t) + e_j(x,t)\delta(\Gamma), \quad j = 1, 2, \dots, n, \tag{2.2}$$

where $\widehat{u}_j(x,t), e_j(x,t)$ are regular functions, Γ is the discontinuity curve in the half-plane $\{(x,t) : x \in \mathbb{R}, t \geq 0\}$, $\delta(\Gamma)$ is the delta function concentrated on Γ (see (2.9)). Thus, we must discover *in which sense* distributions (2.2) may satisfy a nonlinear system (1.1). Thereto, similarly to [5], we use some *inexact reasoning* leading to the integral identities (2.6) which generalize the classical L^∞-integral identities and allow a definition of δ-shocks for system (1.1).

Let us assume that system (1.1) admits a δ-shock wave type solution of the form (2.2), where $\widehat{u}_j(x,t)$ is a piecewise smooth function (for example, $u_{j0}(x,t) + u_{j1}(x,t)H(-S(x,t))$) and $e_j(x,t)$ is continuous, $j = 1, \dots, n$, where $H(x)$ is the Heaviside function. Since we seek a solution in the space of Schwartzian distributions $\mathcal{D}'(\mathbb{R} \times [0,\infty))$ and $(u_j)_x$ is a distribution, according to (1.1), flux-functions $u_j\, f_j(u_1 + \cdots + u_n)$ will necessarily be distributions. Suppose for the moment that $u_j\, f_j(\mu_1 u_1 + \cdots + \mu_n u_n) = A_j(\widehat{u}) + B_j(\widehat{u}, e)\delta(\Gamma)$, where $A_j(\widehat{u})$ is a piecewise smooth function and $B_j(\widehat{u}, e)$ is continuous, $j = 1, \dots, n$. Substituting the last relation and (2.2) into system (1.1) and using formulas [8], we find that the coefficient of $\delta'(\Gamma)$ is $e_j \frac{S_t}{|S_x|} + B_j \frac{S_x}{|S_x|} = 0$, i.e., $B_j = -\frac{S_t}{S_x} e_j$, $j = 1, \dots, n$. Thus

$$\begin{gathered} u_j\, f_j(\mu_1 u_1 + \cdots + \mu_n u_n) = A_j(\widehat{u}) + u_\delta(x,t) e_j \delta(\Gamma), \\ j = 1, \dots, n, \end{gathered} \tag{2.3}$$

where $u_\delta(x,t) = -\frac{S_t}{S_x}$ is the *velocity* of the δ-shock (see [8], (2.8), and [13]).

Let us consider the following expression $\big\langle (u_j)_t + \big(u_j\, f_j(\mu_1 u_1 + \cdots + \mu_n u_n)\big)_x, \varphi \big\rangle = 0$, $\varphi \in \mathcal{D}(\mathbb{R} \times [0,\infty))$. Integrating it by parts and taking into account (2.2), (2.3) we obtain $\int_0^\infty \int \big(\widehat{u}_j \varphi_t + A_j(\widehat{u})\varphi_x\big)\, dx\, dt + \int \widehat{u}_j^0(x)\varphi(x,0)\, dx + \big\langle \delta(\Gamma), e_j\, (\varphi_t + u_\delta \varphi_x) \big\rangle = 0$. Next, using (2.9) and (2.7), we derive the integral identities (2.6).

In order to define the term $f_j(\mu_1 u_1 + \cdots + \mu_n u_n)$ in the sense of Schwartzian distributions, we need to assume that relation (2.5) holds.

Using the above reasoning, similarly to definitions of δ-shocks from the paper [13], we introduce the following definition.

Definition 2.1. A vector-distribution $u(x,t) = (u_1(x,t),\dots,u_n(x,t))$ and a set of curves Γ, where

$$u_j(x,t) = \widehat{u}_j(x,t) + e_j(x,t)\delta(\Gamma), \tag{2.4}$$

$\widehat{u}_j \in L^\infty(\mathbb{R}\times(0,\infty);\mathbb{R})$, $e_j(x,t)\delta(\Gamma) \stackrel{def}{=} \sum_{i\in I} e_{j;i}(x,t)\delta(\gamma_i)$, $e_{j;i} \in C(\gamma_i)$, $i \in I$, $j = 1,2,\dots,n$, and

$$\sum_{j=1}^{n} \mu_j e_j(x,t) = 0 \tag{2.5}$$

is called a *δ-shock wave type solution* of the Cauchy problem (1.1), (2.1) if the integral identities

$$\int_0^\infty \int \widehat{u}_j\Big(\varphi_t + f_j(\mu_1\widehat{u}_1 + \cdots + \mu_n\widehat{u}_n)\varphi_x\Big)\,dx\,dt + \int \widehat{u}_j^0(x)\varphi(x,0)\,dx$$
$$+\sum_{i\in I}\int_{\gamma_i} e_{j;i}(x,t)\frac{\delta\varphi(x,t)}{\delta t}\frac{dl}{\sqrt{1+G^2}} + \sum_{k\in I_0} e_{j;k}^0\varphi(x_k^0,0) = 0, \tag{2.6}$$

hold for all test functions $\varphi \in \mathcal{D}(\mathbb{R}\times[0,\infty))$, $j = 1,2,\dots,n$, where

$$\frac{\delta\varphi}{\delta t}\bigg|_{\gamma_i} = \left(\frac{\partial\varphi}{\partial t} - \frac{(S_i)_t}{(S_i)_x}\frac{\partial\varphi}{\partial x}\right)\bigg|_{S_i(x,t)=0} \tag{2.7}$$

is the *δ-derivative* with respect to time [8],

$$u_\delta(x,t)\big|_{\gamma_i} = G\nu = -\frac{(S_i)_t}{(S_i)_x}\bigg|_{\gamma_i} \tag{2.8}$$

is the velocity of a δ-shock on γ_i, $i \in I$; $\int_{\gamma_i} \cdot\, dl$ is the line integral over the arc γ_i.

Here the δ-derivative (2.7) coincides with the Lagrangian derivative $\frac{D\varphi}{Dt} = \big(\frac{\partial\varphi}{\partial t} + u_\delta\frac{\partial\varphi}{\partial x}\big)\big|_{\gamma_i}$; the delta function $\delta(\gamma_i)$ on the curve γ_i is defined in [8] as

$$\big\langle\delta(S),\ \varphi(x,t)\big\rangle = \int_{\gamma_i}\varphi(x,t)\frac{dl}{\sqrt{1+G^2}}, \quad \forall\,\varphi \in \mathcal{D}(\mathbb{R}\times[0,\ \infty)). \tag{2.9}$$

If $\Gamma = \{\gamma_i : i \in I\}$, where $\gamma_i = \{(x,t) : x = \phi_i(t)\}$, $\phi_i(t) \in C^1$, $i \in I$, and $\dot{(\cdot)} = \frac{d}{dt}(\cdot)$, then

$$\frac{\delta\varphi}{\delta t}\bigg|_{\gamma_i} = \varphi_t(\phi_i(t),t) + \dot\phi_i(t)\varphi_x(\phi_i(t),t) = \frac{d\varphi(\phi_i(t),t)}{dt}. \tag{2.10}$$

The Rankine–Hugoniot conditions. By using Definition 2.1, we derive the *δ-shock Rankine–Hugoniot conditions* for system (1.1).

Theorem 2.2. *Let us assume that $\Omega \subset \mathbb{R}_+ \times (0,\infty)$ is a region cut by a smooth curve $\Gamma = \{(x,t) : S(x,t) = 0\}$ into the left- and right-hand parts $\Omega_\mp$. Let $u = (u_1,\dots,u_n)$ and let Γ be a δ-shock wave type solution of system (1.1) such that $u_j(x,t) = \widehat{u}_j(x,t) + e_j(x,t)\delta(\Gamma)$ are smooth in $\Omega_\pm$ and have one-sided limits $\widehat{u}_{j\pm}$ on Γ, $j = 1,\dots,n$. Then the* Rankine–Hugoniot conditions for the δ-shock

$$\begin{aligned}\frac{\delta e_j(x,t)}{\delta t} &= \Big([u_j\, f_j(\mu_1 u_1 + \cdots + \mu_n u_n)]_\Gamma - [u_j]_\Gamma u_\delta\Big) \frac{S_x}{|S_x|}\Big|_\Gamma, \\ u_\delta(x,t) &= \frac{\sum_{j=1}^n \mu_j [u_j\, f_j(\mu_1 u_1 + \cdots + \mu_n u_n)]}{\sum_{j=1}^n \mu_j [u_j]}\Big|_\Gamma, \quad j = 1,\dots,n,\end{aligned} \tag{2.11}$$

hold along Γ, where $u_\delta(x,t)$ is the velocity (2.8) of a δ-shock.

Proof. Let $n = (\nu_1,\nu_2) = \frac{(S_x,S_t)}{|\nabla_{(x,t)} S|}$ be a unit normal to the curve Γ oriented from Ω_- to Ω_+, $\nabla_{(x,t)}S = (S_x,S_t)$. For any test function $\varphi \in \mathcal{D}(\Omega)$, we have $\varphi(x,t) = 0$, if $(x,t) \notin E$, $\overline{E} \subset \Omega$. Selecting a test function $\varphi(x,t)$ with support in $\Omega_\pm$, by integrating by part we deduce from (2.6) that $(u_j)_t + \big(u_j\, f_j(\mu_1 u_1 + \cdots + \mu_n u_n)\big)_x = 0$ for $(x,t) \in \Omega_\pm$, respectively. Taking into account the last relation and integrating by parts, similarly to the proof of [13, Theorem 2.1.] we obtain

$$\begin{aligned}&\iint\limits_{\Omega_\pm \cap E} u_j\big(\varphi_t + f_j(\mu_1 u_1 + \cdots + \mu_n u_n)\varphi_x\big)\, dx\, dt \\ &= \mp \int\limits_\Gamma u_{j\pm}\big(\nu_2 + \nu_1 f_j(\mu_1 u_{1\pm} + \cdots + \mu_n u_{n\pm})\big)\varphi\, dl \\ &\quad - \int\limits_{\Omega_\pm \cap E \cap \mathbb{R}} u_j^0(x)\varphi(x,0)\, dx.\end{aligned}$$

Adding the latter relations, we have

$$\begin{aligned}&\int_0^\infty \int \widehat{u}_j\big(\varphi_t + f_j(\mu_1 \widehat{u}_1 + \cdots + \mu_n \widehat{u}_n)\varphi_x\big)\, dx\, dt + \int \widehat{u}_j^0(x)\varphi(x,0)\, dx \\ &= \int_\Gamma \big([u_j\, f_j(\mu_1 u_1 + \cdots + \mu_n u_n)]\nu_1 + [u_j]\nu_2\big)\varphi(x,t)\, dl, \ \forall \varphi \in \mathcal{D}(\Omega). \end{aligned} \tag{2.12}$$

Next, using the formula for integration by parts [13], we obtain

$$\begin{aligned}&\int_\Gamma e_j(x,t)\frac{\delta \varphi(x,t)}{\delta t}\frac{dl}{\sqrt{1+G^2}} \\ &= -\int_\Gamma \frac{\delta e_j(x,t)}{\delta t}\varphi(x,t)\frac{dl}{\sqrt{1+G^2}} - e_j^0\varphi(x,0)\big|_{S(x,0)=0},\end{aligned} \tag{2.13}$$

where the δ-derivative $\frac{\delta\varphi}{\delta t}$ is defined in (2.7). Adding (2.12) and (2.13), we obtain

$$\int_{\Gamma}\Big([u_j f_j(\mu_1 u_1+\cdots+\mu_n u_n)]\nu_1+[u_j]\nu_2-\frac{\delta e_j}{\delta t}\frac{1}{\sqrt{1+G^2}}\Big)\varphi(x,t)\,dl=0,$$

for any $\varphi\in\mathcal{D}(\Omega)$. Thus, due to (2.8), the first relation in (2.11) holds. This relation and (2.5) imply the second relation in (2.11). □

If $\Gamma=\{(x,t): x=\phi(t)\}$, $\phi(t)\in C^1(0,+\infty)$, in view of (2.10), the relations (2.11) take the form

$$\begin{aligned}\dot e_j(t)&=\big([u_j f_j(\mu_1 u_1+\cdots+\mu_n u_n)]-[u_j]\dot\phi(t)\big)\big|_{x=\phi(t)},\\ \dot\phi(t)&=\frac{\sum_{j=1}^n \mu_j[u_j f_j(\mu_1 u_1+\cdots+\mu_n u_n)]}{\sum_{j=1}^n \mu_j[u_j]}\Big|_{x=\phi(t)},\quad j=1,\dots,n,\end{aligned}\qquad(2.14)$$

where $e_j(t)\stackrel{def}{=}e_j(\phi(t),t)$.

The right-hand side of relation (2.11) (or (2.14)) is called the *Rankine–Hugoniot deficit* in u_j, $j=1,\dots,n$.

Propagation of δ-shock. In order to describe the propagation of a singular wave front (point) starting from the initial position $x_0=0$, we need to solve the Cauchy problem for system (1.1) with the initial data

$$u^0=(u_1^0,\dots,u_n^0),\quad u_j^0(x)=u_{j+}^0(x)+[u_j^0(x)]H(-x)+e_j^0\delta(-x),\quad(2.15)$$

where $[u_j^0]=u_{j-}^0-u_{j+}^0$ is the jump of the function $u_j^0(x)$ across the discontinuity point $x_0=0$; $u_j^0(x)=u_{j-}^0(x)$ if $x<0$, and $u_j^0(x)=u_{j+}^0(x)$ if $x>0$; $u_{j\pm}^0(x)$ are given smooth functions; e_j^0 are given constants, $j=1,2,\dots,n$.

We will seek a δ-shock wave type solution of the Cauchy problem (1.1), (2.15) in the form of a vector-distribution $u=(u_1,\dots,u_n)$, where

$$u_j(x,t)=u_{j+}(x,t)+[u_j(x,t)]H(-S(x,t))+e_j(x,t)\delta(-S(x,t)),\quad(2.16)$$

and the vector-function $u_\pm$ and functions e_j, S are to be found, $[u_j]=u_{j-}-u_{j+}$, $j=1,2,\dots,n$.

To solve the Cauchy problem (1.1), (2.15) we can easily adapt our results [3, 5, 13].

3 Transport processes related with δ-shocks

It is well known that δ-shocks are connected with *transport and concentration processes* (see [13] and the references therein).

Let $\Gamma = \{(x,t) : x = \phi(t)\}$ be a discontinuity curve in the upper half-plane $\{(x,t) : x \in \mathbb{R},\ t \in [0,\infty)\}$ issued from $\phi(0) = 0$. Denote by

$$S_j(t) = \int_{-\infty}^{\phi(t)} u_j(x,t)\,dx + \int_{\phi(t)}^{+\infty} u_j(x,t)\,dx, \quad t \geq 0, \tag{3.1}$$

the area under the graph $y = \widehat{u}_j(x,t)$, $j = 1,\dots,n$.

Theorem 3.1. *Let $u = (u_1,\dots,u_n)$ and a discontinuity curve $\Gamma = \{(x,t) : x = \phi(t)\}$ be a δ-shock wave type solution of the Cauchy problem (1.1), (2.1), where $u_j(x,t) = \widehat{u}_j(x,t) + e_j(t)\delta(\Gamma)$, $\widehat{u}_j(x,t)$ are compactly supported with respect to x. Then the following balance relations hold:*

$$\begin{aligned} \dot{S}_j(t) &= -\dot{e}_j(t) \\ &= -\big([u_j\, f_j(\mu_1 u_1 + \cdots + \mu_n u_n)] - [u_j]\dot{\phi}(t)\big)\big|_{x=\phi(t)}, \end{aligned} \tag{3.2}$$

$$\begin{aligned} &\int_{-\infty}^{\phi(t)} u_j(x,t)\,dx + \int_{\phi(t)}^{+\infty} u_j(x,t)\,dx + e_j(t) \\ &= \int_{-\infty}^{0} u_j^0(x)\,dx + \int_{0}^{+\infty} u_j^0(x)\,dx + e_j(0). \end{aligned} \tag{3.3}$$

Proof. Differentiating (3.1) and using (1.1), we obtain

$$\begin{aligned} \dot{S}_j(t) &= u_{j-}\dot{\phi}(t) - v_{j+}\dot{\phi}(t) + \int_{-\infty}^{\phi(t)} (u_j)_t(x,t)\,dx + \int_{\phi(t)}^{+\infty} (u_j)_t(x,t)\,dx \\ &= [u_j]\big|_{x=\phi(t)}\dot{\phi}(t) - [u_j\, f_j(\mu_1 u_1 + \cdots + \mu_n u_n)]\big|_{x=\phi(t)} \\ &\quad + u_j\, f_j(\mu_1 u_1 + \cdots + \mu_n u_n)\big|_{x=-\infty} - u_j\, f_j(\mu_1 u_1 + \cdots + \mu_n u_n)\big|_{x=+\infty}. \end{aligned}$$

Since $\widehat{u}_j(x,t)$ are compactly supported with respect to x, $u_j\, f_j(u_1 + \cdots + u_n)\big|_{x=-\infty} = u_j\, f_j(u_1 + \cdots + u_n)\big|_{x=+\infty} = 0$. Now using (2.14), we conclude that (3.2) holds. Integrating (3.2), we obtain (3.3). □

References

[1] S. I. Agafonov and E. V. Ferapontov, Systems of conservation laws from the point of view of the projective theory of congruences. (Russian) Izv. Ross. Akad. Nauk Ser. Mat. 60 (1996), no. 6, 3–30; translation in Izv. Math. 60 (1996), no. 6, 1097–1122.

[2] T. V. Alekseyevskaya, Study of the system of quasilinear equations for lsotachophoresis. Advances in Applied Mathematics 11 (1990), 63–107.

[3] V. G. Danilov and V. M. Shelkovich, Delta-shock wave type solution of hyperbolic systems of conservation laws. Quarterly of Applied Mathematics 63 (2005), no. 3, 401–427.

[4] C. M. Edwards, S. D. Howinson, H. Ockendon and J. R. Ockendon, Nonclassical shallow water flows. Journal of Applied Mathematics 73 (2008), 137–157.

[5] L. C. Evans, Partial Differential Equations. Amer. Math. Soc. Providence, Road Island, 1998.

[6] I. Fouxon, B. Meerson, M. Assaf and E. Livne, Formation of density singularities in hydrodynamics of inelastic gases. Phys. Review E 75 (2007), 050301(R) (4 pages).

[7] F. Huang and Z. Wang, Well posedness for pressureless flow. Commun. Math. Phys. 222 (2001), 117–146.

[8] R. P. Kanwal, Generalized Functions: Theory and Technique. Birkhäuser Boston-Basel-Berlin, 1998.

[9] B. L. Keyfitz and H. C. Kranzer, Spaces of weighted measures for conservation laws with singular shock solutions. Journal of Differential Equations 118 (1995), 420–451.

[10] A. N. Kraiko, Discontinuity surfaces in medium without self-pressure. Prikladnaia Matematika i Mekhanika 43 (1979), 539–449. (In Russian)

[11] M. Mazzotti, Nonclassical composition fronts in nonlinear chromatography: delta-shock. Ind. Eng. Chem. Res. 48, (2009), 7733-7752.

[12] E. Y. Panov and V. M. Shelkovich, δ'-Shock waves as a new type of solutions to systems of conservation laws. Journal of Differential Equations 228 (2006), 49–86.

[13] V. M. Shelkovich, δ- and δ'-shock types of singular solutions to systems of conservation laws and the transport and concentration processes. Uspekhi Mat. Nauk 63:3(381), (2008), 73–146. English transl. in Russian Math. Surveys 63:3, (2008), 473–546.

[14] W. Shen and T. Zhang, The Riemann problem for the transportaion equations in gas dynamics. Memoirs of the Amer. Math. Soc. 137 (1999), no. 654, 1–77.

[15] B. Temple, Systems of conservations laws with invariant submanifolds. Transactions of the American Mathematical Society 280 (1983), no. 2, 781–795.

[16] H. Yang, Generalized plane delta-shock waves for n-dimensional zero-pressure gas dynamics. Journal of Mathematical Analysisi and Applications 260 (2001), 18–35.

A Conservation Law with Nonlocal Flux for Erosion Profiles

Wen Shen, Tianyou Zhang
Department of Mathematics, Penn State University, USA
E-mail: shen_w@math.psu.edu, zhang_t@math.psu.edu

Abstract

In this paper we review some results on a model for the erosion of mountain profile caused by small avalanches. The equation is a scalar conservation law with a non-local flux. Under suitable assumptions on the erosion rate, the mountain profile develops three types of singularities, namely kinks, shocks and hyper-kinks. Entropy weak solutions to the Cauchy problem can be constructed globally in time, taking limits of piecewise affine approximate solutions. A one-sided Oleinik type entropy inequality is also established.

1 Introduction and main results

Let $u(t,x)$ describe the height of a mountain profile at time t, depending on a one-dimensional space variable x. Assuming that the slope does not change sign, say $u_x > 0$, the following model describing the erosion of the mountain profile by small avalanches was proposed in [19]

$$u_x - F_x = 0\,, \quad F = \exp \int_x^\infty f(u_x(y))\,dy\,, \qquad u(0,x) = \bar{u}(x)\,. \tag{1.1}$$

Here the function f denotes the erosion rate per unit length of distance covered by a unit size avalanche. The basic assumptions on f are

$$f(1) = 0, \quad f' \geq \eta_0 > 0, \quad f'' < 0, \quad \lim_{w\to 0+} f(w) = -\infty\,. \tag{1.2}$$

Then, f is positive if the slope $u_x > 1$, and negative if $u_x < 1$. The model (1.1) describes the following phenomenon. Granular matter is poured at unit rate from $x = \infty$ (or from an uphill location outside the interval of interest), and slides downhill along the mountain slope, from right to left. As it flows downhill, if the slope u_x is less than 1, part of the granular matter is deposited. If the slope is greater than 1, then the size of the avalanche grows, putting more granular matter in motion. The (normalized) critical slope is 1, where no erosion or deposition occurs.

The model can be derived formally as follows. Let $m(a)$ be the size of the avalanche as it reaches $x = a$. We have

$$m(a) = \exp \int_a^\infty f(u_x(t,y))\,dy\,.$$

Hence, for any interval $[a, b]$, we have

$$\begin{aligned}\frac{d}{dt}\int_a^b u(t,x)\,dx &= m(b) - m(a)\\ &= \exp \int_b^\infty f(u_x(t,y))\,dy - \exp \int_a^\infty f(u_x(t,y))\,dy\,.\end{aligned}$$

This formally leads to equation (1.1).

The model (1.1) was first derived in [2], with

$$f(w) = (w-1)/w\,, \tag{1.3}$$

as the slow erosion limit of a 2×2 model proposed by [13]

$$\begin{cases} h_t = (hu_x)_x - (1-u_x)h\,,\\ u_t = (1-u_x)h\,. \end{cases} \tag{1.4}$$

Here u and h are the heights of the standing and moving layers, respectively. Note that the f in (1.3) does not satisfy the assumption (1.2), because $\lim_{w\to+\infty} f'(w) = 0$. In this case, the slope u_x remains bounded globally. The only type of singularities are the kinks in the mountain profile (jumps in the slope). Well-posedness of solutions with such functions f is studied in [3].

In [19] we study a wider class of functions $f(u_x)$, leading to the formation of different types of singularities. The assumptions (1.2) imply

$$\lim_{w\to+\infty} f'(w) = \eta_0\,. \tag{1.5}$$

Therefore $f(w)$ grows at a linear rate for large w, and the following holds

$$\int_2^\infty f^{-2}(w)\,dw \;<\; \infty\,. \tag{1.6}$$

Besides the assumptions (1.2), we assume further the following decay property for f'':

$$\limsup_{w\geq\kappa_0} |f''(w)|\,w^3 \;<\; \infty\,. \tag{1.7}$$

This rapid decay of f'' implies that f approaches an asymptote as $w \to +\infty$,

$$\lim_{w\to+\infty} |f(w) - w\eta_0| < \infty .$$

Under the assumptions (1.2) and (1.7), the mountain profile $u(\cdot)$ can exhibit three types of singularities:

- *Kinks*: where the mountain profile is continuous but its slope has distinct right and left limits: $0 < u_x(x-) < u_x(x+) < \infty$.
- *Hyperkinks*: where the mountain profile u is continuous but its slope has an infinite limit from the right: $u_x(x+) = +\infty$.
- *Jumps*: where the mountain profile has an upward discontinuity: $u(x-) < u(x+)$.

We seek solutions in a suitable class of strictly increasing functions. It is important to control the total variation of the characteristic speed $f'(u_x)F$, or $f'(u_x)$. Observe that $f'(u_x)$ changes very little for large u_x and approaches constant η_0 as $u_x \to +\infty$. This motivates the definition of the following auxiliary function

$$\zeta(w) \doteq \begin{cases} (w-2)/f^2(2), & \text{if} \quad w < 2\,, \\ \int_2^w f^{-2}(s)\,ds, & \text{if} \quad w \ge 2\,. \end{cases} \tag{1.8}$$

We see that $\zeta(w)$ is a strictly increasing function, but remains bounded because of (1.6), and $\zeta'(w) \to 0$ as $w \to +\infty$.

Solutions of the Cauchy problem will be obtained within the class $\mathcal{W}$ consisting of all functions $u : \mathbb{R} \mapsto \mathbb{R}$ satisfying the following properties:

(W_1) There exists a constant $\kappa_0 > 0$ such that

$$u(x) - u(y) \;\ge\; \kappa_0(x-y), \quad \forall x > y\,. \tag{1.9}$$

(W_2) There exists an interval $I = [a, b]$ such that $u_x = 1$ outside I.

(W_3) The function $U(x) \doteq u(x) - x$ has bounded variation.

(W_4) By setting

$$\zeta(u_x(x)) \;=\; \zeta(+\infty) \doteq \lim_{w\to+\infty} \zeta(w)$$

for x in the support of the singular part of the measure $D_x u$, the function $x \mapsto \zeta(u_x(x))$ has bounded variation.

A definition of entropy weak solutions for (1.1) is now given.

Definition 1.1. A function $u : [0,T] \times \mathbb{R} \mapsto \mathbb{R}$ is called an *entropy weak solution* for (1.1) with initial data $\bar{u}(x) \in \mathcal{W}$ if $u(0,x) = \bar{u}(x)$, and

- $u(t,\cdot) \in \mathcal{W}$ for every $t \in [0,T]$.

• For every test function $\phi \in \mathcal{C}_c^\infty(\mathbb{R}^2)$, one has the integral identity

$$\int_0^T \int_{\mathbb{R}} (u\phi_t - F\phi_x)\,dx\,dt = \int_{\mathbb{R}} [u(T,x)\phi(T,x) - \bar{u}(x)\phi(0,x)]\,dx.$$

• There exist some constants C_1, C_2 (that do not depend on x, y), such that for any $x < y$ the characteristic speed $f'(u_x)F$ satisfies the one-sided Oleinik type entropy inequality

$$f'(u_x(t,y))F(t,y) - f'(u_x(t,x))F(t,x) \le \max\{C_1/t, C_2\}\,(y-x). \quad (1.10)$$

The main result of [19] is summarized in next Theorem.

Theorem 1.2. *Let $T > 0$ and an initial data $\bar{u} \in \mathcal{W}$ be given. Then the Cauchy problem* (1.1) *admits an entropy weak solution $u = u(t,x)$ defined for all $t \in [0,T]$.*

Other PDE models for granular flows can be found in [4, 17] and in the book [11]. A mathematical analysis of steady state solutions for (1.4) was carried out in [9, 10]; a numerical study has been performed in [12]. The analytical properties of time dependent solutions of (1.4) were first studied in [18], where an existence result for global smooth solutions was proved. The paper [1] established the global existence to the Cauchy problem for (1.4) with large data, within a class of functions with bounded variation. Furthermore, in [2] the authors proved the global existence of large BV solutions for an initial-boundary value problem for (1.4).

Other examples of scalar conservation laws with integro-differential terms (involving integral of u_x) have been considered in the literature, and are worth mentioning here. In particular, the variational wave equation

$$u_t + f(u)_x = \int_0^x f''(u)u_x^2\,dy \quad (x > 0)$$

was studied in [7, 14], while the Camassa-Holm equation

$$u_t + \left(\frac{u^2}{2}\right)_x + \left[\frac{1}{2}\int_{-\infty}^{\infty} e^{-|x-y|}\left(u^2(y) + \frac{u_x^2(y)}{2}\right)dy\right]_x = 0$$

was analyzed in a large number of papers, e.g., [5, 6, 8]. In both cases, writing a balance law for the quantity u_x^2 one obtains the a priori bound

$$\|u_x(t,\cdot)\|_{\mathbf{L}^2} \le C.$$

In turn, this implies that the solution $u(t,\cdot)$ remains Hölder continuous at all times. No shocks occur in the solutions. This is the main difference in our case: for (1.1) the solutions develop shocks, and the distributional derivative u_x contains point masses.

2 Basic analysis on singularities

Let $w \doteq u_x$ denote the slope. For smooth solutions, by the method of characteristics we have $\dot{x} = Ff'(w)$ and

$$\dot{u} = (f'(w)w - f(w))F\,, \tag{2.1}$$
$$\dot{w} = Ff^2(w)\,. \tag{2.2}$$

We observe that the slope w is actually a conserved quantity,

$$w_t - (f(w)F)_x = 0. \tag{2.3}$$

Since f is a non-linear function, different characteristics can meet at a same point, producing jumps in w. These are kinks in u, and we call them *weak singularities*.

Furthermore, by (2.2) w is increasing along a characteristic. Assuming $F \geq F_0 > 0$, if $w(0) > 1$, w can blow up in finite time because of (1.6). Shocks will form in the solutions of u, which we refer to as the *strong singularities*.

Wave speeds follow from the Rankine-Hugonoit conditions. We have

$$\lambda_k = \frac{f(w^+) - f(w^-)}{w^+ - w^-}F\,, \qquad \lambda_s = \frac{\exp\{\eta_0(u^+ - u^-)\} - 1}{u^+ - u^-}F^+, \tag{2.4}$$

for kinks and shocks respectively, at a point x_0. Here $u^\pm$, $w^\pm$, and $F^\pm$ denote the limits of $u(x)$, $w(x)$, and $F(x)$ as $x \to x_0\pm$, respectively. Notice that the flux F is continuous at kinks, hence at these points one has $F^+ = F^- = F$.

By the concavity of f, only convex kinks are admissible. In the case of shocks, the standard Lax admissibility condition [15] requires that characteristics must flow into the shock. This leads to a constraint on the relation between the shock size and the slope on the right

$$\frac{\exp\{\eta_0(u^+ - u^-)\} - 1}{u^+ - u^-} \geq f'(w(x_0^+))\,. \tag{2.5}$$

A third type of singularity, bridging the previous two, is the hyperkink. This is a convex kink with infinite slope on the right, and it travels with the speed

$$\lambda_{hk} = \eta_0 F(x_0)\,.$$

When two singularities interact, if they are of the same type, they will merge into one; if they are of different types, the weaker one will merge into the stronger one. No new waves are generated, which is similar to a scalar conservation law with convex flux.

3 BV solutions for the Cauchy problem

Existence of solutions is achieved through piecewise affine approximate solutions that allow discontinuities. Approximate solutions are constructed by an algorithm similar to wave front tracking. Each nodal point x_i is treated as a singularity (kink or shock), and it travels with the corresponding speed. In the end, the algorithm determines a set of ODEs for the nodal points $x_i(t)$ and for the nodal values $u_i(t) = u(x_i(t))$. Suitable a priori estimates for the approximate solution are established. As customary, the limit of a convergent subsequence then provides an entropy weak solution.

To see that the solutions remain in the space $\mathcal{W}$, we give some formal arguments, assuming that solutions remain smooth.

(W_1) By (2.2), w is increasing, which gives the lower bound for w.

(W_2) The finite wave speed leads to the bound on the support of $u_x - 1$.

(W_3) Setting $U(x) \doteq u(x) - x$, the quantity $q \doteq U_x = u_x - 1$ satisfies

$$q_t + (f(q+1)F)_x = 0 .$$

Since $\text{sign}(q)f(q+1) = |f(q+1)|$ and $F > 0$, then $\|q(t)\|_{\mathbf{L}^1}$ is non-increasing, leading to the BV bound for U.

(W_4) The evolution of the function ζ along a characteristic follows

$$\dot{\zeta} = \zeta_t + f'F\zeta_x = \zeta'(w)f^2(w)F = \begin{cases} f^2(w)F/f^2(2), & w < 2, \\ F, & w \geq 2. \end{cases}$$

This provides a bound on the total variation for ζ, since F has bounded variation thanks to (W_3). We then have, for some constants C_1, C_2,

$$\frac{d}{dt}\text{TV}\{\zeta(w)\} \leq C_1 + C_2\text{TV}\{\zeta(w)\} .$$

TV$\{\zeta\}$ can grow exponentially, but remains finite for finite time.

The total variation of $x \mapsto f'(u_x)$ can be controlled by the total variation of $x \mapsto \zeta$. Indeed, consider f' as a function of ζ. By the Chain Rule, we have

$$\frac{d(f')}{d(\zeta)} = \frac{f''(w)}{\zeta'(w)} = \begin{cases} f''(w)f^2(2), & w \leq 2, \\ f''(w)f^2(w), & w > 2, \end{cases}$$

Thanks to the decay of f'' in (1.7), this is negative and uniformly bounded for $w \geq \kappa_0$. By the Mean Value Theorem, we get

$$f'(\bar{w}) - f'(\tilde{w}) \leq C(\zeta(\tilde{w}) - \zeta(\bar{w})), \quad \bar{w} \leq \tilde{w} . \tag{3.1}$$

Therefore, the bounded variation for $x \mapsto \zeta(u_x)$ yields the same bound for $x \mapsto f'(u_x)$.

To get a formal argument for the Oleinik type inequality, we let $v \doteq \zeta_x$, and consider its evolution along a characteristic

$$\frac{d}{dt}v(t,x(t)) = \begin{cases} -f^2(2)f''Fv^2 + 3f'fFv - f^3F/f^2(2)\,, & \kappa_0 \le w \le 2\,, \\ -f''f^2Fv^2 + f'fFv - fF\,, & w > 2\,. \end{cases}$$

For large values of $|v|$, the first term that contains v^2 dominates. The coefficient for this term is $-f''f^2F$, which is strictly positive. This implies that v can blow up only to $+\infty$. A standard comparison argument gives

$$\zeta(w(t,x)) - \zeta(w(t,y)) \le \max\{C_1/t, C_2\}\,(y-x)\,, \quad \text{for} \quad x \le y\,.$$

By using (3.1), we obtain a similar inequality for f'

$$f'(w(t,y)) - f'(w(t,x)) \le \max\{C_3/t, C_4\}\,(y-x)\,, \quad \text{for} \quad x \le y\,.$$

Combining this with the lower bound for F_x and the admissibility condition for shocks, one gets the desired inequality (1.10).

Let u^ε be the approximate solutions and F^ε be the corresponding approximate flux. Convergence of u^ε and F^ε follows from compactness. However, to achieve the existence of a weak solution, F^ε must converge to a correct limit, i.e., as $u^\varepsilon \to u$, we must have

$$\exp\left\{\int_x^{+\infty} f(u_x^\varepsilon(y))\,dy\right\} \;\to\; \exp\left\{\int_x^{+\infty} f(u_x(y))\,dy\right\}.$$

This is not straight forward because $x \mapsto u_x^\varepsilon(t,x)$ is a measure-valued function.

To overcome this technical difficulty, we make a variable change, and consider x as function of (t,u). By condition (W_1), this variable change is well-defined, and the map $u \to x^\varepsilon(t,u)$ is uniformly Lipschitz for all t. Then, the inverse slope function $u \to x_u^\varepsilon(t,u)$ is uniformly bounded with bounded variation. We define the function

$$\mathcal{G}(s) \doteq sf\left(\frac{1}{s}\right). \tag{3.2}$$

This function is Lipschitz for $\frac{1}{s} \ge \kappa_0$ because

$$\mathcal{G}'(s) = f\left(\frac{1}{s}\right) - \frac{1}{s}f'\left(\frac{1}{s}\right) < \infty\,,$$

Then, the flux F^ε can be redefined as

$$F^\varepsilon(t,u) \doteq \exp\left\{\int_u^{\infty} \mathcal{G}(x_u^\varepsilon(t,v))\,dv\right\}. \tag{3.3}$$

The map $u \mapsto F^\varepsilon(t,u)$ is uniformly Lipschitz continuous. The a priori BV estimates on $\mathcal{G}(u_x^\varepsilon)$ and F^ε can then be used to establish the compactness of a subsequence of approximate solutions towards the correct limits, achieving existence of weak solutions.

We refer to [19] for details of the proof.

References

[1] D. Amadori and W. Shen, Global existence of large BV solutions in a model of granular flow. Comm. PDEs 34 (2009), 1003–1040.

[2] D. Amadori and W. Shen, The slow erosion limit in a model of granular Flow. Arch. Ration. Mech. Anal. 199 (2011), 1–31.

[3] D. Amadori and W. Shen, An Integro-Differential Conservation Law arising in a Model of Granular Flow. Preprint, 2011.

[4] T. Boutreux and P. G. Gennes, Surface flows of granular mixtures, I. General principles and minimal model. J. Phys. I France 6 (1996), 1295–1304.

[5] A. Bressan and A. Constantin, Global dissipative solutions of the Camassa-Holm equation. Anal. Appl. (Singap.) 5 (2007), 1–27.

[6] A. Bressan and A. Constantin, Global conservative solutions of the Camassa-Holm equation. Arch. Ration. Mech. Anal. 183 (2007), 215–239.

[7] A. Bressan, P. Zhang and Y. Zheng, Asymptotic variational wave equations. Arch. Ration. Mech. Anal. 183 (2007), 163–185.

[8] R. Camassa and D. Holm, An integrable shallow water equation with peaked solitons. Phys. Rev. Lett. 71 (1993), 1661–1664.

[9] P. Cannarsa and P. Cardaliaguet, Representation of equilibrium solutions to the table problem for growing sandpiles. J. Eur. Math. Soc. (JEMS) 6 (2004), 435–464.

[10] P. Cannarsa, P. Cardaliaguet, G. Crasta and E. Giorgieri, A boundary value problem for a PDE model in mass transfer theory: representation of solutions and applications. Calc. Var. PDEs 24 (2005), 431–457.

[11] J. Duran, Sands, Powders, and Grains, An Introduction to the Physics of Granular Materials. Springer-Verlag, 2000.

[12] M. Falcone and S. F. Vita, A finite-difference approximation of a two-layer system for growing sandpiles. SIAM J. Sci. Comput. 28 (2006), 1120–1132.

[13] K. P. Hadeler and C. Kuttler, Dynamical models for granular matter. Granular Matter 2 (1999), 9–18.

[14] J. K. Hunter and Y. Zheng, On a nonlinear hyperbolic variational equation I and II. Arch. Ration. Mech. Anal. 129 (195), 305–383.

[15] P. Lax, Hyperbolic systems of conservation laws II. Commun. Pure Appl. Math. 10 (1957), 537–566.

[16] T. P. Liu, Admissible solutions of hyperbolic conservation laws. Mem. Amer. Math. Soc. 30 (1981), no. 240.

[17] S. B. Savage and K. Hutter, The dynamics of avalanches of granular materials from initiation to runout. I. Analysis. Acta Mech. 86 (1991), 201–223.

[18] W. Shen, On the shape of avalanches. J. Math. Anal. Appl. 339 (2008), 828–838.

[19] W. Shen and T. Y. Zhang, Erosion profile by a global model for granular flow. Accepted for publication in Arch. Ration. Mech. Anal.

A Moving-mesh Relaxation Scheme for One-dimensional Barotropic Two-phase Flow with Cavitation*

Keh-Ming Shyue
Department of Mathematics, Taiwan University
Taipei 10617, China
E-mail: shyue@math.ntu.edu.tw

1 Introduction

Cavitation is commonly defined as a phenomenon in a liquid-flowing system when the pressure of the liquid falls sufficiently low in some region of the flow so that vapor bubbles are formed. The study of the dynamics of cavitation is an active research in many fields of science and engineering. Typical examples in relation to various features and characteristics of cavitating flows can be found, for example, at the website: http://en.wikipedia.org/wiki/Cavitation, and references therein.

To compute cavitating flow numerically, one popular approach is to use a barotropic two-phase model (cf. [8]) in that, if we ignore the physical effects such as mass transfer, surface tension, and viscosity, the Eulerian formulation of the basic conservation laws in one space dimension takes the form

$$\frac{\partial}{\partial t}\begin{pmatrix}\alpha_1\rho_1\\ \alpha_2\rho_2\\ \rho u\end{pmatrix}+\frac{\partial}{\partial x}\begin{pmatrix}\alpha_1\rho_1 u\\ \alpha_2\rho_2 u\\ \rho u^2+p\end{pmatrix}=0. \tag{1.1}$$

Here α_k, ρ_k, and p_k are the volume fraction, density, and pressure for the fluid phase k, $k=1,2$, respectively. The notations $\rho=\alpha_1\rho_1+\alpha_2\rho_2$, $p=\alpha_1 p_1+\alpha_2 p_2$, and u denote in turn the total density, pressure, and particle velocity of the flow.

To close the system, the phasic pressure $p_k(\rho)$ is assumed to be a one-to-one function of the density (this should be true locally at the

*The author was supported in part by "National Science Council of Taiwan Grant" #96-2115-M-002-008-MY3.

least), and so we may use the saturation condition $\alpha_1 + \alpha_2 = 1$ directly, yielding a nonlinear algebraic equation to be solved for the pressure p,

$$g(p) = \frac{\alpha_1\rho_1}{\rho_1(p)} + \frac{\alpha_2\rho_2}{\rho_2(p)} - 1 = 0, \tag{1.2}$$

where $\alpha_1\rho_1$ and $\alpha_2\rho_2$ are known a priori. With that, there is no difficulty to compute the remaining flow variables such as ρ_1, ρ_2, α_1, and α_2.

In this case, it is known that combining (1.1) with (1.2) gives a hyperbolic model that is viable for a class of barotropic homogeneous two-phase flow problems with and without cavitation. However, due to the non-monotonic behavior of the mixture sound speed (denoted by c) versus the volume fraction, $1/\rho c^2 = \alpha_1/\rho_1 c_1^2 + \alpha_2/\rho_2 c_2^2$, it poses a major difficulty to maintain a suitable stability condition when the model is discretized by an explicit diffuse-interface type method.

To overcome this numerical difficulty, we are interested in a relaxation approach proposed by Saurel et al. [5] in that in addition to (1.1) a transport equation with a stiff relaxation source term is included in the model for the volume fraction, such as α_1, of the form

$$\frac{\partial \alpha_1}{\partial t} + u\frac{\partial \alpha_1}{\partial x} = \mu\left(p_1\left(\rho_1\right) - p_2\left(\rho_2\right)\right), \tag{1.3}$$

where μ is a relaxation parameter. In contrast with the aforementioned conventional model, here the equilibrium pressure p is obtained by taking the limit of infinite relaxation $\mu \to \infty$ to the solution of (1.3), yielding $p = p_1(\rho_1) = p_2(\rho_2)$, and so an algebraic equation for the relaxed volume fraction α_1,

$$g\left(\alpha_1\right) = p_1\left(\frac{\alpha_1\rho_1}{\alpha_1}\right) - p_2\left(\frac{\alpha_2\rho_2}{1-\alpha_1}\right) = 0. \tag{1.4}$$

It is important to note that since this relaxation model will be solved by a fractional-step method, see Section 3, in the zero relaxation limit $\mu \to 0$, it possesses a nice monotonic behavior of the frozen speed of sound versus the mass fractions, $c_f^2 = Y_1 c_1^2 + Y_2 c_2^2$, and so is an easier one to use as compared to the conventional model for numerical approximation. Here $Y_k = \rho_k/\rho$ is the mass fraction for the fluid phase k.

Our goal in this work is to employ a state-of-the-art moving-mesh method for the efficient numerical resolution of this relaxation model for barotropic two-phase flow with cavitation in one space dimension. This is a fundamental step in our further development of the method to problems in multiple dimensions and also more complicated cavitating flows of practical importance (cf. [3, 4, 5]).

2 Relaxation model in moving coordinates

For the ease of the latter discussion, it is useful to write the relaxation model described above into a more compact expression by

$$\frac{\partial q}{\partial t} + \frac{\partial f(q)}{\partial x} + B\frac{\partial q}{\partial x} = \mu\psi(q), \tag{2.1}$$

where

$$\begin{aligned} q &= \left(\alpha_1\rho_1,\ \alpha_2\rho_2,\ \rho u,\ \alpha_1\right)^T, \\ f &= \left(\alpha_1\rho_1 u,\ \alpha_2\rho_2 u,\ \rho u^2 + p,\ 0\right)^T, \\ B &= \operatorname{diag}\left(0,\ 0,\ 0,\ u\right), \\ \psi &= \operatorname{diag}\left(0,\ 0,\ 0,\ p_1\left(\rho_1\right) - p_2\left(\rho_2\right)\right). \end{aligned} \tag{2.2}$$

Then, as a preliminary, we introduce a coordinate mapping from the physical domain (t, x) to the computational domain (τ, ξ) as

$$dt = d\tau, \qquad dx = \upsilon d\tau + J d\xi,$$

where $\upsilon = \partial x/\partial\tau$ is the mesh velocity, and $J = \partial x/\partial\xi$ is the metric of the mapping. With that, (2.1) can be transformed into

$$J\frac{\partial q}{\partial\tau} + \frac{\partial f(q)}{\partial\xi} + (B - \upsilon I)\frac{\partial q}{\partial\xi} = J\mu\psi(q), \tag{2.3}$$

where I denotes a 4×4 identity matrix. If the geometric compatibility condition (or called the geometric conservation law)

$$\frac{\partial J}{\partial\tau} - \frac{\partial\upsilon}{\partial\xi} = 0 \tag{2.4}$$

is used in the derivation, we have an alternative form of the equations

$$\frac{\partial(Jq)}{\partial\tau} + \frac{\partial}{\partial\xi}\left(f(q) - \upsilon q\right) + B\frac{\partial q}{\partial\xi} = J\mu\psi(q), \tag{2.5}$$

that are essential to many existing moving mesh methods (cf. [2] and references therein).

3 Relaxation solver on moving meshes

In each time step, our numerical method for solving barotropic two-phase flows on a moving mesh consists of the following steps:

(1) (Moving mesh step) Determine the cell-interface velocity and the cell-interface location in the physical space over a time step.

(2) (Frozen step $\mu \to 0$) Solve the homogeneous part of the relaxation model described in Section 2 on a moving mapped grid over the same time step as in step 1.

(3) (Relaxation step $\mu \to \infty$) Solve the model system with only the source terms in the infinite relaxation limit.

Note that, in step 1, we may use a variational approach as proposed by Tang and Tang [6] for the determination of the cell-interface location from the current time at x_i^n to the next at x_i^{n+1}. With that, if the motion of x_i^n is assumed to move at a constant speed over each time interval $[t_n, t_{n+1}]$, the cell-interface velocity can be computed by $v_i^n = (x_i^{n+1} - x_i^n)/\Delta t$, where $\Delta t = t_{n+1} - t_n$, see Figure 3.1 for an illustration.

It is clear that, alternatively, we may also use a classical Lagrangian approach in this moving-mesh step in that v_i^n is taken simply as the particle velocity of the local Riemann problem at the cell interface x_i^n. We then set x_i^{n+1} to be $x_i^{n+1} = x_i^n + v_i^n \Delta t$ over a time step Δt.

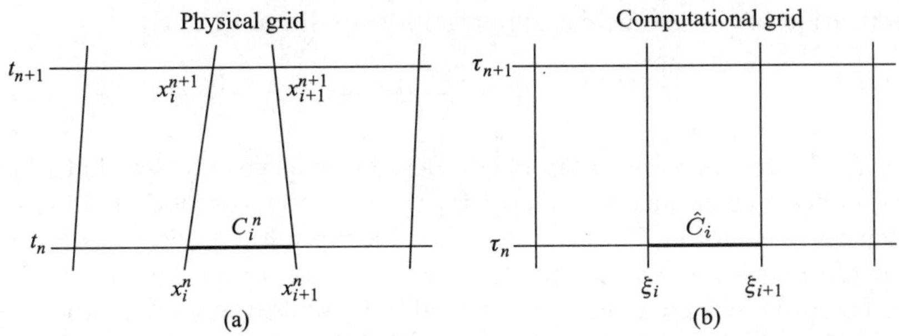

Figure 3.1 A sample grid system in our one-dimensional moving mesh method. The motion of the cell-interface location x_i^n at a time t_n is assumed to move at a constant speed v_i^n over each time step $[t_n, t_{n+1}]$.

Next, in step 2 of the method, we use a finite-volume formulation of the solution in that the approximate value of the cell average of the solution q over the ith grid cell at a time t_n can be written as

$$Q_i^n \approx \frac{1}{\mathcal{M}(C_i^n)} \int_{C_i^n} q(x, t_n)\, dx = \frac{1}{J_i^n \Delta\xi} \int_{\hat{C}_i} q(\xi, \tau_n)\, d\xi,$$

where C_i^n and $\hat{C}_i$ represent the regions occupied by the ith grid cell in the physical and computational domains, respectively, and $\mathcal{M}(C_i^n) = x_{i+1}^n - x_i^n = J_i^n \Delta\xi$ is the measure (length) of C_i^n. Here we define $J_i^n = J(C_i^n)$ as being the metric of the mapping of this cell, and assume a uniform mesh size $\Delta\xi = \xi_{i+1} - \xi_i$ in the computational domain; the time step from the current time τ_n to the next τ_{n+1} is denoted by $\Delta\tau = \Delta t$.

In this setup, the first order version of the wave propagation method for the equations (2.5) without the source terms is a Godunov-type scheme on a mapped grid that can be written in the fluctuation form as

$$J_i^{n+1}Q_i^{n+1} = J_i^n Q_i^n - \frac{\Delta\tau}{\Delta\xi}\left(\mathcal{A}^+\Delta Q_i + \mathcal{A}^-\Delta Q_{i+1}\right), \tag{3.1}$$

where $\mathcal{A}^+\Delta Q_i$ and $\mathcal{A}^-\Delta Q_{i+1}$ are the right- and left-moving fluctuations, respectively, which are defined as

$$\begin{aligned}\mathcal{A}^+\Delta Q_i &= f\left(Q_i^n\right) - f\left(Q_i^*\right) + \upsilon_i^n Q_i^* + B_i^+\left(Q_i^n - Q_{i-1}^n\right),\\ \mathcal{A}^-\Delta Q_{i+1} &= f\left(Q_{i+1}^*\right) - f\left(Q_i^n\right) - \upsilon_{i+1}^n Q_{i+1}^* + B_{i+1}^-\left(Q_{i+1}^n - Q_i^n\right).\end{aligned}$$

Here the values Q_ι^* for $\iota = i,\ i+1$ are assigned to be the state obtained in the solution of the Riemann problem between the grid cells with $Q_{\iota-1}^n$ and Q_ι^n along the ray $(x - x_\iota^n)/(t - t_n) = \upsilon_\iota^n$ (cf. [1]). As usual, we set $B^+ = \max\{B, 0\}$ and $B^- = \min\{B, 0\}$.

Note that, to ensure the free-stream (constant states) preservation, which is one of the basic properties that a moving mesh method should have, it is necessary to supplement the method with a consistent discretization of the geometric compatibility condition (3.2) as

$$J_i^{n+1} = J_i^n - \frac{\Delta\tau}{\Delta\xi}\left(\upsilon_i^n - \upsilon_{i+1}^n\right) \tag{3.2}$$

for J_i^{n+1} with all i and n. In one dimension, it is easy to see that (3.2) is the direct consequence of step 1 for the cell-interface motion. In more than one dimension, however, it requires a careful numerical discretization of the method on a 2D quadrilateral or 3D hexahedral mesh (cf. [7]).

For practical purposes, it is desirable to write the method that retains the free-stream preservation in any space dimension, when the associated geometric conservation law is satisfied only approximately. One approach to do this is discussed in [1] for one-dimensional hyperbolic balance laws in that (3.2) is substituted to (3.1), yielding easily

$$Q_i^{n+1} = Q_i^n - \frac{1}{J_i^{n+1}}\frac{\Delta\tau}{\Delta\xi}\left(\mathcal{A}^+\Delta Q_i + \mathcal{A}^-\Delta Q_{i+1}\right) \tag{3.3}$$

with

$$\begin{aligned}\mathcal{A}^+\Delta Q_i &= f\left(Q_i^n\right) - f\left(Q_i^*\right) - \upsilon_i^n\left(Q_i^n - Q_i^*\right) + B_i^+\left(Q_i^n - Q_{i-1}^n\right),\\ \mathcal{A}^-\Delta Q_{i+1} &= f\left(Q_{i+1}^*\right) - f\left(Q_i^n\right) - \upsilon_{i+1}^n\left(Q_{i+1}^* - Q_i^n\right) + B_{i+1}^-\left(Q_{i+1}^n - Q_i^n\right),\end{aligned}$$

which is a consistent discretization of equations (2.3). To achieve high resolution, it is a standard matter to include correction terms to the method, see [1] for the details.

Finally, in step 3, using the solution obtained in step 2 as the initial condition, we solve the model system with the source term of the form

$$\frac{\partial(Jq)}{\partial\tau} = J\mu\psi(q),$$

which, in the infinite relaxation limit, leads to

$$g\left(\alpha_1^{n+1}\right) = p_1\left(\frac{(\alpha_1\rho_1)^{n+1}}{\alpha_1^{n+1}}\right) - p_2\left(\frac{(\alpha_2\rho_2)^{n+1}}{1-\alpha_1^{n+1}}\right) = 0$$

for the relaxed volume fraction α_1^{n+1} which can be solved by a standard iterative root-finding solver such as the secant method.

4 Numerical results

We now present preliminary results obtained using our moving-mesh method described in Section 3 for barotropic two-phase flow problems with and without cavitation in one space dimension. For simplicity, we assume that the constitutive law for each of the fluid phases satisfies the Tait equation of state of the form

$$p_k(\rho) = (p_{0k} + \mathcal{B}_k)\left(\frac{\rho}{\rho_{0k}}\right)^{\gamma_k} - \mathcal{B}_k \qquad \text{for } k = 1, 2. \tag{4.1}$$

The set of parameters we take in the tests considered here are, for phase 1 (the liquid phase), $(\gamma, \mathcal{B}, \rho_0, p_0)_1 = \left(7, 3000 \text{ bar}, 10^3 \text{ kg/m}^3, 1 \text{ bar}\right)$, and for phase 2 (the gas phase), $(\gamma, \mathcal{B}, \rho_0, p_0)_2 = \left(1.4, 0, 1 \text{ kg/m}^3, 1 \text{ bar}\right)$. In carrying out the runs, the basic grid motion is governed by the quasi-Lagrangian approach as $\partial x/\partial\tau = (1-h)u$ with $h = 10^{-6}$, and the high resolution version of the method is used throughout.

Example 4.1. Our first example is a cavitation-free two-phase (gas-liquid) Riemann problem. Initially, in a region where $x \in [-2, 0)m$, we have the state variables

$$(\rho_1,\ \rho_2,\ u,\ \alpha_1)_L = \left(10^3 \text{ kg/m}^3,\ 719.686 \text{ kg/m}^3,\ 0,\ \epsilon\right),$$

while in a region where $x \in [0, 2]m$, we have the state variables

$$(\rho 1,\ \rho_2,\ u,\ \alpha_1)_R = \left(10^3 \text{ kg/m}^3,\ 1 \text{ kg/m}^3,\ 0,\ 1-\epsilon\right),$$

where $\epsilon = 10^{-6}$.

Figure 4.1 shows contours of the computed density and pressure in the x-t plane up to time $t = 4$ ms using 100 mesh points, where solid wall boundary conditions are used at the both end. It is clear that breaking of the gas-liquid membrane results in a self-similar solution that consists of a left-moving rarefaction wave, a right-moving contact discontinuity, and a shock wave. Complicated wave interactions occur when the waves are reflecting back from the boundaries. From the figure, sharpness of

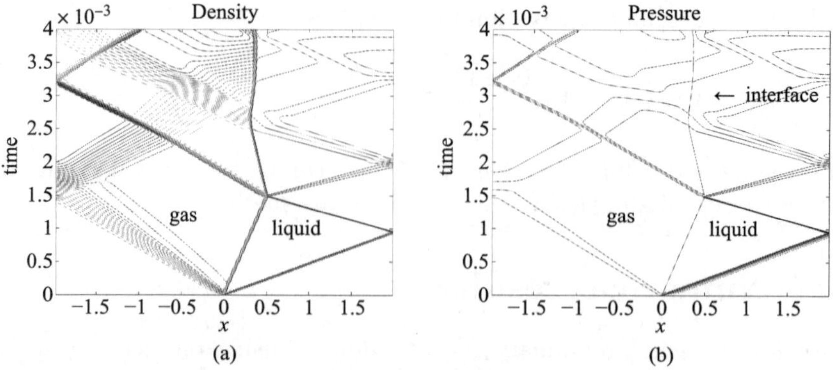

Figure 4.1 Moving-mesh results for a two-phase Riemann problem. Contours of the density and pressure are shown in the x-t plane up to 4ms.

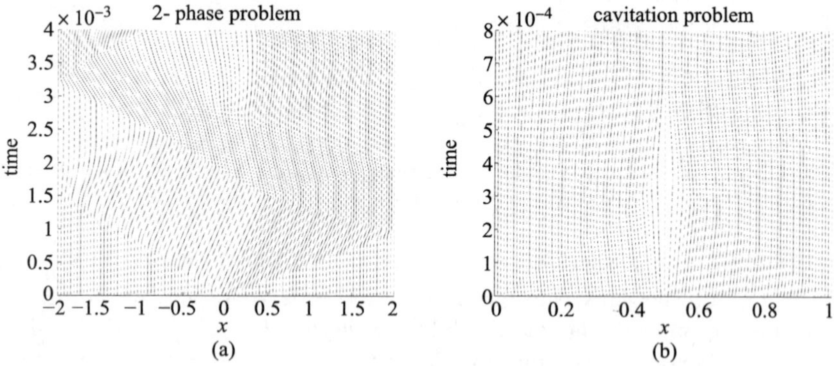

Figure 4.2 Moving-meshes in the x-t plane for the two-phase Riemann problem (a) and cavitation problem (b).

the density and also free of spurious oscillation in the pressure near the contact discontinuity are easily seen. The physical grid system obtained by the method is shown in Figure 4.2.

Example 4.2. We are next concerned with a water-vapor cavitation problem in that inside a shock tube of one-meter length with closed ends the fluid is a homogeneous air-water mixture at the standard atmospheric condition $(\rho_1, \rho_2, \alpha_1) = \left(10^3 \text{ kg/m}^3, 1 \text{ kg/m}^3, 0, 1-\epsilon\right)$, where $\epsilon = 10^{-6}$. Initially, inside the tube, there is a jump on the velocity at $x = 1/2$ m with speed $u = -100$ m/s on the left and $u = 100$ m/s on the right of the tube. In addition to that, since it is a closed tube, there are jumps also on the velocity at the both ends.

With this condition, as times go on, two rarefaction waves are formed, causing the decrease of the pressure and the formation of the cavitation zone inside the tube. In the meantime, there are inward-moving shock waves propagating from the boundaries, yielding the collapse of the region of cavitation due to the shock-cavitation interaction. Figure 4.3 shows pseudo-colors of the volume fraction and pressure in the x-t plane up to time 0.8 ms, see Figure 4.2 for the meshes over time. We observe reasonable resolution of the results.

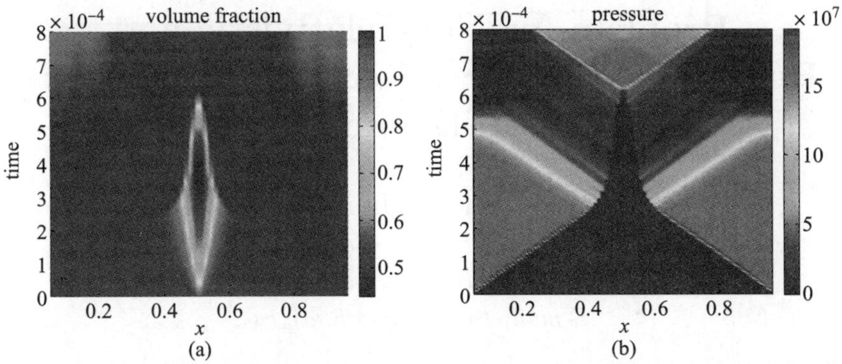

Figure 4.3 Moving-mesh results for a water-vapor cavitation problem. Pseudo-colors of the volume fraction and pressure are shown in the x-t plane up to time 0.8 ms.

References

[1] R. Fazio and R. J. LeVeque, Moving-mesh methods for one-dimensional hyperbolic problems using CLAWPACK. Comput. Math. with Appl. 45 (2003), 273–298.

[2] W. H. Hui, The unified coordinate system in computational fluid dynamics. Comm. Comput. Phys. 2(2007), 577–610.

[3] F. Petitpas, J. Massoni, R. Saurel, E. Lapebie and L. Munier, Diffuse interface model for high speed cavitating underwater systems. Int. J. Multiphase Flow 35 (2009), 747–759.

[4] R. Saurel, F. Petitpas and R. Abgrall. Modelling phase transition in metastable liquids: application to cavitating and flashing flows. J. Fluid. Mech. 607 (2008), 313–350.

[5] R. Saurel, F. Petitpas and R. A. Berry, Simple and efficient relaxation methods for interfaces separating compressible fluids, cavitating flows and shocks in multiphase mixtures. J. Comput. Phys. 228 (2009), 1678–1712.

[6] H. Tang and T. Tang, Adaptive mesh methods for one- and two-dimensional hyperbolic conservation laws. SIAM J. Numer. Anal. 41 (2003), 487–515.

[7] P. D. Thomas and C. K. Lombard, Geometric conservation law and its application to flow computations on moving grids. AIAA J. 17 (1979), 1030–1037.

[8] S. Venkateswaran, J. W. Lindau, R. F. Kunz and C. L. Merkle, Computation of multiphase mixture flows with compressibility effects. J. Comput. Phys. 180 (2002), 54–77.

Well-posedness of the Free Boundary Problem for Non-relativistic and Relativistic Compressible Euler Equations with a Vacuum Boundary Condition*

Yuri Trakhinin
Sobolev Institute of Mathematics
Novosibirsk, Koptyug av. 4, Russia
E-mail: trakhin@math.nsc.ru

Abstract

We survey recent results [7] for the free boundary problem for the equations of compressible Euler equations with a vacuum boundary condition. We consider the case when the density does not go to zero continuously, but jumps. We recover in Eulerian coordinates the earlier well-posedness result obtained by Lindblad [4] for the isentropic Euler equations and extend it to the case of full gas dynamics. Our approach is still directly applicable to the relativistic version of the problem in the setting of special relativity. We also briefly discuss open problems for the general relativistic case and their solution in the framework of relativistic magnetohydrodynamics (plasma-vacuum interface problem [9]).

1 Introduction

We are interested in the motion of an ideal compressible fluid in vacuum described by the Euler equations

$$\partial_t \rho + \operatorname{div}(\rho v) = 0, \tag{1.1}$$

$$\partial_t(\rho v) + \operatorname{div}(\rho v \otimes v) + \nabla p = \rho \mathcal{G}, \tag{1.2}$$

$$\partial_t \left(\rho \left(e + \frac{1}{2}|v|^2 \right) \right) + \operatorname{div} \left(\left(\rho \left(e + \frac{1}{2}|v|^2 \right) + p \right) v \right) = \rho(v, \mathcal{G}), \tag{1.3}$$

*The author is supported by RFBR (Russian Foundation for Basic Research, grant No. 10-01-00320-a) and the programme "Development of scientific potential of the Russian Higher School (2009–2010)" (grant No. 2.1.1/4591).

in a space-time domain $\Omega(t)$ which boundary $\Sigma(t) = \{F(t,x) = 0\}$ is to be determined and moves with the velocity of the gas particles at the boundary:

$$\frac{dF}{dt} = 0, \qquad p = 0 \quad \text{on} \quad \Sigma(t) \tag{1.4}$$

(for all $t \in [0,T]$), where $d/dt = \partial_t + (v, \nabla)$; $v \in \mathbb{R}^3$ denotes velocity, $p = p(\rho, S)$ pressure, ρ density, S entropy, $e = e(\rho, S)$ internal energy, and $\mathcal{G} \in \mathbb{R}^3$ gravitational field. This free boundary problem can be used for modeling the motion of the ocean or a star.

A local-in-time existence theorem was recently proved by Lindblad [4] for the free boundary problem (1.1)–(1.4) with *non-vanishing density* on the boundary (the case of *compressible liquid*) for the isentropic Eulér equations. Namely, the local-in-time existence of smooth solutions was shown in [4] under the natural physical assumption

$$\frac{\partial p}{\partial N} \leq -\epsilon < 0 \qquad \text{on} \quad \Sigma(0), \tag{1.5}$$

where $\partial/\partial N = (\nabla F, \nabla)$, together with the hyperbolicity conditions $\rho(0,x) > 0$ and $(\partial p/\partial \rho)(0,x) > 0$ ($x \in \overline{\Omega(0)}$), provided that the initial domain $\Omega(0)$ is diffeomorphic to a ball. The main tool in [4] is the passage to the Lagrangian coordinates for reducing the original problem to that in a fixed domain. Such a technique seems most natural for free boundary problems with boundary conditions like (1.4). At the same time, for compressible fluids it is connected with a lot of technical difficulties and it is not quite clear how to extend the results to similar problems for more complicated fluid dynamics models like, for example, relativistic gas dynamics or magnetohydrodynamics (MHD). Even the extension of the existence theorem in [4] to full gas dynamics does not seem to be just a technical matter.

We propose another approach to studying the well-posedness of the mentioned free boundary problem (for the case of liquid) and similar free boundary problems for other systems of hyperbolic conservation laws. Our main goal is to recover in Eulerian coordinates the result in [4] for the isentropic Euler equations and extend it to the case of full gas dynamics. For technical simplicity we consider the case of an unbounded domain whose boundary has the form of a graph:

$$\Omega(t) = \{x_1 > \varphi(t, x')\}, \tag{1.6}$$

where the function $\varphi(t, x')$ is to be determined (we refer to [7] for short comments about the case of a bounded domain). We prove the local-in-time existence in Sobolev spaces [7] by the technique applied earlier to weakly stable shock waves and characteristic discontinuities [2, 6]. It contains, in particular, the reduction to a fixed domain, using the "good unknown" of Alinhac [1], and a suitable Nash-Moser-type iteration scheme. A certain modification of such an approach is caused by the fact that the symbol associated to the free surface is *not elliptic*, i.e., we are not able to resolve our boundary conditions for $(\partial_t F, \nabla F)$.

Our approach is still directly applicable to the relativistic version of the problem in the setting of special relativity:

$$\partial_t(\rho\Gamma) + \mathrm{div}\,(\rho u) = 0, \tag{1.7}$$

$$\partial_t(\rho h\Gamma u) + \mathrm{div}\,(\rho h u \otimes u) + \nabla p = \rho \mathcal{G}, \tag{1.8}$$

$$\partial_t(\rho h\Gamma^2 - p) + \mathrm{div}\,(\rho h\Gamma u) = \rho(u, \mathcal{G}), \tag{1.9}$$

where $h = 1 + e + (p/\rho)$ is the specific enthalpy, p is the pressure, $e = e(\rho, S)$ is the specific internal energy per particle, S is the entropy per particle, $u = (u^1, u^2, u^3)$, $v = u/\Gamma$; $\Gamma = u^0 = (1 - |v|^2)^{-1/2}$ is the Lorentz factor, and the speed of the light is equal to unity. The proof of the local-in-time existence theorem for problem (1.7)–(1.9), (1.4) is absolutely the same as for the non-relativistic case. The only important point is that system (1.7)–(1.9) can be symmetrized (see Section 2).

In the end of this paper we briefly discuss open problems for the case of general relativity. Surprisingly, the difficulties appearing for the general relativistic version of our vacuum problem can be easily overcome in the framework of MHD, i.e., for the relativistic plasma-vacuum interface problem [9]. This is so because in MHD the magnetic field can make the symbol associated to the free surface *elliptic*. In Section 3 we briefly announce first results in this direction (yet for special relativistic MHD) and refer to [9] for more details.

2 Main theorem and sketch of the proof

We can easily symmetrize system (1.1)–(1.3) by rewriting it in the non-conservative form

$$\frac{1}{\rho c^2}\frac{\mathrm{d}p}{\mathrm{d}t} + \mathrm{div}\, v = 0, \quad \rho\frac{\mathrm{d}v}{\mathrm{d}t} + \nabla p = \rho\mathcal{G}, \quad \frac{\mathrm{d}S}{\mathrm{d}t} = 0, \tag{2.1}$$

where $c^2 = p_\rho(\rho, S)$ is the square of the sound velocity. Equations (2.1) read as the symmetric quasilinear system

$$A_0(U)\partial_t U + \sum_{j=1}^{3} A_j(U)\partial_j U + Q(U) = 0, \tag{2.2}$$

where $Q(U) = (0, -\rho\mathcal{G}, 0)$, and the symmetric matrices A_α can be easily written down. System (2.2) is symmetric hyperbolic if the the hyperbolicity condition $A_0 > 0$ holds:

$$\rho > 0, \quad p_\rho > 0. \tag{2.3}$$

One can alternatively consider the *isentropic* Euler equations, i.e., system (1.1) and (1.2) for the same variables except for the entropy S.

For domain (1.6) the boundary conditions (1.4) take the form

$$\partial_t \varphi = v_N, \quad p = 0, \tag{2.4}$$

on $\Sigma(t)$, and the gravitational field $\mathcal{G} = (G, 0, 0)$, where $v_N = (v, N)$, $N = (1, -\partial_2\varphi, -\partial_2\varphi)$, and G denotes Newton's gravitational constant.

To reduce the free boundary value problem (2.2) and (2.4) to that in a fixed domain we straighten, as usual, the unknown free surface Σ. That is, the unknown U being smooth in $\Omega(t)$ is replaced by the vector-function $\widetilde{U}(t,x) := U(t, \Phi(t,x), x')$, which is smooth in the fixed domain $\mathbb{R}^3_+ = \{x_1 > 0,\ x' \in \mathbb{R}^2\}$, where $\Phi(t, 0, x') = \varphi(t, x')$ and $\partial_1\Phi > 0$. As in [6], to avoid assumptions about compact support of the initial data in the nonlinear existence theorem and work globally in $\mathbb{R}^3_+$ we use the following choice of $\Phi(t,x)$:

$$\Phi(t,x) := x_1 + \Psi(t,x), \quad \Psi(t,x) := \chi(x_1)\varphi(t,x'),$$

where $\chi \in C_0^\infty(\mathbb{R})$ equals to 1 on $[0,1]$, and $\|\chi'\|_{L_\infty(\mathbb{R})} < 1/2$.

Dropping for convenience tildes in $\widetilde{U}$, we reduce (2.2) and (2.4) to the initial boundary value problem

$$\mathbb{L}(U,\Psi) = 0 \quad \text{in } [0,T] \times \mathbb{R}^3_+, \tag{2.5}$$

$$\mathbb{B}(U,\varphi) = 0 \quad \text{on } [0,T] \times \{x_1 = 0\} \times \mathbb{R}^2, \tag{2.6}$$

$$U|_{t=0} = U_0 \quad \text{in } \mathbb{R}^3_+, \quad \varphi|_{t=0} = \varphi_0 \quad \text{in } \mathbb{R}^2, \tag{2.7}$$

where $\mathbb{L}(U,\Psi) = L(U,\Psi)U + Q(U)$, $L(U,\Psi) = A_0(U)\partial_t + \widetilde{A}_1(U,\Psi)\partial_1 + \sum_{k=2}^3 A_k(U)\partial_k$,

$$\widetilde{A}_1(U,\Psi) = \frac{1}{\partial_1\Phi}\Big(A_1(U) - A_0(U)\partial_t\Psi - \sum_{k=2}^3 A_k(U)\partial_k\Psi\Big),$$

and (2.6) is the compact form of the boundary conditions (2.4).

We are now in a position to state the local-in-time existence theorem for problem (2.5)–(2.7).

Theorem 2.1 ([7]). *Let $m \in \mathbb{N}$ and $m \geq 6$. Suppose the initial data (2.7), with*

$$(U_0 - \check{U}, \varphi_0) \in H^{m+7}(\mathbb{R}^3_+) \times H^{m+7}(\mathbb{R}^2) \text{ and } \rho(p_0, S_0) - \epsilon_1 \in H^{m+7}(\mathbb{R}^3_+),$$

satisfy the hyperbolicity condition (2.3) for all $x \in \overline{\mathbb{R}^3_+}$ and are compatible up to order $m+7$ in a suitable sense (see [7]). Here

$$\check{U} = (2\epsilon x_1, 0, 0, 0, 0), \quad \epsilon_1 = 2\epsilon/G \quad (\epsilon = \text{const} > 0).$$

Let also the initial data satisfy the physical condition (*cf.* (1.5))

$$\partial_1 p \geq \epsilon > 0 \qquad \text{at} \quad x_1 = 0 \tag{2.8}$$

for all $x' \in \mathbb{R}^2$. *Then, there exists a sufficiently short time* $T > 0$ *such that problem* (2.5)–(2.7) *has a unique solution*

$$(U, \varphi) \in \{\check{U} + H^m([0,T] \times \mathbb{R}^3_+)\} \times H^m([0,T] \times \mathbb{R}^2).$$

Moreover, $\rho - \epsilon_1 \in H^m([0,T] \times \mathbb{R}^3_+)$.

Below we will briefly discuss main steps of the proof of Theorem 2.1 (see [7] for more details). For the relativistic version of problem (2.5)–(2.7) we obtain the local-in-time existence (and uniqueness) theorem in the form of Theorem 2.1. We should only supplement (2.3) by the relativistic causality condition

$$0 < c_s^2 < 1, \tag{2.9}$$

where c_s is the relativistic speed of sound, $c_s^2 = c^2/h$. Indeed, equations (1.7)–(1.9) can be rewritten in the nonconservative form

$$\begin{aligned}
&\frac{\Gamma}{\rho c^2}\frac{\mathrm{d}p}{\mathrm{d}t} + (v, \partial_t u) + \operatorname{div} u = 0, \\
&(\rho h \Gamma)\left(\frac{\mathrm{d}u}{\mathrm{d}t} - v\left(v, \frac{\mathrm{d}u}{\mathrm{d}t}\right)\right) + (\partial_t p)v + \nabla p = \rho \mathcal{G}, \qquad \frac{\mathrm{d}S}{\mathrm{d}t} = 0.
\end{aligned} \tag{2.10}$$

System (2.10) being written in the quasilinear form (2.2) is already symmetric with

$$A_0 = \begin{pmatrix} \dfrac{\Gamma}{\rho c^2} & v^{\mathsf{T}} & 0 \\ v & \rho h \Gamma \mathcal{B} & 0 \\ 0 & 0 & 1 \end{pmatrix}, \qquad A_j = \begin{pmatrix} \dfrac{u^j}{\rho c^2} & e_j^{\mathsf{T}} & 0 \\ e_j & \rho h u^j \mathcal{B} & 0 \\ 0 & 0 & v^j \end{pmatrix}, \tag{2.11}$$

where $\mathcal{B} = I - v \otimes v$, $\quad e_j = (\delta_{1j}, \delta_{2j}, \delta_{3j})$. The matrix $A_0 > 0$ provided that inequalities (2.3) are satisfied together with (2.9).

Sketch of the proof of Theorem 2.1 (and its relativistic version).

1) *Linearization.* We linearize problem (2.5)–(2.7) about a given sufficiently smooth vector-function $(\widehat{U}(t,x), \hat{\varphi}(t,x'))$ (basic state).

2) *Passage to the "good unknown".* The linearized interior equations contain the term

$$\left\{L\left(\widehat{U}, \widehat{\Psi}\right)\Psi\right\}\frac{\partial_1(\widehat{U} + \check{U})}{\partial_1 \widehat{\Phi}}$$

which is a first-order operator applied to φ, where U and φ are now perturbations. Following Alinhac [1], we pass to the "good unknown"

$$\dot{U} := U - \frac{\Psi}{\partial_1 \widehat{\Phi}} \partial_1 \left(\widehat{U} + \check{U} \right)$$

in terms of which the differential operator for the linearized interior equations becomes a zero-order operator in φ.

3) *Basic a priori estimate.* Assuming that (2.8) is satisfied for the basic state and using carefully the linearized boundary conditions, we derive the basic a priori estimate [7]

$$\|\dot{U}\|_{L_2(\Omega_T)} + \|\varphi\|_{L_2(\partial\Omega_T)} \leq C \left\{ \|f\|_{L_2(\Omega_T)} + \|g\|_{H^1(\partial\Omega_T)} \right\},$$

where $\dot{U}|_{t=0} = 0$, $\varphi|_{t=0} = 0$ for $t < 0$, $\Omega_T = (-\infty, T] \times \mathbb{R}^3_+$, $\partial\Omega_T = (-\infty, T] \times \{x_1 = 0\} \times \mathbb{R}^2$, and f and g are source terms in the linearized interior equations and the linearized boundary conditions respectively (f and g vanish in the past).

4) *Tame estimate.* The crucial role in the proof of convergence of Nash-Moser iterations is played by a so-called tame estimate for solutions of the linearized problem in H^s for s large enough. This estimate is, roughly speaking, linear in high norms (that are multiplied by low norms) and we refer to [7] for more details.

5) *Nash-Moser iterations.* The realization of the Nash-Moser procedure is very close to that in [2] for 2D vortex sheets (the main difference is that we lose "one derivative from the front") and it is much simpler as that in [6] for current-vortex sheets.

3 Open problems

The general relativistic version of our fluid-vacuum problem is in fact an interface problem. Indeed, we have the vacuum Einstein equations for the metric g in the vacuum domain whereas in the fluid domain we have the relativistic Euler equations written with respect to the metric g whose components $g_{\alpha\beta}$ satisfy the Einstein equations for the fluid tensor $T^{\alpha\beta}$. Moreover, in addition to the boundary conditions (1.4) we have the smoothness conditions

$$[g_{\alpha\beta}] = 0, \quad [g_{\alpha\beta\gamma}] = 0 \quad \text{on } \Sigma(t)$$

(see [3] and references therein) for $g_{\alpha\beta}$ and their first-order derivatives $g_{\alpha\beta\gamma} = \partial_\gamma g_{\alpha\beta}$. As for shock waves in general relativity studied in [3], we can work in harmonic natural coordinates. Then the Einstein equations form a symmetric hyperbolic system for the vector W whose components are $g_{\alpha\beta}$ and $g_{\alpha\beta\gamma}$ (see [3] and references therein).

To cover general relativity we could try to follow a scheme proposed in [7]. Namely, in [7] we proposed to resolve the Euler equations by Nash-Moser iterations whereas at each Nash-Moser iteration step we find the metric from the Einstein equations by Picard iterations. It is quite possible that such a scheme of the proof of the existence of solutions to the nonlinear problem could work (however, we are not yet absolutely sure in success), but it is still unclear how to prove the uniqueness of a solution. The point is that to prove uniqueness we have to derive an a priori estimate for the linearized problem with all the lower-order terms. It means that we have to pass to the Alinhac's "good unknown" even in the linearized Einstein equations. That is the linear boundary conditions $[g_{\alpha\beta\gamma}] = 0$ after the passage to the "good unknown" $\dot{g}_{\alpha\beta}$ become

$$[\dot{g}_{\alpha\beta\gamma}] = -\varphi[\partial_1 \hat{g}_{\alpha\beta\gamma}],$$

where $\hat{g}_{\alpha\beta}$ is an "unperturbed" metric and $[\cdot]$ denotes a jump at the straightened interface $x_1 = 0$. The right-hand sides in the above boundary conditions can be nonzero and the fact that the interface symbol is not elliptic prevents obtaining an a priori estimate.

It is astonishing that for the general relativistic plasma-vacuum interface problem (see [9] for its special relativistic version) we do not have such a difficulty at all because there is a natural stabilizing physical mechanism. This mechanism is the magnetic field (!). Namely, under the assumption that the plasma and vacuum magnetic fields are non-zero and non-parallel to each other everywhere on the interface (analogous assumption appears for the magnetic fields on a current-vortex sheet [6]) the symbol associated to the interface is *elliptic* and, therefore, the lower-order terms like $\varphi[\partial_1 \hat{g}_{\alpha\beta\gamma}]$ are not dangerous anymore.

Here we just very briefly announce our recent results obtained in [9] for the linearized relativistic plasma-vacuum interface problem (in the setting of special relativity). First of all, unlike the non-relativistic version of the plasma-vacuum problem [8], we have to assume that the plasma expands into the vacuum. Actually, there appears a condition [9] on the interface speed which for the case of a bounded plasma domain should be interpreted in the above sense. If this condition is violated then the linearized problem is underdetermined. We show that even if this necessary condition is satisfied the planar interface can be still violently unstable. By using a suitable secondary symmetrization of the Maxwell equations in vacuum, we find a sufficient condition that precludes violent instabilities. Under this condition we derive a basic a priori estimate in the anisotropic weighted Sobolev space H^1_* for the variable coefficients linearized problem for nonplanar plasma-vacuum interfaces and prove the well-posedness of this problem (see [9] for more details).

In the basic a priori estimate obtained in [9] for the linearized relativistic plasma-vacuum problem we have a loss of derivatives from the source terms to the solution. Therefore, we again plan to prove the existence of solutions to the original nonlinear problem by a suitable Nash-Moser-type iteration scheme. We suppose that the scheme of the proof will be really similar to that in [2, 6, 7]. We do not see any principal difficulties in this direction and the corresponding work is already in progress. After that one can also cover the general relativistic case under the mentioned crucial assumption for the plasma and vacuum magnetic fields on the interface.

References

[1] S. Alinhac, Existence d'ondes de raréfaction pour des systèmes quasi-linéaires hyperboliques multidimensionnels. Commun. Partial Differential Equations 14 (1989), no. 2, 173–230.

[2] J. F. Coulombel and P. Secchi, Nonlinear compressible vortex sheets in two space dimensions. Ann. Sci. Ecole Norm. Sup. 41 (2008), no. 1, 85–139.

[3] H. Freistühler and M. Raoofi, Stability of perfect-fluid shock waves in special and general relativity. Classical Quantum Gravity 24 (2007), no. 17, 4439–4455.

[4] H. Lindblad, Well posedness for the motion of a compressible liquid with free surface boundary. Commun. Math. Phys. 260 (2005), no. 2, 319–392.

[5] A. Majda, The existence of multi-dimensional shock fronts. Mem. Amer. Math. Soc. 43 (1983), no. 281.

[6] Y. Trakhinin, The existence of current-vortex sheets in ideal compressible magnetohydrodynamics. Arch. Ration. Mech. Anal. 191 (2009), no. 2, 245–310.

[7] Y. Trakhinin, Local existence for the free boundary problem for non-relativistic and relativistic compressible Euler equations with a vacuum boundary condition. Commun. Pure and Appl. Math. 62 (2009), no. 11, 1551–1594.

[8] Y. Trakhinin, On the well-posedness of a linearized plasma-vacuum interface problem in ideal compressible MHD. J. Differ. Equ., To appear.

[9] Y. Trakhinin, Stability of relativistic plasma-vacuum interfaces. http://arxiv.org/abs/1006.1089

Convergence Rate to the Stationary Waves for Viscous Conservation Laws with Non-convexity on the Half Space*

Yoshihiro Ueda
Mathematical Institute, Tohoku University
Sendai 980-8578, Japan
E-mail: ueda@math.tohoku.ac.jp

Abstract

We study the initial-boundary value problem for scalar viscous conservation laws with non-convexity on the half line. Hashimoto-Matsumura [1] had already shown the asymptotic stability result toward the stationary solution. In this paper, we derive the corresponding convergence rate for the stationary solution constructed in [1]. Our proof is based on the time-spatial weighted L^2 energy method.

1 Introduction

In this paper, we consider the following scalar viscous conservation laws on the half line:

$$\begin{cases} u_t + f(u)_x = u_{xx}, & t > 0,\ x > 0, \\ u(t,0) = u_-, & t > 0, \\ \lim_{x\to\infty} u(t,x) = u_+, & t > 0, \\ u(x,0) = u_0(x), & x > 0, \end{cases} \tag{1.1}$$

where the flux $f = f(u)$ is a given smooth function of u satisfying $f(0) = 0$ and $u_\pm$ are given constants with $u_+ = 0$. In this problem, we assume that the initial function $u_0(x)$ satisfies $u_0(0) = u_-$ and $\lim_{x\to\infty} u_0(x) = u_+$ as the compatibility conditions. Throughout this paper, we impose the following non-convex condition on the flux $f(u)$:

$$f''(0) > 0, \qquad f(u) > 0 \quad \text{for} \quad u \in [u_-, 0). \tag{1.2}$$

*The author is partially supported Grant-in-Aid for Young Scientists (B) No. 21740111 from Japan Society for the Promotion of Science.

Then, it is known that the solution of the problem (1.1) converges to the corresponding stationary solution under the condition $u_- < 0$. Here the stationary solution $\phi(x)$ is defined by the solution of the stationary problem corresponding to (1.1):

$$\begin{cases} f(\phi)_x = \phi_{xx}, & x > 0, \\ \phi(0) = u_-, & \lim_{x\to\infty} \phi(x) = 0. \end{cases} \tag{1.3}$$

More precisely, Hashimoto-Matsumura [1] showed the following result.

Theorem 1.1 ([1]). *Suppose that* (1.2), $u_- < 0$ *and* $u_0 - \phi(x) \in H^1$. *Then there is a positive constant* ε_0 *such that, if* $\| u_0 - \phi\|_{H^1} \le \varepsilon_0$, *then the initial-boundary value problem* (1.1) *has a unique global solution* $u(x,t)$ *with* $u - \phi \in C([0,\infty); H^1)$. *Moreover, the solution satisfies the asymptotic relation:*

$$\lim_{t\to\infty} \sup_{x>0} |u(x,t) - \phi(x)| = 0. \tag{1.4}$$

Here $\phi(x)$ *is the stationary solution satisfying* (1.3).

We remark that Hashimoto-Matsumura [1] derived the asymptotic stability of the superposition of rarefaction wave and stationary solution. Furthermore, Hashimoto-Ueda-Kawashima [2] obtained the convergence rate of solutions constructed in [1] by assuming that the initial data belong to $H^1 \cap L^1$. The main purpose of the present paper is to obtain the convergence rate under the assumption $u_0 - \phi \in H^1_\alpha$ and the flux condition (1.2). Namely, we derive the following theorem.

Theorem 1.2 (Main Theorem). *In addition to the conditions in Theorem* 1.1, *we assume that* $f'(0) < 0$ *and* $u_0 - \phi \in H^1_\alpha$ *for* $\alpha > 0$. *Then there is a positive constant* ε_1 *such that, if* $\| u_0 - \phi\|_{H^1} \le \varepsilon_1$, *then the solution of the initial-boundary value problem* (1.1), *which is constructed in Theorem* 1.1, *satisfies the following quantitative estimates:*

$$\|(u-\phi)(t)\|_{H^1} \le C(1+t)^{-\alpha/2}, \tag{1.5}$$

where C *is a positive constant. Here* $\phi(x)$ *is the same as in Theorem* 1.1.

Remark 1.3. We can also obtain the exponential convergence rate by assuming that the initial perturbation $u_0 - \phi$ belongs to the exponentially weighted space $L^2_{\alpha,\mathrm{exp}} = \{u;\ e^{(\alpha/2)x_1}u \in L^2\}$. The proof of this result is similar to the derivation of Theorem 1.2, so that the details are omitted here. For its proof, we refer the readers to [7].

Notations. We denote by $L^2 = L^2(\mathbb{R}_+)$ the usual Lebesgue space over $\mathbb{R}_+$ with the norm $\|\cdot\|_{L^2}$, and $H^1 = H^1(\mathbb{R}_+)$ the corresponding first order Sobolev space with the norm $\|\cdot\|_{H^1}$. For $\alpha > 0$, $L^2_\alpha = L^2_\alpha(\mathbb{R}_+)$ denotes the polynomially weighted L^2 space with the norm

$$\|f\|_{L^2_\alpha} := \left(\int_0^\infty (1+x)^\alpha |f(x)|^2 dx\right)^{1/2}.$$

Similarly, we define the corresponding weighted Sobolev spaces $H^1_\alpha = H^1_\alpha(\mathbb{R}_+)$. For an interval I and a Banach space X, $C(I;X)$ denotes the space of continuous functions on I with values in X. Finally, letters C and c in this paper are positive generic constants which may vary from line to line.

2 Preliminaries

To complete this procedure, we first review the basic properties of the stationary solution $\phi(x)$. For the details, see [4, 5, 7].

Lemma 2.1. *Assume that* (1.2). *Then the stationary problem* (1.3) *with* $u_+ = 0$ *has a unique smooth solution* $\phi(x)$ *satisfying* $u_- < \phi(x) < 0$ *and* $\phi_x(x) > 0$ *for* $x > 0$. *Moreover, for the non-degenerate case* $f'(0) < 0$, *we have*

$$|\partial_x^k \phi(x)| \le Ce^{-cx}, \quad x \ge 0$$

for each nonnegative integer k.

We next give some preparations for a weight function which is useful for the weighted energy method. Since $f''(0) > 0$ by (1.2), there exist positive constants r and ν such that $f''(u) \ge \nu$ for $|u| \le r$. In this situation, we choose the weight function as

$$w(u) = f(u) + \delta g(u) \quad \text{for} \quad u \in [u_-, 0], \tag{2.1}$$

where $g(u)$ is defined by $g(u) = -u^{2m} + r^{2m}$, and δ and m are positive constants determined later. For the weight function (2.1), we obtain the following lemmas.

Lemma 2.2 (Hashimoto-Matsumura [1]). *Suppose that* $f(u)$ *satisfies* (1.2). *Let* $w(u)$ *be the weight function defined in* (2.1). *Then, for suitably small* $\delta > 0$ *and suitably large integer* m, *there exist positive constants* c *and* C *such that*

$$c \le w(u) \le C, \quad (f''w - fw'')(u) \ge c \tag{2.2}$$

for $u \in [u_-, 0]$.

Lemma 2.3 (Hashimoto-Ueda [3]). *Suppose that $f(u)$ satisfies (1.2) and $f'(0) < 0$. Let $w(u)$ be the weight function defined by (2.1). Then, for suitably large integer m, there exists a positive constant c such that*

$$(fw' - f'w)(u) \geq c \tag{2.3}$$

for $u \in [u_-, 0]$.

For the proof of Lemma 2.2 and 2.3, readers are referred to [1] and [3], respectively. We will use these lemmas in Section 3 to prove our main theorem.

3 Proof of Theorem 1.2

In the final section, we prove Theorem 1.2. Before proceeding the proof, we reformulate our problem. Let $\phi(x)$ be the stationary solution satisfying (1.3). As in the previous works, we introduce a new function

$$u(x,t) = \phi(x,t) + v(x,t). \tag{3.1}$$

Then, by using the stationary problem (1.3), we reformulate our problem (1.1) as follows:

$$\begin{cases} v_t + (f(\phi + v) - f(\phi))_x = v_{xx}, & x > 0,\ t > 0, \\ v(0,t) = 0, & t > 0, \\ v(x,0) = v_0(x), & x > 0, \end{cases} \tag{3.2}$$

where we put $v_0(x) := u_0(x) - \phi_0(x)$. It is the standard strategy for solving our stability problem to treat the problem (3.2). Indeed, for the problem (1.1) with convex function, Liu-Matsumura-Nishihara [4] showed the asymptotic stability result by using the reformulated problem (3.2). Furthermore, when Hashimoto-Matsumuta [1] derived Theorem 1.1, they introduced a new weight function denied in (2.1) and new function $\tilde{v}$ such that

$$v(x,t) = w(\phi(x,t))\tilde{v}(x,t). \tag{3.3}$$

Then, (3.2) is rewritten as

$$\begin{cases} (w\tilde{v})_t + \big(f(\phi + w\tilde{v}) - f(\phi)\big)_x = (w\tilde{v})_{xx}, & x > 0,\ t > 0, \\ \tilde{v}(0,t) = 0, & t > 0, \\ \tilde{v}(x,0) = \tilde{v}_0(x), & x > 0, \end{cases} \tag{3.4}$$

where we put $\tilde{v}_0(x) = v_0(x)/w(\phi_0(x))$. For this problem, we apply the time-spatial weighted energy method to obtain Theorem 1.2.

Before deriving the L^2 decay estimate for (3.4), we recall the uniform L^2 estimate for (3.2) which was established in [1] under the assumptions in Theorem 1.1:

$$\|v(t)\|_{H^1}^2 + \int_0^t (\|v_x(\tau)\|_{H^1}^2 + \|\sqrt{\phi_x} v(\tau)\|_{L^2}^2)\, d\tau \le C\|v_0\|_{H^1}^2. \tag{3.5}$$

This uniform estimate implies that the function $\tilde{v}(v,x)$ defined by (3.3) is also small in L^∞, provided that $\|v_0\|_{H^1}$ is suitably small. This fact will be used in showing the following L^2 decay estimate for (3.4).

Proof of Theorem 1.2. We multiply the equation in (3.4) by $\tilde{v}$. Then, by the basic computations, we have

$$\frac{1}{2}w(\tilde{v}^2)_t + \Big\{F - \frac{1}{2}w_x\tilde{v}^2 - w\tilde{v}\tilde{v}_x\Big\}_x + w\tilde{v}_x^2 - \frac{1}{2}w_{xx}\tilde{v}^2 + G = 0, \tag{3.6}$$

where

$$F = (f(\phi + w\tilde{v}) - f(\phi))\tilde{v} - \int_0^{\tilde{v}} f(\phi + w\eta) - f(\phi)\, d\eta,$$

$$G = \int_0^{\tilde{v}} \{f'(\phi + w\eta)(\phi_x + w_x\eta) - f'(\phi)\phi_x\}\, d\eta.$$

We now derive a simpler expression of the term F and G in (3.6). For the term F, by the straightforward computation, we have

$$\begin{aligned} F &= (f(\phi + w\tilde{v}) - f(\phi))\tilde{v} - \int_0^{\tilde{v}} f(\phi + w\eta) - f(\phi)\, d\eta, \\ &= (f'(\phi)w\tilde{v} + O(\tilde{v}^2))\tilde{v} - \int_0^{\tilde{v}} f'(\phi)w\eta + O(\eta^2)\, d\eta \\ &= \frac{1}{2}f'(\phi)w\tilde{v}^2 + O(|\tilde{v}|^3) \end{aligned} \tag{3.7}$$

On the other hand, a simple calculation gives

$$\begin{aligned} & f'(\phi + w\eta)(\phi_x + w_x\eta) - f'(\phi)\phi_x \\ &= (f'(\phi + w\eta) - f'(\phi))\phi_x + f'(\phi + w\eta)w_x\eta \\ &= (f''(\phi)w\eta + O(\eta^2))\phi_x + (f'(\phi) + O(\eta))w_x\eta \\ &= (f'(\phi)_x w + f'(\phi)w_x)\eta + O(\eta^2)\phi_x = (f'(\phi)w)_x\eta + O(\eta^2)\phi_x, \end{aligned}$$

where we have used the fact that $w_x = w'(\phi)\phi_x = O(1)\phi_x$. Therefore we have the expression of G as

$$G = \int_0^{\tilde{v}} \{(f'(\phi)w)_x\eta + O(\eta^2)\phi_x\}\, d\eta = \frac{1}{2}(f'(\phi)w)_x\tilde{v}^2 + O(|\tilde{v}|^3)\phi_x. \tag{3.8}$$

By using (3.7) and (3.8), we obtain

$$F - \frac{1}{2}w_x\tilde{v}^2 - w\tilde{v}\tilde{v}_x = \frac{1}{2}(f'w - fw')\tilde{v}^2 + O(|\tilde{v}||\tilde{v}_x|) + O(|\tilde{v}|^3),$$
$$-\frac{1}{2}w_{xx}\tilde{v}^2 + G = \frac{1}{2}(f''w - fw'')\phi_x\tilde{v}^2 + O(|\tilde{v}|^3)\phi_x,$$

where we use $\phi_x = f(\phi)$ obtained in (1.3). Consequently, substituting the above two equalities into (3.6), we arrive at the energy equality

$$\begin{aligned}\frac{1}{2}(w\tilde{v}^2)_t &+ \left\{\frac{1}{2}(f'w - fw')\tilde{v}^2 + O(|\tilde{v}||\tilde{v}_x|) + O(|\tilde{v}|^3)\right\}_x \\ &+ w\tilde{v}_x^2 + \frac{1}{2}(f''w - fw'')\phi_x\tilde{v}^2 + O(|\tilde{v}|^3)\phi_x = 0.\end{aligned} \tag{3.9}$$

Let be γ and β be arbitrary positive constants satisfying $0 \le \gamma, \beta \le \alpha$. Multiplying the equation (3.9) by $(1+t)^\gamma(1+x)^\beta$ and integrating the resultant equality with respect to t and x over $(0,t) \times \mathbb{R}_+$, this yields

$$\begin{aligned}(1+t)^\gamma\|\tilde{v}\|^2_{L^2_\beta} &+ \int_0^t (1+t)^\gamma\left(\|(\tilde{v}_x, \sqrt{\phi_x}\tilde{v})\|^2_{L^2_\beta} + \beta\|\tilde{v}\|^2_{L^2_{\beta-1}}\right) d\tau \\ &\le C\|\tilde{v}_0\|^2_{L^2_\beta} + \gamma C\int_0^t (1+t)^{\gamma-1}\|\tilde{v}\|^2_{L^2_\beta}\, d\tau + \beta C\int_0^t (1+t)^\gamma\|\tilde{v}_x\|^2_{L^2_{\beta-1}}\, d\tau \\ &\quad + CN(t)\int_0^t (1+t)^\gamma\left(\|\sqrt{\phi_x}\tilde{v}\|^2_{L^2_\beta} + \beta\|\tilde{v}\|^2_{L^2_{\beta-1}}\right) d\tau,\end{aligned}$$

where we define that $N(t) = \sup_{0\le\tau\le t}\|\tilde{v}(\tau)\|_{L^\infty}$. Here, we use the inequalities (2.2) and (2.3). Moreover, by employing (3.5), the third and forth terms on the right hand side of the above inequality are estimated. (For the detail, see [7, 8].) Namely, we obtain

$$\begin{aligned}(1+t)^\gamma\|\tilde{v}\|^2_{L^2_\beta} &+ \int_0^t (1+t)^\gamma\left(\|(\tilde{v}_x, \sqrt{\phi_x}\tilde{v})\|^2_{L^2_\beta} + \beta\|\tilde{v}\|^2_{L^2_{\beta-1}}\right) d\tau \\ &\le C\|\tilde{v}_0\|^2_{L^2_\beta} + \gamma C\int_0^t (1+t)^{\gamma-1}\|\tilde{v}\|^2_{L^2_\beta}\, d\tau,\end{aligned} \tag{3.10}$$

where C is a constant independent of γ and β. For the energy inequality (3.10), applying the induction argument with respect to γ and β, we can obtain $\|(u-\phi)(t)\|_{L^2} \le C(1+t)^{-\alpha/2}$. Furthermore, employing the same argument, we get the decay estimate of the v_x. Hence, we obtain the desired estimate (1.5) in Theorem 1.2. For the details, we refer the readers to [6, 7, 8]. This completes the proof of Theorem 1.2. □

References

[1] I. Hashimoto and A. Matsumura, Large time behavior of solutions to an initial boundary value problem on the half line for scalar viscous conservation law. Methods Appl. Anal. 14 (2007), 45–60.

[2] I. Hashimoto, Y. Ueda and S. Kawashima, Large time behavior of solutions to an initial boundary value problem on the half line for scalar viscous conservation law. To appear in Methods Appl. Anal.

[3] I. Hashimoto and Y. Ueda, Asymptotic behavior of solutions for damped wave equations with non-convex convection term on the half line. Submitted for publication.

[4] T. P. Liu, A. Matsumura and K. Nishihara, Behaviors of solutions for the Burgers equation with boundary corresponding to rarefaction waves. SIAM J. Math. Anal. 29 (1998), 293–308.

[5] T. P. Liu and K. Nishihara, Asymptotic behavior for scalar viscous conservation laws with boundary effect. J. Differential Equations 133 (1997), 296–320.

[6] M. Nishikawa, Convergence rate to the traveling wave for viscous conservation laws. Funkcial. Ekvac. 41 (1998), 107–132.

[7] Y. Ueda, Asymptotic stability of stationary waves for damped wave equations with a nonlinear convection term. Adv. Math. Sci. Appl. 18 (2008), 329–343.

[8] Y. Ueda, Stability of travelling wave solutions to a semilinear hyperbolic system with relaxation. Math. Meth. Appl. Sci. 32 (2009), 419–434.

Numerical Simulation of Shock-bubble Interactions Using Conservative Front Tracking Method*

M. A. Ullah, Wenbin Gao, De-Kang Mao
Department of Mathematics, Shanghai University
Shanghai 200444, China
E-mail: dkmao@staff.shu.edu.cn

Abstract

In this paper, we present the numerical simulation of two shock-bubble interaction problems using the conservative front-tracking method developed in [D. Mao, Towards front-tracking based on conservation in two space dimensions, tracking discontinuities in capturing fashion, J. Comput. Phys. 226 (2007), 1550–1588] and modified recently, and compare the results with the experiment presented in [J. F. Hass and B. Sturtevant, Interaction of weak shock waves with cylindrical and spherical gas inhomogeneities, J. Fluid Mech., 181 (1987), 41–76]. Our simulation well captured the early phases of the interaction with the numerical results in good agreement with the experiment. Moreover, our simulation also gave sound results for the largely deformed interfaces in the late phases of the interaction.

1 Introduction

In [3], Hass and Sturtevant presented an experimental study of interaction of a weak shock with cylindrical bubbles filled with either helium or $R22$ gas. Since then these two shock-bubble interaction problems have been widely used as benchmark problems to test efficiency of numerical methods, see [1, 5, 6, 10, 11] and the references cited therein. Quite a lot of numerical methods can well capture the early phases of the interaction process with the numerical results generally in good agreement with the experiment. However, it seems that few of them gives satisfied result for the largely deformed interface in the late phases of the interaction

*The authors are supported by NSFC (No. 10971132) .

precess. This is because the Euler system, the model used for the simulations, is inviscid; therefore, the numerical diffusions of the methods used for simulation will produce numerical artifacts and then severely pollute the interfacial structures

For many years, Mao and his coworkers have been developing a conservative front-tracking method for the 2D Euler system, see [7] and [8]. Like all the other front-tacking methods, say [2, 11], it is almost free of numerical dissipation. However, the following two features distinguish it from the other front-tracking methods: 1) The interfaces are tracked by enforcing the conservation properties of the flow rather than by the propagation speed obtained by solving Riemann problem on the interfaces. 2) The movement of the interfaces is locally described by 1D PDE's derived from the Euler system, and the tracking is realized by numerically solving these 1D PDE's on the Cartesian sub-grid. Designed in such a way, Mao's front-tracking method is much simpler than the other front-tracking methods, it runs on Cartesian grid and uses no adaptive grid.

Recently, we designed a numerical surface tension for this conservative front-tracking method to stabilize the tracked interfaces, and then generalized the method to multi-component flows. In this paper we present our numerical simulation of Hass and Sturtevant's shock-bubble interaction problems with this generalized method. It will be seen that our simulation well captured the early phases of the interactions as the other methods did. Moreover, it also gave sound results for the largely deformed interfaces in the late phases of the interaction.

The paper is organized in the following fashion: Section 1 is the introduction, Section 2 briefly describes the conservative front-tracking method, Section 3 presents the numerical simulation, and Section 4 is the conclusion.

2 Conservative front-tracking method

In this section, we are going to conceptually describe the conservative front-tracking method for tracking an interface, and the details of the method can be found in the above cited papers.

1) If locally we view a segment of an interface as either $y = s(x,t)$ or $x = w(y,t)$, then the motion of the segment can be locally described by 1D PDE's with either s as the unknown and x and t as the independent variables or with w as the unknown and y and t as the independent variables. These 1D PDE's can be easily obtained by integrating the 2D Euler system in either y- or x-direction and they are in conservation forms with s or w appearing in some nonlinear functions related to the mass, momentum, and energy. Mathematically, these 1D PDE's are

equivalent to the Hugoniot conditions satisfied by the mass, momentum, and energy across the interface. Our conservative front-tracking is then accomplished by numerically solving these 1D PDE's on either x- or y-sub-grid.

2) The numerical solution is defined on a 2D Cartesian grid. The interface is described by a list of grid cells that are crossed by the interface. The cells crossed by the interface are called discontinuity cells and the cells away from the interface are called smooth cells. In a smooth cell, the numerical solution is a cell-average approximation to the true solution. However, in a discontinuity cell the numerical solution has three cell-average approximations, one is the cell-average approximation in the ordinary sense, called ordinary cell-average, and the other twos are the cell-average approximations to the true solution on the two sides of the interface, called the left and right cell-averages. Data-structured in this way, no irregular grid cells are used in the numerical solution. The left and right cell-averages in a discontinuity cell are "ghost" states, and our conservative front-tracking method will maintain the conservation of cell-averages in smooth cells and ordinary cell-averages in discontinuity cells.

3) A finite-volume scheme is used as the underlying scheme. In the current algorithm, the 5th-order WENO scheme developed in [4] is adopted as the underlying scheme. Cell-averages in all smooth cells and left and right cell-averages in all discontinuity cells are computed with this underlying scheme. Whenever stencils of the underlying scheme go across the interface and data on the other side of the interface are requested in the computation, extrapolation data from the present side are substituted.

4) The ordinary cell-average in a discontinuity cell is computed by numerically solving the 1D conservative PDE's obtained in 1) with conservative discretization. In the current algorithm, they are solved by a Lax-Wendroff-like scheme with a properly designed numerical diffusion, which acts on the interface (not on the flow) to stabilize it and is thus called numerical surface tension. Whether the 1D PDE's in the x-direction (with w as the unknown) or in the y-direction (with s as the unknown) are solved is determined based on a geometric observation on the interface in the vicinity of the discontinuity cell. The numerical fluxes in solving the 1D PDE's are evaluated using the numerical fluxes used in computing the nearby smooth solution on the two sides of the interface, so that the fluxes between the smooth region and the discontinuity cell are balanced, and thus the local conservation is maintained. Once the ordinary cell-averages are computed, the interface segment in the discontinuity cell is recovered by enforcing the conservation of mass in the cell.

5) When tracking an interface, waves associated with other charac-

teristic fields should propagate across the interface. This is accomplished by replacing the velocity and pressure in the left and right cell-averages in each discontinuity cell with uniform velocity and pressure at the end of a time step. The uniform velocity is computed as the quotient of the ordinary momentums and mass. The uniform pressure is computed as a linear combination of the left and right pressures, and the coefficients are so chosen that the ordinary internal energy in the cell is conserved. Thus, we complete a time step of computation.

The 2D conservative front-tracking method for single-component flows has been validated by numerical experiments in [7] and [8] and showed a good quality in tracking shocks and contact discontinuities.

3 Numerical simulation

As described in [3], the set-up of the experiment is as follows: A cylindrical bubble with 5.0 cm in diameter is initially placed in the middle of a long square tube with 8.9 cm in width and filled with air. The bubble is filled with either helium (lighter than air) or R22 gas (heavier than air). A planar shock, with Mach number of 1.22, propagates through air from the right to the left, collides with the bubble and causes the latter to deform.

The solution region is set to be $(-L \text{ cm}, L \text{ cm}) \times (-4.45 \text{ cm}, 4.45 \text{ cm})$ with reflecting boundary conditions on the top and bottom and outflow boundary conditions on left and right ends of the tube. The number L is so chosen that the left-moving transmitted shock will not reach the the left end at the end of the simulation. We use a grid with 160 cells in the vertical direction for the simulation. Here we compute the flow in the whole region, rather than compute it only in a half region and use the symmetry to find the other half as many other people do.

1. *Helium bubble case.* In this case we choose $L = 62.3$. Figure 3.1 presents our numerical results compared with the experiment at times 52 μs, 72 μs, 102 μs, 674 μs, and 983 μs. It is seen from the frames at 52 μs, 72 μs, and 102 μs that our simulation well captured the early phases of the interaction process; the patterns of the reflecting, refracting, and diffracting waves in the numerical solution are in good agreement with the experiment. At the latest two times, especially at 983 μs, our numerical simulation still gave sound pictures of the deformed interface. Though at these moments, the interface in the experiment became severely smeared because of the mass and thermal diffusion and viscosity across it, the agreement of our numerical result with the experiment can still be vaguely seen. We note that there were quite a lot of numerical simulations of this problem, but few of them gave good result at 983 μs.

2. *R22 bubble case.* In this case we choose $L = 89$. Figure 3.2

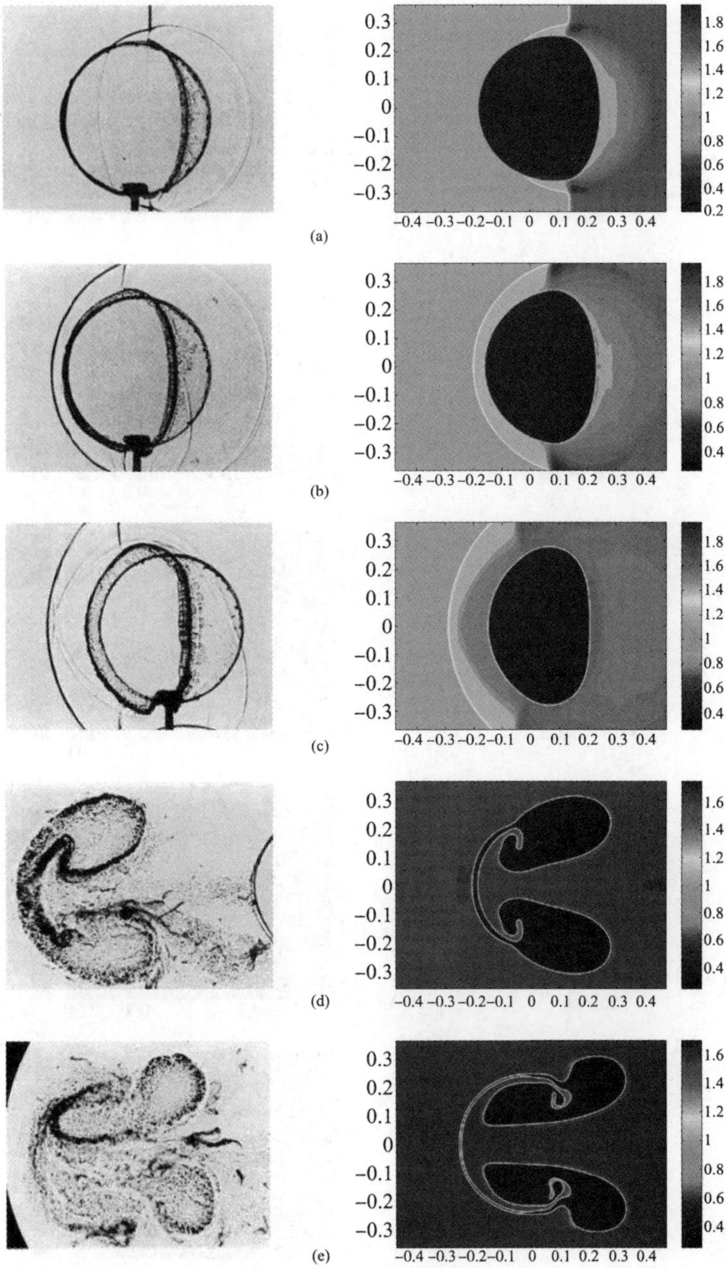

Figure 3.1 Comparison between the numerical and experimental results at times: (a) 52 μs, (b) 72 μs, (c) 102 μs, (d) 674 μs, and (e) 983 μs. Left column is for Hass and Sturtevant experimental shadow-photographs and right column is for our numerical results.

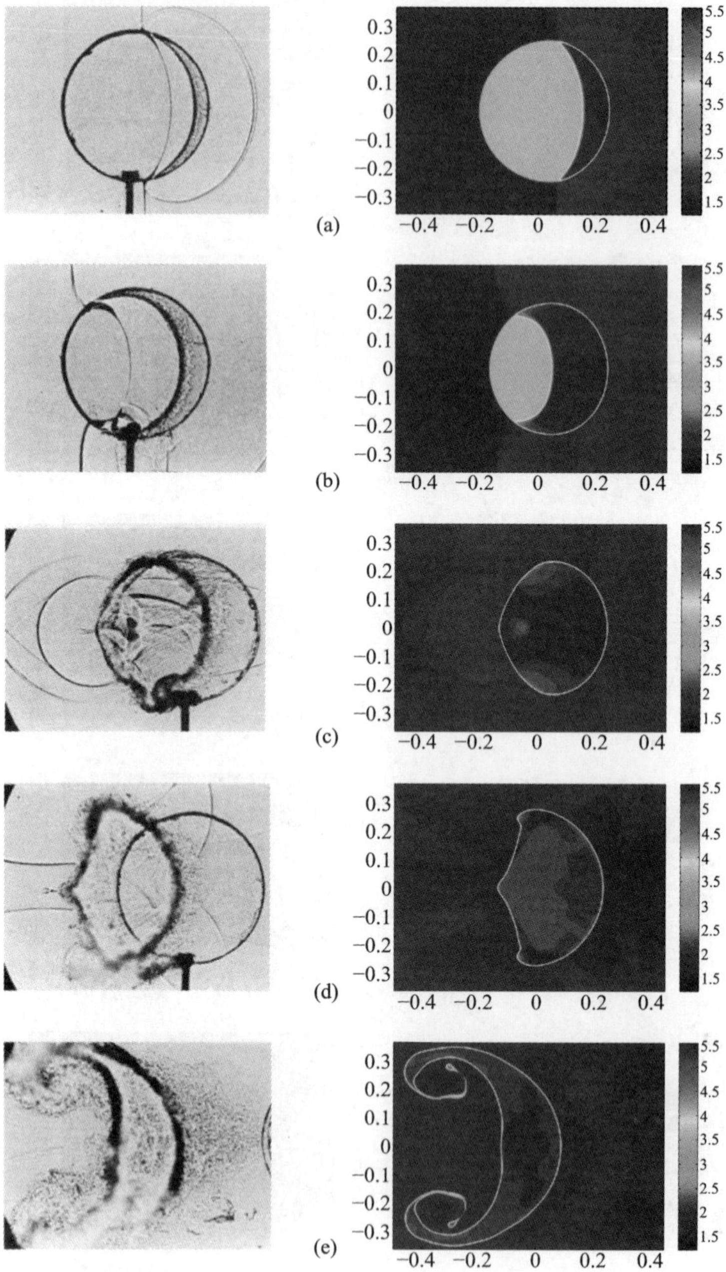

Figure 3.2 Comparison between the numerical and experimental results for air-R_{22} at times: (a) 55 μs, (b) 115 μs, (c) 247 μs, (d) 417 μs, and (e) 1020 μs. Left column is for the experimental shadow-photographs and right column is for numerical results.

presents our numerical results compared with the experiment at times 55 μs, 115 μs, 247 μs, 417 μs, and 1020 μs. It is also seen from the pictures at 55 μs, 115 μs, 247 μs, and 417 μs that our simulation well captured the early phases of the interaction process; the patterns of the reflecting, refracting, and diffracting waves in the numerical solution are in good agreement with the experiment. At the latest time 1020 μs, our numerical simulation still gave a sound picture of the deformed interface. Though at this moment, the interface in the experiment was also severely smeared because of the mass and thermal diffusion and viscosity across it, the agreement of our numerical result with the experiment can still be vaguely seen. We also note that there were quite a lot of numerical simulations of this problem, but only a few of them gave good results at 1020 μs.

4 Conclusion

We have used our conservative front-tracking method to numerically simulate the shock-bubble interaction problems. Our simulations well captured the early phases of the interaction processes in the two examples with the wave patterns in good agreement with the experiment. Our simulations also gave sound results for the later phases of the interactions in the two examples with vague agreement with the experiments.

References

[1] S. Galera, P. H. Maire and J. Breil, A two-dimensional unstructured cell-centered multi-material ALE scheme using VOF interface reconstruction. J. Comput. Phys., 229 (2010), 5755–5787.

[2] J. Glimm, X. L. Li, Y. J. Liu, Z. L. Xu and N. Zhao, Conservative front-tracking with improve accuracy. SIAM. J. Numer. Anal., 41(5) (2003), 1926–1947.

[3] J. F. Hass and B. Sturtevant, Interaction of weak shock waves with cylindrical and spherical gas inhomogeneities. J. Fluid Mech., 181 (1987), 41–76, Cambridge University Press.

[4] G. S. Jiang and C. W. Shu, Efficient Implementation of Weighted ENO Schemes. J. Comput. Phys., 126 (1996), 202–228.

[5] S. Kokh and F. Lagoutiere, An anti-diffusive numerical scheme for the simulation of interfaces between compressible fluids by means of a five-equation model. J. Comput. Phys., 229 (2010), 2773–2809.

[6] J. J. Kreeft and B. Koren, A new formulation of Kapila's five-equation model for compressible two-fluid flow, and its numerical treatment. J. Comput. Phys., 229 (2010), 6220–6242.

[7] D. K. Mao, Towards front-tracking based on conservation in two space dimensions II, tracking discontinuities in capturing fashion. J. Comput. Phys., 226 (2007), 1550–1588.

[8] D. K. Mao, Towards front tracking based on conservation in two space dimensions. SIAM. J. Sci. Comput., 22(1) (2000), 113–151.

[9] J. J. Quirk and S. Karni, On the dynamics of shock-bubble interaction. J. Fluid Mech., 318 (1996), 129–163, Cambridge University Press.

[10] K. M. Shyue, A wave-propagation based volume tracking method for compressible multicomponent flow in two space dimensions. J. Comput. Phys., 215 (2006), 219–244.

[11] H. Terashima and G. Tryggvason, A front-tracking/ghost-fluid method for fluid interfaces in compressible flows. J. Comput. Phys., 228 (2009), 4012–4037.

Asymptotic Stability of Viscous Contact Wave for the 1D Radiation Hydrodynamics System*

Jing Wang
Department of Mathematics, Shanghai Normal University
Shanghai 200433, China
E-mail: franklyme53@gmail.com

Feng Xie
Department of Mathematics, Shanghai Jiao Tong University
Shanghai 200240, China
E-mail: tzxief@sjtu.edu.cn

Abstract

This paper is concerned with the large time behavior of solutions to a radiating gas model, which is represented mathematically as a Cauchy problem for a one-dimensional hyperbolic-elliptic coupled system, with suitably given far field states. Suppose the corresponding Riemann problem for the Euler system admits a contact discontinuity wave, then we can construct a "viscous contact wave" for such a hyperbolic-elliptic system. Based on the energy methods and the ellipticity of the radiation flux equation, we prove that the "viscous contact wave" is asymptotically stable provided the strength of contact discontinuity and the perturbation of the initial data are suitably small.

1 Introduction

The fundamental system of equations in Lagrange coordinates, describing the one-dimensional motion of a radiating gas, can be written in the

*The authors are partly supported by Science and Technology Commission of Shanghai Municipality (09PJ1408200), NSFC (No.10926163, No.10801016), and Shanghai Jiao Tong University ChenXing Project.

following form (see [14, 15, 16, 17, 21]):

$$\begin{cases} v_t - u_x = 0, \\ u_t + P_x = 0, \\ (e + \frac{u^2}{2})_t + (Pu)_x + q_x = 0, \\ -(\frac{q_x}{v})_x + avq + b(\theta^4)_x = 0, \end{cases} \tag{1.1}$$

where $x \in \mathbb{R}, t > 0$, and a and b are two positive constants related to the absorption coefficient and the Stefan-Boltzmann constant. The unknown is (v, u, θ, e, P, q) with $v > 0$ the specific volume, u the velocity, $\theta > 0$ the absolute temperature, $e > 0$ the inertial energy, P the pressure, and q the radiation flux. Here we focus on the ideal polytropic gas, that is, the state equations take the form

$$P = R\frac{\theta}{v} = Av^{-\gamma}e^{\frac{\gamma-1}{R}s}, e = \frac{R}{\gamma - 1}\theta,$$

where s is the entropy, $\gamma > 1$ is the adiabatic exponent, and A, R are two positive constants. We impose the following initial data and far fields states

$$\begin{cases} (v, u, \theta)(x, 0) = (v_0, u_0, \theta_0), & x \in \mathbb{R}, \\ (v, u, \theta, q)(\pm\infty, t) = (v_\pm, u_\pm, \theta_\pm, 0), & t > 0. \end{cases} \tag{1.2}$$

We are concerned with the global existence of solutions of the initial value problem (1.1)–(1.2) and their large time behavior in relation with the spatial asymptotic states $(v_\pm, u_\pm, \theta_\pm)$.

If the radiation effect is neglected in the fluids, the system (1.1) is reduced to the following compressible Euler system

$$\begin{cases} v_t - u_x = 0, \\ u_t + P_x = 0, \\ (e + \frac{u^2}{2})_t + (Pu)_x = 0, \end{cases} \tag{1.3}$$

with the asymptotic states $(v_\pm, u_\pm, \theta_\pm)$ being the Riemann initial data

$$(v, u, \theta)(x, 0) = \begin{cases} (v_-, u_-, \theta_-), & x < 0, \\ (v_+, u_+, \theta_+), & x > 0. \end{cases} \tag{1.4}$$

It is known that the contact discontinuity solution of the Riemann problem (1.3)–(1.4) takes the form [18, 19]

$$(\bar{V}, \bar{U}, \bar{\Theta})(x, t) = \begin{cases} (v_-, u_-, \theta_-), & x < 0, \quad t > 0, \\ (v_+, u_+, \theta_+), & x > 0, \quad t > 0, \end{cases} \tag{1.5}$$

provided that

$$u_- = u_+, \quad P_- \triangleq \frac{R\theta_-}{v_-} = \frac{R\theta_+}{v_+} \triangleq P_+. \tag{1.6}$$

In the setting of the hyperbolic-elliptic coupled system (1.1), the corresponding wave to the contact discontinuity becomes smooth and behaves like a diffusion wave due to the radiation effect. We call this wave "viscous contact wave". Now, we construct the "viscous contact wave" (V, U, Θ, Q) in this paper as follows.

Due to (1.6), the pressure of "viscous contact wave" (V, U, Θ, Q) is expected to be almost constant, that is,

$$\bar{P} \equiv \frac{R\Theta}{V} \approx P_+, \tag{1.7}$$

and the radiation flux Q is expected to act as a diffusion term

$$Q \approx -\frac{b}{aV}\left(\Theta^4\right)_x. \tag{1.8}$$

Then the leading order of the energy equation $(1.1)_3$ is

$$\frac{R}{\gamma-1}\Theta_t + P_+ U_x = \left(\frac{4b}{a}\frac{\Theta^3\Theta_x}{V}\right)_x. \tag{1.9}$$

Then using $(1.1)_1$ and (1.7) we have the following nonlinear diffusion equation:

$$\Theta_t = \frac{\gamma-1}{R\gamma}\left(\frac{4bP_+}{aR}\Theta^2\Theta_x\right)_x, \quad \Theta(\pm\infty, t) = \theta_\pm, \tag{1.10}$$

which has a unique self-similarity solution $\Theta(x,t) = \Theta(\xi), \xi = \frac{x}{\sqrt{1+t}}$ due to [1, 4, 20]. Furthermore, $\Theta(\xi)$ is a monotone function, increasing if $\theta_+ > \theta_-$ and decreasing if $\theta_+ < \theta_-$. Let $\delta = |\theta_+ - \theta_-|$, then $\Theta(\xi)$ satisfies

$$(1+t)^{l/2}|\partial_x^l \Theta| + |\Theta - \theta_\pm| \le c_1 \delta e^{-c_2 x^2/(1+t)}, \quad |x| \to \infty, \quad l \ge 1, \tag{1.11}$$

where c_1, c_2 are two positive constants depending only on θ_+ and δ. Once Θ is determined, then the "viscous contact wave" profile $(V, U, \Theta, Q)(x,t)$ is then defined as

$$V = \frac{R}{P_+}\Theta, \quad U = u_- + \frac{(\gamma-1)}{R\gamma}\frac{4b}{a}\Theta^2\Theta_x, \quad \Theta = \Theta, \quad Q = -\frac{4bP_+}{aR}\Theta^2\Theta_x. \tag{1.12}$$

Our main goal is to show that the solution to (1.1)–(1.2) tends asymptotically to the "viscous contact wave" as time tends to $+\infty$, provided that the perturbation of initial data and strength of corresponding contact wave are small.

We now recall some related topics in this aspect and make some remarks on the analysis in this paper. It is known that there are many interesting results on the asymptotic behavior of solutions to the Navier-Stokes system, which are characterized by those of the solutions of the corresponding Riemann problems for the Euler system. Here, we only mention the contact discontinuity wave case for the Euler system (see [3, 5, 6, 25] and references therein). In [3, 5], the authors studied the asymptotic behavior of solutions to the Navier-Stokes system by the viscous contact wave constructed there. Furthermore, the decay estimates are given by the anti-derivative methods in [5]. In our case, there are no viscosity and heat-conductivity terms but an additional term q_x in the energy equation $(1.1)_3$, so the system is of less dissipation. Thus, how to construct an ansatz and how to derive the suitable a priori estimates will be worth investigating. Since the higher order derivatives of q are expected to decay faster than q itself, thus, the approximation (1.8) is proposed here. This is different from the construction of the "viscous contact wave" in [3, 5]. Then, in order to show the stability of our "viscous contact wave" defined in (1.12), we need to prove the global existence of solutions to (1.1)–(1.2) near the "viscous contact wave".

We should remark that for the study of the non-equilibrium radiation hydrodynamics, most papers are concerned with the "baby model" which is the third order approximation of (1.1) (see [2, 8, 11, 13, 22] and references therein). Recently, many mathematicians have started to study the hyperbolic-elliptic coupled system (1.1) [9, 10, 12, 14, 15]. In [9, 10] Kawashima, Nikkuni, and Nishibata established the global existence of classical solutions for the initial value problem, which is the small perturbation problem of constant states. And in [14] the authors proved the global existence of shock profile for (1.1), and then Lattanzio, Mascia, and Serre [12] obtained the global existence of shock profile for a general hyperbolic-elliptic coupled system. In [15], Lin, Coulombel, and Goudon studied the stability of shock profile obtained in [14]. But the results of the stability of "viscous contact discontinuity" for the hyperbolic-elliptic coupled system (1.1) are much more limited, which is also one motivation of this paper.

In addition, we should remark that the methods in this paper can also be used to deal with singular limit problem considered in [23]. Moreover, when the strength of the contact discontinuity wave is small, we can obtain the global convergence to the contact discontinuity wave as the Bouguer number tends to infinity, and the convergence rates are the same as those in [23].

The remainder part of this paper is organized as the follows:

1. Formulation of the problem and a priori estimates;

2. Main theorem and its proof.

2 Reformulation of the problem and a priori estimates

To establish the global existence of solutions to the system (1.1) with the asymptotic behavior (3.1), it suffices to show that there exists an exact solution to (1.1) in a neighborhood of the approximate solution (V, U, Θ, Q). Suppose (v, u, θ, q) is the exact solution to (1.1) with the initial data (v_0, u_0, θ_0). We decompose the solution as

$$\phi = v - V, \quad \psi = u - U, \quad \zeta = \theta - \Theta, \quad \omega = q - Q. \tag{2.1}$$

Thus, we obtain

$$\begin{cases} \phi_t - \psi_x = 0, \\ \psi_t + \left(\dfrac{R\zeta - P_+\phi}{v}\right)_x = F, \\ \dfrac{R}{\gamma - 1}\zeta_t + Pu_x - P_+U_x = -\omega_x, \\ -\left(\dfrac{\zeta}{\theta}\right)_x = -\dfrac{\zeta\Theta_x}{\theta\Theta} - \dfrac{\Theta}{4b\theta^5}\left(\dfrac{q_x}{v}\right)_x + \dfrac{a\Theta}{b}\left(\dfrac{vq}{4\theta^5} - \dfrac{VQ}{4\Theta^5}\right), \\ (\phi, \psi, \zeta, \omega)(\pm\infty, t) = 0, \\ (\phi, \psi, \zeta)(x, 0) = (\phi_0, \psi_0, \zeta_0), \end{cases} \tag{2.2}$$

where

$$F = -U_t. \tag{2.3}$$

The local existence of the solution to (2.2) is similar as in [7, 15], which relies on the linear iteration scheme and fixed point theorem. To prove the main theorem, it suffices for us to close the following a priori estimates. For any time T

$$N(T) = \sup_{0 \le t \le T} \|(\phi, \psi, \zeta, \omega, \omega_x)(\cdot, t)\|_2 \le \varepsilon_0, \tag{2.4}$$

$$M^2(T) = \int_0^t (\|(\phi_x, \psi_x, \zeta_x)(\cdot, s)\|_1^2 + \|\omega(\cdot, s)\|_3^2) ds \le \varepsilon_0^2, \tag{2.5}$$

where $\varepsilon_0 < 1$ is a small number independent of T, which is to be determined later. Below, we will use the following function space:

$$\begin{aligned} E = &\Big\{(\phi, \psi, \zeta, \omega) | (\phi, \psi, \zeta, \omega, \omega_x) \in L^\infty(0, T; H^2(\mathbb{R})), \\ &(\phi_x, \psi_x, \zeta_x) \in L^2(0, T; H^1(\mathbb{R})), \omega \in L^2(0, T; H^3(\mathbb{R}))\Big\}. \end{aligned}$$

By the methods used in [24], we have the following a priori estimates.

Proposition 2.1. *For any $T > 0$ and $(\phi, \psi, \zeta, \omega) \in E$ satisfying* (2.2) *with suitably small ε_0, δ_0, we have for $0 < \delta < \delta_0$ and $t \in [0, T]$ that*

$$\sup_{0\le t\le T} \|(\phi, \psi, \zeta, \omega, \omega_x)(\cdot, t)\|_2^2 + \int_0^t (\|\omega\|_3^2 + \|\phi_x\|_1^2 + \|\psi_x\|_1^2 + \|\zeta_x\|_1^2) ds \\ \le C\|(\phi_0, \psi_0, \zeta_0)\|_2^2 + C\delta. \tag{2.6}$$

3 Main theorem and its proof

Theorem 3.1. *For any given (v_-, u_-, θ_-), suppose that (v_+, u_+, θ_+) satisfies* (1.6). *Let (V, U, Θ, Q) be the "viscous contact wave" defined in* (1.12) *with strength $\delta = |\theta_+ - \theta_-|$. Then there exist suitably small positive constants δ_0 and σ_0, such that if $0 < \delta < \delta_0$ and the initial data (v_0, u_0, θ_0) satisfies*

$$\|(v_0(\cdot) - V(\cdot, 0), u_0(\cdot) - U(\cdot, 0), \theta_0(\cdot) - \Theta(\cdot, 0))\|_{H^2} \le \sigma_0,$$

then the Cauchy problem (1.1)–(1.2) *admits a unique global solution (v, u, θ, q) satisfying $(v - V, u - U, \theta - \Theta, q - Q) \in E$. Moreover,*

$$\lim_{t\to\infty} \sup_{x\in\mathbb{R}} |(v - V, u - u_-, \theta - \Theta, q)(x, t)| = 0. \tag{3.1}$$

Proof. Once Proposition 2.1 is proved, we can choose the initial perturbation $\|(\phi_0, \psi_0, \zeta_0)\|_2$ and the strength of the contact discontinuity δ suitably small. Then we can close the uniform a priori estimates (2.4) and (2.5). In this way, we can extend the local solution to $T = \infty$.

Moreover, the estimate (2.6) and (2.2) imply that

$$\int_0^\infty \left(\|(\phi_x, \psi_x, \zeta_x, \omega_x)(t)\|^2 + \left|\frac{d}{dt}\|(\phi_x, \psi_x, \zeta_x, \omega_x)(t)\|^2\right| \right) dt < \infty,$$

which as well as (2.6) and the Sobolev inequality lead to the following asymptotic behavior of solutions:

$$\lim_{t\to\infty} \sup_{x\in\mathbb{R}} |(v - V, u - U, \theta - \Theta, q - Q)(x, t)| = 0. \tag{3.2}$$

From (3.2), (1.11), and (1.12), we have (3.1) easily. Thus we finish the proof of the main theorem.

References

[1] D. G. Aronson, The porous media equations. In: Nonlinear Diffusion Problem. Lecture Notes in Math., Vol. 1224, A. Fasano and M. Primicerio eds., Springer-Verlag, Berlin, 1986.

[2] W. Gao and C. J. Zhu, Asymptotic decay toward the planar rarefaction waves for a model system of the radiating gas in two dimensions. Math. Models Methods Appl. Sci. 18 (2008), no. 4, 511–541.

[3] F. M. Huang, J. Li and A. Matsumura, Asymptotic stability of combination of viscous contact wave with rarefaction waves for one-dimensional compressible Navier-Stokes system. Arch. Rational Mech. Anal. 197 (2010) 89–116.

[4] F. M. Huang and R. H. Pan, Convergence rate for compressible Euler equations with damping and vacuum. Arch. Ration. Mech. Anal. 166 (2003), no. 4, 359–376.

[5] F. M. Huang, A. Matsumura and Z. P. Xin, Stability of contact discontinuities for the 1-D compressible Navier-Stokes equations. Arch. Rational. Mech. Anal. 179 (2006), no. 1, 55–77.

[6] F. M. Huang, Z. P. Xin and T. Yang, Contact discontinuity with general perturbations for gas motions. Adv. Math. 219 (2008), no. 4, 1246–1297.

[7] S. Kawashima and A. Matsumura, Asymptotic stability of traveling wave solutions of systems for one-dimensional gas motion. Comm. Math. Phys. 101 (1985), no. 1, 97–127.

[8] S. Kawashima and S. Nishibata, Shock waves for a model system of the radiating gas. SIAM J. Math. Anal. 30 (1999), no. 1, 95–117.

[9] S. Kawashima, Y. Nikkuni and S. Nishibata, The initial value problem for hyperbolic-elliptic coupled systems and applications to radiation hydrodynamics. Analysis of systems of conservation laws, Aachen (1997), 87–127.

[10] S. Kawashima, Y. Nikkuni and S. Nishibata, Large-time behavior of solutions to hyperbolic-elliptic coupled systems. Arch. Ration. Mech. Anal. 170 (2003), no. 4, 297–329.

[11] S. Kawashima and Y. Tanaka, Stability of rarefaction waves for a model system of a radiating gas. Kyushu J. Math. 58 (2004), no. 2, 211–250.

[12] C. Lattanzio, C. Mascia and D. Serre, Shock waves for radiative hyperbolic-elliptic systems. Indiana Univ. Math. J. 56 (2007), no. 5, 2601–2640.

[13] C. Lattanzio and P. Marcati, Global well-posedness and relaxation limits of a model for radiating gas. J. Differential Equations 190 (2003), no. 2, 439–465.

[14] C. J. Lin, J. F. Coulombel and T. Goudon, Shock profiles for non-equilibrium radiating gases. Phys. D 218 (2006), no. 1, 83–94.

[15] C. J. Lin, J. F. Coulombel and T. Goudon, Asymptotic stability of shock profiles in radiative hydrodynamics. C. R. Math. Acad. Sci. Paris 345 (2007), no. 11, 625–628.

[16] D. Mihalas and B. Mihalas, Foundation of Radiation Hydrodynamics, Oxford University Press, 1984.

[17] G. Pamraning, The Equations of Radiation Hydrodynamics, Pergamon Press, 1973.

[18] D. Serre, Systems of Conservation Laws. 1. Cambridge University Press, 1999.

[19] J. Smoller, Shock Waves and Reaction-Diffusion Equations. Springer-Verlag, 1994

[20] C. J. Van Duyn and L. A. Peletier, A class of similarity solutions of the nonlinear diffusion equation. Nonlinear Anal. 1 (1976/77), no. 3, 223–233.

[21] W. Vincenti and C. Kruger, Introduction to Physical Gas Dynamics, Wiley, New York, 1965.

[22] W. K. Wang and W. J. Wang, The pointwise estimates of solutions for a model system of the radiating gas in multi-dimensions. Nonlinear Anal. 71 (2009), no. 3–4, 1180–1195.

[23] J. Wang and F. Xie, Singular limit to strong contact discontinuity for a 1D compressible radiation hydrodynamics model (2010). Submitted.

[24] J. Wang and F. Xie, Asymptotic stability of viscous contact wave for the 1D radiation hydrodynamics system (2010). Submitted.

[25] Z. P. Xin, On nonlinear stability of contact discontinuities. Hyperbolic problems: theory, numerics, applications (Stony Brook, NY, 1994), 249–257, World Sci. Publ. River Edge, NJ, 1996.

Dispersive Limits of the Nonlinear Klein-Gordon Equation*

Kung-Chien Wu
Department of Applied Mathematics and
Center of Mathematical Modeling and Scientific Computing
Chiao Tung University
Hsinchu 30010, China
E-mail: kjwu.am94g@nctu.edu.tw

Chi-Kun Lin
Department of Applied Mathematics and
Center of Mathematical Modeling and Scientific Computing
Chiao Tung University
Hsinchu 30010, China
E-mail: cklin@math.nctu.edu.tw

Abstract

We present several dispersive limits of the modulated nonlinear Klein-Gordon equation. First we perform the mathematical derivation of the compressible Euler equation. Second, we consider the semiclassical limit directly from the modulated Klein-Gordon equation.

1 Introduction

The relativistic quantum mechanic equation for a free particle can be derived by writing

$$E^2 = c^2p^2 + m^2c^4,$$

where E is energy, p is momentum, m is mass, and c is the speed of light. The quantum mechanical description of a relativistic free particle results from applying the *corresponding principle*, which allows one to replace classical observable by quantum mechanical operators acting on wave

*The authors are partly supported by NCTS, Mathematics Division, Taipei, China.

functions [5, 6]. Let $\hbar$ denote the Planck constant, then the Schrödinger correspondence principle given by

$$E \to i\hbar\partial_t, \quad p \to -i\hbar\nabla,$$

will result in the Klein-Gordon equation

$$-\hbar^2\partial_t^2\Psi = -c^2\hbar^2\Delta\Psi + m^2c^4\Psi$$

for wave function Ψ. The nonlinear Klein-Gordon equation is easily obtained by adding the term $V'(|\Psi|^2)\Psi$, where V is the potential energy density of the fields;

$$\frac{\hbar^2}{2mc^2}\partial_t^2\Psi - \frac{\hbar^2}{2m}\Delta\Psi + \frac{mc^2}{2}\Psi + V'(|\Psi|^2)\Psi = 0\,. \tag{1.1}$$

Since mc^2t and $\hbar$ have the same dimension of action, $[mc^2t] = [\hbar] =$ [action], and we may consider the modulated wave function [4]

$$\psi(x,t) = \Psi(x,t)\exp(imc^2t/\hbar)\,, \tag{1.2}$$

where the factor $\exp(imc^2t/\hbar)$ describes the oscillations of the wave function, then ψ satisfies the modulated nonlinear Klein-Gordon equation

$$i\hbar\partial_t\psi + \frac{\hbar^2}{2m}\Delta\psi - V'(|\psi|^2)\psi = \frac{\hbar^2}{2mc^2}\partial_t^2\psi\,. \tag{1.3}$$

The relations between different terms in (1.3) are best seen when the equation is written in terms of dimensionless variables, which will be adorned with carets. The dimensionless independent variables are given by

$$x = L\hat{x}\,, \quad t = T\hat{t}\,,$$

where L and T denote the reference length and time respectively. We also define the reference velocity by $U = L/T$ and rescale the potential energy as

$$V' = mU^2\hat{V}'\,.$$

Substituting all of these rescaled quantities into the original equation (1.3), and dropping all carets, yields

$$i\varepsilon\partial_t\psi + \frac{1}{2}\varepsilon^2\Delta\psi - V'(|\psi|^2)\psi = \frac{1}{2}\varepsilon^2\nu^2\partial_t^2\psi\,. \tag{1.4}$$

Note that the first important dimensionless parameter ν is given by the ratio of reference velocity and speed of light, $\nu = U/c$, and the scaled Planck constant $\varepsilon = \frac{\hbar}{mU^2T}$ is the second important dimensionless parameter. The two dimensionless parameters ν and ε show the relativistic and quantum effects respectively.

In addition to the introduction this paper is decomposed into two parts. In Section 2, we discuss the hydrodynamic limit of the Klein-Gordon equation, we consider the nonrelativistic-semiclassical limit, i.e., the two parameters $\varepsilon, \nu \to 0$ simultaneously, of the modulated nonlinear Klein-Gordon equation (1.4). To avoid carrying out a double limit, we restrict the case when ν and ε are related. In this situation, the compressible Euler equations are recovered as the nonrelativistic-semiclassical limit.

In Section 3, we show that the semiclassical limit of the modulated cubic nonlinear Klein-Gordon equation is a relativistic wave map and the associated phase function satisfies a linear relativistic wave equation.

2 Hydrodynamic Limit

In this section, we consider the convergence towards the compressible Euler equations. In fact, we will consider the so called nonrelativistic-semiclassical limit, i.e., $\nu \to 0$ and $\varepsilon \to 0$ simultaneously. In order to avoid carrying out a double limits the parameters ν and ε must be related. For convenience we set $\nu = \varepsilon^\kappa$ for some $\kappa > 0$, $0 < \varepsilon \ll 1$ and assume the potential energy $V'(|\psi^\varepsilon|^2) = |\psi^\varepsilon|^{2(\gamma-1)}$. Indeed we consider the modulated nonlinear Klein-Gordon equation

$$i\varepsilon\partial_t\psi^\varepsilon - \frac{1}{2}\varepsilon^{2+2\kappa}\partial_t^2\psi^\varepsilon + \frac{1}{2}\varepsilon^2\Delta\psi^\varepsilon - |\psi^\varepsilon|^{2(\gamma-1)}\psi^\varepsilon = 0\,, \tag{2.1}$$

supplemented with the initial conditions

$$\psi^\varepsilon(x,0) = \psi_0^\varepsilon(x)\,, \quad \partial_t\psi^\varepsilon(x,0) = \psi_1^\varepsilon(x)\,, \quad x \in \mathbb{T}^n\,, \tag{2.2}$$

satisfying the energy bound

$$\int_{\mathbb{T}^n} \frac{1}{2}\varepsilon^{2+2\kappa}|\psi_1^\varepsilon|^2 + \frac{1}{2}\varepsilon^2|\nabla\psi_0^\varepsilon|^2 + \frac{1}{\gamma}|\psi_0^\varepsilon|^{2\gamma}dx \le C\,. \tag{2.3}$$

Here C denotes various positive constants independent of ε.

Associated with (2.1) are the local conservation laws corresponding to charge, momentum (current) and energy conservation. In fact, we have the hydrodynamic variables: Schrödinger part charge ρ_S^ε, relativistic part charge ρ_K^ε, Schrödinger part momentum (current) J_S^ε, relativistic part momentum (current) J_K^ε, and energy e^ε given as follows:

$$\begin{aligned}
&\rho_S^\varepsilon = |\psi^\varepsilon|^2, \qquad \rho_K^\varepsilon = \tfrac{i}{2}\varepsilon^{1+2\kappa}\left(\psi^\varepsilon\partial_t\overline{\psi^\varepsilon} - \overline{\psi^\varepsilon}\partial_t\psi^\varepsilon\right),\\
&J_S^\varepsilon = (J_{S,1}^\varepsilon, J_{S,2}^\varepsilon, \dots, J_{S,n}^\varepsilon) = \tfrac{i}{2}\varepsilon\left(\psi^\varepsilon\nabla\overline{\psi^\varepsilon} - \overline{\psi^\varepsilon}\nabla\psi^\varepsilon\right),\\
&J_K^\varepsilon = (J_{K,1}^\varepsilon, J_{K,2}^\varepsilon, \dots, J_{K,n}^\varepsilon) = \tfrac{1}{2}\varepsilon^{2+2\kappa}\Big(\partial_t\psi^\varepsilon\nabla\overline{\psi^\varepsilon} + \partial_t\overline{\psi^\varepsilon}\nabla\psi^\varepsilon\Big),\\
&e^\varepsilon = \tfrac{1}{2}\varepsilon^{2+2\kappa}|\partial_t\psi^\varepsilon|^2 + \tfrac{1}{2}\varepsilon^2|\nabla\psi^\varepsilon|^2 + \tfrac{1}{\gamma}|\psi^\varepsilon|^{2\gamma}.
\end{aligned} \tag{2.4}$$

The local conservation laws of the modulated Klein-Gordon equation (2.1) are the charge, momentum (current), and energy given below:
(A) Conservation of charge

$$\frac{\partial}{\partial t}(\rho_S^\varepsilon - \rho_K^\varepsilon) + \nabla\cdot J_S^\varepsilon = 0\,, \tag{2.5}$$

(B) Conservation of momentum (current)

$$\begin{aligned}
&\frac{\partial}{\partial t}(J_S^\varepsilon - J_K^\varepsilon) + \frac{1}{4}\varepsilon^2\nabla\cdot\Big[2(\nabla\psi^\varepsilon\otimes\nabla\overline{\psi^\varepsilon} + \nabla\overline{\psi^\varepsilon}\otimes\nabla\psi^\varepsilon) - \nabla^2(|\psi^\varepsilon|^2)\Big]\\
&\quad + \frac{1}{4}\varepsilon^{2+2\kappa}\nabla\partial_t\Big(\psi^\varepsilon\partial_t\overline{\psi^\varepsilon} + \overline{\psi^\varepsilon}\partial_t\psi^\varepsilon\Big) + \frac{\gamma-1}{\gamma}\nabla|\psi^\varepsilon|^{2\gamma} = 0\,,
\end{aligned} \tag{2.6}$$

(C) Conservation of energy

$$\frac{\partial}{\partial t}e^\varepsilon - \nabla\cdot\left[\frac{1}{2}\varepsilon^2(\nabla\psi^\varepsilon\partial_t\overline{\psi^\varepsilon} + \nabla\overline{\psi^\varepsilon}\partial_t\psi^\varepsilon)\right] = 0\,. \tag{2.7}$$

They play the crucial roles of the proof of the hydrodynamic limit. Motivated by Brenier's pioneer work [1], we can define the modulated energy of (2.1) as

$$\begin{aligned}
H^\varepsilon(t) = &\frac{\varepsilon^2}{2}\int_{\mathbb{T}^n}|(\nabla - i\varepsilon^{-1}u)\psi^\varepsilon|^2dx\\
&+\frac{1}{2}\int_{\mathbb{T}^n}|\varepsilon^{1+\kappa}\partial_t\psi^\varepsilon|^2dx + \int_{\mathbb{T}^n}\Theta(\rho_S^\varepsilon,\rho)dx,
\end{aligned} \tag{2.8}$$

where

$$\Theta(\rho_S^\varepsilon,\rho) = \frac{1}{\gamma}\Big((\rho_S^\varepsilon)^\gamma - \rho^\gamma\Big) - \rho^{\gamma-1}(\rho_S^\varepsilon - \rho) \tag{2.9}$$

is a convex function, minimum occurs at $\rho_S^\varepsilon = \rho$ and satisfies $\Theta(\rho_S^\varepsilon,\rho) \geq 0$. We also assume

$$\begin{aligned}
H^\varepsilon(0) = &\frac{\varepsilon^2}{2}\int_{\mathbb{T}^n}|(\nabla - i\varepsilon^{-1}u_0)\psi_0^\varepsilon|^2dx + \frac{1}{2}\int_{\mathbb{T}^n}|\varepsilon^{1+\kappa}\psi_1^\varepsilon|^2dx\\
&+\int_{\mathbb{T}^n}\Theta(|\psi_0^\varepsilon|^2,\rho_0)dx \to 0 \quad \text{as} \quad \varepsilon\to 0\,,
\end{aligned} \tag{2.10}$$

i.e., we consider the well-prepared initial data. Therefore to obtain the hydrodynamic limit, we only need to show that the modulated energy $H^\varepsilon(t)$ tends to zero as $\varepsilon \to 0$. Indeed, we have the following theorem [3].

Theorem 2.1. *Let $\gamma \geq 2$ and let ψ^ε be the solution of the modulated nonlinear Klein-Gordon equation (2.1)–(2.2) with initial condition $(\psi_0^\varepsilon, \psi_1^\varepsilon) \in H^{s+1}(\mathbb{T}^n) \oplus H^s(\mathbb{T}^n)$, $s > \frac{n}{2} + 1$, satisfying (2.3) and (2.10). Then there exist $T_* > 0$ such that*

$$\|(\rho_S^\varepsilon - \rho)(\cdot,t)\|_{L^\gamma(\mathbb{T}^n)} \to 0\,, \qquad \|\rho_K^\varepsilon(\cdot,t)\|_{L^{\frac{2\gamma}{\gamma+1}}(\mathbb{T}^n)} \to 0\,, \tag{2.11}$$

$$\|(J_S^\varepsilon - \rho u)(\cdot,t)\|_{L^{\frac{2\gamma}{\gamma+1}}(\mathbb{T}^n)} \to 0\,, \qquad \|J_K^\varepsilon(\cdot,t)\|_{L^1(\mathbb{T}^n)} \to 0\,, \tag{2.12}$$

for $t \in [0, T_*)$ as $\varepsilon \downarrow 0$, where $(\rho, u) \in C\big([0,T_*); H^s(\mathbb{T}^n)\big)$ is the unique local smooth solution of the γ-law compressible Euler equations

$$\begin{cases} \partial_t \rho + \nabla\cdot(\rho u) = 0\,, & x \in \mathbb{T}^n, \quad t \in [0, T_*)\,, \\ \partial_t(\rho u) + \nabla\cdot(\rho u \otimes u) + \nabla P(\rho) = 0\,, & \\ \rho(x,0) = \rho_0(x), \quad u(x,0) = u_0(x)\,, & x \in \mathbb{T}^n\,, \end{cases} \tag{2.13}$$

where $0 < \rho_0 \in H^s(\mathbb{T}^n)$, $u_0 \in H^s(\mathbb{T}^n)$ and the equation of states is given by $P(\rho) = \frac{\gamma-1}{\gamma}\rho^\gamma$.

3 Semiclassical limit

We discuss the semiclassical limit of the modulated cubicnonlinear Klein-Gordon equation [2]

$$i\partial_t\psi^\varepsilon - \frac{1}{2}\varepsilon\nu^2\partial_t^2\psi^\varepsilon + \frac{\varepsilon}{2}\Delta\psi^\varepsilon - \left(\frac{|\psi^\varepsilon|^2 - 1}{\varepsilon}\right)\psi^\varepsilon = 0\,, \tag{3.1}$$

supplemented with initial conditions

$$\psi^\varepsilon(x,0) = \psi_0^\varepsilon(x)\,, \quad \partial_t\psi^\varepsilon(x,0) = \psi_1^\varepsilon(x)\,, \quad x \in \mathbb{T}^n\,, \tag{3.2}$$

where $(\psi_0^\varepsilon, \psi_1^\varepsilon) \in H^1(\mathbb{T}^n) \oplus L^2(\mathbb{T}^n)$ and $\frac{|\psi_0^\varepsilon|^2 - 1}{\varepsilon} \in L^2(\mathbb{T}^n)$.

The main idea is based on the conservation laws of charge and energy;

$$\frac{\partial}{\partial t}\left[|\psi|^2 + \frac{i}{2}\varepsilon\nu^2(\overline{\psi}\partial_t\psi - \psi\partial_t\overline{\psi})\right] + \nabla\cdot\left[\frac{i}{2}\varepsilon(\psi\nabla\overline{\psi} - \overline{\psi}\nabla\psi)\right] = 0\,, \tag{3.3}$$

$$\frac{\partial}{\partial t}\left[\left(\nu^2|\partial_t\psi|^2 + |\nabla\psi|^2\right) + \frac{2}{\varepsilon^2}V\right] - \nabla\cdot\left[(\nabla\psi\partial_t\overline{\psi} + \nabla\overline{\psi}\partial_t\psi)\right] = 0\,. \tag{3.4}$$

For the model (3.1)–(3.2) we have the following weak formulation:

$$0 = i\Big\langle \psi^\varepsilon(\cdot,t_2) - \psi^\varepsilon(\cdot,t_1), \varphi\Big\rangle - \frac{1}{2}\varepsilon\nu^2\Big\langle \partial_t\psi^\varepsilon(\cdot,t_2) - \partial_t\psi^\varepsilon(\cdot,t_1), \varphi\Big\rangle$$

$$-\frac{\varepsilon}{2}\int_{t_1}^{t_2}\Big\langle \nabla\psi^\varepsilon(\cdot,\tau), \nabla\varphi\Big\rangle d\tau - \int_{t_1}^{t_2}\Big\langle\Big(\frac{|\psi^\varepsilon|^2-1}{\varepsilon}\Big)\psi^\varepsilon(\cdot,\tau), \varphi\Big\rangle\, d\tau\,, \tag{3.5}$$

for every $[t_1, t_2] \subset [0, T]$ and for all $\varphi \in C_0^\infty(\mathbb{T}^n)$. Moreover, for all $t \in [0, T]$, it satisfies the charge-energy inequality

$$\int_{\mathbb{T}^n} |\psi^\varepsilon|^2 + \nu^2|\partial_t\psi^\varepsilon|^2 + |\nabla\psi^\varepsilon|^2 + \frac{1}{2}\Big(\frac{|\psi^\varepsilon|^2-1}{\varepsilon}\Big)^2 dx \le C\,. \tag{3.6}$$

We assume the initial datum satisfy $|\psi_0^\varepsilon| = 1$ almost everywhere and $(\psi_0^\varepsilon, \psi_1^\varepsilon) \to (\psi_0, 0)$ in $H^1(\mathbb{T}^n) \oplus L^2(\mathbb{T}^n)$ as $\varepsilon \to 0$, thus $|\psi_0| = 1$ almost everywhere. First we deduce from the charge-energy inequality (3.6) that

$$\{\psi^\varepsilon\}_\varepsilon \quad \text{is bounded in} \quad L^\infty([0,T]; H^1(\mathbb{T}^n))\,, \tag{3.7}$$

$$\{\partial_t\psi^\varepsilon\}_\varepsilon \quad \text{is bounded in} \quad L^\infty([0,T]; L^2(\mathbb{T}^n))\,, \tag{3.8}$$

$$\Big\{\frac{|\psi^\varepsilon|^2-1}{\varepsilon}\Big\}_\varepsilon \quad \text{is bounded in} \quad L^\infty([0,T]; L^2(\mathbb{T}^n))\,, \tag{3.9}$$

then the classical compactness argument shows that there exists a subsequence still denoted by $\{\psi^\varepsilon\}_\varepsilon$ and a function ψ satisfying

$$\psi \in L^\infty([0,T]; H^1(\mathbb{T}^n)), \qquad \partial_t\psi \in L^\infty([0,T]; L^2(\mathbb{T}^n))$$

such that

$$\psi^\varepsilon \rightharpoonup \psi \quad \text{weakly} * \text{in} \quad L^\infty([0,T]; H^1(\mathbb{T}^n))\,, \tag{3.10}$$

$$\partial_t\psi^\varepsilon \rightharpoonup \partial_t\psi \quad \text{weakly} * \text{in} \quad L^\infty([0,T]; L^2(\mathbb{T}^n))\,. \tag{3.11}$$

Also, from (3.9), we have

$$|\psi^\varepsilon|^2 \to 1 \quad \text{a.e. and strongly in } L^2(\mathbb{T}^n). \tag{3.12}$$

Note that (3.9) only shows that $\{\frac{|\psi^\varepsilon|^2-1}{\varepsilon}\}_\varepsilon$ is a weakly relative compact set in $L^\infty([0,T]; L^2(\mathbb{T}^n))$. In order to overcome the difficulty caused by nonlinearity, the fourth term on the right hand side of (3.5), we use the compactness argument to prove $\psi^\varepsilon \to \psi$ strongly in $C([0,T]; L^2(\mathbb{T}^n))$. Moreover, the quantity $\frac{|\psi^\varepsilon(x,t)|^2-1}{\varepsilon}$ is bounded in $L^\infty([0,T]; L^2(\mathbb{T}^n))$, and hence it converges weakly $*$ to some function $w \in L^\infty([0,T]; L^2(\mathbb{T}^n))$.

To find the explicit form of w, we define two functions $W(\psi^\varepsilon)$ and $Z(\psi^\varepsilon)$ respectively by

$$W(\psi^\varepsilon) = \frac{i}{2}\left(\psi^\varepsilon \nabla \overline{\psi^\varepsilon} - \overline{\psi^\varepsilon}\nabla \psi^\varepsilon\right), \quad Z(\psi^\varepsilon) = \frac{i}{2}\nu^2\left(\overline{\psi^\varepsilon}\partial_t \psi^\varepsilon - \psi^\varepsilon \partial_t \overline{\psi^\varepsilon}\right). \tag{3.13}$$

We rewrite the conservation of charge (3.3) as

$$\frac{|\psi^\varepsilon(x,t)|^2 - 1}{\varepsilon} = -Z(\psi^\varepsilon) + Z(\psi^\varepsilon(x,0)) - \int_0^t \text{div } W(\psi^\varepsilon) d\tau \,. \tag{3.14}$$

To obtain the compactness of the sequence $\left\{\frac{|\psi^\varepsilon(x,t)|^2-1}{\varepsilon}\right\}_\varepsilon$, we have to treat the compactness of $\{Z(\psi^\varepsilon)\}_\varepsilon$ and $\{W(\psi^\varepsilon)\}_\varepsilon$ separately. In fact, we have $Z(\psi^\varepsilon) \rightharpoonup Z(\psi)$, $Z(\psi^\varepsilon(x,0)) \rightharpoonup 0$ and

$$\int_0^t \text{div } W(\psi^\varepsilon) d\tau \rightharpoonup \int_0^t \text{div } W(\psi) d\tau$$

in $\mathcal{D}'((0,T) \times \mathbb{T}^n)$, thus

$$\frac{|\psi^\varepsilon(x,t)|^2 - 1}{\varepsilon} \rightharpoonup -Z(\psi) - \int_0^t \text{div } W(\psi) d\tau \tag{3.15}$$

in $\mathcal{D}'((0,T) \times \mathbb{T}^n)$, and the limit function w is given explicitly by

$$w = -Z(\psi) - \int_0^t \text{div } W(\psi) d\tau \,.$$

Letting $\varepsilon \to 0$ we can show that the limit wave function ψ satisfies

$$i\partial_t \psi + \left[Z(\psi) + \int_0^t \text{div } W(\psi) d\tau\right]\psi = 0 \tag{3.16}$$

in the sense of distribution. Using $|\psi|^2 = 1$ and differentiating (3.16) with respect to time variable t, one can show ψ satisfies the relativistic wave map equation

$$(1+\nu^2)\partial_t^2 \psi - \Delta\psi = \left[|\nabla\psi|^2 - (1+\nu^2)|\partial_t \psi|^2\right]\psi, \quad |\psi| = 1 \quad \text{a.e.} \tag{3.17}$$

supplemented with the initial conditions

$$\psi(x,0) = \psi_0(x), \quad \partial_t \psi(x,0) = 0, \quad x \in \mathbb{T}^n, \quad |\psi_0| = 1 \quad \text{a.e.} \tag{3.18}$$

Using the fact $|\psi| = 1$ and writing $\psi = e^{i\theta}$ shows

$$(1+\nu^2)\partial_t^2 \theta = \Delta\theta \,, \quad \theta(x,0) = \arg \psi_0 \,, \quad \partial_t \theta(x,0) = 0 \,, \tag{3.19}$$

i.e., θ is a distribution solution of the linear relativistic wave equation, the ν^2 terms in (3.17) and (3.19) show the relativistic effect.

Theorem 3.1. *Let* $(\psi_0^\varepsilon, \psi_1^\varepsilon) \in H^1(\mathbb{T}^n) \oplus L^2(\mathbb{T}^n)$, $|\psi_0^\varepsilon| = 1$ *a.e. and* $(\psi_0^\varepsilon, \psi_1^\varepsilon) \to (\psi_0, 0)$ *in* $H^1(\mathbb{T}^n) \oplus L^2(\mathbb{T}^n)$, $|\psi_0| = 1$ *a.e., and let* ψ^ε *be the corresponding weak solution of the modulated cubic nonlinear Klein-Gordon equation* (3.1)–(3.2). *Then the weak limit* ψ, *satisfying* $|\psi| = 1$ *a.e., solves the relativistic wave map* (3.17)–(3.18). *Moreover, let* $\psi = e^{i\theta}$ *then the phase function* θ *satisfies the relativistic wave equation* (3.19).

References

[1] Y. Brenier, Convergence of the Vlasov-Poisson system to the incompressible Euler equations. Commun. in Partial Differential Equations. 25 (2000), 737–754.

[2] C. K. Lin and K. C. Wu, Singular limits of the Klein-Gordon equation. Archive for Rational Mechanics and Analysis. 197 (2010), 689–711.

[3] C. K. Lin and K. C. Wu, Hydrodynamic limits of the nonlinear Klein-Gordon equation. Submitted for publication, 2009.

[4] S. Machihara, K. Nakanishi and T. Ozawa, Nonrelativistic limit in the energy space for nonlinear Klein-Gordon equations. Mathematische Annalen. 322 (2002), 603–621.

[5] A. Messiah, Quantum Mechanics, Vol. 1 & 2 (two volumes bound as one). Dover Publications Inc., 1999.

[6] H. M. Pilkuhn, Relativistic Quantum Mechanics. Texts and Monographs in Physics, Springer-Verlag, 2003.

Semiclassical Limit of Quantum Semiconductor Models*

Hao Yan
Zhou Peiyuan Center For Applied Mathematics, Tsinghua University
Beijing 100084, China
E-mail: yanhao06@mails.tsinghua.edu.cn

Li Chen
Department of Mathematical Sciences, Tsinghua University
Beijing 100084, China
E-mail: lchen@math.tsinghua.edu.cn

Abstract

In this contributed paper, we first briefly review some macroscopic quantum semiconductor models and then focus on the semiclassical limit problem of the quantum drift diffusion model and a simplified quantum energy transport model. In the second section, some of the known semiclassical limit results with positive boundary condition and the main technical tools will be reviewed. In the last section, we will introduce an interesting problem, i.e., the semiclassical limit problem with vanishing density on the boundary, which comes from the inversion layer in MOSFET devices.

1 Quantum semiconductor models

In [28], the authors explained the reason to introduce macroscopic quantum models. The reason is that the semiconductor devices, the building block of almost all electronic devices, are now reaching to ultrasmall scale. So in the simulation, the classical models couldn't work well. On the other hand, the simulation of the microscopic models based on Schrödinger-Poisson needs a lot of resources. People are trying to find some reasonable macroscopic quantum models for application.

Quantum Drift Diffusion (or Density Gradient) model was first introduced in 1987 by M. G. Ancona et al. In order to find a macroscopic

*Supported by the National Natural Science Foundation of China (No. 10871112, No. 10910257)

description of the strong inversion layer with the scale of 10 nm, Ancona and his coauthors applied this theory to 1-d steady state MIM tunneling in [2] and other quantum transport in semiconductor in [3, 4, 5], to analysis the tunneling phenomena in ultrathin oxide MOS capacitors in [8]. They also analyzed the legitimacy, physical fidelity, and limitations by using detailed comparisons with nonequilibrium Green's function simulations in 1-d in [6]. All the results show that the density gradient approach is quite accurate and can be of the great value for engineering oriented device analysis in quantum regimes.

Applied mathematicians derived the macroscopic quantum models formally from Schrödinger-Poisson(or Wigner-Poisson) system by asymptotic analysis in 1990s, [17, 18, 27]. The picture of fluid dynamic quantum models, as their classical counterparts, contains quantum hydrodynamic models, quantum energy transport models, and quantum drift diffusion models. One can check the whole hierarchy in P. Markowich et al.'s book [27] or A. Jüngel's book [19]. So far, some mathematical analysis has been done on those macroscopic quantum models, including some works on quantum hydrodynamic (or the quantum moment equations) models, quantum drift diffusion models, and the simplified quantum energy transport models.

Following A. Jüngel's lecture notes [20] and P. Markowich et al.'s book [27] on the model derivation, the so called scaled transient quantum drift diffusion (QDD) model can be derived. The QDD model is

$$\begin{aligned} \partial_t n &= \operatorname{div}\Big[-\varepsilon^2 n \nabla\Big(\frac{\triangle\sqrt{n}}{\sqrt{n}}\Big) + \nabla(\theta n) - n\nabla V\Big], \\ \lambda^2 \triangle V &= n - C(x), \end{aligned} \tag{1.1}$$

where the electron density n and the electrostatic potential V are unknown variables; the doping profile $C(x)$ which is representing the distribution of charged background ions is supposed to be independent of time t; θ is the constant lattice temperature. The scaled Planck constant $\varepsilon > 0$ and scaled Debye length $\lambda > 0$ are parameters.

Recently, A. Jüngel and J-P. Milišić derived a so called simplified quantum energy transport (sQET) model formally from quantum hydrodynamic model in the large time and small regime.

$$\begin{aligned} \partial_t n &= \operatorname{div}\Big[-\varepsilon^2 n \nabla\Big(\frac{\triangle\sqrt{n}}{\sqrt{n}}\Big) + \nabla(n\theta) - n\nabla V\Big], \\ -\operatorname{div}(n\nabla\theta) &= \frac{n}{\tau_e}(\theta_L(x) - \theta), \\ \lambda^2 \triangle V &= n - C(x), \end{aligned} \tag{1.2}$$

It is easy to see that the main difference between sQET and QDD is the equation for temperature. If one takes the relaxation limit $\tau_e \to 0$ in (1.2), QDD (1.1) could be obtained.

So mathematically, in studying the QDD and sQET models, the main difficult term comes from the forth order term, i.e., $-\varepsilon^2 \text{div}\left[n\nabla\left(\frac{\triangle\sqrt{n}}{\sqrt{n}}\right)\right]$. In both of the quantum models (1.1) and (1.2), the semiclassical limit problem means to study the limit of the solution when $\varepsilon \to 0$. The key point in getting the semiclassical limit is to send the forth order term goes to 0 in some sense. In all the discussions, sQET requires more estimates on the temperature θ which is not an easy task since usually we don't have uniform lower bound for electron density n.

2 Semiclassical limit

Semiclassical limit is one of the most important topics in the study of semiconductor models. It shows the connection between classical and quantum models. We will explain briefly in this part the main idea in proving the semiclassical limit in QDD and sQET models. The key step is to have a uniform bound on the highest order term with respect to ε.

A series of results have be obtained including the existence of weak solution, its long time behavior, quasi-neutral limit and semiclassical limit with different boundary conditions, [10, 12, 13, 14, 16, 22, 23, 24, 25, 26].

We take the QDD model with a mixed Dirichlet-Neumann boundary condition in one dimension as an example. The multi dimension case with periodic boundary conditions can be discussed similarly. In 1-d case, the quantum drift diffusion model is written in the following form with a new variable $\rho = \sqrt{n}$

$$(\rho^2)_t = -\varepsilon^2\Big(\rho^2\big(\frac{\rho_{xx}}{\rho}\big)_x\Big)_x + (2\rho\rho_x - \rho^2 V_x)_x,$$
$$\lambda^2 V_{xx} = \rho^2 - C(x),$$

where $(x,t) \in (0,1) \times [0,T) = \Omega \times [0,T) = Q_T$. We impose the initial condition for (1.1) and (1.2)

$$\rho|_{t=0} = \rho_0 \geq 0. \tag{2.1}$$

The mixed Dirichlet-Neumann boundary conditions is

$$\rho = 1, \rho_x = 0, V = V_\Omega \text{on} \partial\Omega \tag{2.2}$$

In [16], the authors prove the global existence of weak solution to the above problem. Based on the existence results, a next reasonable problem that one should prove is the semiclassical limit, i.e., let ρ_ε be the solution of (1.1) for fixed ε, study whether the limit of the sequence $\{\rho_\varepsilon\}$ exists and which kind of equation the limit satisfies. The first semi-classical limit was obtained by Chen and Ju [9] with periodic boundary condition. Later on, by similar argument, the entropy based compactness argument, a series of results were obtained with different boundary conditions [12, 13, 15]. Recently, a semiclassical limit result was obtained in sQET model in [11].

To briefly introduce the method, we still use the above mixed Dirichlet-Neumann boundary problem as an example. The semiclassical limit result is

Theorem 2.1. [16] *Let $\{(\rho_\varepsilon, V_\varepsilon)\}_{\varepsilon>0}$ be weak solution, then as $\varepsilon \to 0$, there exists a subsequence which is not relabeled, such that*

$$\begin{aligned} \rho_\varepsilon^2 &\rightharpoonup n && \text{in } L^4(0,T;L^2(\Omega)) \cap L^{\frac{4}{3}}(0,T;H^1(\Omega)), \\ V_\varepsilon &\rightharpoonup V && \text{in } L^2(0,T;H^2(\Omega)), \\ \rho_\varepsilon^2 &\to n && \text{in } L^{\frac{4}{3}}(0,T;C^{0,\alpha}(\bar{\Omega}))\ (0<\alpha<\frac{1}{2}), \\ V_\varepsilon &\to V && \text{in } L^{\frac{4}{3}}(0,T;H^2(\Omega)), \\ \partial_t\rho_\varepsilon^2 &\rightharpoonup \partial_t n && \text{in } L^{\frac{4}{3}}(0,T;H^{-2}(\Omega)), \\ \varepsilon^2\rho_\varepsilon^2(\ln\rho_\varepsilon^2)_{xx} &\to 0 && \text{in } L^{\frac{8}{5}}(0,T;L^2(\Omega)), \end{aligned}$$

and

$$\int_0^T \langle \partial_t n, \psi\rangle_{\langle H^{-2},H_0^2\rangle} dt = -\int_0^T\int_\Omega (n_x + nV_x)\psi_x dxdt,\ \forall\ \psi \in L^4(0,T;H_0^2(\Omega))$$

$$-\int_0^T\int_\Omega V_{xx}\phi dx = \int_0^T\int_\Omega [n - C(x)]\phi dx,\quad \forall\ \phi \in L^2(Q_T).$$

Here $n \geq 0$.

In the proof of this theorem, it was shown that $\|\varepsilon^{\frac{1}{4}}\rho_\varepsilon\|_{L^8(0,T;L^\infty(\Omega))} \leq C$ where C is independent of ε, so that $\varepsilon^{\frac{5}{4}}\rho_\varepsilon^2(\ln\rho_\varepsilon)_{xx}$ is bounded in $L^{\frac{8}{5}}(0,T;L^2(\Omega))$ uniformly in ε. The main *a priori* estimate to obtain this uniform estimate is from entropy inequalities and a very careful Sobolev type interpolation. In the flowing part of this section, we will describe this idea.

The entropy inequalities used can be derived formally by taking $\ln\rho_\varepsilon^2$ $(n = \rho_\varepsilon^2)$ and $1 - 1/\rho_\varepsilon^2$ as a test function, in which the terms including electrostatic potential V could be dealt by using delicate calculations,

$$\begin{aligned} \frac{d}{dt}\int_\Omega [\rho_\varepsilon^2(\ln\rho_\varepsilon^2 - 1) + 1] dx + 2\varepsilon^2\int_\Omega |(\rho_\varepsilon)_{xx}|^2 dx + \frac{2}{3}\varepsilon^2\int_\Omega \left|\frac{(\rho_\varepsilon)_x^2}{\rho_\varepsilon}\right|^2 dx \\ +4\int_\Omega |(\rho_\varepsilon)_x|^2 dx + \int_\Omega \rho_\varepsilon^4 dx \leq C\int_\Omega \rho_\varepsilon^2 dx + C, \end{aligned} \tag{2.3}$$

$$\frac{d}{dt}\int_\Omega (\rho_\varepsilon^2 - \ln \rho_\varepsilon^2)dx + 2\varepsilon^2 \int_\Omega |(\ln \rho_\varepsilon)_{xx}|^2 dx + 4\int_\Omega |(\ln \rho_\varepsilon)_x|^2 dx$$
$$\le C\int_\Omega |\ln \rho_\varepsilon| dx + C. \tag{2.4}$$

In the compactness argument, the key points are

- $(\rho_\varepsilon^2)_x$ be bounded in some sense (to make sure that the limit of ρ_ε^2 satisfy the classical drift diffusion equation);
- In the semiclassical discussion, $\varepsilon^2\rho_\varepsilon^2\left(\frac{(\rho_\varepsilon)_{xx}}{\rho_\varepsilon}\right)_x$ goes to zero in some sense.

The following estimates are obtained by Gagliardo-Nirenberg inequalities and the known a priori estimates from entropy inequalities shown above.

$$\varepsilon \left\|(\rho_\varepsilon)_x\right\|_{L^{\frac{8}{3}}(0,T;L^\infty(\Omega))} \le C\varepsilon \left\|\rho_\varepsilon\right\|^{\frac{3}{4}}_{L^2(0,T;H^2(\Omega))} \left\|\rho_\varepsilon\right\|^{\frac{1}{4}}_{L^\infty(0,T;L^2(\Omega))} \le C.$$

Separating $\varepsilon\rho_\varepsilon(\ln \rho_\varepsilon)_{xx}$ into two terms $\varepsilon(\rho_\varepsilon)_{xx}$ and $-\varepsilon\frac{(\rho_\varepsilon)_x^2}{\rho_\varepsilon}$ we get the uniform estimate

$$\varepsilon \parallel \rho_\varepsilon(\ln \rho_\varepsilon)_{xx} \parallel_{L^2(Q_T)} \le C.$$

So we have the estimates for quantum term on the right hand side of the equation,

$$\parallel \rho_\varepsilon^2(\ln \rho_\varepsilon)_{xx} \parallel_{L^{\frac{8}{5}}(0,T;L^2(\Omega))} \le \parallel \rho_\varepsilon \parallel_{L^8(0,T;L^\infty(\Omega))} \parallel \rho_\varepsilon(\ln \rho_\varepsilon)_{xx} \parallel_{L^2(Q_T)} \le C_\varepsilon,$$

which gives the estimates for time derivative of ρ_ε^2. Furthermore,

$$\left\|\varepsilon^{\frac{5}{4}}\rho_\varepsilon^2(\ln \rho_\varepsilon^2)_{xx}\right\|_{L^{\frac{8}{5}}(0,T;L^2(\Omega))}$$
$$\le \varepsilon^{\frac{1}{4}} \left\|\rho_\varepsilon\right\|_{L^8(0,T;L^\infty(\Omega))} \left\|\varepsilon\rho_\varepsilon(\ln \rho_\varepsilon)_{xx}\right\|_{L^2(Q_T)} \le C,$$

so that the quantum term vanish as $\varepsilon \to 0$, i.e.,

$$\left\|\varepsilon^2\rho_\varepsilon^2\left(\frac{(\rho_\varepsilon)_{xx}}{\rho_\varepsilon}\right)_x\right\|_{L^{\frac{8}{5}}(0,T;L^2(\Omega))} \le \varepsilon^{\frac{3}{4}} \left\|\varepsilon^{\frac{5}{4}}\rho_\varepsilon^2(\ln \rho_\varepsilon^2)_{xx}\right\|_{L^{\frac{8}{5}}(0,T;L^2(\Omega))} \le C\varepsilon^{\frac{3}{4}}.$$

The convergence of other terms can be obtained by Aubin's lemma and routine argument. We omit them here.

For the sQET model with periodic boundary conditions, with the assumption that lattice temperature $\theta_L(x)$ is bounded from above and below, i.e., $0 < m_L \le \theta_L(x) \le M_L$, one have the same bound for temperature θ by maximum principle,

$$0 < m_L \le \theta(x) \le M_L.$$

Notice that we don't have lower bound for the electron density n. To have a uniform estimate for $\nabla\theta$, a new test function θ/ρ^2 can be used,

$$\int_{\mathbb{T}^d} \rho^2 \nabla\theta \cdot \nabla(\frac{\theta}{\rho^2}) dx = \int_{\mathbb{T}^d} \theta(\theta_L(x) - \theta) dx$$

by a similar entropy estimate in (2.4), one gets

$$\begin{aligned}\int_{\mathbb{T}^d} |\nabla\theta|^2 dx &= \int_{\mathbb{T}^d} \theta \nabla\theta \nabla \ln \rho dx + \int_{\mathbb{T}^d} \theta(\theta_L(x) - \theta) dx \\ &\leq M_L \Big| \int_{\mathbb{T}^d} \nabla\theta \cdot \nabla \ln \rho dx \Big| + C \leq C.\end{aligned}$$

3 Inversion layer problem

In the discussion of semiclassical limit above, one needs the boundary condition of electron density to be positive. While in some real world applications, for example, in MOSFET, the boundary condition for electron density near Oxide must be 0. Physically, the boundary layer of this problem is called inversion layer. For simplicity, we set up the problem by a 1-d stationary QDD model with vanishing density on the boundary

$$\begin{aligned} w_{xx} &= n - e^{w/T_P} - C(x), \\ w - \varphi &= \ln n - \varepsilon^2 \frac{(\sqrt{n})_{xx}}{\sqrt{n}}, \end{aligned}$$

where n is the electron density and w is the electrostatic potential. φ is a constant Fermi level, T_p is the lattice temperature, $C(x)$ is the doping profile of a semiconductor. The reasonable boundary conditions are

$$\begin{aligned} w_x(0) = c(w(0) - V_{gs}),\ & w(1) = \varphi + \ln n_i = V_B, \\ n(0) = 0,\ & n(1) = n_i, \end{aligned}$$

where V_{gs} is the scaled gate voltage , n_i is the scaled equilibrium density, and the constants satisfies $0 = n_i - e^{V_B/T_P} - C(1)$. Note that the boundary condition at $x = 1$ are chosen so that there is no boundary layer at $x = 1$. This will be studied in our coming research paper.

The idea in studying the inversion layer problem should start from the existence of solution n_ε (using similar variational method as in [1] with different boundary conditions). After that by using matched asymptotic expansion, one need to find an approximate solution n_{app}, where one part of the approximate solution is the classical solution. And the last step is to compare n_ε and n_{app}. This is the strategy to study semiclassical limit of the inversion layer problem.

This set up of the problem is the simplest among all macroscopic quantum semiconductor models. One can also use the sQET model and vQHD model to study the same problem with temperature effect and momentum effect.

References

[1] N. B. Abdallah and A. Unterreiter, On the stationatry quantum drift diffusion model. Z. Angew. Math. Phys. 49 (2) (1998), 251–275.

[2] M. G. Ancona, Macroscopic description of quantum-mechanical tunneling. Physical Review B 42 (2) (1990), 1222–1233.

[3] M. G. Ancona, Density-gradient analysis of field emission from metals. Physical Review B 46 (8) (1992), 4874–4883.

[4] M. G. Ancona and G. J. Iafrate, Quantum correction of the equation of state of an electron gas in a semiconductor. Physical Review B 39 (19) (1989), 9536–9540.

[5] M. G. Ancona and K. Lilja, Multi-dimensional tunneling in density-gradient theory. J. comput. Electron. 3 (3–4) (2004), 189–192.

[6] M. G. Ancona and A. Svizhenko, Density-gradient theory of tunneling: Physics and verification in one dimension. J. Appl. Phys. (104) 073726, (2008).

[7] M. G. Ancona and H. F. Tiersten, Macroscopic physics of the silicon inversion layer. Physical Review B 35 (15) (1987), 7959–7965.

[8] M. G. Ancona, Z. Yu, R. W. Dutton, P. J. Vande Voorde, M. Cao and D. Vook, Density-gradient analysis of MOS tunneling. IEEE Trans. Electron Devices 47 (12) (2000), 2310–2318.

[9] L. Chen and Q. C. Ju, Existence of weak solution and semiclassical limit for quantum drift-diffusion model. Z. Angew. Math. Phys. 58 (2007), no. 1, 1–15.

[10] L. Chen and Q. C. Ju, The semiclassical limit in the quantum drift-diffusion equations with isentrophic pressure. Chin. Ann. Math. 29B (4) (2008) , 369–384.

[11] X. Q. Chen and L. Chen, The Semiclassical Limit of Quantum Energy-Transport Model. Priprint.

[12] X. Q. Chen, L. Chen and H. Y. Jian, Existence, semiclassical limit and long-time behavior of weak solution to quantum drift-diffusion model. Nonlinear Analysis RWA 10 (2009), 1321–1342.

[13] X. Q. Chen, L. Chen and H. Y. Jian, The Existence and long-time behavior of weak solution to bipolar quantum drift-diffusion model. Chinese Annals of Math. B 28 (2007), no. 6, 651–664.

[14] X. Q. Chen, L. Chen and H. Y. Jian, The Dirichlet problem of the quantum drift-diffusion model. Nonlinear Analysis 69 (2008), 3084–3092.

[15] X. Q. Chen and L. Chen, Initial time layer problem for quantum drift-diffusion model. J. Math. Anal. Appl. 343 (2008), no. 1, 64–80.

[16] X. Q. Chen and L. Chen, The Dirichlet-Neumann problem of quantum drift-diffusion model. Tsinghua Science and Technology. 13 (4) (2008), 560–569.

[17] C. L. Gardner, The quantum hydrodynamic model for semidoncuctor devices. SIAM J. Appl. Math. 54 (2) (1994), 409–427.

[18] I. Gasser and P. A. Markowich, Quantum hydrodynamics, Wigner transforms and the classical limit. Asympt. Anal. 14(97–116) (1997).

[19] A. Jünger, Quasi-hydrodynamic Semiconductor Equations. Birkhäuser, PNLDE 41, 2001.

[20] A. Jüngel, Transport Equation for Semiconductors. Preliminary Lecture Notes, 2005.

[21] A. Jüngel and H. L. Li, Quantum Euler-Poisson systems: global existence and exponential decay. Quart. Appl. Math.

[22] A. Jüngel, M. C. Mariani and D. Rial, Local existence of solutions to the transient quantum hydrodynamic equations. Math. Models Methods Appl. Sci. 12 (4) (2002), 485–495.

[23] A. Jüngel and R. Pinnau, Global non-negative solutions of a nonlinear fourth order parabolic equation for quantum systems. SIAM J. Math. Anal. 32 (4) (2000), 760–777.

[24] A. Jüngel and R. Pinnau, A positivity preserving numerical scheme for a nonlinear fourth-order parabolic system, SIAM J. Num. Anal. 39 (2) (2001), 385–406.

[25] A. Jüngel and R. Pinnau, Convergent semidiscretization of a nonlinear fourth order parabolic system. Math. Mod. Num. Anal. 37 (2) (2003), 277–289.

[26] A. Jüngel and I. violet, First-order entropies for the Derrida-Lebowitz-Speer-Spohn equation. Discrete Cont. Dyn. Sys. B 8 (2007), 861-877.

[27] P. Markowich, C. Ringhofer and C. Schmeiser. Semiconductor Equations. Springer, Vienna, 1990.

[28] R. Pinnau, A review on the quantum drift diffusion model. Transp. Theory Stat. Phys. 31 (4–6) (2002), 367–395.

[29] J. Simon, Compact sets in the space $L^p(0,T;B)$. Ann. Mat. Pura Appl. 146 (1987), 65–96.

A Result on Global Solutions to 3D Complex Ginzburg-Landau Equation*

Penghong Zhong
Department of Applied Mathematics, Beijing University of Technology
Beijing 100124, China
E-mail: zhongph@emails.bjut.edu.cn

Shu Wang
Department of Applied Mathematics, Beijing University of Technology
Beijing 100124, China
E-mail: wangshu@bjut.edu.cn

Ke Wang
Department of Applied Mathematics, Beijing University of Technology
Beijing 100124, China
E-mail: coco@emails.bjut.edu.cn

Ronghui Yang
Department of mathematics, Yunnan Nationalities University
Kunming 650031, China
E-mail: yangronghui1978@yahoo.cn

Abstract

In this study, we prove the existence of the global solution of the initial value problem for the 3D complex Ginzburg-Landau equation with cubic term under periodic boundary condition by using the theory of operator semigroup and the method of priori estimates in Sobolev spaces.

*This work is supported by NSFC (Grant No. 10771009 and No. 10861014), BSFC (Grant No. 1085001) of China, Funding Project for Academic Human Resources Development in Institutions of Higher Leading Under the Jurisdiction of Beijing Municipality (PHR-IHLB 200906103), Beijing Education Committee Funds and Ph.D. Innovation Project of Beijing University of Technology (Grant No. ykj-2010-3418 and No. ykj-2010-4151).

1 Introduction

The complex Ginzburg-Landau equation (CGLE) with cubic term is one of the most important nonlinear equation in physics. In the present paper, we consider the CGLE as following:

$$u_t = \rho u + (a + i\gamma)\Delta u - (b + i\mu)\left|u\right|^2 u, \tag{1.1}$$

where u is a complex valued function defined in $n+1$ dimensional space-time R^{n+1}. Δ is the Laplacian in R^n and $\rho,\ \gamma,\ \mu,\ a,\ b$ are real parameters with $\rho,\ a,\ b > 0$. The equation arises in physics in particular as a "modulational" (or "envelope" or "amplitude") equation. There are many important papers on CGLE. In [2], Wang and Zhan established the well-posedness of solutions in L^2 in space dimension 2 and the existence of weak solutions in L^p with $p \geq 4$ for case of space dimension 3. In [3], Wang and Su improved this result to L^p with $p \geq 3$ when space dimension is 3. For the global solution of CGLE, Ghidaglia and Héron [4], Guo and Wang [5], Mielke and Schneider [6], etc., studied the finite dimensional global attractor and related dynamical issues for one or two spatial dimensional CGLE with cubic nonlinearity. In [7] Bu obtained the global existence of the Cauchy problem of the 2D GLE.

In this study, we concentrate on proving the existence of the global solution for the 3D complex Ginzburg-Landau equation with cubic term. In Section 3, we combine the results of Section 2 and the technical part of operator semigroup and the method of priori estimates to derive the existence and uniqueness of solution globally defined in time.

2 Preliminary estimates

We consider the $\Omega-$periodic function spaces $L^p_{per}(\Omega)$ and $H^k_{per}(\Omega)$, where $\Omega = (0,\ L) \times (0,\ L) \times (0,\ L) \in R^3$, with the norms denoted respectively by $\|\cdot\|_p$ ($\|\cdot\|_2 = \|\cdot\|$) and $\|\cdot\|_{H^k(\Omega)}$.

We borrow from [[8], Theorem 7.2.5, Theorem 7.2.7] the following result:

Lemma 2.1. Defining the linear operator A on H by $\mathrm{A}u = -(a + i\gamma)\Delta u$ for $u \in H$. If $0 < a \leq 1$ then A is a strongly elliptic operator and the operator $-A$ is the infinitesimal generator of an analytic semigroup of contractions on $H = L^2(\Omega)$.

From [[8], Theorem 6.1.4, Theorem 6.3.1] we borrow the following results, in which the Banach space is denoted by X.

Lemma 2.2. Let $F:\ [0,\ \infty] \times X \to X$ be continuous in t for $t \geq 0$ and locally Lipschitz continuous in u, uniformly in t on bounded intervals. If $-A$ is the infinitesimal generator of a C_0 semigroup $S(t)$ on

X then for every $u_0 \in X$, there is a $t_{\max} \le \infty$, such that the initial value problem

$$\begin{cases} u_t + Au(t) = F(t,\ u(t)), t \ge 0, \\ u(0) = u_0 \end{cases} \tag{2.1}$$

has a unique mild solution u on $[0,\ t_{\max}]$. Moreover, if $t_{\max} < \infty$ then

$$\lim_{t \to t_{\max}} \|u(t)\| = \infty.$$

Lemma 2.3. Let $-A$ be the infinitesimal generator of an analytic semigroup $S(t)$, $S(t) \le M$ and assume further that $0 \in \rho(-A)$. If F is locally Lipschitz continuous in X, then for every initial data $(t_0,\ u_0) \in [0,\ \infty] \times X$ the initial value problem (2.1) has a unique local solution

$$u \in C([t_0,\ T) :\ X) \cap C^1((t_0,\ T) :\ X),$$

where $T(t_0,\ u_0) > t_0$.

Remark 2.4. If $F : [t_0,\ T] \times X \to X$ is locally Lipschitz continuous in X uniformly in $[t_0,\ T]$, it follows from Lemma 2.2 that for every $u_0 \in D(A)$ the initial value problem possesses a classical solution on a maximal interval $[t_0,\ t_{\max}]$. Additionally, if $t_{\max} < T$, then

$$\lim_{t \to t_{\max}} (\|u(t)\| + \|Au(t)\|) = \infty.$$

We note that in this situation it may well be that $\|u(t)\|$ is bounded on $[t_0,\ t_{\max}]$ and $\|Au(t)\| \to \infty$ (if and only if $t \to t_{\max}$).

3 The existence of global solutions

3.1 Local well posedness

Defining the linear operator A on H by $D(A) \in H^2$ and $Au = -(a + i\gamma)\Delta u$ for $u \in D(A)$. The initial value problem of CGLE (1.1) can be rewritten as

$$\begin{cases} u_t + Au = F(u), \\ u(0) = u_0, \end{cases} \tag{3.1}$$

where $F(u) = \rho u + (b + i\gamma) |u|^2 u$.

From Lemma 2.1 it follows that A is a strongly elliptic operator for $0 < a \le 1$ and the operator $-A$ is the infinitesimal generator of an analytic semigroup of contractions on H.

In order to prove the existence of a local solution of the initial value problem (3.1) for every $u_0 \in H^2$, we will use Lemma 2.3 and the remark 2.4. Next we prove the needed properties of the nonlinear operator F.

Lemma 3.1. The nonlinear mapping $F(u) = \rho u + (a + i\gamma)|u|^2 u$ maps H^2 into H and satisfies for $u, v \in H^2$,

$$\|F(u)\| \le c\|u\|_\infty^2 \|u\|_{H^2}, \tag{3.2}$$

$$\|F(u) - F(v)\| \le c(\|u\|_{H^2}, \|v\|_{H^2})\|u - v\|_{H^2}. \tag{3.3}$$

We give the brief proof by using the standard methods given by A. Pazy [8].

Proof. From Sobolev's theorem in R^3, it follows that $H^2(R^3) \subset L^\infty(R^3)$. And there is a constant c such that

$$\|u\|_\infty \le c\|u\|_{H^2} \quad (u \in H^2(R^3)). \tag{3.4}$$

Denoting by D any first order differential operator, we have for every $u \in H^2(R^3)$,

$$\left|D^2(u + |u|^2 u)\right| = \left|D^2 u + D^2(|u|^2 u)\right| \le c((|u|^2 + 1)|D^2 u| + |u||Du|^2)$$

and therefore

$$\left\|(u + |u|^2 u)\right\| \le c(\|u\|_\infty^2 \|u\|_{H^2} + \|u\|_\infty \|u\|_{1,4}^2). \tag{3.5}$$

From Gagliardo-Nirenberg inequality we have

$$\|u\|_{1,4} \le c\|u\|_\infty^{1/2} \|u\|_{H^2}^{1/2}, \tag{3.6}$$

and combining (3.4), (3.5), and (3.6) we obtain (3.2).

Therefore, from the Lemma 3.1 and Lemma 2.3, it implies:

Lemma 3.2. For every $u_0 \in H^2$ there exists a unique solution u of the initial value problem (3.1) defined for $t \in [0, T_{\max}]$ such that

$$u \in C^1((0,\ T_{\max});\ H^2) \cap C([0,\ T_{\max});\ H^2),$$

where $\lim_{t \to T_{\max}} \|u(t)\|_{H^2} = \infty$ with the property that either $T_{\max} = \infty$ or $T_{\max} < \infty$.

3.2 The fundamental priori estimates

In order to get the results we follow the approach in [8]. We will first show $\|u(t)\|$ is bounded on $[0, T]$, i.e., $u(t)$ is bounded in H.

Lemma 3.3. Let $u_0 \in H$ and let $u(t)$ be the solution of the initial value problem (3.1) on $[0, T]$, then $\|u(t)\|$ is bounded on $[0, T]$, i.e.,

$$\|u(t)\| < C_1(\|u_0\|, T), \forall t \in [0,\ T],$$

where C_1 depends on the initial value and T, moreover, if $\|u_0\| \le K$ then T depends on K.

Proof. By taking the inner product of equation (1.1) with u in H and considering the real part of the result, we see that u satisfies the L^2 energy-type equation

$$\frac{1}{2}\frac{d}{dt}\|u\|^2 = \rho\int |u|^2 - a\|\nabla u\|^2 - b\int |u|^4, \tag{3.7}$$

for $t \in [0, T]$, where ∇ is the gradient operator in R^3. By applying the Young's inequality we can get that

$$\rho\int |u|^2 = \int |u|^2 \cdot \rho \le \int \left(\frac{(\sqrt{b}\,|u|^2)^2}{2} + \frac{(\sqrt{\rho}/\sqrt{b})^2}{2} \right). \tag{3.8}$$

Substituting (3.8) into (3.7), we obtain

$$\frac{d}{dt}\|u\|^2 + b\int |u|^4 \le c, \tag{3.9}$$

where c is a constant and depends on ρ, γ, μ, Ω, but is indepent of the initial value.

By using Young's inequality again, we have

$$\|u\|^2 = \int |u|^2 \le \int \frac{(\sqrt{2b}\,|u|^2)^2}{2} + c. \tag{3.10}$$

From (3.9) and (3.10), by using Gronwall's inequality, we know that

$$\|u\| \le \|u_0\| e^t + c(1 - e^{-t}). \tag{3.11}$$

Therefore, there exists $C_1 = C_1(\|u_0\|, T)$ such that

$$\|u(t)\| < C_1(\|u_0\|, T), \quad \forall t \in [0, T].$$

Thus, the proof is complete.

Lemma 3.4. The solution $u(t)$ of the initial value problem (3.1) satisfies

$$\begin{aligned} \frac{1}{4}\frac{d}{dt}\int |u|^4 \le -\frac{b}{2}\int |u|^6 - \frac{1}{4}\int (3a\left|\nabla |u|^2\right|^2 \\ -2\gamma\nabla |u|^2 \cdot i(u\nabla u^* - u^*\nabla u) \\ +a\,|u\nabla u^* - u^*\nabla u|^2) + c. \end{aligned} \tag{3.12}$$

Proof. By taking the inner product of equation (1.1) with $|u|^2 u$ in H and considering the real part of the result we find

$$\frac{1}{4}\frac{d}{dt}\int |u|^4 = \rho\int |u|^4 + Re(a + i\gamma)\int \Delta u\,|u|^2 u^* - b\int |u|^6. \tag{3.13}$$

By a careful calculating which according to Young's inequality and Lemma 3.3, we point out here that Lemma 3.4 is hold.

Lemma 3.5. Let $p \leq \frac{2(a^2+\gamma^2)+2a\sqrt{a^2+\gamma^2}}{\gamma^2}$, then the solution $u(t)$ of the initial value problem (3.1) satisfies

$$\|u\|_p^p \leq C_2, \ \forall t > t_2,$$

where C_2 is indepent of the initial value, t_2 depends on K where $\|u_0\|_p \leq K$.

Proof. By calculating we have

$$\begin{aligned}
\frac{1}{p}\frac{d}{dt}\|u\|_p^p &= \frac{1}{p}\int \frac{d}{dt}|u|^p = \frac{1}{p}\int p\cdot |u|^{p-2}\cdot Re(u^*\cdot u_t) \\
&= Re \int |u|^{p-2} \cdot (u^* \cdot u_t) \\
&= Re \int |u|^{p-2} u^*(\rho u + (a+i\gamma)\Delta u - (b+i\mu)|u|^2 u) \\
&= \rho \int |u|^p - b\int |u|^{p-2}|u|^2 u^* u + Re\int (a+i\gamma)|u|^{p-2} u^*\Delta u \\
&= \rho\|u\|_p^p - b\|u\|_{p+2}^{p+2} + Re(a+i\gamma)\int |u|^{p-2} u^*\Delta u. \qquad (3.14)
\end{aligned}$$

Considering the last term of (3.14), we can deduce some estimates about it by applying the Young' inequality and Gronwall's inequality. A careful calculation about (3.14) show that for some C_2, the Lemma 3.5 holds.

Lemma 3.6. If there exists $\delta > 0$, such that the matrix B shown below is nonnegative definite

$$B = \begin{pmatrix} 3(b+a\delta) & \mu - \gamma\delta \\ \mu - \gamma\delta & b + a\delta \end{pmatrix}.$$

Then, the solution $u(t)$ of the initial value problem (3.1) satisfies

$$\|\nabla u\|^2 \leq C_3, \ \forall t \geq t_3,$$

where C_3 is indepent of the initial value, moreover, if $\|u_0\|_4 \leq K$, $\|u_0\|_{H^1} \leq K$, then t_3 depends on K.

Proof. By taking the inner product of equation (1.1) with Δu in H and considering the real part of the result we see that

$$\frac{1}{2}\frac{d}{dt}\|\nabla u\|^2 + a\|\Delta u\|^2 = \rho\|\nabla u\|^2 + Re(b+i\mu)\int |u|^2 u\Delta u^*. \qquad (3.15)$$

From Lemma 3.3 and by applying the Cauchy-Schwarz inequality we obtain

$$\|\nabla u\|^2 = (-\Delta u, u) \leq \|\Delta u\| \|u\| \leq c\|\Delta u\|.$$

Thanks to the Young's inequality, we can get that

$$(\rho+\frac{1}{2})\left\|\nabla u\right\|^2 \le (\rho+\frac{1}{2})c\left\|\Delta u\right\| \le a\left\|\Delta u\right\|^2 + \frac{1}{4a}(\rho+\frac{1}{2})^2c^2.$$

From Lemma 3.3 and by applying the Young's inequality we obtain

$$\frac{\delta}{4}\int |u|^4 = \int \sqrt{\delta b}\,|u|^3 \cdot \frac{1}{4\sqrt{\delta b}}\,|u| \le \frac{\delta b}{2}\int |u|^6 + c\int |u|^2 \le \frac{\delta b}{2}\int |u|^6 + c.$$

Whence we deduce that

$$\frac{d}{dt}V_\delta(u(t)) + V_\delta(u(t)) \le c.$$

Thanks to Gronwall inequality, we have

$$V_\delta(u(t)) = \left\|\nabla u\right\|^2 + \frac{\delta}{2}\int |u|^4 \le V_\delta(u(0))e^{-t} + c(1-e^{-t}).$$

The proof of the Lemma 3.6 is complete.

Lemma 3.7. If $a^2 + a\sqrt{a^2+\gamma^2} \ge 3\gamma^2$, i.e., $|\gamma| \le \frac{\sqrt{7}}{3}a$ then the solution $u(t)$ of the initial value problem (3.1) satisfies

$$\left\|\Delta u\right\|^2 \le C_4(T, \left\|u_0\right\|_{H^2}^2),\ t\in[0,T],$$

where C_4 depends on the initial value and T, T depends on K, $\left\|u_0\right\|_{H^2} \le K$.

Proof. By multiplying equation (1.1) by $\Delta^2 u$ in H and considering the real part of the result we see that

$$\frac{1}{2}\frac{d}{dt}\left\|\Delta u\right\|^2 = \rho\left\|\Delta u\right\|^2 - \left\|\nabla\Delta u\right\|^2 - Re(b+i\mu)\int |u|^2 u\Delta^2 u^*. \quad (3.16)$$

Now we consider the last term on the right-hand side of (3.16),

$$\begin{aligned} Re(b+i\mu)\int |u|^2 u\Delta^2 u^* &= -Re(b+i\mu)\int \nabla(|u|^2 u)\cdot\nabla\Delta u^* \\ &= -Re(b+i\mu)\int |u|^2\,\nabla u\cdot\nabla\Delta u^* \\ &\quad -Re(b+i\mu)\int \nabla|u|^2\cdot u\nabla\Delta u^* \\ &= Re(b+i\mu)\int |u|^2\,\nabla u\cdot\nabla\Delta u^* \\ &\quad +Re(b+i\mu)\int (u\nabla u^* + u^*\nabla u)u\nabla\Delta u^*. \end{aligned} \quad (3.17)$$

By applying Lemma 3.5 we know that if $8 \leq \frac{2(a^2+\gamma^2)+2a\sqrt{a^2+\gamma^2}}{\gamma^2}$, i.e., $|\gamma| \leq \frac{\sqrt{7}}{3}a$, then $\|u\|_8^8 < c$. So we can estimate the first term of (3.17) to find that

$$\begin{aligned}
&Re(b+i\mu)\int |u|^2 \nabla u\cdot\nabla\Delta u^* \\
&\leq \sqrt{b^2+\mu^2}\int |u|^2 |\nabla u| |\nabla\Delta u| \\
&\leq \sqrt{b^2+\mu^2}\,\|u\|_8^2 \|\nabla u\|_{L^4} \|\nabla\Delta u\| \\
&\leq \frac{1}{12}\|\nabla\Delta u\|^2 + 3(b+\mu)\|\nabla u\|_{L^4}^2 \|u\|_8^4 \\
&\leq \frac{1}{12}\|\nabla\Delta u\|^2 + \frac{1}{12}\|\nabla\Delta u\|^2 + c\|\nabla u\|^2\cdot(\|u\|_8^8)^2 \\
&\leq \frac{1}{6}\|\nabla\Delta u\|^2 + c.
\end{aligned} \tag{3.18}$$

Similarly, estimating the last term of (3.17) we can get

$$Re(b+i\mu)\int (u\nabla u^* + u^*\nabla u)u\nabla\Delta u^* \leq \frac{1}{6}\|\nabla\Delta u\|^2 + c. \tag{3.19}$$

Form (3.17)–(3.19) it follows that

$$\frac{d}{dt}\|\Delta u\|^2 + \|\nabla\Delta u\|^2 \leq 2\rho\|\Delta u\|^2 + c. \tag{3.20}$$

We omit the term $\|\nabla\Delta u\|^2$ to get the result

$$\frac{d}{dt}\|\Delta u\|^2 \leq 2\rho\|\Delta u\|^2 + c.$$

Then, by using the Gronwall inequality we obtain

$$\|\Delta u\|^2 \leq C_4(T, \|u_0\|_{H^2}^2), t \in [0, T].$$

From the Lemma 3.3–3.7, we know that $u(t)$ is bounded in H^2, i.e., there exists $C > 0$, such that $\|u(t)\|_{H^2} \leq C$.

Theorem 3.8. If the parameters γ, μ, a, b satisfy

(i) $|\gamma| \leq \frac{\sqrt{7}}{3}a$.

(ii) There is a $\delta > 0$ such that $|\mu - \gamma\delta| \leq \sqrt{3}|b + a\delta|$. Then for $\forall T > 0$, $u_0 \in H^2$, the initial value problem for the 3D cubic complex Ginzburg-Landau equation

$$\begin{cases} u_t = \rho u + (a+i\gamma)\Delta u - (b+i\mu)|u|^2 u, \\ u(0,x) = u_0(x),\ x \in \Omega \end{cases} \tag{3.21}$$

has a unique global solution $u \in C^1((0,\ T];\ H^2) \cap C([0,\ T];\ H^2)$.

References

[1] N. N. Akhmediev and A. Ankiewicz. Solitions: Nonlinear Pulses and Beams. Chapman and Hall, London, 1997.

[2] S. Wang and M. Zhan, L^p solutions to the time dependent Ginzburg-Landau equations of superconductivity. Nonlinear Anal. 36 (1999), 661–677.

[3] B. Wang and N. Su, Existence of the solutions for the Ginzburg-Landau equations of superconductivity in three spatial dimensions. Proc. Roy. Soc. Edinburg Sect. A 129 (1999), 627–639.

[4] J. M. Ghidaglia and B. Héron, Dimension of the attractor associated to the Ginzburg-Landau equation. Physica D 28 (1987), 282–304.

[5] B. Guo and B. Wang, Finite dimensional behavior for the derivative Ginzburg-Landau equation in two spatial dimensions. Physica D 89 (1995), 83–99.

[6] A. Mielke and G. Schneider, Attractors for modulation equation on unbounded domains, existence and comparison. Nonlinearity 8 (1995), 743–768.

[7] C. Bu, On the Cauchy problem for the 1+2 complex Ginzburg-Landau equation. J. Austral Math. Soc. Ser. B 36 (1994), 313–324.

[8] A. Pazy, Semigroup of Linear Operators and Applications to PDE. Springer, New York, 1983.

Exact Configuration for 3D Ginzburg-Landau Equation Based on Some ODEs*

Penghong Zhong

Department of Applied Mathematics, Beijing University of Technology
Beijing 100124, China
E-mail: zhongph@emails.bjut.edu.cn

Shu Wang

Department of Applied Mathematics, Beijing University of Technology
Beijing 100124, China
E-mail: wangshu@bjut.edu.cn

Ke Wang

Department of Applied Mathematics, Beijing University of Technology
Beijing 100124, China
E-mail: coco@emails.bjut.edu.cn

Ronghui Yang

Department of mathematics, Yunnan Nationalities University
Kunming 650031, China
E-mail: yangronghui1978@yahoo.cn

Abstract

In this paper, two different ordinary differential equations (ODE) which deduce the Jacobi elliptic function method, the extended tanh-function method are used to construct exact solutions of 3D Complex Ginzburg-Landau equation. Exact configurations about the exact solutions are obtained according to these two different methods.

*This work is supported by NSFC (Grant No. 10771009 and No. 10861014), BSFC (Grant No. 1085001) of China, Funding Project for Academic Human Resources Development in Institutions of Higher Leading Under the Jurisdiction of Beijing Municipality (PHR-IHLB 200906103), and Beijing Education Committee Funds.

1 Introduction

As mathematical models of the nonlinear phenomena, the investigation of exact solutions of nonlinear partial differential equations will help one to understand the mechanism that governs these various physical models or to better provide knowledge of the physical problem and possible applications. In these nonlinear models, the investigation of exact solutions of complex Ginzburg-Landau equation (CGLE)[1] plays an important role in the study of nonlinear physical phenomena. Ordinarily, the 3D cubic CGLE is given by

$$u_t = \rho u + (1 + i\gamma)\Delta u - (1 + i\mu)\left|u\right|^2 u, \tag{1.1}$$

where u is a complex valued function defined on $n+1$ dimensional space-time R^{n+1}. Δ is the Laplacian operator in R^n and ρ, γ, μ are real parameters with $\rho > 0$. This equation describes on a qualitative, and often even on a quantitative, level a vast variety of phenomena from nonlinear waves to second-order phase transitions, from superconductivity, superfluidity, and Bose-Einstein condensation to liquid, crystals and strings in field theory. Some similar anamorphosis about this equation has been studied extensively in various aspects. The existence and stability of the periodic solution for the one-dimensional case are proven in[2, 3]. One-dimensional chirped bright and dark soliton-like solutions of cubic CGLE were investigated [4]. The new exact homoclinic wave and soliton solutions for the 2D CGLE were obtained [7]. Existence of the global attractor is proven in [6] for the n-dimensional space case.

In this paper by using the Jacobi elliptic function method [7], the extended tanh-function [5] method together with homogeneous balance principle [8], we obtain some exact solutions for 3D CGLE including periodic wave and singularity solutions.

2 The auxiliary function method

In this section, by use of Jacobi elliptic function method together with homogeneous balance principle we obtain the new periodic wave solutions. Let

$$u = e^{i\omega t}\varphi(\xi). \tag{2.1}$$

Substituting (2.1) into equation (1.1) and canceling $e^{i\omega t}$, we get

$$i\omega\varphi + \varphi_t - \rho\varphi - (1 + i\gamma)(\varphi_{xx} + \varphi_{yy} + \varphi_{zz}) + (1 + i\mu)\varphi^3 = 0. \tag{2.2}$$

Separating the real part and imaginary part of resulting we have

$$\varphi_t - \rho\varphi - (\varphi_{xx} + \varphi_{yy} + \varphi_{zz}) + \varphi^3 = 0, \tag{2.3}$$

$$\omega\varphi - \gamma(\varphi_{xx} + \varphi_{yy} + \varphi_{zz}) + \mu\varphi^3 = 0. \tag{2.4}$$

Suppose

$$\varphi = f(\xi), \tag{2.5}$$

where $\xi = \alpha_1 x + \alpha_2 y + \alpha_3 z + \xi_0$, and $\alpha_i (i = 1, 2, 3), \xi_0$ are constants to be determined.

Substituting (2.5) into equation (2.3)–(2.4), we get the nonlinear ODES:

$$-Af'' + f^3 - \rho f = 0, \tag{2.6}$$

$$-\gamma Af'' + \mu f^3 + \omega f = 0, \tag{2.7}$$

where $A = \alpha_1^2 + \alpha_2^2 + \alpha_3^2$.

Introducing an auxiliary function

$$f(\xi) = a_0 + a_1 F(\xi) + a_2 F^2(\xi) + \cdots + a_m F^m(\xi),$$

where $F(\xi)$ satisfies following equation

$$F'^2(\xi) = q_0 + q_2 F^2(\xi) + q_4 F^4(\xi). \tag{2.8}$$

Considering the balance of the highest order of f'' andf^3 on F, we use the homogenous balance principle to get that $m + 2 = 3m$, i.e., $m = 1$, so we have

$$f(\xi) = a_0 + a_1 F(\xi), a_1 \neq 0, \tag{2.9}$$

where a_0, a_1 are constants to be determined.

Substituting (2.9) into (2.6)–(2.7)

$$-Aa_1(q_2 F + 2q_4 F^3) + (a_0 + a_1 F)^3 - \rho(a_0 + a_1 F) = 0, \tag{2.10}$$

$$-\gamma Aa_1(q_2 F + 2q_4 F^3) + \mu(a_0 + a_1 F)^3 + \omega(a_0 + a_1 F) = 0. \tag{2.11}$$

Taking the coefficients of different powers of F in (2.10) and (2.11) as 0, we obtain a set of algebraic equations:

$$F^0 : a_0^3 - \rho a_0 = 0, \quad \mu a_0^3 + \omega a_0 = 0, \tag{2.12}$$

$$F^1 : -Aa_1 q_2 + 3a_0^2 a_1 - \rho a_1 = 0, \quad -\gamma Aa_1 q_2 + 3\mu a_0^2 a_1 + \omega a_1 = 0, \tag{2.13}$$

$$F^2 : 3a_0 a_1^2 = 0, \quad 3\mu a_0 a_1^2 = 0, \tag{2.14}$$

$$F^3 : -2\gamma Aa_1 q_4 + \mu a_1^3 = 0, \quad -2Aa_1 q_4 + a_1^3 = 0. \tag{2.15}$$

Solving above equations we have

$$a_0 = 0, \quad \rho = -Aq_2, \quad \omega = \gamma Aq_2,$$
$$-2\gamma Aq_4 + \mu a_1^2 = 0, \quad \cdot\, a_1^2 = 2Aq_4,$$

i.e., $A = -\frac{\rho}{q_2}$, $\omega = -\gamma\rho$, $a_1 = \pm\sqrt{-\frac{2\rho q_4}{q_2}}$, and a constraint condition of coefficents in equation $\gamma = \mu$. Substituting all these results into (2.1) we get the solutions as follows

$$u = \pm e^{-i\gamma\rho t}\sqrt{-\frac{2\rho q_4}{q_2}}F(\xi), \tag{2.16}$$

where $a_1 = \pm\sqrt{-\frac{2\rho q_4}{q_2}}$, a_1 is real if $\frac{q_4}{q_2} < 0$.

From (2.8) associated with various values of q_0, q_2, q_4 we obtain solutions (Jacobi elliptic function) of Eq. (1.1) as follows:

(i) If $q_0 = 1$, $q_2 = -(1+R^2)$, $q_4 = R^2$, it follows from (2.8) $F(\xi) = \text{sn}\xi$. Then

$$u(x,y,z,t) = u = \pm e^{-i\gamma\rho t}\sqrt{\frac{2\rho R^2}{1+R^2}}\,\text{sn}\xi,$$

where $\xi = \alpha_1 x + \alpha_2 y + \alpha_3 z + \xi_0, \alpha_1^2 + \alpha_2^2 + \alpha_3^2 = \frac{\rho}{1+m^2}$.

In the limit case when $R \to 1$, $\text{sn}\xi \to \tanh\xi$, thus

$$u(x,y,z,t) = u = \pm e^{-i\gamma\rho t}\sqrt{\rho}\tanh\xi.$$

(ii) Here we give the table 2.1 to show the details about the other solutions.

Table 2.1. Relations about the solutions (2.16).

$F(\xi,R)$	q_4	q_2	q_0	$F(R\to 0)$	$F(R\to 1)$
ns	R^2	$-(1+R^2)$	1	csc	cth
dn	$-(1-R^2)$	$2-R^2$	-1	1	sech
cn	$1-R^2$	$2R^2-1$	$-R^2$	cos	sech
nd	-1	$2-R^2$	$-(1-R^2)$	1	ch
sc	$1-R^2$	$2-R^2$	1	tan	sh
sd	$-R^2(1-R^2)$	$2R^2-1$	1	sin	sh
cs	1	$2-R^2$	$1-R^2$	cot	ch
cd	R^2	$-(1+R^2)$	1	cos	1
ds	1	$2R^2-1$	$-R^2(1-R^2)$	csc	csch

According to (2.16) and Table 2.1, the other solutions can be deduced easily.

3 The extended tanh-function method

In this section, more details about the other solutions which include the singularity solutions are obtained by using the extended tanh-function method.

We can consider a similar 3D CGLE (Obviously, (1.1) is a subcase of it)

$$u_t = \rho u + (a + i\gamma)\Delta u - (b + i\mu)\left|u\right|^2 u. \tag{3.1}$$

Suppose

$$u = e^{i\eta} f(\xi), \tag{3.2}$$

where $\eta = k_1 x + k_2 y + k_3 z + \omega t + \eta_0$, $\xi = \alpha_1 x + \alpha_2 y + \alpha_3 z + \beta t + \xi_0$ and β, ω, k_i, α_i $(i = 1, 2, 3)$ are constants to be determined.

Substituting (3.2) into (3.1) and separating the real part and the imaginary part, we have

$$f'' + \frac{2aB}{\gamma A} f' - \frac{\omega + \gamma C}{\gamma A} f - \frac{\mu}{\gamma A} f^3 = 0 \tag{3.3}$$

and

$$f'' - \frac{2\gamma B - \beta}{aA} f' - \frac{C - \rho}{\gamma A} f - \frac{b}{aA} f^3 = 0, \tag{3.4}$$

where $A = \alpha_1^2 + \alpha_2^2 + \alpha_3^2$, $B = k_1\alpha_1 + k_2\alpha_2 + k_2\alpha_2$, $C = k_1^2 + k_2^2 + k_3^2$.

If the coefficients in (3.3) and (3.4) satisfy

$$\frac{2aB}{\gamma A} = -\frac{\beta + 2\gamma B}{aA}, \ \frac{\omega + \gamma C}{\gamma A} = \frac{C - \rho}{\gamma A}, \ \frac{\mu}{\gamma A} = \frac{b}{aA}, \text{ i.e.}$$

$$a\mu = b\gamma, \ \omega = \frac{\gamma(C - \rho - aC)}{a}, \ \beta = -\frac{2(a^2 + \gamma^2)B}{\gamma},$$

then we can transform (3.3) and (3.4) into the equation

$$f'' - Mf' + Nf + Pf^3 = 0, \tag{3.5}$$

where $M = \frac{-2aB}{\gamma A}$, $N = \frac{\rho - C}{aA}$, $P = \frac{-b}{aA}$. We suppose that the solution of equation (3.5) is

$$f(\xi) = \sum_{i=1}^{n} \sigma^{i-1}(\xi)\left[A_i\sigma(\xi) + B_i\tau(\xi)\right] + A_0, \tag{3.6}$$

where $(A_i(i = 0,\ 1,\ \cdots,\ n))$, $(B_i(i = 0,\ 1,\ \cdots,\ n))$ are constants to be determined. In addition, σ, τ satisfy

$$\sigma'(\xi) = \varepsilon\sigma(\xi)\tau(\xi), \tag{3.7}$$

$$\tau'(\xi) = R + \varepsilon\tau^2(\xi) - m\sigma(\xi), \ \varepsilon = \pm 1, \tag{3.8}$$

$$\tau^2(\xi) = -\varepsilon\left[R - 2m\sigma(\xi) + \frac{(m^2 - 1)}{R}\sigma^2(\xi)\right], \tag{3.9}$$

where R, m are constants to be determined and $R \neq 0$. Then we have the following solutions of equation (3.8).

For $\varepsilon = 1$

$$\begin{aligned}\sigma_1(\xi) &= \frac{R\sec(\sqrt{R}\xi)}{m\sec(\sqrt{R}\xi)+1}, \quad \tau_1(\xi) = \frac{\sqrt{R}\tan(\sqrt{R}\xi)}{m\tan(\sqrt{R}\xi)+1}, \\ \sigma_2(\xi) &= \frac{R\csc(\sqrt{R}\xi)}{m\csc(\sqrt{R}\xi)+1}, \quad \tau_2(\xi) = \frac{\sqrt{R}\cot(\sqrt{R}\xi)}{m\cot(\sqrt{R}\xi)+1}.\end{aligned} \tag{3.10}$$

For $\varepsilon = -1$

$$\begin{aligned}\sigma_3(\xi) &= \frac{R\,\mathrm{sech}(\sqrt{R}\xi)}{m\,\mathrm{sech}(\sqrt{R}\xi)+1}, \quad \tau_3(\xi) = \frac{\sqrt{R}\tanh(\sqrt{R}\xi)}{m\tanh(\sqrt{R}\xi)+1}, \\ \sigma_4(\xi) &= \frac{R\,\mathrm{csch}(\sqrt{R}\xi)}{m\,\mathrm{csch}(\sqrt{R}\xi)+1}, \quad \tau_4(\xi) = \frac{\sqrt{R}\coth(\sqrt{R}\xi)}{m\coth(\sqrt{R}\xi)+1}.\end{aligned} \tag{3.11}$$

Considering the balance of the highest order of f'' and f^3 on f in (3.6), we use the homogenous balance principle to get that $n = 1$, so we have

$$f = A_0 + A_1\sigma + B_1\tau. \tag{3.12}$$

Substituting (3.8), (3.9) and (3.12) into (3.5) and equating the coefficients of different powers of $\sigma^j\tau^i (i = 0, 1,\ j = 0, 1, 2, 3)$ to 0, we get

$$\sigma^3 : -\frac{2A_1\varepsilon^3(m^2-1)}{R} + PA_1^3 - \frac{3PA_1B_1^2\varepsilon(m^2-1)}{R} = 0, \tag{3.13}$$

$$\sigma^2\tau : -\frac{2B_1\varepsilon^3(m^2-1)}{R} + 3PA_1^2B_1 - \frac{PB_1^3\varepsilon(m^2-1)}{R} = 0, \tag{3.14}$$

$$\begin{aligned}\tau^2 : 4A_1\varepsilon^3 m - A_1\varepsilon m + \frac{MB_1\varepsilon^2(m^2-1)}{R} + 3PA_0A_1^2 \\ - \frac{3PA_0B_1^2\varepsilon(m^2-1)}{R} + 6PA_1B_1^2\varepsilon m = 0,\end{aligned} \tag{3.15}$$

$$\sigma\tau : 2B_1\varepsilon^3 m - B_1\varepsilon m - MA_1\varepsilon + 2PB_1^3\varepsilon m + 6PA_0A_1B_1 = 0, \tag{3.16}$$

$$\begin{aligned}\sigma : -2A_1\varepsilon^3 R + A_1\varepsilon R - 2MB_1\varepsilon^2 m + MB_1 m + 3PA_0^2A_1 \\ +6PA_0B_1^2\varepsilon m - 3PA_1B_1^2\varepsilon R + NA_1 = 0,\end{aligned} \tag{3.17}$$

$$\tau : NB_1 - PB_1^3\varepsilon R + 3PA_0^2B_1 = 0, \tag{3.18}$$

$$\sigma^0\tau^0 : -MB_1R + MB_1\varepsilon^2 R + NA_0 + PA_0^3 - 3PA_0B_1^2\varepsilon R = 0. \tag{3.19}$$

Solving the above equations, we can get the following results.

(1) For $\varepsilon = 1$, $M = 0$, and $m = 0$, we have

(i) $A_0 = B_1 = 0$, $\beta = 0$, $A_1^2 = \frac{2a^2A^2}{b(\rho-C)}$, $R = \frac{\rho-C}{aA}$. Then we obtain the following solutions:

$$u_1 = \sqrt{\frac{2(\rho-C)}{b}} \cdot \sec\left(\sqrt{\frac{\rho-C}{aA}}\xi\right) \cdot e^{i\eta}, \tag{3.20}$$

$$u_2 = \sqrt{\frac{2(\rho - C)}{b}} \cdot \csc\left(\sqrt{\frac{\rho - C}{aA}}\xi\right) \cdot e^{i\eta}. \tag{3.21}$$

(ii) $A_0 = A_1 = 0$, $\beta = 0$, $B_1^2 = \frac{2aA}{b}$, $R = \frac{C-\rho}{2aA}$. We have

$$u_3 = \sqrt{\frac{C - \rho}{b}} \cdot \tan\left(\sqrt{\frac{C - \rho}{2aA}}\xi\right) \cdot e^{i\eta}, \tag{3.22}$$

$$u_4 = \sqrt{\frac{C - \rho}{b}} \cdot \cot\left(\sqrt{\frac{C - \rho}{2aA}}\xi\right) \cdot e^{i\eta}, \tag{3.23}$$

where $B = k_1\alpha_1 + k_2\alpha_2 + k_2\alpha_2 = 0$, $\omega = \frac{\gamma(C-\rho-aC)}{a}$.

(2) For $\varepsilon = 1$, $M = 0$ $(B = 0)$, we have $A_0 = 0$, $A_1^2 = \frac{a^2A^2(m^2-1)}{4b(\rho-C)}$, $B_1^2 = \frac{aA}{2b}$, $R = \frac{2(C-\rho)}{aA}$, $\beta = 0$,

$$u_5 = \left[\frac{\sqrt{\frac{a^2A^2(m^2-1)}{4b(\rho-C)}}\frac{2(C-\rho)}{aA}\sec\left(\sqrt{\frac{2(C-\rho)}{aA}}\xi\right)}{m\sec\left(\sqrt{\frac{2(C-\rho)}{aA}}\xi\right)+1} + \frac{\sqrt{\frac{C-\rho}{b}}\tan\left(\sqrt{\frac{2(C-\rho)}{aA}}\xi\right)}{m\tan\left(\sqrt{\frac{2(C-\rho)}{aA}}\xi\right)+1}\right] e^{i\eta}, \tag{3.24}$$

$$u_6 = \left[\frac{\sqrt{\frac{a^2A^2(m^2-1)}{4b(\rho-C)}}\frac{2(C-\rho)}{aA}\csc\left(\sqrt{\frac{2(C-\rho)}{aA}}\xi\right)}{m\csc\left(\sqrt{\frac{2(C-\rho)}{aA}}\xi\right)+1} + \frac{\sqrt{\frac{C-\rho}{b}}\cot\left(\sqrt{\frac{2(C-\rho)}{aA}}\xi\right)}{m\cot\left(\sqrt{\frac{2(C-\rho)}{aA}}\xi\right)+1}\right] e^{i\eta}. \tag{3.25}$$

(3) For $\varepsilon = -1$, $M = 0$ $(B = 0)$, and $m = 0$, we have

(i) $A_0 = B_1 = 0$, $\beta = 0$, $A_1^2 = \frac{2a^2A^2}{b(\rho-C)}$, $R = \frac{C-\rho}{aA}$,

$$u_7 = \sqrt{\frac{2a^2A^2}{b(\rho - C)} \cdot \frac{C - \rho}{aA}}\,\mathrm{sech}\left(\sqrt{\frac{C - \rho}{aA}}\xi\right) \cdot e^{i\eta}, \tag{3.26}$$

$$u_8 = \sqrt{\frac{2a^2A^2}{b(\rho - C)} \cdot \frac{C - \rho}{aA}}\,\mathrm{csch}\left(\sqrt{\frac{C - \rho}{aA}}\xi\right) \cdot e^{i\eta}. \tag{3.27}$$

(ii) $A_0 = A_1 = 0$, $\beta = 0$, $B_1^2 = \frac{2aA}{b}$, $R = \frac{\rho - C}{2aA}$,

$$u_9 = \sqrt{\frac{\rho - C}{b}} \cdot \tanh\left(\sqrt{\frac{\rho - C}{2aA}}\xi\right) \cdot e^{i\eta}, \tag{3.28}$$

$$u_{10} = \sqrt{\frac{\rho - C}{b}} \cdot \coth\left(\sqrt{\frac{\rho - C}{2aA}}\xi\right) \cdot e^{i\eta}. \tag{3.29}$$

(4) For $\varepsilon = -1$, $M = 0$, we have $A_0 = 0$, $A_1^2 = \frac{a^2A^2(m^2-1)}{4b(C-\rho)}$, $B_1^2 = \frac{aA}{2b}$, $R = \frac{2(\rho - C)}{aA}$, $\beta = 0$,

$$u_{11} = \left[\frac{\sqrt{\frac{a^2A^2(m^2-1)}{4b(C-\rho)}}\frac{2(\rho - C)}{aA}\operatorname{sech}\left(\sqrt{\frac{2(\rho-C)}{aA}}\xi\right)}{m\operatorname{sech}\left(\sqrt{\frac{2(\rho-C)}{aA}}\xi\right) + 1} + \frac{\sqrt{\frac{\rho-C}{b}}\tanh\left(\sqrt{\frac{2(\rho-C)}{aA}}\xi\right)}{m\tanh\left(\sqrt{\frac{2(\rho-C)}{aA}}\xi\right) + 1}\right] e^{i\eta}, \tag{3.30}$$

$$u_{12} = \left[\frac{\sqrt{\frac{a^2A^2(m^2-1)}{4b(C-\rho)}}\frac{2(\rho-C)}{aA}\operatorname{csch}\left(\sqrt{\frac{2(\rho-C)}{aA}}\xi\right)}{m\operatorname{csch}\left(\sqrt{\frac{2(\rho-C)}{aA}}\xi\right) + 1} + \frac{\sqrt{\frac{\rho-C}{b}}\coth\left(\sqrt{\frac{2(\rho-C)}{aA}}\xi\right)}{m}\coth\left(\sqrt{\frac{2(\rho - C)}{aA}}\xi\right) + 1\right] e^{i\eta}, \tag{3.31}$$

where $B = k_1\alpha_1 + k_2\alpha_2 + k_2\alpha_2 = 0$, $\omega = \frac{\gamma(C - \rho - aC)}{a}$.

4 Conclusions

In this paper, by using the Jacobi elliptic function method together with homogeneous balance principle and the extended tanh-function method, we obtain exact solutions of 3D complex Ginzburg-Landau equation under the constraint condition $\gamma = \mu$. As a result, we obtain periodic wave and kink wave solutions.

References

[1] N. N. Akhmediev and A. Ankiewicz, Solitons: Nonlinear Pulses and Beams. Chapman and Hall London, (1997), 267–311.

[2] B. L. Guo and R. Yuan, Existence of periodic solutions of the generalized Ginzburg-Landau equation with periodic boundary condition. Communications in Nonlinear Science and Numerical Simulation. 5 (2000), no. 2 69–73.

[3] B. L. Guo, B. N. Lu and G. H. Wan, Stability of travelling wave solutions of the derivative Ginzburg-Landau equations. Communications in Nonlinear Science and Numerical Simulation. 3 (1997), no. 2 150–156.

[4] H. M. Fu and Z. D. Dai, Exact chirped solitary-wave solutions for Ginzburg-Landau equation. Communications in Nonlinear Science and Numerical Simulation. 15 (2010), no. 6, 1462-1465.

[5] E. G. Fan, Extended tanh-function method and its applications to nonlinear equations. Phys. Lett. A. 277 (2000), 212–218.

[6] H. Y. Huang, On the higher-order nonlinear Ginzburg-Landau system in n spatial dimensions. Communications in Nonlinear Science and Numerical Simulation. 3 (1998), no. 1, 15–19.

[7] P. H. Zhong, R. H. Yang and G. S. Yang, Exact periodic and blow up solutions for 2D Ginzburg-Landau equation. Phys. Lett. A. 373 (2008), no. 1, 19–22.

[8] M. L. Wang, Solitary wave solution for Boussinesq equations. Phys. Lett. A. 199 (1995), 162–172.

Author Index

Hyperbolic Problems

Theory, Numerics and Applications

Volume 1

Series in Contemporary Applied Mathematics CAM

---

1 Jiongmin Yong, Rama Cont, eds., *Mathematical Finance: Theory and Practice* (2000)

2 F. Dubois, Huamo Wu, eds., *New Advances in Computational Fluid Dynamics: Theory, Methods and Applications* (2001)

3 Hanji Shang, Alain Tosseti, eds., *Actuarial Science: Theory and Practice* (2002)

4 Alexandre Ern, Weiping Liu, eds., *Mathematical Problems in Environmental Science and Engineering* (2002)

5 Haïm Brezis, Tatsien Li, eds., *Ginzburg-Landau Vortices* (2005)

6 Tatsien Li, Pingwen Zhang, eds., *Frontiers and Prospects of Contemporary Applied Mathematics* (2006)

7 Deguan Wang, Christian Duquennoi, Alexandre Ern, eds., *Mathematical Methods for Surface and Subsurface Hydrosystems* (2007)

8 Rolf Jeltsch, Tatsien Li, Ian H. Sloan, eds., *Some Topics in Industrial and Applied Mathematics* (2007)

9 Philippe G. Ciarlet, Tatsien Li, eds., *Differential Geometry: Theory and Applications* (2008)

10 Tatsien Li, Pingwen Zhang, eds., *Industrial and Applied Mathematics in China* (2009)

11 Zhien Ma, Yicang Zhou, Jianhong Wu, eds., *Modeling and Dynamics of Infectious Diseases* (2009)

12 Tomas Y. Hou, Chun Liu, Jianguo Liu, eds., *Multi-Scale Phenomena in Complex Fluids: Modeling, Analysis and Numerical Simulation* (2010)

13 Guiqiang Chen, Tatsien Li, Chun Liu, eds., *Nonlinear Conservation Laws, Fluid Systems and Related Topics* (2010)

14 A. Damlamian, S. Jaffard, eds., *Wavelet Methods in Mathematical Analysis and Engineering* (2010)

15 Tatsien Li, Yuejun Peng, Bopeng Rao, eds., *Some Problems on Nonlinear Hyperbolic Equations and Applications* (2010)

16 A. Damlamian, B. Miara, Tatsien Li, eds., *Multiscale Problems: Theory, Numerical Approximation and Applications* (2011)

17 Tatsien Li, Song Jiang, eds., *Hyperbolic Problems: Theory, Numerics and Applications*, Vol. 1 (2012)

18 Tatsien Li, Song Jiang, eds., *Hyperbolic Problems: Theory, Numerics and Applications*, Vol. 2 (2012)

Series in Contemporary Applied Mathematics CAM 17

Hyperbolic Problems

Theory, Numerics and Applications

Volume 1

editors

Tatsien Li
Fudan University, China

Song Jiang
IAPCM, China

Higher Education Press

World Scientific

NEW JERSEY • LONDON • SINGAPORE • BEIJING • SHANGHAI • HONG KONG • TAIPEI • CHENNAI

Tatsien Li
School of Mathematical Sciences
Fudan University
Shanghai 200433, China
Email: dqli@fudan.edu.cn

Song Jiang
Institute of Applied Physics and
Computational Mathematics
Beijing 100088, China
Email: jiang@iapcm.ac.cn

Editorial Assistant: Chunlian Zhou

Higher Education Press Limited Company
4 Dewai Dajie, Beijing 100120, P. R. China and
World Scientific Publishing Co Pte Ltd
5 Toh Tuch Link, Singapore 596224

ISBN: 978-981-4417-06-8 (set)
ISBN: 978-981-4417-07-5 (v1)

Printed in P. R. China

Scientific Committee

Yann Brenier (France)
Alberto Bressan (USA)
Constantine Dafermos (USA)
Xiaqi Ding (China)
Ling Hsiao (China)
Rolf Jeltsch (Switzerland)
Song Jiang (China)
Kenneth Karlsen (Norway)
Shuichi Kawashima (Japan)
Dietmar Kroener (Germany)
Randall LeVeque (USA)
Tatsien Li (China)
Taiping Liu (USA)
H. J. Nussenzveig Lopes (Brazil)
Pierangelo Marcati (Italy)
Denis Serre(France)
Chi-Wang Shu (USA)
Eitan Tadmor (USA)
Zhouping Xin (China)

Local Organizing Committee

Tatsien Li (Co-Chair)
Song Jiang(Co-Chair)

Feimin Huang
Hailiang Li
Yaguang Wang
Huijiang Zhao
Changjiang Zhu

Preface

The 13th International Conference on Hyperbolic Problems: Theory, Numerics and Applications (HYP 2010) was held in Beijing, China, from June 15 to 19, 2010. Over 200 participants attended the conference and 162 among them came from abroad. There were 10 plenary lectures, 18 invited talks and around 140 contributed talks in parallel sessions.

The objective of the conference is to bring together experts, researchers and students with interest in theoretical analysis, numerical simulations, and applications of hyperbolic partial differential equations and related mathematical models appearing in applied sciences. The conference keeps the traditional balance of the HYP series, blending theory, numerics and applications, and an emphasis is put on nonlinear problems and applications in various fields such as fluid mechanics, elasticity, astrophysics, biomathematics, traffic flow, etc. As has been done in the past, a special effort has been made it possible for young scientists to attend this conference and to promote their interaction with the more senior researchers.

The biannual HYP series of international conferences on Hyperbolic Problems was initiated by C. Carasso, P.-A. Raviart and D. Serre with the first conference held in Saint-Étienne, France in 1986. Since then it has been organized in Aachen (Germany, 1988), Uppsala (Sweden, 1990), Taormina (Italy, 1992), Stony Brook (USA, 1994), Hong Kong (China, 1996), Zurich (Swizerland, 1998), Magdeburg (Germany, 2000), Pasadena (USA, 2002), Osaka (Japan, 2004), Lyon (France, 2006), and Maryland (USA, 2008). Throughout these years, it has become one of the highest quality and most successful conference series in applied mathematics.

Hyperbolic problems, which are probably originated from Euler's study on the acoustic wave in 1755, not only have a long history but also have extremely rich physical background. The development is highly stimulated by their applications to Physics, Biology, and Engineering Sciences, in particular by the design of effective numerical algorithms. Due to recent rapid development of computers, more and more scientists use hyperbolic partial differential equations and related evolutionary equations as basic tools when propose new mathematical models of various phenomena and related numerical algorithms. We believe that various fields in science and engineering will bring us further into future interests in hyperbolic problems.

The scientific committee of this conference consists of Yann Brenier (France), Alberto Bressan (USA), Constantine Dafermos (USA), Xiaqi Ding (China), Ling Hsiao (China), Rolf Jeltsch (Switzerland), Song Jiang (China), Kenneth Karlsen (Norway), Shuichi Kawashima (Japan), Dietmar Kroener (Germany), Randall LeVeque (USA), Tatsien Li (China), Tai-Ping Liu (USA), Helena J. Nussenzveig Lopes (Brazil), Pierangelo Marcati (Italy), Denis Serre (France), Chi-Wang Shu (USA), Eitan Tadmor (USA) and Zhouping Xin (China). We thank all members of the scientific committee for recommending plenary and invited speakers and promoting this conference.

This conference is supported by the Academy of Mathematics and Systems Science, the Institute of Applied Physics and Computational Mathematics, Capital Normal University, Central China Normal University, Fudan University, Shanghai Jiaotong University, and Wuhan University. It is also sponsored by Shanghai Key Laboratory for Contemporary Applied Mathematics, Wuhan Institute of Physics and Mathematics, Chinese Academy of Sciences, National Natural Science Foundation of China, and Tianyuan Foundation for Mathematics.

These two volumes contain 80 original research and review papers which are written by leading researchers and promising young scientists, and cover a diverse range of multi-disciplinary topics addressing theoretical, modeling and computational issues arising under the umbrella of "Hyperbolic Partial Differential Equations". All the articles are peer-reviewed. We would like to thank all the referees for their efforts done for reading and judging the submitted articles. Special thanks go to Dr. Tao Wang for his laborious job arranging the final layout of articles.

Finally, we are extremely thankful to our colleagues Professors Feimin Huang, Hailiang Li, Yaguang Wang, Huijiang Zhao and Changjiang Zhu, and Dr. Xiaoyun Zhai, Dr. Yi Wang, and many graduate students from the Academy of Mathematics and Systems Science, the Institute of Applied Physics and Computational Mathematics, Capital Normal University, Central China Normal University, Fudan University, Shanghai Jiaotong University and Wuhan University, for their assistance in making this conference successful and comfortable.

June 2011
Tatsien Li, Song Jiang

Contents

Volume 1

Plenary Talks

Invited Talks

Volume 2

Contributed Talks

Plenary Talks

SBV Regularity for Scalar Conservation Laws*

Stefano Bianchini
SISSA
Trieste I-34136, Italy
E-mail: bianchin@sissa.it

Abstract

We outline a short proof of SBV regularity which can be extended to systems of conservation laws. The fundamental idea is the construction of an interaction measure which controls the creation of jumps.

1 Introduction

In recent years regularity estimates for nonlinear equation have received a lot of attention. A natural regularity question in hyperbolic conservation laws is the following: being the solution u in general BV, under which conditions it belongs to the smaller space Special BV (SBV)? We recall that the space SBV is the space of all BV functions for which the measure derivative u_x does not contain a Cantor part.

Since solution u to Hamilton-Jacobi equations with uniformly convex Hamiltonian are known to be semiconcave, then a similar question can be stated in this cases: we would like to prove that the measure second derivative of u does not have a Cantor part.

In this introduction we first review the most important results concerning SBV regularity.

The first positive result has been given in [1], where it is shown that the solution $u(t)$ of a genuinely nonlinear scalar conservation law in one space dimension is SBV up to countably many times. In that paper, the authors considers the characteristic lines

$$\dot{x} = f'(u(t,x)), \quad u(0,x) = y,$$

*This work is supported by ERC Grant CONSLAW.

and prove the following: every time a Cantor part in $u_x(t)$ appears, then there is a set of positive measure A such that all the characteristics starting from $y \in A$ are defined in the interval $[0, t]$ but cannot be prolonged more than t. By the σ-finiteness of $\mathcal{L}^1$, one can apply the same observation used to prove that the positive part of $u_x(t)$ is absolutely continuous up to countably many times, and deduce that up to countably many times the solution $u(t)$ is SBV.

The use of the measure of the set $A(t)$ of initial points for characteristics which can be prolonged up to time t has been applied to obtain extension of the above result: in [10] the SBV estimate is used for scalar balance laws, later extended to Temple systems in [3] and in [5] to the case of Hamilton-Jacobi equation in several space dimension with uniformly convex Hamiltonian.

In [4] a different approach is introduced. The authors show that there is a bounded measure μ^{jump} which controls the creation of jumps in the entropic solution u to a genuinely nonlinear hyperbolic system of conservation laws. The idea is particularly simple to understand: the genuinely nonlinearity makes easy to create jumps, but only by interaction and cancellation one can remove them. This measure allows to make balances concerning only the continuous part of the measure u_x, and thus to recover SBV regularity as a consequence of decay properties of the solution: in fact, for Burgers equation, the fundamental estimate

$$-\frac{\mathcal{L}^1(B)}{t-T} + \mu^{\text{jump}}(T, t) \leq u_x(T, B) \leq \frac{\mathcal{L}^1(B)}{T} \tag{1.1}$$

resembles the decay estimate for positive waves. In (1.1), B is a Borel set and $0 \leq T \leq t$.

In this short paper we give the fundamental ideas in the particularly simple case of a scalar conservation law with convex flux, and we show how the quantity $f'(u)$ is in SBV, i.e. $\sigma := D_x f(u)$ does not have a Cantor part. The paper is organized as follows. After recalling some basic fact of wavefront tracking approximation, we introduce the wave balances of the quantities $\sigma = D_x f'(u)$ and it positive, negative, continuous and jump part. These balances allows to define the new interaction measure μ^{jump}, and to prove that it is a finite measure on $\mathbb{R}^+ \times \mathbb{R}$. Once this measure has been introduced, we use this to compute the wave balance in regions bounded by characteristics, i.e. solution to the differential inclusion $\dot{x} \in f'(u(t, x))$. The argument at this point is standard, the new ingredient is the use of the measure μ^{jump} to control the continuous part of σ.

2 Preliminaries

We consider the scalar conservation law

$$u_t + f(u)_x = 0, \qquad u(0,x) = u_0(x), \tag{2.1}$$

with $u_0(x) \in \mathrm{BV}(\mathbb{R})$: for simplicity we will assume $|u_0(x)| = 1$. The flux f is assumed smooth, at least $C^2(\mathbb{R})$, and we will denote with f', f'' its first and second derivatives.

A standard method for constructing a sequence of approximate solutions converging to the unique entropy solution of (2.1) is given in [8]: let $\nu \in \mathbb{N}$ and consider the points $\{z2^{-\nu}\}_{z\in\mathbb{Z}}$. Let f^ν be the piecewise linear function whose nodes are at the points $\{(z2^{-\nu}, f(z2^{-\nu}))\}_{z\in\mathbb{Z}}$:

$$f^\nu(u) := (1-\alpha)f\big(z2^{-\nu}\big)+\alpha f\big((z+1)2^{-\nu}\big), \quad u = (1-\alpha)z2^{-\nu}+\alpha(z+1)2^{-\nu}.$$

The standard Riemann solver for f^ν takes values only in $\{z2^{-\nu}\}_{z\in\mathbb{Z}}$ if u^-, u^+ belongs to the same set $\{z2^{-\nu}\}_{z\in\mathbb{Z}}$, so that a solution can be constructed by starting with $u_0^\nu \in BV(\mathbb{R}, \{z2^{-\nu}\}_{z\in\mathbb{Z}})$. In fact, one solves the initial Riemann problem (finitely many!), and then each time two or more waves collide (an interaction point) the procedure starts again by solving this newly generated Riemann problem.

The key argument is that at each interaction either the Glimm interaction functional decreases (or equivalently the number of waves), or the total variation decreases, and in both cases of a fixed positive quantity depending only on ν. Hence the number of interaction is finite, and one has not to worry about the cases of infinite interaction points in finite time.

The convergence to the entropy solution follows because the approximated solutions generate a 1-Lipschitz semigroup in L^1, and the set $BV(\mathbb{R}, \{z2^{-\nu}\}_{z\in\mathbb{Z}})$ is dense in $L^\infty(\mathbb{R}, [0,1])$ as $\nu \to +\infty$ w.r.t. the L^1-norm.

Since the number of interactions is countable, we can perturb a little bit the speed of the waves in order to require that no multiple interaction occurs: only two waves at a time interact. This will simplify the computations in the next sections.

The fundamental assumption on the flux f is the following.

Assumption 2.1. *The flux f is a convex function.*

This implies that the entropy admissible jumps $[u^-, u^+]$ satisfy $u^+ < u^-$. As for notation, by C we will denote a sufficiently large constant.

3 Wave balances

In this section we develop some balances for particular waves measures. These are nonlinear functions of the derivative Du, which is a signed measure on $\mathbb{R}$.

Definition 3.1. We define the *wave measure* σ by $\sigma(t) = D_x f'(u(t))$, i.e.

$$\forall \phi \in C_c(\mathbb{R}) \left(\int \phi\sigma := -\int \frac{d}{dx}\phi(x) f'(u(t,x))\mathcal{L}^1(dx) \right).$$

Since $t \mapsto u(t)$ is a Lipschitz function in $L^1_{\rm loc}(\mathbb{R})$, it follows easily that σ is continuous w.r.t. the weak* topology.

Similarly, we can consider the Jordan decomposition

$$\sigma(t) = \sigma^+(t) + \sigma^-(t), \qquad \sigma^+, -\sigma^- \geq 0,\ \sigma^+ \perp \sigma^-,$$

and the maps $t \mapsto \sigma^+(t), \sigma^-(t)$. It is a general fact [4] that these maps are universally measurable if $t \mapsto \sigma(t)$ is continuous, but simple examples show that the continuity is lost.

Define the speed of the each jump by the Rankine-Hugoniot conditions: outside of interaction times,

$$\tilde{\lambda}(t,x) := \begin{cases} \dfrac{f(u^+) - f(u^-)}{u^+ - u^-} & u \text{ has the jump } [u^-, u^+] \text{ in } (t,x), \\ f'(u(t,x)) & u \text{ is continuous in } (t,x). \end{cases}$$

Since the set of times where $\tilde{\lambda}$ is not defined in finite, then this set is negligible for the measures

$$\int \sigma(t)\mathcal{L}^1(dt), \quad \int \sigma^+(t)\mathcal{L}^1(dt), \quad \int \sigma^-(t)\mathcal{L}^1(dt)$$

(it is negligible when integrating in times), so that the divergence forms

$$\partial_t\sigma(t) + \partial_x\big(\tilde{\lambda}\sigma(t)\big),\ \partial_t\sigma^+(t) + \partial_x\big(\tilde{\lambda}\sigma^+(t)\big),\ \partial_t\sigma^-(t) + \partial_x\big(\tilde{\lambda}\sigma^-(t)\big) \quad (3.1)$$

are distributions in $\mathbb{R}^2$.

Another way of interpreting (3.1) is that we ask if the maps $t \mapsto \sigma, \sigma^+, \sigma^-$ satisfy some PDE in the weak (distributional) sense.

In all three cases, it follows that the set where these divergence forms are not 0 are the interaction points $\{(t_i, x_i)\}_{i=1}^I$: in fact, the speed $\tilde{\lambda}$ has been chosen exactly to be the speed of the jump. Hence, a direct computation yields for example

$$\partial_t\sigma + \partial_x\big(\tilde{\lambda}\sigma\big) = \sum_i p_i \delta_{(t_i,x_i)},$$

where

$$p_i = \sum_{\text{exiting}} \sigma - \sum_{\text{entering}} \sigma,$$

i.e. the difference of the strength (with sign) of the waves after the interaction and before the interaction.

The same formula holds also for σ^+, σ^-, the difference being in the count of the entering and exiting waves in the interaction point: more precisely,

$$\partial_t \sigma^\pm + \partial_x \big(\tilde{\lambda}\sigma^\pm\big) = \sum_i p_i^\pm \delta_{(t_i, x_i)},$$

with

$$p_i^\pm = \sum_{\text{exiting}} \sigma^\pm - \sum_{\text{entering}} \sigma^\pm.$$

A final observation is that $p_i = p_i^+ + p_i^-$, simply because $\sigma = \sigma^+ + \sigma^-$.

3.1 Balance for σ

Using the fact that in the approximated solution u^ν only jumps are present, by the definition of σ we have that

$$\sigma(t) = \sum_{j \text{ jumps}} \Big(f'\big(u(t, x_j(t)+\big) - f'\big(u(t, x_j(t-))\big)\Big)\delta_{x_j(t)}$$

where $x_j(t)$ are piecewise linear functions defined in some time interval.

For the interaction point (t_i, x_i) we obtain

$$\sum_{\text{entering}} \sigma = f\big(u(t_i, x_i-)\big) - f\big(u(t_i, x_i+)\big) = \sum_{\text{exiting}} \sigma,$$

and hence the measure $\sigma(t)$ satisfies the conservation equation in divergence form

$$\partial_t \sigma + \partial_x \big(\tilde{\lambda}\sigma\big) = 0.$$

We note here that this equation has many more solutions, and we are selecting one particular solution, namely $\sigma(t) = D_x f'(u(t))$.

3.2 Balance for $\sigma^\pm$

Recall that by our assumption on the approximation, only two waves collide at a time. We thus consider two cases. For shortness we will remove the index i from the quantity p.

3.2.1 Waves with the same sign

This case can happen only for σ^-, and by the assumption on convexity from the interaction one single wave exist which is the sum of the two entering: hence $p^+ = p^- = 0$. Another way of seeing it is that since the waves are negative, we conclude that $p^+ = 0$ and thus

$$p^- = p - p^+ = 0.$$

3.2.2 Waves with different sign

In this case a positive wave is cancelled by a negative one, so that the quantity $2^{1-\nu}$ of total variation disappears. The computation for p^+ is thus

$$p^+ = -\sigma^+, \quad \sigma^+ \text{ entering.}$$

If we denote with u^-, u^m, u^+ the values of u before the interaction, then using the regularity of f we have that for

$$\sigma^+ = f'(u^+) - f'(u^m) \leq \int_{u^m}^{u^+} f''(s)ds \leq C(u^+ - u^-) = C2^{-\nu}.$$

This shows that the measure $\sum_i p^+\delta_{(t_i,x_i)}$ is bounded by

$$\left\| \sum_i p^+\delta_{(t_i,x_i)} \right\| = \sum_i |p_i| \leq C \text{ Tot.Var.}(u).$$

We call it the *cancellation measure* $\mu^{C,\nu}$.

Since $p_i^- = -p_i^+$, we recover the same bound for $\sum_i p^-\delta_{(t_i,x_i)}$.

4 The jump measure

Fix $0 < \epsilon_0 = k_0 2^{-\nu} \leq \epsilon_1/2 = k_1 2^{-\nu-1}$. We assume that ϵ_0 is a multiple of ν.

Definition 4.1. A Lipschitz curve $\gamma : [t_1, t_2] \to \mathbb{R}^2$, $\gamma(t) = (t, x(t))$, is an (ϵ_0, ϵ_1)-shock if

1. $\sigma(\gamma(t)) \geq \epsilon_0$ for all $t \in [t_1, t_2]$;
2. $\sigma(\gamma(\bar{t})) \geq \epsilon_1$ for at least one $\bar{t} \in [t_1, t_2]$.

Since the flux function f is smooth, one has

$$\sigma(t,x) = f'(u(t,x+)) - f'(u(t,x-)) = \int_{u^-}^{u+} f''(s)ds \leq C|u^+ - u^-|,$$

and from $u \in BV$ it follows that there are at most finitely many (ϵ_0, ϵ_1)-jumps.

Definition 4.2. The set $J^{\epsilon_0,\epsilon_1}$ of the (ϵ_0, ϵ_1)-shocks is

$$J^{\epsilon_0,\epsilon_1} := \bigcup \Big\{ \gamma_i([t_{1,i}, t_{2,i}]) : \gamma_i \ (\epsilon_0, \epsilon_1)\text{-shock in } [t_{1,i}, t_{2,i}] \Big\}.$$

Denote with σ^ϵ the measure

$$\sigma^\epsilon := \sigma\big|_{J^{\epsilon_0,\epsilon_1}}.$$

We compute the distribution

$$\partial_t \sigma^\epsilon + \partial_x\big(\tilde{\lambda}\sigma^\epsilon\big) = \sum_i p_i^\epsilon \delta_{(t_i,x_i)}.$$

At each interaction, the following cases happen:

1. two negative waves collide, generating the beginning point of an (ϵ_0,ϵ_1)-shock: then $p_i^\epsilon \leq 0$;
2. two (ϵ_0,ϵ_1)-jumps collinde: hence $p_i^\epsilon = 0$;
3. an (ϵ_0,ϵ_1)-jump of size $> k_0 2^{-\nu}$ collides with a positive wave of size $2^{-\nu}$: then its size remains above ϵ_0 and $p_i^\epsilon = C2^{-\nu}$;
4. an (ϵ_0,ϵ_1)-jump of size $k_0 2^{-\nu}$ collides with a positive wave of size $2^{-\nu}$, and thus it disappears as an (ϵ_0,ϵ_1)-shock: then $p_i^\epsilon = k_0 2^{-\nu}$. Then along the shock at least a cancellation of order $\epsilon_1 - \epsilon_0$ occurred, and thus

$$\begin{aligned}\epsilon_0 \sharp\Big\{\text{number of terminal points}\Big\} &\leq \frac{\epsilon_0}{\epsilon_1-\epsilon_0}\Big\{\text{cancellation measure}\Big\}\\ &\leq \frac{k_0}{k_1-k_0}\text{Tot.Var.}(u(0)).\end{aligned}$$

We conclude that the measure

$$\mu^{\text{jump},\epsilon,\nu} := \partial_t\sigma^\epsilon + \partial_x\big(\tilde{\lambda}\sigma^\epsilon\big)$$

is a uniformly bounded measure, because its positive part is uniformly bounded and

$$\begin{aligned}\mu^-(\mathbb{R}^+\times\mathbb{R}) &= \mu^+(\mathbb{R}^+\times\mathbb{R}) - \mu(\mathbb{R}^+\times\mathbb{R})\\ &\leq \left(1+\frac{k_0}{k_1-k_0}\right)\mu^{C,\nu}(\mathbb{R}^+\times\mathbb{R})\\ &\quad + \int_0^{+\infty}\int_{\mathbb{R}} \partial_t\sigma^\epsilon + \partial_x\big(\tilde{\lambda}\sigma^\epsilon\big)dxdt\\ &\leq \left(1+\frac{k_0}{k_1-k_0}\right)\mu^{C,\nu}(\mathbb{R}^+\times\mathbb{R}) + \text{Tot.Var.}(u(0)).\end{aligned}$$

5 Decay estimates

The key argument here is the balance of positive waves in backward cones and the balance of negative waves in forward cones. For simplicity we will consider these balances for the entropic solution and not for the approximated solution: in the latter case one has to add an $\mathcal{O}(\nu+\epsilon_1)$ to the estimates.

5.1 Decay of positive waves

Consider an interval $I = [a, b]$, and let $a(t)$, $b(t)$ be the minimal backward characteristics such that $a(T) = a$, $b(T) = b$. The definition of wavefront solution implies that no waves can enter in the interval $I(t) = [a(t), b(t)]$, so that the definition of σ implies

$$\frac{d}{dt}(b(t) - a(t)) = \sum \sigma(t, I(t)) = \sigma(T, I).$$

Since $b(0) - a(0) = \sigma(T, I)T \geq 0$, we deduce immediately that $\sigma(T, I) \leq \frac{b-a}{T}$.

5.2 Decay of negative waves

In this case the analysis is similar, but negative waves can enter in the future, and we need only to consider the continuous part: the balance is measured by μ^{jump}. For this, we have that the difference in speed satisfies

$$\frac{dz}{dt} \leq \sigma^{\text{cont}}(t, I(t)) = \sigma^{\text{cont}}(T, I) - \mu^{\text{jump}}(T, t).$$

Integrating in time and using that $z(t) > 0$ we obtain

$$\sigma^{\text{cont}}(T, I) \geq -\frac{z(T)}{t - T} + \mu^{\text{jump}}(T, t).$$

Both estimates can be extended to finitely many intervals, and with standard methods to Borel subsets. Collecting the two estimates we arrive to the following proposition.

Proposition 5.1. *If u is the entropy BV solution to a scalar conservation law with convex flux, then the measure $\sigma = f'(u)_x$ satisfies the estimate*

$$-\frac{\mathcal{L}^1}{t - T} + \mu^{\text{jump}}(T, t) \leq \sigma^{\text{cont}}(t, B) \leq \frac{\mathcal{L}^1(B)}{t}$$

for all B Borel.

In particular, if B is of Lebesgue measure 0 but $\sigma^{\text{cont}}(B) > 0$, we obtain that

$$\mu^{\text{jump}}(t) \leq \sigma^{\text{cont}}(t, B) \leq 0,$$

from which we deduce that the measure $\mu^{\text{jump}}(t)$ is not 0 every time a Cantor part appears in σ^{cont}: this means that immediately this Cantor measure is transformed into jumps.

Corollary 5.2. *Up to countably many times, the function $f'(u(t))$ is SBV.*

References

[1] L. Ambrosio and C. De Lellis, A note on admissible solutions of 1D scalar conservation laws and 2D Hamilton-Jacobi equations. J. Hyperbolic Differential Equations. 31 (2004), 813–826.

[2] L. Ambrosio, N. Fusco and D. Pallara, Functions of Bounded Variation and Free Discontinuity Problems. Oxford Clarendon Press, 2000.

[3] F. Ancona and K. T. Nguyen, SBV regularity of L^∞ solutions to genuinely nonlinear temple systems of balance laws. In preparation.

[4] S. Bianchini and L. Caravenna, SBV regularity for genuinely nonlinear, strictly hyperbolic systems of conservation laws in one space dimension. Submitted.

[5] S. Bianchini, C. De Lellis and R. Robyr, SBV regularity for Hamilton-Jacobi equations in R^n. Preprint.

[6] A. Bressan, Hyperbolic Systems of Conservation Laws: The One-Dimensional Cauchy Problem. Oxford University Press, 2000.

[7] C. M. Dafermos, Hyperbolic Conservation Laws in Continuous Physics. Springer, 2000.

[8] C. M. Dafermos, Polygonal approximations of solutions of the initial value problem for a conservation law. J. Math. Anal. Appl. 38 (1972), 202–212.

[9] C. M. Dafermos, Wave fans are special. Acta Mathematicae Applicatae Sinica, English Series. 24 (2008), 369–374.

[10] R. Robyr, SBV regularity of entropy solutions for a class of genuinely nonlinear scalar balance laws with non-convex flux function. J. Hyperbolic Differential Equations. 5 (2008), 449–475.

Entropy Stable ENO Scheme*

Ulrik S. Fjordholm
Seminar for Applied Mathematics, ETH Zürich
HG J. 48, Rämistrasse 101, Zürich, Switzerland
E-mail: ulrikf@sam.math.ethz.ch

Siddhartha Mishra
Seminar for Applied Mathematics, ETH Zürich
HG G. 57.2, Rämistrasse 101, Zürich, Switzerland
E-mail: smishra@sam.math.ethz.ch

Eitan Tadmor
Center for Scientific Computation and Mathematical Modeling
&
Department of Mathematics, Institute for Physical Science & Technology
University of Maryland, College Park, MD 20742, USA
E-mail: tadmor@cscamm.umd.edu

Abstract

We prove the first stability estimates for the ENO reconstruction procedure. They take the form of a sign property: we show that the jump in the reconstructed pointvalues at each cell interface has the same sign as the jump in underlying cell averages (cell centered values) across the interface. Moreover their ratio is upper bounded. These estimates hold for arbitrary orders of accuracy of the reconstruction as well as for non-uniform meshes. We then combine the ENO reconstruction together with entropy conservative fluxes to construct new entropy stable ENO schemes of arbitrary order.

1 Introduction

The Essentially Non-Oscillatory (ENO) method was designed by Harten et al. in [8] for approximating solutions of nonlinear hyperbolic conservation laws. Let $\{x_{i+1/2}\}_{i\in\mathbb{Z}}$ be a given mesh, and define the cells

*Partially supported by NSF grants DMS07-07949, DMS10-08397 and ONR grant N000140910385.

$\mathcal{I}_i := [x_{i-1/2}, x_{i+1/2})$ and their sizes $h_i = x_{i+1/2} - x_{i-1/2}$ and $h = \max_i h_i$. The starting point is the collection $\{\overline{v}_i\}_{i\in\mathbb{Z}}$ of cell averages of a given function $v(x)$,

$$\overline{v}_i := \frac{1}{h_i}\int_{\mathcal{I}_i} v(x)dx, \tag{1.1}$$

from which we get the piecewise constant approximation

$$\mathcal{A}v(x) = \sum_{i\in\mathbb{Z}} \overline{v}_i \mathbb{1}_{\mathcal{I}_i}(x)$$

of v, which, when v is piecewise continuous, is accurate to first order. The role of the *ENO reconstruction method* is to construct a p-th ($p \in \mathbb{N}$) order accurate approximation

$$\mathcal{RA}v(x) = \sum_{i\in\mathbb{Z}} f_i(x)\mathbb{1}_{\mathcal{I}_i}(x)$$

of v, where $f_i \in \Pi_{p-1}$ is a polynomial recovered from neighboring cell averages. Each interpolant f_i is p-th order accurate in the sense that

$$f_i(x) = v(x) + \mathcal{O}(h^p) \qquad \text{for } x \in \mathcal{I}_i. \qquad (\textbf{Accuracy})$$

The function f_i is an interpolant in the sense that it preserves cell averages,

$$\frac{1}{h_i}\int_{\mathcal{I}_i} f_i(x)dx = \overline{v}_i, \qquad (\textbf{Conservation})$$

that is, $\mathcal{ARA} = \mathcal{A}$. To ensure the conservation property, we consider the primitive of v,

$$V(x) := \int_{-\infty}^{x} v(s)ds.$$

Letting $V_{j+1/2} = V(x_{j+1/2})$, it follows from (1.1) that

$$V_{j+1/2} := \int_{-\infty}^{x_{j+1/2}} v(s)ds = \sum_{i=-\infty}^{j} \int_{x_{i-1/2}}^{x_{i+1/2}} v(s)ds = \sum_{i=-\infty}^{j} h_i\overline{v}_i. \tag{1.2}$$

Therefore, given the cell averages $\{\overline{v}_i\}_i$, point values of the primitive V at the cell interfaces $x_{i+1/2}$ can be explicitly evaluated, up to an additive constant. Now, let $\mathbf{V}(x)$ be the unique p-th order polynomial interpolating the $p+1$ point values $V_{i-r-1/2}, \ldots, V_{i-r+p+1/2}$. We define the $(p-1)$-th order polynomial $f(x) = \mathbf{V}'(x)$. It is a simple consequence of (1.2) that f satisfies (**Accuracy**) and (**Conservation**). Hence, a *reconstruction* from cell averages requires an *interpolation* of the primitive.

Both the accuracy and conservation requirements can be met by interpolating the primitive V on any of the p stencils

$$\mathcal{S}_i^r = \{x_{i-r-1/2}, \ldots, x_{i-r+p+1/2}\},$$

where $r \in \{0, \dots, p-1\}$ is the *stencil offset*. As we are interested in approximating piecewise smooth functions, this stencil has to be selected carefully to avoid spurious oscillations in the reconstructed function. The principal idea of the ENO procedure is to choose the stencil offset in an iterative manner, based on divided differences. The ENO algorithm is defined as follows.

Algorithm 1.1 (ENO stencil selection procedure). *Let point values of the primitive $V_{i-p-1/2}, \dots, V_{i+p+1/2}$ be given.*

- *Let $r_1 = 0$.*
- *For each $j = 1, \dots, p-1$: If*

$$\left|V\left[x_{i-r_j-3/2}, \dots, x_{i-r_j+j-1/2}\right]\right| < \left|V\left[x_{i-r_j-1/2}, \dots, x_{i-r_j+j+1/2}\right]\right|$$

  then let $r_{j+1} = r_j + 1$; otherwise, let $r_{j+1} = r_j$.
- *Let $r = r_p$ and interpolate V over the stencil $\mathcal{S}_i^r$.*

As the j-th order divided difference of a function is a good measure of its j-th derivative, we obtain an interpolating polynomial with the smallest possible derivatives. Thus, the ENO procedure involves a *nonlinear data dependent* selection of the stencil in the *direction of smoothness* of the underlying piecewise smooth function.

Given the reconstructed function $\mathcal{RA}v = \sum_i f_i \mathbb{1}_{\mathcal{I}_i}$, we obtain at each cell interface p-th order accurate approximations

$$v_i^- = f_i(x_{j-1/2}), \qquad v_i^+ = f_i(x_{i+1/2}) \tag{1.3}$$

such that

$$v_i^- = v(x_{i-1/2}) + \mathcal{O}(h^p), \qquad v_i^+ = v(x_{i+1/2}) + \mathcal{O}(h^p). \tag{1.4}$$

These values are the inputs to high-order accurate finite volume schemes to approximate nonlinear conservation laws (see Shu [17]), and the relation between these values and the cell averages will be the main point of study in this paper.

Despite of the existence of the ENO reconstruction procedure for the last 25 years, no global (mesh independent) stability results are available. Given this background, we present the following stability theorem.

Theorem 1.1 (the sign property of ENO reconstruction [5]). *Let $p \in \mathbb{N}$, and let v_i^+, v_{i+1}^- be left and right values at a cell interface $x_{i+1/2}$, obtained through a p-th order ENO reconstruction of cell averages $\{\overline{v}_i\}_i$. Then*

$$\operatorname{sgn}\left(v_{i+1}^- - v_i^+\right) = \operatorname{sgn}\left(\overline{v}_{i+1} - \overline{v}_i\right) \qquad (i \in \mathbb{Z}). \tag{1.5}$$

Moreover, there is a constant $\mathcal{C}_p$, depending only on p and on the ratios of neighboring grid cells, $\frac{h_{j+1}}{h_j}$, such that

$$0 \leq \frac{v_{i+1}^- - v_i^+}{\overline{v}_{i+1} - \overline{v}_i} \leq \mathcal{C}_p \qquad (i \in \mathbb{Z}). \tag{1.6}$$

The sign property says that at each cell interface, the jump in reconstructed values has the same sign as the jump in the underlying cell averages.

We remark that the above theorem is valid for any order of reconstruction and for any mesh size. It is valid for non-uniform meshes and makes no assumptions on the function v, other than that the cell averages $\overline{v}_i$ must be well-defined, which is guaranteed if e.g. $v \in L^1_{\text{loc}}(\mathbb{R})$.

The sign property is illustrated in Figure 1.1, where we show a third-, fourth- and fifth-order ENO reconstruction of randomly chosen cell averages. Even though the reconstructed polynomial can have large variations within a cell, its jumps at cell interfaces always have the same sign as the jumps of the cell averages.

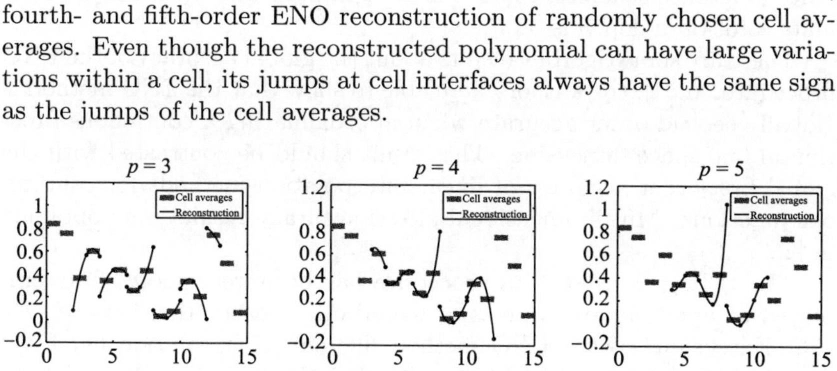

Figure 1.1 ENO reconstruction of randomly chosen cell averages.

The ENO procedure for reconstructing piecewise polynomials, given the point values of a piecewise smooth function $v_i = v(x_i)$ can be defined analogously, see [17]. This procedure is also stable in the following sense,

Theorem 1.2 ([5]). *Let $p \in \mathbb{N}$, and let v_i^+, v_{i+1}^- be left and right values at a cell interface $x_{i+1/2}$, obtained through a p-th order ENO reconstruction of point values $\{v_i\}_i$. Then*

$$\text{sgn}\big(v_{i+1}^- - v_i^+\big) = \text{sgn}\big(v_{i+1} - v_i\big) \qquad (i \in \mathbb{Z}). \tag{1.7}$$

Moreover, there is a constant $\widehat{\mathcal{C}}_p$, depending only on p and on the ratios of neighboring grid cells, $\frac{h_{j+1}}{h_j}$, such that

$$0 \leq \frac{v_{i+1}^- - v_i^+}{v_{i+1} - v_i} \leq \widehat{\mathcal{C}}_p \qquad (i \in \mathbb{Z}). \tag{1.8}$$

This sign property for the ENO reconstruction from point values will be used to construct arbitrarily high-order accurate *entropy stable* ENO schemes for systems of conservation laws in section 2 below.

Our stability results should be compared with that of existing *local* stability results for the ENO procedure like the stability result of [7] that for a given function v, there is an upper mesh size $h_0 > 0$ and a function $z(x)$, *both dependent on* v, such that

$$f(x) = z(x) + \mathcal{O}(h^p)$$

and

$$\mathrm{TV}(z) \leq \mathrm{TV}(v)$$

whenever the mesh size h is less than h_0. What is more, z is monotone inside cells where v has a discontinuity. For ENO-based finite volume or finite difference schemes, this result is of limited utility, as the approximate solution v depends on h.

There are some rigorous results about the global accuracy of the ENO procedure. In [1], the authors were able to show that the ENO method is globally second-order accurate when approximating a continuous function in one space dimension. This result should be contrasted with the global first-order accuracy of linear interpolation methods for continuous functions. Multi-dimensional global accuracy results were obtained in [2].

The ENO reconstruction procedure has been very successfully employed to approximate systems of hyperbolic conservation laws, [8, 15, 17]. Extensions of the ENO method like subcell resolution [9], ENO with biasing [16] and Weighted ENO (WENO) schemes [15, 17] have also been employed in computational fluid dynamics with considerable success. See the excellent review article [17] of Shu for an overview.

However, there are no rigorous stability proofs for any of the above ENO schemes. In the following, we will present a conservative finite difference scheme that approximates systems of conservation laws to arbitrarily high order of accuracy and is entropy stable.

2 Entropy stable ENO schemes

2.1 Continuous setting

In one space dimension, a system of conservation laws takes the form,

$$\begin{aligned} \mathbf{U}_t + \mathbf{f}(\mathbf{U})_x &= 0, \quad \forall\ (x,t) \in \mathbb{R} \times \mathbb{R}_+, \\ \mathbf{U}(x,0) &= \mathbf{U}_0(x), \quad \forall\ x \in \mathbb{R}. \end{aligned} \tag{2.1}$$

Here, $\mathbf{U} : \mathbb{R} \times \mathbb{R}_+ \mapsto \mathbb{R}^m$ is the vector of unknowns and $\mathbf{f}$ is the (nonlinear) flux vector.

It is well known ([3]) that solutions of (2.1) contain discontinuities in form of *shock waves*, even for smooth initial data. Hence, solutions of (2.1) are sought in a weak sense. Weak solutions may not be unique and need to be supplemented with extra admissibility criteria, termed as *entropy conditions*, in order to single out a physically relevant solution [3]. Assume that there exists a convex function $E : \mathbb{R}^m \mapsto \mathbb{R}$ and functions $\mathbf{V}, Q$ such that

$$\mathbf{V} = \partial_{\mathbf{U}} E, \quad \partial_{\mathbf{U}} Q = \langle \mathbf{V}, \partial_{\mathbf{U}} \mathbf{f} \rangle. \tag{2.2}$$

Then, a straightforward calculation using (2.1) and (2.2) shows that *smooth* solutions of (2.1) satisfy the *entropy identity*:

$$E_t + Q_x = 0. \tag{2.3}$$

However, the solutions of (2.1) are not smooth in general and the entropy has to be dissipated at shocks. Hence, we obtain an *entropy inequality*,

$$E_t + Q_x \leq 0, \tag{2.4}$$

that holds in the sense of distributions. We term Q and $\mathbf{V}$ in (2.2) as the entropy flux function and the vector of entropy variables, respectively. Integrating (2.4) over space results in the bound:

$$\frac{d}{dt} \int_{\mathbb{R}} E dx \leq 0 \quad \Rightarrow \quad \int_{\mathbb{R}} E\left(\mathbf{U}(x,T)\right) dx \leq \int_{\mathbb{R}} E(\mathbf{U}_0(x)) dx, \quad \forall T > 0. \tag{2.5}$$

As E is convex, the above entropy bound can be converted into an a priori estimate on the solution of (2.1) in suitable L^p spaces [3].

2.2 Conservative finite difference schemes

For simplicity, we consider a uniform Cartesian mesh $\{(x_i)\}$ in $\mathbb{R}$ with mesh size $x_{i+1} - x_i = \Delta x$. The midpoint values are $x_{i+1/2} = \frac{x_i + x_{i+1}}{2}$ and the domain is partitioned into intervals $I_i = [x_{i-1/2}, x_{i+1/2}]$. The conservative finite difference (finite volume) method updates point values (cell averages in I_i) of the solution $\mathbf{U}$ resulting in the semi-discrete scheme:

$$\frac{d}{dt}\mathbf{U}_i(t) = -\frac{1}{\Delta x}\left(\mathbf{F}_{i+1/2}(t) - \mathbf{F}_{i-1/2}(t)\right), \tag{2.6}$$

with numerical flux $\mathbf{F}_{i+1/2} = \mathbf{F}(\mathbf{U}_i(t), \mathbf{U}_{i+1}(t))$ computed from the (approximate) solution of the Riemann problem for (2.1) at the interface $x_{i+1/2}$, [13]. Second order spatial accuracy can be obtained with non-oscillatory TVD [11] and even higher order of accuracy can be obtained with ENO [8] and WENO [15] piecewise polynomial reconstructions.

However, these high-order schemes are not rigorously shown to be stable. As mentioned before, we will construct stable arbitrarily high-order conservative finite difference schemes. We do so in two stages.

2.3 Entropy conservative schemes

First, we construct entropy conservative schemes i.e., finite difference schemes of the form (2.6) which satisfy a discrete version of the entropy identity (2.3). Introducing the notation,

$$[\![a]\!]_{i+1/2} = a_{i+1} - a_i, \quad \bar{a}_{i+1/2} = \frac{1}{2}(a_i + a_{i+1}),$$

we follow the general procedure developed in [19] to define two-point entropy conservative fluxes:

Theorem 2.1 (Tadmor, [19]). *Consider the one dimensional system of conservation laws* (2.1) *with entropy function* E, *entropy variables* $\mathbf{V}$, *entropy flux* Q *and define the entropy potential* $\psi := \langle \mathbf{V}, \mathbf{f} \rangle - Q$. *Let the finite difference scheme* (2.6) *approximate* (2.1) *and the numerical flux be given by* $\mathbf{F}_{i+1/2} = F^*_{i+1/2} = \mathbf{F}^*(\mathbf{U}_i, \mathbf{U}_{i+1})$. *We assume that it is consistent with* $\mathbf{f}$ *i.e.* $\mathbf{F}^*(\mathbf{U}, \mathbf{U}) = \mathbf{f}(\mathbf{U})$ *and satisfies the following,*

$$\langle [\![\mathbf{V}]\!]_{i+1/2}, \mathbf{F}^*_{i+1/2} \rangle = [\![\psi]\!]_{i+1/2}, \quad \forall i. \tag{2.7}$$

Then the scheme (2.6) *with numerical flux* (2.7) *satisfies the discrete entropy identity,*

$$\begin{aligned} \frac{d}{dt}(E(\mathbf{U}_i)(t)) &= -\frac{1}{\Delta x}(\tilde{Q}_{i+1/2} - \tilde{Q}_{i-1/2}), \\ \tilde{Q}_{i+1/2} &= \left\langle \frac{(\mathbf{V}_i + \mathbf{V}_{i+1})}{2}, \mathbf{F}^*_{i+1/2} \right\rangle - \frac{(\psi_i + \psi_{i+1})}{2}. \end{aligned} \tag{2.8}$$

Summing over all i, *we obtain entropy conservation:*

$$\frac{d}{dt} \sum_i E_i \equiv 0.$$

Furthermore, the scheme (2.6) *with flux* $\mathbf{F}^*$ *is second-order accurate.*

We note that the condition (2.7) provides a single algebraic equation for m unknowns. In general, it is not clear whether a solution of (2.7) exists. Furthermore, the solutions of (2.7) will not be unique except in the case of scalar equations i.e., $m = 1$. In [19], Tadmor showed the existence of at least one solution of (2.7) for any system of conservation laws. Explicit solutions were constructed in [20]. However, the entropy conservative fluxes of [20] are computationally expensive, [4]. Instead, we follow recent papers [4, 14] to obtain algebraically simple and computational inexpensive solutions of (2.7). As an example, we consider the Euler equations of gas dynamics i.e., (2.1) with

$$\mathbf{U} = (\rho, \rho u, \mathcal{E})^\top, \quad \mathbf{f} = (\rho u, \rho u^2 + p, (\mathcal{E} + p)u)^\top \tag{2.9}$$

Here, ρ, u and p are the density, velocity and pressure of the fluid. The total energy $\mathcal{E}$ is related to other variables by the equation of state:

$$\mathcal{E} = \frac{p}{\gamma - 1} + \frac{1}{2}\rho u^2, \tag{2.10}$$

with γ being the gas constant. Let $s = \log(p) - \gamma \log(\rho)$ be the thermodynamic entropy. An entropy-entropy flux pair for the Euler equations is

$$E := \frac{-\rho s}{\gamma - 1}, \quad Q := \frac{-\rho u s}{\gamma - 1}. \tag{2.11}$$

The corresponding entropy variables and potential are

$$\mathbf{V} = \left\{ \frac{\gamma - s}{\gamma - 1} - \frac{\rho u^2}{2p}, \frac{\rho u}{p}, -\frac{\rho}{p} \right\}, \qquad \psi = \rho u. \tag{2.12}$$

An explicit entropy conservative flux for the Euler equations (see [14]) is given by

$$\begin{aligned}
\mathbf{F}^*_{i+1/2} &= \{\mathbf{F}^{1,*}_{i+1/2}, \mathbf{F}^{2,*}_{i+1/2}, \mathbf{F}^{3,*}_{i+1/2}\}, \quad \mathbf{F}^{1,*}_{i+1/2} = \overline{z}_{2,i+1/2}(z_3)^{\ln}_{i+1/2}, \\
\mathbf{F}^{2,*}_{i+1/2} &= \frac{\overline{z}_{3,i+1/2}}{\overline{z}_{1,i+1/2}} + \frac{\overline{z}_{2,i+1/2}}{\overline{z}_{1,i+1/2}} \mathbf{F}^{1,*}_{i+1/2}, \\
\mathbf{F}^{3,*}_{i+1/2} &= \frac{1}{2} \frac{\overline{z}_{2,i+1/2}}{\overline{z}_{1,i+1/2}} \left(\frac{\gamma + 1}{\gamma - 1} \frac{(z_3)^{\ln}_{i+1/2}}{(z_1)^{\ln}_{i+1/2}} + \mathbf{F}^{2,*}_{i+1/2} \right).
\end{aligned} \tag{2.13}$$

Here, the parameter vectors are defined as

$$\mathbf{z} = (z_1, z_2, z_3) = \left\{ \sqrt{\frac{\rho}{p}}, \sqrt{\frac{\rho}{p}} u, \sqrt{\rho p} \right\}, \tag{2.14}$$

and $a^{\ln}$ is the logarithmic mean defined as

$$(a)^{\ln}_{i+1/2} := \frac{[\![a]\!]_{i+1/2}}{[\![\log(a)]\!]_{i+1/2}}. \tag{2.15}$$

Other examples of explicit two-point entropy conservative fluxes are described in [6].

2.3.1 High-order entropy conservative fluxes

We follow the procedure of LeFloch, Mercier and Rohde [12] in order to obtain arbitrary order accurate entropy conservative fluxes.

Theorem 2.2. *Consider the system* (2.1), *equipped with the entropy function E. Assume that there exists the two-point entropy conservative* *flux*, $\mathbf{F}^*_{i+1/2} = \mathbf{F}^*(\mathbf{U}_i, \mathbf{U}_{i+1})$ *satisfying* (2.7). *For any integer* $p \geq 1$, *define the numerical flux*,

$$\mathbf{F}^{2p,*}_{i+1/2} = \sum_{\{0 \leq r,s \leq p\}} \alpha_{rs} \mathbf{F}^*(\mathbf{U}_{i-r}, \mathbf{U}_{i+s}). \tag{2.16}$$

The co-efficients α_{rs} *satisfy the consistency condition:* $\sum_{r,s} \alpha_{rs} = 1$, *and are computed using a Taylor expansion as in [12]. Then, the finite difference scheme* (2.6) *with numerical flux* (2.16) *satisfies the following properties:*

(i) *The scheme is* $2p$*-th order accurate i.e., the truncation error of the scheme* (2.6) *satisfies,*

$$\mathcal{T}_i(t) = \mathcal{O}(\Delta x^{2p}), \quad \forall t > 0,$$

provided that the solutions of (2.1) *are sufficiently smooth.*

(ii) *The scheme* (2.6) *with numerical flux* (2.16) *is entropy conservative as it satisfies the discrete entropy identity:*

$$\begin{aligned}
\frac{d}{dt}(E(\mathbf{U}_i)(t)) &= -\frac{1}{\Delta x}(\tilde{Q}^p_{i+1/2} - \tilde{Q}^p_{i-1/2}), \\
\tilde{Q}^p_{i+1/2} &= \frac{1}{2} \sum_{r,s} \alpha_{rs} (\langle (\mathbf{V}_{i-r} + \mathbf{V}_{i+s}), \mathbf{F}^*(\mathbf{U}_{i-r}, \mathbf{U}_{i+s}) \rangle \\
&\quad - (\psi_{i-r} + \psi_{i+s})).
\end{aligned} \tag{2.17}$$

Some explicit examples of (2.16) are described in [6]. We remark that the arbitrarily high-order entropy conservative fluxes are linear combinations of the two point entropy conservative flux (2.7). Hence, computationally inexpensive forms of (2.7) like (2.13) are absolutely essential to design a computationally tractable high-order entropy conservative scheme.

2.4 ENO based numerical diffusion operators

Entropy is only conserved if the solutions (2.1) are smooth. However, the solutions of (2.1) develop discontinuities and entropy is *dissipated* at shocks resulting in the entropy inequality (2.4). Hence, entropy conservative schemes described in the previous section will lead to high-frequency oscillations near shocks (see [4] for numerical examples). Consequently, we need to add some dissipative mechanism to ensure that entropy is dissipated near shocks. This is achieved by designing suitable entropy stable numerical diffusion operators.

The resulting scheme (2.6) will have the numerical flux,

$$\mathbf{F}^k_{i+1/2} = F^{2p,*}_{i+1/2} - \frac{1}{2}\mathbf{D}_{i+1/2}\langle \mathbf{V}\rangle_{i+1/2}. \tag{2.18}$$

Here, $k \geq 1$ and $p = k/2$ if k is even or $p = (k+1)/2$ if k is odd. The flux $\mathbf{F}^{2p,*}$ is the high order entropy conservative flux given by (2.16). The matrix $\mathbf{D}$ is any symmetric positive definite matrix. We perform a *suitable* reconstruction of the entropy variables $\mathbf{V}$ and obtain some piecewise polynomial function $\mathbf{V}_i(x)$ of degree $k-1$ for $x \in I_i$. Denoting

$$\mathbf{V}_i^- = \mathbf{V}_i(x_{i-1/2}), \quad \mathbf{V}_i^+ = \mathbf{V}_i(x_{i+1/2}), \quad \text{and} \quad \langle \mathbf{V}\rangle_{i+1/2} = \mathbf{V}_{i+1}^- - \mathbf{V}_i^+, \tag{2.19}$$

completes the description of the numerical flux (2.18). We need to modify the reconstruction procedure in order to ensure entropy stability. If $B = \operatorname{diag}(b_1, \ldots, b_n)$ is a diagonal matrix and $s \in \mathbb{R}$, then we write

$$B \leq s$$

provided $b_j \leq s$ for all j, and similarly for $B \geq s$.

We present sufficient conditions for the reconstruction to be entropy stable below.

Lemma 2.1. *Consider the one dimensional system* (2.1) *with entropy variables* $\mathbf{V}$. *Let* $\{\widehat{r}^l_{i+1/2}\}_{l=1}^m$ *be a basis of* $\mathbb{R}^m$ *for each* i. *Define the matrix*

$$\widehat{R}_{i+1/2} = \left[\widehat{r}^1_{i+1/2} | \widehat{r}^2_{i+1/2} | \cdots | \widehat{r}^m_{i+1/2}\right].$$

Let $\widehat{\Lambda}_{i+1/2}$ *be any diagonal matrix with non-negative entries. Define the numerical diffusion matrix:*

$$\widehat{\mathbf{D}}_{i+1/2} = \widehat{R}_{i+1/2}\widehat{\Lambda}_{i+1/2}\widehat{R}^{-1}_{i+1/2}. \tag{2.20}$$

Let $\mathbf{V}_i(x)$ *be a polynomial reconstruction of the entropy variables in the cell* I_i. *Assume that for each* i, *there exists a diagonal matrix* $B_{i+1/2} \geq 0$ *such that the following holds:*

$$\langle V\rangle_{i+1/2} = \widehat{R}_{i+1/2} B_{i+1/2} \widehat{R}^{\top}_{i+1/2} [\![\mathbf{V}]\!]_{i+1/2}. \tag{2.21}$$

Then the numerical scheme (2.6) *with numerical flux,*

$$\mathbf{F}^k_{i+1/2} = F^{2p,*}_{i+1/2} - \frac{1}{2}\widehat{\mathbf{D}}_{i+1/2}\langle \mathbf{V}\rangle_{i+1/2} \tag{2.22}$$

is entropy stable, i.e. it satisfies the entropy dissipation estimate:

$$\frac{d}{dt}E(\mathbf{U}_i) + \frac{1}{\Delta x}\left(\widehat{Q}_{i+1/2} - \widehat{Q}_{i-1/2}\right) \leq 0, \tag{2.23}$$

where the numerical entropy flux function $\widehat{Q}$ is defined as

$$\widehat{Q}_i = \widetilde{Q}^{2p}_{i+1/2} - \frac{1}{2}\left\langle \overline{\mathbf{V}}_{i+1/2}\widehat{\mathbf{D}}_{i+1/2}\langle \mathbf{V}\rangle_{i+1/2}\right\rangle .$$

Here, $\widetilde{Q}^{2p}$ is defined by replacing $\mathbf{D}$ *by* $\widehat{\mathbf{D}}$ *in* (2.17).

The proof of this lemma is presented in [6]. We need to find a reconstruction procedure which satisfies the crucial requirement that the matrix B in (2.21) is positive. Let $\mathbf{V}_i, \mathbf{V}_{i+1}, \mathbf{V}_i^+, \mathbf{V}_{i+1}^-$ be given. Define the *scaled entropy variables*,

$$\mathbf{W}_i^{\pm} = R_{i\pm 1/2}^{\top}\mathbf{V}_i, \quad \widetilde{\mathbf{W}}_i^{\pm} = R_{i\pm 1/2}^{\top}\mathbf{V}_i^{\pm}.$$

Dropping the $\pm$ superscript for notational convenience, the condition (2.21) reads as

$$\langle \widetilde{\mathbf{W}}\rangle_{i+1/2} = B_{i+1/2}\langle \mathbf{W}\rangle_{i+1/2}.$$

This is a component-wise condition; denoting the l-th component of $\mathbf{W}_i$ and $\widetilde{\mathbf{W}}_i$ by w_i^l and $\widetilde{w}_i^l$, respectively, the above condition is equivalent to

$$\operatorname{sgn}\langle \widetilde{w}^l\rangle_{i+1/2} = \operatorname{sgn}\langle w^l\rangle_{i+1/2}, \quad \forall i, l. \tag{2.24}$$

Note that the above condition amounts to the sign property in theorem 1.1. As the ENO reconstruction procedure satisfies the sign property, we can use it to define an entropy stable arbitrarily high order numerical diffusion operator. We obtain an arbitrarily high-order entropy stable scheme by using the ENO reconstruction procedure to reconstruct w^l for each l. The properties of the scheme are summarized below:

Theorem 2.3. *For any $k \geq 1$, let $2p = k$ (if k is even) or $2p = k+1$ (if k is odd). Define the entropy conservative flux $\mathbf{F}^{2p,*}$ by* (2.16)*. Let $\langle \mathbf{V}\rangle$ in* (2.22) *be defined by the k-th order accurate ENO reconstruction procedure. Then the finite difference scheme* (2.6) *with numerical flux* (2.22) *satisfies the following properties,*

(i) The scheme is k-th order accurate for smooth solutions of (2.1).

(ii) The scheme satisfies the discrete entropy inequality (2.23)*. Hence, it is entropy stable.*

We term the arbitrarily high-order accurate entropy stable schemes as *ECENO* schemes as they combine entropy conservative (EC) and ENO schemes.

Although any basis $\{r^l\}_{l=1}^m$ of $\mathbb{R}^m$ and any diagonal matrix $\widehat{\Lambda}$ will suffice for constructing an entropy stable scheme, we choose the following,

$$\widehat{R} = R.$$

Here, R is the matrix of eigenvectors of the flux Jacobian, $\mathbf{f_U}$ and Λ is a positive diagonal matrix that depends on the eigenvalues of the flux Jacobian. Two examples of the matrix $\widehat{\Lambda}$ are

(i) **Roe type diffusion operator:**

$$\widehat{\Lambda} = |\Lambda|. \tag{2.25}$$

Here, Λ denotes the matrix of eigenvalues of $\mathbf{f_U}$.

(ii) **Rusanov type diffusion operator:**

$$\widehat{\Lambda} = |\lambda^{\max}_{i+1/2}|\mathbf{ID}, \tag{2.26}$$

with $\lambda^{\max}$ denoting the largest eigenvalue of $\mathbf{f_U}$.

3 Numerical experiments

We test the following schemes:

ENOk: k-th order accurate standard ENO scheme in the MUSCL formulation [8].

ECENOk: k-th order accurate entropy stable scheme (2.6) with numerical flux (2.22),

for $k = 3, 4$ and 5 on a suite of numerical experiments. The ENO-MUSCL and ECENO schemes are semi-discrete and are integrated in time with a standard Runge-Kutta method.

We consider the Euler equations i.e., (2.1) with (2.9). The entropy function and entropy variables are given in (2.11), (2.12). We define the ECENO scheme with entropy conservative flux given by (2.13) and diffusion matrix being of the Roe type (2.25). The eigenvalues and eigenvectors of the Jacobian are computed at the average of the left and right states. The ENO-MUSCL schemes use the standard Roe numerical flux.

3.1 Sod shock tube

The Sod shock tube experiment considers the Euler equations with Riemann initial data,

$$\mathbf{U}(x,0) = \begin{cases} \mathbf{U}_L & \text{if } x < 0 \\ \mathbf{U}_R & \text{otherwise,} \end{cases} \tag{3.1}$$

with

$$\begin{bmatrix} \rho_L \\ u_L \\ p_L \end{bmatrix} = \begin{bmatrix} 1 \\ 0 \\ 1 \end{bmatrix}, \quad \begin{bmatrix} \rho_R \\ u_R \\ p_R \end{bmatrix} = \begin{bmatrix} 0.125 \\ 0 \\ 0.1 \end{bmatrix},$$

in the computational domain $[-5, 5]$. The initial discontinuity breaks into a left going rarefaction wave, a right going shock wave and a right going contact discontinuity. The computed density with the ENO3, ENO4, ECENO3 and ECENO4 schemes at time $T = 1.3$ on a mesh of 100 points is shown in Figure 3.1. The results show that the ENO and ECENO schemes are quite good at resolving the waves. The ENO4 scheme is slightly oscillatory behind the contact whereas the ECENO3 and ECENO4 schemes resolve all the waves without any noticeable oscillations.

3.2 Shock-Entropy wave interaction

This numerical example was proposed by Shu and Osher in [15] and is a good test of a scheme's ability in resolving a complex solution with both strong and weak shocks and highly oscillatory but smooth waves. The Euler equations are considered in the computational domain $[-5, 5]$ with initial data,

$$\mathbf{U}(x, 0) = \begin{cases} \mathbf{U}_L & \text{if } x < -4 \\ \mathbf{U}_R & \text{otherwise,} \end{cases}$$

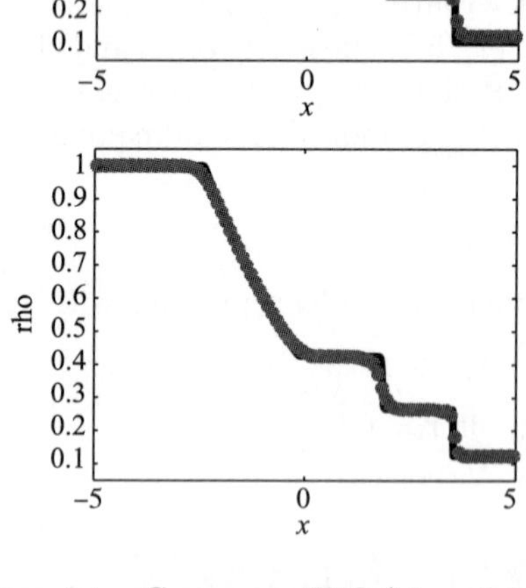

Figure 3.1 Comparing ENO (blue circles) and ECENO (red circles) with the reference solution (black line) for the Sod shock tube. Density at $T = 1.3$ on a mesh of 100 points is plotted.

with

$$\begin{bmatrix}\rho_L\\ u_L\\ p_L\end{bmatrix}=\begin{bmatrix}3.857143\\ 2.629369\\ 10.33333\end{bmatrix},\qquad \begin{bmatrix}\rho_R\\ u_R\\ p_R\end{bmatrix}=\begin{bmatrix}1+\varepsilon\sin(5x)\\ 0\\ 1\end{bmatrix}.$$

We compute up to $t = 1.8$ using a 3rd order Runge-Kutta integration method for the 3rd-order schemes, and a 4th order Runge-Kutta method for the 4th and 5th order schemes. The CFL number is set to 0.5. As a reference solution, we compute with the ENO3 scheme on a mesh of 1600 grid points. The approximate solutions are computed on a mesh of 200 grid points, corresponding to about 7 grid points for each period of the entropy waves. The solution computed by the ENO and ECENO schemes are displayed in Figure 3.2. There are very minor differences

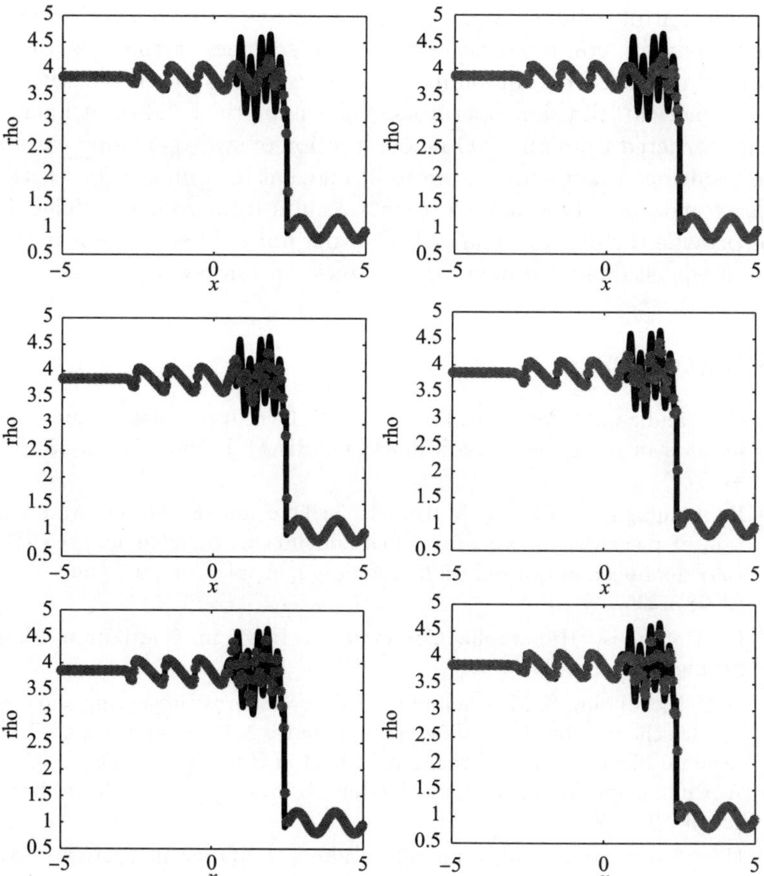

Figure 3.2 Comparing ENO (blue circles) and ECENO (red circles) with a reference solution (black line) on the Shu-Osher shock-entropy wave interaction problem. The plotted quantity is the density at time $T = 1.8$ on a mesh of 200 points.

between the ENO and ECENO schemes of the same order. The test also illustrates that the higher order schemes perform better than a low order scheme.

4 Conclusion

We prove a global stability property for the ENO reconstruction procedure. This property termed as the *sign property* says that the jump in the reconstructed values at each interface has the same sign as the jump in the underlying cell averages (cell centered values). We also obtain an upper bound of the jump in the reconstructed values in terms of the jump in the underlying values. The sign property is crucial in the construction of arbitrarily high-order accurate entropy stable schemes to approximate conservation laws. These schemes, termed as *ECENO* schemes are based on combining entropy conservative fluxes with suitable numerical diffusion operators. The numerical diffusion operators are constructed from an ENO reconstruction of scaled entropy variables. The resulting schemes are shown to be efficient in computations with the Euler equations. The paper presents results from recent articles [5, 6] that provide the first arbitrarily high-order finite difference schemes that are entropy stable for any system of conservation laws.

References

[1] F. Arandiga, A. Cohen, R. Donat and N. Dyn, Interpolation and approximation of piecewise smooth functions. SIAM J. Num. Anal. 43 (2005), 41–57.

[2] F. Arandiga, A. Cohen, R. Donat, N. Dyn and B. Matei, Approximation of piecewise smooth functions and images by edge-adapted (ENO-EA) nonlinear multiresolution techniques. Appl. Comput. Har. Anal. 24 (2008), 225–250.

[3] C. Dafermos, Hyperbolic Conservation Laws in Continuum Physics. Springer, Berlin, 2000.

[4] U. S. Fjordholm, S. Mishra and E. Tadmor, Energy preserving and energy stable schemes for the shallow water equations. "Foundations of Computational Mathematics", Proc. FoCM held in Hong Kong 2008 (F. Cucker, A. Pinkus and M. Todd, eds). London Math. Soc. Lecture Notes Ser. 363 (2009), 93–139.

[5] U. S. Fjordholm, S. Mishra and E. Tadmor, ENO reconstruction is stable. Preprint 2010.

[6] U. S. Fjordholm, S. Mishra and E. Tadmor, Arbitrarily high-order accurate entropy stable essentially non-oscillatory schemes for systems of conservation laws. In preparation 2010.

[7] A. Harten, B. Engquist, S. Osher and S. R. Chakravarty, Some results on high-order accurate essentially non-oscillatory schemes. Appl. Num. Math. 2 (1986), 347–377.

[8] A. Harten, B. Engquist, S. Osher and S. R. Chakravarty, Uniformly high order accurate essentially non-oscillatory schemes. J. Comput. Phys. (1987), 231–303.

[9] A. Harten, ENO schemes with subcell resolution. J. Comput. Phys. 83 (1989), 148–184.

[10] G. Jiang and C-W. Shu, Efficient implementation of weighted ENO schemes. J. Comput. Phys. 126 (1996), 202–226.

[11] B. VanLeer, Towards the ultimate conservative scheme V. A second-order sequel to Godunov's method. J. Comput. Phys. 32 (1979), 101–136.

[12] P. G. LeFloch, J. M. Mercier and C. Rohde, Fully discrete entropy conservative schemes of arbitraty order. SIAM J. Numer. Anal. 40 (2002), 1968–1992.

[13] R. J. LeVeque, Finite Volume Methods for Hyperbolic Problems. Cambridge University Press, Cambridge, 2002.

[14] P. L. Roe, Entropy conservative schemes for Euler equations. Talk at HYP 2006, Lyon, France. Unpublished, Lecture available from http://math.univ-lyon1.fr/~hyp2006.

[15] C. W. Shu and S. Osher, Efficient implementation of essentially non-oscillatory schemes - II. J. Comput. Phys. 83 (1989), 32–78.

[16] C. W. Shu, Numerical experiments on the accuracy of ENO and modified ENO schemes. J. Sci. Comput. 5 (2), 127–149.

[17] C. W. Shu, Essentially non-oscillatory and weighted essentially non-oscillatory schemes for hyperbolic conservation laws. ICASE Technical report, NASA, 1997.

[18] P. Sweby, High resolution schemes using flux limiters for hyperbolic conservation laws. SIAM J. Num. Anal. 21 (1984), 995–1011.

[19] E. Tadmor, The numerical viscosity of entropy stable schemes for systems of conservation laws, I. Math. Comp. 49 (1987), 91–103.

[20] E. Tadmor, Entropy stability theory for difference approximations of nonlinear conservation laws and related time-dependent problems. Act. Numerica. (2004), 451–512.

Large Time Behavior of Solutions for Compressible Navier-Stokes Equations*

Feimin Huang
Institute of Applied Mathematics, AMSS
Chinese Academy of Sciences, China
E-mail: fhuang@amt.ac.cn

Abstract

In this note, I will present recent progress on the large time behavior of solutions for one-dimensional compressible Navier-Stokes equations toward the viscous version of Riemann solutions of compressible Euler equations.

1 Introduction

The 1-dimensional compressible Navier-Stokes equations in *Lagrangian* coordinates reads

$$\begin{cases} v_t - u_x = 0, \\ u_t + p_x = \mu(\frac{u_x}{v})_x, \\ (e + \frac{u^2}{2})_t + (pu)_x = (\kappa\frac{\theta_x}{v} + \mu\frac{uu_x}{v})_x, \end{cases} \tag{1.1}$$

where $x \in \mathbb{R}^1$, $t > 0$, $v(x,t) > 0$, $u(x,t)$, $\theta(x,t) > 0$, $e(x,t) > 0$ and $p(x,t)$ are the specific volume, velocity, internal energy, temperature, and pressure respectively, while $\mu > 0$ and $\kappa > 0$ denote the viscosity and heat conduction coefficients respectively. Here we study the polytropic gas so that p and e are given by

$$p = \frac{R\theta}{v}, \quad e = \frac{R}{\gamma - 1}\theta + const. \tag{1.2}$$

*The research of FMH was supported in part by NSFC Grant No. 10825102 for distinguished youth scholar, and National Basic Research Program of China (973 Program) under Grant No. 2011CB808002.

where $\gamma > 1$ is the adiabatic exponent and $R > 0$ is the gas constant.

When neglecting the effects of viscosity and heat conductivity, i.e., $\kappa = \mu = 0$, (1.1) becomes the well-known compressible Euler system:

$$\begin{cases} v_t - u_x = 0, \\ u_t + p_x = 0, \\ (e + \frac{u^2}{2})_t + (pu)_x = 0, \end{cases} \tag{1.3}$$

which is one of the most important nonlinear strictly hyperbolic system of conservation laws. The basic waves for the system (1.3) are dilation invariant solutions: shock waves, rarefaction waves (which are nonlinear waves), and contact discontinuities, [23], [2], and the linear combinations of these basic waves, called **Riemann solutions**, govern both local and large time asymptotic behavior of general solutions to the inviscid Euler system (1.3) [14]. Since the inviscid system (1.3) is an idealization when the dissipative effects are neglected, thus it is of great importance to study the large time asymptotic behavior of solutions to the corresponding viscous systems, such as (1.1), toward the viscous version of **Riemann solutions**. In the beginning, it is very interesting to study the stability of the viscous version of these basic waves. Indeed, there have been intensive studies in this respect in the development of theory of viscous conservation laws since 1985, started with studies on the nonlinear stability of viscous shock profiles by Goodman [3] and Matsumura-Nishihara [19]. Deeper understanding has been achieved on the asymptotic stability toward nonlinear waves, viscous shock profiles and viscous rarefaction waves, which have been shown to be nonlinearly stable for quite general perturbation for the compressible Navier-Stokes system (1.1) and more general system of viscous strictly hyperbolic conservation laws, and new phenomena have been discovered and new techniques, such as weighted characteristic energy methods and uniform approximate Green's functions, have been developed based on the intrinsic properties of the underlying nonlinear waves, see [11], [12], [17], [24], [15], [16], [25],[28] and the references therein.

However, the problem of stability of contact discontinuities is more subtle. One of the main reasons is the contact discontinuities are associated with linear degenerate fields and are less stable compared with the nonlinear waves for the inviscid system (1.3), [14]. Thus the stabilizing effects around a damped contact wave pattern for Navier-Stokes equations should come mainly from the viscosity and heat conductivity. A general perturbation of a contact wave may introduce waves in the nonlinear sound wave families, and interactions of these waves with the linear contact wave are some of the major difficulties to overcome, see [9]. The mathematical aspect on the stability toward contact waves for solutions to systems of viscous conservation laws was first studied by Xin in

[26], where the metastability of a weak contact discontinuity for the compressible Euler equations with uniform viscosity, was proved by showing that although a contact discontinuity is not an asymptotic attractor for the viscous system, yet a viscous contact wave, which approximates the contact discontinuity on any finite time interval, is asymptotically nonlinear stable for small generic perturbations. This was later generalized by Liu-Xin in [18] to show the metastability of contact discontinuities for a class of general systems of nonlinear conservation laws with uniform viscosity.

The first stability result of viscous contact wave for compressible Navier-Stokes equations was due to Huang-Matsumura-Shi in [6] for a free boundary value problem with a particle path as free boundary. This result was extended in [10] for the global stability which admits the large initial perturbations. Later, the stability of viscous contact wave for the Cauchy problem was obtained in [7] under small initial perturbation with zero mass condition which excludes the possible presence of diffusion waves in the sound wave families. The zero mass assumption in [7], which is not physical, was successfully removed in [9]. Furthermore, a convergence rate is also obtained. The idea is based on the key observations that even though the energy estimate involving the lower order may grow at the rate $(1+t)^{\frac{1}{2}}$, it can be compensated by the decay in the energy estimate for derivatives which is of the order of $(1+t)^{-\frac{1}{2}}$. Thus, these reciprocal order of decay rates for the time evolution of the perturbation are essential to close the priori estimate containing the uniform bounds of the L^∞ norm on the lower order estimate and then it gives the decay of the solution to the contact wave pattern.

We note that the above important progress is made only for the single wave pattern, i.e., viscous shock wave, viscous rarefaction wave or viscous contact wave. However, few results are known for the composite wave, which consists of at least two wave patterns, due to various difficulties. As a starting point, we in [5] studied the asymptotic stability of a composite wave consisting of 1-viscous shock wave and 3-viscous shock wave under *general* initial perturbations for the full Navier-Stokes system (1.1). It is worthy to point out that the stability of above composite wave with *zero* integral perturbation is straightforward by a similar method of [11]. The main contribution of [5] is to overcome the difficulty arising from the *general* initial perturbation by introducing a new linear diffusion wave which will be shown in the next section. Later, we further in [4] showed the nonlinear stability of the linear superposition of rarefaction wave and viscous contact wave. As a byproduct, a weighted estimate for a Sobolev space was derived, see [4] for details. This kind of weighted estimate has various applications in various partial differential equations with dissipative structure like viscous conservation laws [27], inflow problem for compressible Navier-Stokes equations [21], [22] and hydrodynamic model for semiconductor devices [8]. In the rest of the present note, we mainly introduce the stability result on the composite waves for compressible Navier-Stokes equations.

2 Two viscous shock waves

Before stating the main results, we recall the Riemann problem for the compressible Euler equation (1.3) with the Riemann initial data

$$(v,u,\theta)(x,0)=\begin{cases}(v_-,u_-,\theta_-), & x<0,\\ (v_+,u_+,\theta_+), & x>0.\end{cases} \tag{2.1}$$

It is known that the system (1.3) has three eigenvalues: $\lambda_1=-\sqrt{\gamma p/v}<0, \lambda_2=0, \lambda_3=-\lambda_1>0$ where the second characteristic field is linear degenerate and the others are genuinely nonlinear. In the present paper, we focus our attention on the situation where the Riemann solution of (1.3),(2.1) consists of two shock waves (and three constant states), that is, there exists an intermediate state (v_m,u_m,θ_m) such that (v_-,u_-,θ_-) connects with (v_m,u_m,θ_m) by the 1-shock wave with the shock speed $s_1<0$ and (v_m,u_m,θ_m) connects with (v_+,u_+,θ_+) by the 3-shock wave with the shock speed $s_3>0$. Here the shock speeds s_1 and s_3 are constants determined by RH condition and satisfy entropy conditions

$$\lambda_1(v_-,u_-,\theta_-)>s_1>\lambda_1(v_m,u_m,\theta_m),$$
$$\lambda_3(v_m,u_m,\theta_m)>s_3>\lambda_3(v_+,u_+,\theta_+).$$

By the standard arguments (e.g. [23]), for each (v_-,u_-,θ_-) we can see our situation takes place provided (v_+,u_+,θ_+) is located on a curved surface in a neighborhood of (v_-,u_-,θ_-). In what follows, the neighborhood of (v_-,u_-,θ_-) is denoted by Ω_-. To describe the strengths of the shock waves for later use, we set

$$\delta_1=|v_m-v_-|+|u_m-u_-|+|\theta_m-\theta_-|, \quad \delta_3=|v_m-v_+|+|u_m-u_+|+|\theta_m-\theta_+|$$

and $\delta=\min(\delta_1,\delta_3)$. When we choose $|(v_+-v_-,u_+-u_-,\theta_+-\theta_-)|$ small in our situation for the fixed (v_-,u_-,θ_-), we note that it holds

$$\delta_1+\delta_3\le C|(v_+-v_-,u_+-u_-,\theta_+-\theta_-)|, \tag{2.2}$$

where C is a positive constant depending only on (v_-,u_-,θ_-). Then, if it also holds

$$\delta_1+\delta_3\le C\delta, \quad \delta_1+\delta_3\to 0 \tag{2.3}$$

for a positive constant C, we call the strengths of the shock waves are "small with same order". In what follows, we always assume (2.3).

Next we recall the definitions of viscous shock waves of (1.1) which correspond to the above shock waves. We see that the 1-viscous shock wave which corresponds to the 1-shock wave is a traveling wave solution of (1.1) with the formula $(V_1, U_1, \Theta_1)(x - s_1 t)$ which is determined by

$$\begin{cases} -s_1 V_1' - U_1' = 0, \\ -s_1 U_1' + p(V_1, \Theta_1)' = \mu(\frac{U_1'}{V_1})', \\ -s_1(e(V_1, \Theta_1) + \frac{U_1^2}{2})' + (p(V_1, \Theta_1)U_1)' = (\kappa\frac{\Theta_1'}{V_1} + \mu\frac{U_1 U_1'}{V_1})', \\ (V_1, U_1, \Theta_1)(-\infty) = (v_-, u_-, \theta_-), \\ (V_1, U_1, \Theta_1)(+\infty) = (v_m, u_m, \theta_m) \end{cases} \tag{2.4}$$

where $' = \frac{d}{d\xi}$, $\xi = x - s_1 t$. Integrating (2.4) with respect to x, we have

$$\mu\frac{U_{1,x}}{V_1} = P_1 + s_1^2 V_1 - b_1, \;\; b_1 = p_- + s_1^2 v_- = p_m + s_1^2 v_m, \;\; P_1 = \frac{R\Theta_1}{V_1}. \tag{2.5}$$

Similarly, the 3-viscous shock wave $(V_3, U_3, \Theta_3)(x - s_3 t)$ is defined by

$$\begin{cases} -s_3 V_3' - U_3' = 0, \\ -s_3 U_3' + p(V_3, \Theta_3)' = \mu(\frac{U_3'}{V_3})', \\ -s_3(e(V_3, \Theta_3) + \frac{U_3^2}{2})' + (p(V_3, \Theta_3)U_3)' = (\kappa\frac{\Theta_3'}{V_3} + \mu\frac{U_3 U_3'}{V_3})', \\ (V_3, U_3, \Theta_3)(-\infty) = (v_m, u_m, \theta_m), \\ (V_3, U_3, \Theta_3)(+\infty) = (v_+, u_+, \theta_+) \end{cases} \tag{2.6}$$

where $' = \frac{d}{d\eta}$, $\eta = x - s_3 t$ and

$$\mu\frac{U_{3,x}}{V_3} = P_3 + s_3^2 V_3 - b_3, \;\; b_3 = p_+ + s_3^2 v_+ = p_m + s_3^2 v_m, \;\; P_3 = \frac{R\Theta_3}{V_3}. \tag{2.7}$$

By the results in [11], we have

Proposition 2.1. *For each (v_-, u_-, θ_-), suppose $(v_+, u_+, \theta_+) \in \Omega_-$ and the Riemann solution of (1.3),(2.1) consists of two shock waves. Also, assume $1 < \gamma \le 2$. Then there exist positive constants c and C which depend only on (v_-, u_-, θ_-) such that, for $i = 1, 3$,*

$$\begin{aligned} &|(V_1 - v_m, U_1 - u_m, \Theta_1 - \theta_m)(x,t)| \le C\delta_1 e^{-c\delta_1|x - s_1 t|}, \quad x > s_1 t, \; t \ge 0, \\ &|(V_3 - v_m, U_3 - u_m, \Theta_3 - \theta_m)(x,t)| \le C\delta_3 e^{-c\delta_3|x - s_3 t|}, \quad x < s_3 t, \; t \ge 0, \\ &U_{i,x}(x,t) < 0, \quad |(V_{i,x}, U_{i,x}, \Theta_{i,x})(x,t)| \le C\delta_i^2 e^{-c\delta_i|x - s_i t|}, \quad x \in R, \; t \ge 0, \\ &|\Theta_{i,x}(x,t)| \le C|V_{i,x}(x,t)|, \quad x \in R, \; t \ge 0, \\ &|s_i^2 - \tfrac{\gamma p_m}{v_m}| \le C\delta_i. \end{aligned} \tag{2.8}$$

Let $E=\theta+\frac{\gamma-1}{2R}u^2$ and $m(x,t)={}^t(v,u,E), \bar{m}(x,t)={}^t(\bar{v},\bar{u},\bar{E})$, with

$$\begin{cases} \bar{v}=V_1(x-s_1t)+V_3(x-s_3t)-v_m, \\ \bar{u}=U_1(x-s_1t)+U_3(x-s_3t)-u_m, \\ \bar{E}=E_1(x-s_1t)+E_3(x-s_3t)-E_m,\ E_m=\theta_m+\frac{\gamma-1}{2R}u_m^2. \end{cases} \tag{2.9}$$

Since the present paper is concerning with a general initial perturbation, the integral $\int_{-\infty}^{\infty}(m-\bar{m})(x,0)\,dx$ is in general not zero, to use the anti-derivative technique, we need to find an ansatz $\tilde{m}$ such that $\int_{-\infty}^{\infty}m(x,t)-\tilde{m}(x,t)\,dx=0$, meanwhile $|\tilde{m}-\bar{m}|$ tends to zero as $t\to\infty$. This idea was first applied in [15] to study the asymptotic stability of single viscous shock wave for general viscous conservation laws with uniformly artificial viscosities by introducing linear or nonlinear diffusion waves. Here we will borrow the idea of [15] to construct the desired ansatz $\tilde{m}$ by introducing a new linear diffusion wave. In the following, we shall show how to construct the linear diffusion wave with details. Let

$$A(v,u,\theta)=\begin{pmatrix} 0 & -1 & 0 \\ -\frac{p}{v} & 0 & \frac{R}{v} \\ 0 & \frac{\gamma-1}{R}p & 0 \end{pmatrix}$$

be the Jacobi matrix of the flux ${}^t(-u,p,\frac{\gamma-1}{R}pu)$ with respect to (v,u,θ). Then the second right eigenvector of the matrix $A(v,u,\theta)$ at the point (v_m,u_m,θ_m) is

$$r_2={}^t(1,0,\frac{p_m}{R}).$$

Furthermore, a direct computation yields that the three vectors $r_1={}^t(v_m-v_-,u_m-u_-,E_m-E_-)$, r_2 and $r_3={}^t(v_+-v_m,u_+-u_m,E_+-E_m)$ are linearly independent in R^3 if $\delta_1+\delta_3$ is small. So if the initial mass $\int m(x,0)-\bar{m}(x,0)\,dx$ is not zero, we can distribute the initial mass along the three independent directions r_1,r_2 and r_3, that is,

$$\int(m(x,0)-\bar{m}(x,0))\,dx=\sum_{i=1}^{3}\alpha_i r_i, \tag{2.10}$$

where $\alpha_i, i=1,2,3$, are constants uniquely determined by the initial data. The excessive mass $\alpha_1 r_1$ in the first characteristic field can be eliminated by the translated 1-viscous shock wave with a shift α_1, i.e., $V_1(x-s_1t+\alpha_1)$. Similarly, we can eliminate $\alpha_3 r_3$ by replacing $V_3(x-s_3t)$ by $V_3(x-s_3t+\alpha_3)$. We remove the excessive mass in the second characteristic field, i.e., $\alpha_2 r_2$ by introducing a linear diffusion wave, that is,

$$\Theta(x,t)=\frac{\alpha_2}{\sqrt{4\pi a(1+t)}}e^{-\frac{x^2}{4a(1+t)}},\quad a=\frac{(\gamma-1)\kappa}{\gamma R v_m}>0 \tag{2.11}$$

which is the unique solution of

$$\Theta_t = a\Theta_{xx}, \quad \Theta|_{t=-1} = \alpha_2\delta(x), \quad \int_{-\infty}^{\infty} \Theta(x,t)\,dx = \alpha_2. \tag{2.12}$$

Thus we define the ansatz $\tilde{m} = {}^t(\tilde{v}, \tilde{u}, \tilde{E})$ as

$$\begin{aligned}
\tilde{v} &= V_1(x - s_1t + \alpha_1) + V_3(x - s_3t + \alpha_3) - v_m + \Theta, \\
\tilde{u} &= U_1(x - s_1t + \alpha_1) + U_3(x - s_3t + \alpha_3) - u_m + a\Theta_x, \\
\tilde{E} &= E_1(x - s_1t + \alpha_1) + E_3(x - s_3t + \alpha_3) - E_m + \tfrac{p_m}{R}\Theta + \tfrac{\gamma-1}{R}u_m a\Theta_x,
\end{aligned} \tag{2.13}$$

which approximates the Navier-Stokes system (1.1) in the form

$$\begin{cases}
\tilde{v}_t - \tilde{u}_x = 0, \\
\tilde{u}_t + \tilde{p}_x = \mu(\dfrac{\tilde{u}_x}{\tilde{v}})_x + \tilde{R}_{1,x}, \\
(\tilde{e} + \dfrac{\tilde{u}^2}{2})_t + (\tilde{p}\tilde{u})_x = \kappa(\dfrac{\tilde{\theta}_x}{\tilde{v}})_x + (\dfrac{\mu}{\tilde{v}}\tilde{u}\tilde{u}_x)_x + \tilde{R}_{2,x},
\end{cases} \tag{2.14}$$

where $\tilde{R}_i$ $(i = 1, 2)$ satisfy

$$\begin{aligned}
|\tilde{R}_i|,\ |\tilde{R}_{i,x}| \ &\le C(\delta^2 + |\alpha_2|\delta^{\frac{3}{2}})e^{-c\delta(|x|+t)} \\
&\quad + C\frac{|\alpha_2|}{(1+t)^{\frac{3}{2}}}e^{-\frac{cx^2}{1+t}} + C(\delta + |\alpha_2|)e^{-c(|x|+t)}.
\end{aligned} \tag{2.15}$$

Furthermore,

$$\begin{aligned}
&\int_{-\infty}^{\infty} (m(x,0) - \tilde{m}(x,0))\,dx \\
&\quad = \int_{-\infty}^{\infty} (m(x,0) - \bar{m}(x,0))\,dx + \int_{-\infty}^{\infty} (\bar{m}(x,0) - \tilde{m}(x,0))\,dx \\
&\quad = \sum_{i=1}^{3} \alpha_i r_i + \sum_{i\neq 2} \int_{-\infty}^{\infty} {}^t(V_i(x) - V_i(x+\alpha_i), U_i(x) - U_i(x+\alpha_i), \\
&\qquad\qquad E_i(x) - E_i(x+\alpha_i))\,dx - \alpha_2 r_2 = 0,
\end{aligned} \tag{2.16}$$

where we have used the fact that $\int_{-\infty}^{\infty} \Theta\,dx = \alpha_2$. Thus $\tilde{m}(x,t)$ is the desired ansatz. The exact derivation of the linear diffusion wave $\Theta(x,t)$ can be found in [5].

Remark 2.1. It is worthy to recall that in the study of single viscous shock wave for viscous conservation laws with artificial viscosities, to derive good error terms like $\tilde{R}_1$ and $\tilde{R}_2$, besides the nonlinear or linear diffusion waves of Liu [15], Szepessy and Xin [24] introduced additional coupled linear diffusion waves which do not carry any mass. Here our linear diffusion wave Θ is somehow like that of Liu [15] and $a\Theta_x$ is like the coupled diffusion wave of Szepessy and Xin [24]. Furthermore, in this particular system of compressible Navier-Stokes equations, we get an explicit formula of the coupled diffusion wave, i.e., $a\Theta_x$, which is a higher order correction corresponding to the linear diffusion wave Θ.

Under the above preparations, we are now on the stage to state our main results. We first fix any (v_-, u_-, θ_-), and assume that $(v_+, u_+, \theta_+) \in \Omega_-$ and the Riemann solution of (1.3),(2.1) consists of two shock waves. Then the composite wave $(\bar{v}, \bar{u}, \bar{\theta})(x,t)$ alluded above is well defined, and we assume the initial data satisfy

$$(v_0 - \bar{v}(\cdot, 0), u_0 - \bar{u}(\cdot, 0), \theta_0 - \bar{\theta}(\cdot, 0)) \in H^1 \cap L^1. \tag{2.17}$$

Under the assumption (2.17), $\tilde{m} = (\tilde{v}, \tilde{u}, \tilde{E})$ in (2.13) is well defined and the shifts α_1 and α_3 are uniquely determined by (2.10) so that $\int (m(x,0) - \tilde{m}(x,0))\,dx = 0$. Thus, if we set

$$\begin{aligned}
&\Phi_0(x) = -\int_x^\infty (v_0(y) - \tilde{v}(y,0))\,dy, \quad \Psi_0(x) = -\int_x^\infty (u_0(y) - \tilde{u}(y,0))\,dy, \\
&\bar{W}_0(x) = -\int_x^\infty ((e_0 + \frac{u_0^2}{2})(y) - (\tilde{e} + \frac{\tilde{u}^2}{2})(y,0))\,dy,
\end{aligned} \tag{2.18}$$

we can assume the initial data further satisfy

$$(\Phi_0, \Psi_0, \bar{W}_0) \in L^2 \tag{2.19}$$

where it is noted that (2.17) and (2.19) imply $(\Phi_0, \Psi_0, \bar{W}_0) \in H^2$. To state the results precisely, we also set

$$\begin{aligned}
\bar{v}_{\alpha_1,\alpha_3}(x,t) &= V_1(x - s_1 t + \alpha_1) + V_3(x - s_3 t + \alpha_3) - v_m, \\
\bar{u}_{\alpha_1,\alpha_3}(x,t) &= U_1(x - s_1 t + \alpha_1) + U_3(x - s_3 t + \alpha_3) - u_m, \\
\bar{\theta}_{\alpha_1,\alpha_3}(x,t) &= \Theta_1(x - s_1 t + \alpha_1) + \Theta_3(x - s_3 t + \alpha_3) - \theta_m,
\end{aligned} \tag{2.20}$$

where we note $(\bar{v}, \bar{u}, \bar{\theta})_{0,0} = (\bar{v}, \bar{u}, \bar{\theta})$. We have

Theorem 1. ([5]) *For each (v_-, u_-, θ_-), assume that $(v_+, u_+, \theta_+) \in \Omega_-$ and the Riemann solution of* (1.3),(2.1) *consists of two shock waves whose strengths satisfy (2.3). Further assume that the initial data satisfy* (2.17) *and* (2,19), *and also* $1 < \gamma \le 2$. *Then there exist positive constants δ_0 and ϵ_0, such that if $|(v_+ - v_-, u_+ - u_-, \theta_+ - \theta_-)| \le \delta_0$ and*

$$\|(v_0 - \bar{v}(\cdot,0), u_0 - \bar{u}(\cdot,0), \theta_0 - \bar{\theta}(\cdot,0))\|_{H^1 \cap L^1} + \|(\Phi_0, \Psi_0, \bar{W}_0)\| \le \epsilon_0,$$

then the Cauchy problem of (1.1)-(1.2) *of the compressible Navier-Stokes system admits a unique global solution in time (v, u, θ) satisfying*

$$\begin{aligned}
&(v - \tilde{v}, u - \tilde{u}, \theta - \tilde{\theta}) \in C([0, +\infty); H^1), \quad v - \tilde{v} \in L^2(0, \infty; H^1), \\
&(u - \tilde{u}, \theta - \tilde{\theta}) \in L^2(0, \infty; H^2),
\end{aligned}$$

where $(\tilde{v},\tilde{u},\tilde{\theta})$ *is defined by* (2.13)*, and the shifts* α_1,α_3 *in* $(\tilde{v},\tilde{u},\tilde{\theta})$ *are defined by* (2.10)*. Furthermore, it holds*

$$\sup_{x\in R}|(v-\bar{v}_{\alpha_1,\alpha_3},u-\bar{u}_{\alpha_1,\alpha_3},\theta-\bar{\theta}_{\alpha_1,\alpha_3})(x,t)| \ \to\ 0,\ t\to\infty.$$

3 Two rarefaction waves and one viscous contact wave

To state our main results, we first recall the viscous contact wave (V,U,Θ) for the compressible Navier-Stokes system (1.1) defined in [7]. For the Riemann problem (1.3)(2.1), it is known that the contact discontinuity solution takes the form

$$(\tilde{V},\tilde{U},\tilde{\Theta})(x,t)=\begin{cases}(v_-,u_-,\theta_-),\ x<0,t>0,\\(v_+,u_+,\theta_+),\ x>0,t>0,\end{cases} \tag{3.1}$$

provided that

$$u_-=u_+,\quad p_-\triangleq\frac{R\theta_-}{v_-}=p_+\triangleq\frac{R\theta_+}{v_+}. \tag{3.2}$$

In the setting of the compressible Navier-Stokes system (1.1), the corresponding wave (V,U,Θ) to the contact discontinuity $(\tilde{V},\tilde{U},\tilde{\Theta})$ becomes smooth and behaves as a diffusion wave due to the dissipation effect. We call this wave "viscous contact wave". The viscous contact wave (V,U,Θ) can be constructed as follows. Since the pressure for the profile (V,U,Θ) is expected to be almost constant corresponding to the arguments in [6], i.e., we set

$$\frac{R\Theta}{V}=p_+,$$

which indicates the leading part of the energy equation $(1.1)_3$ is

$$\frac{R}{\gamma-1}\Theta_t+p_+U_x=\kappa\left(\frac{\Theta_x}{V}\right)_x. \tag{3.3}$$

The equations (3.3) and $(1.1)_1$ lead to a nonlinear diffusion equation,

$$\Theta_t=a\left(\frac{\Theta_x}{\Theta}\right)_x,\quad \Theta(\pm\infty,t)=\theta_\pm,\quad a=\frac{\kappa p_+(\gamma-1)}{\gamma R^2}>0, \tag{3.4}$$

which has a unique self similarity solution $\Theta(x,t)=\Theta(\xi),\xi=\frac{x}{\sqrt{1+t}}$ due to [1] and [7]. Furthermore, $\Theta(\xi)$ is a monotone function, increasing if

$\theta_+ > \theta_-$ and decreasing if $\theta_+ < \theta_-$; On the other hand, there exists some positive constant $\overline{\delta}$, such that for $\delta = |\theta_+ - \theta_-| \leq \overline{\delta}$, Θ satisfies

$$(1+t)|\Theta_{xx}| + (1+t)^{1/2}|\Theta_x| + |\Theta - \theta_\pm| \leq c_1\delta e^{-\frac{c_2x^2}{1+t}} \text{ as } |x| \to \infty, \quad (3.5)$$

where c_1 and c_2 are both positive constants depending only on θ_- and $\overline{\delta}$. Once Θ is determined, the contact wave profile $(V, U, \Theta)(x, t)$ is then defined as follows:

$$V = \frac{R}{p_+}\Theta, \quad U = u_- + \frac{\kappa(\gamma-1)}{\gamma R}\frac{\Theta_x}{\Theta}, \quad \Theta = \Theta. \quad (3.6)$$

It is straightforward to check that (V, U, Θ) satisfies

$$\|V - \tilde{V}, U - \tilde{U}, \Theta - \tilde{\Theta}\|_{L^p} = O\left(\kappa^{1/(2p)}\right)(1+t)^{1/(2p)}, p \geq 1,$$

which means the nonlinear diffusion wave $(V, U, \Theta)(x, t)$ approximates the contact discontinuity $(\tilde{V}, \tilde{U}, \tilde{\Theta})(x, t)$ to the Euler system (1.3) in L^p norm, $p \geq 1$ on any finite time interval as the heat conductivity coefficient κ tends to zero. More importantly, the contact wave (V, U, Θ) solves the compressible Navier-Stokes system (1.1) time asymptotically, ie.,

$$\begin{cases} V_t - U_x = 0, \\ U_t + (\frac{R\Theta}{V})_x = \mu\left(\frac{U_x}{V}\right)_x + R_1, \\ \left(e(V,\Theta) + \frac{U^2}{2}\right)_t + (p(V,\Theta)U)_x = \left(\kappa\frac{\Theta_x}{V} + \mu\frac{UU_x}{V}\right)_x + R_2, \end{cases}$$

where

$$\begin{aligned} R_1 &= \frac{\kappa(\gamma-1)}{\gamma R}\left((\ln\Theta)_{xt} - \mu\left(\frac{p_+}{R\Theta}(\ln\Theta)_{xx}\right)_x\right) \\ &= O(\delta)(1+t)^{-3/2}e^{-\frac{c_2x^2}{1+t}} \text{ as } |x| \to \infty, \\ R_2 &= \left(\frac{\kappa(\gamma-1)}{\gamma R}\right)^2\left((\ln\Theta)_x(\ln\Theta)_{xt} - \mu\left(\frac{p_+}{R\Theta}(\ln\Theta)_x(\ln\Theta)_{xx}\right)_x\right) \\ &= O(\delta)(1+t)^{-2}e^{-\frac{c_2x^2}{1+t}} \text{ as } |x| \to \infty. \end{aligned}$$

Here we consider the stability of the superposition of viscous contact wave and rarefaction ones. In this situation, we assume that

$$(v_+, u_+, \theta_+) \in R_1CR_3(v_-, u_-, \theta_-) \subset \Omega_-$$

where

$$R_1CR_3(v_-, u_-, \theta_-) \triangleq \Big\{ (u, v, \theta) \in \Omega(v_-, u_-, \theta_-) \Big| \, s \neq s_-,$$

$$u \geq u_- - \int_{v_-}^{e^{\frac{\gamma-1}{R\gamma}(s_- - s)}v} \lambda_-(\eta, s_-)d\eta, \, u \geq u_- - \int_{e^{\frac{\gamma-1}{R\gamma}(s - s_-)}v_-}^{v} \lambda_+(\eta, s)d\eta \Big\}$$

with

$$s=\frac{R}{\gamma-1}\ln\frac{R\theta}{A}+R\ln v,\quad s_\pm=\frac{R}{\gamma-1}\ln\frac{R\theta_\pm}{A}+R\ln v_\pm,$$

and

$$\lambda_\pm(v,s)=\pm\sqrt{A\gamma v^{-\gamma-1}e^{(\gamma-1)s/R}}.$$

It is well-known ([23]) that there exists some suitably small $\delta_1>0$ such that for

$$(v_+,u_+,\theta_+)\in R_1CR_3(v_-,u_-,\theta_-),\quad |\theta_--\theta_+|\le\delta_1, \tag{3.7}$$

there exist a positive constant $C=C(\theta_-,\delta_1)$ and a unique pair of points (v_-^m,u^m,θ_-^m) and (v_+^m,u^m,θ_+^m) in $\Omega(v_-,u_-,\theta_-)$ satisfying

$$\frac{R\theta_-^m}{v_-^m}=\frac{R\theta_+^m}{v_+^m}\triangleq p^m,$$

and

$$|v_\pm^m-v_\pm|+|u^m-u_\pm|+|\theta_\pm^m-\theta_\pm|\le C|\theta_--\theta_+|;$$

Moreover, the points (v_-^m,u^m,θ_-^m) and (v_+^m,u^m,θ_+^m) belong to the 1-rarefaction wave curve $R_-(v_-,u_-,\theta_-)$ and the 3-rarefaction wave curve $R_+(v_+,u_+,\theta_+)$ respectively, where

$$R_\pm(v_\pm,u_\pm,\theta_\pm)=\left\{(v,u,\theta)\,\middle|\,s=s_\pm,u=u_\pm-\int_{v_\pm}^{v}\lambda_\pm(\eta,s_\pm)d\eta,v>v_\pm\right\}.$$

The 1-rarefaction wave $(v_-^r,u_-^r,\theta_-^r)(\frac{x}{t})$ (resp. the 3-rarefaction wave $(v_+^r,u_+^r,\theta_+^r)(\frac{x}{t})$) connecting (v_-,u_-,θ_-) and (v_-^m,u^m,θ_-^m) (resp. $(v_+^m,$ u^m, $\theta_+^m)$ and (v_+,u_+,θ_+)) is the weak solution of the Riemann problem of the Euler system (1.3) with the following initial Riemann data

$$(v_\pm^r,u_\pm^r,\theta_\pm^r)(x,0)=\begin{cases}(v_\pm^m,u^m,\theta_\pm^m), & \pm x<0,\\ (v_\pm,u_\pm,\theta_\pm), & \pm x>0.\end{cases} \tag{3.8}$$

Since the rarefaction waves $(v_\pm^r,u_\pm^r,\theta_\pm^r)$ are not smooth enough solutions, it is convenient to construct smooth approximate ones of them. Motivated by [20], the smooth solutions of Euler system (1.3), $(V_\pm^r,U_\pm^r,\Theta_\pm^r)$, which approximate $(v_\pm^r,u_\pm^r,\theta_\pm^r)$, are given by

$$\begin{cases}\lambda_\pm(V_\pm^r(x,t),s_\pm)=w_\pm(x,t),\\ U_\pm^r=u_\pm-\displaystyle\int_{v_\pm}^{V_\pm^r(x,t)}\lambda_\pm(\eta,s_\pm)d\eta,\\ \Theta_\pm^r=\theta_\pm(v_\pm)^{\gamma-1}(V_\pm^r)^{1-\gamma},\end{cases} \tag{3.9}$$

where w_- (resp. w_+) is the solution of the initial problem for the typical Burgers equation:

$$\begin{cases} w_t + ww_x = 0, & (x,t) \in \mathbb{R} \times (0,\infty), \\ w(x,0) = (w_r + w_l)/2 + ((w_r - w_l)/2)\tanh x, \end{cases} \tag{3.10}$$

with $w_l = \lambda_-(v_-, s_-), w_r = \lambda_+(v_-^m, s_-)$ (resp. $w_l = \lambda_+(v_+^m, s_+), w_r = \lambda_+(v_+, s_+)$).

Let $(V^{cd}, U^{cd}, \Theta^{cd})(x,t)$ be the viscous contact wave constructed in (3.6) with $(v_\pm, u_\pm, \theta_\pm)$ replaced by $(v_\pm^m, 0, \theta_\pm^m)$ respectively. We define

$$\begin{pmatrix} V \\ U \\ \Theta \end{pmatrix}(x,t) = \begin{pmatrix} V^{cd} + V_-^r + V_+^r \\ U^{cd} + U_-^r + U_+^r \\ \Theta^{cd} + \Theta_-^r + \Theta_+^r \end{pmatrix}(x,t) - \begin{pmatrix} v_-^m + v_+^m \\ u^m \\ \theta_-^m + \theta_+^m \end{pmatrix}, \tag{3.11}$$

and

$$(\phi, \psi, \zeta)(x,t) = (v - V, u - U, \theta - \Theta)(x,t).$$

Then we have

Theorem 2. ([4]) For any given (v_-, u_-, θ_-), suppose that (3.7) holds for some small $\delta_1 > 0$. Let (V, U, Θ) be as in (3.11) with strength $\delta = |\theta_+ - \theta_-| \le \bar{\delta}$. Then there exist positive constants $\delta_0(\le \min\{\delta_1, \bar{\delta}\})$ and ε_0, such that if $\delta < \delta_0$ and the initial data (v_0, u_0, θ_0) satisfies

$$\|(v_0(\cdot) - V(\cdot,0), u_0(\cdot) - U(\cdot,0), \theta_0(\cdot) - \Theta(\cdot,0))\|_{H^1} \le \varepsilon_0,$$

then the Cauchy problem of (1.1) admits a unique global solution (v, u, θ) satisfying $(v - V, u - U, \theta - \Theta) \in X([0,\infty))$ and

$$\lim_{t\to\infty} \sup_{x\in\mathbb{R}} \begin{pmatrix} |(v - v_-^r - V^{cd} - v_+^r + v_-^m + v_+^m)(x,t)| \\ |(u - u_-^r - u_+^r + u^m)(x,t)| \\ |(\theta - \theta_-^r - \Theta^{cd} - \theta_+^r + \theta_-^m + \theta_+^m)(x,t)| \end{pmatrix} = 0, \tag{3.12}$$

where the $(v_-^r, u_-^r, \theta_-^r)(x,t)$ and $(v_+^r, u_+^r, \theta_+^r)(x,t)$ are the 1-rarefaction wave and 3-rarefaction one uniquely determined by (3.9) respectively.

4 Key lemma

In this section, we derive an elementary inequality concerning the heat kernel which will is very useful. For $\alpha > 0$, we define

$$\omega(x,t) = (1+t)^{-1/2} \exp\left\{-\frac{\alpha x^2}{1+t}\right\}, \quad g(x,t) = \int_{-\infty}^{x} \omega(y,t)dy. \tag{4.1}$$

It is easy to check that

$$4\alpha g_t = g_{xx} = \omega_x, \quad \|g(\cdot,t)\|_{L^\infty} = \sqrt{\pi}\alpha^{-1/2}. \tag{4.2}$$

Lemma 4.1. ([4]) For $0 < T \leq +\infty$, suppose that $h(x,t)$ satisfies

$$h \in L^\infty(0,T;L^2(\mathbb{R})), \quad h_x \in L^2(0,T;L^2(\mathbb{R})), \quad h_t \in L^2(0,T;H^{-1}(\mathbb{R})).$$

Then the following estimate holds:

$$\begin{aligned}&\int_0^T \int_{\mathbb{R}} h^2\omega^2 dxdt \\ &\leq 4\pi\|h(0)\|^2 + 4\pi\alpha^{-1}\int_0^T \|h_x(t)\|^2 dt + 8\alpha \int_0^T \left\langle h_t, hg^2 \right\rangle_{H^{-1}\times H^1} dt.\end{aligned} \tag{4.3}$$

Proof. We get by standard arguments and using (4.2) that

$$\begin{aligned}&2\int_0^T \left\langle h_t, hg^2\right\rangle_{H^{-1}\times H^1} dt \\ &= \int_{\mathbb{R}} (hg)^2(x,T)dx - \int_{\mathbb{R}} (hg)^2(x,0)dx - \frac{1}{2\alpha}\int_0^T\int_{\mathbb{R}} h^2 g\omega_x dxdt \\ &\geq -\int_{\mathbb{R}} h^2(x,0)g^2(x,0)dx + \frac{1}{2\alpha}\int_0^T\int_{\mathbb{R}} h^2\omega^2 dxds \\ &\quad + \frac{1}{\alpha}\int_0^T\int_{\mathbb{R}} hh_x g\omega dxdt \\ &\geq -\frac{\pi}{\alpha}\int_{\mathbb{R}} h^2(x,0)dx + \frac{1}{4\alpha}\int_0^T\int_{\mathbb{R}} h^2\omega^2 dxdt - \frac{\pi}{\alpha^2}\int_0^T \|h_x(t)\|^2 dt,\end{aligned}$$

which yields (4.3) directly.

Corollary 4.1. In addition to the condition of Lemma 4.1, assume that h satisfies

$$h_t = h_{xx}, \quad h(x,0) = h_0(x) \in L^2(\mathbb{R}).$$

Then there exists some positive constant $C(\alpha)$ depending only on α such that

$$\int_0^T \int_{\mathbb{R}} h^2\omega^2 dxdt \leq C(\alpha)\|h_0\|^2, \tag{4.4}$$

i.e.,

$$\int_0^T (1+t)^{-1}\int_{\mathbb{R}} h^2 e^{-\frac{\alpha x^2}{1+t}} dxdt \leq C(\alpha)\|h_0\|^2.$$

Proof. Energy inequality of heat equation leads to

$$\frac{1}{2}\sup_{0\leq t\leq T}\|h(t)\|^2 + \int_0^T \|h_x\|^2 dt \leq \frac{1}{2}\|h_0\|^2,$$

which together with Lemma 4.1 yields (4.4) directly due to the following simple fact:

$$\begin{aligned}
&\int_0^T \langle h_t, hg^2\rangle_{H^{-1}\times H^1} dt \\
&= -\int_0^T\int_{\mathbb{R}} h_x^2 g^2 dxdt - 2\int_0^T\int_{\mathbb{R}} h_x hg\omega dxdt \\
&\leq \frac{1}{16\alpha}\int_0^T\int_{\mathbb{R}} h^2\omega^2 dxdt + 16\pi\int_0^T \|h_x\|^2 dt.
\end{aligned}$$

Remark 4.1. The proof of estimate (4.3) is quite simple and elementary. However, it seems that this estimate can be used in more complicated systems and therefore has extensive applications. For example, it has been used in viscous conservation laws [27], inflow problem for compressible Navier-Stokes equations [21], [22] and hydrodynamical model for semiconductor devices [8].

5 Concluding remarks

To complete the program, that is, to prove the nonlinear stability of the viscous version of Riemann solutions for the compressible Navier-Stokes equations, the remaining thing is to solve the following two cases:

- 1-shock wave+2-contact discontinuity+3-shock wave:
- 1-rarefaction wave+2-contact discontinuity+3-shock wave:

Since in the study of single viscous contact wave [9], although the L^2 norm of the antiderivative for the perturbation admits growth at some rate, it can be compensated by the decay in the derivatives so that the L^∞ norm is uniformly bounded. When considering the combination of viscous shock waves and viscous contact wave, it is difficult to get the decay rate for the L^2 norm of the perturbation. Thus the uniform bound of the L^2 norm of the antiderivative for the perturbation is the main difficulty to overcome in the first case. On the other hand, we regard the second case as a very important open problem because the framework for the study on the stability on shock wave is quite different from that for rarefaction wave, how to treat the stability of the shock wave and rarefaction in a uniform way is a big challenging open problem.

References

[1] F. V. Atkinson and L. A. Peletier, Similarity solutions of the nonlinear diffusion equation. Arch. rational Mech. Anal. 54 (1974), 373–392.

[2] R. Courant and K. O. Friedrichs, Supersonic Flows and Shock Waves. Wiley-Interscience: New York, 1948.

[3] J. Goodman, Nonlinear asymptotic stability of viscous shock profiles for conservation laws. Arch. Rational Mech. Anal. 95 (4) (1986), 325–344.

[4] F. M. Huang, J. Li and A. Matsumura, Asymptotic stability of combination of viscous contact wave with rarefaction waves for 1-D compressible Navier-Stokes system. Arch. Rat. Mech. Anal. 197 (2010), 89–116.

[5] F. M. Huang and A. Matsumura, Stability of a composite wave of two viscous shockwaves for the full compressible Navier-Stokes equations. Comm. Math. Phys. 289 (2009), 841–861.

[6] F. M. Huang, A. Matsumura and X. Shi, On the stability of contact discontinuity for compressible Navier-Stokes equations with free boundary. Osaka J. Math. 41 (2004), no. 1, 193–210.

[7] F. M. Huang, A. Matsumura and Z. P. Xin, Stability of contact discontinuities for the 1-D compressible Navier-Stokes equations. Arch. Rat. Mech. Anal. 179 (2005), 55–77.

[8] F. M. Huang, M. Mei and Y. Wang, Large time behavior of solutions to n-dimensional bipolar hydrodynamic model for semiconductors. SIAM J. Math. Anal. in press.

[9] F. M. Huang, Z. P. Xin and T. Yang, Contact discontinuity with general perturbations for gas motions. Adv. Math., 219 (4) (2008), 1246–1297.

[10] F. M. Huang and H. J. Zhao, On the global stability of contact discontinuity for compressible Navier-Stokes equations. Rend. Sem. Mat. Univ. Padova. 109 (2003), 283–305.

[11] S. Kawashima and A. Matsumura, Asymptotic stability of traveling wave solutions of systems for one-dimensional gas motion. Comm. Math. Phys. 101 (1) (1985), 97–127.

[12] S. Kawashima, A. Matsumura and K. Nishihara, Asymptotic behavior of solutions for the equations of a viscous heat-conductive gas. Proc. Japan Acad. 62 Ser.A (1986), 249–252.

[13] T.-P. Liu, Decay to N-waves of solutions of general systems of nonlinear hyperbolic conservation laws. Comm. Pure Appl. Math. 30 (1977), no. 5, 586–611.

[14] T.-P. Liu, Linear and nonlinear large time behavior of general systems of hyperbolic conservation laws. Comm. Pure Appl. Math. 30 (1977), 767–796.

[15] T.-P. Liu, Nonlinear stability of shock waves for viscous conservation laws. Mem. Amer. Math. Soc. 56 (329) (1985), 1–108.

[16] T.-P. Liu and Z. P. Xin, Nonlinear stability of rarefaction waves for compressible Navier-Stokes equations. Comm. Math. Phys. 118 (1988), no. 3, 451–465.

[17] T.-P. Liu, Pointwise convergence to shock waves for viscous conservation laws. Comm. Pure Appl. Math. 50 (1997), no. 11, 1113–1182.

[18] T.-P. Liu and Z. P. Xin, Pointwise decay to contact discontinuities for systems of viscous conservation laws. Asian J. Math. 1 (1997), 34–84.

[19] A. Matsumura and K. Nishihara, On the stability of traveling wave solutions of a one-dimensional model system for compressible viscous gas. Japan J. Appl. Math. 2 (1) (1985), 17–25.

[20] A. Matsumura and K. Nishihara, Asymptotics toward the rarefaction waves of a one-dimensional model system for compressible viscous gas. Japan J. Appl. Math. 3 (1) (1985), 3–13.

[21] X. H. Qin and Y. Wang, Stability of wave patterns to the inflow problem of full compressible Navier-Stokes equations. SIAM J. Math. Anal., 41 (2009), no. 5, 2057–2087.

[22] X. H. Qin and Y. Wang, Large time behavior of solutions to the inflow problem of full compressible Navier-Stokes equations. SIAM J. Math. Anal. 43 (2011), 341–366.

[23] J. Smoller, Shock Waves and Reaction-diffusion Equations. Springer, New York, 1994.

[24] A. Szepessy and Z. P. Xin, Nonlinear stability of viscous shock waves. Arch. Rat. Mech. Anal. 122 (1993), 53–103.

[25] A. Szepessy and K. Zumbrun, Stability of rarefaction waves in viscous media. Arch. Rational Mech. Anal. 133 (1996), no. 3, 249–298.

[26] Z. P. Xin, On nonlinear stability of contact discontinuities. Hyperbolic problems: theory, numerics, applications (Stony Brook, NY, 1994), 249–257, World Sci. Publishing, River Edge, NJ, 1996.

[27] H. Zeng, Stability of a superposition of shock waves with contact discontinuities for systems of viscous conservation laws. J. Diff. Eqns. 246 (2009), 2081–2102.

[28] K. Zumbrun, Stability of large-amplitude shock waves of compressible Navier-Stokes euations. Handbook of Mathematical Fluid Dynamics, Vol.III, 311–533, North-Holland, Amsterdam, 2004.

On Asymptotic Behavior of Solutions of the Compressible Navier-Stokes Equation around a Parallel Flow

Yoshiyuki Kagei
Faculty of Mathematics, Kyushu University
Fukuoka 819-0395, Japan
E-mail: kagei@math.kyushu-u.ac.jp

Abstract

Large time behavior of solutions to the compressible Navier-Stokes equation is considered in an infinite layer of $\mathbf{R}^n$ around stationary parallel flows. It is shown that the parallel flow is asymptotically stable under sufficiently small perturbations, if the Reynolds and Mach numbers are sufficiently small. The asymptotic behavior of the perturbation is described by a function satisfying an $n-1$ dimensional heat equation when $n \geq 3$ and by a function satisfying a one dimensional viscous Burgers equation when $n = 2$.

1 Introduction

We consider the compressible Navier-Stokes equation

$$\partial_t \rho + \mathrm{div}\,(\rho v) = 0, \tag{1.1}$$

$$\rho(\partial_t v + v \cdot \nabla v) - \mu \Delta v - (\mu + \mu')\nabla \mathrm{div}\, v + \nabla P(\rho) = \rho \boldsymbol{g} \tag{1.2}$$

in an n dimensional infinite layer $\Omega_\ell = \mathbf{R}^{n-1} \times (0, \ell)$:

$$\Omega = \{x = (x', x_n)\, ;\, x' = (x_1, \cdots, x_{n-1}) \in \mathbf{R}^{n-1}, 0 < x_n < \ell\} \quad (n \geq 2).$$

Here $\rho = \rho(x,t)$ and $v = {}^\top(v^1(x,t), \cdots, v^n(x,t))$ denote the unknown density and velocity at time $t \geq 0$ and position $x \in \Omega$, respectively; $P = P(\rho)$ is the pressure that is assumed to be a smooth function of ρ satisfying

$$P'(\rho_*) > 0$$

for a given constant $\rho_* > 0$; μ and μ' are the viscosity coefficients that are assumed to be constants and satisfy $\mu > 0$, $\frac{2}{n}\mu + \mu' \geq 0$; and $\boldsymbol{g} = \boldsymbol{g}(x_n)$ is an external force which has the form

$$\boldsymbol{g} = {}^\top(g^1(x_n), 0, \cdots, 0, g^n(x_n)),$$

where g^j $(j = 1, n)$ are given smooth function of x_n. Here and in what follows the superscript ${}^\top\cdot$ stands for the transposition.

The system (1.1)–(1.2) is considered under the boundary condition

$$v|_{x_n=0} = 0, \quad v|_{x_n=\ell} = V^1\boldsymbol{e}_1, \tag{1.3}$$

and the initial condition

$$^\top(\rho, v)|_{t=0} = {}^\top(\rho_0, v_0). \tag{1.4}$$

Here V^1 is a given constant and $\boldsymbol{e}_1$ is the unit vector $\boldsymbol{e}_1 = {}^\top(1, 0, \cdots, 0) \in \mathbf{R}^n$.

Under suitable smallness conditions on g^n, problem (1.1)–(1.3) has a stationary solution ${}^\top(\overline{\rho}_s, \overline{v}_s)$ of parallel flow with properties:

$$\overline{\rho}_s = \overline{\rho}_s(x_n), \quad \overline{v}_s = {}^\top(\overline{v}_s^1(x_n), 0, \cdots, 0),$$

$$\|\overline{\rho}_s - \rho_*\|_{C^0[0,\ell]} \leq C\frac{\ell\,\|g^n\|_{C^0[0,\ell]}}{P'(\rho_*)},$$

$$\|\overline{v}_s^1\|_{C^0[0,\ell]} \leq C\left(\frac{\rho_*\ell^2\|g^1\|_{C^0[0,\ell]}}{\mu} + |V^1|\right).$$

Here and in what follows $\|\cdot\|_{C^k[a,b]}$ denotes the usual C^k norm for functions on the interval $[a, b]$.

In this paper we survey the results in [1, 2, 4] on large time behavior of solutions to (1.1)–(1.4) when the initial value ${}^\top(\rho_0, v_0)$ is sufficiently close to a parallel flow ${}^\top(\rho_s, v_s)$. We also announce a result on two dimensional problem in [3].

A non-dimensional form of the problem for the perturbation is written as follows. We set $V = \frac{\rho_*\ell^2\|g^1\|_{C^0[0,\ell]}}{\mu} + |V^1| > 0$ and introduce the parameters:

$$\nu = \frac{\mu}{\rho_*\ell V}, \quad \nu' = \frac{\mu'}{\rho_*\ell V}, \quad \gamma = \frac{\sqrt{P'(\rho_*)}}{V},$$

and

$$\tilde{\omega}_{m+1} = \frac{\ell(1+\ell^m)\|g^n\|_{C^m[0,\ell]}}{V^2}$$

for some $m \in \mathbf{N}$ satisfying $m \geq [n/2] + 1$. We note that $Re = 1/\nu$ and $Ma = 1/\gamma$ are the Reynolds and Mach numbers. Then the perturbation ${}^\top(\phi, w) = {}^\top(\frac{\gamma^2}{\rho_*}(\rho - \overline{\rho}_s), \frac{1}{V}(v - \overline{v}_s))$ is governed by

$$\partial_t\phi + v_s \cdot \nabla\phi + \gamma^2 \mathrm{div}\,(\rho_s w) = f^0, \tag{1.5}$$

$$\begin{aligned}&\partial_t w - \tfrac{\nu}{\rho_s}\Delta w - \tfrac{\tilde{\nu}}{\rho_s}\nabla \mathrm{div}\, w + \nabla\left(\tfrac{\tilde{P}'(\rho_s)}{\gamma^2\rho_s}\phi\right) \\ &\quad + \left(\tfrac{\nu}{\gamma^2\rho_s^2}\Delta v_s\right)\phi + v_s\cdot\nabla w + w\cdot\nabla v_s = \boldsymbol{f},\end{aligned} \tag{1.6}$$

$$w|_{x_n=0,1} = 0, \tag{1.7}$$

$$^\top(\phi, w)|_{t=0} = {}^\top(\phi_0, w_0). \tag{1.8}$$

Here $\tilde{\nu} = \nu + \nu'$; f^0 and $\boldsymbol{f}$ denote the nonlinearities; the domain Ω_ℓ is transformed into Ω_1. We write (1.5)–(1.8) in the form

$$\begin{cases}\partial_t u + Lu = \boldsymbol{F}, \quad w|_{x_n=0,1} = 0, \\ u|_{t=0} = u_0,\end{cases} \tag{1.9}$$

where $u = {}^\top(\phi, w)$; L is the linearized operator; $\boldsymbol{F}$ denotes the nonlinearity; and $u_0 = {}^\top(\phi_0, w_0)$.

To see large time behavior of small solutions to (1.9) let us consider the linearized problem, i.e., (1.9) with $\boldsymbol{F} = 0$. The Fourier transform of (1.9) with $\boldsymbol{F} = 0$ in $x' \in \mathbf{R}^{n-1}$ can be written as

$$\begin{cases}\partial_t \hat{u} + \hat{L}_{\xi'}\hat{u} = 0, \quad \hat{w}|_{x_n=0,1} = 0, \\ \hat{u}|_{t=0} = \hat{u}_0,\end{cases} \tag{1.10}$$

where $\xi' = (\xi_1, \cdots, \xi_{n-1}) \in \mathbf{R}^{n-1}$ is the dual variable. The operator $\hat{L}_{\xi'}$ has different characters between the cases $|\xi'| \ll 1$ and $|\xi'| \gg 1$. We thus decompose the semigroup e^{-tL} associated with (1.9) into two parts: $e^{-tL} = \mathscr{F}^{-1}[e^{-t\hat{L}_{\xi'}}|_{|\xi'|\le r}] + \mathscr{F}^{-1}[e^{-t\hat{L}_{\xi'}}|_{|\xi'|\ge r}]$ for some small $r > 0$, where $\mathscr{F}^{-1}$ denotes the inverse Fourier transform. If $|\xi'| \ll 1$, then $\hat{L}_{\xi'}$ can be regarded as a perturbation from $\hat{L}_0$ and we will find that the spectrum of $-L$ near the imaginary axis is given by that of $-\hat{L}_{\xi'}$ with $|\xi'| \ll 1$, which is parameterized as $-ia_1\xi_1 - \kappa_1\xi_1^2 - \kappa''|\xi''|^2 + O(|\xi'|^3)$ with some $a_1 \in \mathbf{R}$, $\kappa_1 > 0$ and $\kappa'' > 0$, where $\xi' = (\xi_1, \xi'')$, $\xi'' = (\xi_2, \cdots, \xi_{n-1})$, provided that $\nu \gg 1$, $\gamma \gg 1$ and $\tilde{\omega}_{m+1} \ll 1$. On the other hand, if $|\xi'| \gg 1$, we can derive the exponential decay property of the corresponding part of the semigroup e^{-tL} by the Fourier transformed version of Matsumura-Nishida's energy method, provided that $\nu \gg 1$, $\gamma \gg 1$ and $\tilde{\omega}_{m+1} \ll 1$. As a result, one can see that the solution of the linearized problem behaves as a function $\sigma(x', t) = \mathscr{F}^{-1}(e^{-(ia_1\xi_1 + \kappa_1\xi_1^2 + \kappa''|\xi''|^2)t})$, provided that $\nu \gg 1$, $\gamma \gg 1$ and $\tilde{\omega}_{m+1} \ll 1$ ([1, 4]).

Concerning the nonlinear problem, the large time behavior of the solution to (1.9) is described by the same function as in the case of the linearized problem when $n \geq 3$ ([1, 2]). But, if the space dimension n is 2, then the asymptotic behavior for the nonlinear problem is no longer described by solutions of the linearized problem; and the asymptotic leading part is given by a solution of a one dimensional viscous Burgers equation ([3]).

This paper is organized as follows. In section 2 we rewrite the problem into a non-dimensional one and show the existence of stationary parallel flows. In section 3 we consider the linearized problem at a parallel flow. In section 4 we discuss the nonlinear problem for $n \geq 3$. In section 5 we consider the two dimensional problem and briefly explain how a one dimensional viscous Burgers equation appears in the asymptotic leading part.

2 Parallel flows and equation for the perturbation

In this section we rewrite the problem into a non-dimensional form and show the existence of stationary parallel flows. We then rewrite the equation into the one for the perturbation. In the end of this section we introduce notations using in this paper.

2.1 Non-dimensional form

We introduce the following dimensionless variables:

$$x = \ell \widetilde{x}, \quad t = \frac{\ell}{V}\widetilde{t}, \quad v = V\widetilde{v}, \quad \rho = \rho_* \widetilde{\rho}, \quad P = \rho_* V^2 \widetilde{P}$$

with

$$V = \frac{\rho_* \ell^2 |g^1|_\infty}{\mu} + |V^1| > 0.$$

Then the problem (1.1)–(1.4) is transformed into the following dimensionless problem on the layer $\Omega \equiv \Omega_1 = \mathbf{R}^{n-1} \times (0,1)$. After omitting "tildes" in $\tilde{x}$, $\tilde{t}$, $\tilde{\rho}$ and $\tilde{v}$, we have

$$\partial_t \rho + \operatorname{div}(\rho v) = 0, \tag{2.1}$$

$$\rho(\partial_t v + v \cdot \nabla v) - \nu \Delta v - (\nu + \nu')\nabla \operatorname{div} v + \tilde{P}'(\rho)\nabla \rho = \frac{\ell}{V^2}\rho \tilde{\boldsymbol{g}}, \tag{2.2}$$

$$v|_{x_n=0} = 0, \quad v|_{x_n=1} = \frac{V^1}{V}\boldsymbol{e}^1, \tag{2.3}$$

$$^\top(\rho, v)|_{t=0} = {}^\top(\rho_0, v_0). \tag{2.4}$$

Here $\tilde{\boldsymbol{g}}(x_n) = \boldsymbol{g}(\ell x_n)$; and ν, ν' and γ are the non-dimensional parameters:

$$\nu = \frac{\mu}{\rho_* \ell V}, \quad \nu' = \frac{\mu'}{\rho_* \ell V}, \quad \gamma = \sqrt{\widetilde{P}'(1)} = \frac{\sqrt{P'(\rho_*)}}{V}.$$

Remark. The Reynolds number Re and Mach number Ma are given by $Re = \nu^{-1}$ and $Ma = \gamma^{-1}$, respectively.

2.2 Parallel flows

One can see that if $\|g^n\|_{C^0[0,\ell]}$ is small enough, then a stationary solution of the form ${}^\top(\rho_s, v_s) = {}^\top(\rho_s(x_n), v_s^1(x_n)\boldsymbol{e}_1)$ exists. More precisely, we will look for solutions of (2.1)–(2.3) of the form ${}^\top(\rho_s, v_s) = {}^\top(\rho_s(x_n), v_s^1(x_n)\boldsymbol{e}_1)$ with

$$\int_0^1 \rho_s(x_n)\, dx_n = 1. \tag{2.5}$$

Proposition 2.1. ([2]) *Assume that $\tilde{P}'(\rho) > 0$ for $\rho_1 \le \rho \le \rho_2$ with some $0 < \rho_1 < 1 < \rho_2$. Let $\Phi(\rho) = \int_1^\rho \frac{\tilde{P}'(\eta)}{\eta}\, d\eta$ fot $\rho_1 \le \rho \le \rho_2$ and let $\Psi(r) = \Phi^{-1}(r)$ for $r_1 \le r \le r_2$. Here Φ^{-1} denotes the inverse function of Φ and $r_j = \Phi(\rho_j)$ $(j = 1, 2)$.*

If

$$\omega_0 \equiv \frac{\ell}{V^2}\|g^n\|_{C^0[0,\ell]} \le C \min\left\{|r_1|, r_2, \frac{1}{4\gamma^2\|\Psi''\|_{C^0[r_1,r_2]}}\right\},$$

then there exists a smooth stationary solution $(\rho_s, v_s) = (\rho_s(x_n), v_s^1(x_n)\boldsymbol{e}_1)$ of (2.1)–(2.3) with (2.5) satisfying

$$\rho_1 \le \rho_s(x_n) \le \rho_2,$$

$$|\rho_s - 1|_\infty \le C\frac{\omega_0}{\gamma^2}, \quad |v_s|_\infty \le C.$$

Furthermore, for each $k \in \mathbf{N}$ there exists a $\delta_k = \delta_k(\ell, \gamma^2, \Psi) > 0$ such that if $\omega_k \equiv \frac{1}{V^2}\|g^n\|_{C^{k-1}[0,\ell]} \le \delta_k$, then

$$|\partial_{x_n}^k \rho_s|_\infty \le C_k\omega_k, \quad |\partial_{x_n}^k v_s|_\infty \le C_k \quad \textit{for} \quad k = 0, 1, 2, \cdots.$$

Here C_k is a positive constant depending on k, δ_k and $\|\Psi\|_{C^k[r_1,r_2]}$.

2.3 Equation for perturbation

We now rewrite the system (2.1)–(2.4) into the one for the perturbation. We set $\rho = \rho_s + \gamma^{-2}\phi$ and $v = v_s + w$ in (2.1)–(2.4). Then we arrive at the initial boundary value problem for the perturbation $u = {}^\top(\phi, w)$:

$$\partial_t \phi + v_s \cdot \nabla\phi + \gamma^2 \mathrm{div}\,(\rho_s w) = f^0, \tag{2.6}$$

$$\begin{aligned}(2.7)\qquad &\partial_t w - \frac{\nu}{\rho_s}\Delta w - \frac{\tilde{\nu}}{\rho_s}\nabla \mathrm{div}\, w + \nabla\left(\frac{\tilde{P}'(\rho_s)}{\gamma^2\rho_s}\phi\right)\\ &\quad + \left(\frac{\nu}{\gamma^2\rho_s^2}\Delta v_s\right)\phi + v_s\cdot\nabla w + w\cdot\nabla v_s = \boldsymbol{f},\end{aligned}$$

$$(2.8)\qquad w|_{x_n=0,1} = 0,$$

$$(2.9)\qquad u|_{t=0} = u_0 = {}^\top(\phi_0, w_0).$$

Here $\tilde{\nu} = \nu + \nu'$; and f^0 and $\boldsymbol{f}$ denote the nonlinearities:

$$f^0 = -\mathrm{div}\,(\phi w),$$

$$\begin{aligned}\boldsymbol{f} = &-w\cdot\nabla w + \frac{\nu\phi}{(\phi+\gamma^2\rho_s)\rho_s}\left(-\Delta w + \left(\frac{1}{\gamma^2\rho_s}\Delta v_s\right)\phi\right)\\ &-\frac{\tilde{\nu}\phi}{(\phi+\gamma^2\rho_s)\rho_s}\nabla \mathrm{div}\, w\\ &+\frac{\phi}{\gamma^2\rho_s}\nabla\left(\frac{\tilde{P}'(\rho_s)}{\gamma^2\rho_s}\phi\right) - \frac{1}{2\gamma^2\rho_s}\nabla\left(\tilde{P}''(\rho_s)\phi^2\right)\\ &+\tilde{P}^3(\rho_s,\phi,\partial_x\phi),\end{aligned}$$

where

$$\begin{aligned}\tilde{P}_3 = &\frac{\phi^3}{\gamma^4(\phi+\gamma^2\rho_s)\rho_s^3}\nabla\tilde{P}(\rho_s) - \frac{1}{2\gamma^6\rho_s}\nabla\left(\phi^3 P_3(\rho_s,\phi)\right)\\ &+\frac{\phi}{2\gamma^6\rho_s^2}\nabla\left(\tilde{P}''(\rho_s)\phi^2 + \frac{1}{\gamma^2}\phi^3 P_3(\rho,\phi)\right)\\ &-\frac{\phi^2}{\gamma^2(\phi+\gamma^2\rho_s)\rho_s}\nabla\left(\frac{\tilde{P}'(\rho_s)}{\gamma^2}\phi + \frac{1}{2\gamma^4}\tilde{P}''(\rho_s)\phi^2 + \frac{1}{2\gamma^6}\phi^3 P_3(\rho_s,\phi)\right)\end{aligned}$$

with

$$P_3(\rho_s,\phi) = \int_0^1 (1-\theta)^2 P'''(\theta\gamma^{-2}\phi + \rho_s)\, d\theta.$$

2.4 Notation

We finally introduce some notation which will be used throughout the paper. For a domain D and $1 \le p \le \infty$ we denote by $L^p(D)$ the usual Lebesgue space on D and its norm is denoted by $\|\cdot\|_{L^p(D)}$. Let m be a nonnegative integer. The symbol $H^m(D)$ denotes the m-th order L^2 Sobolev space on D with norm $\|\cdot\|_{H^m(D)}$. $C_0^m(D)$ stands for the set of all C^m functions which have compact support in D. We denote by $H_0^1(D)$ the completion of $C_0^1(D)$ in $H^1(D)$.

We simply denote by $L^p(D)$ (resp., $H^m(D)$) the set of all vector fields $w = {}^\top(w^1, \cdots, w^n)$ on D with $w^j \in L^p(D)$ (resp., $H^m(D)$), $j = 1, \cdots, n$, and its norm is also denoted by $\|\cdot\|_{L^p(D)}$ (resp., $\|\cdot\|_{H^m(D)}$). For $u = {}^\top(\phi, w)$ with $\phi \in H^k(D)$ and $w = {}^\top(w^1, \cdots, w^n) \in H^m(D)$, we define $\|u\|_{H^k(D)\times H^m(D)}$ by $\|u\|_{H^k(D)\times W^m(D)} = \|\phi\|_{H^k(D)} + \|w\|_{H^m(D)}$. When $k = m$, we simply write $\|u\|_{H^k(D)\times H^k(D)} = \|u\|_{H^k(D)}$.

In case $D = \Omega$ we abbreviate $L^p(\Omega)$ (resp., $H^m(\Omega)$) as L^p (resp., H^m). In particular, the norm $\|\cdot\|_{L^p(\Omega)} = \|\cdot\|_{L^p}$ is denoted by $\|\cdot\|_p$.

In case D is the interval $(0,1)$ we denote the norm of $L^p(0,1)$ by $|\cdot|_p$. The norm of $H^m(0,1)$ is denoted by $|\cdot|_{H^m}$.

We denote the inner product of $L^2(0,1)$ by

$$(f, g) = \int_0^1 f(x_n)g(x_n)\,dx_n, \quad f, g \in L^2(0,1).$$

We also introduce a weighted inner product $\langle\cdot,\cdot\rangle$ defined by

$$\langle u_1, u_2\rangle = \int_0^1 \phi_1\phi_2 \frac{\tilde{P}'(\rho_s)}{\gamma^2\rho_s}\,dx_n + \int_0^1 w_1 w_2 \rho_s\,dx_n$$

for $u_j = {}^\top(\phi_j, w_j) \in L^2(0,1)^{n+1}$ $(j = 1, 2)$. Note that $\langle\cdot,\cdot\rangle$ defines an inner product of $L^2(0,1)^{n+1}$, since $0 < \rho_1 \le \rho_s \le \rho_1$ and $\tilde{P}'(\rho_s) > 0$ for $\rho_1 \le \rho_s \le \rho_1$ from Proposition 2.1. Furthermore, we denote the mean value of $f \in L^1(0,1)$ by $\langle\cdot\rangle$:

$$\langle f\rangle = \int_0^1 f(x_n)\,dx_n.$$

For $u = {}^\top(\phi, w) \in L^1(0,1)$ with $w = {}^\top(w^1, \cdots, w^n)$ we define $\langle u\rangle$ by

$$\langle u\rangle = \langle\phi\rangle + \langle w^1\rangle + \cdots + \langle w^n\rangle.$$

We often write $x \in \Omega$ as

$$x = {}^\top(x', x_n), \quad x' = {}^\top(x_1, \cdots, x_{n-1}) \in \mathbf{R}^{n-1}.$$

Partial derivatives of a function u in x, x', x_n and t are denoted by $\partial_x u$, $\partial_{x'} u$, $\partial_{x_n} u$ and $\partial_t u$, respectively. We also write higher order partial derivatives of u in x as $\partial_x^k u = (\partial_x^\alpha u;\, |\alpha| = k)$.

We denote the $k \times k$ identity matrix by I_k. We also define $(n+1)\times(n+1)$ diagonal matrices Q_0, Q_n and $\widetilde{Q}$ by

$$Q_0 = \mathrm{diag}\,(1, 0, \cdots, 0), \quad Q_n = \mathrm{diag}\,(0, \cdots, 0, 1)$$

and

$$\widetilde{Q} = \mathrm{diag}\,(0, 1, \cdots, 1).$$

Note that

$$\langle Q_0 u\rangle = \langle \phi\rangle \quad \text{for} \quad u = {}^\top(\phi, w).$$

For a function $f = f(x')$ ($x' \in \mathbf{R}^{n-1}$), we denote its Fourier transform by $\widehat{f}$ or $\mathscr{F} f$:

$$\widehat{f}(\xi') = (\mathscr{F} f)(\xi') = \int_{\mathbf{R}^{n-1}} f(x') e^{-i\xi'\cdot x'}\, dx'.$$

The inverse Fourier transform is denoted by $\mathscr{F}^{-1}$:

$$(\mathscr{F}^{-1} f)(x) = (2\pi)^{-(n-1)} \int_{\mathbf{R}^{n-1}} f(\xi') e^{i\xi'\cdot x'}\, d\xi'.$$

3 Linearized problem

We write (2.6)–(2.9) in the form

(3.1) $$\partial_t u + Lu = \boldsymbol{F}, \quad w|_{x_n=0,1} = 0, \quad u|_{t=0} = u_0,$$

where $u = {}^\top(\phi, w)$ and L is the operator of the form

$$L = \begin{pmatrix} v_s^1 \partial_{x_1} & \gamma^2 \mathrm{div}\,(\rho_s\,\cdot) \\ \nabla\left(\frac{\tilde{P}'(\rho_s)}{\gamma^2\rho_s}\,\cdot\right) & -\frac{\nu}{\rho_s}\Delta I_n - \frac{\tilde{\nu}}{\rho_s}\nabla\mathrm{div} \end{pmatrix} + \begin{pmatrix} 0 & 0 \\ \frac{\nu\Delta v_s^1}{\gamma^2\rho_s^2}\boldsymbol{e}_1 & v_s^1\partial_{x_1} I_n + (\partial_{x_n} v_s^1)\boldsymbol{e}_1{}^\top\boldsymbol{e}_n \end{pmatrix},$$

and $\boldsymbol{F} = {}^\top(f^0, \boldsymbol{f})$ is the nonlinearity.

To investigate the linearized problem, we consider the Fourier transform of (2.6)–(2.9) with $f^0 = 0$ and $\boldsymbol{f} = \boldsymbol{0}$, that is written as

(3.2) $$\partial_t\hat{\phi} + i\xi_1 v_s^1\hat{\phi} + i\gamma^2\xi'\cdot(\rho_s\hat{w}') + \gamma^2\partial_{x_n}(\rho_s\hat{w}^n) = 0,$$

(3.3) $$\begin{aligned} &\partial_t\hat{w}' + \tfrac{\nu}{\rho_s}(|\xi'|^2 - \partial_{x_n}^2)\hat{w}' - i\tfrac{\tilde{\nu}}{\rho_s}\xi'(i\xi'\cdot\hat{w}' + \partial_{x_n}\hat{w}^n) + i\xi'(\tfrac{\tilde{P}'(\rho_s)}{\gamma^2\rho_s}\hat{\phi}) \\ &+\tfrac{\nu\Delta v_s^1}{\gamma^2\rho_s^2}\hat{\phi}\boldsymbol{e}_1' + i\xi_1 v_s^1\hat{w}' + (\partial_{x_n} v_s^1)\hat{w}^n\boldsymbol{e}_1' = 0, \end{aligned}$$

(3.4) $$\begin{aligned} &\partial_t\hat{w}^n + \tfrac{\nu}{\rho_s}(|\xi'|^2 - \partial_{x_n}^2)\hat{w}^n - \tfrac{\tilde{\nu}}{\rho_s}\partial_{x_n}(i\xi'\cdot\hat{w}' + \partial_{x_n}\hat{w}^n) \\ &+\partial_{x_n}(\tfrac{\tilde{P}'(\rho_s)}{\gamma^2\rho_s}\hat{\phi}) + i\xi_1 v_s^1\hat{w}^n = 0, \end{aligned}$$

$$\hat{w}|_{x_n=0,1} = 0 \tag{3.5}$$

$$\hat{u}|_{t=0} = \hat{u}_0 = (\hat{\phi}_0, \hat{w}_0). \tag{3.6}$$

Here $\hat{\phi} = \hat{\phi}(\xi', x_n, t)$ and $\hat{w} = \hat{w}(\xi', x_n, t)$ are the Fourier transform of $\phi = \phi(x', x_n, t)$ and $w = w(x', x_n, t)$ in $x' \in \mathbf{R}^{n-1}$ with $\xi' \in \mathbf{R}^{n-1}$ being the dual variable. We thus arrive at the following problem

$$\partial_t u + \hat{L}_{\xi'} u = 0, \quad u|_{t=0} = u_0 \tag{3.7}$$

with a parameter $\xi' \in \mathbf{R}^{n-1}$. Here $u = {}^\top(\phi(x_n, t), w(x_n, t))$ ($x_n \in [0,1], t \geq 0$) and $\hat{L}_{\xi'}$ is the operator on $H^1(0,1) \times L^2(0,1)$ with domain $D(\hat{L}_{\xi'}) = H^1(0,1) \times (H^2(0,1) \cap H_0^1(0,1))$ defined by

$$\hat{L}_{\xi'} u = \hat{A}_{\xi'} u + \hat{B}_{\xi'} u + \hat{C}_0 u \quad \text{for} \quad u \in D(\hat{L}_{\xi'}),$$

where

$$\hat{A}_{\xi'} = \begin{pmatrix} 0 & 0 & 0 \\ 0 & \frac{\nu}{\rho_s}(|\xi'|^2 - \partial_{x_n}^2) I_{n-1} + \frac{\tilde{\nu}}{\rho_s} \xi'^\top \xi' & -i\frac{\tilde{\nu}}{\rho_s} \xi' \partial_{x_n} \\ 0 & -i\frac{\tilde{\nu}}{\rho_s} {}^\top\xi' \partial_{x_n} & \frac{\nu}{\rho_s}(|\xi'|^2 - \partial_{x_n}^2) - \frac{\tilde{\nu}}{\rho_s} \partial_{x_n}^2 \end{pmatrix},$$

$$\hat{B}_{\xi'} = \begin{pmatrix} i\xi_1 v_s^1 & i\gamma^2 \rho_s {}^\top\xi' & \gamma^2 \partial_{x_n}(\rho_s \,\cdot) \\ i\xi' \frac{\tilde{P}'(\rho_s)}{\gamma^2 \rho_s} & i\xi_1 v_s^1 I_{n-1} & 0 \\ \partial_{x_n}\left(\frac{\tilde{P}'(\rho_s)}{\gamma^2 \rho_s} \,\cdot\right) & 0 & i\xi_1 v_s^1 \end{pmatrix},$$

$$\hat{C}_0 = \begin{pmatrix} 0 & 0 & 0 \\ \frac{\nu \Delta v_s^1}{\gamma^2 \rho_s^2} \mathbf{e}_1' & 0 & (\partial_{x_n} v_s^1) \mathbf{e}_1' \\ 0 & 0 & 0 \end{pmatrix}.$$

We note that, for each fixed ξ', $-\hat{L}_{\xi'}$ generates an analytic semigroup $e^{-t\hat{L}_{\xi'}}$ in $H^1(0,1) \times L^2(0,1)$. Therefore, for each fixed ξ', problem (3.7) has a unique solution $\hat{u}(t) = e^{-t\hat{L}_{\xi'}} \hat{u}_0 \in C([0,\infty); H^1(0,1) \times L^2(0,1))$ with $\hat{u}(t) \in C((0,\infty); D(\hat{L}_{\xi'}))$. The solution $\mathscr{U}(t) u_0$ of (3.1) with $\boldsymbol{F} = \mathbf{0}$ is then given by $\mathscr{U}(t) u_0 = \mathscr{F}^{-1}(e^{-t\hat{L}_{\xi'}} \hat{u}_0)$.

We begin with a lemma on the eigenvalue problem for $\hat{L}_0$. We denote by $\hat{L}_0^*$ the formal adjoint of $\hat{L}_0$ with respect to the weighted inner product $\langle \cdot, \cdot \rangle$.

Lemma 3.1. ([4]) (i) *$\lambda = 0$ is a simple eigenvalue of $\hat{L}_0$ and $\hat{L}_0^*$.*

(ii) *The eigenspaces for* $\lambda = 0$ *of* $\hat{L}_0$ *and* $\hat{L}_0^*$ *are spanned by* $u^{(0)}$ *and* $u^{(0)*}$ *respectively, where*

$$u^{(0)} = {}^\top(\phi^{(0)}, w^{(0),1}\mathbf{e}_1', 0), \quad u^{(0)*} = {}^\top(\phi^{(0)*}, 0, 0)$$

with

$$\phi^{(0)}(x_n) = \alpha_0 \tfrac{\gamma^2 \rho_s(x_n)}{\tilde{P}'(\rho_s(x_n))}, \quad \alpha_0 = \left(\int_0^1 \tfrac{\gamma^2 \rho_s}{\tilde{P}'(\rho_s)}\, dx_n\right)^{-1},$$

$$w^{(0),1}(x_n) = \tfrac{\ell}{\gamma^2 V^2} \int_0^1 G(x_n, y_n)\tilde{g}^1(y_n)\phi^{(0)}(y_n)\, dy_n,$$

$$\phi^{(0)*}(x_n) = \tfrac{\gamma^2}{\alpha_0}\phi^{(0)}(x_n).$$

Here

$$G(x_n, y_n) = \begin{cases} (1 - x_n)y_n & (0 < y_n < x_n), \\ x_n(1 - y_n) & (x_n < y_n < 1). \end{cases}$$

(iii) *The eigenprojection* $\hat{\Pi}^{(0)}$ *for* $\lambda = 0$ *of* $\hat{L}_0$ *is given by*

$$\hat{\Pi}^{(0)} u = \langle u, u^{(0)*}\rangle u^{(0)} = \langle \phi \rangle u^{(0)} \quad \text{for} \quad u = {}^\top(\phi, w).$$

Remark. In [4] only a special case of a Poiseuille type flow is considered. But one can easily see that the argument in [4] is valid for general parallel flows given in Proposition 2.1 if ω_0 is sufficiently small.

We will treat $\hat{L}_{\xi'}$ as a perturbation from $\hat{L}_0$. As for the case $\xi' = 0$ we have the following result.

Lemma 3.2. ([4]) *There exists a constant* $\tilde{\omega} > 0$ *such that if* $\omega_3 \le \tilde{\omega}$, *then the following assertions hold.*

(i) *There exist positive numbers* η_0 *and* θ_0 *with* $\theta_0 \in (\frac{\pi}{2}, \pi)$ *such that* $\Sigma(-\eta_0, \theta_0) \setminus \{0\} \subset \rho(-\hat{L}_0)$. *Furthermore, the following estimates hold uniformly for* $\lambda \in \rho(-\hat{L}_0) \cap \Sigma(-\eta_0, \theta_0) \setminus \{0\}$:

$$\left|(\lambda + \hat{L}_0)^{-1} f\right|_{H^1 \times L^2} \le \tfrac{C}{|\lambda|} |f|_{H^1 \times L^2},$$

$$\left|\partial_{x_n}^l \widetilde{Q}(\lambda + \hat{L}_0)^{-1} f\right|_2 \le C\left(\tfrac{1}{|\lambda|} + \tfrac{1}{(|\lambda|+1)^{1-\frac{l}{2}}}\right) |f|_{H^1 \times L^2}$$

for $l = 1, 2$,

$$\left|\partial_{x_n}^2 Q_0(\lambda + \hat{L}_0)^{-1} f\right|_2 \le C\left(\tfrac{1}{|\lambda|} + \tfrac{1}{(|\lambda|+1)^{\frac{1}{2}}}\right) |f|_{H^2 \times H^1}.$$

(ii) $\lambda = 0$ *is a simple eigenvalue of* $-\hat{L}_0$*; and the associated eigenprojection is given by the one given in Lemma* 3.1 (iii).

Based on Lemma 3.2, we can obtain the following estimates on $(\lambda + \hat{L}_{\xi'})^{-1}$ for $|\xi'| \ll 1$, by changing η_0 and θ_0 suitably if necessary.

Lemma 3.3. ([4]) *Let* $\tilde{\omega}$ *be the number given in Lemma* 3.2. *Then if* $\omega_3 \le \tilde{\omega}$, *then there exists a positive number* $r_0 = r_0(\eta_0, \theta_0)$ *such that the set* $\Sigma(-\eta_0, \theta_0) \cap \{\lambda;\ |\lambda| \ge \frac{\eta_0}{2}\}$ *is in* $\rho(-\hat{L}_{\xi'})$ *for* $|\xi'| \le r_0$. *Furthermore, the following estimates hold uniformly in* $\lambda \in \Sigma(-\eta_0, \theta_0) \cap \{\lambda;\ |\lambda| \ge \frac{\eta_0}{2}\}$ *and* ξ' *with* $|\xi'| \le r_0$*:*

$$\left|(\lambda + \hat{L}_{\xi'})^{-1} f\right|_{H^1 \times L^2} \le \frac{C}{|\lambda|} |f|_{H^1 \times L^2},$$

$$\left|\partial_{x_n}^l \widetilde{Q}(\lambda + \hat{L}_{\xi'})^{-1} f\right|_2 \le C\left(\frac{1}{|\lambda|} + \frac{1}{(|\lambda|+1)^{1-\frac{l}{2}}}\right) |f|_{H^1 \times L^2}$$

for $l = 1, 2$,

$$\left|\partial_{x_n}^2 Q_0(\lambda + \hat{L}_{\xi'})^{-1} f\right|_2 \le C\left(\frac{1}{|\lambda|} + \frac{1}{(|\lambda|+1)^{\frac{1}{2}}}\right) |f|_{H^2 \times H^1}.$$

As for the spectrum of $-\hat{L}_{\xi'}$ near $\lambda = 0$, we have the following result. We write $\hat{L}_{\xi'}$ in the following form:

$$\hat{L}_{\xi'} = \hat{L}_0 + \sum_{j=1}^{n-1} \xi_j \hat{L}_j^{(1)} + \sum_{j,k=1}^{n-1} \xi_j \xi_k \hat{L}_{jk}^{(2)},$$

where $\xi' = {}^\top(\xi_1, \cdots, \xi_{n-1})$, and

$$\hat{L}_0 = \begin{pmatrix} 0 & 0 & \gamma^2 \partial_{x_n}(\rho_s \,\cdot\,) \\ \frac{\nu \Delta v_s^1}{\gamma^2 \rho_s^2} \boldsymbol{e}_1' & -\frac{\nu}{\rho_s} \partial_{x_n}^2 I_{n-1} & (\partial_{x_n} v_s^1) \boldsymbol{e}_1' \\ \partial_{x_n}\left(\frac{\tilde{P}'(\rho_s)}{\gamma^2 \rho_s} \,\cdot\,\right) & 0 & -\frac{\nu + \tilde{\nu}}{\rho_s} \partial_{x_n}^2 \end{pmatrix},$$

$$\hat{L}_j^{(1)} = \begin{pmatrix} i v_s^1 \delta_{1j} & i\gamma^2 \rho_s \,{}^\top \boldsymbol{e}_j' & 0 \\ i \frac{\tilde{P}'(\rho_s)}{\gamma^2 \rho_s} \boldsymbol{e}_j' & i v_s^1 \delta_{1j} I_{n-1} & -i \frac{\tilde{\nu}}{\rho_s} \boldsymbol{e}_j' \partial_{x_n} \\ 0 & -i \frac{\tilde{\nu}}{\rho_s} {}^\top \boldsymbol{e}_j' \partial_{x_n} & i v_s^1 \delta_{1j} \end{pmatrix},$$

$$\hat{L}_{jk}^{(2)} = \begin{pmatrix} 0 & 0 & 0 \\ 0 & \frac{\nu}{\rho_s} \delta_{jk} I_{n-1} + \frac{\tilde{\nu}}{\rho_s} \boldsymbol{e}_j' \,{}^\top \boldsymbol{e}_k' & 0 \\ 0 & 0 & \frac{\nu}{\rho_s} \delta_{jk} \end{pmatrix}.$$

Here $\boldsymbol{e}'_j$ denotes the unit vector of $\mathbf{R}^{n-1}$ in ξ_j-direction.

By using the analytic perturbation theory [5], we have the following

Lemma 3.4. ([4]) *There exist positive numbers ν_0, γ_0, $\tilde{\omega}$ and r_0 such that if $\nu \ge \nu_0$, $\gamma^2/(\nu+\tilde{\nu}) \ge \gamma_0^2$ and $\omega_3 \le \tilde{\omega}$, then for each ξ' with $|\xi'| \le r_0$ it holds that*

$$\sigma(-\hat{L}_{\xi'}) \cap \{\lambda; |\lambda| < \frac{\eta_0}{2}\} = \{\hat{\lambda}_0(\xi')\},$$

where $\hat{\lambda}_0(\xi')$ is a simple eigenvalue of $-\hat{L}_{\xi'}$ that has the form

$$\hat{\lambda}_0(\xi') = -ia_1\xi_1 - \kappa_1\xi_1^2 - \kappa''|\xi''|^2 + O(|\xi'|^3)$$

as $|\xi'| \to 0$ with some constants $a_1 \in \mathbf{R}$, $\kappa_1 > 0$ and $\kappa'' > 0$. Here $\xi'' = (\xi_2, \cdots, \xi_{n-1})$.

The associated eigenprojection $\hat{\Pi}(\xi')$ for $\hat{\lambda}_0(\xi')$ is given by

$$\hat{\Pi}(\xi') = \hat{\Pi}^{(0)} + \sum_{j=1}^{n-1} \xi_j \hat{\Pi}_j^{(1)} + O(|\xi'|^2),$$

where $\hat{\Pi}^{(0)}$ is the projection given in Lemma 3.1 and

$$\hat{\Pi}_j^{(1)} = -\hat{\Pi}^{(0)}\hat{L}_j^{(1)}\hat{S} - \hat{S}\hat{L}_j^{(1)}\hat{\Pi}^{(0)}$$

with $\hat{S} = \Big((I - \hat{\Pi}^{(0)})\hat{L}_0(I - \hat{\Pi}^{(0)})\Big)^{-1}$.

In view of these lemmas, one can see that the low frequency part of the semigroup behaves as a solution of an $n-1$ dimensional heat equation. The high frequency part $|\xi'| \gg 1$ can be analyzed by a Fourier transformed version of the Matsumura-Nishida energy method; and one can obtain the exponentially decaying estimates for the corresponding part of the semigroup ([1]). We thus obtain the following result.

Theorem 3.5.([1, 4]) *There exist positive numbers ν_0, γ_0, $\tilde{\omega}$ and r_0 such that if $\nu \ge \nu_0$, $\gamma^2/(\nu+\tilde{\nu})$ and $\omega_3 \le \tilde{\omega}$, then $\mathscr{U}(t)$ is decomposed as*

$$\mathscr{U}(t) = U_0(t) + U_\infty(t).$$

Here

$$U_0(t) = \mathscr{F}^{-1}[e^{-t\hat{L}_{\xi'}}|_{|\xi'|\le r_0}], \quad U_\infty(t) = \mathscr{F}^{-1}[e^{-t\hat{L}_{\xi'}}|_{|\xi'|\ge r_0}]$$

and each part has the following properties. Suppose that $u_0 \in L^1(\Omega) \cap \big[H^1(\Omega) \times L^2(\Omega)\big]$ with $\partial_{x'} w_0 \in L^2(\Omega)$. Then the following estimates hold.

(i) *$U_0(t)u_0$ is written in the form*

$$U_0(t)u_0 = \mathscr{U}^{(0)}(t)u_0 + \mathscr{R}^{(0)}(t)u_0 + \mathscr{R}^{(\infty)}(t)u_0.$$

Here

$$\mathscr{U}^{(0)}(t)u_0 = \mathscr{F}^{-1}(e^{-(ia_0\xi_1+\kappa_1\xi_1^2+\kappa''|\xi''|^2)t}\hat{\Pi}_0\hat{u}_0).$$

The function $\mathscr{R}^{(0)}(t)u_0$ satisfies the following estimate. For $m = 0, 1$, there exists a positive constant C such that

$$\|\partial_x^m \mathscr{R}^{(0)}(t)u_0\|_2 \le Ct^{-\frac{n-1}{4}-\frac{1}{2}}\|u_0\|_1$$

holds for $t \ge 1$.

There exists a positive constant d such that $\mathscr{R}^{(\infty)}(t)u_0$ satisfies

$$\|\mathscr{R}^{(\infty)}(t)u_0\|_{H^1\times L^2} \le Ce^{-dt}\|u_0\|_{H^1\times L^2}$$

for all $t \ge 1$.

(ii) *$U_\infty(t)u_0$ satisfies*

$$\|U_\infty(t)u_0\|_{H^1} \le Ce^{-dt}\|u_0\|_{H^1}$$

for all $t \ge 1$.

4 Nonlinear problem: the case $n \ge 3$

In this section we consider the nonlinear problem for $n \ge 3$. In case $n \ge 3$, the global existence of small solutions can be shown in principle by the Matsumura-Nishida energy method [6]. However, due to the lower order term $\left(\frac{\nu}{\gamma^2\rho_s^2}\Delta v_s\right)\phi$ in (2.7), a direct application of the Matsumura-Nishida energy method does not work even though the coefficient $\left(\frac{\nu}{\gamma^2\rho_s^2}\Delta v_s\right)$ is assumed to be sufficiently small. We note that the Poincaré inequality holds for w but not for ϕ in general. To overcome this, we employ a decomposition of solutions.

As was shown in Lemma 3.1, the operator $\hat{L}_0$ (as an operator on $H^1(0,1) \times L^2(0,1)$) has a one dimensional kernel spanned by a function $u^{(0)} = {}^\top(\phi^{(0)}(x_n), w^{(0)}(x_n))$ with $\int_0^1 \phi^{(0)}(x_n)\,dx_n = 1$. We now define a projection P_1 by

$$P_1 u = \mathscr{F}^{-1}\left(\hat{\chi}_1(\xi')\hat{\Pi}^{(0)}\hat{u}\right) = \mathscr{F}^{-1}\left(\hat{\chi}_1(\xi')\langle\hat{\phi}\rangle\right)u^{(0)}(x_n)$$

for $u = {}^\top(\phi, w)$. Here $\hat{\chi}_1$ is a cut off function: $\hat{\chi}_1(\xi) = 1$ for $|\xi| \le 1$ and $\hat{\chi}_1(\xi) = 0$ for $|\xi| > 1$. Decomposing the solution $u(t)$ of (3.1) into

$$u(t) = P_1u(t) + P_\infty u(t) \equiv (\sigma_1 u^{(0)})(t) + u_\infty(t), \tag{4.1}$$

where $P_\infty = I - P_1$ and $P_1 u(t) = (\sigma_1 u^{(0)})(t)$ with

$$\sigma_1 = \sigma_1(x', t) = \mathscr{F}^{-1}\left(\hat{\chi}_1(\xi')\langle\hat{\phi}(t)\rangle\right).$$

We note that the Poincaré inequality $\|u_\infty\|_{L^2} \le C\|\partial_x u_\infty\|_{L^2}$ holds for u_∞-part and differentiations of any order in x variable are bounded operators on the subspace Range (P_1), i.e., $\|\partial_x^k(\sigma_1 u^{(0)})\|_{L^2} \le C_k\|\sigma_1 u^{(0)}\|_{L^2}$ for any $k = 1, 2, \cdots$.

Using this decomposition, we can show the following result by a variant of the Matsumura-Nishida energy method.

Theorem 4.1. ([2]) *Assume that* $n \ge 3$.

(i) *Let* m *be an integer satisfying* $m \ge [n/2] + 1$. *Then there are positive numbers* ν_0, γ_0 *and* $\tilde{\omega}$ *such that if* $\nu \ge \nu_0$, $\gamma^2/(\nu + \tilde{\nu}) \ge \gamma_0^2$ *and* $\omega_{m+1} \le \tilde{\omega}$, *then the following assertion holds. There is a positive number* ε_0 *such that if* $u_0 = {}^\top(\phi_0, w_0) \in H^m$ *satisfies* $\|u_0\|_{H^m} \le \varepsilon_0$ *and a suitable compatibility condition, then there exists a unique global solution* $u(t) = {}^\top(\phi(t), w(t))$ *of* (3.1) *in* $\cap_{j=0}^{[\frac{m}{2}]} C^j([0,\infty); H^{m-2j})$ *which satisfies*

$$\|u(t)\|_{H^m}^2 + \int_0^t (\|\partial_{x'}\sigma_1\|_2^2 + \|\partial_x \phi_\infty\|_{H^{m-1}}^2 + \|\partial_x w_\infty\|_{H^m}^2)\, d\tau \le C\|u_0\|_{H^m}^2 \tag{4.2}$$

uniformly for $t \ge 0$. *Here* $\sigma_1 = \sigma_1(x', t) = \mathscr{F}^{-1}(\hat{\chi}_1\langle\hat{\phi}(t)\rangle)$; $\hat{\chi}_1(\xi') = 1$ *for* $|\xi'| \le 1$ *and* $\hat{\chi}_1(\xi') = 0$ *for* $|\xi'| > 1$; $u^{(0)} = u^{(0)}(x_n)$ *is the function given in Lemma* 3.1; *and* $u_\infty = {}^\top(\phi_\infty, w_\infty) = u - \sigma_1 u^{(0)}$.

(ii) *Furthermore, let* $m \ge [n/2] + 2$. *Then*

$$\|u(t)\|_2 = O(t^{-\frac{n-1}{4}}),$$

$$\|u(t) - (\sigma u^{(0)})(t)\|_2 = O(t^{-\frac{n-1}{4} - \frac{1}{2}} L(t))$$

as $t \to \infty$, *provided that* $u_0 = (\phi_0, w_0) \in H^m \cap L^1$ *with* $\|u_0\|_{H^m \cap L^1} \ll 1$. *Here* $\sigma = \sigma(x', t) = \mathscr{F}^{-1}(e^{-(ia_1\xi_1 + \kappa_1\xi_1^2 + \kappa''|\xi''|^2)t}\langle\hat{\phi}_0\rangle)$ *with some constants* $a_1 \in \mathbf{R}$, $\kappa_1 > 0$ *and* $\kappa'' > 0$; *and* $L(t) = 1$ *when* $n \ge 4$ *and* $L(t) = \log(1 + t)$ *when* $n = 3$.

The assertion (ii) can be proved in a standard manner by using (4.2) and the decay estimates for the linearized problem given in Theorem 3.5.

5 Nonlinear problem: the case $n = 2$

In this section we consider the case $n = 2$. In the case of the two dimensional problem, we need a more detailed decomposition, since the nonlinearity $\boldsymbol{F}$ has quadratic nonlinearities. In this case the asymptotic leading part $\sigma(x_1,t)$ is given by a solution of a one dimensional viscous Burgers equation

$$\partial_t\sigma - \kappa_1\partial_{x_1}^2\sigma + a_1\partial_{x_1}\sigma + a_2\partial_{x_1}(\sigma^2) = 0. \tag{5.1}$$

Theorem 5.1. ([3]) *Assume that* $n = 2$. *Let* m *be an integer satisfying* $m \geq 3 = [n/2]+2$. *Then there are positive numbers* ν_0, γ_0 *and* $\tilde{\omega}$ *such that if* $\nu \geq \nu_0$, $\gamma^2/(\nu+\tilde{\nu}) \geq \gamma_0^2$ *and* $\omega_{m+1} \leq \tilde{\omega}$, *then the following assertion holds. For any* $u_0 = {}^\top(\phi_0, w_0) \in H^m \cap L^1$ *satisfying a suitable compatibility condition with* $\|u_0\|_{H^m\cap L^1} \ll 1$ *the perturbation* $u(t)$ *uniquely exists in* $\cap_{j=0}^{[\frac{m}{2}]} C^j([0,\infty); H^{m-2j})$ *which satisfies*

$$\|u(t)\|_2 = O(t^{-\frac{1}{4}}) \quad (t\to\infty)$$

and

$$\|u(t) - (\sigma u^{(0)})(t)\|_2 = O(t^{-\frac{3}{4}+\varepsilon}) \quad (t\to\infty)$$

with any $\varepsilon > 0$. *Here* $\sigma = \sigma(x_1,t)$ *satisfies the one dimensional viscous Burgers equation* (5.1) *with* $\sigma|_{t=0} = \langle\phi_0\rangle$.

We here try to explain how the nonlinear term $a_2\partial_{x_1}(\sigma^2)$ in (5.1) appears in the asymptotic leading part of $u(t)$. By the Duhamel principle, $\hat{u}(\xi_1;t)$ is given by

$$\hat{u}(\xi_1;t) = e^{-t\hat{L}_{\xi_1}}\hat{u}_0 + \int_0^t e^{-(t-\tau)\hat{L}_{\xi_1}}\hat{\boldsymbol{F}}(\tau)\,d\tau. \tag{5.2}$$

In view of the linearized analysis, the most slowly decaying part of $u(t)$ will be given by $\hat{\Pi}^{(0)}\hat{u}(\xi';t)$ with $|\xi'| \ll 1$. So we apply $\hat{\Pi}^{(0)}\hat{\Pi}(\xi')$ to (5.2) and obtain

(5.3)

$$\hat{\Pi}^{(0)}\hat{\Pi}(\xi')\hat{u}(\xi_1;t) = e^{-\hat{\lambda}_0(\xi')t}\hat{\Pi}^{(0)}\hat{\Pi}(\xi')\hat{u}_0 + \int_0^t e^{-\hat{\lambda}_0(\xi')(t-\tau)}\hat{\Pi}^{(0)}\hat{\Pi}(\xi')\hat{\boldsymbol{F}}(\tau)\,d\tau$$

with $\hat{\lambda}_0(\xi') = -(ia_1\xi_1 + \kappa_1\xi_1^2) + O(|\xi_1|^3)$. Consider the inhomogeneous term in (5.3). Substituting $u = \sigma_1 u^{(0)} + u_\infty$ into $\boldsymbol{F}$, we have

$$\begin{aligned}\boldsymbol{F} &= \begin{pmatrix} -\operatorname{div}(\phi w) \\ f^1 \\ f^n \end{pmatrix} \\ &= \begin{pmatrix} -\operatorname{div}(\phi w) \\ f^1 \\ \tilde{f}^n \end{pmatrix} + \sigma_1^2 \begin{pmatrix} 0 \\ 0 \\ \frac{1}{2\rho_s\gamma^4}\partial_{x_n}(P''(\rho_s)\{\phi^{(0)}\}^2) \end{pmatrix} \equiv \tilde{\boldsymbol{F}} + \sigma_1^2\boldsymbol{F}_1.\end{aligned}$$

Note that $\boldsymbol{F}_1 = \boldsymbol{F}_1(x_n)$ and that $\tilde{\boldsymbol{F}}$ does not include the terms $O(\sigma_1^2)$, namely includes the terms decaying faster such as $O(\sigma_1^3)$, $O(\sigma_1 \partial_{x_1} \sigma_1)$ and so on. Furthermore, $\hat{\Pi}^{(0)} \boldsymbol{F}_1 = 0$ and $\hat{\Pi}^{(0)} \boldsymbol{F} = \hat{\Pi}^{(0)} \tilde{\boldsymbol{F}} = -\partial_{x_1} \langle \phi w^1 \rangle u^{(0)}$.

Recall that $\hat{\Pi}(\xi_1)$ is expanded as

$$\hat{\Pi}(\xi_1) = \Pi^{(0)} + \xi_1 \Pi_1^{(1)} + \Pi^{(2)}(\xi_1) \quad \text{for} \quad |\xi_1| \ll 1$$

with $\Pi_1^{(1)} = -\hat{\Pi}^{(0)} \hat{L}_1^{(1)} \hat{S} - \hat{S} \hat{L}_1^{(1)} \hat{\Pi}^{(0)}$ (see Lemma 3.4). Noting that $\hat{\Pi}^{(0)} \hat{S} = \hat{S} \hat{\Pi}^{(0)} = 0$, we see that for $|\xi_1| \ll 1$

$$\begin{aligned}
\hat{\Pi}^{(0)} \hat{\Pi}(\xi_1) \hat{\boldsymbol{F}} &= \hat{\Pi}^{(0)} (\tilde{\boldsymbol{F}})\hat{} + \xi_1 \hat{\Pi}^{(0)} \hat{\Pi}_1^{(1)} \hat{\boldsymbol{F}} + \hat{\Pi}^{(0)} \hat{\Pi}^{(2)}(\xi_1) \hat{\boldsymbol{F}} \\
&\sim -i\xi_1 \langle \widehat{(\phi w^1)} \rangle u^{(0)} - \xi_1 \langle \hat{L}^{(1)} \hat{S} (\widehat{(\sigma_1^2)} \boldsymbol{F}_1) \rangle u^{(0)} \\
&\sim -\langle \phi^{(0)} w^{(0),1} \rangle i\xi_1 \widehat{(\sigma_1^2)} u^{(0)} + i \langle \hat{L}^{(1)} \hat{S} \boldsymbol{F}_1 \rangle i\xi_1 \widehat{(\sigma_1^2)} u^{(0)}, \\
&\sim a_2 \mathscr{F} \left(\partial_{x_1} (\sigma_1^2) \right) u^{(0)}
\end{aligned}$$

with $a_2 = -\langle \phi^{(0)} w^{(0),1} - i\hat{L}^{(1)} \hat{S} \boldsymbol{F}_1 \rangle$, which gives $a_2 \partial_{x_1} (\sigma^2)$ in (5.1). A detailed argument will be give in [3].

References

[1] Y. Kagei, Asymptotic behavior of solutions of the compressible Navier-Stokes equation around the plane Couette flow. MI Preprint Series 2009-7, Kyushu University, accepted for publication in J. Math. Fluid Mech. 2009.

[2] Y. Kagei, Global existence of solutions to the compressible Navier-Stokes equation around parallel flows. Preprint.

[3] Y. Kagei, Asymptotic behavior of solutions of the compressible Navier-Stokes equation around parallel flows. In preparation.

[4] Y. Kagei, Y. Nagafuchi and T. Sudo, Decay estimates on solutions of the linearized compressible Navier-Stokes equation around a Poiseuille type flow. J. Math-for-Ind. 2A (2010), 39–56.

[5] T. Kato, Perturbation Theory for Linear Operators. Springer-Verlag, Berlin, Heidelberg, New York, 1980.

[6] A. Matsumura and T. Nishida, Initial boundary value problems for the equations of motion of compressible viscous and heat-conductive fluids. Comm. Math. Phys. 89 (1983), 445–464.

On Nonlinear Stability of Viscous Shock Waves with Physical Viscosity*

Tai-Ping Liu
Department of Mathematics, Stanford University
Stanford, CA 94305, USA
E-mail: liu@math.stanford.edu

Yanni Zeng
Department of Mathematics, University of Alabama at Birmingham
Birmingham, AL 35294, USA
E-mail: zeng@math.uab.edu

Abstract

We study the nonlinear stability of viscous shock waves and large time behavior for physical conservation laws. Under a generic perturbation of a weak viscous shock profile, the solution to the Cauchy problem of a one-dimensional $n \times n$ system approaches to a shifted shock profile time asymptotically. The details of the convergence are obtained. The result is an extension of our recent result on conservation laws with artificial viscosity, and applies to the Navier-Stokes equations for the compressible fluids and the equations of Magnetohydrodynamics.

1 Introduction

We study the Cauchy problem of an $n \times n$ system of viscous conservation laws

$$u_t + f(u)_x = (B(u)u_x)_x, \qquad x \in \mathbb{R}, \quad t > 0, \tag{1.1}$$

$$u(x,0) = u_0(x). \tag{1.2}$$

Here $u(x,t) \in \mathbb{R}^n$ is the density function, $f(u) \in \mathbb{R}^n$ is the flux function, and $B(u) \in \mathbb{R}^{n\times n}$ is the viscosity matrix. For physical models, B is

*The research of Tai-Ping Liu is supported in part by the National NSC Grant 96-2628-M-001-011 and NSF Grant DMS-0709248, and Yanni Zeng by NSF Grant DMS-0207154.

rank deficient, hence (1.1) is hyperbolic-parabolic rather than uniformly parabolic. Important examples of (1.1) include the Navier-Stokes equations for compressible fluids and the system of magnetohydrodynamics. In Lagrangian coordinates, the Navier-Stokes equations read:

$$\begin{cases} (1/\rho)_t - v_x = 0, \\ v_t + p_x = (\mu\rho v_x)_x, \\ (e + \frac{1}{2}v^2)_t + (pv)_x = (\kappa\rho\theta_x + \mu\rho v v_x)_x, \end{cases} \tag{1.3}$$

where ρ, v, p, e and θ are, respectively, the density, velocity, pressure, internal energy, and temperature of the fluid. For simplicity we assume that the viscosity coefficient $\mu \geq 0$ and the heat conductivity $\kappa > 0$ are constants. Similarly, in Lagrangian coordinates the equations of magnetohydrodynamics are

$$\begin{cases} (1/\rho)_t - v_{1x} = 0, \\ v_{1t} + \left[p + \frac{1}{2\mu_0}\left(b_2^2 + b_3^2\right)\right]_x = (\nu\rho v_{1x})_x, \\ v_{2t} - \left(\frac{1}{\mu_0} b_1^* b_2\right)_x = (\mu\rho v_{2x})_x, \\ v_{3t} - \left(\frac{1}{\mu_0} b_1^* b_3\right)_x = (\mu\rho v_{3x})_x, \\ (b_2/\rho)_t - (b_1^* v_2)_x = \left(\frac{1}{\sigma\mu_0}\rho b_{2x}\right)_x, \\ (b_3\rho)_t - (b_1^* v_3)_x = \left(\frac{1}{\sigma\mu_0}\rho b_{3x}\right)_x, \\ \left[e + \frac{1}{2}\left(v_1^2 + v_2^2 + v_3^2\right) + \frac{1}{2\mu_0\rho}\left(b_2^2 + b_3^2\right)\right]_t \\ \quad + \left[\left(p + \frac{1}{2\mu_0}\left(b_2^2 + b_3^2\right)\right) v_1 - \frac{1}{\mu_0} b_1^* \left(b_2 v_2 + b_3 v_3\right)\right]_x \\ \quad = \left[\nu\rho v_1 v_{1x} + \mu\rho(v_2 v_{2x} + v_3 v_{3x}) + \kappa\rho\theta_x + \frac{1}{\sigma\mu_0^2}\rho(b_2 b_{2x} + b_3 b_{3x})\right]_x, \end{cases} \tag{1.4}$$

where ρ, $v = (v_1, v_2, v_3^t)$, p, $b = (b_1^*, b_2, b_3)^t$, e and θ represent, respectively, the density, velocity, pressure, magnetic induction, internal energy and temperature. In one space dimension, the first component of the magnetic field is a constant, denoted as b_1^*. For simplicity we take the following as constant parameters: μ and ν are the two viscosity coefficients; μ_0 is the magnetic permeability; κ is the heat conductivity; and $1/\sigma$ is the electrical resistivity. For those parameters we set $\mu_0 > 0$ and $\kappa > 0$, while all others are nonnegative.

We are interested in the nonlinear behavior of solutions to (1.1). This is closely related to the nonlinear wave of its inviscid counterpart, the hyperbolic conservation laws

$$u_t + f(u)_x = 0, \qquad x \in \mathbb{R}, \quad t > 0. \tag{1.5}$$

It is well-known that there are several primary waves for (1.5): the compressive shock wave, the expansive rarefaction wave, and the linear contact discontinuity. Correspondingly, we have similar wave phenomena for (1.1). That is, we have the viscous shock wave, which is a traveling wave solution to (1.1) connecting two different end-states, and the end-states and the wave speed form a shock wave for (1.5). We then consider a perturbation to the viscous shock profile, and study the convergence of solution to the shock at large time and the details of the convergence. The case of rarefaction wave is different. For a rarefaction wave of the inviscid system (1.5) there is no exact explicit representation of viscous rarefaction wave. However, there is an accurate approximation using the Burgers rarefaction wave solution. That is, one can construct a solution to (1.1) that approaches the inviscid rarefaction wave time asymptotically, [13]. The stability of such solution may then be considered. For a contact discontinuity of (1.5), although it is not a time asymptotic attractor of (1.1), there is a viscous contact wave that is a smooth solution to (1.1) and approximates the inviscid one at a finite time interval as the viscosity tends to zero, [7] and references therein. The stability of the viscous contact wave and the large time behavior of solution can also been considered. To study the wave behavior, including various combination of these waves, the estimates in the analysis are necessarily "local".

A common approach to study stability for viscous conservation laws is the energy method, [5, 20, 10, 23], etc. By choosing appropriate weights, it may yield local estimates. For instance, we can use weighted energy methods to study the stability of viscous shock wave of a scalar conservation law and obtain pointwise estimates in space and in time, see [14]. For systems of viscous conservation laws, there is strong coupling of waves pertaining to distinct characteristic fields. An effective method to handle the coupling is the "Green's function approach", [15, 12, 16, 17]. This is to linearized the system along the background wave and obtain estimates on the Green's function. Then we use Duhamel's principle to perform a priori estimates. Here there are several key steps. The first one is the estimate on the Green's function. We may perform an estimate on the exact Green's function, or we may construct an approximate one that agrees with the exact one in the leading order. However, the estimate needs to be accurate and able to provide all vital information. The second step is to assess different wave interactions, [11]. This step is important since it leads to the correct ansatz. Another step is the a priori estimate that closes the nonlinear analysis. For system (1.1), which has physical viscosity, energy type estimates are needed as well to close the nonlinear analysis, [15, 16]. This is due to the singularities in the Green's function as to be further discussed.

For the Green's function of a scalar conservation law, the heat kernel and the Burgers Green's function yield all the vital information. This further generalizes to systems with artificial viscosity, $B = I$ in (1.1), [12, 17]. In this case the corresponding linear system is in fact a system of n decoupled equations, and the Green's function consists of those for scalar equations. For the general case of physical viscosity the corresponding linear system is a true system, and the Green's function is a non-diagonal matrix. The off-diagonal entries represent the coupling of different equations in the linear level. The rank deficiency of viscosity matrix B further complicates the situation: The singularity from initial data propagates into the solution in several directions. The singularities exist all the time though decay exponentially. The Green's function for (1.1) linearized around a constant state is studied using the Fourier analysis and complex analysis, [24, 15]. In an upcoming monograph, [19], the Green's function for (1.1) linearized around a weak viscous shock profile is done by interpolating the Green's functions for the systems at the two end-states. This requires deep understanding on the coupling in the Green's function for systems with constant coefficients. In a recent paper, [18], the authors refine the secondary waves obtained in [15] to reveal the nature of the coupling: The coupling is weak not just in the strength but also by a special block structure. Besides, there is continuous differentiability with respect to the coefficients of the system.

On the level of local estimates, the study of nonlinear stability of weak viscous shock waves with artificial viscosity is done completely only recently, [17]. The result involves subtle ansatz of various scales, induced by the nonlinear coupling. In particular, it shows explicitly the interwinding of the shock strength with the decay of solution in space and in time. Both the perturbation and the shock strength are weak but independent. As the shock strength tends to zero, the optimal result for constant state perturbation obtained in [15] is recovered. The independence of the perturbation and the wave strength is an important feature needed for the studying of more complex wave patterns.

In this note we announce a result for nonlinear stability of viscous shock waves for (1.1). It is similar to the one for the case of artificial viscosity obtained in [17], and applies to both Navier-stokes equations and the equations for magnetohydrodynamics, (1.2) and (1.3). For the latter we allow multiple eigenvalues in the transversal fields as long as the shock is classical. In next section we formulate the problem. The main result is stated in the third section, while the proof is given in an upcoming research monograph, [19].

2 Nonlinear stability of viscous shock wave

Consider system (1.1) in a small open set $\mathbb{O} \in \mathbb{R}^n$ and assume that $f(u)$ and $B(u)$ are sufficiently smooth in $\mathbb{O}$. Recall that the viscosity matrix B is rank deficient. However, there are physical assumptions that guarantee the well-posedness of (1.1). These assumptions may have different mathematical forms. Here we use Kawashima-Shizuta assumptions, [21], in $\mathbb{O}$:

Assumption 2.1. *System* (1.1) *has a strictly convex entropy* η.

Assumption 2.1 means that there exist an entropy pair $\eta(u)$ and $F(u)$, such that η is strictly convex, $(\nabla\eta)f' = \nabla F$, and $(\nabla^2\eta)B$ is symmetric and semi-positive definite, [22, 8]. We assume that η and F are sufficiently smooth. To simplify our notation we let

$$A_0(u) = \nabla^2\eta(u). \tag{2.1}$$

Clearly A_0 is symmetric, positive definite.

Assumption 2.2. $B(u) \neq 0$. *There exists a smooth one-to-one mapping*

$$u = f_0(\tilde{u}) \tag{2.2}$$

such that the null space $\mathbb{N}$ *of*

$$\tilde{B}(\tilde{u}) \equiv B(f_0(\tilde{u}))f_0'(\tilde{u}) \tag{2.3}$$

is independent of $\tilde{u}$. *Also,* $\mathbb{N}^\perp$ *is invariant under* $f_0'(\tilde{u})^t A_0(f_0(\tilde{u}))$, *and* $\tilde{B}(\tilde{u})$ *maps* $\mathbb{R}^n$ *into* $\mathbb{N}^\perp$.

Assumption 2.2 means that by change of variables a new viscosity matrix $\tilde{B}$ has a special block structure.

Assumption 2.3. *Any eigenvector of* $f'(u)$ *is not in the null space of* $B(u)$.

Assumption 2.3, together with Assumption 2.1, implies that (1.1) is fully dissipative, even though B is rank deficient.

We now consider a viscous shock wave of (1.1), which is a traveling wave solutions connecting two different end states

$$u(x,t) = \phi(x - st), \qquad \phi(\mp\infty) = u^{\mp}. \tag{2.4}$$

Substituting (2.4) into (1.1) we have

$$-s\phi'(x) + f(\phi(x))' = (B(\phi(x))\phi'(x))'.$$

Therefore, the wave speed s and the end states $u^{\mp}$ need to satisfy the Rankine-Hugoniot condition

$$s(u^+ - u^-) = f(u^+) - f(u^-) \tag{2.5}$$

as for the hyperbolic conservation laws

$$u_t + f(u)_x = 0. \tag{2.6}$$

It is a classical result that under Assumption 2.1, $A_0(u)f'(u)$ is symmetric, [4]. Therefore, (2.6) is completely hyperbolic. That is, $f'(u)$ has n real eigenvalues $l_1(u) \leq l_2(u) \leq \cdots \leq l_n(u)$, and a complete set of eigenvectors:

$$\begin{gathered} f'(u)r_i(u) = \lambda_i(u)r_i(u), \; l_i^t(u)f'(u) = \lambda_i(u)l_i^t(u), \\ l_i^t(u)r_j(u) = \delta_{ij}, \; 1 \leq i,j \leq n, \end{gathered} \tag{2.7}$$

where δ_{ij} is the Kronecker δ. We assume that each characteristic field is either genuinely nonlinear,

$$\nabla l_i(u) \cdot r_i(u) \neq 0,$$

or linearly degenerate,

$$\nabla l_i(u) \cdot r_i(u) \equiv 0$$

in $\mathbb{O}$, [9]. We further assume that for some p, $1 \leq p \leq n$, $\lambda_p(u)$ is simple in $\mathbb{O}$, and the p-field is genuinely nonlinear. Then for the inviscid system (2.6) there is a classical p-shock wave, which satisfies (2.5) and the Lax entropy condition,

$$\lambda_p(u^+) < s < \lambda_p(u^-), \tag{2.8}$$

provided that the shock strength is sufficiently small,

$$\varepsilon \equiv |u^+ - u^-| \ll 1, \tag{2.9}$$

see [9]. Consequently, a viscous p-shock wave exists for (1.1), [3].

In this note we study the nonlinear stability of the viscous p-shock wave introduced above. We allow eigenvalues of $f'(u)$ other than $\lambda_p(u)$ to have multiplicities greater then one. However, we need to restrict the splitting of eigenvalues along the Rankine-Hugoniot curve of the p-family and along the viscous p-shock wave.

Assumption 2.4. *In the open set $\mathbb{O}$, along the Rankine-Hugoniot curve*

$$s(u - u^-) = f(u) - f(u^-) \tag{2.10}$$

and along the viscous p-shock wave $\phi(x)$, each distinct eigenvalue $\lambda_i(u)$ of $f'(u)$ has a constant multiplicity. For each distinct eigenvalue $\lambda_i(u)$ let

$l_i(u)$ and $r_i(u)$ be the sub-matrices formed by all the left and all the right corresponding eigenvectors, respectively. Then along the same curves each distinct eigenvalue of $l_i(u)^t B(u) r_i(u)$ has a constant multiplicity as well.

Under Assumptions 2.1 – 2.4, we consider the nonlinear stability of the viscous p-shock wave. Consider the Cauchy problem (1.1), (1.2) when the solution is a small perturbation of ϕ:

$$u(x,t) = \phi(x - st) + \bar{u}(x,t). \tag{2.11}$$

Our goal is to obtain a precise estimates on $\bar{u}$, with optimal rates in space, in time, and in the shock strength ε, which is sufficiently small.

The p-shock wave ϕ is in fact a one-parameter family since any translation of it is still a p-shock wave. A generic initial perturbation pushes the shock to a new location, and generates diffusion waves in the transversal field directions. The limit profile and the diffusion waves have been identified in [10], by the decomposition of the excess mass:

$$\int_{-1}^{1} [u_0(x) - \phi(x)]\, dx = x_0 \left(u^+ - u^-\right) + \sum_{i<p} c_i r_i(u^-) + \sum_{i>p} c_i r_i(u^+). \tag{2.12}$$

The coefficient x_0 gives the particular profile in the time-asymptotic limit, while the other coefficients c_i are the mass carried by the diffusion waves in the transversal field directions. That is, replacing (2.11), a more precise description of u is

$$u(x,t) = \phi(x - st + x_0) + \sum_{i<p} \theta_i(x,t) r_i(u^-) + \sum_{i>p} \theta_i(x,t) r_i(u^+) + v(x,t),$$

where θ_i is the diffusion wave carrying mass c_i to be defined below, and v is a higher order term in time decay rate.

Without loss of generality we may set $x_0 = 0$ and $s = 0$ (stationary shock) to simplify our notations. This can always be done by change of variables. To further simplify our notations we introduce

$$u_i^0 = \begin{cases} u^-, & \text{if } i < p, \\ u^+, & \text{if } i > p, \end{cases} \tag{2.13}$$

and set $\lambda_i^0 = \lambda_i(u_i^0)$, $r_i^0 = r_i(u_i^0)$, $l_i^0 = l_i(u_i^0)$, etc. Therefore, we arrive at

$$u(x,t) = \phi(x) + \sum_{i \neq p} \theta_i(x,t) r_i^0 + v(x,t). \tag{2.14}$$

We now define the diffusion waves. If λ_i^0 is simple, θ_i is defined as the self-similar solution to

$$\begin{aligned} &\theta_{it} + \lambda_i^0\theta_{ix} + C_{ii}(\theta_i^2)_x = (l_i^t B r_i)(u_i^0)\theta_{ixx}, \\ &\int_{-\infty}^{\infty} \theta_i(x,t)\,dx = c_i, \\ &C_{ii} = \frac{1}{2}(l_i^0)^t f''(u_i^0)(r_i^0, r_i^0), \end{aligned} \tag{2.15}$$

where c_i is determined by (2.12).

By differentiating (2.7) we have $l_i^t(u)f''(u)(r_i(u), r_i(u)) = \nabla\lambda_i(u) \cdot r_i(u)$. Therefore, C_{ii} is zero or not, depending on whether the i-field is linearly degenerate or genuinely nonlinear. Consequently, (2.15) is either the heat equation or the Burgers equation, and θ_i is the heat kernel or the Burgers kernel, carrying the mass c_i. If λ_i^0 is of multiplicity $m_i > 1$, and $l_{i-1}^0 < l_i^0 = \cdots = l_{i+m_i-1}^0 < l_{i+m_i}^0$. By our assumption $i \neq p$. Let $\Theta_i = (\theta_i, \cdots, \theta_{i+m_i-1})^t$. Then similar to (2.15) we define Θ_i as the self-similar solution to the $m_i \times m_i$ generalized Burgers equation

$$\begin{aligned} &\Theta_{it} + \lambda_i^0\Theta_{ix} + \frac{1}{2}(L_i^0)^t f''(u_i^0)(R_i^0\Theta_i, R_i^0\Theta_i)_x = (L_i^t B R_i)(u_i^0)\Theta_{ixx}, \\ &\int_{-\infty}^{\infty} \Theta_i(x,t)dx = (c_i, \cdots, c_{i+m_i-1})^t, \end{aligned} \tag{2.16}$$

where

$$L_i = (l_i, \cdots, l_{i+m_i-1}), \qquad R_i = (r_i, \cdots, r_{i+m_i-1})$$

are $n \times m_i$ matrices. For simple λ_i^0, θ_i has explicit expression as a heat kernel or solved by the Hopf-Cole transformation, [6, 2]. For multiple λ_i^0, Θ_i has been studied in [1], and detailed estimates similar to heat kernels or Burgers waves are given.

From (2.12) – (2.16) and by our notation $x_0 = 0$, we have

$$\int_{-\infty}^{\infty} v(x,t)\,dx = \int_{-\infty}^{\infty} [u(x,t) - \phi(x)]\,dx - \sum_{i\neq p} c_i r_i^0 = 0, \tag{2.17}$$

using the fact that u is conserved. Because v has zero mass, we expect it to decay faster the θ_i. Our main result in next section is a detailed estimate on v.

3 Main result

Note that in (2.14) the leading terms of the perturbation are the diffusion waves, the heat kernels or the Burgers waves. These decay as

$(t+1)^{-1/2}$. Due to (2.17) we expect v decays faster in time. However, v decays algebraically rather than exponentially in space. To describe v we introduce the following waves of algebraic types.

$$\begin{aligned}
&\psi_i(x,t) \equiv [(x-\lambda_i^0(t+1))^2+t+1]^{-\frac{1}{2}}, \qquad i \neq p,\\
&\bar{\psi}_i(x,t) \equiv [|x-\lambda_i^0(t+1)|^3+(t+1)^2]^{-\frac{1}{3}}, \qquad i \neq p,\\
&\chi_i(x,t) \equiv \min\{\varepsilon^{-\frac{1}{2}}\psi_i^{\frac{3}{2}}(x,t), \varepsilon^{\frac{1}{2}}(t+1)^{-\frac{1}{2}}\psi_i^{\frac{1}{2}}(x,t)\}\mathrm{char}_i(x,t), \qquad i \neq p,\\
&\mathrm{char}_i(x,t) \equiv \begin{cases} 1, \text{ if } 0 < x < \lambda_i^0(t+1) \text{ and } i > p,\\ 1, \text{ if } \lambda_i^0(t+1) < x < 0 \text{ and } i < p,\\ 0, \text{ otherwise,}\end{cases}\\
&\psi_p(x,t) \equiv [(|x|+\varepsilon(t+1))^2+t+1]^{-\frac{1}{2}},\\
&\bar{\psi}_p(x,t) \equiv [(|x|+\varepsilon(t+1))^3+(t+1)^2]^{-\frac{1}{3}},\\
&\eta_1(x,t) \equiv \min\{\varepsilon(x^2+1)^{-\frac{1}{2}}(|x|+t+1)^{-\frac{1}{2}}, (x^2+1)^{-\frac{3}{4}}(|x|+t+1)^{-\frac{1}{2}}\},\\
&\eta_2(x,t) \equiv \min\{\varepsilon^2(x^2+1)^{-1}(t+1)^{-1}, (x^2+1)^{-\frac{3}{2}}(t+1)^{-1}\}. \qquad (3.1)
\end{aligned}$$

Here we recall that ε is the shock strength defined by (2.9). Also recall that the viscous shock wave has been set stationary, hence the entropy condition (2.8) becomes

$$\lambda_p^+ \equiv \lambda_p(u^+) < 0 < \lambda_p(u^-) \equiv \lambda_p^-. \qquad (3.2)$$

Our main result can now be stated as the following.

Theorem 3.1. *Suppose that the initial perturbation* $\bar{u}(x,0) = u_0(x) - \phi(x) \in H^{13}(\mathbb{R})$ *satisfies*

$$\begin{aligned}
&\sum_{j=0}^{3}\left|\frac{\partial^j}{\partial x^j}\bar{u}(x,0)\right| = O(1)(|x|+1)^{-\frac{3}{2}},\\
&\left|\frac{\partial^4}{\partial x^4}\bar{u}(x,0)\right| + \left|\frac{\partial^5}{\partial x^5}\bar{u}(x,0)\right| = O(1)(|x|+1)^{-1}, \qquad (3.3)\\
&\left|\frac{\partial^6}{\partial x^6}\bar{u}(x,0)\right| + \left|\frac{\partial^7}{\partial x^7}\bar{u}(x,0)\right| = O(1)(|x|+1)^{-\frac{1}{2}}.
\end{aligned}$$

Let

$$\begin{aligned}
&\sup_{x\in\mathbb{R}}\left\{(|x|+1)^{\frac{3}{2}}\sum_{j=0}^{3}\left|\frac{\partial^j}{\partial x^j}\bar{u}(x,0)\right| + (|x|+1)\sum_{j=4}^{5}\left|\frac{\partial^j}{\partial x^j}\bar{u}(x,0)\right|\right. \qquad (3.4)\\
&\left. +(|x|+1)^{\frac{1}{2}}\sum_{j=6}^{7}\left|\frac{\partial^j}{\partial x^j}\bar{u}(x,0)\right|\right\} + \|\bar{u}(x,0)\|_{H^{13}} = \delta_0.
\end{aligned}$$

Then there exist $\bar{\delta}_0 > 0$ *and* $\bar{\varepsilon} > 0$ *such that if* $\delta_0 < \bar{\delta}_0$ *and* $\varepsilon < \bar{\varepsilon}$*, the Cauchy problem* (1.1), (1.2) *has a unique global solution* $u(x,t)$*, which* *tends to the viscous shock wave and diffusion waves, (2.14) – (2.16), as following: For* $-\infty < x < \infty$, $t \geq 0$,

$$
\begin{aligned}
v(x,t) =& \sum_{i=1}^{n} v_i(x,t) r_i(\phi(x)), \\
v_i(x,t) =& O(1)\delta_0 \Big[\psi_i^{\frac{3}{2}}(x,t) + \sum_{k\neq i} \bar{\psi}_k^{\frac{3}{2}}(x,t) + \chi_i(x,t) \\
& + \varepsilon^2 e^{-|\lambda_p||x|/\mu} \psi_p^{\frac{1}{2}}(x,t) + \eta_2(x,t) \Big], \quad i \neq p, \qquad (3.5) \\
v_p(x,t) =& O(1)\delta_0 \Big[\psi_p^{\frac{3}{2}}(x,t) + \sum_{k\neq p} \bar{\psi}_k^{\frac{3}{2}}(x,t) \\
& + \varepsilon e^{-|\lambda_p||x|/\mu} \psi_p^{\frac{1}{2}}(x,t) + \eta_1(x,t) \Big],
\end{aligned}
$$

where $|\lambda_p| = \min\{\lambda_p^-, -\lambda_p^+\}$, *and* $\mu > 0$ *is a constant.*

Corollary 3.2. *Under the assumptions of Theorem 3.1, we have*

$$
\begin{aligned}
&\|v(\cdot,t)\|_{L^1} = O(1)\delta_0 (t+1)^{-\frac{1}{4}}, \\
&\|v(\cdot,t)\|_{L^2} = O(1)\delta_0 (t+1)^{-\frac{1}{2}}, \qquad (3.6) \\
&\|v(\cdot,t)\|_{L^\infty} = O(1)\delta_0 \big[(t+1)^{-\frac{3}{4}} + \varepsilon^{\frac{1}{2}} (t+1)^{-\frac{1}{2}} \big].
\end{aligned}
$$

To finish this section we make a few remarks. From (3) both $\psi_i(x,t)$ and $\bar{\psi}_i(x,t)$ center at $x - \lambda_i^0(t+1)$. Along the center line, $\psi_i(x,t)$ decays at the rate $(t+1)^{-\frac{1}{2}}$ while $\bar{\psi}_i(x,t)$ decays faster at $(t+1)^{-\frac{2}{3}}$. However, the support of $\psi_i(x,t)$ has size $(t+1)^{\frac{1}{2}}$, while $\bar{\psi}_i(x,t)$ has a wider support in the size $(t+1)^{\frac{2}{3}}$. In (3.5) the primary wave $\psi_i^{\frac{3}{2}}$ or $\psi_p^{\frac{3}{2}}$ decays at $(t+1)^{-\frac{3}{4}}$ along the main characteristic direction. The term $\sum_{k\neq i} \bar{\psi}_k^{\frac{3}{2}}$ or $\sum_{k\neq p} \bar{\psi}_k^{\frac{3}{2}}$ represents waves along the secondary characteristic directions. These waves decay faster at $(t+1)^{-1}$. They should have supports of the same size as the primary wave as $(t+1)^{\frac{1}{2}}$. Here we have introduced these waves to simplify the description of v. While keeping the optimal decay rate, they artificially enlarge the supports. The accurate support size $(t+1)^{\frac{1}{2}}$ is in fact used in the proof of Theorem 3.1.

The main difference between waves in the p-direction and the waves in the i-directions, $i \neq p$, is the effect of compression. For instance, although $\psi_p^{\frac{3}{2}}$ has decay rate $(t+1)^{-\frac{3}{4}}$, when multiplied by a positive power of ε, the power can be use to trade for decay rates up to $(t+1)^{-\frac{3}{2}}$.

Comparing with the special case of artificial viscosity, $B = I$, studied in [17], here we have the extra terms for the case of the physical viscosity as following: $\bar{\psi}_p^{\frac{3}{2}} + \eta_2(x,t)$ in $v_i(x,t)$, $i \neq p$, and $\eta_1(x,t)$ in $v_p(x,t)$. The term $\bar{\psi}_p^{\frac{3}{2}}$ in v_i comes from the linear coupling of the p-characteristic family into the i-characteristic family. In the case of artificial viscosity all the coupling is nonlinear, hence there is no such a term. The term η_1 in v_p comes from the linear coupling of the transversal characteristic families into the p-characteristic family. In particular, in the Green's function such a coupling cannot be expressed as derivatives of heat kernels for higher order terms, where there are bounded functions expressed as $O(1)$ in the estimate. Because of these unknown bounded functions, we cannot use differential equations to achieve cancelation. The convolution of these higher order terms in the Green's function with slow waves in the source results in the slow wave η_1. The term η_2 in v_i arrives for the same reason.

References

[1] I-L. Chern, Multiple-mode diffusion waves for viscous nonstrictly hyperbolic conservation laws. Comm. Math. Phys. 138 (1991), 51–61.

[2] J. D. Cole, On a quasi-linear parabolic equation occurring in aerodynamics. Quart. Appl. Math. 9 (1951), 225–236.

[3] H. Freistühler, C. Fries and C. Rohde, Existence, bifurcation, and stability of profiles for classical and non-classical shock waves. Ergodic Theory, Analysis, and Efficient Simulation of Dynamical Systems, Springer, Berlin, 814 (2001), 287–309.

[4] K. O. Friedrichs and P. D. Lax, Systems of conservation equations with a convex extension. Proc. Nat. Acad. Sci. U.S.A. 68 (1971), 1686–1688.

[5] J. Goodman, Nonlinear asymptotic stability of viscous shock profiles for conservation laws. Arch. Rational Mech. Anal. 95 (1986), 325–344.

[6] E. Hopf, The partial differential equation $u_t + uu_x = \mu u_{xx}$. Comm. Pure Appl. Math. 3 (1950), 201–230.

[7] F. M. Huang, Z. P. Xin and T. Yang, Contact discontinuity with general perturbations for gas motions. Adv. Math. 219 (2008), 1246–1297.

[8] S. Kawashima, Systems of a hyperbolic-parabolic composite type, with applications to the equations of magnetohydrodynamics. Doctoral thesis, Kyoto University, 1983.

[9] P. D. Lax, Hyperbolic systems of conservation laws, II. Comm. Pure Appl. Math. 10 (1957), 537–566.

[10] T.-P. Liu, Nonlinear stability of shock waves for viscous conservation laws. Mem. Amer. Math. Soc. 56 (1985), no.328, v+108 pp.

[11] T.-P. Liu, Interactions of nonlinear hyperbolic waves. Nonlinear Analysis, F.-C. Liu and T.-P. Liu, World Scientific, Singapore, (1991), 171–184.

[12] T.-P. Liu, Pointwise convergence to shock waves for viscous conservation laws. Comm. Pure Appl. Math. 50 (1997), 1113–1182.

[13] T.-P. Liu and S.-H Yu, Viscous rarefaction waves. Bulletin. Inst. Math., Academia Sinica (N.S.), 5 (2010), 123-179.

[14] T.-P. Liu, S.-H Yu and Y. Zeng, Viscous conservation laws, part I: scalar laws. Bulletin, Inst. Math. Academia Sinica, to appear.

[15] T.-P. Liu and Y. Zeng, Large time behavior of solutions for general quasilinear hyperbolic-parabolic systems of conservation laws. Mem. Amer. Math. Soc. 125 (1997), no. 599, viii+120 pp.

[16] T.-P. Liu and Y. Zeng, Compressible Navier-Stokes equations with zero heat conductivity. J. Differential Equations 153 (1999), 225–291.

[17] T.-P. Liu and Y. Zeng, Time-asymptotic behavior of wave propagation around a viscous shock profile. Commun. Math. Phys. 290 (2009), 23–82.

[18] T.-P. Liu and Y. Zeng, On Green's function for hyperbolic-parabolic systems. Acta Mathematica Scientia. 29 (2009), 1556–1572.

[19] T.-P. Liu and Y. Zeng, Nonlinear stability of viscous shock waves with physical viscosity. Preprint.

[20] A. Matsumura and K. Nishihara, On a stability of travelling wave solution of a one-dimensional model system for compressible viscous gas. Japan J. Appl. Math. 3 (1986), 1–13.

[21] Y. Shizuta and S. Kawashima, Systems of equations of hyperbolic-parabolic type with applications to the discrete Boltzmann equation. Hokkaido Math. J. 14 (1985), 249–275.

[22] J. Smoller, Shock waves and reaction-diffusion equations, 2nd ed. Springer-Verlag, New York, 1994.

[23] A. Szepessy and Z. P. Xin, Nonlinear stability of viscous shock waves. Arch. Rational Mech. Anal. 122 (1993), 53–103.

[24] Y. Zeng, L^1 asymptotic behavior of compressible, isentropic, viscous 1-D flow. Comm. Pure Appl. Math. 47 (1994), 1053–1082.

Stellar Structure, Dynamics and Stability*

Tao Luo
Department of Mathematics and Statistics, Georgetown University
Washington, DC 20057, USA
E-mail: tl48@georgetown.edu

Joel Smoller
Department of Mathematics, University of Michigan
Ann Arbor, MI 48109, USA
E-mail: smoller@umich.edu

Abstract

We survey some results on the equations for Newtonian and General Relativistic equations of stars of perfect fluids with the emphasis on the dynamical stability of the Newtonian white dwarfs.

1 Introduction

In the theory of general relativity, properties of the gravitational field are determined by a Lorentzian metric g (cf. [45]). In a coordinate system $x = (x^0,\ x^1, x^2, x^3)$, $x^0 = ct$ (c: speed of light), the metric $g_{ij} = g_{ij}(x)$ $i, j = 0 - 3$ is a symmetric tensor defined on a 4-dimensional space-time manifold. The celebrated Einstein equation is given by

$$R_{ij} - \frac{1}{2}Rg_{ij} = kT_{ij}, \tag{1.1}$$

for $i, j = 0,\ 1, \cdots, 3$, where R_{ij} is the Ricci curvature tensor, R stands for the scalar curvature , $R = g^{ij}R_{ij}$ ($(g^{ij})=(g_{ij})^{-1}$). T_{ij} is the energy-momentum tensor, $k = \frac{8\pi G}{c^4}$ (G is the gravitational constant). In the case of a perfect fluid, the energy momentum tensor is given by

$$T_{ij} = (\rho + p)u_i u_j + pg_{ij}, i, j = 0,\ 1, \cdots, 3,$$

where p denotes the pressure, ρ is the mass-energy density, u represents the 4-velocity of the fluid, satisfying $g_{ij}u^i u^j = -1$. (Here and in the

*Luo was supported in part by the National Science Foundation under Grants DMS-0839864. Smoller was supported in part by the National Science Foundation under Grant DMS-0603754.

following the Einstein summation convention is used). From the Bianchi identity of geometry, we have the following conservation laws of energy and momentum:

$$0 = divT = T^{ij}_{;i} = T^{ij}_{,i} + \Gamma^{i}_{jk}T^{kj} + \Gamma^{j}_{ik}T^{ik}.$$

(The semicolon denotes covariant derivative, and the comma denotes ordinary partial derivative.)

In the Newtonian approximation ($g_{00} \simeq -(1+2\phi), T_{00} \simeq \rho$), we have the following Euler-Poisson equations:

$$\begin{cases} \rho_t + \nabla \cdot (\rho \mathbf{v}) = 0, \\ (\rho \mathbf{v})_t + \nabla \cdot (\rho \mathbf{v} \otimes \mathbf{v}) + \nabla p(\rho) = -\rho \nabla \Phi, \\ \Delta \Phi = 4\pi G \rho. \end{cases} \tag{1.2}$$

In this paper, we survey some results in the study of non-rotating stars and rotating stars in both General Relativity and Newtonian settings. The emphasis will be on the dynamical stability in the Newtonian setting. Some related open problems concerning the uniqueness of rotating star solutions, instability for the non-rotating star solutions for the polytropic star when $6/5 < \gamma < 4/3$, and free boundary problems to capture the physical vacuum boundary can be found in [30].

2 Non-rotating and rotating stars

We begin this section with the problems of stars of perfect fluids in the General Relativity setting.

2.1 Problems in General Relativity setting

In the general relativistic setting in astrophysics, the Tolman-Oppenheimer-Volkoff (TOV) equations are used to model non-rotating stars: the structure of a spherically symmetric star of isotropic material which is in static gravitational equilibrium. Let $(x_0, x_1, x_2, x_3) = (t, r, \theta, \phi)$, and consider the spherically symmetric time-independent metric:

$$ds^2 = -B(r)c^2dt^2 + A(r)^{-1}dr^2 + r^2(d\theta^2 + sin^2\theta d\phi^2).$$

In the energy-momentum tensor, the fluid 4- velocity is given by $u_0 = \sqrt{B}$, $u_r = u_\theta = u_\phi = 0$ (the fluid is fixed in the (t, r, θ, ϕ)-coordinate system. Then

$$\begin{cases} \frac{B'}{B} = -2\frac{p'}{p+\rho c^2}, A(r) = 1 - 2\frac{GM(r)}{c^2 r}, M(r) = 4\pi \int_0^r \rho(s)s^2 ds, \\ -r^2 \frac{dp}{dr} = GM(r)\rho(1 + \frac{p}{\rho c^2})(1 + \frac{4\pi r^3 p}{M(r)c^2})A^{-1}. \end{cases} \tag{2.1}$$

Weinberg ([45]), refers to this system as *the fundamental equation of Newtonian astrophysics, with general-relativistic corrections supplied by the last three factors.*

The exterior solutions (the stationary spherical symmetric solution of vacuum Einstein equation $R_{ij} = 0$, which models the solution outside stars) are given by the Schwarzschild solution ([1]):

$$ds^2 = -(1 - \frac{2GM_0}{c^2 r})c^2 dt^2 + (1 - \frac{2GM_0}{c^2 r})^{-1} dr^2 + r^2(d\theta^2 + sin^2\theta d\phi^2).$$

$r_s = \frac{2GM_0}{c^2}$ is called the Schwarzschild radius. The interior solution and the exterior solution are matched as follows: Let R be the radius of the star, then we require:

$$\begin{cases} p(R) = \rho(R) = 0, \\ B(R) = (1 - \frac{2GM_0}{c^2 R}), A(R) = (1 - \frac{2GM_0}{c^2 R}), \end{cases} \tag{2.2}$$

where M_0 is the total mass of the star.

If $R < \frac{9}{8} r_s$, it was found by Buchdahl ([4]) that the star is unstable to gravitational collapse. The refined result by Smoller and Temple (cf. [40] and [41]) showed that if the radius of the star is less that 9/8 of the Schwarzschild radius, then the metric must have a singularity at $r = 0$ ($B(r) = O(r^{-1})$ and the scalar curvature $R \geq Cr^{-6}$ as $r \to 0$). Moreover, for such solutions the density and pressure always have smooth profiles that decrease to zero as the radius $r \to 0$, and this implies that the gravitational field becomes repulsive near $r = 0$. In order to allow the possibility of a star whose surface lies inside 9/8ths of the Schwarzschild radius, Smoller and Temple [42] placed a shock wave inside a surface layer with the idea that a shock wave could supply the pressure required to hold up a surface layer when it is inside 9/8ths of the Schwarzschild radius. Also, it was introduced ([41]) a new formulation of the TOV equations, which was used in ([41]) to show that, under mild assumptions on the equation of state, black holes never form in solutions of the TOV equations.

Other studies on TOV solutions can be found in [32] and [33].

The Kerr metric of vacuum Eistein equation, describing the geometry of spacetime around a rotating massive body with mass M and rotating with angular momentum J, is given by (cf. [1])

$$\begin{aligned} ds^2 =& -(1 - \frac{r_s r}{\eta^2})c^2 dt^2 + \frac{\eta^2}{\Delta} dr^2 + \eta^2 d\theta^2 \\ &+ (r^2 + \alpha^2 + \frac{r_s r \alpha^2}{\eta^2} sin^2\theta) sin^2\theta) d\phi^2 - \frac{2 r_s r \alpha sin^2\theta}{\eta^2} c dt d\phi. \end{aligned}$$

where the coordinates r, θ, ϕ are standard spherical coordinate system, and r_s is the Schwarzschild radius,

$$\alpha = \frac{J}{Mc}, \eta^2 = r^2 + \alpha^2 cos^2\theta, \Delta = r^2 - r_s r + \alpha^2.$$

There have been almost no results on the interior solution to match the exterior kerr metric.

2.2 Newtonian non-rotating and rotating stars

In Newtonian setting, the compressible Euler-Poisson equations in three spatial dimensions (1.2) are used to model the evolution of a gaseous star (cf. [8]),in which ρ is the density, $\mathbf{v} = (v_1, v_2, v_3)$ is the velocity, p is the pressure and Φ is the gravitational potential, and G is the Newtonian gravitational constant. For the isentropic fluids the pressure function is given by $p = p(\rho)$ satisfying $p'(\rho) > 0$ for $\rho > 0$. The pressure function for the polytropic stars are given by $p(\rho) = K\rho^\gamma$, $K > 0$ and $\gamma > 1$ are constants. When $\gamma = 4/3$, a star is called supermassive ([45]). For a white dwarf star (a star in which gravity is balanced by electron degeneracy pressure), the pressure function $p(\rho)$ obeys the following asymptotics ([8], Chapter 10):

$$\begin{cases} p(\rho) = c_1\rho^{4/3} - c_2\rho^{2/3} + \cdots, & \rho \to \infty, \\ p(\rho) = d_1\rho^{5/3} - d_2\rho^{7/3} + O(\rho^3), & \rho \to 0, \end{cases} \tag{2.3}$$

where c_1, c_2, d_1 and d_2 are positive constants.

In the study of time-independent solutions of system (1.2), there are two cases, non-rotating stars and rotating stars.

Non-rotating star solutions: A non-rotating star solution is a time -independent radial symmetric solution of form: $(\rho, 0, \Phi)(|x|)$ to (1.2) satisfying (cf [8] and [45]),

$$-\eta^2\frac{dp(\rho)}{d\eta} = GM(\eta)\rho(\eta), M(\eta) = 4\pi\int_0^\eta s^2\rho^2(s)ds, \tag{2.4}$$

where $\eta = |x| = \sqrt{x_1^2 + x_2^2 + x_3^2}$. For a polytropic star, $p(\rho) = K\rho^\gamma$ $(\gamma > 1)$, let

$$\eta = \left(\frac{K\gamma}{4\pi G(\gamma - 1)}\right)^{1/2} \rho(0)^{\frac{\gamma-2}{2}}\xi, \qquad \rho = \rho(0)\theta^{\frac{1}{\gamma-1}}.$$

Then

$$\frac{d}{d\xi}(\xi^2\frac{d\theta}{d\xi}) + \xi^2\theta^{\frac{1}{\gamma-1}} = 0. \tag{2.5}$$

Boundary conditions are

$$\theta(0) = 1, \quad \theta'(0) = 0. \tag{2.6}$$

The function $\theta(\xi)$ defined by (2.5) and (2.6) is called *Lane-Emden function.* For $\gamma > 6/5$, $\theta(\xi)$ vanishes at some finite ξ_1:

$$\theta(\xi_1) = 0, \quad \theta'(\xi_1) < 0.$$

The relation between the radius of the star and central density is

$$R = \left(\frac{K\gamma}{4\pi G(\gamma-1)}\right)^{1/2} \rho(0)^{\frac{\gamma-2}{2}} \xi_1.$$

The relation between the total mass and central density of the star is given by

$$M = 4\pi\rho(0)^{\frac{3\gamma-4}{2}} \left(\frac{K\gamma}{4\pi G(\gamma-1)}\right)^{3/2} \xi_1^2 |\theta'(\xi)|. \tag{2.7}$$

So

$$M = 4\pi R^{\frac{3\gamma-4}{\gamma-2}} \left(\frac{K\gamma}{4\pi G(\gamma-1)}\right)^{-\frac{1}{\gamma-2}} \xi_1^{\frac{3\gamma-4}{\gamma-2}} \xi^2 |\theta'(\xi)|. \tag{2.8}$$

When $\gamma = 4/3$, $\xi_1 \approx 6.89685, \xi_1^2|\theta'(\xi_1)| \approx= 2.01824$, and the total mass M is independent of the central density,

$$M_c = \frac{4}{\sqrt{\pi}} K^{3/2} \xi^2 |\theta'(\xi)|.$$

This mass is called *Chandrasekhar limit* for a white dwarf star. The above analysis can be found in [45]. The gravitational collapse for non-rotating white dwarfs was discovered in [7] and [8] if the total mass exceeds the Chandrasekhar limit . This gravitational collapse was also verified by Lieb and Yau ([24]) as the limit of Quantum Mechanics.

Rotating star solutions: We introduce some notation. We use $\int$ to denote $\int_{\mathbf{R}^3}$, and use $||\cdot||_q$ to denote $||\cdot||_{L^q(\mathbf{R}^3)}$. For any point $x = (x_1, x_2, x_3) \in \mathbf{R}^3$, let

$$r(x) = \sqrt{x_1^2 + x_2^2},\ z(x) = x_3,\ B_R(x) = \{y \in \mathbf{R}^3,\ |y - x| < R\}. \tag{2.9}$$

For any function $f \in L^1(\mathbf{R}^3)$, we define the operator B by

$$Bf(x) = \int \frac{f(y)}{|x-y|} dy = f * \frac{1}{|x|}. \tag{2.10}$$

Also, we use ∇ to denote the spatial gradient, i.e., $\nabla = \nabla_x = (\partial_{x_1}, \partial_{x_2}, \partial_{x_3})$. C will denote a generic positive constant.

A rotating star solution $(\tilde{\rho}, \tilde{\mathbf{v}}, \tilde{\Phi})(r, z)$, where $r = \sqrt{x_1^2 + x_2^2}$ and $z = x_3$, $x = (x_1, x_2, x_3) \in \mathbf{R}^3$, is an *axi-symmetric* time-independent solution of system (1.2), which models a star rotating about the x_3-axis. Suppose the angular momentum (per unit mass), $J(m_{\tilde{\rho}}(r))$ is prescribed, where

$$m_{\tilde{\rho}}(r) = \int_{\sqrt{x_1^2+x_2^2}<r} \tilde{\rho}(x)dx = \int_0^r 2\pi s \int_{-\infty}^{+\infty} \tilde{\rho}(s, z)dsdz \qquad (2.11)$$

is the mass in the cylinder $\{x = (x_1, x_2, x_3) : \sqrt{x_1^2 + x_2^2} < r\}$, and J is a given function. In this case, the velocity field $\tilde{\mathbf{v}}(x) = (v_1, v_2, v_3)$ takes the form

$$\tilde{\mathbf{v}}(x) = (-\frac{x_2 J(m_{\tilde{\rho}}(r))}{r^2}, \frac{x_1 J(m_{\tilde{\rho}}(r))}{r^2}, 0).$$

Substituting this in (1.2), we find that $\tilde{\rho}(r, z)$ satisfies the following two equations:

$$\begin{cases} \partial_r p(\tilde{\rho}) = \tilde{\rho}\partial_r(B\tilde{\rho}) + \tilde{\rho}L(m_{\tilde{\rho}}(r))r^{-3}, \\ \partial_z p(\tilde{\rho}) = \tilde{\rho}\partial_z(B\tilde{\rho}), \end{cases} \qquad (2.12)$$

where the operator B is defined in (2.10), and

$$L(m_{\tilde{\rho}}) = J^2(m_{\tilde{\rho}})$$

is the square of the angular momentum. We define

$$A(\rho) = \rho \int_0^\rho \frac{p(s)}{s^2} ds. \qquad (2.13)$$

It is easy to verify that (cf. [2]) (2.12) is equivalent to

$$A'(\tilde{\rho}(x)) + \int_{r(x)}^\infty L(m_{\tilde{\rho}}(s)s^{-3}ds - B\tilde{\rho}(x) = \lambda, \quad \text{where } \tilde{\rho}(x) > 0, \qquad (2.14)$$

for some constant λ. Here $r(x)$ and $z(x)$ are as in (2.9).

2.3 Existence and dynamical stability of Newtonian rotating and non-rotating star solutions via the energy minimizing approach

The existence theory for non-rotating star solutions including white dwarf stars for which the total mass M of the star is not greater than a critical mass M_c ($M \leq M_c$)is classical([8]). For rotating white dwarf stars with prescribed total mass and angular momentum distribution, Auchumuty and Beals ([2]) proved that if the angular momentum distribution is nonnegative, then existence holds if $M \leq M_c$. Friedman and

Turkington ([13]) proved existence for any mass provided that the angular momentum distribution is everywhere positive; see Li ([23]), Chanillo & Li ([9]) and Luo & Smoller ([27]) for related results for rotating star solutions with prescribed constant angular velocity. For non-rotating stars, G. Rein ([35]) proved nonlinear dynamical stability for general perturbations. However, his result does not apply to white dwarf stars. Other related results besides those mentioned above for compressible fluid rotating stars can be found in [3], [5], [12], and [27].

Let M be a positive constant and let W_M be the set of functions ρ defined by

$$W_M = \Big\{\rho : \mathbf{R}^3 \to \mathbf{R},\ \rho \text{ is axisymmetric},\ \rho \geq 0, a.e., \int \rho(x)dx = M,$$
$$\int \left(A(\rho(x)) + \frac{\rho(x)L(m_\rho(r(x)))}{r(x)^2} + \rho(x)B\rho(x)\right) dx < +\infty \Big\}.$$

For $\rho \in W_M$, we define the **energy functional** F by

$$F(\rho) = \int [A(\rho(x)) + \frac{1}{2}\frac{\rho(x)L(m_\rho(r(x)))}{r(x)^2} - \frac{1}{2}\rho(x)B\rho(x)]dx. \tag{2.15}$$

In (2.15), the first term denotes the potential energy, the middle term denotes the rotational kinetic energy and the third term is the gravitational energy.

For a white dwarf star, the pressure function $p(\rho)$ satisfies the following conditions:

$$\lim_{\rho\to 0+} \frac{p(\rho)}{\rho^{4/3}} = 0,\ \lim_{\rho\to\infty} \frac{p(\rho)}{\rho^{4/3}} = \mathfrak{K},\ p'(\rho) > 0 \text{ as } \rho > 0, \tag{2.16}$$

where $\mathfrak{K}$ is a finite positive constant. Assuming that the function $L \in C^1[0, M]$ and satisfies

$$L(0) = 0,\ L(m) \geq 0,\ \text{for } 0 \leq m \leq M, \tag{2.17}$$

Auchmuty and Beals (cf. [2]) proved the existence of a minimizer of the functional $F(\rho)$ in the class of functions $W_{M,S} = W_M \cap W_{sym}$, where

$$W_{sym} = \{\rho : \mathbf{R}^3 \to \mathbf{R},\ \rho(x_1, x_2, -x_3) = \rho(x_1, x_2, x_3),\ x_i \in \mathbf{R}, i = 1, 2, 3\} \tag{2.18}$$

for white dwarfs when the total mass is below a critical mass.

We are interested in minimizers of functional F in the *larger* class W_M and its dynamical stability. First, we give the following existence and compactness result. For the angular momentum, besides the condition (2.17), we also assume that it satisfies the following conditions:

$$L(am) \geq a^{4/3}L(m),\ 0 < a \leq 1,\ 0 \leq m \leq M, \tag{2.19}$$

$$L'(m) \geq 0, \qquad 0 \leq m \leq M. \tag{2.20}$$

Condition (2.20) is called the Sölberg stability criterion ([43]).
Compactness of Minimizing Sequence: The following compactness result for the minimizing sequences of the functional F is crucial for the existence and stability analysis. The proof can be found in ([29]).

Theorem 2.1 ([29]). *Suppose that the square of the angular momentum L satisfies (2.17), (2.19) and (2.20), and the pressure function p satisfies the following conditions*

$$\begin{aligned} &p \in C^1[0,+\infty),\ \lim_{\rho\to 0+} \frac{p(\rho)}{\rho^{4/3}} = 0,\ \lim_{\rho\to\infty} \frac{p(\rho)}{\rho^\gamma} = K, \\ &p(\rho) \geq 0, p'(\rho) > 0 \text{ for } \rho > 0, \end{aligned} \tag{2.21}$$

where $0 < K < +\infty$ and $\gamma \geq 4/3$. Then there exists a constant $\mathfrak{M}_c$ satisfying $0 < \mathfrak{M}_c < \infty$ if $\gamma = 4/3$ and $\mathfrak{M}_c = \infty$ if $\gamma > 4/3$, such that if $M < \mathfrak{M}_c$, then,
(1)

$$\inf_{\rho \in W_M} F(\rho) < 0, \tag{2.22}$$

(2) for $\rho \in W_M$,

$$\int [A(\rho)(x) + \frac{1}{2}\frac{\rho(x)L(m_\rho(r(x)))}{r(x)^2}]dx \leq C_1 F(\rho) + C_2, \tag{2.23}$$

for some positive constants C_1 and C_2;
(3) If $\{\rho^i\} \subset W_M$ is a minimizing sequence for the functional F, then there exist a sequence of vertical shifts $a_i\mathbf{e_3}$ ($a_i \in \mathbf{R}$, $\mathbf{e_3} = (0,0,1)$), a subsequence of $\{\rho^i\}$ (still labeled $\{\rho^i\}$), and a function $\tilde\rho \in W_M$, such that for any $\epsilon > 0$ there exists $R > 0$ with

$$\int_{|x|\geq R} T\rho^i(x)dx \leq \epsilon, \quad i \in \mathbb{N}, \tag{2.24}$$

and

$$T\rho^i(x) \rightharpoonup \tilde\rho, \text{ weakly in } L^\gamma(\mathbf{R}^3), \text{ as } i \to \infty, \tag{2.25}$$

where $T\rho^i(x) := \rho^i(x + a_i\mathbf{e_3})$;
Moreover
(4)

$$\nabla B(T\rho^i) \to \nabla B(\tilde\rho) \text{ strongly in } L^2(\mathbf{R}^3), \text{ as } i \to \infty, \tag{2.26}$$

and $\tilde\rho$ is a minimizer of F in W_M.

We call such a minimizer $\tilde{\rho}$ a *rotating star* solution with total mass M and angular momentum $\sqrt{L(m)}$.

With the help of this theorem, the nonlinear dynamical stability results of the rotating star solutions including white dwarfs were established by Luo and Smoller for the axi-symmetric perturbations. Please refer to [28] and [29] for details. In the following, we give the nonlinear dynamical stability of non-rotating white dwarf stars with general perturbations, which extends the results in [35] (The result in [35] does not apply to the white dwarf case).

2.4 Nonlinear dynamical stability of non-rotating white dwarf stars with general perturbations

For white dwarf stars, as mentioned before, the pressure function satisfies

$$p \in C^1[0,+\infty), \ \lim_{\rho\to 0+}\frac{p(\rho)}{\rho^{\gamma_1}} = \beta, \ \lim_{\rho\to\infty}\frac{p(\rho)}{\rho^{\gamma}} = K, \ p'(\rho) > 0 \text{ for } \rho > 0, \tag{2.27}$$

where $\gamma_1 > 4/3$, $0 < \beta < +\infty$ and $0 < K < +\infty$ are constants. In this section, we always assume that the pressure function satisfies (2.27). First, we define for $0 < M < +\infty$,

$$X_M = \left\{\rho : \mathbf{R}^3 \to \mathbf{R}, \rho \geq 0, a.e., \int \rho(x)dx = M, \right. \\ \left. \int [A(\rho(x)) + \frac{1}{2}\rho(x)B\rho(x)]dx < +\infty\right\}, \tag{2.28}$$

where $A(\rho)$ is the function given in (2.13). For $\rho \in X_M$, we define the **energy functional** G for non-rotating stars by

$$G(\rho) = \int [A(\rho(x)) - \frac{1}{2}\rho(x)B\rho(x)]dx. \tag{2.29}$$

Then we have the the following theorem.

Theorem 2.2. *Suppose that the pressure function p satisfies (2.27). Let $\tilde{\rho}_N$ be a minimizer of the energy functional G in X_M and let*

$$\Gamma_N = \{x \in \mathbf{R}^3 : \ \tilde{\rho}_N(x) > 0\}, \tag{2.30}$$

then there exists a constant λ_N such that

$$\begin{cases} A'(\tilde{\rho}_N(x)) - B\tilde{\rho}_N(x) = \lambda_N, & x \in \Gamma_N, \\ -B\tilde{\rho}_N(x) \geq \lambda_N, & x \in \mathbf{R}^3 - \Gamma_N. \end{cases} \tag{2.31}$$

The proof of this theorem is well-known, cf. [35] or [2].

Remark 2.3. 1) We call the minimizer $\tilde{\rho}_N$ of the functional G in X_M a non-rotating star solution.
2) It follows from [24] that the minimizer $\tilde{\rho}_N$ of the functional G in X_M is actually radial, and has a compact support.

Similar to Theorem 2.1, we have the following compactness theorem.

Theorem 2.4 (Luo-Smoller [29]). *Suppose that the pressure function p satisfies (2.27). There exists a constant M^c $(0 < M^c < \infty)$ such that if $M < M^c$, then the following hold:*
(1)

$$\inf_{\rho \in X_M} G(\rho) < 0, \tag{2.32}$$

(2) for $\rho \in X_M$,

$$\int A(\rho)(x)dx \leq C_1 G(\rho) + C_2, \tag{2.33}$$

for some positive constants C_1 and C_2,
(3) if $\{\rho^i\} \subset X_M$ is a minimizing sequence for the functional G, then there exist a sequence of translations $\{x^i\} \subset \mathbf{R}^3$, a subsequence of $\{\rho^i\}$, (still labeled $\{\rho^i\}$), and a function $\tilde{\rho}_N \in X_M$, such that for any $\epsilon > 0$ there exists $R > 0$ with

$$\int_{|x| \geq R} T\rho^i(x)dx \leq \epsilon, \quad i \in \mathbb{N}, \tag{2.34}$$

and

$$T\rho^i(x) \rightharpoonup \tilde{\rho}_N, \text{ weakly in } L^{4/3}(\mathbf{R}^3), \text{ as } i \to \infty, \tag{2.35}$$

where $T\rho^i(x) := \rho^i(x + x^i)$.
Moreover
(4)

$$\nabla B(T\rho^i) \to \nabla B(\tilde{\rho}_N) \text{ strongly in } L^2(\mathbf{R}^3), \text{ as } i \to \infty, \tag{2.36}$$

and
(5) $\tilde{\rho}_N$ is a minimizer of G in X_M.
(6) The minimizers of G in X_M are unique up to a translation $\rho_N(x) \to \rho_N(x + y)$.

The proof of this theorem can be found in [29].

Let $\tilde{\rho}_N$ be the minimizer of G on X_M and λ_N be the constant in (2.31). For $\rho \in X_M$, we define

$$\begin{aligned} d(\rho, \tilde{\rho}_N) &= \int \{[A(\rho) - A(\tilde{\rho}_N)] - (\rho - \tilde{\rho}_N)(\lambda_N + B\tilde{\rho}_N\}dx, \\ &= \int \{[A(\rho) - A(\tilde{\rho}_N)] - B\tilde{\rho}_N(\rho - \tilde{\rho}_N)\}dx, \end{aligned} \tag{2.37}$$

where we have used the identity

$$\int \rho dx = \int \tilde{\rho}_N dx = M,$$

for $\rho \in X_M$. It is easy to see that due to the convexity of A,

$$d(\rho, \tilde{\rho}_N) \geq 0, \tag{2.38}$$

for any $\rho \in X_M$. Our nonlinear stability theorem of non-rotating white dwarf star solutions is the following theorem, which extends the results in [35].

Theorem 2.5 (Luo-Smoller [29]). *Suppose that the pressure function satisfies (2.27). Let $\tilde{\rho}_N$ the minimizer of the functional G in X_M. Let $(\rho, \mathbf{v}, \Phi)(x,t)$ be an entropy weak solution of the Cauchy problem (1.2) for $(x,t) \in \mathbf{R}^3 \times [0,T)$ with the initial data $\rho(x,0) = \rho_0(x)$ and $\mathbf{v}(x,0) = \mathbf{v}_0(x)$ satisfying $\int_{\mathbf{R}^3} \rho(x,t)dx = \int_{\mathbf{R}^3} \rho_0(x)dx$ and $E(t) \leq E(0)$, where*

$$E(t) = E(\rho(t), \mathbf{v}(t)) = \int \left(A(\rho) + \frac{1}{2}\rho|\mathbf{v}|^2 \right)(x,t)dx - \frac{1}{8\pi}\int |\nabla\Phi|^2(x,t)dx. \tag{2.39}$$

If the initial data satisfies

$$\int \rho_0(x) = \int \rho_N(x)dx = M,$$

then there exists a constant M^c $(0 < M^c < \infty)$ such that if $M < M^c$, then for every $\epsilon > 0$, there exists a number $\delta > 0$ independent of T such that if

$$d(\rho_0, \tilde{\rho}_N) + \frac{1}{8\pi}||\nabla B\rho_0 - \nabla B\tilde{\rho}_N||_2^2 + \frac{1}{2}\int \rho_0(x)(|v_0|^2)(x)dx < \delta, \tag{2.40}$$

then for every $t \in (0,T)$, there is a translation $y(t) \in \mathbf{R}^3$ such that,

$$d(\rho(t), T^{y(t)}\tilde{\rho}_N) + \frac{1}{8\pi}||\nabla B\rho(t) - \nabla BT^{y(t)}\tilde{\rho}_N||_2^2 + \frac{1}{2}\int \rho(x,t)|v(x,t)|^2 dx < \epsilon \tag{2.41}$$

for $t \in (0,T)$, where $T^{y(t)}\tilde{\rho}_N(x) =: \tilde{\rho}_N(x + y(t))$.

The proof of this theorem can be found in [29].

References

[1] R. Adler, M. Bazin and M. Schiffer, Introduction to General Relativity. McGraw-Hill Book Co., New York-Toronto-London-Sydney, 1965.

[2] G. Auchmuty and R. Beals, Variational solutions of some nonlinear free boundary problems. Arch. Rat. Mech. Anal. 43 (1971), 255–271.

[3] G. Auchmuty, The global branching of rotating stars. Arch. Rat. Mech. Anal. 114 (1991), 179–194.

[4] W. A. Buchdahl, Phys. Rev. 116 (1956), 1027.

[5] L. Caffarelli and A. Friedman, The shape of axi-symmetric rotating fluid. J. Funct. Anal. 694 (1980), 109–142.

[6] D. Coutand and S. Shkoller, Well-posedness in smooth function spaces for the moving-boundary 1-D compressible Euler equations in physical vacuum. arXiv:0910.3136v1 [math.AP].

[7] S. Chandrasekhar, Phil. Mag. 11 (1931), 592; Astrophys. J. 74 (1931), 81; Monthly Notices Roy. Astron. Soc. 91 (1931), 456; Rev. Mod. Phys. 56 (1984), 137.

[8] S. Chandrasekhar, Introduction to the Stellar Structure. University of Chicago Press, 1939.

[9] S. Chanillo and Y. Y. Li, On diameters of uniformly rotating stars. Comm. Math. Phys. 166 (1994), 417–430.

[10] Y. Deng, T. P. Liu, T. Yang and Z. Yao, Solutions of Euler-Poisson equations for gaseous stars. Arch. Rat. Mech. Anal. 164 (2002), 261–285.

[11] R. H. Fowler, Monthly Notices Roy. Astron. Soc. 87 (1926), 114.

[12] A. Friedman and B. Turkington, Asymptotic estimates for an axi-symmetric rotating fluid. J. Func. Anal. 37 (1980), 136–163.

[13] A. Friedman and B. Turkington, Existence and dimensions of a rotating white dwarf. J. Diff. Eqns. 42 (1981), 414–437.

[14] J. Jang, Nonlinear instability in gravitational Euler-Poisson system for $\gamma = \frac{6}{5}$. To appear in Arch. Rat. Mech. Anal.

[15] J. Jang and N. Masmoudi, Well-posedness for compressible Euler equations with physical vacuum singularity. Comm. Pure Appl. Math. 62 (2009), no. 10, 1327–1385.

[16] Y. Guo and G. Rein, Stable steady states in stellar dynamics. Arch. Rat. Mech. Anal. 147 (1999), 225–243.

[17] Y. Guo and G. Rein, Stable models of elliptical galaxies. Mon. Not. R. Astron. Soc. 344 (2002), 1296–1306.

[18] D. Gilbarg and N. Trudinger, Elliptic Partial Differentail Equations of Second Order (2nd ed). Springer, 1983.

[19] L. Landau, Phys. Z. Sowjetunion 1 (1932), 285.

[20] P. Lax, Shock waves and entropy. Contribution to Nonlinear Functional Analysis, E. A. Zarantonello, ed., Academic Press, New York 603–634, 1971.

[21] N. R. Lebovitz, The virial tensor and its application to self-gravitating fluids. Astrophys. J. 134 (1961), 500–536.

[22] N. R. Lebovitz and A. Lifschitz, Short-wavelength instabilities of Riemann ellipsoids. Philos. Trans. Roy. Soc. London Ser. A 354 (1996), 927–950.

[23] Y. Y. Li, On uniformly rotating stars. Arch. Rat. Mech. Anal. 115 (1991), 367–393.

[24] E. H. Lieb and H. T. Yau, The Chandrasekhar theory of stellar collapse as the limit of quantum mechanics. Comm. Math. Phys. 112 (1987), 147–174.

[25] S. S. Lin, Stability of gaseous stars in spherically symmetric motions. SIAM J. Math. Anal. 28 (1997), 539–569.

[26] P. L. Lions, The concentration-compactness principle in the calculus of variations. The locally compact case. Part I, Ann. Inst. H. Poincaré Anal. Non Linéaire, 1 (1984), 109–145.

[27] T. Luo and J. Smoller, Rotating fluids with self-gravitation in bounded domains. Arch. Rat. Mech. Anal. 173 (2004), 345–377.

[28] T. Luo and J. Smoller, Existence and nonlinear stability of rotating star solutions of the compressible Euler-Poisson equations. Arch. Rational Mech. Anal. 191 (2009), 447–496.

[29] T. Luo and J. Smoller, Nonlinear dynamical stability of Newtonian rotating and non-rotating white dwarfs and rotating supermassive stars. Comm. Math. Phys. 284 (2008), 425–457.

[30] T. Luo and J. Smoller, On the Euler-Poisson equations of self-gravitating compressible fluids. Accepted for publication in Proceedings of the IMA summer school on Nonlinear Conservation Laws and Applications.

[31] T. Makino, Blowing up of the Euler-Poisson equation for the evolution of gaseous star. Transport Theory and Statistical Physics, 21 (1992), 615–624.

[32] T. Makino, On spherically symmetric stellar models in general relativity. J. Math. Kyoto Univ. 38 (1998), 55–69.

[33] A. D. Rendall and B. G. Schmidt, Existence and properties of spherically symmetric static fluid bodies with a given equation of state. Classical Quantum Gravity 8 (1991), 985–1000.

[34] G. Rein, Reduction and a concentration-compactness principle for energy-casimir functionals. SIAM J. Math. Anal. 33 (2001), 896–912.

[35] G. Rein, Non-linear stability of gaseous stars. Arch. Rat. Mech. Anal. 168 (2003), 115–130.

[36] P. Secchi, On the evolution equations of viscous gaseous stars. Ann. Scuola Norm. Sup. Pisa Cl. Sci. (4) 18 (1991), 295–318.

[37] P. Secchi, On the uniqueness of motion of viscous gaseous stars. Math. Methods Appl. Sci. 13 (1990), no. 5, 391–404.

[38] S. H. Shapiro and S. A. Teukolsky, Black Holes, White Dwarfs, and Neutron Stars. WILEY-VCH, 2004.

[39] J. Smoller, Shock Waves and Reaction-Diffusion Equations (2nd Ed). Springer, Berlin, New York, 1994.

[40] J. Smoller and B. Temple, Solutions of the Oppenheimer-Volkoff equations inside 9/8ths of the Schwarzschild radius. Comm. Math. Phys. 184 (1997), no. 3, 597–617.

[41] J. Smoller and B. Temple, On the Oppenheimer-Volkoff equations in general relativity. Arch. Rational Mech. Anal. 142 (1998), no. 2, 177–191.

[42] J. Smoller and B. Temple, Shock waves near the Schwarzschild radius and stability limits for stars. Phys. Rev. D (3) 55 (1997), no. 12, 7518–7528.

[43] J. L. Tassoul, Theory of Rotating Stars. Princeton University Press, Princeton, 1978.

[44] D. Wang, Global solutions and stability for self-gravitating isentropic gases. J. of Math. Anal. Appl. 229 (1999), 530–542.

[45] S. Weinberg, Gravitation and Cosmology. John Wiley and Sons, New York, 1972.

On Some Non-strictly Hyperbolic Systems of Conservation Laws*

Evgueni Yu. Panov
Novgorod State University
Veliky Novgorod, Russia
E-mail: Eugeny.Panov@novsu.ru

Abstract

We study special hyperbolic systems of conservation laws, which can be written as single conservation laws on matrix algebras and include, in particular, the known systems of Keyfitz-Kranzer type. The theory of strong generalized entropy solutions of the Cauchy problem is developed.

1 Introduction

The natural generalization of a scalar conservation law is the following system

$$U_t + (\tilde{f}(U))_x = 0, \tag{1.1}$$

where $U = U(t,x) \in X$, $(t,x) \in \Pi = \mathbb{R}_+ \times \mathbb{R}$, X is a linear subspace of some finite-dimensional algebra such that $U^2 \in X \ \forall U \in X$, $U \mapsto \tilde{f}(U)$ is a functional calculus operator corresponding to some scalar function f. As follows from results of [16], the system under consideration is (non-strictly) hyperbolic if and only if it can be reduced to the model case when X is a space of symmetric (or Hermitian) matrixes of some order n. Study of generalized entropy solutions to the Cauchy problem for system (1.1) with an initial condition $U(0,x) = U_0(x) \in L^\infty(\mathbb{R}, X)$ was begun in papers [12, 13], where the case $n = 2$ was completely investigated. We consider here other special case when the matrix $U = u \otimes u$, $u = u(t,x) \in \mathbb{R}^n$. In this case the problem is reduced to the Cauchy problem for the well-known system of Keyfitz-Kranzer type, introduced in [8]:

$$u_t + (\varphi(|u|)u)_x = 0, \ u = u(t,x) \in \mathbb{R}^n; \ |u| = (u_1^2 + \cdots + u_n^2)^{1/2} \tag{1.2}$$

*The author is partly supported by the Russian Foundation for Basic Research (grant No. 09-01-00490-a) and the Deutsche Forschungsgemeinschaft (DFG project No. 436 RUS 113/895/0-1).

in a half-plane $(t,x) \in \Pi = \mathbb{R}_+ \times \mathbb{R}$, $\mathbb{R}_+ = (0,+\infty)$ with an initial condition

$$u(0,x) = u_0(x) \in L^\infty(\mathbb{R},\mathbb{R}^n). \tag{1.3}$$

The study of problem (1.2), (1.3) is the main subject of this paper. Only in the last section 6 we return to the general case of systems (1.1). The presentation is based on the results of papers [12, 13, 14, 15, 17, 18].

2 Hyperbolicity, entropies

We assume that the function $\varphi(r) \in C(\mathbb{R}_+)$ in (1.2) is only continuous and satisfies the condition

$$r\varphi(r) \underset{r\to 0+}{\to} 0, \tag{2.1}$$

i.e. the function $f(z) = \varphi(|z|)z$ is continuous on $\mathbb{R}$, here $f(0) = 0$ according to (2.1).

Suppose firstly that $f(z) \in C^1(\mathbb{R})$. Then the map $F(u) = \varphi(|u|)u$ is C^1-smooth on $\mathbb{R}^n$ and system (1.2) can be written in the quasilinear form

$$u_t + A(u)u_x = 0, \tag{2.2}$$

$$A(u) = dF(u) = \varphi(|u|)E + \varphi'(|u|)u \otimes u/|u|,$$

E is the identity matrix.

By symmetricity of the matrix $A(u)$ system (1.2)-(2.2) is non-strictly hyperbolic, that is, all its eigenvalues are real and there is a basis of corresponding eigenvectors. These eigenvalues are exactly

$$\lambda_i = \varphi(|u|),\ i = 1,\dots,n-1; \quad \lambda_n = \varphi(|u|) + \varphi'(|u|)|u| = f'(|u|).$$

We see that there are coinciding eigenvalues, therefore the system is not strictly hyperbolic. Remark also that the characteristic field $\lambda(u) = \varphi(|u|)$ is linearly degenerate, i.e. $(\nabla\lambda(u), r(u)) \equiv 0$ for any eigenvector $r(u)$ of the matrix $A(u)$ corresponding to $\lambda(u)$. On the other hand the characteristic field $\lambda_n(u)$ is genuinely nonlinear if the function $f(r)$ is convex for $r > 0$.

Denote by $\mathcal{F}$ the set of functions $p(u) \in C^1(\mathbb{R}^n \setminus \{0\})$, which are locally Lipschitz in $\mathbb{R}^n$.

Definition 2.1. A function $p(u) \in \mathcal{F}$ is called an entropy of system (1.2) if there exists a function $q(u) \in \mathcal{F}$, called the entropy flux, such that for all $u \in \mathbb{R}^n \setminus \{0\}$ $\nabla q(u) = A(u)\nabla p(u)$.

In the case then $p, q \in C^1(\mathbb{R}^n)$ Definition 2.1 is the classical definition of entropy pairs. If $u = u(t,x)$ is a C^1-smooth solution of system (1.2), $p(u)$ is an entropy and $q(u)$ is the corresponding entropy flux then after multiplying equality (2.2) by $\nabla p(u)$ we obtain the relation

$$p(u)_t + q(u)_x = 0 \tag{2.3}$$

in the sense of distributions (in $\mathcal{D}'(\Pi)$).

We describe the set of entropies of system (1.2).

Theorem 2.2. *Let $\eta(r) \in C^1(\mathbb{R}_+)$ and be locally Lipschitz on $[0,\infty)$, $h(u) \in C^1(\mathbb{R}^n \setminus \{0\})$ be a homogeneous (of degree 1) function:*

$$h(ru) = rh(u), \quad \forall r > 0.$$

Then

$$p(u) = \eta(|u|) + h(u) \tag{2.4}$$

is an entropy of system (1.2) with a flux

$$q(u) = \psi(|u|) + \varphi(|u|)h(u),$$

where $\psi(r)$ is defined up to an additive constant by the identity $\psi'(r) = \eta'(r)f'(r)$. If $\varphi'(r) \neq 0$ on a dense set of values $r \in \mathbb{R}_+$ then any entropy $p(u)$ of system (1.2) has the form (2.4).

Entropies of the form (2.4) are called admissible. Remark that an admissible entropy $p(u) \in C^1(\mathbb{R}^n)$ if and only if $p(u) = \eta(|u|) + h(u)$ for some function $\eta(r) \in C^1([0,+\infty))$, $\eta'(0) = 0$, and a linear function $h(u)$.

In the following proposition we give conditions of convexity for admissible entropies.

Proposition 2.3. 1) *If an admissible entropy is convex on $\mathbb{R}^n$ then the function $\eta(r)$ is convex on $\mathbb{R}_+$;* 2) *An entropy $p(u) = \eta(|u|)$ is convex on $\mathbb{R}^n$ if and only if the function $\eta(r)$ is convex and nondecreasing on $\mathbb{R}_+$;* 3) *A homogeneous entropy $p(u) = h(u)$ is convex on $\mathbb{R}^n$ if and only if the function $h(u)$ is sub-linear.*

3 Generalized entropy solutions

Definition 3.1. A function $u = u(t,x) \in L^\infty(\Pi, \mathbb{R}^n)$ is called a generalized entropy solution (g.e.s. for short) of problem (1.2), (1.3) if the Kruzhkov-Lax entropy condition is satisfied

$$p(u)_t + q(u)_x \le 0 \quad \text{in } \mathcal{D}'(\Pi) \tag{3.1}$$

for any admissible convex entropy $p(u) \in \mathcal{F}$ and the corresponding flux $q(u)$, and

$$\operatorname*{ess\,lim}_{t\to 0+} u(t,\cdot) = u_0 \quad \text{in} \quad L^1_{loc}(\mathbb{R}, \mathbb{R}^n). \tag{3.2}$$

Remark that Definition 3.1 is correct even in the case of only continuous $f(r)$, we need only redefine the function $\psi(r)$ in the expression for $q(u)$, setting

$$\psi(r) = f(r)\eta'(r) - \int_0^r f(s)d\eta'(s)$$

(by Proposition 2.3 the function η is convex and $d\eta'(s)$ is a Borel measure). Taking in (3.1) linear entropies $p(u) = \pm(c, u)$, $c \in \mathbb{R}^n$ we derive that g.e.s. $u(t, x)$ satisfies system (1.2) in the sense of distributions. Remark also that by relation (2.3) smooth solutions satisfy condition (3.1) with the equality sign.

It is well known that for smooth entropies condition (3.1) is suggested by the idea of "vanishing viscosity" method. We claim that it remains true even in the case when entropy $p(u) \in \mathcal{F}$. Consider the parabolic approximation of (1.2):

$$u_t + (\varphi(|u|)u)_x = u_t + A(u)u_x = \varepsilon u_{xx}, \quad \varepsilon > 0. \tag{3.3}$$

Proposition 3.2. *Assume that for some subsequence $\varepsilon_k \to 0$ there is a solution $u^k = u^k(t, x) \in C^2(\Pi, \mathbb{R}^n)$ of (3.3) with $\varepsilon = \varepsilon_k$ and the sequence u^k converges in $L^1_{loc}(\Pi, \mathbb{R}^n)$ to a vector-function $u = u(t, x)$ as $k \to \infty$. Then $u(t, x)$ satisfies condition (3.1) for any convex entropy $p(u) \in \mathcal{F}$ and the corresponding flux $q(u)$.*

Taking in (3.1) convex admissible entropies $p(u) = \eta(|u|)$ with convex and nondecreasing functions $\eta(r)$, we establish the following result.

Theorem 3.3. *If $u = u(t, x)$ is a g.e.s. of (1.2), (1.3) then $r = |u(t, x)|$ is a generalized entropy sub-solution to the scalar Cauchy problem*

$$r_t + f(r)_x = 0, \quad r(0, x) = r_0(x) \tag{3.4}$$

with the corresponding initial function $r_0(x) = |u_0(x)|$.

Applying comparison principle from [3] or [10], we obtain the following *maximum principle.*

Corollary 3.4. *If $u(t, x)$ is a g.e.s. of (1.2), (1.3) then*

$$|u(t, x)| \le \operatorname{ess\,sup} |u_0(x)| \quad \text{a.e. on } \Pi.$$

Let $e \in \mathbb{R}^n$ be a unit vector, $0 \le \delta \le 1$. Define cones

$$\Gamma(e, \delta) = \{\ u \in \mathbb{R}^n \mid (u, e) \ge \delta|u|\ \},$$
$$\dot{\Gamma}(e, \delta) = \{\ u \in \mathbb{R}^n \mid (u, e) > \delta|u|\ \} \cup \{0\}.$$

Using the entropy condition (3.1) with special homogeneous entropies, we prove that these cones are invariant regions for system (1.2).

Theorem 3.5. *Suppose that $u_0(x) \in \Gamma(e, \delta)$ ($u_0(x) \in \dot{\Gamma}(e, \delta)$) for a.e. $x \in \mathbb{R}$. If $u = u(t, x)$ is a g.e.s. of (1.2), (1.3) then $u(t, x) \in \Gamma(e, \delta)$ ($u(t, x) \in \dot{\Gamma}(e, \delta)$ respectively) a.e. on Π.*

4 Strong generalized entropy solutions

In spite of the fact that condition (3.1) is rather restrictive problem (1.2), (1.3) can have more than one g.e.s. (see Example 4.8 below). In this connection we define a notion of a strong g.e.s.

Definition 4.1. A function $u = u(t,x) \in L^\infty(\Pi, \mathbb{R}^n)$ is called a strong g.e.s. of problem (1.2), (1.3) if the function $r = |u(t,x)|$ is a g.e.s. of scalar problem (3.4) with the initial value $r_0 = |u_0|$, $u_t + (\varphi(r)u)_x = 0$ in $\mathcal{D}'(\Pi, \mathbb{R}^n)$, and initial condition (3.2) is satisfied.

One can prove that smooth solutions of (1.2), (1.3) are strong g.e.s. as well. If $u(t,x)$ is a strong g.e.s. of problem (1.2), (1.3) then the function $r = |u(t,x)|$ is uniquely defined as a g.e.s. of scalar problem (3.4) (see [9] and also [10] for the case of only continuous flux) and the functions $v_i = u_i/r$, $i = 1, \ldots, n$ (if $r = 0$ we can choose values of these functions arbitrarily) must satisfy in $\mathcal{D}'(\Pi)$ the linear equation

$$(Av)_t + (Bv)_x = 0, \tag{4.1}$$

where $A = A(t,x) = r(t,x)$, $B = B(t,x) = f(r(t,x))$, and the initial condition

$$v(0,x) = v_0(x), \tag{4.2}$$

with the corresponding data $v_0(x) = u_{0i}/r_0$. We assume that the coefficients $A, B \in L^\infty(\Pi)$ satisfy the conditions

$$\forall \varepsilon > 0 \ |B| \le N(\varepsilon) \cdot (A + \varepsilon) \ \text{ a.e. on } \Pi, \ \varepsilon N(\varepsilon) \underset{\varepsilon \to 0+}{\to} 0; \tag{4.3}$$

$$A_t + B_x = 0 \ \text{ in } \mathcal{D}'(\Pi). \tag{4.4}$$

As follows from (4.4), the map $t \to A(t,\cdot)$, $t \ge 0$, is weakly-$*$ continuous in $L^\infty(\mathbb{R})$ and, in particular, there exists a weak trace $A(0,x) \in L^\infty(\mathbb{R})$. We will assume that $v_0(x)$ is merely measurable function.

Definition 4.2. A measurable function $v = v(t,x) \in L^\infty(\Pi)$ is called a generalized solution (g.s. for short) of (4.1), (4.2) if $A(0,x)v_0(x) \in L^1_{loc}(\mathbb{R})$, $Av, Bv \in L^1_{loc}(\bar\Pi)$, where $\bar\Pi = [0, +\infty) \times \mathbb{R}$, and

$$\int_\Pi [Avh_t + Bvh_x] dtdx + \int_{\mathbb{R}} A(0,x)v_0(x)h(0,x)dx = 0$$

for any test function $h = h(t,x) \in C_0^\infty(\bar\Pi)$;

A measurable function $v = v(t,x)$ is called a renormalized solution (r.s.) of (4.1), (4.2) if for each bounded function $p(u) \in C(\mathbb{R})$ the function $p(v(t,x))$ is a g.s. of this problem with initial data $p(v_0(x))$.

Proposition 4.3. (i) *There exists a r.s.* $v(t,x)$ *of (4.1), (4.2);*

(ii) *if* $v_0(x)A(0,x) = 0$ *a.e. in* $\mathbb{R}$ *then* $v(t,x)A(t,x) = 0$ *a.e. in* Π *(uniqueness);*

(iii) *if* $v_i = v_i(t,x)$, $i = 1,\dots,m$, *are r.s. of (4.1), (4.2) with initial functions* $v_{0i}(x)$ *then for any Borel function* $p(u)$ *on* $\mathbb{R}^m$ *the function* $p(v_1,\dots,v_m)$ *is a r.s. of (4.1), (4.2) with the initial function* $p(v_{01}(x),\dots,v_{0m}(x))$*;*

(iv) *any g.s.* $v(t,x)$ *of (4.1), (4.2) is also a r.s. of this problem. Conversely, if* $v = v(t,x)$ *is a r.s. of (4.1), (4.2), and* $A(0,x)v_0(x) \in L^1_{loc}(\mathbb{R})$, $Av, Bv \in L^1_{loc}(\bar{\Pi})$ *then* $v(t,x)$ *is a g.s. of (4.1), (4.2);*

(v) *assume that* $v_0(x) \in L^\infty(\mathbb{R}, A(0,x)dx)$ *and* v *is a r.s. of (4.1), (4.2). Then* $v \in L^\infty(\Pi, Adtdx)$, *and* $\|v\|_\infty \le \|v_0\|_\infty$*;*

(vi) *if* $v = v(t,x)$ *is a g.s. of (4.1), (4.2) then*

$$\operatorname*{ess\,lim}_{t\to 0+} A(t,\cdot)|v(t,\cdot) - v_0(x)| = 0 \quad in\ L^1_{loc}(\mathbb{R}).$$

In particular, if $A(0,x)$ *is a strong trace of* A, *i.e.* $\operatorname*{ess\,lim}_{t\to 0+} A(t,\cdot) = A(0,x)$ *in* $L^1_{loc}(\mathbb{R})$, *then* $A(0,x)v_0(x)$ *is a strong trace of* Av.

It follows from assertion (iv),(v) that for a bounded initial function there always exists a g.s. of problem (4.1), (4.2). For unbounded v_0 this is not true (even if $A(0,x)v_0(x)$ is bounded), see [15].

Remark 4.4. For a multidimensional transport equation

$$v_t + (b, \nabla_x v) = 0, \quad b = b(t,x) \in L^\infty(\mathbb{R}^+ \times \mathbb{R}^n, \mathbb{R}^n)$$

the Cauchy problem is generally ill-posed, a g.s. can be nonunique, see examples in [4, 5, 15]. For such the equations the additional regularity assumptions are necessary for the well-posedness, cf. [6, 1]. As was firstly mentioned in [19], in the case of one space variable, no regularity assumptions are needed. Problem (4.1), (4.2) was systematically investigated in papers [12, 13, 14, 15], where the theory of generalized and renormalized solutions was developed (in [15] the case of transport equations with source functions was also included) and, in particular, the statements of Proposition 4.3 were established.

Using Proposition 4.3, we now can prove the existence and the uniqueness of strong g.e.s. to problem (1.2), (1.3).

Theorem 4.5. *Let* $A = r$, $B = f(r)$; $v_i = v_i(t,x)$ *be a g.s. of problem (4.1), (4.2) with the initial data* $v_{0i} = u_{0i}/r_0$, $i = 1,\dots,n$. *Then the vector* $u = rv$, $v = (v_1,\dots,v_n)$ *is the unique strong g.e.s. of problem (1.2), (1.3) and it is also a g.e.s. of this problem. Besides for any homogeneous function* $h(u)$

$$h(u)_t + (\varphi(|u|)h(u))_x = 0 \quad in\ \mathcal{D}'(\Pi). \tag{4.5}$$

Proof. By Proposition 4.3,(iii),(iv) the function $|v|$ is a r.s. of (4.1), (4.2) with initial data $|v_0| = 1$. By assertion (ii) of this Proposition $Av = r|v| = r$. Hence, $|u| = r$ is a g.e.s. of scalar problem (3.4). In view of Proposition 4.3,(vi) initial requirement (3.2) is also satisfied. Therefore, $u = rv$ is a strong g.e.s. of (1.2), (1.3). The uniqueness of a strong g.e.s. readily follows from the uniqueness of g.e.s. for problem (3.4) and Proposition 4.3,(ii). Equality (4.5) follows from the relations $h(u) = Ah(v)$, $\varphi(|u|)h(u) = Bh(v)$ and Proposition 4.3,(iii). For any entropy (2.4) the function $r = |u|$ satisfies the entropy relation $\eta(r)_t + \psi(r)_x \le 0$ in $\mathcal{D}'(\Pi)$, as a g.e.s. of scalar problem (3.4). This relation and (4.5) readily imply (3.1). Hence, u is a g.e.s. of (1.2), (1.3). □

Remark 4.6. As was shown in [15], Theorem 4.5 remains valid also for inhomogeneous systems of Keyfitz-Kranzer type $u_t+(\varphi(|u|)u)_x = S(t,x)u$, where $S(t,x)$ is a bounded skew-symmetric matrix.

Under the following condition

$$\text{for some unit vector } e \in \mathbb{R}^n \ u_0(x) \in \dot{\Gamma}(e,0) \text{ a.e. on } \mathbb{R} \tag{4.6}$$

there is the uniqueness even in the class of g.e.s.

Theorem 4.7. *Under the condition (4.6) a g.e.s. $u(t,x)$ of problem (1.2), (1.3) is strong and, therefore, is unique.*

Proof. Without loss of generality we may assume that $e = e_1 = (1,0,\ldots,0)$. By Theorem 3.5 we have $u(t,x) \in \dot{\Gamma}(e,0)$ a.e. on Π. This means that $u_1(t,x) > 0$ a.e. on the set where $r = |u(t,x)| > 0$. Let $A = u_1(t,x)$, $B = \varphi(r)u_1(t,x)$. Then the pair (A,B) satisfies (4.3), (4.4). Clearly, $v_i = u_i(t,x)/u_1(t,x)$ (we may set $v_i = 0$ on the set where $u_1 = 0$) are g.s. to problem (4.1), (4.2) with the corresponding initial data. By Proposition 4.3,(iii) $|v| = r/u_1$ is a g.s. as well. Hence, $r_t + f(r)_x = (A|v|)_t + (B|v|)_x = 0$ in $\mathcal{D}'(\Pi)$. This, together with the fact that r is a g.e.sub-s. of (3.4) implies that actually $r = |u|$ is a g.e.s. of (3.4), i.e. u is a strong g.e.s. of (1.2), (1.3). □

Emphasize that the cone $\dot{\Gamma}(e,0)$ in condition (4.6) cannot be replaced by $\Gamma(e,0)$. We confirm this by the following example.

Example 4.8. Assume that $n = 2$, $\varphi(r) = r$. Then system (1.2) has the form

$$u_t + (ru)_x = 0, \ u = (u_1,u_2), \ r = |u| = (u_1^2 + u_2^2)^{1/2}. \tag{4.7}$$

We set the following initial condition

$$u(0,x) = u_0(x) = (0, \operatorname{sign} x). \tag{4.8}$$

Remark that $u_0 \in \Gamma(e,0)$ with $e = (1,0)$. Here $r_0 = |u_0| \equiv 1$ and if $u = u(t,x)$ is a strong g.e.s. of (4.7), (4.8) then $r = |u| \equiv 1$ (as a g.e.s.

to the corresponding scalar problem with the initial data $r_0 \equiv 1$) and $u_t + u_x = 0$ in $\mathcal{D}'(\Pi, \mathbb{R}^2)$. Therefore

$$u(t,x) = u_0(x-t) = (0, \operatorname{sign}(x-t))$$

is the unique strong g.e.s. of problem (4.7), (4.8).

Let $v = v(t,x)$ be a g.e.s. to the scalar Cauchy problem $v_t + (|v|v)_x = 0$ with initial function $v_0(x) = \operatorname{sign} x$. It is easy to verify that

$$v(t,x) = \begin{cases} -1, & x < 2(\sqrt{2}-1)t, \\ x/(2t), & 2(\sqrt{2}-1)t \le x \le 2t, \\ 1, & x > 2t, \end{cases}$$

and $v(t,x)$ is a limit in $L^1_{loc}(\Pi)$ of classical solutions $v^\varepsilon(t,x) \in C^2(\Pi)$ to the problem $v_t + (|v|v)_x = \varepsilon v_{xx}$ as "viscosity" $\varepsilon \to 0+$. Denote $\bar{u}(t,x) = (0, v(t,x))$, $u^\varepsilon(t,x) = (0, v^\varepsilon(t,x))$. Then the vector $u^\varepsilon \in C^2$ is a solution to parabolic system (3.3) and $u^\varepsilon(t,x) \to \bar{u}(t,x)$ in $L^1_{loc}(\Pi, \mathbb{R}^2)$ as $\varepsilon \to 0+$.

By Proposition 3.2 the limit function $\bar{u}(t,x)$ satisfies entropy condition (3.1) for all convex entropies $p(u) \in \mathcal{F}$ and evidently it satisfies initial condition (3.2) as well. So $\bar{u}(t,x)$ is a g.e.s. to problem (4.7), (4.8), which is different from $u(t,x)$.

Let $u = u(t,x)$ be a strong g.e.s. of problem (1.2), (1.3), $v = h(u) \in \mathbb{R}^m$ be a homogeneous map of degree one, such that $r = |u|$ is a function of new variables: $r = r(v)$. Then by equality (4.5) the vector $v = h(u(t,x))$ satisfies the system

$$v_t + (\varphi(r)v)_x = 0, \ \ r = r(v) \tag{4.9}$$

in $\mathcal{D}'(\Pi, \mathbb{R}^m)$, and the initial condition

$$v(0,x) = v_0(x) = h(u_0(x)). \tag{4.10}$$

One can define a strong g.e.s. of problem (4.9), (4.10) in the same way as for problem (1.2), (1.3) and prove its existence and uniqueness. We conclude that for strong g.e.s. $u = u(t,x)$ of (1.2), (1.3) the vector-function $v = h(u)$ is a strong g.e.s. of problem (4.9), (4.10). Consider some particular cases.

1) Let $\|\cdot\|$ be a norm in $\mathbb{R}^n$, $v = h(u) = |u| \cdot u/\|u\|$ (for $u = 0$ we set $v = 0$). Then in (4.9) $r = \|v\|$ and this is system (1.2) but with arbitrary norm instead of $|\cdot|$. As was proved in [17], the main results remain valid even for the infinite-dimensional Keyfitz-Kranzer systems $u_t + (\varphi(\|u\|)u)_x = 0$ when the unknown function $u = u(t,x)$ takes its values in an arbitrary Banach space $(X, \|\cdot\|)$. For example, if $X = L^p(\mathbb{R})$, $p \ge 1$ then the Keyfitz-Kranzer system has the form of the integro-differential equation

$$\frac{\partial}{\partial t} u(t,x,\lambda) + \frac{\partial}{\partial x}\left[h\left(\int |u(t,x,\mu)|^p d\mu\right) u(t,x,\lambda)\right] = 0.$$

2) Let $v = h(u) = (u_1^2/r, \dots, u_n^2/r)$ ($h(u) = 0$ for $r = 0$), $\varphi(r) = 1/(1+r)$. Then $r = v_1 + \cdots + v_n$ and system (4.9) coincides the following system of multicomponent chromatography

$$\begin{gathered} v_t + (v/(1+r))_x = 0, \\ r = v_1 + \cdots + v_n,\ v_i \geq 0,\ i = 1, \dots, n. \end{gathered} \tag{4.11}$$

Observe that the map $h(u)$ may be restricted on the set $\{\, u \in \mathbb{R}^n \mid u_i \geq 0,\ i = 1, \dots, n \,\}$, which is invariant domain for system (1.2) (by Theorem 3.5), and it is contained in some cone $\dot{\Gamma}(e, 0)$. Hence, it follows from Theorem 4.7 that the Cauchy problem for (4.11) has a unique g.e.s. for any non-negative initial data $v_{0i} \in L^\infty(\mathbb{R})$, $i = 1, \dots, n$.

5 Justification of "vanishing viscosity" method

Suppose that the function $f(u)$ is not linear on non-degenerate intervals and the initial data $u_0(x)$ satisfies condition (4.6). We consider the Cauchy problem for parabolic equation (3.3) with the initial condition

$$u(0, x) = u_{0,\varepsilon}(x), \tag{5.1}$$

where the functions $u_{0,\varepsilon}(x)$ are uniformly bounded in $L^\infty(\mathbb{R}, \mathbb{R}^n)$ and $u_{0,\varepsilon} \to u_0$ in $L^1_{loc}(\mathbb{R}, \mathbb{R}^n)$ as $\varepsilon \to 0+$.

Theorem 5.1. *Let $u_\varepsilon(t, x) \in C^2(\Pi, \mathbb{R}^n)$ be a solution of problem (3.3), (5.1). Then $u_\varepsilon(t, x) \to u(t, x)$ in $L^1_{loc}(\Pi, \mathbb{R}^n)$ as $\varepsilon \to 0$ where $u(t, x)$ is the unique g.e.s. of problem (1.2), (1.3).*

For the proof of Theorem 5.1 we firstly establish the following maximum principle.

Proposition 5.2. *Let $\varepsilon > 0$, $u = u(t, x) \in C^2(\Pi, \mathbb{R}^n) \cap L^\infty(\Pi, \mathbb{R}^n)$ be the solution of the Cauchy problem for equation (3.3) with an initial vector $u_0(x) \in L^\infty(\mathbb{R}, \mathbb{R}^n)$. Then for all $(t, x) \in \Pi$ $|u(t, x)| \leq M = \operatorname*{ess\,sup}_{x \in \mathbb{R}} |u_0(x)|$.*

Proposition 5.2 enables us to pass to the limit along a subsequence of solutions $u_k = u_\varepsilon(t, x)$, $\varepsilon = \varepsilon_k \to 0$ and obtain a bounded *measure valued solution* $\nu_{t,x}(u)$ to problem (1.2), (1.3). This solution takes its values in the space of Borel probability measures with compact support in $\mathbb{R}^n$. To prove Theorem 5.1 it is sufficient to establish that $\nu_{t,x}(u)$ are Dirac measures. Using compensated compactness theory we derive the

Tartar-Murat commutation relations: for almost all $(t,x) \in \Pi$

$$\int_{\mathbb{R}^n} [p_1(u)q_2(u) - p_2(u)q_1(u)]d\nu_{t,x}(u) = \int_{\mathbb{R}^n} p_1(u)d\nu_{t,x}(u) \int_{\mathbb{R}^n} q_2(u)d\nu_{t,x}(u) - \int_{\mathbb{R}^n} p_2(u)d\nu_{t,x}(u) \int_{\mathbb{R}^n} q_1(u)d\nu_{t,x}(u) \tag{5.2}$$

for any pair of convex entropies p_1, p_2 and the corresponding fluxes q_1, q_2. Taking in (5.2) $p_i(u) = \varphi_i(r)$, where $r = |u|$, $i = 1, 2$ and arguing as in [20], we derive that $r = r(t,x)$ on $\operatorname{supp} \nu_{t,x}$, that is, the measures $\nu_{t,x}(u)$ are supported on spheres in $\mathbb{R}^n$.

Further, condition (4.6) and the measure-valued analogue of Theorem 3.5 yield the inclusion $\operatorname{supp} \nu_{t,x}(u) \in \dot{\Gamma}(e,0)$ for a.e. $(t,x) \in \Pi$. We may assume, without loss of generality, that $e = e_1$. Then this inclusion means that for a.e. (t,x) either $\bar{u}_1 = \int u_1 d\nu_{t,x}(u) > 0$ or $r = r(t,x) = 0$. Denote $A = \bar{u}_1$, $B = \varphi(r)\bar{u}_1$. Then these coefficients satisfies assumptions (4.3), (4.4). Further, $v_i = \bar{u}_i/\bar{u}_1$, where $\bar{u}_i = \int u_i d\nu_{t,x}(u)$ are g.s. to problem (4.1), (4.2) and, by renormalization property (iii) in Proposition 4.3, $|v| = |\bar{u}|/\bar{u}_1$ is a g.s. of this problem with initial data r_0/u_{01}. This means that $|\bar{u}|_t + (\varphi(r)|\bar{u}|)_x = 0$. In view of entropy inequality with the entropy $p(u) = r$ we have $r_t + (\varphi(r)r)_x \le 0$. Hence,

$$(r - |\bar{u}|)_t + (\varphi(r)(r - |\bar{u}|))_x \le 0. \tag{5.3}$$

Since $r - |\bar{u}| = \int p(u)d\nu_{t,x}(u) - p\left(\int u d\nu_{t,x}(u)\right) \ge 0$ and $(r - |\bar{u}|)|_{t=0} = 0$ one can easily derive from (5.3) that $r - |\bar{u}| = 0$ a.e. on Π_T or $|\bar{u}|^2 = r^2$. This implies that $\int |u - \bar{u}|^2 d\nu_{t,x}(u) = 0$. Therefore, $\nu_{t,x}(u) = \delta(u - \bar{u}(t,x))$, and $\bar{u}(t,x)$ is a g.e.s. of (1.2), (1.3). We conclude that $u_k(t,x) \to \bar{u}(t,x)$ in $L^1_{loc}(\Pi, \mathbb{R}^n)$. By Theorem 4.7, the solution $\bar{u}(t,x)$ is unique and therefore does not depend on the choice of the subsequence u_k. Hence the original family $u_\varepsilon(t,x)$ converges to $\bar{u}(t,x)$, which completes the proof of Theorem 5.1.

Remark 5.3. The described above theory of the problem (1.2), (1.3) is essentially one-dimensional. To treat multidimensional Keyfitz-Kranzer systems, additional regularity assumptions are necessary, see [2, 7].

6 Hyperbolic conservation laws on matrix algebras

Let X be the space H_n of Hermitian or the space S_n of symmetric matrices of order n. We consider system (1.1)

$$U_t + \left(\tilde{f}(U)\right)_x = 0, \tag{6.1}$$

where $U = U(t,x) \in X$, $(t,x) \in \Pi = (0,+\infty) \times \mathbb{R}$, $\tilde{f} : X \mapsto X$, $U \mapsto \tilde{f}(U)$ being the functional calculus operator corresponding to some real function $f(u) \in C^1(\mathbb{R})$.

The operator $A(U) = d\tilde{f}(U)$ can be written explicitly and it is a symmetric operator on X with respect to the scalar multiplication $(U,V) = \operatorname{Tr} UV$. The eigenvalues of $A(U)$ are given by the formula

$$\mu_{kl} = \mu_{kl}(U) = \begin{cases} \dfrac{f(\lambda_k) - f(\lambda_l)}{\lambda_k - \lambda_l}, & \lambda_k \neq \lambda_l, \\ f'(\lambda_k), & \lambda_k = \lambda_l. \end{cases}$$

Now we describe entropy pairs of system (6.1).

Theorem 6.1. *Assume that $\varphi(u) \in C^1(\mathbb{R})$. Then the function $p(U) = \operatorname{Tr}\tilde{\varphi}(U)$ is an entropy of system (6.1) with the flux $q(U) = \operatorname{Tr}\tilde{\psi}(U)$, where the function $\psi(u) \in C^1(\mathbb{R})$ is defined up to an additive constant by the identity $\psi'(u) = \varphi'(u)f'(u)$, that is, $\psi(u)$ is the entropy flux corresponding to the entropy $\varphi(u)$ for the scalar equation*

$$u_t + f(u)_x = 0. \tag{6.2}$$

If the function $f(u)$ is not linear on nondegenerate intervals then any entropy of (6.1) has the indicated form, up to a linear term.

Besides, the entropy $p(U) = \operatorname{Tr}\tilde{\varphi}(U)$ is (strictly) convex on X if and only if the function $\varphi(u)$ is (strictly) convex on $\mathbb{R}$.

Remark 6.2. Weakening the regularity requirement, we can extend the class of entropies of (6.1). For instance, the functions

$$p(U) = \sum_{i=1}^{n} \varphi_i(\lambda_i(U)),\ q(U) = \sum_{i=1}^{n} \psi_i(\lambda_i(U))$$

are singular entropy pairs for each functions $\varphi_i(u), \psi_i(u) \in C^1(\mathbb{R})$ such that $\psi_i'(u) = \varphi_i'(u)f'(u)$, $i = 1,\dots,n$. Here $\lambda_1(U) \le \lambda_2(U) \le \cdots \le \lambda_n(U)$ are eigenvalues of U. The entropy $p(U)$ is convex if and only if all functions $\varphi_i(u)$ are convex, and $\varphi_1'(u) \le \varphi_2'(u) \le \cdots \le \varphi_n'(u)$. The function $p(U) \in C^1$ only in the case when the function $\varphi_i(u)$ coincide: $\varphi_i(u) = \varphi(u)$ for all $i = 1,\dots,n$ (then $p(U) = \operatorname{Tr}\tilde{\varphi}(U)$). The family of singular entropies was introduced and completely described in [13] in the case $n = 2$.

Now we consider the Cauchy problem for (6.1) with the initial condition

$$U(0,x) = U_0(x) \in L^\infty(\mathbb{R}, X). \tag{6.3}$$

Definition 6.3. An X-valued function $U = U(t,x) \in L^\infty(\Pi, X)$ is called a g.e.s. of (6.1), (6.3) if for any convex entropy $p(U) \in C^1(X)$

$$p(U)_t + q(U)_x \le 0 \quad \text{in } \mathcal{D}'(\Pi, X), \tag{6.4}$$

and $\operatorname*{ess\,lim}_{t\to 0} U(t,\cdot) = U_0$ in $L^1_{loc}(\mathbb{R}, X)$.

Let $U(t,x)$ be a g.e.s. of (6.1), (6.3), and $\lambda_1(t,x) \le \cdots \le \lambda_n(t,x)$, $\lambda_1^0(x) \le \cdots \le \lambda_n^0(x)$ be eigenvalues of operators $U(t,x)$ and $U_0(x)$, respectively. Then in view of (6.4)

$$\frac{\partial}{\partial t}\sum_{i=1}^n \varphi(\lambda_i(t,x)) + \frac{\partial}{\partial x}\sum_{i=1}^n \psi(\lambda_i(t,x)) \le 0 \quad \text{in } \mathcal{D}'(\Pi). \tag{6.5}$$

for each entropy pair $(\varphi(u), \psi(u))$ for the scalar equation (6.2). Let us define the measure valued functions

$$\nu_x^0(\lambda) = \frac{1}{n}\sum_{i=1}^n \delta(\lambda - \lambda_i^0(x)), \quad \nu_{t,x}(\lambda) = \frac{1}{n}\sum_{i=1}^n \delta(\lambda - \lambda_i(t,x)). \tag{6.6}$$

By (6.5) $\nu_{t,x}$ is a measure valued solution of (6.2) with measure valued initial data

$$\nu_{0,x} = \nu_x^0. \tag{6.7}$$

Using results of the theory of measure valued solutions to problem (6.2), (6.7) developed in [11], we establish the following properties.

Theorem 6.4. *Let $U(t,x)$ be g.e.s. of (6.1), (6.3).* (i) *If $U_0(x) = u_0(x)\cdot E$, $u_0(x) \in L^\infty(\mathbb{R})$, then $U(t,x) = u(t,x)\cdot E$, where $u(t,x)$ – is a unique g.e.s. of the Cauchy problem for (6.2) with initial function $u_0(x)$.* (ii) *Let $|\cdot|$ be the operator norm on X. Then*

$$|U(t,x)| \le M = \operatorname{ess\,sup} |U_0(x)| \quad \textit{for a.e. } (t,x) \in \Pi.$$

(iii) *For a.e. $t > 0$*

$$\int_{\mathbb{R}} \|U(t,x)\|_p^p dx \le \int_{\mathbb{R}} \|U_0(x)\|_p^p dx, \quad \|U\|_p^p \doteq \operatorname{Tr}|U|^p, \ p \ge 1.$$

Example 6.5 (nonuniqueness of g.e.s.). Assume that $f(u) = u^2$, $X = S_2$. We consider the Riemann problem

$$U_t + (U^2)_x = 0, \ U(0,x) = U_0(x), \tag{6.8}$$

where $U_0(x) = \begin{cases} U_+, & x \ge 0, \\ U_-, & x < 0, \end{cases} \quad U_+ = \begin{pmatrix} 1 & 0 \\ 0 & 0 \end{pmatrix}, \quad U_- = \begin{pmatrix} 0 & 0 \\ 0 & 1 \end{pmatrix}.$

Let $a(t,x)$, $b(t,x)$ be g.e.s. for the equation $u_t + (u^2)_x = 0$ with initial functions $a_0(x) = \begin{cases} 1, & x \ge 0, \\ 0, & x < 0, \end{cases}$ $b_0(x) = \begin{cases} 0, & x \ge 0, \\ 1, & x < 0, \end{cases}$ respectively. One can easily compute that

$$a(t,x) = \begin{cases} 0, & x < 0, \\ x/(2t), & 0 \le x < 2t, \\ 1, & x \ge 2t, \end{cases} \qquad b(t,x) = \begin{cases} 1, & x < t, \\ 0, & x \ge t, \end{cases}$$

and the X-valued function

$$U_1(t,x) = \begin{pmatrix} a(t,x) & 0 \\ 0 & b(t,x) \end{pmatrix}$$

is a g.e.s. of (6.8). From the other hand, one can easily verify that the X-valued function

$$U_2(t,x) = \begin{cases} U_+, & x \ge t, \\ U_-, & x < t, \end{cases}$$

is a g.e.s. of (6.8) as well (it is exactly the strong g.e.s., which will be defined soon). Thus, $U_1(t,x)$, $U_2(t,x)$ – are two different g.e.s. of (6.8).

In view of nonuniqueness of g.e.s. we introduce the class of strong g.e.s. of the problem (6.1), (6.3). Let $U \in X$ and $\lambda_i = \lambda_i(U)$, $i = 1, \dots, n$ be eigenvalues of U written in the increasing order: $\lambda_1 \le \lambda_2 \le \dots \le \lambda_n$.

Definition 6.6. An X-valued function $U = U(t,x) \in L^\infty(\Pi, X)$ is called a *strong g.e.s.* to the Cauchy problem (6.1), (6.3) if (1) U satisfies (6.1) in $\mathcal{D}'(\Pi, X)$; (2) the eigenvalues $\lambda_i(t,x) = \lambda_i(U(t,x))$ of $U(t,x)$ are g.e.s. of the scalar problem

$$u_t + (f(u))_x = 0, \quad u(0,x) = u_0(x) \tag{6.9}$$

with $u_0 = \lambda_i^0(x) \doteq \lambda_i(U_0(x))$, $i = 1, \dots, n$; (3) initial condition (6.3) holds.

By the definition, the eigenvalues $\lambda_i = \lambda_i(t,x)$ of a strong g.e.s. $U = U(t,x)$ are uniquely defined as g.e.s. of (6.9). Also observe that by the comparison principle the inequalities $\lambda_1 \le \lambda_2 \le \dots \le \lambda_n$ follows from the corresponding inequalities for the initial functions. Evidently, any strong g.e.s. of (6.1), (6.3) is a g.e.s. of this problem.

The case $n = 2$ was completely investigated in [12, 13], where the existence and uniqueness of strong g.e.s. were established, and vanishing viscosity method was justified. In recent paper [18] the main results was extended to the case when

(C) There are only two different eigenvalues among $\lambda_i^0(x)$, $i = 1, \ldots, n$, i.e., for all $i = 1, \ldots, n$ either $\lambda_i^0(x) = \lambda_1^0(x)$ or $\lambda_i^0(x) = \lambda_n^0(x)$ a.e. on $\mathbb{R}$.

Notice that condition (C) is always satisfied for $n = 2$ and also for the Keyfitz-Kranzer systems (when $U_0 = u_0 \otimes u_0$). Under condition (C) the eigenvalues of strong g.e.s. $U(t,x)$ satisfy the analogous property: for all $i = 1, \ldots, n$ either $\lambda_i(t,x) = \lambda_1(t,x)$ or $\lambda_i(t,x) = \lambda_n(t,x)$, where $\lambda_1(t,x)$, $\lambda_n(t,x)$ – are unique g.e.s. of (6.9) with the initial data $\lambda_1^0(x)$, $\lambda_n^0(x)$, respectively.

Theorem 6.7. *Under condition* (C) *there exists a unique strong g.e.s. of (6.1), (6.3).*

Moreover, as was demonstrated in [18], the strong g.e.s. of (6.1), (6.3) admits the representation $U = \lambda_1 E + (\lambda_n - \lambda_1)V$, where $V = V(t,x)$ is an X-valued g.s. of transport equation (4.1) with coefficients $A = \lambda_n(t,x) - \lambda_1(t,x)$, $B = f(\lambda_n(t,x)) - f(\lambda_1(t,x))$, and initial data $V_0 = V_0(x)$ defined from the representation $U_0 = \lambda_1^0 E + (\lambda_n^0 - \lambda_1^0)V_0$. Remark also that Theorem 6.7 yields the existence and the uniqueness of strong g.e.s. to problem (1.2), (1.3) already established in Theorem 4.5.

Let us show that condition (C) is essential for the uniqueness of strong g.e.s.

Example 6.8 (non-uniqueness of strong g.e.s.). Let us consider the Riemann problem for the system

$$U_t + (U^2)_x = 0, \quad U = U(t,x) \in S_3, \tag{6.10}$$

with the initial matrix

$$U(0,x) = U_0(x) = \begin{cases} \operatorname{diag}\{1,0,0\}, & x < 0, \\ \operatorname{diag}\{0,1,1\}, & x > 0. \end{cases} \tag{6.11}$$

The eigenvalues of $U_0(x)$ take only two values 0 and 1. But the functions

$$\lambda_1^0(x) \equiv 0, \quad \lambda_2^0(x) = \begin{cases} 0, & x < 0, \\ 1, & x > 0, \end{cases} \quad \lambda_3^0(x) \equiv 1$$

are distinct. Therefore, condition (C) fails. One can easily compute the corresponding g.e.s. $\lambda_i(t,x)$, $i = 1,2,3$, of the equation $u_t + (u^2)_x = 0$ with initial functions $\lambda_i^0(x)$:

$$\lambda_1(t,x) \equiv 0, \ \lambda_2(t,x) = \begin{cases} 0, & x < 0, \\ x/(2t), & 0 \le x < 2t, \\ 1, & x \ge 2t, \end{cases} \qquad \lambda_3(t,x) \equiv 1.$$

We define the X-valued functions

$$U_1(t,x) = \begin{cases} \mathrm{diag}\{1,0,0\}, & x<0, \\ \mathrm{diag}\{1,x/(2t),0\}, & 0 \le x < t, \\ \mathrm{diag}\{0,x/(2t),1\}, & t \le x < 2t, \\ \mathrm{diag}\{0,1,1\}, & x \ge 2t, \end{cases}$$

$$U_2(t,x) = \begin{cases} \mathrm{diag}\{1,0,0\}, & x<0, \\ \mathrm{diag}\{1,0,x/(2t)\}, & 0 \le x < t, \\ \mathrm{diag}\{0,1,x/(2t)\}, & t \le x < 2t, \\ \mathrm{diag}\{0,1,1\}, & x \ge 2t. \end{cases}$$

Then $\lambda_i(U_1(t,x)) = \lambda_i(U_2(t,x)) = \lambda_i(t,x)$, $i = 1,2,3$, i.e., $U_1(t,x)$, $U_2(t,x)$ satisfy condition (2) of Definition 6.6. Since both the functions $U = U_1(t,x), U_2(t,x)$ satisfy the Rankine-Hugoniot relation $U^+ - U^- = (U^+)^2 - (U^-)^2$ on the discontinuity line $x = t$, they are g.s. of (6.1) in $\mathcal{D}'(\Pi, X)$. Finally, initial condition (3) is evidently satisfied. Hence, $U_1(t,x)$, $U_2(t,x)$ are strong g.e.s. of (6.10), (6.11).

References

[1] L. Ambrosio, Transport equation and Cauchy problem for BV vector fields. Invent. Math. 158 (2004), 227–260.

[2] L. Ambrosio, F. Bouchut and C. De Lellis, Well-posedness for a class of hyperbolic systems of conservation laws in several space dimensions. Comm. Partial Differential Equations 29 (2004), 1635–1651.

[3] Ph. Benilan and S. N. Kruzhkov, Conservation laws with continuous flux functions. NoDEA. 3 (1996), 395–419.

[4] A. Bressan, An ill-posed Cauchy problem for a hyperbolic system in two space dimensions. Rend. Sem. Mat. Univ. Padova. 110 (2003), 103–117.

[5] F. Colombini, T. Luo and J. Rauch, Uniqueness and nonuniqueness for nonsmooth divergence free transport. Sémin. Équ. Dériv. Partielles, Éc. Polytech., Cent. Math., Palaiseau. 2002–2003, Exp. No. XXII (2003), 21 p.

[6] R. J. DiPerna and P. L. Lions, Ordinary differential equations, transport theory and Sobolev spaces. Invent. Math. 98 (1989), 511–547.

[7] G. Crippa and L. V. Spinolo, An overview on some results concerning the transport equation and its applications to conservation laws. Comm. on Pure and Applied Analysis 9 (2010), no. 5, 1283–1293.

[8] B. L. Keyfitz and H. C. Kranzer, A system of nonstrictly hyperbolic conservation laws arising in elasticity theory. Arch. Rational Mech. Anal. 72 (1980), 219–241.

[9] S. N. Kruzhkov, First order quasilinear equations in several independent variables. Math. USSR Sbornik. 10 (1970), no. 2, 217–243.

[10] S. N. Kruzhkov and E. Yu. Panov, Osgood's type conditions for uniqueness of entropy solutions to Cauchy problem for quasilinear conservation laws of the first order. Annali Univ. Ferrara-Sez. VII-Sc. Mat. 40 (1994), 31–53.

[11] E. Yu. Panov, On measure valued solutions of the Cauchy problem for a first order quasilinear equation. Izvestiya: Mathematics. 60 (1996), no. 2, 335–377.

[12] E. Yu. Panov, A class of systems of quasilinear conservation laws. Sbornik: Mathematics. 188 (1997), no 5, 725–751.

[13] E. Yu. Panov, On nonlocal theory of generalized entropy solutions to Cauchy problem for some hyperbolic systems of conservation laws. Izvestia: Mathematics. 63 (1999), no. 1, 129–179.

[14] E. Yu. Panov, On the theory of generalized entropy solutions of the Cauchy problem for a class of nonstrictly hyperbolic systems of conservation laws. Sbornik: Mathematics. 191 (2000), no. 1, 121–150.

[15] E. Yu. Panov, Generalized Solutions of the Cauchy Problem for a Transport Equation with Discontinuous Coefficients. In the book "Instability in Models Connected with Fluid Flows. II", Claude Bardos & Andrey Fursikov ed., International Mathematical Series, Vol. 7, Springer, 2008, 23–84.

[16] E. Yu. Panov, On symmetrizability of hyperbolic matrix spaces. St. Petersburg Math. J. 20 (2009), 465–471.

[17] E. Yu. Panov, On infinite-dimensional Keyfitz-Kranzer systems of conservation laws. Differential Equations. 45 (2009), no. 2, 274–278.

[18] E. Yu. Panov, On a class of systems of quasilinear conservation laws. J. of Mathematical Sciences. 156 (2009), no. 4, 644–668.

[19] D. Serre, Les ondes planes en electromagnetisme non linéaire. Physica D, 31 (1988), 227–251.

[20] L. Tartar, Compensated compactness and applications to partial differential equations. Nonlinear analysis and mechanics: Heriot. Watt Symposium, vol. 4 (Edinburgh 1979), Res. Notes Math. 39 (1979), 136–212.

Irrotational Flows for Chaplygin Gas: Conical Waves*

Denis Serre
UMPA, UMR CNRS–ENS Lyon # 5669
École Normale Supérieure de Lyon
46, allée d'Italie, F–69364 Lyon, cedex 07, France
E-mail: denis.serre@umpa.ens-lyon.fr

Abstract

The 3-dimensional Riemann problem is too complicated to be analyzed with the existing tools. However, some pieces can be treated rigorously because they are basically 2-D. We investigate the ternary interaction of planar shock waves surrounding a given state. A *conical wave* emerges from the triple point. We prove its existence and uniqueness for a gas obeying to the Chaplygin equation of state.

1 Steady and self-similar flows of a compressible gas

Let $d = 2$ or 3 be the space dimension. The Euler equations of an isentropic compressible fluid governs the mass density $\rho > 0$ and the velocity field u. The pressure p is a prescribed smooth function of ρ. We assume that $\partial p/\partial\rho > 0$, the square root $c(\rho)$ being the sound speed.

$$\partial_t\rho + \operatorname{div}(\rho u) = 0, \tag{1.1}$$

$$\partial_t(\rho u) + \operatorname{div}(\rho u \otimes u) + \nabla p(\rho) = 0. \tag{1.2}$$

Remark that real flows should be non-isentropic. The constancy of the entropy is not compatible once shock waves develop. It is only expected that the amplitude of oscillations for the entropy are of cubic order with

*Part of this work was done while the author was visiting Prof. Yaguang Wang in the Mathematics Department of Jiaotong University in Shanghai. With warm thanks.

respect to those of the flow. There is one notable exception, that of a Chaplygin gas, whose pressure has the form $p(\rho, s) = g(s) - f(s)/\rho$; then the entropy remains constant even beyond shock formation, if it was at initial time. This is proved in [6] for $d = 2$; see [7] for $d = 3$.

Irrotational flows. Assuming that the flow is C^2 and isentropic, the vorticity $\omega := \operatorname{curl} u$ is transported according to

$$(\partial_t + u \cdot \nabla)\frac{\omega}{\rho} = \left(\frac{\omega}{\rho} \cdot \nabla\right) u. \tag{1.3}$$

If the initial velocity is irrotational ($\omega(\cdot, 0) \equiv 0$), the flow remains irrotational. Once again, this conclusion is incorrect for realistic flows once shock waves appear, but it is expected to be a good approximation if the flow has a mild amplitude. But for a *Chaplygin gas*, steady or self-similar irrotational flows are perfectly valid solutions of the full Euler system [6, 7].

For an isentropic, irrotational flow, we introduce the velocity potential ψ by $u = \nabla\psi$. It is governed by the Bernoulli equation

$$\partial_t \psi + \frac{1}{2}|\nabla\psi|^2 + \imath(\rho) = 0, \tag{1.4}$$

where

$$\imath(\rho) := \int^{\rho} \frac{c^2(\mu)}{\mu}\, d\mu$$

is the enthalpy. The flow is determined by initial and/or boundary conditions for ρ and ψ, together with the PDEs (1.1,1.4), where the former can be written in the form

$$\partial_t \rho + \operatorname{div}(\rho \nabla\psi) = 0. \tag{1.5}$$

1.1 Steady flows

For an irrotational steady flow, $\partial_t \psi$ is a constant. We have therefore the Bernoulli equation

$$\frac{1}{2}|\nabla\psi|^2 + \imath(\rho) = \kappa, \tag{1.6}$$

with κ a constant. Since $\imath' = c^2/\rho$ is positive, this can be resolved in terms of the density:

$$\rho = h\left(\frac{1}{2}|\nabla\psi|^2 - \kappa\right).$$

Then the system of PDEs reduces to a single second-order equation

$$\operatorname{div}\left(h\left(\frac{1}{2}|\nabla\psi|^2 - \kappa\right)\nabla\psi\right) = 0. \tag{1.7}$$

Jump relations. Let the flow be discontinuous across a hypersurface with unit normal ν, moving at normal speed σ. The Rankine-Hugoniot condition for mass conservation writes

$$[\rho(u \cdot \nu - \sigma)] = 0. \tag{1.8}$$

Equation (1.8) allows us to define a net flow

$$j := \rho(u \cdot \nu - \sigma),$$

which has the same value on both sides.

The rest of the jump relations, as well as the nature of discontinuities, depends on the model under consideration. In the isentropic case, we have both acoustic waves (shocks) in which $j \neq 0$, and vortex sheets, along which $j = 0$. In the irrotational case, we only have acoustic waves. Across an isentropic shock, we have

$$j[u] + [p]\nu = 0, \tag{1.9}$$

which is equivalent to the combination of

$$[u_T] = 0 \tag{1.10}$$

and

$$j^2 \left[\frac{1}{\rho}\right] + [p] = 0. \tag{1.11}$$

For an irrotational shock, we only have (1.10), which amounts to writing the continuity of ψ.

1.2 Self-similar flows

The Euler system being a system of first-order conservation laws, it is invariant under space-time dilations. Therefore, dilation-invariant initial data (that is data that are positively homogeneous of degree zero) yield dilation-invariant solutions, if the Cauchy problem has a well-posedness property. It is thus interesting to look for solutions that depend only upon the ratio

$$y := \frac{x}{t}\,.$$

The Cauchy problem for a dilation-invariant data $U(\cdot, 0)$ is called the *Riemann problem.*

When acting on a function that depends only upon y, the time derivative is equivalent to the operator $-t^{-1}y \cdot \nabla_y$. The Euler equations takes the new form

$$-y \cdot \nabla_y \rho + \operatorname{div}_y(\rho u) = 0,$$

$$-(y \cdot \nabla_y)(\rho u) + \operatorname{div}_y(\rho u \otimes u) + \nabla p = 0,$$

$$-y \cdot \nabla_y E + \operatorname{div}_y((E + p)u) = 0,$$

This system looks more complicated than the original one, but can be written in a simpler form by employing the so-called *pseudo-velocity*

$$v(y) := u(y) - y,$$

where we notice that both u and y have the dimension of a velocity. We obtain

$$\mathrm{div}_y(\rho v) + d\rho = 0, \tag{1.12}$$

$$\mathrm{div}_y(\rho v \otimes v) + (d+1)\rho v + \nabla p(\rho, s) = 0, \tag{1.13}$$

$$\mathrm{div}_y((E[v] + p)v) + \rho|v|^2 + d(E[v] + p) = 0, \tag{1.14}$$

where d stands for the dimension of the physical space. Its value, an integer between 1 and 3, depends on the applications we have in mind. Equation (1.14) above involves the quantity

$$E[v] := \frac{1}{2}\rho|v|^2 + p(\rho, s)$$

which is obtained by replacing u by v in $E =: E[u]$.

Still, self-similar flows may be isentropic and even irrotational. In the latter situation, the potential ψ of u is replaced by $\Phi = \psi - \frac{t}{2}|y|^2$, that of v,

$$v = \nabla_x \Phi.$$

Remark that Φ is positively homogeneous of degree one. We prefer instead to use $\phi(y) := t^{-1}\Phi$, for which we have $v = \nabla_y \phi$. It satisfies the modified Bernoulli identity

$$\frac{1}{2}|\nabla\phi|^2 + \phi + \imath(\rho) = 0, \tag{1.15}$$

where now the constant of integration has been absorbed into ϕ. We may recover the density in terms of the potential,

$$\rho = h\left(\frac{1}{2}|\nabla\phi|^2 + \phi\right),$$

and rewrite the system as a second-order PDE

$$\mathrm{div}\left(h\left(\frac{1}{2}|\nabla\phi|^2 + \phi\right)\nabla\phi\right) + dh\left(\frac{1}{2}|\nabla\phi|^2 + \phi\right) = 0. \tag{1.16}$$

Initial data. Self-similar flows are generated by self-similar data $U(\cdot, 0) =: U^0$ that are constant along rays around the origin. This yields "boundary conditions at infinity" when solving (1.12, 1.13, 1.14), because of $U(x,t) = U(x/t)$,

$$\lim_{\mu\to+\infty} U(\mu y) = U^0(y). \tag{1.17}$$

Jump relations. Because the principal part of (1.16) is the same as that of (1.7), the jump relations are very similar to the steady irrotational case. They consist on the one hand of the continuity of the potential, from which we derive

$$[v_T] = 0, \tag{1.18}$$

and on the other hand of

$$[\rho v \cdot \nu] = 0. \tag{1.19}$$

1.3 Chaplygin gas

At constant entropy, a Chaplygin gas obeys the equation of state

$$p(\rho, s) = p_\infty - \frac{a^2}{\rho}, \qquad a > 0. \tag{1.20}$$

The sound speed is $c = a/\rho$. The acoustic fields are linearly degenerate, in the terminology of hyperbolic systems. Two extreme regimes are excluded *a priori* for physical reasons. On the one hand, a small density would yield a negative pressure. On the other hand the pressure saturates at very high densities, which allows the mass to concentrate along codimension-one subsets. Therefore the Chaplygin equation of state is used only so far as the density remains in a suitable compact interval of $(0, +\infty)$.

Jump relations. In an acoustic wave, property (1.11) yields $(j^2 - a^2)[\frac{1}{\rho}] = 0$. Since the density is discontinuous across such a wave, we deduce

$$j^2 = a^2. \tag{1.21}$$

It is remarkable that in a Chaplygin gas, the net flow does not depend on the strength of the discontinuity, but always equals $\pm a$. Using the definition of j, this gives on both sides of the wave

$$(u_\pm \cdot \nu - s)^2 = a^2/\rho_\pm^2, \tag{1.22}$$

which tells us that acoustic waves are sonic: $|u \cdot \nu - s| = c$. This property is specific to the Chaplygin gas.

Irrotational flows are true flows. For general flows, there is no reason why an irrotational data would produce an irrotational flow if there are discontinuities. However, the situation is favourable for a steady isentropic flow.

Theorem 1.1 (See [7]). *Consider a steady shock wave in a Chaplygin gas. Let Γ be the surface of discontinuity and ν its unit normal. Let us decompose the vorticity into normal and tangential components along the shock locus:*

$$\omega = \omega_T + (\omega \cdot \nu)\nu.$$

Then we have the jump relations

$$[\omega \cdot \nu] = 0, \qquad \left[\frac{\omega_T}{\rho}\right] = 0. \tag{1.23}$$

In particular, if the flow is irrotational on one side of the shock, it is irrotational on the other side.

In conclusion, piecewise smooth irrotational steady flows are not only good approximations of Euler flows, but they are genuine flows of the full gas dynamics system with a Chaplygin equation of state. According to [7], this is true also for self-similar flows.

Second-order equations for a Chaplygin gas. Because of $\imath = -a^2/(2\rho^2)$, we have

$$h(z) = \frac{a}{\sqrt{2z}}. \tag{1.24}$$

The potential equations (1.7) and (1.16) therefore rewrite, respectively,

$$\operatorname{div} \frac{\nabla \psi}{\sqrt{|\nabla \psi|^2 - 2\kappa}} = 0, \tag{1.25}$$

$$\operatorname{div} \frac{\nabla \phi}{\sqrt{2\phi + |\nabla \phi|^2}} + \frac{d}{\sqrt{2\phi + |\nabla \phi|^2}} = 0. \tag{1.26}$$

2 Second-order equations and their types

For irrotational models, the potential of a steady flow is governed by the second-order equation (1.7), and that of a self-similar flow by (1.16). We now describe a third kind of wave.

2.1 Conical waves

It may well be the case that a flow is simultaneously steady and self-similar. This happens when the initial data (ρ^0, u^0) is homogeneous of degree zero in $\mathbb{R}^d$, if it is a steady solution. Then its potential ψ is a solution of (1.25), homogeneous of degree one. Its pseudo-potential at unit time is

$$\phi(y, 1) = \psi(y) - \frac{1}{2}|y|^2.$$

Let (r,ω) be the spherical coordinates in $\mathbb{R}^3$, with $r>0$, ω in the unit sphere, and $x=r\omega$. Let us write $\psi(x)=r\theta(\omega)$. The domain of ψ is a cone $K=[0,\infty)\times D$. The variational formulation of (1.7) is

$$\int_K h(\frac{1}{2}|\nabla\psi|^2-\kappa)\nabla\psi\cdot\nabla\chi\,dx=0,\qquad\forall\chi\in\mathcal{D}(K).\tag{2.1}$$

Since $\mathcal{D}(K)=\mathcal{D}(]0+\infty[)\otimes\mathcal{D}(D)$, it is enough to choose test functions χ of the form $f(r)g(\omega)$. With $|\nabla\psi|^2=\theta^2+|\nabla_\omega\theta|^2$ and

$$\nabla\psi\cdot\nabla\chi=\theta f'g+\frac{f}{r}\nabla_\omega\theta\cdot\nabla_\omega g,$$

we find

$$\int_0^{+\infty}r^2dr\int_D h\left(\frac{\theta^2+|\nabla_\omega\theta|^2}{2}-\kappa\right)(f'\theta g+\frac{f}{r}\nabla_\omega\theta\cdot\nabla_\omega g)\,d\omega=0$$

for all test f and g. Integrating by parts in r, we see that this formulation does not depend on f at all, and reduces to

$$\int_D h\left(\frac{\theta^2+|\nabla_\omega\theta|^2}{2}-\kappa\right)(-2\theta g+\nabla_\omega\theta\cdot\nabla_\omega g)\,d\omega=0,\qquad\forall g\in\mathcal{D}(D).$$

This yields the following second order PDE over a spherical domain D:

$$\mathrm{div}_\omega(h(\cdots)\nabla_\omega\theta)+2\theta h\left(\frac{\theta^2+|\nabla_\omega\theta|^2}{2}-\kappa\right)=0,\qquad\omega\in D,\tag{2.2}$$

where the divergence operator stands for the adjoint of $-\nabla_\omega$ over the unit sphere.

2.2 The type of the PDEs

The analysis of a Cauchy problem or of a boundary value problem depends crucially on the type of the PDE (or system of) under consideration. We see below that the transition between hyperbolicity and ellipticity occurs at different thresholds for different kind of waves: steady, self-similar and conical ones.

For evolutionary PDEs, hyperbolicity is the algebraic property that makes the Cauchy problem well-posed. The full Euler system is hyperbolic. In every direction $\xi\in\mathbf{S}^{d-1}$, infinitesimal waves propagate at velocities $u\cdot\xi$ and $u\cdot\xi\pm c(\rho,s)$, with $c=a/\rho$ for a Chaplygin gas. The latter are velocities of acoustic waves, whereas $u\cdot\xi$ is the velocity of contact discontinuities and vorticity waves, which are always linearly degenerate. For irrotational flows, there remains only acoustic waves. They are genuinely nonlinear for most equations of state, but are linearly degenerate for a Chaplygin gas.

Steady irrotational flow. Let us devellop (1.7) as a quasilinear equation

$$h\left(\frac{1}{2}|\nabla\psi|^2-\kappa\right)\Delta\psi+h'\left(\frac{1}{2}|\nabla\psi|^2-\kappa\right)\mathrm{D}^2\psi(\nabla\psi,\nabla\psi)=0,$$

whose principal symbol is

$$P(x;\xi)=h(\cdots)|\xi|^2+h'(\cdots)(\xi\cdot\nabla\psi)^2,\qquad \xi\in\mathbb{R}^d. \tag{2.3}$$

The eigenvalues of this quadratic form are h (multiplicity $d-1$) and $h+|\nabla\psi|^2h'$ (simple). Remembering that h is positive (it is the value of a density), we see that the equation is elliptic (respectively hyperbolic) if $h+|\nabla\psi|^2h'$ is positive (resp. negative). From the definition, we have $\imath(h(B))=-B$ and therefore $h'\imath'=-1$, that is $h'c^2=-\rho=-h$. Hence the type depends of the sign of $(|\nabla\psi|^2-c^2)h'$. Thanks to $\imath'>0$, we have $h'<0$ and therefore only the sign of $|\nabla\psi|^2-c^2$ matters. Because of $\nabla\psi=u$, we see that (1.7) is elliptic (resp. hyperbolic) if the flow is subsonic (resp. supersonic), meaning that $|u|<c$ (resp. $|u|>c$).

Self-similar irrotational flow. Again, equation (1.16) is equivalent to a quasilinear one

$$h(\frac{1}{2}|\nabla\phi|^2+\phi)\Delta\phi+h'(\frac{1}{2}|\nabla\phi|^2+\phi)\mathrm{D}^2\phi(\nabla\phi,\nabla\phi)+dh(\cdots)+|\nabla\phi|^2h'(\cdots)=0,$$

whose symbol has a similar expression, but with ψ replaced by ϕ, and $\frac{1}{2}|\nabla\psi|^2-\kappa$ by $\frac{1}{2}|\nabla\phi|^2+\phi$,

$$P(x;\xi)=h(\cdots)|\xi|^2+h'(\cdots)(\xi\cdot\nabla\phi)^2,\qquad \xi\in\mathbb{R}^d.$$

With the same calculation as above, we find that only the sign of $|\nabla\phi|^2-c^2$ matters. This yields the conclusion that the equation is elliptic (resp. hyperbolic) equation if the flow is *pseudo*-subsonic (resp. pseudo-supersonic), meaning that $|v|<c$ (resp. $|v|>c$).

Conical flow. Since a conical flow is both steady and self-similar, whereas the types of equations (1.7) and (1.16) follow distinct rules, we must anticipate that the type for conical waves follows a third one. The symbol of the principal part in (2.2) is nothing but the restriction of the symbol P of (1.7) to the tangent space defined by $\xi\cdot\omega=0$. This gives here

$$P(\omega;\xi)=h(\cdots)|\xi|^2+h'(\cdots)(\xi\cdot\nabla_\omega\theta)^2,$$

where the argument of h is now $\frac{1}{2}(\theta^2+|\nabla_\omega\theta|^2)-\kappa$. The eigenvalues are h and $h+|\nabla_\omega\theta|^2h'$, the latter being $\rho(1-c^{-2}|\nabla_\omega\theta|^2)$. The equation (2.2) is therefore elliptic whenever

$$|\nabla_\omega\theta|<c(\rho). \tag{2.4}$$

In terms of the velocity, this means that $|u_\omega| < c(\rho)$, where $u = u_r\vec{e}_r + u_\omega$ is the decomposition into radial and tangential components. It may happen that $|u_\omega| < c(\rho) < |u|$, in which case the flow is supersonic while equation (2.2) is elliptic !

About the notion of sub/super-sonic flows. The discussion above shows that the notion of sub-/super-sonicity is not universal. On the contrary, it depends a lot upon the context. For instance, the collection of steady flows is stable under the Galilean changes of variables, but their type (subS or superS) is not. For a steady shock, it is natural to choose a frame attached to the shock; then only the normal component of the velocity is Galilean-invariant. It is known that the flow is superS on one side of the shock and subS on the other, except for a Chaplygin gas, where they are both sonic, because of degeneracy. For a self-similar flow, the relevant notion is that of pseudo-sub/super-sonic flow. Finally, for conical flows, the relevant notion is yet another one. This variety of criteria is due to the fact that in the various settings, the relations among the space-time coordinates are different. The moral is that *before speaking about sonicity, one has to* identify these relations, or equivalently to *describe the invariance group of the problem.*

3 The role of special waves

3.1 Self-similar waves

The study of the Riemann problem in one space dimension goes back to Riemann himself in 1860, and a general theory for small data was established by Lax in 1957. The two-dimensional analysis is much less understood, apart from the scalar case and irrotational flows. Let us only mention the breakthrough done by Chen and Feldman [2, 3]. In two space dimensions, they proved the existence of a potential flow in the classical problem of regular reflexion against a solid wedge. One difficulty of this problem is that the (pseudo-)subsonic region is bordered by a reflected shock (which plays the role of a free boundary), and by a sonic line that is characteristic (where ellipticity degenerates).

Chaplygin gas. In this case, the whole boundary is pseudo-sonic; it is characteristic but it is no longer a free boundary. This situation was treated in [6]. The subsonic domain is convex, in the strong sense that its inward curvature is stritly positive. In practice, the boundary is piecewise $\mathcal{C}^2$. The strong convexity turns out to be a necessary and sufficient condition for the existence to the characteristic Dirichlet problem encountered in the construction of the flow.

Theorem 3.1 (D. S. [6]). *Let Ω be a convex open domain in $\mathbb{R}^2$, with piecewise $\mathcal{C}^2$ boundary. Assume that the inward curvature is uniformly positive and bounded. Then there exists a unique positive solution $\phi \in \mathcal{C}^\infty(\Omega) \cap Lip(\overline{\Omega})$ of (1.26) with $d = 2$, with the boundary condition $\phi = 0$.*

Let us recall that $\phi = 0$ is precisely the value at which ellipticity is lost. Whether ϕ is smooth up to the boundary is an open problem .

3.2 Steady waves

Steady waves are governed by the restricted system of conservation laws $\mathrm{Div}_x F(U) = 0$. There does not exist a general theory because such a system may have a moving type, hyperbolic or elliptic according to the solution itself. However, if in gas dynamics we consider flows with a sufficiently high velocity in some direction, say x_1, then the system of the steady Euler equations is hyperbolic in this direction. This happens when the first component of the velocity is larger than the sound speed. Taking x_1 as an artificial time variable, we can construct steady solutions by solving Cauchy problems with data at $x_1 = 0$. Because we wish to treat only forward problems, the best strategy is usually to take an explicit, simple solution for $x_1 \leq 0$, and then to solve the system for $x_1 > 0$. This is an *interaction problem*, where incoming waves are given and outgoing waves are the unknown.

In two space dimensions, an explicit pattern at $x_1 < 0$ is given by two approaching planar shock waves that meet at $x_1 = 0$, say at the origin. The flow at $x_1 = 0$ is made of two constant states $U_\pm$, and there remains to solve a one-dimensional Riemann problem, in the (x_1, x_2)-plane instead of the (t, x)-plane. This is something that we can do, for instance for small incident strengths, thanks to Lax's Theorem. Because the flow is self-similar and steady, this is an example of conical wave, a rather simple one indeed.

If moreover the flow is irrotational, there are *only two* outgoing waves, which are often shocks (for instance if the strengths of the incoming waves are of the same size). Such a flow is piecewise constant and takes four distinct values separated by half-lines merging at the origin. The outgoing shocks are determined by a *shock polar analysis*, for which we refer to [4]. For a Chaplygin gas, the construction of the emerging shocks has a very nice geometrical interpretation [6].

The situation described above does occur in the resolution of 2-dimensional Riemann problems. Let us assume for instance that the initial data $U(x, 0)$ is constant in each coordinate quadrant, the jump of the velocity being purely normal, in order to satisfy irrotationality. Then each initial discontinuity along a half-axis is resolved by a backward and a forward waves, propagating in the normal direction. Up to a Galilean change of variables, we may assume that the forward shock S_{1f+} in the upper half-plane, is a steady shock. Meanwhile, we may assume that the forward shock S_{2f+} in the right half-plane is steady too. Notice that S_{1f+} "moves to the right" and S_{2f+} "moves up". Therefore S_{1f+} and S_{2f+} move towards each other, and an interaction takes place, at the origin since these waves have been made steady. These incoming waves never reach the pseudo-subS zone. Instead the transmitted waves are candidates to reach it, although it is possible that two transmitted waves, produced by distinct wave interactions, interact with each other somewhere in the pseudo-superS zone.

What is important in this description is that for points $y = x/t$ in some conical neighbourhood $\mathcal{V}$ of the origin (the point where the interaction takes place), the dependence cone meets only two initial discontinuities, separating three states. The solution behaves exactly as if $U(x, 0)$ was equal to the planar interaction of two shocks. Therefore the flow in $\mathcal{V}$ is *equal* to this interaction: it is piecewise constant with straight shocks. In particular, the flow is constant behind the interaction.

Remark. When the wave strength in the Riemann data is small, the shock velocity is close to the sound speed $c = c(\rho)$. Therefore, after the shocks $S_{1,2f+}$ have been made steady, the fluid velocity is approximately $(c, c)^T$, whose norm is $c\sqrt{2}$. The flow is thus supersonic in the direction $(1, 1)$, as expected.

3.3 Conical waves in 3-D

The two-dimensional conical waves studied in the previous paragraph were solved by algebraic equations (shock polar analysis). We turn now to the three-dimensional situation, where we encounter a new phenomenon. The conical waves are no longer piecewise constant, hence their construction is much more elaborate.

Following the same strategy as above, we choose in the past ($x_1 < 0$) three incoming steady shock waves, which separate a central state U_0 from three surrounding states U_1, U_2, U_3. We assume that the gas velocity has a dominant component in the x_1 direction, so that the irrotational system is hyperbolic in this direction, and we may expect that the Cauchy problem be solvable. Again, this situation occurs in 3-D Riemann problems, where the data is constant in octants, and x_1 is the diagonal direction in an octant. The state U_0 is the data in the opposite octant. The latter is limited by three planes, which separate U_0 from constant states V_j ($j = 1, 2, 3$). At $t = 0$, each discontinuity $U_0|V_j$ is resolved by a 1-D Riemann problem. This RP produces a backward and a forward waves, with a constant state U_j in between. Up to a Galilean change of frame, we may assume that the backward waves, those between U_0 and the U_j's, are steady. It is their interaction which we are interested in. For data with small strength, the flow velocity is approximately a constant V such that $V \cdot \nu_i \sim c$, where ν_i is the unit normal to the shock betwwen U_0 and U_i. Since the vector $(1,0,0)^T$ is a convex combination of the ν_i's, we deduce $V_1 > c$, hence the flow is supersonic in the x_1 direction. We thus expect that there is a conical neighbourhood $\mathcal{V}$ of the origin in which the solution of the hyperbolic Cauchy problem depends only upon the data in $x_1 < 0$, but not upon that in $x_1 > 0$. Therefore, we do not need to know anything about the data, but the states $U_0, \dots, U_3$ and their separatrices.

We observe that a part of a 3-D RP is made of 2-D RPs, because away from the origin, the three planar shocks $U_0|U_j$ $(j = 1,2,3)$ interact pairwise. We resolve these interactions with the help of the analysis made in Paragraph 3.2. If (i,j,k) denote the indices $(1,2,3)$ in some order, the interaction between $U_0|U_i$ and $U_0|U_j$, produces a transmitted state U^k, and we denote Π_i^k the plane supporting the shock between U_i and U^k.

It is worth to observe that if the system governing the flow was linear, then the Π_i^ks would be the continuations of the incident shock planes. In addition, the state U^k would be given by the formula $U^k = U_i + U_j - U_0$. To have the complete pattern of interaction, there would remain to extend the plane mentionned above, thus dividing the whole space in 8 cells, and to set $\hat{U} := U_1 + U_2 + U_3 - 2U_0$ in the remaining cell. Nothing deep here.

The situation in our fluid problem is much more complicated, due the nonlinearity of the system. Remark that the binary interactions described above take place in $x_1 < 0$, and therefore our "initial data" at $x_1 = 0$ consists of the six constant states U_1, U_2, U_3 and U^1, U^2, U^3. These states are separated by six half-lines emanating from the origin, which are the traces of the planes Π_j^k for $j \neq k$. The state U_0 is supported by a cell contained in $x_1 < 0$, thus we may forget it. What is important for the moment is that nearby states (for instance U_1 and U^2) are linked by a shock wave, thus we do not have to solve a 1-D RP between them. Instead, we remark that the planar shocks between U_1 and its nearby states U^2 and U^3 are approaching each other as x_1 increases. Therefore they interact along some line (the intersection of Π_1^2 and Π_1^3). This interaction produces two emerging shocks, along planes denoted by Π^{21} and Π^{31}, together with a transmitted state $\hat{U}^1$.

In a perfect world (and the linear world is perfect in this respect), the states $\hat{U}^j$ would coincide, and the planes Π^{ij} and Π^{ik} would coincide too, for each i. Instead, we may verify on concrete examples that $\hat{U}^1, \hat{U}^2$ and $\hat{U}^3$ are distinct, and that the Π^{ij} are six distinct planes, instead of only three. This means that gluing the ten states $U_0, U_1, \dots, U^2, \dots, \hat{U}^3$

and the fifteen corresponding shocks does not solve the problem.

The reason of this failure is that we are solving a Cauchy problem for a hyperbolic system in the direction x_1, where the data has a higher-order singularity at the origin. Although the solution is rather simple, piecewise constant, away from the influence cone K of the singularity, there is no reason why it should be so everywhere. Instead, we expect that in K, the potential is a non-trivial solution of the governing PDE (2.2). The problem to solve has a free boundary in general. The cone K is bordered by the states $U^1, U^2, U^3, \hat{U}^1, \hat{U}^2, \hat{U}^3$, and ∂K must be determined with the help of the Rankine–Hugoniot conditions. The boundary is made of shock waves, and/or of pieces of characteristic cones.

For a Chaplygin gas, the acoustic fields are linearly degenerate. These waves are thus supported by the characteristic cones associated with the external states U^j and $\hat{U}^k$. We verify that the cones associated with nearby states (say U^i and $\hat{U}^j$ with $i \neq j$) are interiorly tangent to each other, and the plane Π^{jk} is tangent to them along their intersection line. Thus there exists a unique convex cone K whose boundary is made of pieces of these six cones. It is known *a priori*, and we have to solve an ordinary boundary-value problem, with the difficulty that ellipticity degenerates at the boundary.

For a gas with a general equation of state, say a γ-law, the characteristic cones $K(U^i)$ and $K(\hat{U}^j)$ are not tangent to each other, and not characteristic. Therefore one needs at least three shocks among the six waves to form ∂K. We leave open the question of whether there are actually six shocks or if there are only three, associated with three sonic waves. The example of 2-D shock reflexion against a solid wedge suggests that the waves separating K from the states $\hat{U}^j$ could be pseudo-sonic, whereas those between K and the states U^j would be shocks. If so, the former would be pieces of circular cones, whereas the latter would be free boundaries.

Pictures for a Chaplygin gas. Our first picture is a section in the plane $x_1 + x_2 + x_3 = 1$. It displays the states U_j and U^k. Semi-infinite lines are the traces of the planar shocks Π_j^k. Small segments are the traces of shocks Π^{jk}. The oval is the trace of the "sonic cone"; it is made of six arc of circles, tangent to the previous segments.

The next two pictures display to views of the incoming steady shocks (they form a pyramid) between U_0 and the U_j's, and the outgoing conical wave. The shocks $U_j \mapsto U^k$ and $U^k \mapsto \hat{U}^\ell$ are not shown. In both pictures, the conical wave expands upwards. The x_1 axis passes from the interior of the pyramid to the interior of the piecewise-circular cone.

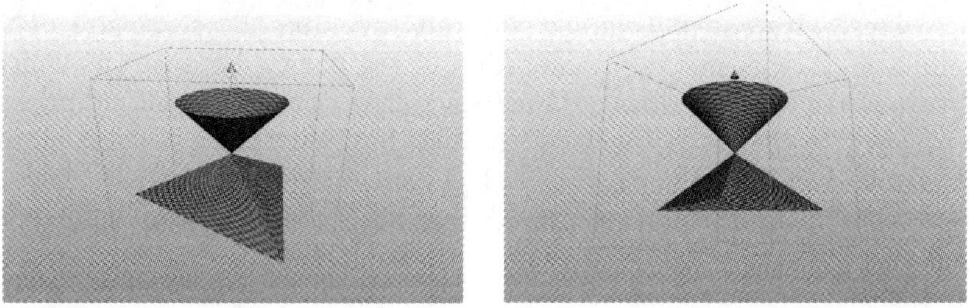

4 Existence results for a Chaplygin gas

Theorems 4.1 and 4.2 are taken from [6]. Other statements of this section are taken from [7].

4.1 Self-similar flows

We are concerned with Equation (1.26), which we rewrite in quasilinear form

$$(2\phi + |\nabla\phi|^2)\Delta\phi - \mathrm{D}^2\phi(\nabla\phi, \nabla\phi) + 2d\phi + (d-1)|\nabla\phi|^2 = 0. \tag{4.1}$$

Recall that it is elliptic whenever $\phi > 0$, and hyperbolic if $\phi < 0$. In the course of the construction of solutions of the d-dimensional Riemann problem, we are interested in the Dirichlet problem in a bounded domain $D \subset \mathbb{R}^d$, the boundary condition being

$$\phi = 0 \qquad \text{on } \partial D. \tag{4.2}$$

Let us pretend that the construction of the outer, hyperbolic solution in $\mathbb{R}^d \setminus D$ is done *a priori*. For the 2-D Riemann problem, $(\rho, u)_{\mathrm{out}}$ is piecewise constant, obtained by resolving binary interactions. In 3-D, it is generally more complicated, because it involves conical waves, as well as other patterns. The boundary of D is the level set $\{\phi_{\mathrm{out}} = 0\}$. The following proposition tells us that D cannot be arbitrary.

Proposition 4.1. *The inward mean curvature H of ∂D is strictly positive. More precisely, it satisfies*

$$H\,\frac{\partial\phi_{\mathrm{out}}}{\partial\nu} = 1,$$

where the derivative is in the inward normal direction to ∂D.

In the 2-D case, we deduce that the connected components of D (D is usually connected) are strictly convex. In particular the boundary is Lipschitz. In 3-D, we say that D is strictly *mean-convex*. Like convexity, mean convexity can be defined without invoking the curvature of the boundary. It does not imply any kind of regularity, because the control of the mean curvature does not mean a control of the second fundamental form, see [1]. At a global level, it may happen that D is not simply connected, is knotted or has two linked connected components. However we are not aware of a Riemann problem for which D would be torus-shaped.

Two-dimensional results.

Theorem 4.1 ($d = 2$). *Let $D \subset \mathbb{R}^2$ be a bounded convex domain, whose boundary is piecewise $\mathcal{C}^2$ and the curvature satisfies $0 < \min\kappa$, $\max\kappa < +\infty$. Then the Dirichlet problem (4.1,4.2) admits a unique positive solution ϕ in $Lip(\bar{D}) \cap \mathcal{C}^\infty(D)$.*

The regularity of ϕ at the boundary remains an open problem.

Theorem 4.1 is used to complete the resolution of 2-D Riemann problems. The data being piecewise constant and separated by rays, the pseudo-supersonic flow can be constructed with the help of binary interactions. It is piecewise constant. This construction may reach an obstruction, because some binary interactions with large incident waves might have no solution. For moderate data however, the solution can be extended until one reaches a pseudo-sonic line, the zero set of ϕ_{out}. This level line is made of arcs of pseudo-sonic circles that are pairwise interiorly tangent. It encloses a strictly convex domain D, in which we apply Theorem 4.1. This yields an existence result.

Theorem 4.2 ($d = 2$, Chaplygin gas). *Given an irrotational Riemann data with discontinuities of moderate strength, the 2-D Riemann problem admits a solution. It consists of finitely many constant states and one non-constant, pseudo-subsonic state. Their domains are separated by straight shocks and by arcs of sonic circles.*

We emphasize that the only obstruction that could be met for strongly discontinuous initial data is that some binary interaction of planar shocks could have no solution. This is a drawback of the Chaplygin equation of state.

Three-dimensional results. In the 3-D case, we need to reformulate the hypothesis of mean convexity. Because ϕ_{out} is not everywhere smooth, ∂D is not $\mathcal{C}^2$, we assume the following.

1. There exists a radius $\rho > 0$ such that, for every point $p \in \partial D$, there is a ball $B(z;\rho) \subset D$, such that $p \in \overline{B(z;\rho)}$. In particular, the sphere $S(z;\rho)$ is tangent to ∂D at p.

2. There is a constant $\alpha > 0$ such that the following holds. For every point $p \in \partial D$, there are orthogonal coordinates $x = (y, x_d)$, an $\epsilon > 0$, and a symmetric matrix S with $\mathrm{Tr}\, S \geq \alpha$, such that $p = (0,0)$, and $D \cap B(p;\epsilon)$ is included in the set defined by

$$x_d > y^T S y.$$

When ∂D is of class C^2, Property 2 is equivalent to the uniform positivity of the mean curvature, whereas Property 1 is associated with an upper bound of the curvature tensor at p.

Theorem 4.3 ($d = 3$). *Let $D \subset \mathbb{R}^3$ be a bounded mean-convex domain, in the sense of the hypotheses 1 and 2. Then the Dirichlet problem (4.1,4.2) admits a unique positive solution ϕ in the class $Lip(\bar{D}) \cap \mathcal{C}^\infty(D)$.*

The proofs of Theorems 4.1 and 4.2 are quite similar. They use the maximum principle satisfied by the auxiliary unknown $w := \sqrt{2\phi}$. We begin with an easy L^∞-estimate based on barrier functions. Being more clever, we establish a Lipschitz estimate at the boundary; this is where we use the mean convexity of D. Then we apply the maximum principle to an auxiliary quantity z to obtain interior Lipschitz estimates. The latter guarranty the uniform ellipticity of the principal part of (4.1), on every compact subdomain. This allows us to apply classical regularity theory. This provides enough compactness to pass to the limit in an approximation procedure. The boundary estimate is easy in 2-D, using the potentials of constant states as barrier functions, whereas it is very technical in 3-D. On the contrary, the interior estimate is easy in 3-D, where we may take $z = |\nabla\phi|^2$, whereas it is complicated in 2-D, where we need a $z = |\nabla\phi|^2 + 2\alpha\phi$ with some parameter $\alpha \in (2,3)$.

The necessary and sufficient condition of mean convexity looks natural once we observe that the principal part of the PDE governing w is nothing but the mean-curvature operator

$$\mathrm{div} \frac{\nabla w}{\sqrt{1 + |\nabla w|^2}}.$$

The mean convexity is known to be sufficient for the solvability of the Plateau problem with Dirichlet boundary condition $w = g$, and it is necessary in order that it be solvable for every continuous g. See [5] for details. Notice also that the above operator is of the form $\mathrm{div} A(\nabla w)$, hence is appropriate for applying the maximum principle, a crucial tool in the proof of the *a priori* estimates.

4.2 Conical waves

As explained in Paragraph 3.3, conical waves are used to resolve ternary interaction of steady planar waves. As such, they arise in some regions of the 3-D Riemann problem. The initial data of this RP consists in six states U^j and U_j ($j = 1, 2, 3$), with U^j and U_k separated by shock waves whenever $j \neq k$. In the x-space, shocks surrounding U_j are approaching each other as x_1 increases, and eventually interact, producing two planar shocks and a transmitted state $\hat{U}^j$. The plane supporting the shock between U^k and $\hat{U}^j$ is characteristic for both states, which means that it is tangent to the cones of equations

$$(|U|^2 - c(\rho)^2)|x|^2 = (U \cdot x)^2.$$

Because the tangential component of the velocity is continuous accross a shock, the plane and both cones are actually tangent along the same ray. We may form a conical surface Γ by gluing one piece of each of the six characteristic cones. This surface is piecewise $\mathcal{C}^2$ (except at the origin, of course) and encloses a convex cone $K \sim \mathbb{R}^+ \times \Omega$. Its curvature vanishes in the radial direction and is piecewise constant (thus uniformly positive) in the orthogonal direction.

The system to solve in Ω is formed of the PDE (2.2) with $h(z) = a/\sqrt{2z}$, together with the boundary condition $|\nabla\theta| = c$. From $h'(z)c^2 = -h(z)$, we deduce (because the gas is Chaplygin) $c^2 = 2z$, that is $c^2 \equiv \theta^2 + |\nabla\theta|^2 - 2\kappa$. Therefore the boundary condition is $\theta^2 = 2\kappa$. Since θ_{out} is positive on $\partial\Omega$, we can recast the boundary condition as

$$\theta = \sqrt{2\kappa} \qquad \text{on } \partial\Omega. \tag{4.3}$$

Inside Ω, the ellipticity tells us $\theta > \sqrt{2\kappa}$. The proof of the following result can be found in [7].

Theorem 4.4. *Let Ω be a spherical domain with piecewise $\mathcal{C}^2$ boundary and uniformly positive inward curvature (in particular, Ω is strictly contained in a half-sphere). Then the BVP (2.2,4.3) with $h(z) \equiv a/\sqrt{2z}$ admits a unique solution $\theta \in Lip(\bar{\Omega}) \cap \mathcal{C}^\infty(\Omega)$.*

As a corollary, the ternary interaction problem between three steady shocks of moderate strengths in a Chaplygin gas admits a solution.

The proof of the solvability of the BVP in every "strongly convex" Ω works more generally for an equation of state $p(\rho) = p_0 - a^2\rho^{-1/\beta}$ with $\beta \in [0, 1]$. The extreme points $\beta = 1$ and $\beta = 0$ correspond to either the Chaplygin gas or the incompressible gas, and we use a homotopy between both. Amazingly, the auxiliary unknown $z := \cosh^{-1}\theta$ satisfies the maximum principle. This is why we are able to establish *a priori* estimates that are uniform in s, ensuring the uniform ellipticity on compact subsets of Ω.

For $\beta = 0$, the PDE reduces to the linear equation $\Delta\theta_0 + 2\theta_0 = 0$. For a half-sphere defined by $\xi \cdot \omega > 0$ with $\xi \in \mathbf{S}^2$, the first eigenvalue of $-\Delta$ is $\lambda_1 = 2$; the corresponding eigenfunction is $\xi \cdot \omega$. Because λ_1 is a decreasing function of the domain, and $\bar{\Omega}$ is contained in a half-sphere, we have $\lambda_1(\Omega) > 2$, and the Dirichlet problem for θ_0 is uniquely solvable.

References

[1] E. Barozzi, E. Gonzalez and U. Massari, Pseudoconvex sets. Annali dell'Università di Ferrara. 55 (2009), 23–35.

[2] G.-Q. Chen and M. Feldman, Potential theory for shock reflection by a large-angle wedge. Proc. Natl. Acad. Sci. USA. 102 (2005), 15368–72.

[3] G.-Q. Chen and M. Feldman, Global solution to shock reflection by large-angle wedges for potential flow. Annals of Maths. 171 (2010), 1067–1182.

[4] R. Courant and K. O. Friedrichs, Supersonic Flow and Shock Waves. Applied Mathematical Sciences 21, Springer-Verlag, New York, 1948.

[5] H. Jenkins and J. Serrin, The Dirichlet problem for the minimal surface equation in higher dimensions. J. für die reine und angewandte Mathematik. 229 (1968), 170–187.

[6] D. Serre, Multi-dimensional shock interaction for a Chaplygin gas. Arch. Rat. Mech. Anal. 191 (2009), 539–577.

[7] D. Serre, Three-dimensional interaction of shocks in irrotational flows. Submitted.

Invited Talks

Dissipative Coupling of Scalar Conservation Laws across an Interface: Theory and Applications

Boris Andreianov

Laboratoire de Mathématiques CNRS UMR 6623
Université de Franche-Comté
16 route de Gray, 25000 Besançon, France
E-mail: boris.andreianov@univ-fcomte.fr

Abstract

We give a brief account on the theory of L^1-contractive solvers of the model conservation law with discontinuous flux:

$$(MP) \quad u_t + (\mathfrak{f}(x,u))_x = 0, \quad \mathfrak{f}(x,\cdot) = f^l(\cdot)\,\mathbb{1}_{x<0} + f^r(\cdot)\,\mathbb{1}_{x>0},$$

constructed in the work [6] of K.H. Karlsen, N.H. Risebro and the author. We discuss the modifications that can be used for extending our approach to the multi-dimensional setting and curved flux discontinuity hypersurfaces; the vanishing viscosity case (see [7]) is presented as an illustration. Applications to a road traffic with point constraint and to a coupled particle-fluid interaction model, coming from the joint works [4] with P. Goatin, N. Seguin and [8, 5] with F. Lagoutière, N. Seguin, T. Takahashi, are presented.

1 Introduction

Well-posedness for conservation laws with discontinuous flux of the form $\mathfrak{f}(t,x;u)$ is an active area of research since the early 1990ies. Several notions of solution were proposed, specifically in the model case (MP). Extensive references on the subject are given in the paper [6] in which a general approach to the Cauchy problem (MP) was proposed; many of the ideas that we used appear in the original contributions to the subject (see in particular [1, 2, 9, 10, 11, 13, 14, 15, 16, 21] and references in [6]).

We describe the notion(s) of entropy solutions to (MP) that lead to an L^1 theory analogous to the classical Kruzhkov theory (namely, with the L^1 contraction principle in the domain of dependence). Each of these notions is fully determined by a subset $\mathcal{G}$ of $\mathbb{R}^2$, called *admissibility germ*;

the role of $\mathcal{G}$ is to describe the coupling, across the interface $\{x=0\}$, of two conservation laws $u_t+(f^{l,r}(u))_x=0$ set up in the domains $\{\mp x>0\}$, respectively. The principal tools we use are the Kato inequalities away from the interface and near the interface, strong traces theory ([19]), adapted entropies ([10, 9, 11]), adapted vanishing viscosities, monotone finite volume schemes with the Godunov solver at the interface, variation bounds away from the interface ([11]), measure-valued solutions and compactification arguments. Section 2 briefly presents the theory of [6].

Establishing general properties of admissibility germs, we were able to classify many of the known uniqueness criteria for particular cases of the problem (MP); in particular, we gave a description of the *vanishing viscosity germ* $\mathcal{G}_{VV}$ which corresponds to solutions constructed by adding the standard viscosity εu_{xx} into the right-hand side of (MP) (see [6, 7]; cf. [13] for an equivalent description of $\mathcal{G}_{VV}$). Further, the theory of [6] (or rather, its straightforward generalizations, cf. Section 3) found new applications, which we briefly describe below (see Section 4 for details).

Namely, in [4] the road traffic model

$$u_t+f(u)_x=0 \text{ with a (formal) constraint } f(u)(t,0)\le F(t) \qquad (1.1)$$

is considered, following [12]. Although the flux here is continuous, the point constraint in (1.1) acts as an interface coupling condition and it may lead to introduction of a non-Kruzhkov shock at the interface. In this setting, we proved well-posedness in L^∞ and constructed a simple but efficient finite volume scheme for the problem (1.1).

Next, following [18], we looked at the Burgers equation

$$u_t+(u^2/2)_x=-u\delta_0(x) \qquad (1.2)$$

(here δ_0 is the Dirac delta function). Formally, the singular source $-u\delta_0(x)$ acts as an absorption term, thus it does not destruct the L^1-dissipativity; but it induces a non-conservative coupling. The coupling is interpreted in [18] using a regularization of δ_0. Dropping the Rankine-Hugoniot coupling restriction (cf. (2.1) below) , in [8, 5] we apply the previous theory to this case and establish well-posedness for solutions defined in [18]; once more, a simple numerical scheme for problem (1.2) is validated. Further, in the work [5] the properties of solutions to (1.2) are exploited in the fixed-point argument that allows to solve the coupled "particle-in-Burgers" problem of [18] (the Burgers equation with singular source $-(u-h'(t))\delta_0(x-h(t))$ for the fluid velocity u is coupled with an ODE for the evolution of the particle position h). In an ongoing work, we eventually achieve a BV well-posedness theory for this free-interface problem with non-conservative interface coupling.

The model case (MP) made apparent a number of ideas that can be now applied to the general setting of multi-dimensional conservation laws with piecewise smooth flux. Actually, references [4, 5, 7, 8] (cf. Section 4) already contain a few extensions. This work is on-going; some generalizations to our definitions and tools are presented in Section 3.

2 Admissibility germs and entropy solutions

A starting assumption is that, away from the flux discontinuities, one uses the Kruzhkov notion of entropy solutions ([17]). Thus, understanding the model case (MP) should reduce to understanding the coupling of two scalar conservation laws across the interface $\Sigma := \{x = 0\}$. We justify this reduction; moreover, thanks to the theory of strong boundary traces for conservation laws (see Panov [19]), the coupling across Σ can be encoded by the set $\mathcal{G}^*$ of the admissible couples $\Big(\gamma^l u\,,\,\gamma^r u\Big)$ of left- and right-sided traces of u at a.e. point of the interface Σ. To be precise, these strong traces do exist if $f^{l,r}(\cdot)$ are non-degenerately nonlinear (see [19]); more generally, introducing the appropriate singular mappings (see [6]), we can always rely on strong traces of $f^{l,r}(u)$ and of the associated Kruzhkov entropy fluxes $q^{l,r}(u,k) := \operatorname{sign}(u-k)(f^{l,r}(u)-f^{l,r}(k))$, $k \in \mathbb{R}$. To simplify the presentation, we assume in the sequel that $\gamma^{l,r}u$ exist.

We make apparent the properties of the set $\mathcal{G}^*$; namely, the coupling is conservative and L^1-dissipative (L^1D, for short) in the sense that

$$\forall (u^l, u^r), (v^l, v^r) \in \mathcal{G}^* \quad \begin{cases} f^l(u^l) = f^r(u^r), \\ q^l(u^l, v^l) \geq q^r(u^r, v^r), \end{cases} \tag{2.1}$$

moreover, the set $\mathcal{G}^* \subset \mathbb{R}^2$ is *maximal* L^1D (i.e. it has no extension that still satisfies the properties (2.1)). Further, in practice it is often enough to describe only a subset $\mathcal{G}$ of $\mathcal{G}^*$ that satisfies properties (2.1) and admits a unique maximal extension. In this situation, we say that $\mathcal{G}$ is a *definite L^1-dissipative (or L^1D) admissibility germ* and $\mathcal{G}^*$ is its maximal L^1D extension. For a definite L^1D germ, we define a $\mathcal{G}$-entropy solution as a juxtaposition of entropy solutions $u|_{x<0}$ (for the conservation law $u_t + f^l(u)_x = 0$) and $u|_{x>0}$ (for $u_t + f^r(u)_x = 0$) coupled via

$$\text{for a.e. } t \in \mathbb{R}^+ \quad \Big((\gamma^l u)(t)\,,\, \gamma^r(u)(t)\Big) \in \mathcal{G}^*. \tag{2.2}$$

In this way, definite germs $\mathcal{G}, \tilde{\mathcal{G}}$ with different maximal L^1D extensions $\mathcal{G}^*, \tilde{\mathcal{G}}^*$ lead to different notions of entropy solution for (MP) (cf. [1, 11]).

We introduce some other notions related to germs; in particular, $\mathcal{G}^*$ is a complete germ if every Riemann problem for (MP) admits a $\mathcal{G}$-entropy solution. The key statement of the theory is the following:

for every definite L^1D germ $\mathcal{G}$ such that $\mathcal{G}^*$ is complete,
problem (MP) is well-posed
in the framework of $\mathcal{G}$-entropy solutions with L^∞ initial data.

More precisely, uniqueness holds for general continuous fluxes $f^{l,r}$; existence is shown under some structure assumptions on $f^{l,r}$ (including e.g. the trivial requirement that the ranges of $f^l(\cdot)$ and $f^r(\cdot)$ intersect); the continuous dependence on the initial datum u_0 can be stated as the L^1 contraction in the domain of dependence (for $f^{l,r}(\cdot)$ locally Lipschitz).

The proof of uniqueness (continuous dependence is similar) for the Cauchy problem for (MP) is very straightforward from this definition. From the Kruzhkov theory away from the interface, the Kato inequality with test functions $\xi\xi_h$, $\xi \geq 0$, $\xi_h := \min\{|x|/h, 1\}$ is derived, namely,

$$\int_0^T \int_{\mathbb{R}} \Big\{ -|u-\hat{u}|(\xi\xi_h)_t - \operatorname{sign}(u-\hat{u})(\mathfrak{f}(x;u) - \mathfrak{f}(x;\hat{u}))\,(\xi\xi_h)_x \Big\} \leq 0 \quad (2.3)$$

for $u, \hat{u}$ two $\mathcal{G}$-entropy solutions associated with the same initial datum. Letting h desrease to zero and using the definition of strong traces, we generate the interface term that takes precisely the form

$$\int_0^T \Big\{ q^l(\gamma^l u, \gamma^l \hat{u}) - q^r(\gamma^r u, \gamma^r \hat{u}) \Big\}(t)\, \xi(t,0)\, dt. \quad (2.4)$$

Now the trace constraint (2.2) and the dissipativity property (2.1) of $\mathcal{G}^*$ make this term non-negative; it can therefore be dropped. Choosing ξ in the classical way of Kruzhkov, we find that $u = \hat{u}$.

Existence arguments require a definition of $\mathcal{G}$-entropy solution that is clearly stable by passage to the limit. We provide the following:

Definition 2.1. A function $u \in L^\infty(\mathbb{R}^+ \times \mathbb{R})$ is a $\mathcal{G}$-entropy solution of (MP) if it is a Kruzhkov entropy solution in the domains $\{\pm x > 0\}$, it is a weak solution in the whole domain (i.e. the Rankine-Hugoniot condition holds), and moreover, the adapted entropy inequalities

$$|u - c(x)|_t + (\mathfrak{q}(x; u, c(x)))_x \leq 0 \text{ in } \mathcal{D}'((0,+\infty) \times \mathbb{R}), \quad (2.5)$$

($\mathfrak{q}(x; \cdot, c(x)) = q^l(\cdot, c^l)\, \mathbb{1}_{x<0} + q^r(\cdot, c^r)\, \mathbb{1}_{x>0}$ being the adapted entropy flux) hold for every function $c(x) = c^l\, \mathbb{1}_{x<0} + c^r\, \mathbb{1}_{x>0}$ with $(c^l, c^r) \in \mathcal{G}$.

Notice that Definition 2.1 admits a straightforward extension to the case of measure-valued ($\mathcal{G}$-entropy-process) solutions. Unfortunately, in our theory we require existence of a $\mathcal{G}$-entropy solution $u(\cdot,\cdot)$ to prove that every $\mathcal{G}$-entropy-process solution $(\hat{u}(\cdot,\cdot,\alpha))_{\alpha\in[0,1]}$ is independent of α and coincides with u. Yet if existence of u can be obtained via some particular strongly convergent approximation procedure, this extension allows to prove convergence of the approximations possessing only a uniform L^∞ bound and compatible with (adapted) entropy inequalities.

Notice that smaller is $\mathcal{G}$, easier it is to check the constraints (2.5); e.g. in the cases considered in [1, 3, 4, 9, 11], one can pick for $\mathcal{G}$ a singleton (A, B), called "(A, B)-connection"; in this way, only one adapted entropy inequality has to be checked. Another advantage of restraining $\mathcal{G}^*$ as much as possible (which leads to the use of definite L^1D germs) is the following. Assume that the approximation procedure in use allows explicitly for some particular stationary solutions of (MP) of the form $u(x) = c^l_{expl}\, 1\!\!1_{x<0} + c^r_{expl}\, 1\!\!1_{x>0}$; assume that we know that the procedure converges, for each fixed initial datum, and that the resulting solver for (MP) is L^1 contractive (this is the case when we start with L^1 contractive approximate solvers; the vanishing viscosity method is one typical example, monotone finite volume methods provide another example). Then we know that the resulting notion of solution is a $\mathcal{G}^*$-entropy solution for some maximal L^1D germ $\mathcal{G}^*$, and $\mathcal{G}^*$ contains the couples (c^l_{expl}, c^r_{expl}) corresponding to the explicit stationary limits of the approximation scheme in use. Eventually, the set of all such couples (c^l_{expl}, c^r_{expl}) may form a definite L^1D germ $\mathcal{G}$ of which $\mathcal{G}^*$ is the unique maximal L^1D extension; $\mathcal{G}^*$ can be calculated starting from the set of all (c^l_{expl}, c^r_{expl}). The proofs of [3, 4, 7, 8] are based on this approach.

Let us explain why, *for a definite L^1D germ $\mathcal{G}$*, Definition 2.1 is equivalent to the previously given definition "with traces" of $\mathcal{G}$-entropy solutions. Firstly, the entropy inequalities (2.5) derive from splitting a test function ξ into $\xi\xi_h + \xi(1-\xi_h)$ with $\xi_h = \min\{|x|/h, 1\}$. Indeed, for the test function $\xi\xi_h$, the Kruzhkov entropy inequalities away from $\{x=0\}$ can be used, with $k = c^{l,r}$; for the family $(\xi(1-\xi_h))_{h>0}$, we pass to the limit as $h \to 0$, use the definition of strong traces and generate the term (2.4) with $\gamma^{l,r}\hat{u} := c^{l,r}$. From the inclusion $\mathcal{G} \subset \mathcal{G}^*$ and the dissipativity property of the germ $\mathcal{G}^*$ (see (2.1)), this term has the good sign and we get (2.5) as $h \to 0$. Secondly, the same calculation made with the test function $\xi(1-\xi_h)$ ($\xi \geq 0$ being arbitrary) allows to go from the adapted entropy inequality (2.5) to the t-a.e. inequalities

$$\forall (c^l, c^r) \in \mathcal{G} \quad q^l((\gamma^l u)(t), c^l) - q^r((\gamma^r u)(t), c^r) \geq 0.$$

By assumption, the unique maximal L^1D extension of $\mathcal{G}$ is $\mathcal{G}^*$; it follows that the couple $\Big((\gamma^l u)(t)\,,\,(\gamma^r u)(t)\Big)$ belongs to $\mathcal{G}^*$. Thus (2.2) holds.

Well-chosen numerical scheme yields a rather general existence result:

Theorem 2.2 (see Andreianov, Karlsen and Risebro [6]).
Let $\mathcal{G}$ be definite L^1D germ with the maximal extension $\mathcal{G}^*$ that is a complete germ. Assume that the functions $f^{l,r}$ are locally Lipschitz continuous on $\mathbb{R}$. Then for any initial function $u_0 \in L^\infty(\mathbb{R})$ there exists a unique $\mathcal{G}$-entropy solution of the Cauchy problem (MP), $u|_{t=0} = u_0$.

Existence results can also derive from different vanishing viscosity approaches: this permits to bypass the tedious check of completeness of germs. We refer to [7, 6] for the standard viscosity case. If $\mathcal{G}$ is a singleton $\{(A, B)\}$, adapted "artificial" viscosities, explicitly allowing for the stationary solution $A\,\mathbb{1}_{x<0} + B\,\mathbb{1}_{x>0}$, can be used (see [6] and Section 4). Also the physically relevant "vanishing capillarity" regularization for the case of Buckley-Leverett equation with discontinuous flux becomes easy to analyze: we refer to the work [3] of the author with C. Cancès.

Let us give an insight into the proof of Theorem 2.2. Uniqueness was already justified, using the definition "with traces"; to prove existence, we construct a family of approximate solutions using a finite volume scheme with any monotone consistent numerical flux away from the interface, and with the Godunov numerical flux at $\{x = 0\}$ (this Godunov flux is constructed from the Riemann solver at the interface $\{x = 0\}$; the Riemann solver is well defined, due to the completeness assumption on $\mathcal{G}^*$). Then we prove that, firstly, the approximate solver is L^1-contractive (more generally, it satisfies approximate Kato inequalities); and secondly, that the solutions $c^l\,\mathbb{1}_{x<0} + c^r\,\mathbb{1}_{x>0}$ with $(c^l, c^r) \in \mathcal{G}^*$ are explicit limits of the scheme (this is evident because the Godunov scheme is well-balanced). If an L^1_{loc} limit u of the approximations $(u_n)_n$ exists, we "inherit" the adapted entropy inequalities (2.5) for u; the Kruzhkov entropy inequalities away from $\{x = 0\}$ and the Rankine-Hugoniot condition are justified in the analogous (a more standard) way. Eventually, we have to justify the L^1_{loc} compactness of $(u_n)_n$. To this end, we use the BV_{loc} estimate device in the way of [11], and a uniform L^∞ estimate on u_n coming from comparison principle with well-chosen stationary solutions (at this point, the completeness of $\mathcal{G}^*$ is used in an indirect way). In this way, solutions with BV initial data are constructed; using localized L^1 contraction for these solutions, we deduce existence for L^∞ data.

It should be noticed that the different properties of L^1D germs are not independent; some relations are listed in [6] and in the Appendix of [7]. To give an example, complete L^1D germs are automatically maximal ones. The slight difference between the definition of the entropy solution in [15] and the Γ-condition of [13] is easily removed making the closure operation on germs (see [6]). The notion of closure of a germ turns out to be useful in the analysis of the vanishing viscosity approximation of (MP), starting from the viscous profiles for (MP) (see [7, 6]). Structural properties of germs may allow to establish that the set of stationary solutions admissible according to some admissibility criteria in not L^1D, which means that the criterion fails. For instance, we show in [6] that the *crossing condition* imposed in the works of [16, 21] is necessary for the uniqueness of the entropy solutions in the sense of inequalities ($\forall k \in \mathbb{R}$)

$$\int_0^T\!\!\int_{\mathbb{R}} \Big\{ |u-k|\,\xi_t + \mathfrak{q}(x,u,k)\,\xi_x \Big\} \geq -\int_0^T |f^r(k) - f^l(k)|\,\xi(\cdot,0). \tag{2.6}$$

Indeed, (2.6) uses the Kruzhkov entropies ([17]), whereas constants are, in general, not solutions to (MP). Thus an error term should be incorporated into the Kato inequality (2.3) (now let us take $\xi_h \equiv 1$ in (2.3)) written for u and $\hat{u} \equiv k$; the right-hand side of (2.6) lower bounds this error term. In Section 3, we use the idea of [21, 16] with k replaced by $c(x) = c^l \mathbb{1}_{x<0} + c^r \mathbb{1}_{x>0}$, but now with couples (c^l, c^r) that are not necessarily in $\mathcal{G}$, and with an *ad hoc* error term in the entropy inequalities.

3 Generalization to variable germs

The theory of Section 2 assumes that the admissibility germ $\mathcal{G}$ governs the interface coupling at a.e. point $(t, 0)$ of the interface $\{x = 0\}$. One could naturally ask what happens if a family $(\mathcal{G}(t))_{t>0}$ (of definite L^1D germs) is given. Such "variable germs" occur for instance in the problem (1.1) considered in [4], and in the free-interface problem of [5] (see Section 4). Moreover, dependence of $\mathcal{G}(\cdot)$ on the point of the interface naturally occurs whenever the interface is curved (indeed, the normal direction to the interface enters the definition of the germ, see [7]).

In these cases, the definition "with traces" of a $\mathcal{G}(\cdot)$-entropy solution carries on without difficulty, provided the interface Σ is regular enough and a non-degeneracy assumption guarantees the existence of one-sided traces $(\gamma^{l,r}u)(\cdot)$ on Σ of an entropy solution u in $(\mathbb{R}^+ \times \mathbb{R}^N) \setminus \Sigma$. We simply have to put $\mathcal{G}^* = \mathcal{G}(t)^*$ in (2.2) (more generally, t has to be replaced with a local coordinate σ on Σ). Clearly, existence of solutions satisfying the so modified interface coupling constraint (2.2) requires some kind of measurability of the family $\mathcal{G}(\cdot)$. The families that appear in practice, such as the vanishing viscosity family $(\mathcal{G}_{VV}(\sigma))_{\sigma \in \Sigma}$ associated with a curved interface Σ, do satisfy this constraint (this can be understood from the fact that the existence of a solution is guaranteed, as a consequence of convergence of the vanishing viscosity method; see [7]).

Let us give a generalization of Definition 2.1 to the case of a variable germ; the idea is to relax the constraint $(c^l, c^r) \in \mathcal{G}$, and to "pay" with a remainder term concentrated on Σ. Assume that Σ is a regular enough hypersurface of $\mathbb{R}^+ \times \mathbb{R}^N$ separating it into two domains $\Omega^{l,r}$, and

$$\mathfrak{f}(t, x; \cdot) = f^l(\cdot)\, \mathbb{1}_{\Omega^l}(t, x) + f^r(\cdot)\, \mathbb{1}_{\Omega^r}(t, x). \tag{3.1}$$

In the definition of the remainder R, we tacitly assume that $(\mathcal{G}(\sigma))_{\sigma \in \Sigma}$ is a measurable family of germs on Σ, in the *ad hoc* sense.

Definition 3.1. A function $u \in L^\infty(\mathbb{R}^+ \times \mathbb{R}^N)$ is a $\mathcal{G}(\cdot)$-entropy solution of $u_t + \mathrm{div}(\mathfrak{f}(t, x; u)) = 0$ (with $\mathfrak{f}$ given by (3.1)) if it is a Kruzhkov entropy solution in the domains $\Omega^{l,r}$, it is a weak solution in the whole domain (i.e. the Rankine-Hugoniot condition holds), and moreover, the adapted entropy inequalities with remainder term $R\left(\cdot; (c^l, c^r)\right)$:

$$|u - c(x)|_t + \mathrm{div}\, \mathfrak{q}(x; u, c(x)) \leq R\left(\sigma; (c^l, c^r)\right) d\mathcal{H}(\sigma) \;\text{ in } \mathcal{D}'((0, +\infty) \times \mathbb{R}^N)$$

hold for every function $c(x) = c^l\,\mathbb{1}_{x<0} + c^r\,\mathbb{1}_{x>0}$ with $(c^l, c^r) \in \mathbb{R}^2$.

Here $d\mathcal{H}(\sigma)$ is the N-dimensional Hausdorff measure on Σ, so that the remainder term R is supported by the interface Σ; moreover, it is assumed that $R : \Sigma \times \mathbb{R}^2 \to \mathbb{R}^+$ is a Carathéodory function such that for a.e. $\sigma \in \Sigma$

$$\forall\ (c^l, c^r) \in \mathcal{G}(\sigma),\ \lim_{r\downarrow 0} \fint_{B_r(\sigma)\cap\Sigma} R(\sigma'; (c^l, c^r))\, d\mathcal{H}(\sigma') = 0 \tag{3.2}$$

and

$$\begin{gathered} \forall\,(c^l, c^r) \in \mathbb{R}^2 \text{ and } \forall\ (a^l, a^r) \in \mathcal{G}(\sigma) \\ q^r(\sigma; a^r, c^r) - q^l(\sigma; a^l, c^l) \leq R(\sigma; (c^l, c^r)). \end{gathered} \tag{3.3}$$

Roughly speaking, property (3.2) means that the remainder vanishes whenever $(c^l, c^r) \in \mathcal{G}(\sigma)$ (thus providing a link with Definition 2.1). Property (3.3) leads to the term R in the entropy inequalities, since R dominates the quantity (2.4) for an entropy solution u and for $\hat{u} = c(x)$.

When $f^{l,r}$ are Lipschitz continuous, the simplest choice of R is

$$R(\sigma; (c^l, c^r)) := \mathit{const}\ \mathrm{dist}\big(\, (c^l, c^r)\ ,\ \mathcal{G}(\sigma)\,\big). \tag{3.4}$$

In the cases considered in [4, 5], the measurability of $R(\cdot, (c^l, c^r))$ given by (3.4) is clear; in a more general setting, we prefer a subtler definition of R based upon the oscillation functions $Osc_{[a^{l,r}, c^{l,r}]} f^{l,r}$ associated with $f^{l,r}|_{[a^{l,r}, c^{l,r}]}$ (see [6]), where (a^l, a^r) are chosen in $\mathcal{G}(\cdot)$, as in (3.3).

Equivalence of the definition "with traces" and Definition 3.1 is established as in the model case (MP). Uniqueness is straightforward. For examples of existence proofs, we refer to [7] and to Section 4.

To end this section, assume that $u, \hat{u}$ are entropy solutions of (MP) corresponding to different families of germs $\mathcal{G}(\cdot)$ et $\hat{\mathcal{G}}(\cdot)$, respectively (this includes the case of two fixed but different germs). Then we can compare $u, \hat{u}$ in terms of an error term containing a "distance" between the two germs. Indeed, we define the "distance" between two elements in $\mathbb{R}^2$ as $\rho\Big(\, (c^l, c^r), (\hat{c}^l, \hat{c}^r)\,\Big) := \Big(q^r(c^r, \hat{c}^r) - q^l(c^l, \hat{c}^l)\Big)^+$; then for $\mathcal{G}, \hat{\mathcal{G}} \subset \mathbb{R}^2$,

$$\rho\Big(\,\mathcal{G}\,,\,\hat{\mathcal{G}}\,\Big) := \sup\Big\{\,\rho\Big(\,(c^l, c^r)\,,\,(\hat{c}^l, \hat{c}^r)\,\Big)\ |\ (c^l, c^r) \in \mathcal{G},\ (\hat{c}^l, \hat{c}^r) \in \hat{\mathcal{G}}\,\Big\}. \tag{3.5}$$

Then, essentially in the same way as the uniqueness of a $\mathcal{G}$-entropy solution is proved, in the case $u(0, \cdot) = \hat{u}(0, \cdot)$ we establish

$$\int_{\mathbb{R}} |u - \hat{u}|(t, \cdot) \leq \int_0^t \rho\Big(\,\mathcal{G}(t)\,,\,\hat{\mathcal{G}}(t)\,\Big)\, dt. \tag{3.6}$$

Notice that definition (3.5) and the associated notion of a ρ-neighbourhood of a given germ $\mathcal{G}$ are also useful to understand the issue of measurability of families of germs $(\mathcal{G}(\sigma))_{\sigma\in\Sigma}$.

4 Applications

One important application was to understand the vanishing viscosity limit for (MP); this was done in [6] (cf. [13]), and extended in [7] to the multi-dimensional setting with a smooth curved interface. The Buckley-Leverett vanishing capillarity limit studied in [3] and the two original applications below allowed, in a sense, to validate the theory (notice that most of the examples given in [6] were already understood in the previous contributions to the subject). The two new applications came from the works [12] and [18], respectively, where the respective notions of admissible solution were already fixed; uncovering the germs underlying these notions, one gets well-posedness quite easily, with the tools of [6]. The second application shows that the theory extends without difficulty to the case of non-conservative (but L^1-dissipative) coupling.

For both applications, we construct convergent "naive" finite volume schemes, and it has to be stressed that the issue of convergence cannot be settled uniquely by the general arguments of Theorem 2.2; additional rather delicate observations and arguments are needed for finer schemes.

• **Road traffic with point constraint**

Consider (1.1) with a flux f with $f(0) = 0 = f(1)$, f nonnegative, Lipschitz, with $f'(\cdot)$ changing sign at $\bar{u} = \operatorname{argmax} f$; take initial data satisfying $0 \le u_0(\cdot) \le 1$. For $F \in [0, f(\bar{u})]$, the F-level set of $f(\cdot)$ consists of two points: $B_F \le \bar{u} \le A_F$. Then we define $\mathcal{G}(t) := \{(A_{F(t)}, B_{F(t)})\}$ and compute that the unique maximal L^1D extension of $\mathcal{G}(t)$ is

$$\mathcal{G}(t)^* := \mathcal{G}(t) \cup \{\text{couples } (c^l, c^r) \text{ such that } f(c^{l,r}) \le F \text{ and } c^l \le c^r\}$$

(in fact, all the couples in $\mathcal{G}(t)^* \setminus \mathcal{G}(t)$ correspond to Kruzhkov admissible stationary solutions of $u_t + f(u)_x = 0$ satisfying $f(u) \le F$, while the stationary solution $c(x) = A_F \mathbb{1}_{x<0} + B_F \mathbb{1}_{x<0}$ is a non-Kruzhkov shock). Thus for all $t > 0$, $\mathcal{G}(t)$ is a definite germ.

Then we define the associated $\mathcal{G}(\cdot)$-entropy solutions (here the family of germs is defined via an L^∞ function $F(\cdot)$, and the measurability of R in Definition 3.1 is straightforward). Uniqueness and L^1 contraction are immediate; for existence, one first works with piecewise constant $F(\cdot)$, then extends the result using (3.6). While Theorem 2.2 applies for $F(t) \equiv const$, let us sketch a proof by converging vanishing viscosity approximation. We study the approximation $u_t + f(u)_x = \varepsilon(k(x,u))_{xx}$ with $k(x, \cdot) := a(\cdot) \mathbb{1}_{x<0} + b(\cdot) \mathbb{1}_{x>0}$ for $a(\cdot)$ defined from A_F by

$$a : r \in [0,1] \mapsto \tfrac{1}{2A_F} \min\{r, A_F\} + \tfrac{1}{2(1-A_F)} (r - A_F)^+,$$

and $b(\cdot)$ defined accordingly from B_F. The point is that $a(\cdot), b(\cdot)$ are increasing, $a(0) = b(0)$, $a(A_F) = b(B_F)$, and $a(1) = b(1)$; therefore

$u_-^\varepsilon := 0$, $u_+^\varepsilon := 1$ and $c^\varepsilon(x) := A_F \mathbb{1}_{x<0} + B_F \mathbb{1}_{x>0}$ are explicit solutions, independent of $\varepsilon > 0$. Moreover, introducing the new unknown $w(t,x) := k(x, u(t,x))$, it is easy to check that the *ad hoc* Kato inequality holds for solutions of the problem. This implies comparison principle that allows to bound the solutions u^ε of the viscous equation (with initial datum u_0) by $0 \equiv u_-^\varepsilon \leq u^\varepsilon \leq u_+^\varepsilon \equiv 1$. Assuming that the level sets of $f'(\cdot)$ do not contain intervals, we can use the compactification results (see e.g. [20]) and deduce that u^ε converge a.e. to a limit u which is a weak solution of $u_t + f(u) = 0$ and also an entropy solution away from the interface. Moreover, the passage to the limit in the Kato inequality written for u^ε and $\hat{u} := c^\varepsilon(x)$ yields the adapted entropy inequality (2.5). Thus u is indeed a $\mathcal{G}$-entropy solution of our problem, because $\{(A_F, B_F)\}$ is a definite germ with the same maximal extension as $\mathcal{G}$.

In [4], we have proved existence for (1.1) by making converge a finite volume scheme with a monotone consistent numerical flux $g(\cdot,\cdot)$: we simply re-define the flux at the interface $\{x = 0\}$ with $g_F := \min\{g, F\}$. Thus, the scheme avoids the intricate Godunov solver used in the proof of Theorem 2.2. Although seemingly naive, the scheme of [4] is partially well-balanced in the sense that it preserves a sufficiently large part of the stationary solutions described by $\mathcal{G}^*$ (namely, constant functions and the key function $c(x) = A_F \mathbb{1}_{x<0} + B_F \mathbb{1}_{x<0}$ are solutions of the scheme). Using the measure-valued (entropy-process) solution techniques (the above existence result is needed), we deduce convegence of the "naive" scheme.

• Particle-in-Burgers model

The interface coupling and the associated Riemann solver for the non-conservative problem (1.2) were fully described in [18]. In [8], we observed that the set of admissible one-sided traces fulfills the "L^1D" inequality of (2.1) (the Rankine-Hugoniot condition in (2.1) has to be dropped); moreover, it admits no L^1D extension, thus it is maximal. As previously, we denote it $\mathcal{G}$ (here $\mathcal{G}^* = \mathcal{G}$). Then $\mathcal{G}$-entropy solutions enjoy uniqueness, in the same way as in Section 2. Also existence can be shown with the Godunov scheme, as in Theorem 2.2, but once more we prefer a "naive" scheme that is by far simpler to implement (indeed, the Riemann solver of [18] involves a bunch of different cases). We have

$$\mathcal{G} = \Big\{(c^l, c^r) \mid c^l - c^r = 1 \text{ or } (\, c^l \geq 0,\ c^r \leq 0,\ -1 \leq c^l + c^r \leq 1)\Big\};$$

it should be stressed that, in a sense, the line $\mathcal{L}$: "$c^l - c^r = 1$" of $\mathbb{R}^2$ is the key part of $\mathcal{G}$. Unfortunately, this part itself is not a definite germ, but adjoining the square $\mathcal{Q} := [0,1] \times [-1, 0]$ to the line $\mathcal{L}$, we get a definite part of the germ $\mathcal{G}$. The finite volume scheme with a consistent monotone numerical flux $g(\cdot,\cdot)$ satisfying some additional constraint (which is nonrestrictive in practice) is constructed by modifying $g(\cdot,\cdot)$ at the interface.

Recall that the conservation at the interface is lost; we define the left interface flux by $g_0^l(u_-,u_+) := g(u_-,u_++1)$ (as if we wanted to connect the right state u_+ to the associated state on the line $\mathcal{L}$), and the right interface flux by $g_0^r(u_-,u_+) := g(u_--1,u_+)$ (as if we wanted to connect the left state u_- to the associated state on the line $\mathcal{L}$). By construction of $g_0^{l,r}$, the scheme is partially well-balanced: it preserves the stationary solutions $c(x) = c^l\, 1\!\!1_{x<0} + c^r\, 1\!\!1_{x>0}$ for $(c^l,c^r) \in \mathcal{L}$. The part $\mathcal{L}$ of $\mathcal{G}$ is not definite, that's why we prove in addition that the stationary solutions with $(c^l,c^r) \in \mathcal{Q}$ are preserved asymptotically (they are obtained at the limit of our "naive" scheme). Then we use the arguments analogous to those of Theorem 2.2 to pass to the limit in the scheme.

Problem (1.2) is a particular case of the Burgers equation with source term $-\lambda(u-h'(t))\,\delta_0(x-h(t))$, where $h \in W^{2,\infty}(0,T)$ defines an internal interface. We first consider the interface $\{x = h(t)\}$ as being fixed; later on, we couple the conservation law to the interface equation that formally reads as $mh''(t) = \lambda(u(t,h(t))-h'(t))$. The latter relation means that the particle of mass m is driven by the difference $u - h'(t)$ of fluid and particle velocities, with a viscosity coefficient λ (see [18, 5] for details); we fix $\lambda = 1$. The rigorous interpretation of the particle-driving equation relates $h''(t)$ to the jump of the normal traces on the interface $\{x = h(t)\}$ of the flux $(u, u^2/2)(t,x)$ of the Burgers equation (see [18] and [5]).

In [8, 5] we start with the case of a straight particle path $h(t)=Vt$; actually, a simple change of variables reduces the problem to the case $V = 0$, i.e. to problem (1.2), and the associated germ $\mathcal{G}_V$ is $\mathcal{G}+(V,V)$ ($\mathcal{G}$ being the germ for problem (1.2)). We get well-posedness for the singular conservation law driven by the source located at $\{x = Vt\}$. Next, we construct solutions to $u_t + (u^2/2)_x = -(u-h'(t))\,\delta_0(x-h(t))$ by approximating a general given path $h(\cdot)$ by piecewise affine continuous functions. These solutions have to be interpreted in the way of Section 3, as $\mathcal{G}(\cdot)$-entropy solutions with the family of germs $\mathcal{G}(t) := \mathcal{G} + (h'(t), h'(t))$. Uniqueness is ensured by the general theory (cf. Section 3). Moreover, in a work in progress of F. Lagoutière, N. Seguin and T. Takahashi and the author it is shown that, constructing solutions with the wave-front tracking algorithm, we can ensure a uniform BV bound on the solution in terms of the variation of the initial datum u_0 and the variation of $h'(\cdot)$ (as the Godunov scheme, the WFT algorithm is more difficult to use in practice because a perfect knowledge of the Riemann solver is required; yet in our case, it yields finer uniform estimates). At this point, fixed-point arguments can be used for the coupled problem of [18]. In [5], we give a sketch of existence proof for L^∞ data. Moreover, exploiting the techniques of continuous dependence of BV solutions of $u_t + f(u)_x = s$ on the flux f, thanks to the Gronwall inequality we deduce uniqueness of BV solutions of the free-interface coupled problem introduced in [18].

References

[1] Adimurthi, S. Mishra and G. D. Veerappa Gowda, Optimal entropy solutions for conservation laws with discontinuous flux-functions. J. Hyp. Diff. Equ. 2 (2005), no. 4, 783–837.

[2] Adimurthi and G. D. Veerappa Gowda, Conservation laws with discontinuous flux. J. Math. Kyoto University, 43 (2003), no. 1, 27–70.

[3] B. Andreianov and C. Cancès, Vanishing capillarity solutions of Buckley-Leverett equation with gravity in two-rocks' medium. In preparation.

[4] B. Andreianov, P. Goatin and N. Seguin, Finite volume schemes for locally constrained conservation laws. Numer. Math. 115 (2010), 609–645.

[5] B. Andreianov, F. Lagoutière, N. Seguin and T. Takahashi, Small solids in an inviscid fluid. Netw. Heter. Media 5 (2010), no. 3, 385–404.

[6] B. Andreianov, K. H. Karlsen and N. H. Risebro, A theory of L^1-dissipative solvers for scalar conservation laws with discontinuous flux. Arch. Rat. Mech. Anal. 2011, DOI: 10.1007/s00205-010-0389-4.

[7] B. Andreianov, K. H. Karlsen and N. H. Risebro, On vanishing viscosity approximation of conservation laws with discontinuous flux. Netw. Heter. Media 5 (2010), no. 3, 617–633.

[8] B. Andreianov and N. Seguin, Well-posedness of a singular balance law. Preprint 2011, http://hal.archives-ouvertes.fr/hal-00576959

[9] E. Audusse and B. Perthame, Uniqueness for scalar conservation laws with discontinuous flux via adapted entropies. Proc. Roy. Soc. Edinburgh A 135 (2005), no. 2, 253–265.

[10] P. Baiti and H. K. Jenssen, Well-posedness for a class of 2×2 conservation laws with L^∞ data. J. Differ. Equ. 140 (1997), no. 1, 161–185.

[11] R. Bürger, K. H. Karlsen and J. Towers, An Engquist-Osher type scheme for conservation laws with discontinuous flux adapted to flux connections. SIAM J. Numer. Anal. 47 (2009), 1684–1712.

[12] R. Colombo and P. Goatin, A well posed conservation law with a variable unilateral constraint. J. Differ. Equ. 234 (2007), no. 2, 654–675.

[13] S. Diehl. A uniqueness condition for nonlinear convection-diffusion equations with discontinuous coefficients. J. Hyp. Diff. Eq. 6 (2009), 127–159.

[14] M. Garavello, R. Natalini, B. Piccoli and A. Terracina, Conservation laws with discontinuous flux. Netw. Heter. Media 2 (2007), 159–179.

[15] T. Gimse and N. H. Risebro, Solution of the Cauchy problem for a conservation law with a discontinuous flux function. SIAM J. Math. Anal. 23 (1992), no. 3, 635–648.

[16] K. H. Karlsen, N. H. Risebro and J. D. Towers, L^1 stability for entropy solutions of nonlinear degenerate parabolic convection-diffusion equations with discontinuous coefficients. Skr. K. Nor. Vidensk. Selsk. 3 (2003).

[17] S. N. Kruzhkov, First order quasilinear equations with several independent variables, Mat. Sb.(N.S) 81(123) (1970), 228–255.

[18] F. Lagoutière, N. Seguin and T. Takahashi, A simple 1D model of inviscid fluid-solid interaction. J. Differ. Equ. 245 (2008), no. 11, 3503–3544.

[19] E. Yu. Panov, Existence of strong traces for quasi-solutions of multidimensional conservation laws. J. Hyp. Diff. Equ. 4 (2007), no. 4, 729–770.

[20] E. Yu. Panov, Existence and strong pre-compactness properties for entropy solutions of a first-order quasilinear equation with discontinuous flux. Arch. Rat. Mech. Anal. 195 (2009), no. 2, 643–673.

[21] J. D. Towers, Convergence of a difference scheme for conservation laws with a discontinuous flux. SIAM J. Numer. An. 38 (2000), 681–698.

Aggregation Equation with Growing at Infinity Attractive-repulsive Potentials*

Daniel Balagué
Departament de Matemàtiques
Universitat Autònoma de Barcelona
Bellaterra E08193, Spain
E-mail: dbalague@mat.uab.cat

José A. Carrillo
ICREA & Departament de Matemàtiques
Universitat Autònoma de Barcelona
Bellaterra E08193, Spain
E-mail: carrillo@mat.uab.cat

Abstract

In this paper, we consider the aggregation equation and global in time solutions for this equation when the potential is radially symmetric, attractive-repulsive with possible polynomial growth at infinity and the initial data is compactly supported. We will discuss certain aspects of the existence of nonsingular steady states of the equation for these type of potentials based on numerical results in the radially symmetric case.

1 Introduction and a short panorama

We will be interested in studying qualitative properties of the solutions to the problem

$$\rho_t = \operatorname{div}[\rho(\rho * \nabla W)] \text{ in } \mathbb{R}^N \times [0,T], \tag{1.1}$$

$$\rho(x,0) = \rho_0(x) \text{ for all } x \in \mathbb{R}^N, \tag{1.2}$$

where W is the so-called interaction potential. These kind of models have been proposed as macroscopic descriptions of large systems of agents interacting by repulsive and attractive forces, see [6] and the references

*The authors acknowledge partial support from DGI-MICINN (Spain) project MTM2008-06349-C03-03 and 2009-SGR-345 from AGAUR-Generalitat de Catalunya.

therein. They are formal gradient flows with respect to the Wasserstein distance as seen in [8] although typically defined by non-convex interaction potentials. This kind of problem have received a lot of attention in the last years due to the possibility of getting blow-up in finite time for fully attractive potentials. Short-range repulsive, long-range attractive potentials are natural candidates to produce equilibration for large times of the density distribution. We will start by doing a small review of the main results in this theory in the last years to concentrate later in a generalization of the well-posedness theory for attractive-repulsive potentials with polynomial growth at infinity for compactly supported initial data. Later, we will discuss some preliminary numerical experiments that indicate the existence of nonsingular radial stationary states for singular repulsive potentials in dimension 2, c.f. [13, 12].

Let us start by quickly reviewing some of the well-posedness results in the literature concerning the Cauchy problem for the aggregation equation (1.1)–(1.2), together with some of the main properties proven on the asymptotic behavior of their solutions. Let us denote by $\mathcal{P}_2(\mathbb{R}^N)$ the space of probability measures with bounded second moment.

Existence and uniqueness of weak solutions for the aggregation equation in $\mathcal{P}_2(\mathbb{R}^N) \cap L^p(\mathbb{R}^N)$ have been given in [14, 2, 3]. The local-in-time existence and uniqueness theory in [3] gives also a control on the evolution of the L^p-norm.

Theorem 1.1 (Local existence). *Consider $1 < q < \infty$ and p its Hölder conjugate. Suppose $\nabla W \in W^{1,q}(\mathbb{R}^N)$ and suppose $\rho_0 \in L^p(\mathbb{R}^N)$ is nonnegative. Then, there exists a time $T^* > 0$ and a nonnegative function $\rho \in \mathcal{C}([0,T^*); L^p(\mathbb{R}^N)) \cap \mathcal{C}^1((0,T^*); W^{-1,p}(\mathbb{R}^N))$ such that*

$$\frac{d\rho}{dt}(t) + \operatorname{div}(\rho(t)u(t)) = 0,$$

in the sense of distributions in $\mathbb{R}^N \times (0,T^*)$ with

$$u(t) = -\rho(t) * \nabla W \quad \forall t \in (0,T^*),$$
$$\rho(0) = \rho_0.$$

Moreover the function $t \to \|\rho(t)\|^p_{L^p(\mathbb{R}^N)}$ is differentiable and satisfies

$$\frac{d}{dt}\{\|\rho(t)\|^p_{L^p(\mathbb{R}^N)}\} = -(p-1)\int_{\mathbb{R}^N} \rho(x,t)^p \operatorname{div} u(x,t)\,dx \quad \forall t \in [0,T^*).$$

Let us remark that the solutions constructed are distributional solutions for which characteristics are well defined flows. The fact that p and q are Hölder conjugates implies that the velocity field $u \in C^1$ in space and continuous in time, see [3]. Uniqueness in this setting is proven following the ideas in [11]. This allows us to extend the local solutions constructed in Theorem 1.1 to a maximal time interval of existence $[0, T_{\max})$. In addition, one can find in [3] some sufficient conditions to get global-in-time existence and uniqueness.

Theorem 1.2 (Global existence: ΔW bounded above). *Under the assumptions of* Theorem 1.1, *and if we assume moreover that* $\rho_0 \in L^1(\mathbb{R}^N)$ *and* $\operatorname{ess\,sup} \Delta W < +\infty$, *then the solution* ρ *exists for all times (i.e.* $T_{\max} = +\infty$*).*

Weak measure solutions to the the Cauchy problem for the aggregation equation (1.1)–(1.2) are constructed in [7] under the condition that the potential is smooth except possibly at the origin, the growth at infinity is no worse than quadratic, and the singularity at the origin of the derivative of the potential is not worse that Lipschitz. A motivating example in [7] is the so-called attractive Morse potential $W(x) = 1 - e^{-|x|}$. One key assumption involves the so-called λ-convexity of the potential $W(x)$ as in [1]. More precise assumptions are to be found in [7]. In fact, it is shown in [7] that the Cauchy problem for the aggregation equation (1.1)–(1.2) is well posed in $\mathcal{P}_2(\mathbb{R}^N)$. This allows to give sense to finite number of particles following the gradient flow dynamics determined by the potential W as a weak-measure solution of (1.1). The velocity field for measure solutions is shown to be well defined as

$$u(t) = -\int_{x \neq y} \nabla W(x-y)\, d\rho(t)$$

and belonging to $L^2(\rho(t))$ for all $t \in (0, \infty)$.

For initial data in $\mathcal{P}_2(\mathbb{R}^N) \cap L^p(\mathbb{R}^N)$, the weak-$L^p$ solutions satisfy equation (1.1) in the distributional sense, and they lead to curves in the space $\mathcal{P}_2(\mathbb{R}^N) \cap L^p(\mathbb{R}^N)$ which are continuous with respect to the strong topology of L^p up to its maximal time existence $T_{\max}$. The velocity field given by $\rho * \nabla W$ belongs to $L^1((0,T), L^2(\rho(t)))$ for all $T \in (0, T_{\max})$. Therefore, weak-L^p solutions with initial data in $\mathcal{P}_2(\mathbb{R}^N) \cap L^p(\mathbb{R}^N)$ are weak measure solutions, up to their maximal time of existence.

The problem is global-in-time well-posed in $L^p(\mathbb{R}^N)$ under Osgood conditions [4, 3] and there is blow-up of the L^p-norm when this condition is violated [5, 4, 7]. To define the Osgood condition we need to talk about natural potentials:

Definition 1.3 (Natural potential). A *natural potential* is a radially symmetric potential $W(x) = k(|x|)$, where $k : (0, +\infty) \to \mathbb{R}$ is smooth function satisfying the following conditions:

(C1) $\sup_{r \in (0,\infty)} |k'(r)| < +\infty$.

(C2) There exists $0 < \alpha < N$ with $k'(r) \simeq k''(r) \simeq O(1/r^\alpha)$ as $r \to \infty$.

(MN1) $\exists \delta_1 > 0$ such that $k''(r)$ is monotonic in $(0, \delta_1)$.

(MN2) $\exists \delta_2 > 0$ such that $rk''(r)$ is monotonic in $(0, \delta_2)$.

Natural potentials satisfy automatically that $\nabla W \in W^{1,q}(\mathbb{R}^N)$, $1 < q < \infty$, see [3, Lemma 5.6]. Moreover, a natural potential is said to be *repulsive in short range* if there exists an interval $(0, \delta)$ on which $k(r)$ is decreasing; and it is said to be *strictly attractive in the short range* if there exists an interval $(0, \delta)$ on which $k(r)$ is strictly increasing. With these concepts, the Osgood condition is given by

Theorem 1.4 (Osgood conditions for global well-posedness in L^p). *Suppose that W is a natural potential.*

(i) *If W is repulsive in the short range, then the aggregation equation is globally well posed in $\mathcal{P}_2(\mathbb{R}^N) \cap L^p(\mathbb{R}^N)$, $p > N/(N-1)$.*

(ii) *If W is strictly attractive in the short range, the aggregation equation is globally well posed in $\mathcal{P}_2(\mathbb{R}^N) \cap L^p(\mathbb{R}^N)$, $p > N/(N-1)$, if and only if*

$$r \mapsto \frac{1}{k'(r)} \quad \textit{is not integrable at 0.}$$

Using results given in [15, 12, 13] and using ideas of Wasserstein distance shown in [10, 9], we can prove the existence and stability or instability of stationary states for the equation (1.1) and compute analytically some stationary states for the aggregation equation in one dimension.

2 Towards attractive-repulsive potentials

In this section, we contribute to extend the global-in-time well posedness L^p-theory to attractive-repulsive potentials under suitable assumptions. Let us remember that for the solutions constructed in [3], the characteristics or flow maps are well defined and the flow maps together with their inverses enjoy uniform Lipschitz bounds in space and time for all bounded time intervals $[0, T]$ with $0 < T < T_{\max}$.

Definition 2.1. We say that W is a *attractive-repulsive potential* if it is a radially symmetric potential $W(x) = k(|x|)$, where $k : (0, +\infty) \to \mathbb{R}$ is a smooth function satisfying (MN1) and (MN2). Moreover we will impose that the potential can be written as $W = W_A + W_R$ with W_A attractive and with $\nabla W_A \in W^{1,q}_{loc}(\mathbb{R}^N)$, $1 < q < \infty$, and W_R compactly supported repulsive in short range. Furthermore, we will assume that $k = k_A + k_R$ satisfies

(C3) $D := \sup_{r \in (0,\infty)} |k'_R(r)| < \infty$.

(C4) There exists $m > 0$ such that $\frac{k_A(r)}{1+r^m}$ is bounded and increasing.

We will restrict to compactly supported initial data in $L^p(\mathbb{R}^N)$. The main novelty we want to include in the L^p-theory is that the growth at infinity of the attractive part of the potential should not affect the global-in-time existence of the solution if the initial data is compactly supported. In fact, the assumption (C4) is not sharp and it can be weakened, but we prefer to deal with polynomial growth at infinity for the sake of simplicity.

Theorem 2.2 (Global existence for attractive-repulsive potentials). *Let $W(x)$ be an attractive-repulsive potential as in Definition* 2.1. *Then there exists a global in time solution for the equation* (1.1) *with compactly supported initial data $\rho_0 \in L^p(\mathbb{R}^N)$.*

Proof. Due to translational invariance we can assume without loss of generality that the center of mass is located at the 0 and the total mass is equal to one. To construct global solutions, we proceed by approximation of the potential. In fact, let us consider a mesa function χ_ε with $0 < \varepsilon < 1$, i.e., a positive C^∞ radially decreasing function, such that $\chi_\varepsilon(r) = 1$ for $0 \le r \le 1$, decaying at infinity such that $(1 + r^m)\chi_\varepsilon(r) = 1$ for $r \ge 1 + 1/\varepsilon$.

Now, we define $W_A^\varepsilon(x) = \chi_\varepsilon(|x|)W_A(x)$ for all $0 < \varepsilon < 1$ and $x \in \mathbb{R}^N$. Finally, we define the regularized potential as $W^\varepsilon(x) = W_A^\varepsilon(x) + W_R(x)$. It is straightforward to check that the regularized potential W^ε is a natural potential in the sense of Definition 1.3, and thus the existence and uniqueness of solution ρ^ε for the equation (1.1) with the regularized potential and initial data compactly supported and $\rho_0 \in L^p(\mathbb{R}^N)$ follows from [3], see our Theorem 1.2. Let us denote by $X^\varepsilon(x_0, t)$, with $x_0 \in \mathbb{R}^N$ and $t \ge 0$, the particle maps associated to the regularized potential. Let us remark that the assumption (C4) allows us to keep the attractive character of the potential W_A^ε.

We will show that we can control uniformly in ε the support of the solutions to the regularized problem in bounded time intervals. Consider any ball $B_\sigma(0)$ containing the support of ρ_0, i.e., $\mathrm{supp}(\rho_0) \subset\subset B_\sigma(0)$. Then, we claim that there exists $\delta > 0$ such that for any point $y \in \mathrm{supp}(\rho_0)$ we have that

$$|X^\varepsilon(y,t)|^2 \le \tilde{\sigma}^2(t) := \sigma^2 e^t + D^2(e^t - 1), \quad t \in [0, \delta]. \tag{2.1}$$

To prove this fact, let us study the flow of any point $x_0 \in \partial B_\sigma(0)$ and let us recall that the particle maps are defined by

$$\frac{d}{dt}X^\varepsilon(x_0,t) = u^\varepsilon(X^\varepsilon(x_0,t),t), \qquad \text{with } u^\varepsilon = -\rho^\varepsilon * \nabla W^\varepsilon,$$
$$X^\varepsilon(x_0,0) = 0,$$

where u^ε is the velocity field. By choosing $\delta > 0$ small enough and using the continuity of the particle map $X^\varepsilon(x_0, t)$ and the compactness of $\partial B_\sigma(0)$, we can deduce that

$$0 < d_1 < |X^\varepsilon(x_0,t)| < d_2, \quad \mathrm{supp}(\rho^\varepsilon(\cdot,t)) \subset B_{d_1}(0)\,,$$

for all $0 \le t \le \delta$ and all $x_0 \in \partial B_\sigma(0)$. Then, for all $t \in (0, \delta)$ we can compute

$$\begin{aligned}\frac{1}{2}\frac{d}{dt}|X^\varepsilon(x_0,t)|^2 &= -\int_{\mathbb{R}^N} k_A^{\varepsilon\prime}(|X^\varepsilon - y|)\,\frac{X^\varepsilon - y}{|X^\varepsilon - y|}\cdot X^\varepsilon\,\rho(y)dy \\ &\quad - \int_{\mathbb{R}^N} k_R'(|X^\varepsilon - y|)\,\frac{X^\varepsilon - y}{|X^\varepsilon - y|}\cdot X^\varepsilon\,\rho(y)dy \qquad (2.2)\end{aligned}$$

where we have omitted the variables for clarity. Notice that $k_A^{\varepsilon\prime}(\eta) \ge 0$, $|(X^\varepsilon - y)/\,|X^\varepsilon - y|\,| = 1$ and $-k_R'(\eta) \ge 0$. Moreover, we have that $(X^\varepsilon - y)\cdot X^\varepsilon \ge 0$ for all $y \in \mathrm{supp}(\rho^\varepsilon)$. This is due to the fact that $\mathrm{supp}(\rho^\varepsilon) \subset B_{d_1}(0)$ which implies that $|y| \le |X^\varepsilon|$ so $(X^\varepsilon - y)\cdot X^\varepsilon = |X^\varepsilon - y|\,(|X^\varepsilon| - |y|\cos\theta) \ge 0$ where θ is the angle between $X^\varepsilon - y$ and X^ε. Then, we can get rid of the first term of the right-hand side in (2.2) to get

$$\begin{aligned}\frac{d}{dt}|X^\varepsilon(x_0,t)|^2 &\le -2\int_{\mathbb{R}^N} k_R'(|X^\varepsilon - y|)\frac{X^\varepsilon - y}{|X^\varepsilon - y|}\cdot X^\varepsilon\rho(y)dy \\ &\le 2D\,|X^\varepsilon(x_0,t)| \qquad (2.3)\end{aligned}$$

which implies for all $0 \le t \le \delta$ and all $x_0 \in \partial B_\sigma(0)$ the following differential inequality

$$\frac{d}{dt}|X^\varepsilon(x_0,t)|^2 \le D^2 + |X^\varepsilon(x_0,t)|^2\,,$$

leading to (2.1).

Now let us take $t = \delta$ and consider $\sigma_1 := \tilde\sigma(\delta)$. Our first claim is that $\mathrm{supp}(\rho^\varepsilon(\cdot,\delta)) \subset\subset B_{\sigma_1}(0)$ and thus, that we are in the same conditions as initially. To show that assume by contradiction that

$$\sup\{\|y\|\ ;\ y \in \mathrm{supp}(\rho^\varepsilon(\cdot,\delta))\} = \sigma_1$$

then, it should hold too that

$$\sup\{\|y\|\ ;\ y \in \mathrm{supp}(\rho^\varepsilon(\cdot,\delta))\} = |X^\varepsilon(x_0,\delta)| \qquad (2.4)$$

for some $x_0 \in \partial B_\sigma(0)$.

Let us denote by $X^\varepsilon_{-1}(y,\delta)$ the inverse map to $X^\varepsilon(y,\delta)$. From the uniform Lipschitz estimates for the particle maps and their inverses in [3], we have that given any $T > \delta$, there exists $K = K(T,\varepsilon)$ such that

$$|X^\varepsilon(x_0,\delta) - y| \ge K\,\left|x_0 - X^\varepsilon_{-1}(y,\delta)\right|$$

for all $x_0 \in \partial B_\sigma(0)$ and $y \in \mathrm{supp}(\rho^\varepsilon(\cdot, \delta))$. Since we obviously have $X^\varepsilon_{-1}(y, \delta) \in \mathrm{supp}(\rho_0)$, then

$$|X^\varepsilon(x_0, \delta) - y| \geq K \mathrm{dist}\,(\partial B_\sigma(0), \mathrm{supp}(\rho_0)) > 0\,,$$

contradicting (2.4).

Therefore, we can repeat the process by choosing $\delta_1 > 0$ small enough to get

$$\frac{d}{dt}\left|X^\varepsilon(x_1, t)\right|^2 \leq D^2 + \left|X^\varepsilon(x_1, t)\right|^2\,,$$

for all $\delta \leq t \leq \delta_1$ and all $x_1 \in \partial B_{\sigma_1}(0)$, from which (2.1) follows. This is due to the fact that

$$\left|X^\varepsilon(x_1, t)\right|^2 \leq f(t)$$

for all $\delta \leq t \leq \delta_1$ where $f(t)$ solves $f'(t) = D^2 + f(t)$ for all $\delta \leq t \leq \delta_1$ with $f(\delta) = \sigma_1^2 = \tilde{\sigma}^2(\delta)$, and thus by uniqueness of this ordinary differential equation we infer that $f(t) = \tilde{\sigma}^2(t)$ for all $\delta \leq t \leq \delta_1$. This finally shows that the procedure can be repeated at all times showing (2.1) for all $t \geq 0$.

Therefore, given any fixed time $T > 0$, we have shown a uniform in ε bound of the support of ρ^ε in $[0, T]$, i.e.

$$\mathrm{diam}\,(\mathrm{supp}(\rho^\varepsilon(\cdot, t))) \leq 2\tilde{\sigma}(T) \quad \forall t \in [0, T]. \tag{2.5}$$

It is trivial then to check that for all $t \in [0, T]$ and all $x \in \mathrm{supp}(\rho^\varepsilon(\cdot, t))$, we have

$$\int_{\mathbb{R}^N} \nabla W_A^\varepsilon(x - y)\rho^\varepsilon(y, t)dy = \int_{B_{\tilde{\sigma}(t)}(0)} \nabla W_A(x - y)\rho^\varepsilon(y, t)dy$$

by choosing ε such that $2\tilde{\sigma}(T)\varepsilon < 1$. Therefore by uniqueness [11, 3], we see that $\rho^{\varepsilon_1}(x, t) = \rho^{\varepsilon_2}(x, t) := \rho(x, t)$ for all $t \in [0, T]$, $x \in \mathbb{R}^N$ provided $2\tilde{\sigma}(T)\max(\varepsilon_1, \varepsilon_2) < 1$, and that ρ is a solution to the Cauchy problem for (1.1) with initial data ρ_0 in the interval $[0, T]$. Due to the uniqueness result in [11, 3], we can perform this procedure for arbitrary $T > 0$. □

3 Numerical results

In this section we find analytically and numerically some stationary states for the aggregation equation in some particular cases. For doing that, ρ is thought as a probability density and we will make use of the inverse distribution function of ρ. Consider $F(x, t)$ the distribution function associated to ρ. Since F is nondecreasing, one defines the pseudo inverse distribution function as

$$\varphi(\eta) = \inf\{x\,;\, F(x) > \eta\} \quad \text{for} \quad \eta \in (0, 1).$$

The aggregation equation in radial coordinates reads [4],

$$\frac{\partial \hat{\rho}}{\partial t} + \frac{\partial}{\partial r}(\hat{\rho}\hat{u}) = 0, \tag{3.1}$$

where $\hat{\rho} = \omega_N r^{N-1}\rho(r,t)$, the velocity field $\hat{u}$ is given by

$$\hat{u}(r,t) = -\int_0^\infty \psi(r,\eta)\hat{\rho}(\eta,t)d\eta,$$

and the function ψ is defined as follows:

$$\begin{aligned}\psi(r,\eta) &= \frac{1}{\omega_N \eta^{N-1}} \int_{\partial B(0,\eta)} \nabla W(re_1 - y)\cdot e_1 d\sigma(y)\\ &= \frac{1}{\omega_N \eta^{N-1}} \int_{\partial B(0,\eta)} k'(|re_1 - y|)\frac{re_1 - y}{|re_1 - y|}\cdot e_1 d\sigma(y).\end{aligned} \tag{3.2}$$

It is a matter of computations to check that the equation for the inverse distribution function is

$$\frac{\partial \varphi}{\partial t}(r,t) + \int_0^1 \psi(\varphi(r,t),\varphi(\xi,t))d\xi = 0. \tag{3.3}$$

We will use equation (3.3) to compute the stationary states in two dimensions.

Total Concentration at the origin: $W(x) = \frac{|x|^2}{2}$

Let us start with the simple case of a fully attractive potential. It is straightforward by characteristics to check that the only stationary state of the equation is a delta Dirac at 0. In terms of φ, we see that from (3.2):

$$\psi(r,\eta) = \frac{1}{2\pi}\int_0^{2\pi} r - \eta\cos\theta d\theta = r.$$

That implies $\frac{\partial \varphi}{\partial t}(r,t) + \varphi(r,t) = 0$, which gives us the result depicted in Figure 3.1.

Uniform distribution on the sphere: $W(x) = |x|^4 - |x|^2$

For a regular attractive-repulsive potential one has the following result. When $W(x) = |x|^4 - |x|^2$, the function ψ has the following expression

$$\psi(r,\eta) = 4r^3 - 2r + 8r\eta^2.$$

To find steady states one can impose $\frac{\partial \varphi}{\partial t}(r,t) = 0$. The equation (3.3) is now

$$0 = -4\varphi(r,t)^3 + 2\varphi(r,t) - 8\varphi(r,t)\int_0^1 \varphi(\eta,t)^2 d\eta. \tag{3.4}$$

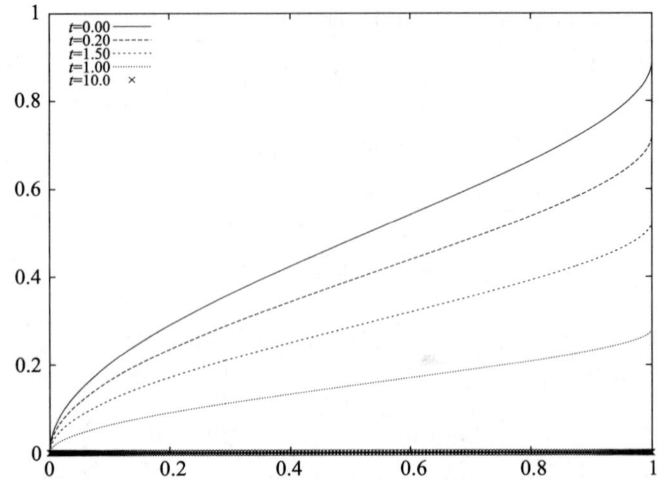

Figure 3.1 Evolution of φ for $W(x) = \frac{|x|^2}{2}$ towards total concentration at 0.

This gives us an equation for finding $M = \int_0^1 \varphi(\eta, t)^2 d\eta = 1/6$. Plugging this value into (3.4) on obtains that the nontrivial steady state for (3.3) is $\varphi(r) \equiv \frac{\sqrt{6}}{6}$. For the probability measure ρ this means that the stationary state is a Dirac measure on a sphere, i.e., uniform distribution on a ring of radius $\frac{\sqrt{6}}{6}$. We show the results for the simulation in Figure 3.2.

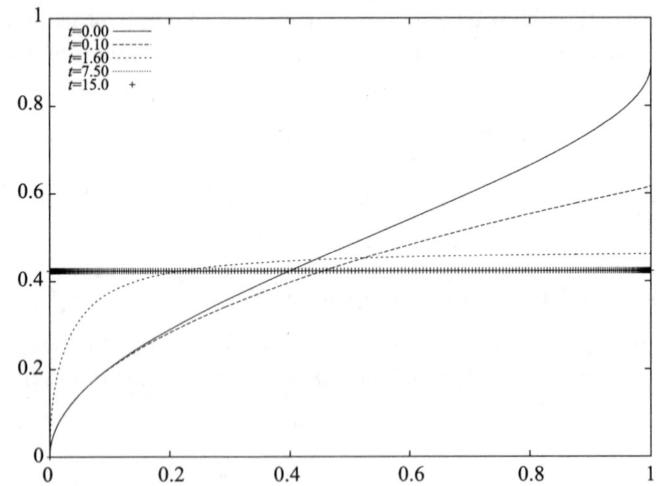

Figure 3.2 Evolution of φ for $W(x) = |x|^4 - |x|^2$ towards the uniform distribution on a ring of radius $\frac{\sqrt{6}}{6}$.

Singular attractive-repulsive potentials: $W(x) = \frac{|x|^2}{2} - |x|$

Since the expression for the inverse equation is difficult to deal with, one chooses the equation for the function $\hat{\rho}$. Imposing $\frac{\partial \hat{\rho}}{\partial t}(r,t) = 0$ one obtains the following equation that $\hat{\rho}$ should satisfy:

$$r = \int_0^\infty \Gamma(r,\eta)\hat{\rho}(r,t)d\eta, \quad \Gamma(r,\eta) = \frac{1}{2\pi}\int_0^{2\pi} \frac{r - \eta\cos\theta}{\sqrt{\eta^2 + r^2 - 2r\eta\cos\theta}}.$$

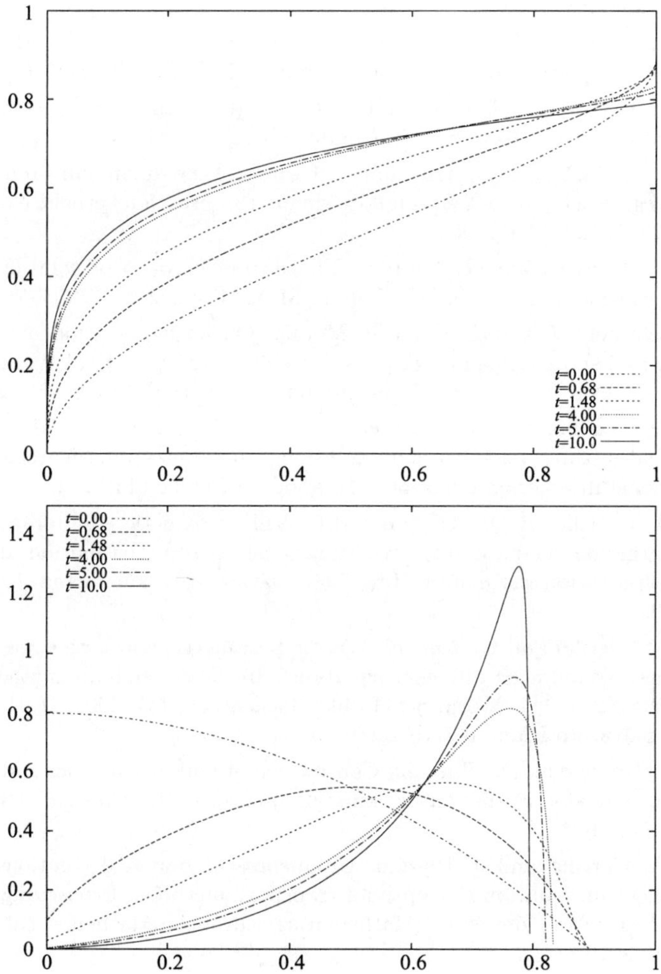

Figure 3.3 Evolution of φ (top) and ρ (below) for $W(x) = \frac{|x|^2}{2} - |x|$. We see the convergence to a profile with no flat parts on φ and thus no concentrations on ρ.

It is an open problem to show the existence of solutions to the previous integrable equation. We conjecture the existence of these integrable stationary states based on the results in one dimension, see [12], and also on the numerical simulations using the inverse distribution function. See Figure 3.3.

References

[1] L. Ambrosio, N. Gigli and G. Savaré, Gradient flows in metric spaces and in the spaces of probability measures. Lectures in Mathematics ETH Zürich, Birkhäuser Verlag, Basel, 2005.

[2] A. L. Bertozzi and J. Brandman, Finite-time blow-up of L^{∞}-weak solutions of an aggregation equation. Comm. Math. Sci. 8(1): 45–65, 2010.

[3] A. L. Bertozzi, T. Laurent and J. Rosado, L^p theory for the aggregation equation, 2010. To appear in Comm. Pur. Appl. Math.

[4] A. L. Bertozzi, J. A. Carrillo and T. Laurent, Blow-up in multidimensional aggregation equations with mildly singular interaction kernels. Nonlinearity 22(3): 683–710, 2009.

[5] A. L. Bertozzi and T. Laurent, Finite-time blow-up of solutions of an aggregation equation in $\mathbf{R}^n$. Comm. Math. Phys. 274(3): 717–735, 2007.

[6] M. Burger, V. Capasso and D. Morale, On an aggregation model with long and short range interactions, Nonlinear Analysis. Real World Applications. An International Multidisciplinary Journal 8: 939–958, 2007.

[7] J. A. Carrillo, M. Di Francesco, A. Figalli, T. Laurent and D. Slepcev, Global-in-time weak measure solutions and finite-time aggregation for nonlocal interaction equations. To appear in Duke Math. J.

[8] J. A. Carrillo, R. J. McCann and C. Villani, Kinetic equilibration rates for granular media and related equations: entropy dissipation and mass transportation estimates. Rev. Matemática Iberoamericana 19: 1–48, 2003.

[9] J. A. Carrillo and G. Toscani, Wasserstein metric and large-time asymptotics of nonlinear diffusion equations. In New trends in mathematical physics, 234–244. World Sci. Publ., Hackensack, NJ, 2004. In Honour of the Salvatore Rionero 70th Birthday.

[10] J. A. Carrillo and G. Toscani, Contractive probability metrics and asymptotic behavior of dissipative kinetic equations. Riv. Mat. di Parma. 6: 75–198, 2007.

[11] J. A. Carrillo and J. Rosado, Uniqueness of bounded solutions to aggregagtion equations by optimal transport methods. Proceedings of the 5th European Congress of Mathematics, European Mathematical Society, 3–16, 2010.

[12] K. Fellner and G. Raoul, Stable stationary states of non-local interaction equations. To appear in Mathematical Models and Methods in Applied Sciences.

[13] K. Fellner and G. Raoul, Stability of stationary states of non-local equations with singular interaction potentials. To appear in Mathematical and Computer Modelling.

[14] T. Laurent, Local and global existence for an aggregation equation. Comm. Partial Differential Equations. 32 (10-12): 1941–1964, 2007.

[15] G. Raoul, Non-local interaction equations: Stationary states and stability analysis. Reprint.

[16] C. Villani, Optimal Transport: Old and New. Volume 338 of Grundlehren der Mathematischen Wissenschaften. Springer-Verlag, Berlin, 2009.

Mixed Type Problems for Transonic Flow in Gas Dynamics and Isometric Embedding in Geometry

Gui-Qiang Chen*
Mathematical Institute, University of Oxford
Oxford, OX1 3LB, UK
E-mail: chengq@maths.ox.ac.uk

Marshall Slemrod†
Department of Mathematics, University of Wisconsin
Madison, WI 53706, USA
E-mail: slemrod@math.wisc.edu

Dehua Wang‡
Department of Mathematics, University of Pittsburgh
Pittsburgh, PA 15260, USA
E-mail: dwang@math.pitt.edu

Abstract

The mixed type problems arising in transonic flows in gas dynamics and isometric embeddings in geometry are considered. The two-dimensional steady Euler equations for the transonic flow past an obstacle such as an airfoil are first discussed. A viscous approximation to the steady transonic flow problem is presented, and

*Also, Department of Mathematics, Northwestern University, Evanston, IL 60208, USA; School of Mathematical Sciences, Fudan University, Shanghai 200433, China. G.-Q. Chen's research was supported in part by the National Science Foundation under Grants DMS-0935967, DMS-0807551, the Natural Science Foundation of China under Grant NSFC-10728101, the Royal Society-Wolfson Research Merit Award (UK), and the UK EPSRC Science and Innovation Award to the Oxford Centre for Nonlinear PDE (EP/E035027/1)

†Also, Department of Mathematics, University of Pittsburgh, Pittsburgh, PA 15260, USA; Department of Mathematical Sciences, Korean Advanced Institute for Science and Technology, Daejeon, Republic of Korea. M. Slemrod's research was supported in part by the National Science Foundation under Grant DMS-0647554.

‡D. Wang's research was supported in part by the National Science Foundation under Grants DMS-0604362 and DMS-0906160, and by the Office of Naval Research under Grant N00014-07-1-0668.

its convergence is obtained by the method of compensated compactness. Then the isometric embedding problem in geometry is discussed. A fluid dynamic formulation of the Gauss-Codazzi system for the isometric embedding of two-dimensional surfaces is provided, and an existence result of isometric immersions with negative Gauss curvature is given. The Div-Curl structure and the weak continuity of the Gauss-Coddazi-Ricci system for the isometric embedding of higher dimensional manifolds are also studied.

1 Introduction

We are concerned with the mixed type problems of partial differential equations arising in the transonic flow past an obstacle in gas dynamics and the isometric embedding in Riemannian geometry. The transonic flow problem past an obstacle arises in many applications of gas dynamics such as an airfoil and is generally of hyperbolic-elliptic mixed type. Shiffman [20] and Bers [2] established the existence of classical subsonic solutions. Morawetz in [15, 16] tackled the problem of transonic flows in a channel or exterior to an airfoil by the method of vanishing viscosity. In Chen-Slemrod-Wang [6], we introduced a new viscous approximation to the transonic flow problem, constructed the invariant regions, and proved the convergence by the method of compensated compactness ([22]). The isometric embedding is a classical problem in differential geometry. The first global existence of smooth C^∞ embeddings was given by Nash [19] (see also [11]). Many other important results have been achieved for the embedding of surfaces with positive Gauss curvature, which can be formulated as an elliptic boundary value problem. For the case of surfaces with negative Gauss curvature, the underlying partial differential equations are hyperbolic. When the Gauss curvature changes signs, the problem of isometric embedding of surfaces is of mixed type. The isometric embedding problem in the cases of hyperbolic and mixed type are much more difficult. See Han-Hong [12] and Yau [23] for related results and discussions. The idea in Chen-Slemrod-Wang [6] on transonic flows was applied to study the isometric embedding problem for surfaces in Chen-Slemrod-Wang [7]. More specifically, in Chen-Slemrod-Wang [7], we introduced a general approach to deal with the isometric immersion problem involving nonlinear partial differential equations of mixed hyperbolic-elliptic type by combining a fluid dynamic formulation of balance laws with a compensated compactness framework. As an example of direct applications of this approach, we show how this idea can be applied to establish an isometric immersion of a two-dimensional Riemannian manifold (or a surface) with negative Gauss curvature in $\mathbb{R}^3$. For the isometric embedding of manifolds with dimensions greater than two, the underlying Gauss-Codazzi-Ricci equations have no type, neither purely hyperbolic nor purely elliptic. In Chen-Slemrod-Wang [8], we explored the Div-Curl structure and established the weak continuity of the Gauss-Coddazi-Ricci system with respect to the uniform L^p-bounded solution sequence for $p > 2$, which implies that the weak limit of the isometric embeddings of the manifold in a fixed coordinate chart is an isometric immersion. More generally, we establish a compensated compactness framework for the Gauss-Codazzi-Ricci system in differential geometry. That is, given any sequence of approximate solutions to this system which is uniformly bounded in $L^p, p > 2$, and has reasonable bounds on the errors made in the approximation (the errors are confined in a compact subset of H^{-1}_{loc}), then the approximating sequence has a weakly convergent subsequence whose limit is a solution of the Gauss-Codazzi-Ricci system. Since the compensated compactness played an important role in our studies, we shall first recall in Section 2 the basic ideas of compensated compactness, particularly the Div-Curl lemma and the weak continuity of 2×2 determinants. In Section 3, we will present a viscous approximation for the transonic flow problem. In Section 4, we discuss a fluid dynamic approach of the isometric immersion problem for two-dimensional surfaces. In Section 5, we show our result on the weak continuity of the Gauss-Coddazi-Ricci system for the isometric embedding of higher dimensional manifolds.

2 Compensated compactness

The methods of compensated compactness, originally introduced by Tartar and Murat [17, 22] and by Ball [1] for a related observation, deal with the behavior of nonlinear functions with respect to weak topology, and are efficient tools for studying the singular limits, especially the compactness of the solution operators, for quasilinear hyperbolic systems of conservation laws. Consider a sequence $\{v^\varepsilon\}$ of uniformly bounded measurable functions. It is possible to extract a convergent subsequence in the weak-star topology of L^∞. Unfortunately, this topology is too weak to guarantee the continuity of nonlinear functions defined on the sequence. In fact, if $v \equiv w^*\text{-}\lim v^\varepsilon$ and f is a generic continuous function, we have $f(v) \neq w^*\text{-}\lim f(v^\varepsilon)$, unless v^ε is strongly convergent, where $w^*\text{-}\lim$ denotes the weak limit in the sense of the weak-star topology in L^∞. However, for any given sequence $v^\varepsilon : \Omega \subseteq \mathbb{R}^n \to \mathbb{R}^m$, uniformly bounded in L^∞, it is possible to associate with a family of probability measures $\{\nu_x\}_{x\in\Omega}$, the Young measures such that, for any continuous function $f : \mathbb{R}^m \to \mathbb{R}^k$, one has $w^*\text{-}\lim f(v^\varepsilon(x)) = \int f(\lambda)d\nu_x(\lambda)$. The strong convergence is then characterized by the Dirac measure. The deviation between weak and strong convergence is measured by the spreading of the support of the Young measure ν_x. To study the behavior of the

Young measure, we need some differential properties of the sequence, especially the H^{-1} compactness of the entropy dissipation measures for conservation laws.

The Div-Curl lemma (cf. [10, 17, 22]) is a basic result in the compensated compactness theory for the weak continuity of the scalar product of two vector fields if we have good enough control on the divergence and curl. The key insight as in Murat [17] is that, if we have enough information concerning various combinations of derivatives, we can sometimes show that certain nonlinear functions are weakly continuous.

For $w = (w_1, w_2, \cdots, w_d) \in L^2$, we define $\operatorname{curl} w \in W^{-1,2}$ by

$$\operatorname{curl} w := (\partial_{x_j} w_i - \partial_{x_i} w_j)_{1\le i,j\le d},$$

which is a $d \times d$ matrix field.

Theorem 2.1 (Div-Curl Lemma). *Let $\Omega \subset \mathbb{R}^d, d \ge 2$, be open bounded. Let $p, q > 1$ such that $\frac{1}{p} + \frac{1}{q} = 1$. Assume that two field sequences $u^\varepsilon \in L^p(\Omega;\mathbb{R}^d)$ and $v^\varepsilon \in L^q(\Omega;\mathbb{R}^d)$ satisfy the following:*

(i) *$u^\varepsilon \rightharpoonup u$ weakly in $L^p(\Omega;\mathbb{R}^d)$ as $\varepsilon \to 0$;*
(ii) *$v^\varepsilon \rightharpoonup v$ weakly in $L^q(\Omega;\mathbb{R}^d)$ as $\varepsilon \to 0$;*
(iii) *$\operatorname{div} u^\varepsilon$ are confined in a compact subset of $W^{-1,p}_{loc}(\Omega;\mathbb{R})$;*
(iv) *$\operatorname{curl} v^\varepsilon$ are confined in a compact subset of $W^{-1,q}_{loc}(\Omega;\mathbb{R}^{d\times d})$.*

Then the scalar product of u^ε and v^ε are weakly continuous:

$$u^\varepsilon \cdot v^\varepsilon \longrightarrow u \cdot v$$

in the sense of distributions.

Using the Div-Curl lemma, one obtains the following weak continuity of 2×2 determinants.

Theorem 2.2 (Weak Continuity of a 2×2 Determinant). *Let $\Omega \subset \mathbb{R} \times \mathbb{R}^+ = \mathbb{R}^2_+$ be a bounded open set and $u^\varepsilon = (u_1^\varepsilon, u_2^\varepsilon, u_3^\varepsilon, u_4^\varepsilon) : \Omega \to \mathbb{R}^4$ be measurable functions satisfying*

$$u^\varepsilon \rightharpoonup u = (u_1, u_2, u_3, u_4) \qquad \text{in } L^2_4(\Omega),$$

and

$$\frac{\partial u_1^\varepsilon}{\partial t} + \frac{\partial u_2^\varepsilon}{\partial x}, \quad \frac{\partial u_3^\varepsilon}{\partial t} + \frac{\partial u_4^\varepsilon}{\partial x} \qquad \text{are compact in } H^{-1}_{loc}(\Omega).$$

Then there exists a subsequence (still labeled) u^ε such that

$$\begin{vmatrix} u_1^\varepsilon & u_2^\varepsilon \\ u_3^\varepsilon & u_4^\varepsilon \end{vmatrix} \rightharpoonup \begin{vmatrix} u_1 & u_2 \\ u_3 & u_4 \end{vmatrix} \qquad \text{in the sense of distributions.}$$

3 Transonic flow in gas dynamics

In gas dynamics, the two-dimensional steady, isentropic transonic flow past an obstacle such as an airfoil is governed by the following steady Euler equations:

$$\begin{cases} (\rho u)_x + (\rho v)_y = 0, \\ (\rho u^2 + p)_x + (\rho uv)_y = 0, \\ (\rho uv)_x + (\rho v^2 + p)_y = 0, \end{cases} \tag{3.1}$$

where (x, y) are the spatial variables, ρ is the density, (u, v) is the velocity, and $p = \rho^\gamma/\gamma$, $\gamma \geq 1$, is the pressure. Under the assumption that the flow is irrotational, (3.1) can be reduced to the following system of irrotationality and conservation of mass:

$$\begin{cases} v_x - u_y = 0, \\ (\rho u)_x + (\rho v)_y = 0, \end{cases} \tag{3.2}$$

together with Bernoulli's law by scaling:

$$\rho = \left(1 - \frac{\gamma - 1}{2} q^2\right)^{\frac{1}{\gamma-1}}, \quad \gamma > 1, \tag{3.3}$$

where q is the flow speed defined by $q^2 = u^2 + v^2$. The sound speed c is defined as $c^2 = p'(\rho) = 1 - \frac{\gamma-1}{2}q^2$. At the cavitation point $\rho = 0$, $q = q_{cav} := \sqrt{\frac{2}{\gamma-1}}$. At the stagnation point $q = 0$, the density reaches its maximum $\rho = 1$. Bernoulli's law (3.3) is valid for $0 \leq q \leq q_{cav}$. Define the critical speed q_{cr} as $q_{cr} := \sqrt{\frac{2}{\gamma+1}}$. We rewrite Bernoulli's law (3.3) in the form:

$$q^2 - q_{cr}^2 = \frac{2}{\gamma + 1}\left(q^2 - c^2\right). \tag{3.4}$$

We see that the flow is subsonic when $q < q_{cr}$, sonic when $q = q_{cr}$, and supersonic when $q > q_{cr}$. Thus, the problem of transonic flows is of mixed type. The isothermal case ($\gamma = 1$) is similar.

When the upstream speed is sufficiently small, the flow remains subsonic and the governing equations are elliptic. Shiffman [20] and Bers [2] proved the existence of a unique subsonic solution. When the upstream speed reaches a certain number, the flow becomes transonic and shock waves form. Morawetz presented a program in her two papers [15, 16] for proving the existence of weak solutions to the transonic flow in a channel or exterior to an airfoil. She proved that, if the solutions of some viscous approximation problem satisfy the compensated compactness framework, then there is a convergent subsequence whose limit is a solution of the transonic flow problem. The viscous approximation problem was not identified in [15, 16]. In Chen-Slemrod-Wang [6], we introduced the polar coordinates in the phase plane:

$$u = q\cos\theta, \quad v = q\sin\theta; \qquad \theta = \tan^{-1}\left(\frac{v}{u}\right),$$

and provided such a viscous approximation problem:

$$\begin{cases} v_x - u_y = R_1 := \varepsilon\Delta\theta, \\ (\rho u)_x + (\rho v)_y = R_2 := \varepsilon\nabla\cdot(\sigma(\rho)\nabla\rho), \end{cases} \tag{3.5}$$

where $\sigma(\rho)$ is any positive smooth function satisfying

$$\sigma(\rho) = 1 - \frac{c^2}{q^2} \qquad \text{for } q > c.$$

Using the Riemann invariants, we found the invariant regions for the viscous solutions and showed that the viscous solutions are bounded and stay uniformly in ε away from cavitation. To show the convergence of the viscous approximate solutions, we introduced the entropy pairs given by the Loewner-Morawetz relation, the corresponding entropy dissipation measures were analyzed so that the viscous approximate solutions satisfy the compensated compactness framework, and finally the argument of Morawetz [16] leads to the convergence. In [6], we showed that the flow is away from cavitation, but assumed that the flow angles remain bounded and the flow is away from the stagnation point.

4 Isometric embedding of surfaces

The isometric embedding of a two-dimensional Riemannian manifold (or a surface) into $\mathbb{R}^3$ can be studied through the Gauss-Codazzi system described as follows. Let $g_{ij}, i, j = 1, 2$, be the given metric of a two-dimensional Riemannian manifold $\mathcal{M}$ parameterized on an open set $\Omega \subset \mathbb{R}^2$. The first fundamental form I for $\mathcal{M}$ on Ω is

$$I := g_{11}(dx)^2 + 2g_{12}dxdy + g_{22}(dy)^2,$$

and the isometric embedding problem is to seek a map $\mathbf{r} : \Omega \to \mathbb{R}^3$ such that $d\mathbf{r}\cdot d\mathbf{r} = I$, that is,

$$\partial_x\mathbf{r}\cdot\partial_x\mathbf{r} = g_{11}, \quad \partial_x\mathbf{r}\cdot\partial_y\mathbf{r} = g_{12}, \quad \partial_y\mathbf{r}\cdot\partial_y\mathbf{r} = g_{22},$$

so that $\{\partial_x\mathbf{r}, \partial_y\mathbf{r}\}$ in $\mathbb{R}^3$ are linearly independent. The corresponding second fundamental form is

$$II := h_{11}(dx)^2 + 2h_{12}dxdy + h_{22}(dy)^2.$$

The fundamental theorem of surface theory (cf. [9, 12]) indicates that there exists a surface in $\mathbb{R}^3$ whose first and second fundamental forms are I and II if the coefficients (g_{ij}) and (h_{ij}) of the two given quadratic forms I and II with $(g_{ij}) > 0$ satisfy the Gauss-Codazzi system. It is indicated in Mardare [14] (Theorem 9) that this theorem holds even when (h_{ij}) is only in L^∞ for given (g_{ij}) in $C^{1,1}$, for which the immersion surface is $C^{1,1}$. This shows that, for the realization of a two-dimensional Riemannian manifold in $\mathbb{R}^3$ with given metric $(g_{ij}) > 0$, it suffices to solve $(h_{ij}) \in L^\infty$ determined by the Gauss-Codazzi system to recover $\mathbf{r}$ a posteriori. The Gauss-Codazzi system (cf. [9, 12]) can be written as

$$\begin{cases} \partial_x M - \partial_y L = \Gamma_{22}^{(2)} L - 2\Gamma_{12}^{(2)} M + \Gamma_{11}^{(2)} N, \\ \partial_x N - \partial_y M = -\Gamma_{22}^{(1)} L + 2\Gamma_{12}^{(1)} M - \Gamma_{11}^{(1)} N, \end{cases} \tag{4.1}$$

with

$$LN - M^2 = \kappa, \tag{4.2}$$

where

$$L = \frac{h_{11}}{\sqrt{|g|}}, \quad M = \frac{h_{12}}{\sqrt{|g|}}, \quad N = \frac{h_{22}}{\sqrt{|g|}}, \quad |g| = \det(g_{ij}) = g_{11}g_{22} - g_{12}^2,$$

$\kappa(x, y)$ is the Gauss curvature that is determined by the relation:

$$\kappa = \frac{R_{1212}}{|g|}, \quad R_{ijkl} = g_{lm}\left(\partial_k \Gamma_{ij}^{(m)} - \partial_j \Gamma_{ik}^{(m)} + \Gamma_{ij}^{(n)}\Gamma_{nk}^{(m)} - \Gamma_{ik}^{(n)}\Gamma_{nj}^{(m)}\right),$$

R_{ijkl} is the curvature tensor and depends on (g_{ij}) and its first and second derivatives, and

$$\Gamma_{ij}^{(k)} = \frac{1}{2} g^{kl} \left(\partial_j g_{il} + \partial_i g_{jl} - \partial_l g_{ij}\right)$$

is the Christoffel symbol and depends on the first derivatives of (g_{ij}), where the summation convention is used, (g^{kl}) denotes the inverse of (g_{ij}), and $(\partial_1, \partial_2) = (\partial_x, \partial_y)$. Therefore, given a positive definite metric $(g_{ij}) \in C^{1,1}$, the Gauss-Codazzi system gives us three equations for the three unknowns (L, M, N) determining the second fundamental form II. From the viewpoint of geometry, the Gauss equation (4.2) is a constraint condition and the Codazzi equations in (4.1) are integrability relations.

Consider (M, N) as the state variables. If $N \neq 0$, the eigenvalues of system (4.1)–(4.2) are

$$\frac{-M \pm \sqrt{-\kappa}}{N}. \tag{4.3}$$

The Gauss-Codazzi system (4.1)–(4.2) is hyperbolic if $\kappa < 0$, elliptic if $\kappa > 0$, and parabolic if $\kappa = 0$. Thus, for a generic surface, the

Gauss curvature κ may change sign, then system (4.1)–(4.2) is of mixed hyperbolic-elliptic type, as in transonic flow (3.2)–(3.3).

Motivated by this similarity of mixed type, we can put the isometric embedding problem into a fluid dynamic formulation so that it may be solved via the approaches for transonic flows of fluid dynamics in Chen-Slemrod-Wang [6]. As in Chen-Slemrod-Wang [7], we set

$$L = \rho v^2 + p, \quad M = -\rho uv, \quad N = \rho u^2 + p,$$

and set $q^2 = u^2 + v^2$ as usual. Then the Codazzi equations (4.1) become the familiar balance laws of momentum:

$$\begin{cases} \partial_x(\rho uv) + \partial_y(\rho v^2 + p) = -(\rho v^2 + p)\Gamma_{22}^{(2)} - 2\rho uv\Gamma_{12}^{(2)} - (\rho u^2 + p)\Gamma_{11}^{(2)}, \\ \partial_x(\rho u^2 + p) + \partial_y(\rho uv) = -(\rho v^2 + p)\Gamma_{22}^{(1)} - 2\rho uv\Gamma_{12}^{(1)} - (\rho u^2 + p)\Gamma_{11}^{(1)}, \end{cases} \tag{4.4}$$

and the Gauss equation (4.2) becomes

$$\rho p q^2 + p^2 = \kappa. \tag{4.5}$$

We choose the pressure p as for the Chaplygin-type gas:

$$p = -\frac{1}{\rho}.$$

Then, from (4.5), we have the "Bernoulli" relation:

$$\rho = \frac{1}{\sqrt{q^2 + \kappa}}. \tag{4.6}$$

This yields

$$p = -\sqrt{q^2 + \kappa}, \tag{4.7}$$

and

$$u^2 = p(p - M), \quad v^2 = p(p - L), \quad M^2 = (N - p)(L - p),$$

which lead to the relations for (u, v) in terms of (L, M, N). We define the "sound" speed as

$$c^2 = p'(\rho) = \frac{1}{\rho^2},$$

then, from our "Bernoulli" relation (4.6), we see

$$c^2 = q^2 + \kappa. \tag{4.8}$$

Hence, under this formulation, when $\kappa > 0$, the "flow" is subsonic, and system (4.4)–(4.5) is elliptic; when $\kappa < 0$, the "flow" is supersonic, and

(4.4)–(4.5) is hyperbolic; and when $\kappa = 0$, the "flow" is sonic, and (4.4)–(4.5) is degenerate. In general, system (4.4)–(4.5) is of mixed hyperbolic-elliptic type, which is consistent with the eigenvalues (4.3) of the original system (4.1)–(4.2).

In Chen-Slemrod-Wang [7], we provided a general framework of weak solutions to the Gauss-Codazzi system (4.1)–(4.2) with an application in the case of negative Gauss curvature $\kappa < 0$. Taking $\kappa = -\gamma^2$ with $\gamma > 0$, we rescale (L, M, N) as

$$\tilde{L} = \frac{L}{\gamma}, \qquad \tilde{M} = \frac{M}{\gamma}, \qquad \tilde{N} = \frac{N}{\gamma},$$

with

$$\tilde{L}\tilde{N} - \tilde{M}^2 = -1.$$

We considered one of the variables, x, as time-like and the other variable y as space-like, introduced a vanishing viscosity method via parabolic regularization:

$$\begin{cases} \partial_x(\rho uv) + \partial_y(\rho v^2 + p) = R_1 + \varepsilon \partial_y^2(\rho v), \\ \partial_x(\rho u^2 + p) + \partial_y(\rho uv) = R_2 + \varepsilon \partial_y^2(\rho u), \end{cases} \tag{4.9}$$

where

$$R_1 := -(\rho v^2 + p)\tilde{\Gamma}_{22}^{(2)} - 2\rho uv\tilde{\Gamma}_{12}^{(2)} - (\rho u^2 + p)\tilde{\Gamma}_{11}^{(2)},$$

$$R_2 := -(\rho v^2 + p)\tilde{\Gamma}_{22}^{(1)} - 2\rho uv\tilde{\Gamma}_{12}^{(1)} - (\rho u^2 + p)\tilde{\Gamma}_{11}^{(1)},$$

$$\tilde{\Gamma}_{11}^{(1)} = \Gamma_{11}^{(1)} + \frac{\gamma_x}{\gamma}, \qquad \tilde{\Gamma}_{12}^{(1)} = \Gamma_{12}^{(1)} + \frac{\gamma_y}{2\gamma}, \qquad \tilde{\Gamma}_{22}^{(1)} = \Gamma_{22}^{(1)},$$

$$\tilde{\Gamma}_{11}^{(2)} = \Gamma_{11}^{(2)}, \qquad \tilde{\Gamma}_{12}^{(2)} = \Gamma_{12}^{(2)} + \frac{\gamma_x}{2\gamma}, \qquad \tilde{\Gamma}_{22}^{(2)} = \Gamma_{22}^{(2)} + \frac{\gamma_y}{\gamma}.$$

For (4.9), we obtained the uniform estimate by identifying invariant regions for the approximate solutions and showed the H_{loc}^{-1}–compactness. Then, as in [6], the compensated compactness framework yields a weak solution to the initial value problem of system (4.4)–(4.5) and thus system (4.1)–(4.2) when the initial data lies in the invariant region for the catenoid with the given metric:

$$g_{11} = g_{22} = (\cosh(cx))^{\frac{2}{\beta^2-1}}, \quad g_{12} = 0, \quad \kappa(x) = -\kappa_0 E(x)^{-\beta^2},$$

with $c \neq 0, \kappa_0 > 0, \beta > \sqrt{2}$. Here we applied the weak continuity of 2×2 determinants in Theorem 2.2 to pass the limit in the constraint equation (4.2). This establishes a $C^{1,1}$-immersion of the Riemannian manifold into $\mathbb{R}^3$. In particular, our existence result asserts the existence of a $C^{1,1}$-surface for the associated metric for a class of non-circular cross-sections prescribed at $x = x_0$ ($\neq 0$) for a catenoid.

We remark that, in [7], we established the existence of $C^{1,1}$-isometric immersions for a class of initial data where the given metric is that of a catenoid. The existence of a global isometric embedding or immersion of a general surface with negative Gauss curvature is still open and is of course generally not possible due to Efimov's theorem (cf. [12]). When the Gauss curvature κ changes sign, the problem becomes transonic, thus mixed hyperbolic-elliptic type. In this mixed-type problem, only special local solutions are known to exist for special data (cf. [13, 12]), and the existence of global solutions is a difficult open problem.

5 Isometric embedding of higher dimensional manifolds

An extremely challenging direction in geometry is the higher dimensional isometric embedding, and all major problems are open. In fact, the isometric embedding problem of higher dimensional manifold $\mathcal{M}^d$, $d \geq 3$, is not elliptic as noted by S.-S. Chern and H. Levy (cf. [3, 21]). In Chen-Slemrod-Wang [8], we explored the structure of the Gauss-Codazzi-Ricci system of higher dimensional isometric embedding and showed the weak continuity of the system by using the compensated compactness argument.

The isometric embedding problems for d-dimensional Riemannian manifolds with given metric g_{ij} into the Euclidean space $\mathbb{R}^N$ can be reduced as the solvability problems of the Gauss-Codazzi-Ricci equations with the following form in local coordinates $g_{ij}dx^i dx^j = ds^2$:

$$\begin{aligned}
&h^a_{ji}h^a_{kl} - h^a_{ki}h^a_{jl} = R_{ijkl},\\
&\frac{\partial h^a_{lj}}{\partial x^k} - \frac{\partial h^a_{kj}}{\partial x^l} + \Gamma^m_{lj}h^a_{km} - \Gamma^m_{kj}h^a_{lm} + \kappa^a_{kb}h^b_{lj} - \kappa^a_{lb}h^b_{kj} = 0,\\
&\frac{\partial \kappa^a_{lb}}{\partial x^k} - \frac{\partial \kappa^a_{kb}}{\partial x^l} - g^{mn}\left(h^a_{ml}h^b_{kn} - h^a_{mk}h^b_{ln}\right) + \kappa^a_{kc}\kappa^c_{lb} - \kappa^a_{lc}\kappa^c_{kb} = 0,
\end{aligned} \tag{5.1}$$

where (g^{mn}) denotes the inverse of (g_{ij}), Γ^k_{ij} the Christoffel symbol, R_{ijkl} the Riemann curvature tensor, h^a_{ij} the coefficients of the second fundamental form, κ^a_{lb} the coefficients of the connection form (torsion coefficients) on the normal bundle, and the indices a, b, c run from 1 to $N-d$, and i, j, k, l, m, n run from 1 to $d \geq 3$. The coefficients of the second fundamental form are symmetric: $h^a_{ij} = h^a_{ji}$, and the coefficients of the connection form are antisymmetric: $\kappa^a_{kb} = -\kappa^b_{ka}$.

In Chen-Slemrod-Wang [8], we presented our main observation on the features of the Codazzi and Ricci equations: the Div-Curl structure, which leads to the weak continuity of the system. Using the Div-Curl lemma (i.e., Theorem 2.1), we established the weak continuity of the Gauss-Codazzi-Ricci system and related compensated compactness framework for approximate solutions to the system. In particular, we have the following weak continuity of the Gauss-Codazzi-Ricci system with respect to the uniform L^p-bounded solution sequence for $p > 2$: *Let* $(h_{ij}^{a,\varepsilon}, \kappa_{lb}^{a,\varepsilon})$ *be a sequence of solutions to the Gauss-Codazzi-Ricci system* (5.1), *which is uniformly bounded in* L^p, $p > 2$. *Then the weak limit vector field* $(h_{ij}^{a}, \kappa_{lb}^{a})$ *of the sequence* $(h_{ij}^{a,\varepsilon}, \kappa_{lb}^{a,\varepsilon})$ *in* L^p *is still a solution to* (5.1). We showed that the solution sequence $(h_{ij}^{a,\varepsilon}, \kappa_{lb}^{a,\varepsilon})$ for the weak continuity can be obtained from a selection criterion. In particular, we wish to select an "energy" minimizer among all embeddings with local coordinates $g_{ij}dx^i dx^j = ds^2$. Non-uniqueness and uniqueness (rigidity) of embeddings was discussed by Berger-Bryant-Griffiths [3, 4]. For the details and more discussions on the higher dimensional isometric embedding, we refer to Chen-Slemrod-Wang [8], as well as [5, 18] for the local existence of embeddings.

References

[1] J. M. Ball, Convexity conditions and existence theorems in nonlinear elasticity. Arch. Rational Mech. Anal. 63 (1977), 337–403.

[2] L. Bers, Mathematical Aspects of Subsonic and Transonic Gas Dynamics. John Wiley & Sons, Inc.: New York; Chapman & Hall, Ltd.: London, 1958.

[3] E. Berger, R. Bryant and P. Griffiths, Some isometric embedding and rigidity results for Riemannian manifolds. Proc. Nat. Acad. Sci. 78 (1981), 4657–4660.

[4] E. Berger, R. Bryant and P. Griffiths, Characteristics and rigidity of isometric embeddings, Duke Math. J. 50 (1983), 803–892.

[5] R. L. Bryant, P. A. Griffiths and D. Yang, Characteristics and existence of isometric embeddings. Duke Math. J. 50 (1983), 893–994.

[6] G.-Q. Chen, M. Slemrod and D. Wang, Vanishing viscosity method for transonic flow. Arch. Rational Mech. Anal. 189 (2008), 159–188.

[7] G.-Q. Chen, M. Slemrod and D. Wang, Isometric immersions and compensated compactness, Commun. Math. Phys. 294 (2010), 411–437.

[8] G.-Q. Chen, M. Slemrod and D. Wang, Weak continuity of the Gauss-Codazzi-Ricci system for isometric embedding, Proc. Amer. Math. Soc. 138 (2010), 1843–1852.

[9] M. P. do Carmo, Riemannian Geometry, Transl. by F. Flaherty. Birkhäuser: Boston, MA, 1992.

[10] L. C. Evans, Weak Convergence Methods for Nonlinear Partial Differential Equations. CBMS-RCSM, Vol. 74. AMS: Providence, 1991.

[11] M. Günther, On the perturbation problem associated to isometric embeddings of Riemannian manifolds. Ann. Global Anal. Geom. 7 (1989), 69–77.

[12] Q. Han and J.-X. Hong, Isometric Embedding of Riemannian Manifolds in Euclidean Spaces. AMS: Providence, RI, 2006.

[13] C.-S. Lin, The local isometric embedding in $\mathbb{R}^3$ of 2-dimensional Riemannian manifolds with Gaussian curvature changing sign cleanly. Comm. Pure Appl. Math. 39 (1986), 867–887.

[14] S. Mardare, On Pfaff systems with L^p coefficients and their applications in differential geometry. J. Math. Pure Appl. 84 (2005), 1659–1692.

[15] C. S. Morawetz, On a weak solution for a transonic flow problem. Comm. Pure Appl. Math. 38 (1985), 797–818.

[16] C. S. Morawetz, On steady transonic flow by compensated compactness. Meth. Appl. Anal. 2 (1995), 257–268.

[17] F. Murat. Compacité par compensation. Ann. Scuola Norm. Sup. Pisa Cl. Sci. 5(4) (1978), 489–507.

[18] G. Nakamura and Y. Maeda, Local isometric embedding problem of Riemannian 3-manifold into $\mathbb{R}^6$. Proc. Japan Acad. Ser. A: Math. Sci. 62 (1986), 257–259.

[19] J. Nash, The imbedding problem for Riemannian manifolds. Ann. Math. (2), 63 (1956), 20–63.

[20] M. Shiffman, On the existence of subsonic flows of a compressible fluid. J. Rational Mech. Anal. 1 (1952), 605–652.

[21] M. Spivak, A Comprehensive Introduction to Differential Geometry. Publish or Perish, Inc., Boston, Mass., Vol. V, 1975.

[22] L. Tartar, Compensated compactness and applications to partial differential equations. In: Nonlinear Analysis and Mechanics: Heriot-Watt Symposium, Vol. IV, pp. 136–212, Res. Notes in Math. 39, Pitman, Boston, Mass.-London, 1979.

[23] S.-T. Yau, Review of geometry and analysis. In: Mathematics: Frontiers and Perspectives, pp. 353–401, International Mathematics Union, Eds. V. Arnold, M. Atiyah, P. Lax and B. Mazur, AMS: Providence, 2000.

Hamiltonian Approximation of Entropy Solutions of the Burgers Equation*

Giuseppe Maria Coclite
Department of Mathematics, University of Bari
Bari 70125, Italy
E-mail: coclitegm@dm.uniba.it

Kenneth Hvistendahl Karlsen
Centre of Mathematics for Applications (CMA), University of Oslo
Oslo 0316, Norway
E-mail: kennethk@math.uio.no

Abstract

We consider a fourth order equation, which contains nonlinear dispersive effects as well as fourth order dissipative effects. We prove that as the diffusion and dispersion parameters tend to zero, with a condition on the relative balance between these two parameters, smooth solutions of the equation converge to discontinuous weak solutions of a scalar conservation law.

1 Introduction

Consider the Cauchy problem for the Burgers equation

$$\begin{cases} \partial_t u + u\partial_x u = 0, & t > 0, \ x \in \mathbb{R}, \\ u(0,\cdot) = u_0 \in L^1(\mathbb{R}) \cap L^\infty(\mathbb{R}), & \text{in } \mathbb{R}. \end{cases} \tag{1.1}$$

In this paper we investigate the convergence of the solution $v_{\varepsilon,\alpha}$ of the elliptic-parabolic system [4]

$$\begin{cases} \partial_t v_{\varepsilon,\alpha} + u_{\varepsilon,\alpha}\partial_x v_{\varepsilon,\alpha} = \varepsilon\partial_{xx}^2 v_{\varepsilon,\alpha}, & t > 0, \ x \in \mathbb{R}, \\ v_{\varepsilon,\alpha} = u_{\varepsilon,\alpha} - \alpha\partial_{xx}^2 u_{\varepsilon,\alpha}, & t > 0, \ x \in \mathbb{R}, \\ v_{\varepsilon,\alpha}(0,x) = u_{0,\varepsilon,\alpha}(x), & x \in \mathbb{R}, \end{cases} \tag{1.2}$$

*This work was initiated while G. M. Coclite visited the Centre of Mathematics for Applications (CMA) at the University of Oslo. He is grateful for CMA financial support and excellent working conditions.

to the unique entropy solution of (1.1), when $\varepsilon, \alpha \to 0$, where $u_{0,\varepsilon,\alpha}$ is a smooth approximation of u_0 (see (1.5)).

In other words we consider the convergence of the solution $u_{\varepsilon,\alpha}$ of the fourth order problem

$$\begin{cases} \partial_t u_{\varepsilon,\alpha} - \alpha \partial^3_{txx} u_{\varepsilon,\alpha} + u_{\varepsilon,\alpha}\partial_x u_{\varepsilon,\alpha} \\ \quad -\alpha u_{\varepsilon,\alpha}\partial^3_{xxx} u_{\varepsilon,\alpha} = \varepsilon \partial^2_{xx} u_{\varepsilon,\alpha} - \varepsilon\alpha \partial^4_{xxxx} u_{\varepsilon,\alpha}, & t > 0,\, x \in \mathbb{R}, \\ u_{\varepsilon,\alpha}(0,x) - \alpha \partial^2_{xx} u_{\varepsilon,\alpha}(0,x) = u_{0,\varepsilon,\alpha}(x), & x \in \mathbb{R}, \end{cases} \tag{1.3}$$

to the unique entropy solution of (1.1), when $\varepsilon, \alpha \to 0$.

The analysis of the case $\varepsilon = 0$ is extremely interesting, indeed due to the maximum principle and the fact that the total variation is bounded one can prove that the family $v_{0,\alpha}$ converges to a weak solution of (1.1), when $\alpha \to 0$ [1]. On the other hand, $v_{0,\alpha}$ does not converge in general to an entropy solution of (1.1), when $\alpha \to 0$, it depends on the structure of the initial condition [2, 14].

In this paper we have very weak assumptions on the initial condition (see (1.1)) and study the convergence of the family $\{v_{\varepsilon,\alpha}\}_{\varepsilon,\alpha}$ to the unique entropy solution of (1.1).

Our starting point is the family of third order nonlinear dispersive equations [7]

$$\partial_t u - \partial^3_{txx} u + (b+1) u \partial_x u = b \partial_x u \partial^2_{xx} u + u \partial^3_{xxx} u, \tag{1.4}$$

indexed over the real constant $b \in \mathbb{R}$. This equation has an Hamiltonian structure for every $b \geq 0$, but it has a bi-Hamiltonian structure and it is completely integrable only in the two cases $b = 2$, $b = 3$ [7]. In other words the unique two integrable equations within that family are the Camassa-Holm [3, 6, 11] and Degasperis-Procesi equations [8], respectively. Choosing $b = 0$, assuming that the dispersive effects are small and the presence of a small viscosity term we get first (1.2) and then the Burgers equation.

Finally, we recall that Coclite and Karlsen [5] and Hwang [10] studied the problem in the case $b = 2$.

Regarding the approximated initial condition $u_{0,\varepsilon,\alpha}$ we assume that

$$\begin{aligned} & u_{0,\varepsilon,\alpha} \to u_0, \qquad \text{a.e. and in } L^p_{\mathrm{loc}}(\mathbb{R}),\ 1 \leq p < \infty, \text{ as } \varepsilon,\ \alpha \to 0, \\ & u_{0,\varepsilon,\alpha} \in C^\infty(\mathbb{R}), \quad \|u_{0,\varepsilon,\alpha}\|_{L^\infty(\mathbb{R})} \leq \|u_0\|_{L^\infty(\mathbb{R})}, \\ & \|u_{0,\varepsilon,\alpha}\|_{L^1(\mathbb{R})} \leq \|u_0\|_{L^1(\mathbb{R})}, \quad \|\partial_x u_{0,\varepsilon,\alpha}\|_{L^1(\mathbb{R})} \leq TV(u_0), \end{aligned} \tag{1.5}$$

for every $\varepsilon,\ \alpha > 0$.

We continue with the defintion of entropy solution to (1.1) [9].

Definition 1.1. *Let* $u\colon [0,\infty) \times \mathbb{R} \to \mathbb{R}$ *be a function. We say that* u *is an entropy solution of the Cauchy problem* (1.1) *if*

(D.1) *$u \in L^\infty((0,\infty)\times\mathbb{R}) \cap L^1((0,T)\times(0,\infty))$, for any $T > 0$;*

(D.2) *u satisfies (1.1) in the sense of distributions on $(0,\infty)\times\mathbb{R}$;*

(D.3) *for each convex map $\eta \in C^2(\mathbb{R})$ the inequality*

$$\partial_t \eta(u) + \partial_x q(u) \leq 0$$

holds in the sense of distributions on $(0,\infty)\times\mathbb{R}$, where q is defined by $q'(u) = u\eta'(u)$.

The Cauchy problem (1.1) admits several distributional solutions but only one entropy solution [9].

The main result of this paper is the following.

Theorem 1.2. *There exist two sequences $\{\varepsilon_n\}_{n\in\mathbb{N}}$, $\{\alpha_n\}_{n\in\mathbb{N}} \subset (0,\infty)$, $\varepsilon_n, \alpha_n \to 0$, and a function $u \in L^\infty((0,\infty)\times\mathbb{R}))\cap L^1((0,T)\times(0,\infty))$, $T > 0$, such that*

$$v_{\varepsilon_n,\alpha_n} \to u \quad \text{a.e. and in } L^p_{loc}((0,\infty)\times\mathbb{R}) \text{ for each } 1 \leq p < \infty, \tag{1.6}$$

and u is a distributional solution of (1.1). If, in addition

$$\alpha = o(\varepsilon^2), \tag{1.7}$$

then u is the unique entropy solution of (1.1).

Our result is based the following lemmas.

Lemma 1.3 (see [15]). *Let $\{v_\nu\}_{\nu>0}$ be a family of functions defined on $(0,\infty)\times\mathbb{R}$ such that*

$$\|v_\nu\|_{L^\infty((0,\infty)\times\mathbb{R})} \leq M, \qquad \nu > 0,$$

and the family

$$\{\partial_t \eta(v_\nu) + \partial_x q(v_\nu)\}_{\nu>0}$$

is compact in $H^{-1}_{loc}((0,\infty)\times\mathbb{R})$, for every $\eta \in C^2(\mathbb{R})$ convex, where $q'(v) = v\eta'(v)$. Then there exist a sequence $\{\nu_n\}_{n\in\mathbb{N}} \subset (0,\infty)$, $\nu_n \to 0$, and a map $v \in L^\infty((0,\infty)\times\mathbb{R})$ such that

$$v_{\nu_n} \to v \quad \text{a.e. and in } L^p_{loc}((0,\infty)\times\mathbb{R}),\ 1 \leq p < \infty.$$

Lemma 1.4 (see [13]). *Let Ω be a bounded open subset of $\mathbb{R}^N$, $N \geq 2$. Suppose the sequence $\{\mathcal{L}_n\}_{n\in\mathbb{N}}$ of distributions is bounded in $W^{-1,\infty}(\Omega)$. Suppose also that*

$$\mathcal{L}_n = \mathcal{L}_{1,n} + \mathcal{L}_{2,n},$$

where $\{\mathcal{L}_{1,n}\}_{n\in\mathbb{N}}$ lies in a compact subset of $H^{-1}_{\rm loc}(\Omega)$ and $\{\mathcal{L}_{2,n}\}_{n\in\mathbb{N}}$ lies in a bounded subset of $L^1_{loc}(\Omega)$. Then $\{\mathcal{L}_n\}_{n\in\mathbb{N}}$ lies in a compact subset of $H^{-1}_{loc}(\Omega)$.

2 Estimates on the elliptic equation

This section is dedicated to some estimates on the second equation in (1.2).

The Green's function of the linear differential operator $1 - \alpha\partial^2_{xx}$ is

$$G_\alpha(x) = \frac{1}{2\sqrt{\alpha}} e^{-|x|/\sqrt{\alpha}}, \qquad \alpha > 0, \qquad x \in \mathbb{R}.$$

Therefore, we have that

$$u_{\varepsilon,\alpha}(t,x) = \big(G_\alpha * v_{\varepsilon,\alpha}(t,\cdot)\big)(x) = \frac{1}{2\sqrt{\alpha}} \int_{\mathbb{R}} e^{-|x-y|/\sqrt{\alpha}} v_{\varepsilon,\alpha}(t,y)dy, \tag{2.1}$$

where the convolution is only in the x variable. Moreover

$$\|G_\alpha\|_{L^1(\mathbb{R})} = 1, \qquad \alpha > 0. \tag{2.2}$$

Lemma 2.1 (L^∞ estimate)**.** *Let $\varepsilon, \alpha > 0$. We have that*

$$\|u_{\varepsilon,\alpha}\|_{L^\infty((0,\infty)\times\mathbb{R})} \leq \|v_{\varepsilon,\alpha}\|_{L^\infty((0,\infty)\times\mathbb{R})}. \tag{2.3}$$

Proof. The Hölder inequality, (2.1), and (2.2) give

$$\begin{aligned}
|u_{\varepsilon,\alpha}(t,x)| &\leq \int_{\mathbb{R}} G_\alpha(x-y)\,|v_{\varepsilon,\alpha}(t,y)|\,dy \\
&\leq \|v_{\varepsilon,\alpha}\|_{L^\infty((0,\infty)\times\mathbb{R})} \int_{\mathbb{R}} G_\alpha(x-y)dy \\
&= \|v_{\varepsilon,\alpha}\|_{L^\infty((0,\infty)\times\mathbb{R})} \|G_\alpha\|_{L^1(\mathbb{R})} = \|v_{\varepsilon,\alpha}\|_{L^\infty((0,\infty)\times\mathbb{R})}
\end{aligned}$$

for every $t \geq 0$, $x \in \mathbb{R}$. □

Lemma 2.2 (L^1 & BV estimates)**.** *Let $\varepsilon, \alpha > 0$. The following estimates hold*

$$\|u_{\varepsilon,\alpha}(t,\cdot)\|_{L^1(\mathbb{R})} \leq \|v_{\varepsilon,\alpha}(t,\cdot)\|_{L^1(\mathbb{R})}, \tag{2.4}$$

$$\|\partial_x u_{\varepsilon,\alpha}(t,\cdot)\|_{L^1(\mathbb{R})} \leq \|\partial_x v_{\varepsilon,\alpha}(t,\cdot)\|_{L^1(\mathbb{R})} \tag{2.5}$$

for every $t \geq 0$.

Proof. The Tonelli theorem, (2.1), and (2.2) give

$$\begin{aligned}
\int_{\mathbb{R}} |u_{\varepsilon,\alpha}(t,x)|dx &\leq \int_{\mathbb{R}}\int_{\mathbb{R}} G_\alpha(x-y)\,|v_{\varepsilon,\alpha}(t,y)|\,dxdy \\
&= \int_{\mathbb{R}} \left(\int_{\mathbb{R}} G_\alpha(x-y)dx\right) |v_{\varepsilon,\alpha}(t,y)|\,dy \\
&= \|G_\alpha\|_{L^1(\mathbb{R})} \int_{\mathbb{R}} |v_{\varepsilon,\alpha}(t,y)|\,dy = \|v_{\varepsilon,\alpha}(t,\cdot)\|_{L^1(\mathbb{R})},
\end{aligned}$$

that is (2.4). Since

$$\partial_x u_{\varepsilon,\alpha}(t,x) = \big(G_\alpha * \partial_x v_{\varepsilon,\alpha}(t,\cdot)\big)(x),$$

using the same argument we get (2.5). □

Lemma 2.3 (L^2 estimates). *Let $\varepsilon, \alpha > 0$. The following identities hold*

$$\begin{aligned} \|u_{\varepsilon,\alpha}(t,\cdot)\|^2_{L^2(\mathbb{R})} &+ 2\alpha \|\partial_x u_{\varepsilon,\alpha}(t,\cdot)\|^2_{L^2(\mathbb{R})} \\ &+ \alpha^2 \left\|\partial^2_{xx} u_{\varepsilon,\alpha}(t,\cdot)\right\|^2_{L^2(\mathbb{R})} = \|v_{\varepsilon,\alpha}(t,\cdot)\|^2_{L^2(\mathbb{R})}, \end{aligned} \tag{2.6}$$

$$\begin{aligned} \|\partial_x u_{\varepsilon,\alpha}(t,\cdot)\|^2_{L^2(\mathbb{R})} &+ 2\alpha \left\|\partial^2_{xx} u_{\varepsilon,\alpha}(t,\cdot)\right\|^2_{L^2(\mathbb{R})} \\ &+ \alpha^2 \left\|\partial^3_{xxx} u_{\varepsilon,\alpha}(t,\cdot)\right\|^2_{L^2(\mathbb{R})} = \|\partial_x v_{\varepsilon,\alpha}(t,\cdot)\|^2_{L^2(\mathbb{R})} \end{aligned} \tag{2.7}$$

for every $t \geq 0$.

Proof. The second equation in (1.2) says

$$v^2_{\varepsilon,\alpha} = u^2_{\varepsilon,\alpha} - 2\alpha u_{\varepsilon,\alpha}\partial^2_{xx} u_{\varepsilon,\alpha} + \alpha^2(\partial^2_{xx} u_{\varepsilon,\alpha})^2.$$

Integrating over $\mathbb{R}$ we get

$$\begin{aligned} \int_{\mathbb{R}} v^2_{\varepsilon,\alpha} dx &= \int_{\mathbb{R}} u^2_{\varepsilon,\alpha} dx - 2\alpha \int_{\mathbb{R}} u_{\varepsilon,\alpha}\partial^2_{xx} u_{\varepsilon,\alpha} dx + \alpha^2 \int_{\mathbb{R}} (\partial^2_{xx} u_{\varepsilon,\alpha})^2 dx \\ &= \int_{\mathbb{R}} u^2_{\varepsilon,\alpha} dx + 2\alpha \int_{\mathbb{R}} (\partial_x u_{\varepsilon,\alpha})^2 dx + \alpha^2 \int_{\mathbb{R}} (\partial^2_{xx} u_{\varepsilon,\alpha})^2 dx, \end{aligned}$$

that is (2.6). Since

$$\partial_x v_{\varepsilon,\alpha} = \partial_x u_{\varepsilon,\alpha} - \alpha\partial^3_{xxx} u_{\varepsilon,\alpha},$$

using the same argument we get (2.7). □

3 Estimates on the transport equation

This section is dedicated to some estimates on the first equation in (1.2).

Lemma 3.1 (L^∞ estimates). *Let ε, $\alpha > 0$. We have that*

$$\|v_{\varepsilon,\alpha}\|_{L^\infty((0,\infty)\times\mathbb{R})} \leq \|u_0\|_{L^\infty(\mathbb{R})}. \tag{3.1}$$

Moreover

$$\|u_{\varepsilon,\alpha}\|_{L^\infty((0,\infty)\times\mathbb{R})}, \ \left\|\alpha\partial^2_{xx} u_{\varepsilon,\alpha}\right\|_{L^\infty((0,\infty)\times\mathbb{R})} \leq \|u_0\|_{L^\infty(\mathbb{R})}. \tag{3.2}$$

Proof. We observe that the maps with constant values $\|u_0\|_{L^\infty(\mathbb{R})}$ and $-\|u_0\|_{L^\infty(\mathbb{R})}$ solve the first equation of (1.2). Hence, (3.1) is consequence of the comparison principle for parabolic equations and (1.5). Finally, (3.2) follows from (3.1) and Lemma 2.1. □

Lemma 3.2 (*BV* estimates). *Let* ε, $\alpha > 0$. *The following estimates hold*

$$\|\partial_x v_{\varepsilon,\alpha}(t,\cdot)\|_{L^1(\mathbb{R})}, \|\partial_x u_{\varepsilon,\alpha}(t,\cdot)\|_{L^1(\mathbb{R})} \leq TV(u_0), \tag{3.3}$$

for every $t \geq 0$.

Proof. Due to the first equation in (1.2)

$$\begin{aligned}
\frac{d}{dt}\int_{\mathbb{R}} |\partial_x v_{\varepsilon,\alpha}|\,dx =& \int_{\mathbb{R}} \operatorname{sign}(\partial_x v_{\varepsilon,\alpha})\,\partial^2_{tx} v_{\varepsilon,\alpha} dx \\
=&\varepsilon \int_{\mathbb{R}} \operatorname{sign}(\partial_x v_{\varepsilon,\alpha})\,\partial^3_{xxx} v_{\varepsilon,\alpha} dx - \int_{\mathbb{R}} \operatorname{sign}(\partial_x v_{\varepsilon,\alpha})\,\partial_x(u_{\varepsilon,\alpha}\partial_x v_{\varepsilon,\alpha})dx \\
=&-\varepsilon \int_{\mathbb{R}} \delta_{\{\partial_x v_{\varepsilon,\alpha}=0\}}(\partial^2_{xx} v_{\varepsilon,\alpha})^2 dx + \int_{\mathbb{R}} \delta_{\{\partial_x v_{\varepsilon,\alpha}=0\}} u_{\varepsilon,\alpha}\partial_x v_{\varepsilon,\alpha}\partial^2_{xx} v_{\varepsilon,\alpha} dx \\
=&-\varepsilon \int_{\mathbb{R}} \delta_{\{\partial_x v_{\varepsilon,\alpha}=0\}}(\partial^2_{xx} v_{\varepsilon,\alpha})^2 dx \leq 0,
\end{aligned}$$

where $\delta_{\{\partial_x v_{\varepsilon,\alpha}=0\}}$ is the Dirac delta concentrated on the set $\{\partial_x v_{\varepsilon,\alpha} = 0\}$. Therefore, we have the bound on $\partial_x v_{\varepsilon,\alpha}$. The estimate on $\partial_x u_{\varepsilon,\alpha}$ is consequence of the one on $\partial_x v_{\varepsilon,\alpha}$ and Lemma 2.2. □

Lemma 3.3 (L^1 estimates). *Let* ε, $\alpha > 0$. *We have that*

$$\begin{aligned}
&\|v_{\varepsilon,\alpha}(t,\cdot)\|_{L^1(\mathbb{R})}, \|u_{\varepsilon,\alpha}(t,\cdot)\|_{L^1(\mathbb{R})} \\
&\qquad\qquad \leq \|u_0\|_{L^1(\mathbb{R})} + TV(u_0)\,\|u_0\|_{L^\infty(\mathbb{R})}\, t,
\end{aligned} \tag{3.4}$$

for every $t \geq 0$.

Proof. Due to the first equation in (1.2)

$$\begin{aligned}
\frac{d}{dt}\int_{\mathbb{R}} |v_{\varepsilon,\alpha}|\,dx =& \int_{\mathbb{R}} \operatorname{sign}(v_{\varepsilon,\alpha})\,\partial_t v_{\varepsilon,\alpha} dx \\
=&\varepsilon \int_{\mathbb{R}} \operatorname{sign}(v_{\varepsilon,\alpha})\,\partial^2_{xx} v_{\varepsilon,\alpha} dx - \int_{\mathbb{R}} \operatorname{sign}(v_{\varepsilon,\alpha})\,u_{\varepsilon,\alpha}\partial_x v_{\varepsilon,\alpha} dx \\
=&-\varepsilon \int_{\mathbb{R}} \delta_{\{v_{\varepsilon,\alpha}=0\}}(\partial_x v_{\varepsilon,\alpha})^2 dx - \int_{\mathbb{R}} \operatorname{sign}(v_{\varepsilon,\alpha})\,u_{\varepsilon,\alpha}\partial_x v_{\varepsilon,\alpha} dx \\
\leq& \|u_{\varepsilon,\alpha}\|_{L^\infty((0,\infty)\times\mathbb{R})}\int_{\mathbb{R}} |\partial_x v_{\varepsilon,\alpha}|\,dx,
\end{aligned}$$

where $\delta_{\{v_{\varepsilon,\alpha}=0\}}$ is the Dirac delta concentrated on the set $\{v_{\varepsilon,\alpha} = 0\}$. Therefore, the estimate on $v_{\varepsilon,\alpha}$ follows from Lemmas 3.1, 3.2. Finally, the bound on $u_{\varepsilon,\alpha}$ is consequence of the one on $v_{\varepsilon,\alpha}$ and Lemma 2.2. □

Lemma 3.4 (L^2 estimates). *Let $\varepsilon, \alpha > 0$. The following inequality holds*

$$\begin{aligned} \|v_{\varepsilon,\alpha}(t,\cdot)\|^2_{L^2(\mathbb{R})} &+ 2\varepsilon \int_0^t \|\partial_x v_{\varepsilon,\alpha}(s,\cdot)\|^2_{L^2(\mathbb{R})} ds \\ &\leq \|u_0\|^2_{L^2(\mathbb{R})} + TV(u_0)\,\|u_0\|^2_{L^\infty(\mathbb{R})}\, t, \end{aligned} \tag{3.5}$$

for every $t \geq 0$. In particular

i) *the families*

$$\{v_{\varepsilon,\alpha}\}_{\varepsilon,\alpha},\ \{u_{\varepsilon,\alpha}\}_{\varepsilon,\alpha},\ \{\sqrt{\alpha}\partial_x u_{\varepsilon,\alpha}\}_{\varepsilon,\alpha},\ \{\alpha\partial^2_{xx} u_{\varepsilon,\alpha}\}_{\varepsilon,\alpha}$$

are bounded in $L^\infty((0,T); L^2(\mathbb{R}))$, for every $T > 0$;

ii) *the families*

$$\{\sqrt{\varepsilon}\partial_x v_{\varepsilon,\alpha}\}_{\varepsilon,\alpha}, \{\sqrt{\varepsilon}\partial_x u_{\varepsilon,\alpha}\}_{\varepsilon,\alpha},\\ \{\sqrt{\varepsilon\alpha}\partial^2_{xx} u_{\varepsilon,\alpha}\}_{\varepsilon,\alpha},\ \{\alpha\sqrt{\varepsilon}\partial^3_{xxx} u_{\varepsilon,\alpha}\}_{\varepsilon,\alpha}$$

are bounded in $L^2((0,T)\times\mathbb{R})$, for every $T > 0$.

Proof. Due to the first equation in (1.2)

$$\begin{aligned} \frac{d}{dt}\int_{\mathbb{R}} \frac{v^2_{\varepsilon,\alpha}}{2} dx &= \int_{\mathbb{R}} v_{\varepsilon,\alpha}\partial_t v_{\varepsilon,\alpha} dx \\ &= \varepsilon \int_{\mathbb{R}} v_{\varepsilon,\alpha}\partial^2_{xx} v_{\varepsilon,\alpha} dx - \int_{\mathbb{R}} v_{\varepsilon,\alpha} u_{\varepsilon,\alpha}\partial_x v_{\varepsilon,\alpha} dx \\ &= -\varepsilon \int_{\mathbb{R}} (\partial_x v_{\varepsilon,\alpha})^2 dx + \frac{1}{2}\int_{\mathbb{R}} v^2_{\varepsilon,\alpha}\partial_x u_{\varepsilon,\alpha} dx \\ &\leq -\varepsilon \int_{\mathbb{R}} (\partial_x v_{\varepsilon,\alpha})^2 dx + \frac{1}{2}\|v_{\varepsilon,\alpha}\|^2_{L^\infty((0,\infty)\times\mathbb{R})} \int_{\mathbb{R}} |\partial_x u_{\varepsilon,\alpha}|\, dx. \end{aligned}$$

Therefore, (3.5) follows from Lemmas 3.1, 3.2.

Finally, *i)* and *ii)* follow from (3.5) and Lemma 2.3. □

4 Proof of Theorem 1.2

In this section we prove Theorem 1.2. We prove the compactness of the family $\{v_{\varepsilon,\alpha}\}_{\varepsilon,\alpha}$ using compensated compactness, then using the assumption (1.7) we show that the limit point of the family is a distributional solution of (1.1), finally we prove that the entropy inequalities hold.

Lemma 4.1 (Compactness). *There exist two sequences $\{\varepsilon_n\}_{n\in\mathbb{N}}, \{\alpha_n\}_{n\in\mathbb{N}}$ of positive numbers and $u \in L^\infty((0,\infty)\times\mathbb{R}) \cap L^1((0,T)\times\mathbb{R})$, $T > 0$, such that $\varepsilon_n, \alpha_n \to 0$, and (1.6) holds.*

Proof of Lemma 4.1. We want to use Lemma 1.3. Let $\eta \in C^2(\mathbb{R})$ be convex. Multiplying the first equation in (1.1) by $\eta'(v_{\varepsilon,\alpha})$ we get

$$\begin{aligned}&\partial_t\eta(v_{\varepsilon,\alpha}) + \partial_x q(v_{\varepsilon,\alpha})\\ =&\varepsilon\eta'(v_{\varepsilon,\alpha})\partial^2_{xx}v_{\varepsilon,\alpha} + (v_{\varepsilon,\alpha} - u_{\varepsilon,\alpha})\partial_x v_{\varepsilon,\alpha}\eta'(v_{\varepsilon,\alpha})\\ =&\varepsilon\partial^2_{xx}\eta(v_{\varepsilon,\alpha}) - \varepsilon\eta''(v_{\varepsilon,\alpha})(\partial_x v_{\varepsilon,\alpha})^2 + (v_{\varepsilon,\alpha} - u_{\varepsilon,\alpha})\partial_x v_{\varepsilon,\alpha}\eta'(v_{\varepsilon,\alpha}),\end{aligned} \tag{4.1}$$

where q satisfies $q'(u) = u\eta'(u)$.

Lemmas 3.1 and 3.2 imply

$$\begin{aligned}&\{(v_{\varepsilon,\alpha} - u_{\varepsilon,\alpha})\partial_x v_{\varepsilon,\alpha}\eta'(v_{\varepsilon,\alpha})\}_{\varepsilon,\alpha}\\ &\qquad\text{is bounded in } L^1((0,T)\times\mathbb{R}) \text{ for each } T>0.\end{aligned} \tag{4.2}$$

Lemmas 3.1 and 3.4 give

$$\begin{aligned}&\{\varepsilon\eta''(v_{\varepsilon,\alpha})(\partial_x v_{\varepsilon,\alpha})^2\}_{\varepsilon,\alpha}\\ &\qquad\text{is bounded in } L^1((0,T)\times\mathbb{R}) \text{ for each } T>0,\end{aligned} \tag{4.3}$$

and

$$\varepsilon\partial_x\eta(v_{\varepsilon,\alpha}) \to 0 \quad \text{in } L^1_{loc}((0,\infty)\times\mathbb{R}). \tag{4.4}$$

Due to Lemma 1.4, (4.2), (4.3), (4.4),

$$\{\partial_t\eta(v_{\varepsilon,\alpha}) + \partial_x q(v_{\varepsilon,\alpha})\}_{\varepsilon,\alpha} \quad \text{is compact in } H^{-1}_{loc}((0,\infty)\times\mathbb{R}). \tag{4.5}$$

Then the claim follows from (4.5) and Lemmas 1.3, 3.1, 3.3. □

Lemma 4.2 (Convergence to weak solution). *The limit function u of Lemma 4.1 is a distributional solution of* (1.1).

Remark 4.3. *It is interesting to notice that in the previous lemmas we require simply the convergence to 0 of the parameters ε and α: we make no use of* (1.7). *This is perfectly coherent with the results of Bhat and Fetecau [1], where ε is set to be 0 from the beginning.*

Proof of Lemma 4.2. Let $\varphi \in C^\infty([0,\infty)\times\mathbb{R})$ be a test function with compact support. Since

$$u_{\varepsilon_n,\alpha_n}\partial^3_{xxx}u_{\varepsilon_n,\alpha_n} = -\frac{1}{4}\partial^3_{xxx}(u_{\varepsilon_n,\alpha_n})^2 + \frac{3}{2}\partial_x(u_{\varepsilon_n,\alpha_n}\partial^2_{xx}u_{\varepsilon_n,\alpha_n}),$$

multiplying (1.3) by φ and integrating by parts, we get

$$\begin{aligned}
-\int_0^\infty\int_{\mathbb{R}}\Big(u_{\varepsilon_n,\alpha_n}\partial_t\varphi+\frac{u^2_{\varepsilon_n,\alpha_n}}{2}\partial_x\varphi\Big)dtdx+\int_{\mathbb{R}}u_{0,\varepsilon_n,\alpha_n}(x)\varphi(0,x)dx\\
=-\alpha_n\int_0^\infty\int_{\mathbb{R}}u_{\varepsilon_n,\alpha_n}\partial^3_{txx}\varphi dtdx+\alpha_n\int_{\mathbb{R}}u_{0,\varepsilon_n,\alpha_n}(x)\partial^2_{xx}\varphi(0,x)dx\\
+\frac{\alpha_n}{4}\int_0^\infty\int_{\mathbb{R}}u^2_{\varepsilon_n,\alpha_n}\partial^3_{xxx}\varphi dtdx\\
-\frac{3\alpha_n}{2}\int_0^\infty\int_{\mathbb{R}}u_{\varepsilon_n,\alpha_n}\partial^2_{xx}u_{\varepsilon_n,\alpha_n}\partial_x\varphi dtdx\\
+\varepsilon_n\int_0^\infty\int_{\mathbb{R}}u_{\varepsilon_n,\alpha_n}\partial^2_{xx}\varphi dtdx-\varepsilon_n\alpha_n\int_0^\infty\int_{\mathbb{R}}u_{\varepsilon_n,\alpha_n}\partial^4_{xxxx}\varphi dtdx.
\end{aligned}\tag{4.6}$$

Lemma 4.1 and (1.5) yield

$$\begin{aligned}
-\int_0^\infty\int_{\mathbb{R}}\Big(u_{\varepsilon_n,\alpha_n}\partial_t\varphi+\frac{u^2_{\varepsilon_n,\alpha_n}}{2}\partial_x\varphi\Big)dtdx\\
+\int_{\mathbb{R}}u_{0,\varepsilon_n,\alpha_n}(x)\varphi(0,x)dx\\
\to-\int_0^\infty\int_{\mathbb{R}}\Big(u\partial_t\varphi+\frac{u^2}{2}\partial_x\varphi\Big)dtdx+\int_{\mathbb{R}}u_0(x)\varphi(0,x)dx
\end{aligned}\tag{4.7}$$

and

$$\begin{aligned}
-\alpha_n\int_0^\infty\int_{\mathbb{R}}u_{\varepsilon_n,\alpha_n}\partial^3_{txx}\varphi dtdx+\alpha_n\int_{\mathbb{R}}u_{0,\varepsilon_n,\alpha_n}(x)\partial^2_{xx}\varphi(0,x)dx\\
+\frac{\alpha_n}{4}\int_0^\infty\int_{\mathbb{R}}u^2_{\varepsilon_n,\alpha_n}\partial^3_{xxx}\varphi dtdx\\
+\varepsilon_n\int_0^\infty\int_{\mathbb{R}}u_{\varepsilon_n,\alpha_n}\partial^2_{xx}\varphi dtdx-\varepsilon_n\alpha_n\int_0^\infty\int_{\mathbb{R}}u_{\varepsilon_n,\alpha_n}\partial^4_{xxxx}\varphi dtdx\to 0.
\end{aligned}\tag{4.8}$$

Finally, we claim that

$$\alpha_n\int_0^\infty\int_{\mathbb{R}}u_{\varepsilon_n,\alpha_n}\partial^2_{xx}u_{\varepsilon_n,\alpha_n}\partial_x\varphi dtdx\to 0.\tag{4.9}$$

Indeed, due to the second equation in (1.1), we have

$$\alpha_n\partial^2_{xx}u_{\varepsilon_n,\alpha_n}=v_{\varepsilon_n,\alpha_n}-u_{\varepsilon_n,\alpha_n}.\tag{4.10}$$

Hence, the Hölder inequality, Lemma 3.4, and (1.7) give

$$\begin{aligned}
\alpha_n\Big|\int_0^\infty\int_{\mathbb{R}}u_{\varepsilon_n,\alpha_n}\partial^2_{xx}u_{\varepsilon_n,\alpha_n}\partial_x\varphi dtdx\Big|\\
\le\|u_{\varepsilon_n,\alpha_n}\|_{L^2(\mathrm{supp}(\varphi))}\|v_{\varepsilon_n,\alpha_n}-u_{\varepsilon_n,\alpha_n}\|_{L^2(\mathrm{supp}(\varphi))}\times\\
\times\|\partial_x\varphi\|_{L^\infty((0,\infty)\times\mathbb{R})}\to 0.
\end{aligned}$$

Therefore, (4.9) is done.

Clearly (4.6), (4.7), (4.8), (4.9) imply

$$\int_0^\infty \int_{\mathbb{R}} \left(u\partial_t \varphi + \frac{u^2}{2}\partial_x \varphi \right) dt dx - \int_{\mathbb{R}} u_0(x)\varphi(0,x)dx = 0. \tag{4.11}$$

□

Lemma 4.4 (Entropy inequalities). *If* (1.7) *holds then the limit function u of Lemma 4.1 is the entropy solution of* (1.1) *in the sense of Definition 1.1.*

Proof. Let $\varphi \in C^\infty([0,\infty)\times\mathbb{R})$ be a positive test function with compact support and $\eta \in C^2(\mathbb{R})$ be a convex entropy. The convexity of η, (4.1), (4.10) give

$$\begin{aligned}
&\partial_t \eta(v_{\varepsilon_n,\alpha_n}) + \partial_x q(v_{\varepsilon_n,\alpha_n}) \\
&= \varepsilon_n \partial_{xx}^2 \eta(v_{\varepsilon_n,\alpha_n}) - \varepsilon_n(\partial_{xx}^2 v_{\varepsilon_n,\alpha_n})^2 \eta''(v_{\varepsilon_n,\alpha_n}) \\
&\quad - (u_{\varepsilon_n,\alpha_n} - v_{\varepsilon_n,\alpha_n})\partial_x v_{\varepsilon_n,\alpha_n} \eta'(v_{\varepsilon_n,\alpha_n}) \\
&\leq \varepsilon_n \partial_{xx}^2 \eta(v_{\varepsilon_n,\alpha_n}) - \alpha_n \partial_{xx}^2 u_{\varepsilon_n,\alpha_n} \partial_x v_{\varepsilon_n,\alpha_n} \eta'(v_{\varepsilon_n,\alpha_n}),
\end{aligned} \tag{4.12}$$

where q satisfies $q'(u) = u\eta'(u)$. Multiplying by φ and integrating by parts

$$\begin{aligned}
-&\int_0^\infty \int_{\mathbb{R}} \left(\eta(v_{\varepsilon_n,\alpha_n})\partial_t \varphi + q(v_{\varepsilon_n,\alpha_n})\partial_x \varphi \right) dt dx \\
&\qquad + \int_{\mathbb{R}} \eta(u_{0,\varepsilon_n,\alpha_n}(x))\varphi(0,x)dx \\
&\leq \varepsilon_n \int_0^\infty \int_{\mathbb{R}} \eta(v_{\varepsilon_n,\alpha_n})\partial_{xx}^2 \varphi dt dx \\
&\quad - \alpha_n \int_0^\infty \int_{\mathbb{R}} \partial_{xx}^2 u_{\varepsilon_n,\alpha_n} \partial_x v_{\varepsilon_n,\alpha_n} \eta'(v_{\varepsilon_n,\alpha_n})\varphi dt dx.
\end{aligned} \tag{4.13}$$

Lemma 4.1 and (1.5) yields

$$\begin{aligned}
-&\int_0^\infty \int_{\mathbb{R}} \left(\eta(v_{\varepsilon_n,\alpha_n})\partial_t \varphi + q(v_{\varepsilon_n,\alpha_n})\partial_x \varphi \right) dt dx \\
+&\int_{\mathbb{R}} \eta(u_{0,\varepsilon_n,\alpha_n}(x))\varphi(0,x)dx \\
&\to -\int_0^\infty \int_{\mathbb{R}} \left(\eta(u)\partial_t \varphi + q(u)\partial_x \varphi \right) dt dx + \int_{\mathbb{R}} \eta(u_0(x))\varphi(0,x)dx
\end{aligned} \tag{4.14}$$

and

$$\varepsilon_n \int_0^\infty \int_{\mathbb{R}} \eta(v_{\varepsilon_n,\alpha_n})\partial_{xx}^2 \varphi dt dx \to 0. \tag{4.15}$$

Finally, we claim that

$$\alpha_n \int_0^\infty \int_{\mathbb{R}} \partial_{xx}^2 u_{\varepsilon_n,\alpha_n} \partial_x v_{\varepsilon_n,\alpha_n} \eta'(v_{\varepsilon_n,\alpha_n}) \varphi dt dx \to 0. \tag{4.16}$$

Indeed, the Hölder inequality, Lemma 3.4, and (1.7) give

$$\begin{aligned}
&\left|\alpha_n \int_0^\infty \int_{\mathbb{R}} \partial_{xx}^2 u_{\varepsilon_n,\alpha_n} \partial_x v_{\varepsilon_n,\alpha_n} \eta'(v_{\varepsilon_n,\alpha_n}) \varphi dt dx\right| \\
&\quad \le \alpha_n \int_0^\infty \int_{\mathbb{R}} |\partial_{xx}^2 u_{\varepsilon_n,\alpha_n}| |\partial_x v_{\varepsilon_n,\alpha_n}| |\eta'(v_{\varepsilon_n,\alpha_n})| \varphi dt dx \\
&\quad \le \alpha_n \left\|\partial_{xx}^2 u_{\varepsilon_n,\alpha_n}\right\|_{L^2(\mathrm{supp}(\varphi))} \left\|\partial_x v_{\varepsilon_n,\alpha_n}\right\|_{L^2(\mathrm{supp}(\varphi))} \times \\
&\qquad\qquad \times \left\|\eta'\right\|_{L^\infty([-\|u_0\|_{L^\infty(\mathbb{R})},\|u_0\|_{L^\infty(\mathbb{R})}])} \left\|\varphi\right\|_{L^\infty((0,\infty)\times\mathbb{R})} \\
&\quad = \frac{\sqrt{\alpha_n}}{\varepsilon_n} \left\|\sqrt{\alpha_n \varepsilon_n} \partial_{xx}^2 u_{\varepsilon_n,\alpha_n}\right\|_{L^2(\mathrm{supp}(\varphi))} \left\|\sqrt{\varepsilon_n} \partial_x v_{\varepsilon_n,\alpha_n}\right\|_{L^2(\mathrm{supp}(\varphi))} \times \\
&\qquad\qquad \times \left\|\eta'\right\|_{L^\infty([-\|u_0\|_{L^\infty(\mathbb{R})},\|u_0\|_{L^\infty(\mathbb{R})}])} \left\|\varphi\right\|_{L^\infty((0,\infty)\times\mathbb{R})} \\
&\quad \le \frac{\sqrt{\alpha_n}}{\varepsilon_n} c \left\|\eta'\right\|_{L^\infty([-\|u_0\|_{L^\infty(\mathbb{R})},\|u_0\|_{L^\infty(\mathbb{R})}])} \left\|\varphi\right\|_{L^\infty((0,\infty)\times\mathbb{R})} \to 0,
\end{aligned}$$

for some constant $c > 0$ independent on α_n and ε_n. Then (4.9) is proved.

Clearly (4.13), (4.14), (4.15), (4.16) imply

$$-\int_0^\infty \int_{\mathbb{R}} \Big(\eta(u)\partial_t \varphi + q(u)\partial_x \varphi\Big) dt dx + \int_{\mathbb{R}} \eta(u_0(x))\varphi(0,x) dx \le 0. \tag{4.17}$$

Due to Lemma 4.2 the proof is concluded. □

Proof of Theorem 1.2. Direct consequence of Lemmas 4.1, 4.2, and 4.4. □

References

[1] H. S. Bhat and R. C. Fetecau, A Hamiltonian regularization of the Burgers equation. J. Nonlinear Sci. 16 (2006), no. 6, 615–638.

[2] H. S. Bhat and R. C. Fetecau, The Riemann problem for the Leray-Burgers equation. J. Differential Equations 246 (2010), no. 10, 3957–3979.

[3] R. Camassa and D. D. Holm, An integrable shallow water equation with peaked solitons. Phys. Rev. Lett., 71 (1993), no. 11, 1661–1664.

[4] G. M. Coclite, H. Holden and K. H. Karlsen, Wellposedness for a parabolic-elliptic system. Discrete Contin. Dyn. Syst., 13 (2005), no. 3, 659–682.

[5] G. M. Coclite and K. H. Karlsen, A singular limit problem for conservation laws related to the Camassa-Holm shallow water equation. Comm. Partial Differential Equations, 31 (2006), no. 7–9, 1253–1272.

[6] H.-H. Dai, Exact travelling-wave solutions of an integrable equation arising in hyperelastic rods. Wave Motion, 28 (1998), no. 4, 367–381.

[7] A. Degasperis, D. D. Holm and A. N. W. Hone, Integrable and non-integrable equations with peakons. In: Nonlinear physics: theory and experiment, II (Gallipoli, 2002), 37–43. World Sci. Publishing, River Edge, NJ, 2003.

[8] A. Degasperis and M. Procesi, Asymptotic integrability. In: Symmetry and perturbation theory (Rome, 1998), 23–37. World Sci. Publishing, River Edge, NJ, 1999.

[9] H. Holden and N. H. Risebro, Front Tracking for Hyperbolic Conservation Laws. Volume 152 of Applied Mathematical Sciences. Springer-Verlag, New York, 2002.

[10] S. Hwang, Singular limit problem of the Camassa-Holm type equation. J. Differential Equations, 235 (2007), no. 1, 74–84.

[11] R. S. Johnson, Camassa-Holm, Korteweg-de Vries and related models for water waves. J. Fluid Mech., 455 (2002) 63–82.

[12] E. H. Lieb and M. Loss, Analysis. Volume 14 of Graduate Studies in Mathematics, American Mathematical Society, Providence, RI, second edition, 2001.

[13] F. Murat, L'injection du cône positif de H^{-1} dans $W^{-1,q}$ est compacte pour tout $q < 2$. J. Math. Pures Appl. (9), 60 (1981), no. 3, 309–322.

[14] G. Norgard and K. Mohseni, On the convergence of the convectively filtered Burgers equation to the entropy solution of the inviscid Burgers equation. Multiscale Model. Simul., 7 (2009), no. 4, 1811–1837.

[15] L. Tartar, Compensated compactness and applications to partial differential equations. In: Nonlinear analysis and mechanics: Heriot-Watt Symposium, Vol. IV, 136–212, Pitman, Boston, 1979.

Coupling Hyperbolic Conservation Laws and Ordinary Differential Equations

Rinaldo M. Colombo
Department of Mathematics
Brescia, Italy
E-mail: rinaldo@ing.unibs.it

Abstract

We present problems based on conservation laws coupled with ordinary differential equations. Basic well posedness results are presented, as well as a few sample applications.

1 Introduction

We are interested in the dynamical systems that result from the coupling of a conservation law and an ordinary differential equation. Several models lead to this analytical framework, they generally consist in a single agent (or a few of them) interacting with a continuum. Examples are the classical piston problem, see [26, §99] or [34, 35], or the situation described in the fairy tale of the pied piper [24], where a musician frees a city from rats using his magic flute. From a deterministic point of view, studying these phenomena leads to a dynamical system consisting of an o.d.e. for the evolution of the agents (the piston or the pied piper) and a partial differential equations for that of the continuum (the fluid or the rats). More precisely, motivated by the above and other applications, we choose conservation laws for the description of the continuum's evolution. In particular, no diffusion is here considered. On one side, this choice makes the analytical treatment technically more difficult, due to the possible singularities arising in the density that describes the continuum. On the other hand, we obtain a framework where all propagation speeds are finite. As a consequence, for instance, a continuum initially confined in a bounded region will remain in a (larger but) bounded region at any positive time, as it is reasonably necessary in the case of the rats.

In the current literature, similar issues have been considered with a great variety of analytical tools. For instance, a traffic problem is considered in [27], while a fire confinement problem modeled through differential inclusions is studied in [9], see [11] for a tumor–induced angiogenesis described through a stochastic geometric model. Further examples are provided by the interaction of a fluid (liquid or gas) with a rigid body or with an elastic structure, like a membrane, see [36, 37]: the evolution of the rigid body is described by a system of o.d.e.s, while the evolution of the fluid is subject to p.d.e.s like Navier-Stokes or Euler equations.

The problems considered below naturally fall in the main directions of the present theory of conservation laws. Indeed, this analytical theory currently appears split in two distinct branches: multi–D scalar equations and 1D systems. Fundamental results for the former branch are Kružkov work [25], in the case of the Cauchy problem, and the paper by Bardos, Leroux and Nédélec [3], in the case with boundary. On the other hand, the first existence result for 1D systems dates back to Glimm's work [22], whereas continuous dependence was proved more recently in [7], in the case of 2 equations, and in the general case in [8]. This approach was then greatly simplified in [10, 29, 30, 31], see also [6, 19].

Throughout, we denote $\mathbb{R}^+ = [0, +\infty[$ and $\mathring{\mathbb{R}}^+ =]0, +\infty[$.

2 Systems in 1D

This section deals with a 1D hyperbolic conservation law on $\mathbb{R}^+$ coupled with an o.d.e. through the boundary condition, i.e.

$$\begin{cases} \partial_t u + \partial_x f(u) = 0 & (t,x) \in \mathbb{R}^+ \times \mathbb{R}^+ \\ b\left(u(t,0)\right) = B\left(t, w(t)\right) & t \in \mathbb{R}^+ \\ \dot{w} = F\left(t, u(t,0), w\right) & t \in \mathbb{R}^+. \end{cases} \tag{2.1}$$

The boundary condition on the second line above is an algebraic relation acting as interface between the p.d.e. and the o.d.e.. Solutions to systems of conservation laws develop jump discontinuities and the sense in which the boundary condition is attained is critical. Here, we restrict our attention to the *non characteristic case*, which allows us to require that the solution to (2.1) attains the value imposed at the boundary in the strong sense of the trace, for a.e. time, see [4, Definition 2.1] and [1, 2, 13, 23].

On system (2.1) we require the following conditions, where we refer to [6, 19] for the standard vocabulary about conservation laws.

(f) f is smooth, $Df(u)$ is strictly hyperbolic for $u \in \Omega$ and each characteristic field is either genuinely nonlinear or linearly degenerate.

Without loss of generality, we may assume that $0 \in \Omega$ and that the n real eigenvalues $\lambda_1(u), \ldots, \lambda_n(u)$ of $Df(u)$ are ordered so that $\lambda_{i-1}(u) < \lambda_i(u)$ for all $u \in \Omega$ and $i = 2, \ldots, n$. Denote the right eigenvectors with $r_1(u), \ldots, r_n(u)$. We also introduce the following conditions.

(NC) There exist a $c > 0$ and $\ell \in \{1, 2, \ldots, n-1\}$ such that for $u \in \Omega$, $\lambda_\ell(u) < -c$ and $\lambda_{\ell+1}(u) > c$.

(b) $b \in \mathbf{C^1}(\Omega; \mathbb{R}^{n-\ell})$ and $\det(D_u b(0)\,[r_{\ell+1}(0)\ \ r_{\ell+2}(0)\ \ \cdots\ \ r_n(0)]) \neq 0$.

(B) $B \in \mathbf{C^1}(\mathbb{R}^+ \times \mathbb{R}^m; \mathbb{R}^{n-\ell})$ is locally Lipschitz, i.e. for every compact $\tilde{K} \subset \mathbb{R}^m$, there exists a $\tilde{C}_{\tilde{K}} > 0$ such that, for $t > 0$ and $w \in \tilde{K}$: $\left\|\partial_t B(t,w)\right\|_{\mathbb{R}^{n-\ell}} + \left\|\partial_w B(t,w)\right\|_{\mathbb{R}^{n-\ell}} \leq \tilde{C}_{\tilde{K}}$.

If the boundary condition is invertible, i.e. $u(t,0) = b^{-1}(B(t, w(t)))$, then (2.1) can be decoupled, solving first $\dot{w} = F(t, b^{-1}(B(t,w)), w)$ and then the remaining balance law. Condition **(b)** above essentially *imposes* b to be *not* invertible.

We impose that the o.d.e. in (2.1) fits into Caratheodory equations.

(F) The map $F \colon \mathbb{R}^+ \times \Omega \times \mathbb{R}^m \longrightarrow \mathbb{R}^m$ is such that

(F.1) For $t > 0$ and $u \in \Omega$, the map $\begin{array}{ccc} \mathbb{R}^m & \longrightarrow & \mathbb{R}^m \\ w & \longmapsto & F(t,u,w) \end{array}$ is continuous.

(F.2) For $t > 0$ and $w \in \mathbb{R}^m$, the map $\begin{array}{ccc} \Omega & \longrightarrow & \mathbb{R}^m \\ u & \longmapsto & F(t,u,w) \end{array}$ is continuous.

(F.3) For $u \in \Omega$ and $w \in \mathbb{R}^m$, the map $\begin{array}{ccc} \mathbb{R}^+ & \longrightarrow & \mathbb{R}^m \\ t & \longmapsto & F(t,u,w) \end{array}$ is measurable.

(F.4) For all compact $K \subset \Omega$, there is a $C_K > 0$ such that if $u \in K$, $w_1, w_2 \in \mathbb{R}^m$, $t \in \mathbb{R}^+$, $\left\|F(t,u,w_1) - F(t,u,w_2)\right\| \leq C_K \left\|w_1 - w_2\right\|$.

(F.5) There exists a $C \in \mathbf{L^1_{loc}}(\mathbb{R}^+; \mathbb{R}^+)$ such that for $t > 0$, $u \in \Omega$, $w \in \mathbb{R}^m$, $\left\|F(t,u,w)\right\|_{\mathbb{R}^m} \leq C(t)\,\left(1 + \|w\|_{\mathbb{R}^m}\right)$.

We are now ready to state the main well posedness result.

Theorem 2.1. [4, Theorem 2.8] *Let **(f)**, **(NC)**, **(b)**, **(B)** and **(F)** hold. Assume that* $0 \in \mathring{\Omega}$ *and* $b(0) = B(0,0)$. *Then, there exist* $\delta, T_\delta > 0$ *with the following property: for every initial data* u_o *and* w_o *satisfying* $\mathrm{TV}(u_o) + \|w_o\|_{\mathbb{R}^m} < \delta$, *problem (2.1) admits a solution on* $[0, T_\delta]$.

By *solution*, we mean here a pair consisting of a weak entropy solution to the conservation law and a Caratheodory solution to the o.d.e., see [4, Definition 2.6] for details.

For uniqueness, continuous dependence from the initial data and stability estimates for (2.1), we defer to [12].

As a first example, consider a rectilinear tube filled with fluid to the right of a piston. In the isentropic, or isothermal, case this system can be described by the p-system coupled with an o.d.e.. More precisely

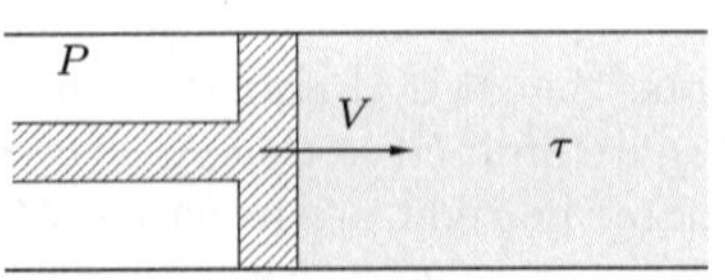

$$\begin{cases} \partial_t \tau - \partial_x v = 0 \\ \partial_t v + \partial_x p(\tau) = 0 \\ V = v(t, 0+) \\ \dot{V} = \alpha \cdot \Big(P(t) - p\left(\tau(t,0+)\right)\Big) \end{cases} \tag{2.2}$$

where t is time, x the Lagrangian mass coordinate, τ the specific volume, v the Lagrangian speed of the flow, p the pressure in the fluid, V the speed of the piston, $P(t)$ the (given) pressure to the left of the piston and α is the ratio between the section of the tube and the mass of the piston. This problem has been widely considered in the literature, mainly with pistons having *assigned* movements, see for instance [34, 35] or [26, §99]. Here, on the contrary, the acceleration of the piston is due to the difference between the pressure of the fluid on its right and that of the outer environment on its left.

Let $n = 2$, $\ell = 1$, $m = 1$ and, with reference to (2.1), fix any $(\bar{\tau}, \bar{v}) \in \mathring{\mathbb{R}}^+ \times \mathbb{R}$. Define

$$u = \begin{bmatrix} \tau - \bar{\tau} \\ v - \bar{v} \end{bmatrix}, \quad f(u) = \begin{bmatrix} -v \\ p(\tau) \end{bmatrix}, \quad \begin{array}{ll} w = V - \bar{v}, & F(t,u,w) = \alpha\left(P(t) - p(\tau)\right), \\ b(u) = v, & B(t,w) = V. \end{array}$$

Then, problem (2.2) is of the type (2.1). Moreover, if $p \in \mathbf{C^4}(\mathring{\mathbb{R}}^+; \mathring{\mathbb{R}}^+)$ with $p'(\tau) < 0$ and $p''(\tau) > 0$ for all $\tau \in \mathring{\mathbb{R}}^+$ and $P \in \mathbf{L^1_{loc}}(\mathbb{R}^+; \mathring{\mathbb{R}}^+)$ then, Theorem 2.1 applies, see [4] for details.

As a second example, consider a single junction in a sewer network. At $x = 0$, a junction joins k horizontal pipes to one vertical manhole. Each pipe is referred to an abscissa $x \in \mathbb{R}^+$. The flow in the i–th tube, for $i = 1, \ldots, k$, can be described by the Saint Venant equations [20], see also [18] or [26, formula (108.1)], while the coupling is obtained imposing the conservation of mass and the equality of the hydraulic heads, see [5, 4, 28] and [33, Chapter 1]. We are thus lead to study the system [5]

$$\begin{cases} \partial_t A_i + \partial_x Q_i = 0 \\ \partial_t Q_i + \partial_x \left(\dfrac{Q_i^2}{A_i} + p_i(A_i) \right) = 0 \\ \hat{h}(t) = \dfrac{1}{2g} \dfrac{Q_i(t,0+)^2}{A_i(t,0+)^2} + h_i\left(A_i(t,0+)\right) \\ h_M(t) = -\dfrac{1}{2g\, A_M^2} \left(\displaystyle\sum_{i=1}^{k} Q_i(t,0+) \right) \left| \displaystyle\sum_{i=1}^{k} Q_i(t,0+) \right| + \hat{h}(t) \\ \dot{h}_M(t) = \dfrac{1}{A_M} \left(Q_{\text{ext}}(t) - \displaystyle\sum_{i=1}^{k} Q_i(t,0+) \right) \end{cases}$$

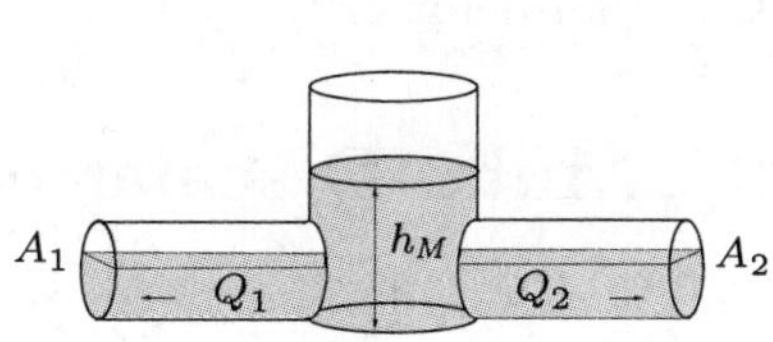

where i ranges over $1, \ldots, k$. This system falls within the class (2.1). Indeed, fix any $k \in \mathbb{N} \setminus \{0\}$ and set $n = 2k$, $\ell = k$, $m = 1$. Fix the areas

$\bar{A}_1, \ldots, \bar{A}_k > 0$, the flows $\bar{Q}_1, \ldots, \bar{Q}_k$ and the height $\bar{h}_M > 0+$. Define

$$u = \begin{bmatrix} A_1 - \bar{A}_1 \\ Q_1 - \bar{Q}_1 \\ \vdots \\ A_k - \bar{A}_k \\ Q_k - \bar{Q}_k \end{bmatrix} \qquad f(u) = \begin{bmatrix} Q_1 \\ \frac{Q_1^2}{A_1} + p_1(A_1) \\ \vdots \\ Q_k \\ \frac{Q_k^2}{A_k} + p_k(A_k) \end{bmatrix} \qquad B(t,w) = \begin{bmatrix} 0 \\ \vdots \\ 0 \\ h_M \end{bmatrix}$$

$$\left.\begin{array}{l} w = h_M - \bar{h}_M \\ F(t,u,w) = \mathrm{sign}(\hat{h} - h_M)\sqrt{2g\left|\hat{h} - h_M\right|} \\ \qquad\qquad + \frac{1}{A_M}\, Q_{\mathrm{ext}}(t) \end{array}\right| \quad b(u) = \begin{bmatrix} \frac{1}{2g}\frac{Q_2^2}{A_2^2} + h_2(A_2) - \hat{h} \\ \vdots \\ \frac{1}{2g}\frac{Q_k^2}{A_k^2} + h_k(A_k) - \hat{h} \\ \frac{-\sum_{i=1}^{k} Q_i}{2g\, A_M^2}\left|\sum_{i=1}^{k} Q_i\right| + \hat{h} \end{bmatrix}$$

where $\hat{h} = \frac{1}{2g}\frac{Q_1^2}{A_1^2} + h_1(A_1)$. Then, Theorem 2.1 applies, provided

$$\frac{1}{2g}\frac{\bar{Q}_1^2}{\bar{A}_1^2} + h_1(\bar{A}_1) = \cdots = \frac{1}{2g}\frac{\bar{Q}_k^2}{\bar{A}_k^2} + h_k(\bar{A}_k) = \bar{h}_M + \frac{\sum_{i=1}^{k} \bar{Q}_i}{2g\, \bar{A}_M^2}\left|\sum_{i=1}^{k} \bar{Q}_i\right|$$

$$|\bar{Q}_i| < \bar{A}_i \sqrt{p_i(\bar{A}_i)} \quad \text{for } i = 1, \ldots, k; \qquad \left|\sum_{i=1}^{k} \bar{Q}_i\right| \sum_{i=1}^{k} \frac{\bar{A}_i}{\sqrt{p_i'(\bar{A}_i)}} \neq 1$$

$$\begin{array}{l} p_i \in \mathbf{C^4}(\mathring{\mathbb{R}}^+; \mathring{\mathbb{R}}^+) \text{ with } p_i' > 0 \text{ and } p_i'' > 0 \text{ for } i = 1, \ldots, k\,, \\ h_i \in \mathbf{C^1}(\mathbb{R}^+; \mathbb{R}^+) \quad \text{and} \quad Q_{\mathrm{ext}} \in \mathbf{L^1_{loc}}(\mathbb{R}^+; \mathbb{R}^+) \end{array}$$

see [4] for details.

3 Multi–D scalar equations

This section is devoted to the problem

$$\begin{cases} \partial_t \rho + \mathrm{div}_x f\left(t, x, \rho, p(t)\right) = 0 & (t,x) \in \mathbb{R}^+ \times \mathbb{R}^{N_x} \\ \dot{p} = \varphi\left(t, p, \left(A\rho(t)\right)(p)\right) & \rho \in \mathbb{R}^+ \\ \rho(0,x) = \bar{\rho}(x) & p \in \mathbb{R}^{N_p} \\ p(0) = \bar{p} \end{cases} \tag{3.1}$$

where the unknowns are $\rho = \rho(t,x)$, i.e. the scalar density describing the macroscopic state of the continuum, and $p = p(t)$, i.e. the state of the individuals. The latter quantity can be the vector of the individuals' positions or of the individuals' positions and speeds. The dynamics of the continuum is described by the flow f, which is in general the product $f = \rho v$ of ρ and a suitable speed $v = v\left(t, x, \rho(t,x), p(t)\right)$. The vector field φ defines the dynamics of the individuals at time t and it depends from ρ through a suitable average $A\left(\rho(t)\right)$. Our driving example below is the convolution in the space variable $A\left(\rho(t)\right) = \rho(t) * \eta$, with a smooth compactly supported kernel η.

Below we present results on the existence and uniqueness of entropy solutions, on their stability with respect the data and the equation. For examples of optimal control problems, see [15]. As usual in this context, see also [14, 16, 17, 25, 32], the hypotheses on f are rather intricate. However, when the continuum can be supposed initially confined in a bounded region, [15, Corollary 2.5] applies and the hypotheses on f are greatly simplified.

Below, $B_{\mathbb{R}^{N_p}}(x, r)$ denotes the closed ball in $\mathbb{R}^{N_p}$ centered at x with radius r. Let $T_{\max} \in [0, +\infty]$ and call $I = [0, T_{\max}]$ if $T_{\max} < +\infty$, while $I = \mathbb{R}^+$ otherwise. The real parameter R, i.e. the maximal possible density is fixed and positive. For a given compact set K in $\mathbb{R}^{N_p}$ and a $T > 0$, we denote $\Omega_T = [0, T] \times \mathbb{R}^{N_x} \times [0, R] \times K$.

(f) The flow $f \colon I \times \mathbb{R}^{N_x} \times [0, R] \times \mathbb{R}^{N_p} \to \mathbb{R}^{N_x}$ is such that

(f.1) $f \in \mathbf{C}^2(I \times \mathbb{R}^{N_x} \times [0, R] \times \mathbb{R}^{N_p}; \mathbb{R}^{N_x})$.

(f.2) For all $(t, x, p) \in I \times \mathbb{R}^{N_x} \times \mathbb{R}^{N_p}$, $f(t, x, 0, p) = f(t, x, R, p) = 0$.

(f.3) For all $T \in I$ and for all compact subsets $K \subset \mathbb{R}^{N_p}$, there exists a C_f such that for $t \in [0, T]$, $x \in \mathbb{R}^{N_x}$, $\rho \in [0, R]$ and $p \in K$, $\left\|\partial_\rho f(t, x, \rho, p)\right\| < C_f$ and $\left|\operatorname{div}_x f(t, x, \rho, p)\right| < C_f$.

(f.4) For all $T \in I$ and for all compact subsets $K \subset \mathbb{R}^{N_p}$, there exists a C_f such that for $t \in [0, T]$, $x \in \mathbb{R}^{N_x}$, $\rho \in [0, R]$ and $p \in K$, $\left\|\nabla_x \partial_\rho f(t, x, \rho, p)\right\| < C_f$.

(f.5) For all compact subsets $K \subset \mathbb{R}^{N_p}$, there exists a C_f such that $\int_I \int_{\mathbb{R}^{N_x}} \sup_{p \in K, \rho \in [0,R]} \left\|\nabla_x \operatorname{div}_x f(t, x, \rho, p)\right\| \mathrm{d}x \, \mathrm{d}t < C_f$.

(f.6) For all compact subsets $K \subset \mathbb{R}^{N_p}$, there exists a C_f such that $\int_I \int_{\mathbb{R}^{N_x}} \sup_{p \in K, \rho \in [0,R]} \left\|\operatorname{div}_x f(t, x, \rho, p)\right\| \mathrm{d}x \, \mathrm{d}t < C_f$.

(f.7) For all $T \in I$ and for all compact subsets $K \subset \mathbb{R}^{N_p}$, there exists a C_f such that for $t \in [0, T]$, $\rho \in [0, R]$, $p \in K$ and $x \in \mathbb{R}^{N_x}$, $\int_{\mathbb{R}^{N_x}} \left\|\nabla_p \operatorname{div}_x f(t, x, \rho, p)\right\| \mathrm{d}x < C_f$ and $\left\|\nabla_p \partial_\rho f(t, x, \rho, p)\right\| < C_f$.

Condition **(f.2)** states that at $\rho = R$, the continuum is at congestion and can not move. The first part of **(f.3)** ensures the finite propagation speed of the solution to the p.d.e., see [15, Proposition 4.2] or [25, Theorem 1]. The second part ensures that the solution is bounded, similarly to **(φ.3)** in the o.d.e.. All these assumptions are satisfied, for instance, by vector fields of the form $u(\rho, x, p) = v(\rho)\, \vec{\mathbf{v}}(x, p)$ with $v \in \mathbf{C}^2([0, R]; \mathbb{R})$ and

$\vec{\mathbf{v}} \in \mathbf{C^2_c}(\mathbb{R}^{N_x} \times \mathbb{R}^{N_x}; \mathbb{R}^{N_x})$. If f does not depend explicitly on t and x, as it is usual when dealing with 1D systems of conservation laws, then the above assumptions reduce to only **(f.1)**, **(f.2)**, the first part of **(f.3)** and the first part of **(f.7)**. Moreover, [15, Corollary 2.5] shows that whenever $\bar\rho$ has compact support, then the requirements on f are reduced, since only **(f.1)**, **(f.2)** and **(f.3)** are necessary.

The individuals' state changes with speed $\varphi = \varphi(t, p(t), A(\rho(t))(p(t)))$ that depends on time t, on the individuals' state p at time t and on an average $A\left(\rho(t)\right)$ of the continuum density ρ evaluated at time t and computed at $p(t)$. On A we require

(A) $A\colon \mathbf{L^1}(\mathbb{R}^{N_x}; \mathbb{R}) \to \mathbf{W^{1,\infty}}(\mathbb{R}^{N_p}; \mathbb{R}^{N_r})$ is linear and continuous.

The operator norm of A is $\|A\|_{\mathcal{L}(\mathbf{L^1}, \mathbf{W^{1,\infty}})}$. If $N_p = N_x$, a typical example is $\left(A(\rho)\right)(p) = (\rho * \eta)(p)$ with $\eta \in \mathbf{C^1_c}(\mathbb{R}^{N_x}; \mathbb{R})$ and $\int_{\mathbb{R}^{N_p}} \eta \, \mathrm{d}x = 1$.

(φ) The vector field $\varphi\colon \mathbb{R}^+ \times \mathbb{R}^{N_p} \times \mathbb{R}^{N_r} \to \mathbb{R}^{N_p}$ is such that

(φ.1) $t \mapsto \varphi(t, p, r)$ is measurable for all $p \in \mathbb{R}^{N_p}$ and all $r \in \mathbb{R}^{N_r}$;

(φ.2) there exists a function $C_\varphi \in \mathbf{L^1}(I; \mathbb{R}^+)$ such that for a.e. $t \in I$, $p_1, p_2 \in \mathbb{R}^{N_p}$ and $r_1, r_2 \in \mathbb{R}^{N_r}$, $\left\|\varphi(t, p_1, r_1) - \varphi(t, p_2, r_2)\right\| \leq C_\varphi(t) \left(\|p_1 - p_2\| + \|r_1 - r_2\|\right)$;

(φ.3) there exists a function $C_\varphi \in \mathbf{L^1}(I; \mathbb{R}^+)$ such that for a.e. $t \in [0, T]$, for all $p \in \mathbb{R}^{N_p}$ and for all $r \in \mathbb{R}^{N_r}$, $\left\|\varphi(t, p, r)\right\| \leq C_\varphi(t) \left(1 + \|p\|\right)$.

We define as *solution* to (3.1) with initial datum $(\bar\rho, \bar p)$ a pair $\rho \in \mathbf{C^0}(I; \mathbf{L^1}(\mathbb{R}^{N_x}; [0, R]))$ and $p \in \mathbf{W^{1,1}}(I; \mathbb{R}^{N_p})$ such that $\rho = \rho(t, x)$ is a Kružkov solution [25, Definition 1] to $\partial_t \rho + \mathrm{div}_x f(t, x, \rho, p(t)) = 0$, $p = p(t)$ is a Caratheodory solution [21, § 1] to $\dot p = \varphi(t, p, A(\rho(t))(p))$ and $\rho(0) = \bar\rho$, $p(0) = \bar p$, see [15, Definition 2.1].

Theorem 3.1. *Under conditions* ***(f)***, ***(φ)*** *and* ***(A)***, *for any initial datum $\bar p \in \mathbb{R}^{N_p}$ and $\bar\rho \in (\mathbf{L^1} \cap \mathbf{BV})(\mathbb{R}^{N_x}; [0, R])$, problem (3.1) admits a unique solution. This solution can be extended to all I.*

Let now f_1, f_2 satisfy ***(f)***; *A_1, A_2 satisfy* ***(A)*** *and φ_1, φ_2 satisfy* ***(φ)***, *for the same interval I and the same parameters or functions R, C_f, C_A, C_φ. Then, for any initial data $(\bar\rho_1, \bar p_1), (\bar\rho_2, \bar p_2) \in (\mathbf{L^1} \cap \mathbf{BV})(\mathbb{R}^{N_x}; [0, R]) \times \mathbb{R}^{N_p}$, the solutions (ρ_1, p_1) and (ρ_2, p_2) to*

$$\begin{cases} \partial_t \rho_1 + \mathrm{div}_x f_1\left(t, x, \rho_1, p_1(t)\right) = 0 \\ \dot p_1 = \varphi_1\left(t, p_1, \left(A_1 \rho_1(t)\right)(p_1)\right) \\ \rho_1(0, x) = \bar\rho_1(x) \\ p_1(0) = \bar p_1 \end{cases} \qquad \begin{cases} \partial_t \rho_2 + \mathrm{div}_x f_2\left(t, x, \rho_2, p_2(t)\right) = 0 \\ \dot p_2 = \varphi_2\left(t, p_2, \left(A_2 \rho_2(t)\right)(p_2)\right) \\ \rho_2(0, x) = \bar\rho_2(x) \\ p_2(0) = \bar p_2 \end{cases}$$

satisfy the inequalities

$$
\begin{aligned}
&\left\|(\rho_1-\rho_2)(t)\right\|_{\mathbf{L^1}}\\
&\le \left(1+\mathcal{K}(t)\right)\|\bar\rho_1-\bar\rho_2\|_{\mathbf{L^1}}\\
&+\mathcal{K}(t)\left(\left\|\partial_\rho(f_1-f_2)\right\|_{\mathbf{L}^\infty(\Omega_t)}+\left\|\mathrm{div}\,(f_1-f_2)\right\|_{\mathbf{L^1}(\mathbb{R}^{N_x})\times\mathbf{L}^\infty([0,t]\times[0,R]\times K_t)}\right)\\
&+\mathcal{K}(t)\left(\|\varphi_1-\varphi_2\|_{\mathbf{L}^\infty([0,t]\times K_t\times[0,C_A])}+\|A_1-A_2\|_{\mathcal{L}(\mathbf{L^1},\mathbf{W}^{1,\infty})}+\|\bar p_1-\bar p_2\|\right)\\
&\left\|(p_1-p_2)(t)\right\|\\
&\le \left(1+\mathcal{K}(t)\right)\|\bar p_1-\bar p_2\|\\
&+\mathcal{K}(t)\left(\left\|\partial_\rho(f_1-f_2)\right\|_{\mathbf{L}^\infty(\Omega_t)}+\left\|\mathrm{div}\,(f_1-f_2)\right\|_{\mathbf{L^1}(\mathbb{R}^{N_x})\times\mathbf{L}^\infty([0,t]\times[0,R]\times K_t)}\right)\\
&+\mathcal{K}(t)\left(\|\varphi_1-\varphi_2\|_{\mathbf{L}^\infty([0,t]\times K_t\times[0,C_A])}+\|A_1-A_2\|_{\mathcal{L}(\mathbf{L^1},\mathbf{W}^{1,\infty})}+\|\bar\rho_1-\bar\rho_2\|_{\mathbf{L^1}}\right)
\end{aligned}
$$

where $\mathcal{K}\in\mathbf{C^0}(I;\mathbb{R}^+)$ *vanishes at* $t=0$.

As a first toy application we consider the situation described in [24, n. 246]. To lure rats away, the city of Hamelin hires a rat-catcher who, playing his magic pipe, attracts all mice out of the city. In this case, $\rho=\rho(t,x)$ is the mice density and $p=p(t)$ is the piper's position. Rats move with a speed $v(\rho)\,\vec{\boldsymbol{v}}(p-x)$, with the scalar v and the vector $\vec{\boldsymbol{v}}$ as in Figure 3.1.

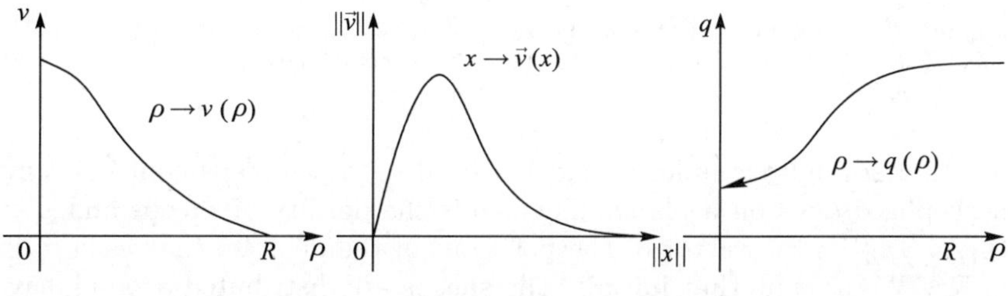

Figure 3.1 Left, v is $\mathbf{C^2}$ and decreasing. Center, $\vec{\boldsymbol{v}}$ is the attraction felt by mice. Right, q is for the acceleration of the piper due to mice density.

More precisely, at density 0 mice have the fastest speed while at density R their speed vanishes. The term $\vec{\boldsymbol{v}}$ accounts for the attraction of the mice due to music. The piper has speed $q(\rho*\eta)\,\vec{\boldsymbol{\psi}}(t)$, i.e. he moves faster when the average mice density around him is higher. On the contrary, when only few rats are near to him, he slows down. This situation fits within the framework of Theorem 3.1, see [15, Lemma 3.1].

To fix a specific situation, in (3.1) we choose: $v(\rho)=V_{\max}\,(1-\rho/R)$, $\vec{\boldsymbol{v}}(x)=x\,e^{-\|x\|^2}$, $q(r)=v_p+(V_p-v_p)r/R$, $\vec{\boldsymbol{\psi}}(t)=\begin{bmatrix}\cos\omega t\\-\sin\omega t\end{bmatrix}$ and $\eta(x)=\left(3/(\pi {r_p}^6)\right)\left(\max\{0,{r_p}^2-\|x\|^2\}\right)^2$ with parameters $V_{\max}=9$, $R=1$, $V_p=7$, $v_p=1$, $\omega=1$ and $r_p=0.15$. At $t=0$, rats are uniformly distributed with density $R = 1$ in $[-0.5, 0] \times [0.35, 0.85]$. The piper starts from $(-1, 0.5)$. See Figure 3.2 for graphs of the resulting solution.

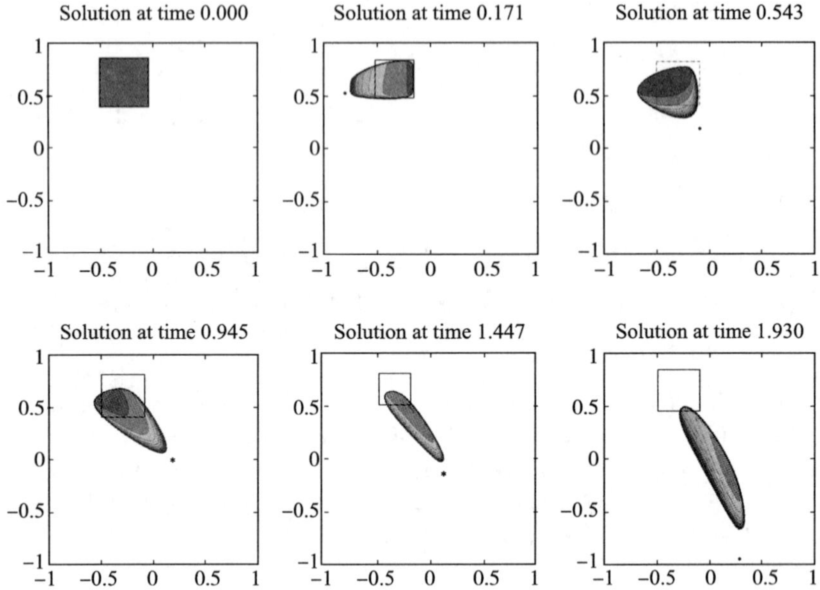

Figure 3.2 The pied piper and the rats, at times 0, when $p = (0, 0.5)$; 0.171; 0.543; 0.945; 1.447 and 1.930, when the rats almost completely left the rectangle and $p = (0.366, -0.983)$.

As a second example, consider a herd of, say, sheeps controlled by n shepherd dogs on a plane. Then, ρ is the density of sheeps and $p \equiv (p_1, \ldots, p_n)$ is the vector of the positions of the dogs, so that each p_i is in $\mathbb{R}^2$. We assume that initially the sheeps are distributed around, say, the origin and tend to disperse moving radially with a speed directed by $\vec{v_r}(x)$. The dogs have to prevent this dispersion and they pursue this goal moving around sheeps or, more precisely, with a speed φ orthogonal to the gradient of the sheeps' density. The sheeps modify their speed escaping from the dogs with a repulsive speed $\vec{v_d}(x, p) = \sum_i \vec{v}(x - p_i)$, where $\vec{v}$ behaves qualitatively as in Figure 3.1, center. Finally, the speed of the sheeps is given by $v(\rho)\,(\vec{v_r}(x) + \sum_{i=1}^{n} \vec{v}(x - p_i))$ where v is maximal at the density zero and vanishes at the maximal density R, which means that the sheeps can not move when their density is maximal.

Under standard regularity assumptions, this situation fits within the framework of Theorem 3.1, see [15, Lemma 3.3].

To fix a specific situation, we choose $n = 2$ and the following functions in (3.1): $v(\rho) = V_{\max}\,(1 - \rho/R)$, $\vec{v}(x) = (\alpha/\sqrt{\ell})\, e^{-\|x\|^2/\ell}\, x$, $\vec{v}_r(x) = (\beta\, x)/(1 + \|x\|^2)$ and $\eta(x) = \big(3/(\pi {r_p}^6)\big)\, (\max\{0, {r_p}^2 - \|x\|^2\})^2$ with parameters $V_{\max} = 1$, $R = 1$, $\alpha = 20$, $\ell = 0.2$, $\beta = 1$, $r_p = 1$ and $V_d = 100$.

At $t = 0$, sheeps are uniformly distributed at density $R = 1$ in the circumference centered at $(0, 0)$ with radius 0.2. Dogs start from (0.7, 0) and (−0.7, 0). Graphs of the corresponding solution are in Figure 3.3.

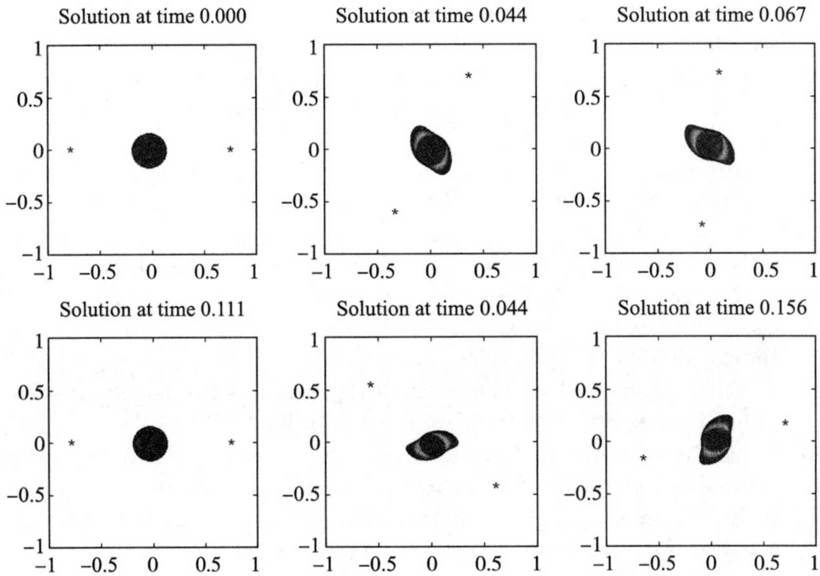

Figure 3.3 Solution to (3.1) in the case of the shepherd dog at times $t = 0$, $t = 0.044$, $t = 0.067$, $t = 0.111$, $t = 0.156$, $t = 0.200$. Sheeps are initially uniformly distributed at the maximal density $R = 1$ in the circumference centered at $(0, 0)$ with radius 0.2. Dogs start moving from (0.7, 0) and (−0.7, 0), they succeed in confining the dispersion of the sheeps, at least for the tie interval considered.

References

[1] D. Amadori, Initial-boundary value problems for nonlinear systems of conservation laws. Nonlinear Differential Equations Appl., 4(1): 1–42, 1997.

[2] D. Amadori and R. M. Colombo, Continuous dependence for 2 × 2 conservation laws with boundary. J. Differential Equations, 138(2): 229–266, 1997.

[3] C. Bardos, A. Y. le Roux and J.-C. Nédélec, First order quasilinear equations with boundary conditions. Comm. Partial Differential Equations, 4(9): 1017–1034, 1979.

[4] R. Borsche, R. M. Colombo and M. Garavello, On the coupling of systems of hyperbolic conservation laws with ordinary differential equations. Preprint, 2010.

[5] R. Borsche and A. Klar, A mixed type model for flooding in urban drainage systems. In preparation, 2010.

[6] A. Bressan, Hyperbolic Systems of Conservation Laws. Volume 20 of Oxford Lecture Series in Mathematics and its Applications. Oxford University Press, Oxford, 2000.

[7] A. Bressan and R. M. Colombo, The semigroup generated by 2×2 conservation laws. Arch. Rational Mech. Anal., 133(1): 1–75, 1995.

[8] A. Bressan, G. Crasta and B. Piccoli, Well-posedness of the Cauchy problem for $n \times n$ systems of conservation laws. Mem. Amer. Math. Soc., 146(694): viii+134, 2000.

[9] A. Bressan and C. De Lellis, Existence of optimal strategies for a fire confinement problem. Comm. Pure Appl. Math., 62(6): 789–830, 2009.

[10] A. Bressan, T.-P. Liu and T. Yang, L^1 stability estimates for $n \times n$ conservation laws. Arch. Ration. Mech. Anal., 149(1): 1–22, 1999.

[11] V. Capasso, A. Micheletti and D. Morale, Stochastic geometric models, and related statistical issues in tumour-induced angiogenesis. Math. Biosci., 214(1-2): 20–31, 2008.

[12] R. M. Colombo and M. Garavello, Well posedness for models based on coupled o.d.e.–conservation laws. In preparation., 2010.

[13] R. M. Colombo and G. Guerra, On general balance laws with boundary. Journal of Differential Equations, 248(5): 1017–1043, 2010.

[14] R. M. Colombo, M. Herty and M. Mercier, Control of the continuity equation with a non-local flow. To appear on ESAIM Control Optim. Calc. Var., 2010.

[15] R. M. Colombo and M. Mercier, An analytical framework to describe the interactions between individuals and a continuum. In preparation., 2010.

[16] R. M. Colombo, M. Mercier and M. D. Rosini, Stability and total variation estimates on general scalar balance laws. Commun. Math. Sci., 7(1): 37–65, 2009.

[17] R. M. Colombo, M. Mercier and M. D. Rosini, Stability estimates on general scalar balance laws. C. R. Math. Acad. Sci. Paris, 347(1-2): 45–48, 2009.

[18] J. Cunge, F. Holly and A. Verwey, Practical Aspects of Computational River Hydraulics. Pitmann Publishing, Bath, 1980.

[19] C. M. Dafermos, Hyperbolic Conservation Laws in Continuum Physics. Volume 325 of Grundlehren der Mathematischen Wissenschaften. Springer-Verlag, Berlin, second edition, 2005.

[20] M. de Saint-Venant, Théorie du mouvement non permanent des eaux, avec application aux crues des riveres et a l'introduction des marees dans leur lit. Acad. Sci. Comptes Rondus, pages 147–154, 237–240, 1871.

[21] A. F. Filippov, Differential Equations with Discontinuous Righthand Sides. Kluwer Academic Publishers Group, Dordrecht, 1988. Translated from the Russian.

[22] J. Glimm, Solutions in the large for nonlinear hyperbolic systems of equations. Comm. Pure Appl. Math., 18: 697–715, 1965.

[23] J. Goodman, Initial Boundary Value Problems for Hyperbolic Systems of Conservation Laws. PhD thesis, California University, 1982.

[24] J. Grimm and W. Grimm, Deutsche Sagen. Nicolaische Verlagsbichhandlung, Berlin, second edition, 1865.

[25] S. N. Kružkov, First order quasilinear equations with several independent variables. Mat. Sb. (N.S.), 81(123): 228–255, 1970.

[26] L. D. Landau and E. M. Lifshitz, Fluid Mechanics. Translated from the Russian by J. B. Sykes and W. H. Reid. Course of Theoretical Physics, Vol. 6. Pergamon Press, London, 1959.

[27] C. Lattanzio, A. Maurizi and B. Piccoli, Moving bottlenecks in car traffic flow: a pde-ode coupled model. Preprint, 2010.

[28] G. Leugering and J. P. G. Schmidt, On the modelling and stabilization of flows in networks of open canals. SIAM Journal on Control and Optimization, 41(1): 164–180, 2002.

[29] T.-P. Liu and T. Yang, L_1 stability for 2×2 systems of hyperbolic conservation laws. J. Amer. Math. Soc., 12(3): 729–774, 1999.

[30] T.-P. Liu and T. Yang, A new entropy functional for a scalar conservation law. Comm. Pure Appl. Math., 52(11): 1427–1442, 1999.

[31] T.-P. Liu and T. Yang, Well-posedness theory for hyperbolic conservation laws. Comm. Pure Appl. Math., 52(12): 1553–1586, 1999.

[32] M. Mercier, Stability estimates on general scalar balance laws. Preprint, 2010.

[33] R. Mulley, Flow of Industrial Fluids: Theory and Equations. CRC Press, 2004.

[34] T. Nishida, Global solution for an initial boundary value problem of a quasilinear hyperbolic system. Proc. Japan Acad., 44: 642–646, 1968.

[35] T. Nishida and J. Smoller, Mixed problems for nonlinear conservation laws. J. Differential Equations, 23(2): 244–269, 1977.

[36] D. Serre, Chute libre d'un solide dans un fluide visqueux incompressible. Existence. Japan J. Appl. Math., 4(1): 99–110, 1987.

[37] J. L. Vázquez and E. Zuazua, Large time behavior for a simplified 1D model of fluid-solid interaction. Comm. Partial Differential Equations, 28(9–10): 1705–1738, 2003.

Stability of Finite Difference Schemes for Hyperbolic Initial Boundary Value Problems*

Jean-François Coulombel
CNRS & Université Lille 1
Laboratoire Paul Painlevé (UMR CNRS 8524)
Cité scientifique, Bâtiment M2
59655 Villeneuve d'Ascq Cedex, France
E-mail: jean-francois.coulombel@math.univ-lille1.fr

Abstract

We review some results on the stability of finite difference schemes for hyperbolic initial boundary value problems. In the case of zero initial data, our main goal is to characterize *strong stability* of the numerical scheme in terms of an appropriate spectral condition that is analogous to the Uniform Kreiss-Lopatinskii Condition for hyperbolic partial differential equations. The last step in our analysis consists in incorporating non-zero initial data and proving *semigroup estimates.*

1 Introduction

A general stability theory for finite difference approximations of hyperbolic initial boundary value problems was initiated in the fundamental contributions [10, 8]. The extension of the results of [8] to the multidimensional case was performed in [15]. The main goal of this theory is to characterize stability of a numerical scheme by means of a certain algebraic spectral condition that is referred to as the Uniform Kreiss-Lopatinskii Condition (UKLC in what follows). The theory is entirely analogous to the characterization of well-posedness for (non-discretized) hyperbolic initial boundary value problems. We refer to [11, 17], see also the recent book [1, chapter 4], for results on such "continuous" problems. We also refer to [9, 18] for some nice connections between these stability theories and the propagation of wave packets. The verification of

*Research of the author was supported by the French Agence Nationale de la Recherche, contract ANR-08-JCJC-0132-01.

the UKLC may be performed either by hand-made calculations or, if the calcutations are too involved, it could be checked numerically with some suitable approximations. Consistency of the numerical scheme usually follows from some Taylor expansions, and together with stability, it yields convergence results, see [6].

Proving that the UKLC is sufficient for stability is the most difficult part of the job. It relies on the construction of suitable symmetrizers where, instead of the theory for the hyperbolic Cauchy problem where one considers propagation of signals in the time direction, here one considers the initial boundary value problem as an evolution equation in the normal variable to the boundary of the space domain. For continuous problems, the symmetrizers construction was performed independently in [11] and [17]. The analysis of numerical schemes is unfortunately more involved for reasons that we explain below. In [8] and [15], the authors made such assumptions on the numerical schemes that in the final part of their analysis, they were able to use the symmetrizers constructed in [11, 17] to derive their results. From a practical point of view, such assumptions on the numerical schemes are not so severe in one space dimension. The dissipation assumption in [15] is much more restrictive in several space dimensions. As a general fact, the simulation of hyperbolic equations should include as less dissipation as possible. It is therefore natural to try to develop a stability theory for discretized initial boundary value problems where the numerical schemes are allowed to have an arbitrarily small dissipation. This was the main motivation for our works [2, 3] where we extended the results of [8] in one space dimension to a large class of numerical schemes with arbitrarily small dissipation. We consider the results of [2, 3] as a natural starting point in view of a future generalization of [15] in the multidimensional case.

This article is organized as follows. In section 2, we review the characterization of stability for zero initial data by means of the UKLC. In section 3, we present the energy method developed in [4] in order to prove semigroup estimates and incorporate non-zero initial data.

2 Strong stability for zero initial data: the normal modes analysis

We consider a hyperbolic initial boundary value problem in one space dimension:

$$\begin{cases} \partial_t u + A\,\partial_x u = F(t,x)\,, & (t,x) \in \mathbb{R}^+ \times \mathbb{R}^+\,, \\ B\,u(t,0) = g(t)\,, & t \in \mathbb{R}^+\,, \\ u(0,x) = f(x)\,, & x \in \mathbb{R}^+\,, \end{cases} \tag{2.1}$$

that we discretize by means of a finite difference scheme. We thus introduce some space and time steps $\Delta x, \Delta t > 0$, where the ratio $\lambda := \Delta t/\Delta x$ is a fixed positive constant, and we let p, q, r, s denote some fixed integers. We keep Δt as a free small parameter while Δx is given by the relation $\Delta x = \Delta t/\lambda$. The solution to (2.1) is approximated by a sequence (U_j^n) defined for $n \in \mathbb{N}$, $j \in 1 - r + \mathbb{N}$, and with values in $\mathbb{R}^N$ with N the size of the system (2.1). For $j = 1-r, \dots, 0$, U_j^n approximates the trace $u(n\,\Delta t, 0)$ on the boundary $\{x = 0\}$, and possibly the trace of normal derivatives. The boundary meshes $[j\,\Delta x, (j+1)\,\Delta x[$, $j = 1-r, \dots, 0$, shrink to $\{0\}$ as the small parameter Δt tends to 0, so the formal continuous limit problem as Δt tends to 0 is set on the half-line $\mathbb{R}^+$. We consider some approximations of (2.1) that read

$$\begin{cases} U_j^{n+1} = \displaystyle\sum_{\sigma=0}^{s} Q_\sigma\, U_j^{n-\sigma} + \Delta t\, F_j^n\,, & j \geq 1\,, \quad n \geq s\,, \\ U_j^{n+1} = \displaystyle\sum_{\sigma=-1}^{s} B_{j,\sigma}\, U_1^{n-\sigma} + g_j^{n+1}\,, & j = 1-r, \dots, 0\,, \quad n \geq s\,, \\ U_j^n = f_j^n\,, & j \geq 1-r\,, \quad n = 0, \dots, s\,, \end{cases} \tag{2.2}$$

where the shift operators Q_σ and $B_{j,\sigma}$ are given by

$$Q_\sigma := \sum_{\ell=-r}^{p} A_{\ell,\sigma}\, T^\ell\,, \quad B_{j,\sigma} := \sum_{\ell=0}^{q} B_{\ell,j,\sigma}\, T^\ell\,, \qquad T^\ell\, U_k^m := U_{k+\ell}^m\,.$$

The $N \times N$ matrices $A_{\ell,\sigma}, B_{\ell,j,\sigma}$ may depend on the CFL parameter λ as well as on A, B in (2.1) but are independent of Δt. Various possibilities for the discretization of the boundary conditions are described in [7, chapter 11]. Here we consider that the discretization of the hyperbolic operator is given, and we wish to characterize the discrete boundary conditions in (2.2) — meaning the operators $B_{j,\sigma}$ — that yield a stable scheme. The first task is to define precisely which notion of stability is considered. Depending on the source terms in (2.2), various possibilities are described in [8]. We shall focus here on *strongly stable* schemes as defined below.

The following notation

$$\|u\|_{m_1,m_2}^2 := \Delta x \sum_{j=m_1}^{m_2} |u_j|^2$$

is used to denote the ℓ^2-norm on the interval $[m_1, m_2]$ (m_1 may equal $-\infty$ and m_2 may equal $+\infty$). We recall the following definition:

Definition 2.1 (Strong stability [8]). The finite difference approximation (2.2) is said to be strongly stable if there exists a constant C_0 such that for all $\gamma > 0$ and all $\Delta t \in]0,1]$, the solution (U_j^n) to (2.2) with vanishing initial data ($f^0 = \cdots = f^s = 0$) satisfies the estimate

$$\frac{\gamma}{\gamma\,\Delta t + 1} \sum_{n \geq s+1} \Delta t\, \mathrm{e}^{-2\,\gamma\, n\, \Delta t} \, \|U^n\|^2_{1-r,+\infty}$$

$$+ \sum_{n \geq s+1} \Delta t\, \mathrm{e}^{-2\,\gamma\, n\, \Delta t} \sum_{j=1-r}^{0} |U_j^n|^2$$

$$\leq C_0 \left\{ \frac{\gamma\,\Delta t + 1}{\gamma} \sum_{n \geq s} \Delta t\, \mathrm{e}^{-2\,\gamma\,(n+1)\,\Delta t} \, \|F^n\|^2_{1,+\infty} \right.$$

$$\left. + \sum_{n \geq s+1} \Delta t\, \mathrm{e}^{-2\,\gamma\, n\, \Delta t} \sum_{j=1-r}^{0} |g_j^n|^2 \right\}.$$

Our goal is to give a "practical" characterization of strong stability in terms of a certain algebraic condition. For later use, we introduce the *amplification matrix* (or symbol) associated with the discretization of the hyperbolic operator:

$$\forall \kappa \in \mathbb{C} \setminus \{0\}, \quad \mathcal{A}(\kappa) := \begin{pmatrix} \widehat{Q_0}(\kappa) & \dots & \dots & \widehat{Q_s}(\kappa) \\ I & 0 & \dots & 0 \\ 0 & \ddots & \ddots & \vdots \\ 0 & 0 & I & 0 \end{pmatrix} \in \mathcal{M}_{N(s+1)}(\mathbb{C}),$$

$$\widehat{Q_\sigma}(\kappa) := \sum_{\ell=-r}^{p} \kappa^\ell \, A_{\ell,\sigma} \,. \tag{2.3}$$

We recall that the uniform power boundedness of $\mathcal{A}(\kappa)$ for $\kappa \in \mathbb{S}^1$ is a necessary and sufficient condition for the ℓ^2-stability of the discretized Cauchy problem, see e.g. [7, chapter 5].

The resolvent equation is formally obtained from (2.2) by applying a Laplace transform in time, see [8]. It reads

$$\begin{cases} w_j - \displaystyle\sum_{\sigma=0}^{s} z^{-\sigma-1}\, Q_\sigma\, w_j = F_j \,, & j \geq 1 \,, \\ w_j - \displaystyle\sum_{\sigma=-1}^{s} z^{-\sigma-1}\, B_{j,\sigma}\, w_1 = g_j \,, & j = 1-r, \dots, 0 \,, \end{cases} \tag{2.4}$$

where $|z| > 1$, $(F_j) \in \ell^2$, and $g_{1-r}, \dots, g_0 \in \mathbb{C}^N$. It is convenient to rewrite the resolvent equation (2.4) as an evolution equation for the sequence (w_j). Assumption **(H1)** below is crucial in order to achieve this transformation. For $\ell = -r, \dots, p$, let us define the matrices

$$\forall\, z \in \mathbb{C} \setminus \{0\}\,, \quad \mathbb{A}_\ell(z) := \delta_{\ell,0}\, I - \sum_{\sigma=0}^{s} z^{-\sigma-1} A_{\ell,\sigma}\,,$$

where δ_{ℓ_1,ℓ_2} is the Kronecker symbol. As in [8], we make the following assumption:

(H1) The matrices $\mathbb{A}_{-r}(z)$ and $\mathbb{A}_p(z)$ are invertible for all $|z| \geq 1$.

Let us assme $q < p$ for simplicity. In that case, all the w_j's involved in the *boundary conditions* for the resolvent equation (2.4) are coordinates of the augmented vector $W_1 := (w_p, \dots, w_{1-r})$. Using Assumption **(H1)**, we can define a matrix $\mathbb{M}(z)$ that is holomorphic on some open neighborhood $\mathcal{V}$ of $\{|z| \geq 1\}$

$$\forall\, z \in \mathcal{V}\,,\ \mathbb{M}(z) := \begin{pmatrix} -\mathbb{A}_p(z)^{-1}\, \mathbb{A}_{p-1}(z) & \dots & \dots & -\mathbb{A}_p(z)^{-1}\, \mathbb{A}_{-r}(z) \\ I & 0 & \dots & 0 \\ 0 & \ddots & \ddots & \vdots \\ 0 & 0 & I & 0 \end{pmatrix} \in \mathcal{M}_{N(p+r)}(\mathbb{C})\,.$$

We can then rewrite the resolvent equation (2.4) as an induction relation for the augmented vector $W_j := (w_{j+p-1}, \dots, w_{j-r})$:

$$\begin{cases} W_{j+1} = \mathbb{M}(z)\, W_j + \widetilde{F_j}\,, & j \geq 1\,, \\ B(z)\, W_1 = g\,, \end{cases} \tag{2.5}$$

with some suitable source terms $(\widetilde{F_j}), g$. It is easy to check that the matrix $B(z) \in \mathcal{M}_{Nr,N(p+r)}$ that encodes the boundary conditions in (2.5) depends holomorphically on $z \in \mathbb{C} \setminus \{0\}$ and has maximal rank $N\,r$ for all z. Our main result is the following:

Theorem 2.2. *Let Assumption* **(H1)** *be satisfied. Assume moreover that the symbol $\mathcal{A}$ defined by (2.3) satisfies the two following conditions*

- Uniform power boundedness: *there exists a constant $C_1 > 0$ such that for all $\kappa \in \mathbb{S}^1$ and all $n \in \mathbb{N}$, $\|\mathcal{A}(\kappa)^n\| \leq C_1$,*
- Geometric regularity of eigenvalues on $\mathbb{S}^1$: *if $\underline{\kappa} \in \mathbb{S}^1$ and $\underline{z} \in \mathbb{S}^1 \cap \mathrm{sp}(\mathcal{A}(\underline{\kappa}))$ has algebraic multiplicity $\underline{\alpha}$, then there exist some functions $\lambda_1(\kappa), \dots, \lambda_{\underline{\alpha}}(\kappa)$ that are holomorphic in a neighborhood $\mathcal{W}$ of $\underline{\kappa}$ in $\mathbb{C}$ and that satisfy*

$$\lambda_1(\underline{\kappa}) = \dots = \lambda_{\underline{\alpha}}(\underline{\kappa}) = \underline{z}\,,\ \det\big(z\, I - \mathcal{A}(\kappa)\big) = \vartheta(\kappa, z) \prod_{j=1}^{\underline{\alpha}} \big(z - \lambda_j(\kappa)\big)\,,$$

with ϑ a holomorphic function of (κ, z) in some neighborhood of $(\underline{\kappa}, \underline{z})$ such that $\vartheta(\underline{\kappa}, \underline{z}) \neq 0$. Furthermore, there exist some vectors $E_1(\kappa), \ldots, E_{\underline{\alpha}}(\kappa) \in \mathbb{C}^{N(s+1)}$ that depend holomorphically on $\kappa \in \mathcal{W}$, that are linearly independent for all $\kappa \in \mathcal{W}$, and that satisfy

$$\forall \kappa \in \mathcal{W}, \quad \forall j = 1, \ldots, \underline{\alpha}, \quad \mathcal{A}(\kappa)\, E_j(\kappa) = \lambda_j(\kappa)\, E_j(\kappa).$$

For $|z| > 1$, we let $\mathbb{E}^s(z)$ denote the generalized eigenspace associated with eigenvalues of $\mathbb{M}(z)$ with modulus less than 1. Then $\mathbb{E}^s(z)$ has constant dimension $N\,r$ for all $|z| > 1$ and $\mathbb{E}^s$ defines a holomorphic vector bundle over $\{|z| > 1\}$. This vector bundle can be extended continuously in a unique way over $\{|z| \geq 1\}$. We still let $\mathbb{E}^s(z)$ denote this continuous extension for $|z| = 1$.

In addition to all previous assumptions, assume that for all $|z| \geq 1$ we have $\mathbb{E}^s(z) \cap \operatorname{Ker} B(z) = \{0\}$. In what follows this condition is referred to as the UKLC. Then the scheme (2.2) is strongly stable.

We refer to [2, 3] for a complete proof. Let us make some comments on the result and on the strategy of the proof. Theorem 2.2 is in the same spirit as the well-posedness results for multidimensional hyperbolic initial boundary value problems. Two main assumptions are made here on the discretization of the hyperbolic operator. The uniform power boundedness of the matrices $\mathcal{A}(\kappa)$ is a *stability assumption* for the discretized Cauchy problem. This condition plays the same role as *hyperbolicity* in the analysis of the continuous problem. Let us observe that for all $\kappa \in \mathbb{S}^1$, the eigenvalues of $\mathcal{A}(\kappa)$ necessarily belong to $\{|z| \leq 1\}$, and eigenvalues on the unit circle are semi-simple. The geometric regularity assumption makes the behavior of eigenvalues and eigenvectors precise near a point $\underline{\kappa} \in \mathbb{S}^1$ where the spectrum of $\mathcal{A}(\underline{\kappa})$ meets $\mathbb{S}^1$. This assumption is similar to the geometric regularity condition that characterizes the so-called block structure for continuous problems, see [14, appendix C].

The assumptions on $\mathcal{A}$ enable us to extend the *stable bundle* $\mathbb{E}^s$ of $\mathbb{M}$ from the open set $\{|z| > 1\}$ to its closure $\{|z| \geq 1\}$ in a unique way. This result is completely independent of the boundary conditions in the scheme (2.2). Once this first result is known, strong stability for (2.2) is encoded in the UKLC. Let us observe that the UKLC is meaningful as far as dimensions of vector spaces are concerned. Indeed, the matrix $B(z)$ has full rank, so its kernel has dimension $N\,p$, and we ask that this kernel does not intersect $\mathbb{E}^s(z)$ which has dimension $N\,r$. In other words, the UKLC is equivalent to requiring the decomposition

$$\mathbb{C}^{N(p+r)} = \mathbb{E}^s(z) \oplus \operatorname{Ker} B(z)$$

for all $|z| \geq 1$.

As described in the introduction, the main result in [8] is similar to Theorem 2.2 — though it is not exactly stated in the same way — but it requires some additional assumptions on the eigenvalues of the amplification matrix $\mathcal{A}$. More precisely, the following behavior is allowed in [8] for the eigenvalues of $\mathcal{A}$ that touch the unit circle: either the eigenvalue λ_j satisfies $\lambda_j'(\underline{\kappa}) \neq 0$ if $\lambda_j(\underline{\kappa}) \in \mathbb{S}^1$ (regular tangency point), or $|\lambda_j(e^{i\,\xi})| = 1$ for all $e^{i\,\xi}$ sufficiently close to $\underline{\kappa}$. The latter behavior occurs for instance with the leap-frog scheme. As described through a series of examples in [2, 3], the only stability assumption on the discretization of the hyperbolic operator, that is, uniform power boundedness of $\mathcal{A}(\kappa)$ for $\kappa \in \mathbb{S}^1$, allows for much more complicated behaviors of the eigenvalues λ_j than the two behaviors allowed in [8]. Taking into account such new behaviors requires a more general analysis that allows for a larger class of numerical schemes. Dropping the technical assumptions of [8] is also of particular importance in view of the analysis of multidimensional discretizations where the behaviors of eigenvalues can be even more complex (in particular, geometric regularity of eigenvalues can be violated).

As a final word on Theorem 2.2, let us sketch the main ideas of the proof. Following the analysis in the continuous case, both conclusions of Theorem 2.2 (extension of the bundle $\mathbb{E}^s$ and sufficiency of the UKLC for strong stability) follow from *one single argument* (!), namely from the existence of a family of K-symmetrizers. This terminology is borrowed from [13] — a great source of inspiration together with [12, 14] for the analysis of [2, 3] — and a precise definition in the context of numerical schemes is [2, Definition 2.5]. In the framework of Theorem 2.2, showing the existence of K-symmetrizers is done on a reduced version of the matrix $\mathbb{M}(z)$. More precisely it can be shown, see [2, section 3], that under the assumptions of Theorem 2.2, the matrix $\mathbb{M}(z)$ satisfies the *discrete block structure condition* defined in [2] and that is analogous to the block structure that appears in the continuous context. The symmetrizers are then constructed separately for each block appearing in the discrete block structure. The main novelty compared to the continuous case is that the numerous possible behaviors of the eigenvalues λ_j of $\mathcal{A}(\kappa)$ yield a huge variety of new blocks in the discrete block structure. In particular, symmetrizing such new blocks requires new arguments since the analysis of [11, 17] can not cover such cases, see [3, appendix B]. A detailed construction of these new symmetrizers was first initiated in [2], and then fully worked out in [3] where a complete classification of blocks was made according to their size and their *dissipation index*. The latter notion is the heart of the argument in [3] since it relates the behaviors of the eigenvalues λ_j for the discrete Cauchy problem to the properties of the blocks in the reduction of $\mathbb{M}(z)$. We hope that this new construction may reveal useful in other contexts such as the analysis of semi-discretized equations or for some other types of partial differential equations.

3 Stability for non-zero initial data: proving semigroup estimates

The complete stability analysis of (2.2) requires considering non-zero initial conditions $f^0, \dots, f^s$. When introducing non-zero initial conditions in some Banach space, say E, it becomes natural to ask for a *semigroup estimate* of the solution (U^n) in $\ell^\infty_n(E)$. For hyperbolic initial boundary value problems, the incorporation of initial conditions was done in [16]. The natural space of initial conditions is L^2 as can be guessed from the theory of dissipative boundary conditions. For numerical schemes, the problem has been understood so far only in the case $s = 0$, that is, when the scheme (2.2) more simply reads

$$\begin{cases} U_j^{n+1} = Q\,U_j^n + \Delta t\,F_j^n\,, & j \geq 1\,, \quad n \geq 0\,, \\ U_j^{n+1} = B_{j,-1}\,U_1^{n+1} + B_{j,0}\,U_1^n + g_j^{n+1}\,, & j = 1-r, \dots, 0\,, \quad n \geq 0\,, \\ U_j^0 = f_j\,, & j \geq 1-r\,. \end{cases} \tag{3.1}$$

In one space dimension, the analysis in [19] reduces the proof of semigroup estimates for initial conditions $f \in \ell^2$ to the verification of strong stability (in the sense of Definition 2.1) for Dirichlet boundary conditions. Hopefully, Dirichlet boundary conditions had been studied in [5] where the authors had checked strong stability by directly computing the stable subspace $\mathbb{E}^s(z)$ and the corresponding Lopatinskii determinant. For scalar equations, or equivalently when all matrices in (2.2) can be simultaneously diagonalized, the Lopatinskii determinant associated with Dirichlet boundary conditions is a Vandermonde determinant that can be proved to be non-vanishing. As a corollary, the main result of [19] establishes stability estimates for (3.1) when all matrices in the definition of the finite difference operator Q are simultaneously diagonalizable.

Simultaneous diagonalization of all matrices happens all the time in one space dimension because the matrices are more or less polynomial functions of $\lambda\,A$. However, this property does not hold anylonger in two or more space dimensions. In the non-commuting case, the computations in [5] for the stable subspace $\mathbb{E}^s(z)$ cease to be valid, and there is therefore little hope to extend the argument in [19] to the framework of multidimensional systems. (There is no doubt that the argument in [19] combined with [5] should work as well for discretized multidimensional scalar problems.) The motivation for our work [4] was to derive semigroup estimates for (3.1) with an argument that would work as well for multidimensional systems. Instead of the simultaneous diagonalizing assumption, we assume the the discretization of the hyperbolic operator does not increase the ℓ^2-norm for the Cauchy problem:

(H2) The operator Q in (3.1) satisfies $\|Q\,v\|_{-\infty,+\infty} \le \|v\|_{-\infty,+\infty}$ for all $v \in \ell^2$.

The first main result of [4] proves semigroup estimates as well as strong stability for (3.1) in the case of Dirichlet boundary conditions, that is for the numerical scheme

$$\begin{cases} V_j^{n+1} = Q\,V_j^n + \Delta t\,F_j^n\,, & j \ge 1\,, \quad n \ge 0\,, \\ V_j^{n+1} = g_j^{n+1}\,, & j = 1-r,\dots,0\,, \quad n \ge 0\,, \\ V_j^0 = f_j\,, & j \ge 1-r\,. \end{cases} \tag{3.2}$$

More precisely, there holds:

Theorem 3.1. *Let Assumptions* **(H1)** *and* **(H2)** *be satisfied. Then there exists a constant C_2 such that for all $\gamma > 0$ and all $\Delta t \in\,]0,1]$, the solution V to* (3.2) *satisfies the estimate*

$$\begin{aligned} \sup_{n\ge 0} \mathrm{e}^{-2\gamma n\Delta t} \|V^n\|_{1-r,+\infty}^2 &+ \frac{\gamma}{\gamma\,\Delta t + 1} \sum_{n\ge 0} \Delta t\, \mathrm{e}^{-2\gamma n\Delta t} \|V^n\|_{1-r,+\infty}^2 \\ &+ \sum_{n\ge 0} \Delta t\, \mathrm{e}^{-2\gamma n\Delta t} \sum_{j=1-r}^{\max(p,q+1)} |V_j^n|^2 \le C_2 \Bigg\{ \|f\|_{1-r,+\infty}^2 \\ &+ \frac{\gamma\,\Delta t + 1}{\gamma} \sum_{n\ge 0} \Delta t\, \mathrm{e}^{-2\gamma(n+1)\Delta t} \|F^n\|_{1,+\infty}^2 \\ &+ \sum_{n\ge 1} \Delta t\, \mathrm{e}^{-2\gamma n\Delta t} \sum_{j=1-r}^{0} |g_j^n|^2 \Bigg\}. \end{aligned} \tag{3.3}$$

In particular, the discretization (3.2) *is strongly stable in the sense of Definition 2.1.*

Theorem 3.1 is an improved version of [5, Lemma 2.3] because we can directly consider non-zero initial data. Moreover we control in (3.3) the $\ell^\infty_n(\ell^2_j)$-norm of the solution V. This property is useful in the analysis of (3.1). One point in Theorem 3.1 is that there is at least one boundary condition that yields a strongly stable scheme (a property that is not so obvious when one considers the characterization of Theorem 2.2).

Theorem 3.1 suggests introducing the Dirichlet boundary conditions as an auxiliary problem in order to incorporate the initial conditions in

(3.1). More precisely, it seems natural to decompose the solution U to (3.1) as $U := V + W$ where V solves (3.2) and W solves

$$\begin{cases} W_j^{n+1} = Q\,W_j^n\,, & j \geq 1\,, \quad n \geq 0\,, \\ W_j^{n+1} = B_{j,-1}\,W_1^{n+1} + B_{j,0}\,W_1^n + \widetilde{g}_j^{n+1}\,, & j = 1-r,\dots,0\,, \quad n \geq 0\,. \\ W_j^0 = 0\,, & j \geq 1-r\,. \end{cases}$$

The source term $\widetilde{g}$ is defined by

$$\forall j = 1-r,\dots,0\,, \quad \forall n \geq 1\,, \quad \widetilde{g}_j^n := B_{j,-1}\,V_1^n + B_{j,0}\,V_1^{n-1}\,.$$

The V-part of the solution U is controlled by (3.3) and it therefore only remains to control the W-part of U (not the most difficult part of the job, though it should be done with care as far as the semigroup estimate is concerned, see [4, Lemma 2.11]). The final main result of [4] in one space dimension is stated as follows.

Theorem 3.2. *Let Assumptions* **(H1)** *and* **(H2)** *be satisfied, and assume that the scheme* (3.1) *is strongly stable in the sense of Definition 2.1. Then there exists a constant* C_3 *such that for all* $\gamma > 0$ *and all* $\Delta t \in]0,1]$, *the solution* U *to* (3.1) *satisfies the estimate*

$$\begin{aligned} &\sup_{n \geq 0} \mathrm{e}^{-2\,\gamma\,n\,\Delta t}\, \|U^n\|_{1-r,+\infty}^2 \\ &+ \frac{\gamma}{\gamma\,\Delta t + 1} \sum_{n \geq 0} \Delta t\, \mathrm{e}^{-2\,\gamma\,n\,\Delta t}\, \|U^n\|_{1-r,+\infty}^2 + \sum_{n \geq 0} \Delta t\, \mathrm{e}^{-2\,\gamma\,n\,\Delta t} \sum_{j=1-r}^{0} |U_j^n|^2 \\ &\qquad \leq C_3 \left\{ \|f\|_{1-r,+\infty}^2 + \frac{\gamma\,\Delta t + 1}{\gamma} \sum_{n \geq 0} \Delta t\, \mathrm{e}^{-2\,\gamma\,(n+1)\,\Delta t}\, \|F^n\|_{1,+\infty}^2 \right. \\ &\qquad\qquad \left. + \sum_{n \geq 1} \Delta t\, \mathrm{e}^{-2\,\gamma\,n\,\Delta t} \sum_{j=1-r}^{0} |g_j^n|^2 \right\}. \end{aligned}$$

The major new point of Theorem 3.2 compared with the main result of [19] is that it only relies on Assumptions **(H1)** and **(H2)**, both of which can be extended in a straightforward way to the multidimensional case. As a matter of fact, all the arguments in the proof of Theorems 3.1 and 3.2 work in exactly the same way for multidimensional problems. Examples of numerical schemes in two space dimensions that satisfy Assumptions **(H1)** and **(H2)** are given in [4], see also the references therein. Another convenient point of [4] compared with [19], at least in the author's opinion, is that the energy method developed in [4] is completely free of the characterization of strong stability by means of the UKLC, while [19] relied on the computations of [5] for the stable subspace. Actually, it is precisely because the energy method in [4] nowhere uses any Laplace transform in time and Kreiss symmetrizers that the method can be extended to multidimensional problems with not too severe assumptions.

Of course, a major restriction of [4], as in [19], is that it only allows numerical schemes with $s = 0$. An extension to the leap-frog scheme and to other discretizations with several time steps is currently in progress.

References

[1] S. Benzoni-Gavage and D. Serre, Multidimensional Hyperbolic Partial Differential Equations. Oxford University Press, Oxford, 2007.

[2] J.-F. Coulombel, Stability of finite difference schemes for hyperbolic initial boundary value problems. SIAM J. Numer. Anal. 47 (2009), no. 4, 2844–2871.

[3] J.-F. Coulombel, Stability of finite difference schemes for hyperbolic initial boundary value problems II. Ann. Scuola Norm. Sup. Pisa Cl. Sci. (4), to appear.

[4] J.-F. Coulombel and A. Gloria, Semigroup stability of finite difference schemes for multidimensional hyperbolic initial boundary value problems. Math. Comp., to appear.

[5] M. Goldberg and E. Tadmor, Scheme-independent stability criteria for difference approximations of hyperbolic initial-boundary value problems II. Math. Comp. 36 (1981), no. 154, 603–626.

[6] B. Gustafsson, The convergence rate for difference approximations to mixed initial boundary value problems. Math. Comp. 29 (1975), 396–406.

[7] B. Gustafsson, H.-O. Kreiss and J. Oliger, Time Dependent Problems and Difference Methods. John Wiley & Sons Inc., New York, 1995.

[8] B. Gustafsson, H.-O. Kreiss and A. Sundström, Stability theory of difference approximations for mixed initial boundary value problems II. Math. Comp. 26 (1972), 649–686.

[9] R. L. Higdon, Initial-boundary value problems for linear hyperbolic systems. SIAM Rev. 28 (1986), no. 2, 177–217.

[10] H.-O. Kreiss, Stability theory of difference approximations for mixed initial boundary value problems. Math. Comp. 22 (1968), 703–714.

[11] H.-O. Kreiss, Initial boundary value problems for hyperbolic systems, Comm. Pure Appl. Math. 23 (1970), 277–298.

[12] G. Métivier, The block structure condition for symmetric hyperbolic systems, Bull. London Math. Soc., 32 (2000), no. 6, 689–702.

[13] G. Métivier and K. Zumbrun, Symmetrizers and continuity of stable subspaces for parabolic-hyperbolic boundary value problems. Discrete Contin. Dyn. Syst., 11 (2004), no. 1, 205–220.

[14] G. Métivier and K. Zumbrun, Hyperbolic boundary value problems for symmetric systems with variable multiplicities. J. Differential Equations, 211 (2005), no. 1, 61–134.

[15] D. Michelson, Stability theory of difference approximations for multidimensional initial boundary value problems. Math. Comp. 40 (1983), no. 161, 1–45.

[16] J. Rauch, $\mathcal{L}_2$ is a continuable initial condition for Kreiss' mixed problems. Comm. Pure Appl. Math. 25 (1972), 265–285.

[17] R. Sakamoto, Mixed problems for hyperbolic equations. I. Energy inequalities. J. Math. Kyoto Univ. 10 (1970), 349–373.

[18] L. N. Trefethen, Group velocity interpretation of the stability theory of Gustafsson, Kreiss, and Sundström. J. Comp. Phys. 49 (1983), 199–217.

[19] L. Wu, The semigroup stability of the difference approximations for initial-boundary value problems. Math. Comp. 64 (1995), no. 209, 71–88.

Hysteresis and Phase Transition in Many-particle Storage Systems

Wolfgang Dreyer
Weierstrass-Institute for Applied Analysis and Stochastics
Mohrenstr. 39, 10117 Berlin, Germany
E-mail: dreyer@wias-berlin.de

Clemens Guhlke
Weierstrass-Institute for Applied Analysis and Stochastics
Mohrenstr. 39, 10117 Berlin, Germany
E-mail: guhlke@wias-berlin.de

Abstract

We study the behavior of systems that can be described as ensembles of interconnected storage particles. Our examples concern the storage of lithium in many-particle electrodes of rechargeable lithium-ion batteries and the storage of air in a system of interconnected rubber balloons. We are particularly interested in those storage systems whose constituents exhibit non-monotone material behavior leading to transitions between two coexisting phases and to hysteresis. In the current study we consider the case that the time to approach equilibrium of a single storage particle is much smaller than the time for full charging of the ensemble. In this regime, the evolution of the probability to find a particle of the ensemble in a certain state may be described by a nonlocal conservation law of Fokker-Planck type. The resulting equation contains two parameters which control whether the ensemble transits the 2-phase region along a Maxwell line or along a hysteresis path, or whether the ensemble shows the same non-monotone behavior as its constituents.

1 Introduction

Reversible storage systems are of key importance in modern energy technologies. They serve to store electrical or chemical energy for later use, in particular in mobile applications. In this study we consider storage systems that consist of an ensemble of many interconnected storage particles. The technological applications concern ensembles of nano-sized crystalline solids, where matter is stored on lattice sites. Modern many-particle electrodes of lithium-ion batteries belong to this class of storage systems. The storage of hydrogen in certain metals provides a further example.

Our first goal is to develop a generic model that is capable of simulating the evolution of many-particle storage systems during slow loading respectively unloading. We apply this model to $FePO_4$ electrodes of lithium-ion batteries. Moreover, in order to illustrate the crucial behavior of many particle systems we also consider a macroscopic system of interconnected rubber balloons. Such a system may serve to store air and exhibits a similar behavior as the ensemble of nano-sized $FePO_4$ particles.

The paper is organized as follows:

In Section 2 we present a general evolution law to describe the evolution of a generic storage system that consists of a large number of storage particles. The evolution law is a non-local and non-linear Fokker-Planck equation.

Section 3 considers two applications: (1) The storage problem for a many-particle electrode of a rechargeable lithium-ion battery. (2) A system of interconnected spherical rubber balloons that serve to store air supplied from a pressure vessel.

In Section 4 we discuss the two physical parameters of the model.

Numerical simulations are found to be in Section 5 where we study a many-particle electrode during loading and unloading.

In Section 6 we illustrate the corresponding behavior of a system of interconnected rubber balloons.

2 The general model

We propose a general model of a storage system that consists of an ensemble of many interconnected storage particles. The thermodynamic state of a single storage particle is described by a single variable $\xi \in [a, b]$. Any particle of the ensemble is equipped with a double well free energy function $F(\xi)$, so that two coexisting phases are possible. The prescribed mean number $q(t)$ of stored molecules in a particle is represented by the mean value of a given strict monotone function $G(\xi)$. The statistical behavior of the ensemble is represented by a probability density $w(t, \xi)$, whose initial- and boundary value problem reads

$$w(0,\xi) = w_0(\xi), \qquad v(t,a)w(t,a) = v(t,b)w(t,b) = 0,$$

$$\frac{\partial w(t,\xi)}{\partial t} + \frac{\partial v(t,\xi)w(t,\xi)}{\partial \xi} = 0, \tag{2.1}$$

$$\tau v(t,\xi) = \big(\Lambda(t)G'(\xi) - F'(\xi)\big) - \nu^2 \frac{\partial \log(w(t,\xi))}{\partial \xi} .$$

The problem is non-local and non-linear because the function $\Lambda(t)$ must be calculated from the side condition

$$q(t) = \int_a^b G(\xi)w(t,\xi)d\xi . \tag{2.2}$$

3 Non-monotone constitutive behavior

Next we will consider two different applications, that have the non-monotonicity of the crucial constitutive function in common. The first example concerns the charging-discharging process of a lithium-ion battery, where lithium atoms are reversibly stored on interstitial lattice sites of iron phosphate particles forming one of the two electrodes of the battery. The chemical potential $\mu = F'(y)$ of the Li_yFePO_4 crystal, where $y \in \{0,1\}$ is the lithium mole fraction of a single $FePO_4$ storage particle, has the non-monotone shape given in Figure 3.1. The generic function $G(y)$ is the identity.

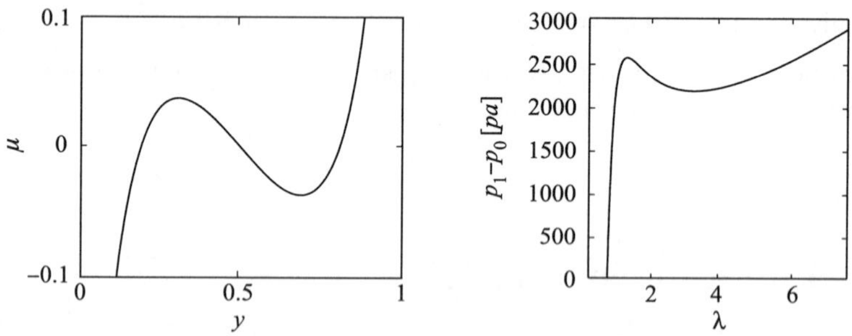

Figure 3.1 Left: Chemical potential of a Li_yFePO_4 storage particle. Right: Pressure jump across a single balloon versus strain λ.

In the second examples we consider a system of interconnected rubber balloons that may serve to store air. In this case the statistical variable ξ is the strain $\lambda = r/r_0$ of a single balloon, where r and r_0 indicate the radii of the actual balloon and the undeformed balloon, respectively. The non-monotonicity here concerns the pressure difference across the membrane of a single balloon as a function of the strain, which qualitatively exhibits the same shape as the chemical potential of Li_yFePO_4, Figure 3.1. The generic function is related to the number $N(\lambda)$ of air molecules by $G(\lambda) = N(\lambda)/\bar{N}$, where $\bar{N}$ is the number of air molecules in the undeformed balloon.

4 Configurational entropy

The properties of solutions of the model are controlled by the two constant positive parameter τ and ν^2. The parameter τ represents the relaxation time of the system and ν^2 controls the influence of the configurational entropy. It describes interaction of the storage particles by exchange of Li atoms and air molecules, respectively. This interaction is important for low charging rates, where it leads to two peculiar phenomena. 1. The storage particles are not loaded at the same time, but according to the rule *one after the other*. 2. The phase transition leading to two coexisting phases does not happen within a single storage particle, rather it is a property of the many-particle ensemble. We observe particles with small filling, forming one phase, and particles with large filling, forming the second phase.

5 Selected simulations for the many-particle electrode

In this section we study a many-particle electrode of a rechargeable lithium-ion battery. We prescribe the history of loading-unloading processes by the filling degree $q(t)$ and solve the evolution equation for the probability density $w(t, y)$.

We solve the equation for $y \in [0, 1]$ and with boundary conditions $v(t, 0)w(t, 0) = v(t, 1)w(t, 1) = 0$. Initially we start the simulation in the 1-phase region, where the storage particles are Gaussian distributed around a small value, $q_0 = 0.1$, for the initial filling degree, viz.

$$w(0, y) = w_0(y) = \sqrt{\frac{\mu'(q_0)}{2\pi\nu^2}} \exp\left(-\frac{\mu'(q_0)}{2\nu^2}(y - q_0)^2\right) . \qquad (5.1)$$

In [2] we have performed simulations for 45 different values of (τ, ν^2) but with the same loading-unloading path: At first we increase the filling degree q linearly from $q = 0.1$ to $q = 0.9$ and afterwards we decrease it from $q = 0.9$ to $q = 0.1$. The time t has been normalized by $|\dot{q}| = 1$, so the parameter $\tau = \tau_D/\tau_L$ gives the ratio between relaxation time τ_D of the storage system and loading time τ_L. In particular, for fixed τ_D

a small τ corresponds to a large loading time. The next Figure shows a typical evolution for parameter τ and ν^2 so that a two phase region appears.

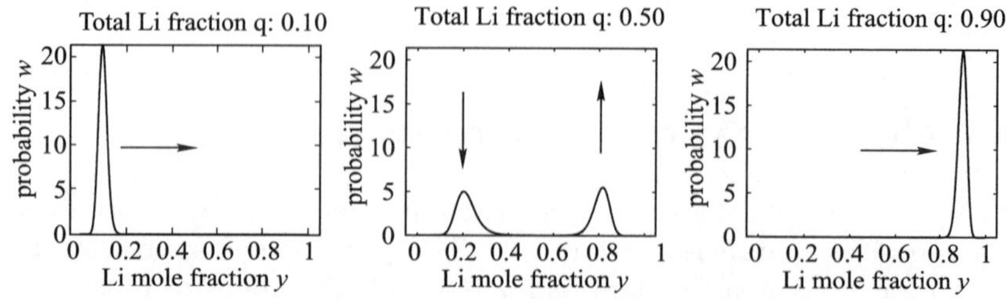

Figure 5.1 Evolution of the probability density for a loading process.

The total lithium mole fraction $q(t)$ is linearly related to the total charge of the battery and the mean chemical potential

$$\langle \mu \rangle (t) = \int_0^1 \mu(y) w(t,y) dy \tag{5.2}$$

is linearly related to the cell voltage of a battery. Thus the crucial behavior of a battery is the dependency of the mean chemical potential on the total lithium mole fraction $q(t)$. Figure 5.2 shows some examples for selected parameters.

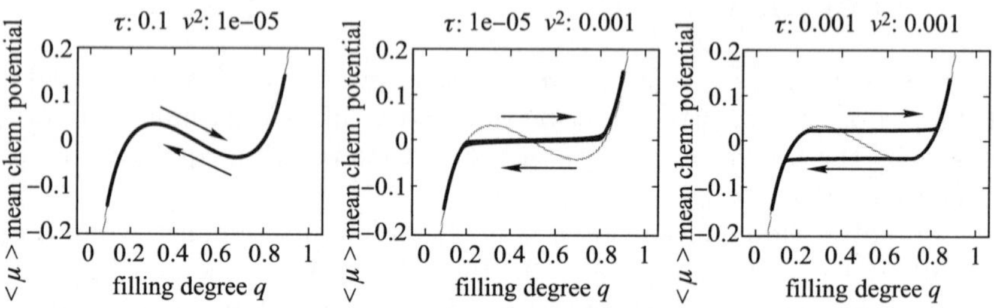

Figure 5.2 Hysteresis loops for various parameter

The location and the shape of the hysteresis is controlled by the two parameter τ and ν^2. The details of the hysteresis curves can be quite involved but we observe three main cases:

1. Figure 5.2 left: The probability density consists of a single pulse only, because there is not sufficient time to interchange molecules between the storage particles according to the configurational entropy effect. The mean chemical potential follows the chemical potential of a single particle.

2. Figure 5.2 middle: The configurational entropy is dominant and the storage system follows the Maxwell line. A hysteresis does not develop here.

3. Figure 5.2 right: The deterministic evolution of the system and interchange of molecules due to the configurational entropy are of the same order. A maximal possible hysteresis is developed.

The diagram on the right hand side of Figure 5.3 concerns a different loading-unloading history that serves to demonstrate that it is possible to change between the loading and unloading plateaus without leaving the two-phase region.

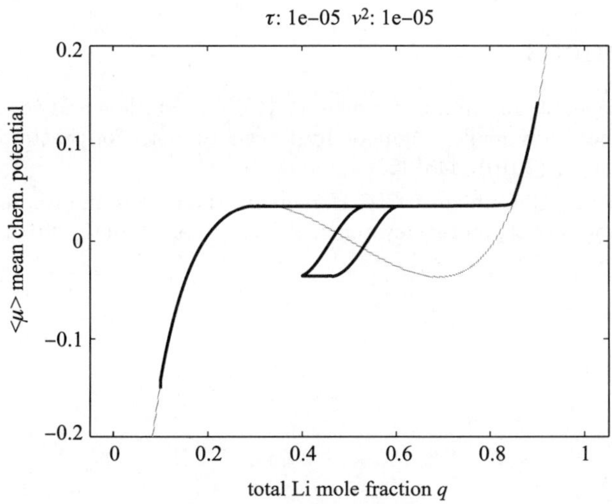

Figure 5.3 Partial loading unloading scenario

6 An experiment with rubber balloons

Finally a system of interconnected rubber balloons serves to demonstrate the rule *one after the other* and to establish the phase separation. The sequence of snapshots in Figure 6.1 are generated for the analog loading-unloading path as above. Initially we observe only balloons with small filling. After a critical filling degree is reached, the transition to a second phase with large balloons sets in. In this period the balloons grow *one after the other* and we have a system of two coexisting phases with small and large balloons.

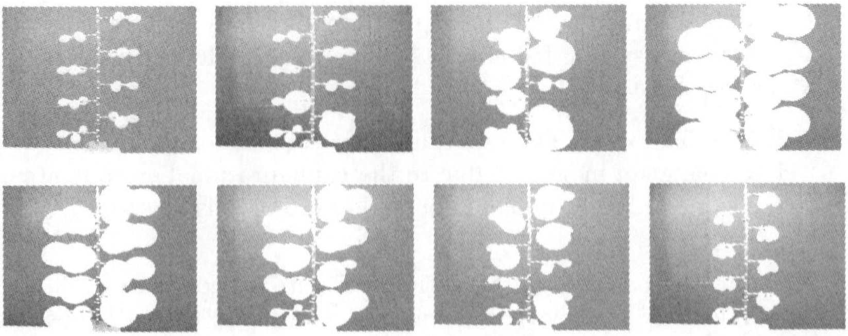

Figure 6.1 Eight different states of interconnected rubber balloons during loading and unloading with air via the pressure vessel. From [2].

References

[1] W. Dreyer, J. Jamnik, C. Guhlke, R. Huth, J. Moškon and M. Gaberšček, The thermodynamic origin of hysteresis in insertion batteries. Nature Materials 9 (2010), 448–453.

[2] W. Dreyer, C. Guhlke and M. Herrmann, Hysteresis and phase transition in many-particle storage systems. WIAS Preprint 1481 (2010).

Existence of Algebraic Vortex Spirals*

Volker Elling
Department of Mathematics, University of Michigan
530 Church Street, Ann Arbor, MI 48109, USA
E-mail: velling@umich.edu

Abstract

We discuss an existence proof for multi-branched algebraic vortex spirals, motivated by the ultimate goal of showing existence of Pullin's separated vortex sheets and other examples of non-uniqueness in the initial-value problem for the Euler equations.

1 Fluids and models

Consider the *compressible Navier-Stokes equations*:

$$\begin{aligned} \frac{\partial}{\partial t}\rho + \nabla\cdot(\rho\vec{v}) &= 0 && \text{[mass]},\\ \frac{\partial}{\partial t}(\rho\vec{v}) + \nabla\cdot(\rho\vec{v}\otimes\vec{v}) + \nabla p &= \nabla\cdot\Sigma && \text{[momentum]},\\ \frac{\partial}{\partial t}(\rho e) + \nabla\cdot(\rho e\vec{v}) + \nabla\cdot(p\vec{v}) &= \nabla\cdot(\Sigma\vec{v}) + \nabla\cdot(\kappa\nabla T) && \text{[energy]}. \end{aligned}$$

Here ρ is mass density, $\vec{v}$ velocity, $e = \frac{1}{2}|\vec{v}|^2 + q$ is specific energy, $q = q(\rho, T)$ specific internal energy, T temperature, $p = p(\rho, T)$ pressure. Σ is the *viscous stress tensor*:

$$\Sigma = \eta(\nabla\vec{v} + (\nabla\vec{v})^T) - \lambda(\nabla\cdot\vec{v})I.$$

The left-hand sides of the Navier-Stokes equations describe transport (convection) and pressure effects. The right-hand sides model heat conduction (coefficient κ) and viscosity (coefficients η, λ), both of which have a smoothing effect on the flow.

In many applications the right-hand side coefficients $\epsilon = (\eta, \lambda, \kappa)$ are very small. Applications rely heavily on the approximation $\epsilon = 0$ which yields the *compressible Euler equations*:

$$\frac{\partial}{\partial t}U + \nabla\cdot(\vec{f}(U)) = 0.$$

*The author is partly supported by NSF grant DMS-0907074.

Solutions of the Euler equations are relevant only as *vanishing viscosity* limits ($\epsilon \downarrow 0$) of Navier-Stokes solutions for the same initial data. This limit, however, is the source of great theoretical and practical difficulties.

It is well-known that solutions of the Euler equations can become discontinuous in finite time, even starting from smooth initial data (Figure 1.1), forming singularities like *shocks* — hypersurfaces across which ρ, T and normal $\vec{v}$ jump, with continuous tangential $\vec{v}$, as well as *contact discontinuities*, in particular *vortex sheets* (Figure 2.1 left) which have continuous ρ, T and normal $\vec{v}$, but jumps in tangential $\vec{v}$. Euler shocks correspond to slightly smooth but sharp transition layers in small-ϵ Navier-Stokes. Hence we must consider *weak (distributional) solutions*: functions U so that

$$\int\int U \frac{\partial}{\partial t}\phi + \vec{f}(U) \cdot \nabla\phi \, dx \, dt = 0 \tag{1.1}$$

for all infinitely often differentiable ϕ with compact support, with additional terms if $\operatorname{supp}\phi$ meets domain boundaries. There are weak solutions, such as *expansion shocks*, that are not vanishing viscosity limits and therefore unphysical. To exclude them it is necessary to impose the *entropy inequality* (second law of thermodynamics)

$$\frac{\partial}{\partial t} S + \nabla \cdot (S\vec{v}) \geq 0 \tag{1.2}$$

which is again understood in a distributional sense; S is the gas-dynamic entropy density.

Vortex sheets are generated by shock waves passing along a concave corner, for example, as observed in experiments. More importantly, even in absence of boundaries several shock waves can interact in some types of *Mach reflection* to generate vortex sheets, some with rollup into single-branch spirals (see Figure 1.1). Hence vortex sheets can arise in finite time from smooth initial data.

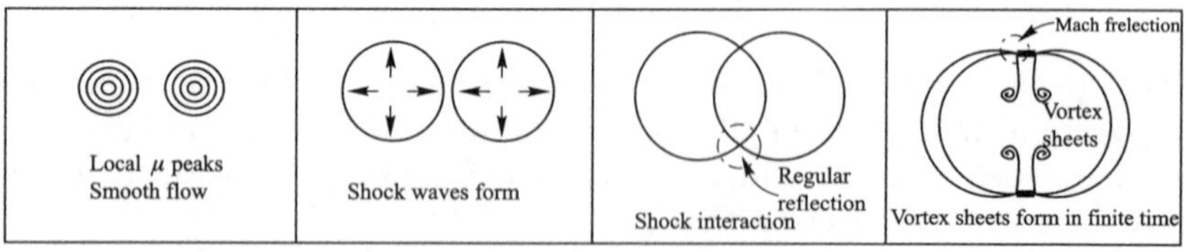

Figure 1.1 Two regions of increased ρ form spherical shock waves that interact, with Mach reflections that generate vortex sheets.

Therefore, a reasonably complete mathematical theory of compressible inviscid flow and its numerics *cannot* avoid vortex sheets and their pathologies.

Vortex sheets also occur in *incompressible flow* which is derived by approximating density ρ and temperature T as constant, an approximation that arises in the limit of low *Mach numbers* $M := |\vec{v}|/c$ (c sound speed), for example. The continuity equation reduces to $\nabla \cdot \vec{v} = 0$, with momentum equations

$$\frac{\partial}{\partial t}\vec{v} + \vec{v} \cdot \nabla \vec{v} + \rho^{-1} \nabla p = 0. \tag{1.3}$$

Taking $\nabla \times$ we obtain the *vorticity equation*

$$\frac{\partial}{\partial t}\omega + \vec{v} \cdot \nabla \omega = 0 \qquad (\Delta \psi = \omega,\ \vec{v} = \nabla^{\perp} \psi). \tag{1.4}$$

Vortex sheets are described by the Birkhoff-Rott equation which is derived as follows: $\omega = 0$ away from the vortex sheets, so ψ is harmonic there. In 2-D it is more convenient to use holomorphic functions: consider the *complex velocity* $W = v^x - iv^y$ as a function of $Z = x + iy$, then the equations $\nabla \cdot \vec{v} = 0$ and $\omega = \nabla \times \vec{v} = 0$ for $\vec{v} = (v^x, v^y)$ are exactly the Cauchy-Riemann equations for W.

In each point of the vortex sheet S, the one-sided limits of the velocity differ by a vector tangential to the vortex sheet: with [ ] denoting jump,

$$[\vec{v}] = \kappa \vec{t}.$$

With s arc length along the sheet, we define the *circulation* Γ:

$$\Gamma = \int \kappa \, ds.$$

$S \ni \Gamma \mapsto Z(t, \Gamma)$ is a parametrization of the vortex sheet curve at time t. For sufficiently smooth Z solutions of the *Birkhoff-Rott equation*

$$\frac{\partial Z}{\partial t}(t, \Gamma) = W(t, \Gamma)^*, \qquad W(t, \Gamma) = \frac{1}{2\pi i}\text{p.v.} \int_S \frac{d\Gamma'}{Z(t, \Gamma) - Z(t, \Gamma')}$$

yield $\vec{v}, p$ that solve the incompressible Euler equation. $*$ is conjugation; "p.v.", *principal value*, defines W as the arithmetic average of W on each side of the sheet.

Apart from our theoretical interest, vortex sheets and spiral rollup are ubiquitous in applications as well. In the "coffee spoon" experiment, a solid line segment is accelerated at initial time in fluid at rest (Figure 1.2 left); a vortex sheet emanates from each end, rolling up into a single-branched spiral (center). The well-known *starting vortex* (Figure 1.2 right) is shed from accelerating aircraft wings, a crucial mechanism in generation of lift and drag forces. Spirals are also found in buoyancy plumes and (more diffused) in turbulence.

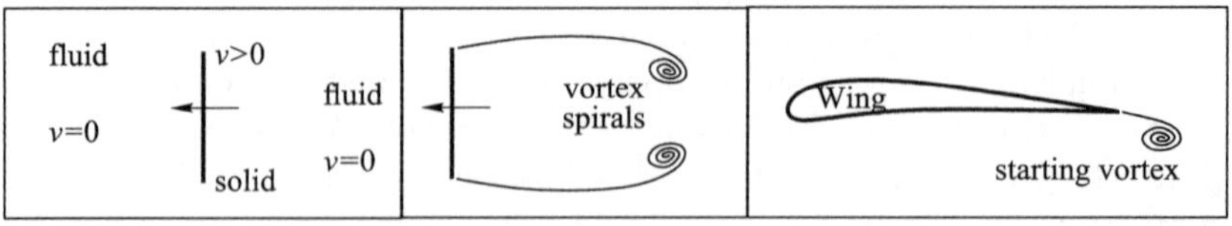

Figure 1.2 Left, center: coffee spoon experiment; right: starting vortex from wing cross-section

2 Pullin's separated sheet

The simplest vortex sheet is straight, with velocity constant on each side and ρ, T constant in the entire 2-D plane (Figure 2.1 left), which is a steady weak solution. Pullin [10] made an intriguing numerical observation: there appear to be additional solutions for the same initial data, with the sheet separating into two halves and each end rolling up into an algebraic spiral (Figure 2.1 center and right).

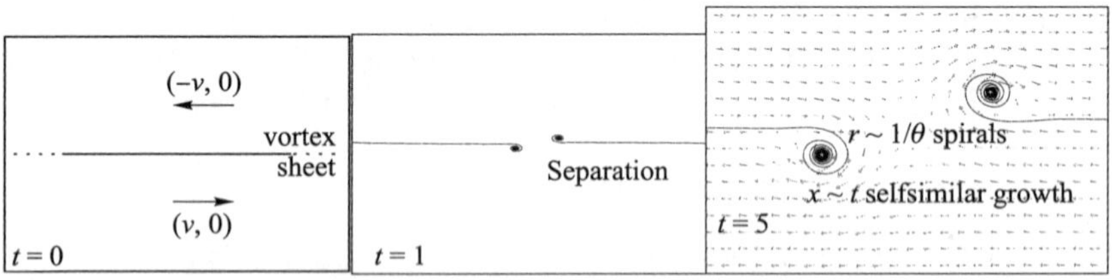

Figure 2.1 Pullin example

The Pullin examples are *self-similar* vortex sheets:

$$Z(t,\Gamma) = t^m z(\gamma), \qquad \gamma := t^{1-2m}\Gamma$$

for a range of m including $m = 1$ which is the hyperbolic scaling of the compressible case and therefore most relevant. The corresponding self-similar Birkhoff-Rott equation is

$$(2m-1)\gamma\frac{\partial z}{\partial \gamma} - mz + \left(\frac{1}{2\pi i}\text{p.v.}\int_S \frac{d\gamma'}{z(\gamma)-z(\gamma')}\right)^* = 0 \qquad (2.1)$$

Vortex sheets and their numerical approximations are notoriously unstable, so a rigorous existence proof is desirable, which is our main motivation.

The linearized Birkhoff-Rott equation is spectrally unstable (Kelvin-Helmholtz). This is not important since we seek evidence of pathologies, not well-posedness, and non-uniqueness is stronger than instability.

3 Separated vortex sheets as singular limits

It is natural to ask how Pullin's separated sheet could arise as a limit of solutions of more complete fluid models like Navier-Stokes, Boltzmann equation or many-particle mechanics. Consider as initial data a straight vortex sheet with a gap (left). In numerical results (Figure 3.1 right) for compressible Euler with added numerical viscosity, we observe the *same* velocity field as in Pullin's separated sheet (Figure 2.1). The pressure (contour lines) has two minima located near the centers of the rotation, counteracting the centrifugal force. A pressure maximum is located in the central saddle-point. For *suffciently large gap* the pressure minima move away to infinity at a constant velocity, exactly as in the incompressible case, resembling Pullin's results. Repeating the calculation with *small* gaps yields convergence to a stable straight sheet.

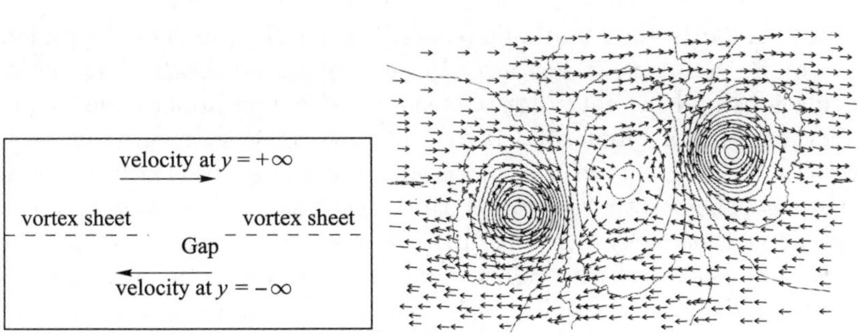

Figure 3.1 Left: initial sheet with gap; right: numerics (contours p, arrows $\vec{v}$).

Assuming existence of a solution of either type for one gap length and one choice of parameters (viscosity/Knudsen numbers/...), we can scale lengths and parameters to obtain a family of solutions, with parameters and gap converging to zero; in particular the initial data is identical in the limit. But there are two families of solutions, one converging to Pullin's separated spirals and one converging to the straight sheet as $t \uparrow \infty$. (Here we use that the inviscid structure, with $x \sim t$ scaling, eventually dominates viscous and other layers with $x \sim \sqrt{t}$ (e.g.) scaling, assuming absence of turbulence and other effects that may change these asymptotics.)

Conclusion: **Non-uniqueness for Euler corresponds to lack of ϵ-uniform stability in Navier-Stokes** (and other models). Pullin's observation should be interpreted as a case of propagation of errors from small to large scales.

In many applications, for example flow around aircraft, viscosity/heat conduction have a significant effect only in turbulent regions, if any, and in boundary or discontinuity layers of width on the order of millimeters. This resolution is prohibitively expensive for 3-D numerical grids that must cover tens of meters sidelength. Instead, simulation and modelling of non-turbulent flows commonly use the Euler equations on coarse grids①, with the crucial assumption/hope that small-scale errors do not propagate to have a significant large-scale effect. This assumption fails more often than previously realized, as shown by Pullin's result and our viscous calculation, as well as observations of our own (below).

Clearly a careful discussion of the basis of contemporary Euler numerics is overdue. Since any formulation of the Euler equations can only be motivated from Navier-Stokes and other limits, we face a theoretical problem as well.

4 The carbuncle phenomenon

Another hope is that non-uniqueness is a purely theoretical problem. But to the contrary, it has already been observed, both in numerical calculations and in reality [4]. Consider a common application: supersonic flow onto a blunt solid body. The expected solution (Figure 4.1 left) is a smooth curved *bow shock* ahead of the body. However, many numerical schemes are susceptible to the *carbuncle phenomenon*, a more well-known pathology that was first observed in [9]: instead of the bow shock, a conical structure called *carbuncle* appears on or near the symmetry axis (Figure 4.1 right). The carbuncle phenomenon is especially common in the Roe and Godunov schemes and generally in schemes that

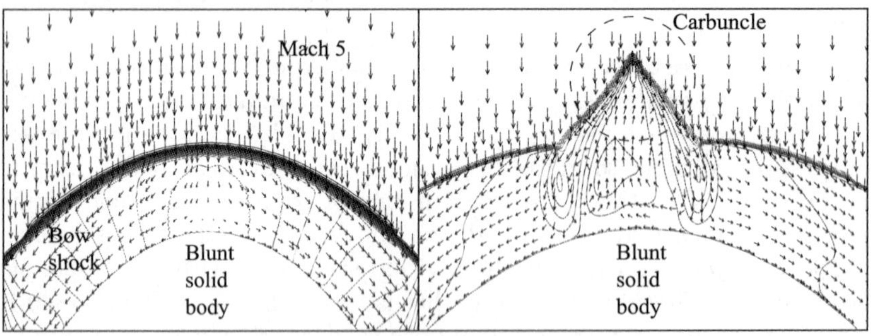

Figure 4.1 Left: bow shock in Mach 5 blunt body flow (Lax-Friedrichs scheme); right: carbuncle (Roe scheme)

attempt to minimize numerical dissipation. On the other hand, more dissipative methods like the Lax-Friedrichs scheme suppress carbuncles

①On coarse grids discretization of small-viscosity Navier-Stokes terms can only be "pro forma". Sometimes subgrid (turbulence) models are added to cover viscous scales at least statistically.

and produce the correct, if diffuse, bow shock (left). The carbuncle phenomenon is more common at large Mach numbers, which are increasingly important in applications (reentry, hypersonic engines/scramjets). Carbuncle-like structures are observed in physical flow [7], hence cannot be dismissed as numerical artefacts.

The carbuncle tip has, up to numerical smoothing and error, the structure of a numerical non-uniqueness example discovered by the author [3]: consider a stationary entropy solution in the entire 2-D plane, with supersonic air from the left, deflected away from the symmetry axis by a conical shock wave (solid lines in Figure 4.2). The deflection leaves room for a cone of $\vec{v} = 0$ air of same pressure as the post-shock air.

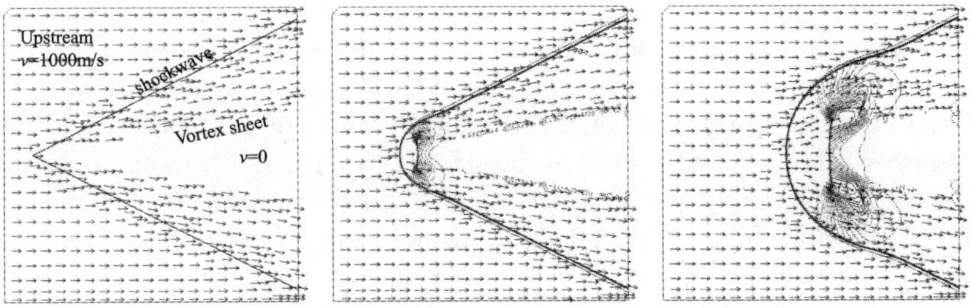

Figure 4.2 Possible non-uniqueness example. Contours: pressure p. Vectors: velocity $\vec{v}$.

But instead, all schemes tested produce another solution: the shock is rounded off at the tip and pushed away from the origin (Figure 4.2 center, right), at a speed of about $350\frac{m}{s}$, for $1000\frac{m}{s}$ flow on the left (the flow appears to be *self-similar*, i.e. $U(t, \vec{x}) = U(\vec{x}/t)$), so the numerical solution is *significantly* different.

5 Program and related work

We consider ideas for proving existence of Pullin's separated spiral. Some existence results for incompressible Euler (e.g. by Delort) allow vortex-sheet initial data. But existence of the steady sheet is trivial, and an abstract compactness-based approach cannot decide existence of a second solution; we need a constructive existence proof that provides detailed structural information.

There are rigorous proofs of different non-uniqueness examples, by Scheffer, Schnirelman for incompressible Euler and most recently de Lellis and Székelyhidi [8] with improved results and including compressible Euler. Their solutions are compactly supported in space, rather than self-similar, have bounded $\vec{v}, T, \rho$ and ρ bounded away from 0. They can be made arbitrarily close (in L^∞) to any constant background solution.

On the other hand, the solutions in any rigorous result are discontinuous and apparently difficult to represent graphically; their physical relevance is unclear. This leaves room for hope that uniqueness can be restored, either by restricting entropy solutions to a smaller function class, or by imposing additional admissibility conditions in the style of the entropy inequality, such as choosing solutions based on the rate of entropy growth.

Our aim is to close the gap between rigorous but irregular and piecewise smooth but unproven examples. We seek an existence proof for solutions that contain only features also observed in physical flow, so that they cannot be explained away by narrowing the function class or augmenting the model. It is likely that Pullin's example is the simplest possible example, perhaps "atomic" in some sense. The simplicity is naturally limited by known *uniqueness* results for the multi-dimensional Euler equations, e.g. Yudovich or Vishik, or Dafermos/DiPerna [1, 2] for compressible flow: whenever a classical (Lipschitz-continuous) solution U exists, it is unique even in the class of L^∞ entropy solutions. Hence the simplest counterexamples can be at most Hölder-continuous; to find the borderline Hölder exponent (if any) is another interesting, but perhaps academic, problem.

6 Kaden-Rott approximation

The principal value integral in (2.1) is in general quite difficult to evaluate exactly. However, Kaden [6] observed that the core of the spiral consists of very dense and almost circular rings, which can be approximated by exact circles. The complex velocity induced by an exact circle enclosing circulation γ is

$$w(z) = \frac{\gamma}{2\pi i z}$$

outside the circle, but $w = 0$ inside. The resulting equation

$$(2m-1)\gamma\frac{\partial z}{\partial \gamma} - mz + \Big(\frac{\gamma}{2\pi i z}\Big)^* = 0$$

has exact solutions

$$z(\gamma) = r(\gamma)e^{i\theta(\gamma)}, \quad r(\gamma) = C\gamma^{\frac{m}{2m-1}}, \quad \theta(\gamma) = \theta_0 + \frac{1}{2\pi C^2}\gamma^{\frac{1}{1-2m}} \tag{6.1}$$

where θ_0, C are free constants. r is distance from the spiral center. θ is the angle around the spiral center, with $\theta \downarrow 0$ corresponding to the infinity end of the sheet whereas $\theta \uparrow \infty$ approaches the spiral center along the infinitely many turns of the spiral. The scaling for $m = 1$ is

$$r, \gamma \sim \theta^{-1}, \qquad |\vec{v}| \sim 1 \qquad (\theta \uparrow \infty).$$

Numerical calculations and observation suggest that the Kaden/Rott approximation is valid and that (6.1) is indeed the correct leading-order behaviour of the spiral core. But although examples of *logarithmic* spirals have been known since the 1920s (Prandtl), no case of *algebraic* spirals has been proven to exist mathematically. In particular, the proof of the famous *Kaden problem* has been open since 1931 [6].

7 Multi-branched algebraic vortex spirals

In [5] we provide the first existence proof of an algebraic vortex spiral in any case. To motivate the work, consider a trivial continuum solution first. Consider self-similar flow:

$$\begin{aligned} &\vec{v}(\vec{x},t) = t^{m-1}\vec{v}(\vec{\xi}), \qquad \vec{\xi} = \frac{\vec{x}}{t^m} \\ \Rightarrow \qquad &\omega(\vec{x},t) = t^{-1}\omega(\vec{\xi}), \qquad \psi(\vec{x},t) = t^{2m-1}\psi(\vec{\xi}) \end{aligned}$$

for some $m > \frac{1}{2}$. The 2-D incompressible Euler equations (1.3) reduce to the form

$$(\vec{v} - m\vec{\xi}) \cdot \nabla\omega - \omega = 0, \qquad \vec{v} = \nabla\psi^{\perp}, \qquad \Delta\psi = \omega.$$

We consider radially symmetric solutions $\psi = \psi(r)$ (where $r = |\vec{\xi}|$) and obtain

$$\psi(r) \sim r^{2-1/m} \qquad (\Rightarrow \quad |\vec{v}(r)| \sim r^{1-1/m},\ \omega(r) \sim r^{-1/m}). \tag{7.1}$$

For this flow the particle paths are *precisely* the Kaden-Rott approximations (6.1). In this continuum problem they are the *exact* solution.

We are motivated to consider the following initial-value problem: instead of a single half-line sheet, as in the Kaden problem, consider n half-lines, meeting in the origin at equal angles at initial time (left, for $n = 4$ and $m = 1$).

This problem can be considered for any $m > \frac{1}{2}$, by changing the distribution of Γ on the initial sheets. Our recent article [5] proves: **for n sufficiently large, there exists such a solution of the Birkhoff-Rott equations,** and each of the n branches has the asymptotic form (6.1) (up to rotation and scaling), i.e. is an *algebraic* spiral. For $n \uparrow \infty$ this can be considered a many-sheet approximation of the radially symmetric continuum vorticity distribution (7.1).

We observe that $n > 1$ is necessary in many cases: for example for $m \geq 1$, including the important acoustic scaling $m = 1$ (Pullin spiral), a single half-line sheet induces infinite velocities. At least two symmetrically opposed sheets are needed to provide a crucial cancellation.

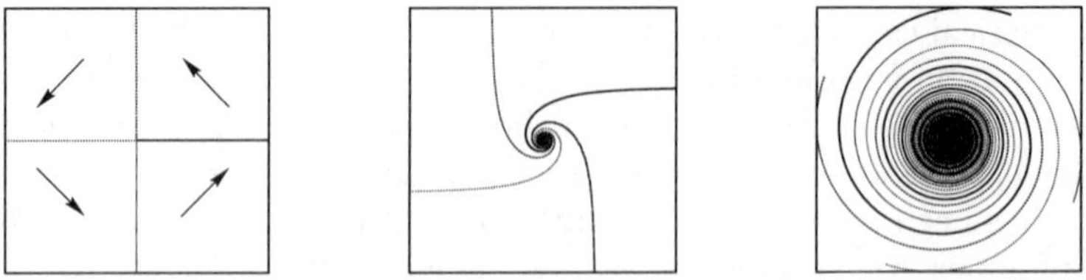

Figure 7.1 Left: initial data ($n = 4, m = 1$); center: rollup; right: core

The main mathematical difficulty is the *spiral* part of the vortex sheets. A standard idea is to construct it as an asymptotic series

$$z(\theta) = (2\pi\theta)^{-m} e^{i\theta} \sum_{k=0}^{\infty} a_k \theta^{-k} \qquad (\theta \approx \infty)$$

where we solve for a_k successively, starting with $k = 0$. This approach is likely to fail for several reasons:

1. For most problems, asymptotic series either do not converge or fail to converge to the correct solution in *any* region, even for relatively simple problems. Consider a separable ODE such as $u_t = u/t^2$, with an irregular singular point $t = 0$. The solutions are multiples of $\exp(-1/t)$. The ansatz

$$u = \sum_{k=0}^{\infty} a_k t^{k+\lambda}$$

yields $a_k = 0$ for all k, which is indeed the correct asymptotic expansion (and obviously convergent), but does not coincide with $\exp(-1/t)$ for any $t \neq 0$.

2. Even if a series converges, it is often only a local solution that cannot be extended to the entire domain.

3. Even if a series converges globally, rigorous proofs are rarely obtained with this approach (ODE with *regular* singular points are one of the few exceptions).

Instead we use the following idea: make an ansatz

$$z(\theta) = (2\pi\theta)^{-m} e^{i\theta} \Big(\sum_{k=0}^{M-1} a_k \theta^{-k} + \tilde{z}(\theta)\theta^{-M} \Big)$$

where the series is *truncated* at some finite M. The leading coefficients a_k are obtained successively. $\tilde{z}$ is sought in a suitable Banach space, using a fixed-point iteration.

The ansatz reduces the problem to an essentially linear one: after fixing a_k the defect is

$$L[\theta^{-M}\tilde{z}] + N_2[\theta^{-M}\tilde{z}, \theta^{-M}\tilde{z}] + N_3[\theta^{-M}\tilde{z}, \theta^{-M}\tilde{z}, \theta^{-M}\tilde{z}] + ...$$

where N_k are k-linear forms representing quadratic and higher self-interactions of the sheet. In the nonlinear terms, θ^{-M} appears at least twice, yielding $O(\theta^{-2M})$ decay (helped by somewhat localized integral kernels). For these terms it is sufficient to use a wasteful analysis based on standard partition of unity arguments combined with Calderon-Zygmund integral operator theory that loses a few orders of decay; we compensate by taking M sufficiently large.

To analyze L we choose $\tilde{z}$ in the Sobolev space $W^{m,2}(\mathbb{R}_+)$, which is a multiplication algebra for $m \geq 2$. Here the largeness of n is crucial: the expansion of the (complex) velocity field of the outer sheet, for example, is a power series

$$w = \sum_{k=0}^{\infty} b_k z^k = \sum_{k=0}^{\infty} b_k r^k e^{ik\theta}.$$

For $n > 1$, $b_0 = 0$; in fact $b_k = 0$ except when $k+1$ is a multiple of n, by *symmetry.* The leading order is $O(r^n)$, i.e. the influence of the outer sheet near the center decays rapidly. On the other hand the outer expansion of the field induced by the *inner* sheet within a certain radius is a *Laurent* series

$$w = \sum_{k=-\infty}^{-1} b_k r^k e^{ik\theta}.$$

The $k = -1$ term is *never* zero — it corresponds to the central vortex of the Kaden-Rott approximation. However, for increasing n, more and more b_k with $k \leq -2$ are zero by symmetry. Therefore, the influence of inner on outer sheet *also* decays; for large n we have a relatively local problem, with the interaction of very dense vortex rings posing the main difficulty.

Even this interaction decreases in size when n grows, but the reasons are much more subtle; the proof requires a combination of delicate techniques from the theory of singular integral operators and some entirely new ideas for integral operators that are locally Calderon-Zygmund, but degenerate as a domain boundary is approached.

We obtain an approximate inverse G for L. In the next step, we show that the Kaden-Rott approximation z_0 is a good initial approximation, with small defect $N(z_0)$. Here we exploit the analogy to (7.1): for large n, the multibranched spiral approaches a radially symmetric vorticity field for which z_0 represents the *exact* particle paths. (This key idea was the first step towards the solution of the problem, reducing the difficulty of a particular algebraic rollup problem just enough to make it tractable.)

Finally a quasi-Newton iteration utilizing G converges to an exact solution of the problem.

In a last step the paper asserts:

1. We have obtained a weak solution of the incompressible Euler equations as well.
2. The vortex sheets are analytic except in the center.

References

[1] C. Dafermos, The second law of thermodynamics and stability. Arch. Rat. Mech. Anal. 70: 167–179, 1979.

[2] R. DiPerna, Uniqueness of solutions to hyperbolic conservation laws. Indiana Univ. Math. J. 28(1): 137–188, 1979.

[3] V. Elling, A possible counterexample to well-posedness of entropy solutions and to Godunov scheme convergence. Math. Comp. 75: 1721–1733, 2006.

[4] V. Elling, The carbuncle phenomenon is incurable. Acta Math. Sci. (ser. B) 29(6): 1647–1656, 2009.

[5] V. Elling, Existence of multi-branched algebraic vortex spirals. Submitted to Arch. Rat. Mech. Anal., 2010.

[6] H. Kaden, Aufwicklung einer unstabilen Unstetigkeitsfläche. Ingenieur-Archiv 2: 140–168, 1931.

[7] I. M. Kalkhoran, P. M. Sforza and F. Y. Wang, Experimental study of shock-vortex interaction in a Mach 3 stream. Technical report, 1991. AIAA Paper 1991–3270.

[8] C. De Lellis and L. Székelyhidi Jr, On admissibility criteria for weak solutions of the Euler equations. Arch. Rat. Mech. Anal. 195(1): 225–260, Jan. 2010.

[9] K. M. Peery and S. T. Imlay, Blunt-body flow simulations. AIAA paper 88–2904, 1988.

[10] D. Pullin, On similarity flows containing two-branched vortex sheets. In R. Caflisch, editor, Mathematical Aspect of Vortex Dynamics, 97–106. SIAM, 1989.

Robust Finite Volume Schemes for Simulating Waves in the Solar Atmosphere

Franz G. Fuchs, Andrew D. McMurry
Center of mathematics for applications, University of Oslo
Oslo, Norway
E-mail: franzf@math.uio.no, a.d.mcmurry@ifi.uio.no

Siddhartha Mishra
Seminar for Applied Mathematics, ETH Zurich
CH-8092, Switzerland
E-mail: siddhartha.mishra@sam.math.ethz.ch

Knut Waagan
CSCAMM, University of Maryland
College Park, USA
E-mail: kwaagan@gmail.com

Abstract

We present well-balanced high-resolution finite volume schemes for simulating waves in the outer solar atmosphere. The schemes approximate the stratified MHD equations with an upwind discretization of the Godunov-Powell source term and a locally hydrostatic pressure reconstruction that preserves discrete steady states. The paper summarizes recent articles [4, 5, 6].

1 Introduction

The study of the sun is of paramount importance. In particular, the solar climate is a key determinant of climate on earth. Solar weather impacts us directly in the form of magnetic storms that affect the electrical grid and indirectly through solar flares and prominences that pose a hazard for satellites and space vehicles. The sun (see Figure 1.1) consists of the following parts: an inner core where most of the solar energy is generated, a radiative zone, a convective zone and the *outer* solar atmosphere. In turn, the solar atmosphere consists of the photosphere (the visible part of the sun), the chromosphere and the corona. It is well known that there is a considerable variation in the temperature of the outer solar atmosphere. The temperature of the photosphere is around 6000 K and the chromosphere is almost isothermal at about 10000 K. However, the temperature in the corona is about 10^6 K (see Figure 1.1). This massive variation in temperature takes place in a very narrow layer (as compared to the typical solar length scale) termed as the *transition region.*

Many interesting questions about the solar atmosphere remain unanswered. Examples include how the solar energy is transferred to the overlaying chromospheric and coronal plasmas and heats them. The above issues are related to the role played by waves and oscillations in the solar atmosphere. It is well known ([1]) that waves carry energy from the base of the solar atmosphere (in the convection zone) upto the chromosphere and the corona. While traveling up the atmosphere, these waves interact with the magnetic field of the sun. This complex interaction between waves and the magnetic field is a possible candidate for explaining the intricate dynamics of the solar atmosphere [1].

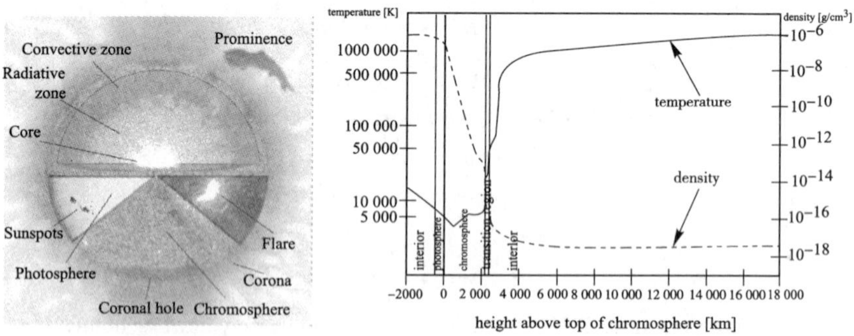

(a) Parts of the sun

(b) Temperature distribution

Figure 1.1 The sun.

1.1 Models of the outer solar atmosphere

Matter in the sun exists in the form of *plasma* i.e., magnetized fluid. The outer solar atmosphere can be modeled by the equations of *stratified Magnetohydrodynamics* (sMHD):

$$
\begin{aligned}
\rho_t + \operatorname{div}(\rho \mathbf{u}) &= 0, \\
(\rho \mathbf{u})_t + \operatorname{div}\left(\rho \mathbf{u} \otimes \mathbf{u} + \left(p + \frac{1}{2}|\overline{\mathbf{B}}|^2\right) I - \overline{\mathbf{B}} \otimes \overline{\mathbf{B}}\right) &= -\rho g \mathbf{e}_3, \\
\overline{\mathbf{B}}_t + \operatorname{div}\left(\mathbf{u} \otimes \overline{\mathbf{B}} - \overline{\mathbf{B}} \otimes \mathbf{u}\right) &= 0, \\
E_t + \operatorname{div}\left(\left(E + p + \frac{1}{2}|\overline{\mathbf{B}}|^2\right) \mathbf{u} - (\mathbf{u} \cdot \overline{\mathbf{B}})\, \overline{\mathbf{B}}\right) &= -\rho g\, (\mathbf{u} \cdot \mathbf{e}_3), \\
\operatorname{div}(\overline{\mathbf{B}}) &= 0,
\end{aligned}
\tag{1.1}
$$

where ρ is the density, $\mathbf{u} = \{u_1, u_2, u_3\}$ and $\overline{\mathbf{B}} = \{\overline{B}_1, \overline{B}_2, \overline{B}_3\}$ are the velocity and magnetic fields respectively, p is the thermal pressure, g is the constant acceleration due to gravity , $\mathbf{e}_3$ represents the unit vector in the vertical (z-) direction. E is the total energy, for simplicity determined by the ideal gas equation of state:

$$E = \frac{p}{\gamma - 1} + \frac{1}{2}\rho |\mathbf{u}|^2 + \frac{1}{2}\left|\overline{\mathbf{B}}\right|^2, \tag{1.2}$$

where $\gamma > 1$ is the adiabatic gas constant. The above equations need to be augmented with suitable initial and boundary conditions.

Our primary objective is to simulate waves in the solar atmosphere. We will present robust and highly efficient numerical schemes to simulate waves in the outer solar atmosphere in this paper. We summarize some recent results ([4],[5],[6]) in this direction.

2 The ideal MHD equations

The first step in designing efficient schemes for the stratified MHD equations is to discretize the ideal MHD equations obtained from (1.1) by setting the gravity $g = 0$. The resulting equations are a system of conservation laws of the form,

$$\mathbf{U}_t + \mathbf{F}_x + \mathbf{G}_y + \mathbf{H}_z = 0. \tag{2.1}$$

Here $\mathbf{U} = \{\rho, \mathbf{u}, \overline{\mathbf{B}}, E\}$ and the fluxes $\mathbf{F}, \mathbf{G}, \mathbf{H}$ can be read from (1.1). The MHD equations are hyperbolic but not strictly hyperbolic. Furthermore, the shock structure of MHD equations is quite complicated on account of the non-strict hyperbolicity and the non-convexity of these equations.

The absence of explicit solution formulas or theoretical results implies that numerical simulations are the main tools for studying MHD and related equations. Finite volume methods ([8]) are among the most popular frameworks for approximating conservation laws like (2.1). We consider (2.1) in $\mathbf{x} = (x, y, z) \in [X_l, X_r] \times [Y_l, Y_r] \times [Z_b, Z_t]$ and discretize it by a uniform grid in all directions with the grid spacing $\Delta x, \Delta y$ and Δz. We set $x_i = X_l + i\Delta x$, $y_j = Y_l + j\Delta y$ and $z_k = Z_b + k\Delta z$. The indices are $0 \le i \le N_x$, $0 \le j \le N_y$ and $0 \le k \le N_z$. Set $x_{i+1/2} = x_i + \Delta x/2$, $y_{j+1/2} = y_j + \Delta y/2$ and $z_{k+1/2} = z_k + \Delta z/2$, and let $\mathcal{C}_{i,j,k} = [x_{i-1/2}, x_{i+1/2}) \times [y_{j-1/2}, y_{j+1/2}) \times [z_{k-1/2}, z_{k+1/2})$ denote a typical cell. The cell average of the unknown state vector $\mathbf{W}$ (approximating $\mathbf{U}$) over $\mathcal{C}_{i,j,k}$ at time t^n is denoted $\mathbf{W}^n_{i,j,k}$. By integrating the conservation law over the cell $\mathcal{C}_{i,j,k}$ and the time interval $[t^n, t^{n+1})$ with $t^{n+1} = t^n + \Delta t^n$, where the time-step Δt^n is determined by a suitable CFL condition results in a standard (first-order in space and time) finite volume scheme written down as

$$
\begin{aligned}
\mathbf{W}^{n+1}_{i,j,k} = \mathbf{W}^{n}_{i,j,k} - \frac{\Delta t^n}{\Delta z}(\mathbf{H}^n_{i,j,k+1/2} - \mathbf{H}^n_{i,j,k-1/2}) \\
- \Delta t^n \left(\frac{\mathbf{F}^n_{i+1/2,j,k} - \mathbf{F}^n_{i-1/2,j,k}}{\Delta x} + \frac{\mathbf{G}^n_{i,j+1/2,k} - \mathbf{G}^n_{i,j-1/2,k}}{\Delta y} \right).
\end{aligned} \tag{2.2}
$$

The numerical fluxes $\mathbf{F}$, $\mathbf{G}$ and $\mathbf{H}$ are determined in terms of (approximate) Riemann solvers. A wide variety of approximate Riemann solvers are available for the MHD system; see [7, 3] and references therein.

A crucial issue that arises in approximating the MHD equations is a discretization of the divergence constraint in (1.1). The standard finite volume scheme (2.2) *will not* preserve a discrete version of the constraint, leading to numerical instabilities and oscillations, see [10]. Recipes for treating the divergence constraint include the projection method [2] and methods based on staggering [10] (and references therein).

An alternative involves adding the Godunov-Powell source term to the MHD equations resulting in the balance law,

$$
\begin{aligned}
&\mathbf{U}_t + \mathbf{F}_x + \mathbf{G}_y + \mathbf{H}_z = \mathbf{s}. \\
&\mathbf{s} = \{-\overline{\mathbf{B}}(\mathrm{div}\overline{\mathbf{B}}), -\mathbf{u}(\mathrm{div}\overline{\mathbf{B}}), -(\mathbf{u}\cdot\overline{\mathbf{B}})(\mathrm{div}\overline{\mathbf{B}})\}.
\end{aligned} \tag{2.3}
$$

The above equations are (formally) equivalent to the standard form of MHD equations (1.1) without gravity. The added advantages of the Godunov-Powell form (2.3) are its Galilean invariance and symmetrizability. Furthermore, the magnetic field in (2.3) satisfies the divergence transport equation:

$$
(\mathrm{div}\overline{\mathbf{B}})_t + \mathrm{div}(\mathbf{u}(\mathrm{div}\overline{\mathbf{B}})) = 0. \tag{2.4}
$$

Hence, initial divergence errors are transported out of the domain enabling us to treat the divergence constraint approximately in a stable manner. The method was proposed in [9] but recent papers ([3]) show that the additional source terms in (2.3) need to discretized in a suitable manner for numerical stability.

2.1 HLL three wave solvers and upwind discretizations of the Godunov-Powell source terms.

Following [4], we present a judicious combination of an approximate Riemann solver and upwind source discretization for the Godunov-Powell MHD system. For simplicity, we consider (2.3) in one space dimension and write the resulting system as,

$$
\mathbf{U}_t + \mathbf{F}_x = \mathbf{s}^{\mathbf{1}}. \tag{2.5}
$$

Here, the source $\mathbf{s^1} = \mathbf{s^1}(\mathbf{U}, (\overline{B}_1)_x)$ involves a non-conservative product and needs to be discretized carefully. The corresponding first-order finite volume scheme for (2.3) is

$$\mathbf{W}_i^{n+1} = \mathbf{W}_i^n - \frac{\Delta t^n}{\Delta x}(\mathbf{F}_{i+1/2}^n - \mathbf{F}_{i-1/2}^n) + \Delta t^n \mathbf{S}_i^{1,n}. \tag{2.6}$$

We will approximate the eight waves in the MHD Riemann problem with three waves, i.e, two representing the outermost fast waves and a middle wave approximating the material contact discontinuity. The approximate solution and fluxes are given by

$$\mathbf{W}^{H_3} = \begin{cases} \mathbf{W}_L & \text{if } \frac{x}{t} \le s_L, \\ \mathbf{W}_L^* & \text{if } s_L < \frac{x}{t} < s_M, \\ \mathbf{W}_R^* & \text{if } s_M < \frac{x}{t} < s_R, \\ \mathbf{W}_R & \text{if } s_R \le \frac{x}{t}, \end{cases} \qquad \mathbf{F}^{H_3} = \begin{cases} \mathbf{F}_L & \text{if } \frac{x}{t} \le s_L, \\ \mathbf{F}_L^* & \text{if } s_L < \frac{x}{t} < s_M, \\ \mathbf{F}_R^* & \text{if } s_M < \frac{x}{t} < s_R, \\ \mathbf{F}_R & \text{if } s_R \le \frac{x}{t}. \end{cases} \tag{2.7}$$

We set $\pi_1 = p + \frac{\overline{B}_2^2 + \overline{B}_3^2}{2}$. The outer wave speeds s_L and s_R model the fast magneto-sonic waves and are defined as in [7]. In order to describe the solver, we need to determine the speed of the middle wave s_M and the intermediate states $\mathbf{W}_L^*, \mathbf{W}_R^*$. The middle wave models a material contact discontinuity. Hence, the velocity field and the tangential magnetic fields are assumed to be constant across the middle wave. This allows us to define $\mathbf{u}^* = \mathbf{u}_L^* = \mathbf{u}_R^*$, $\overline{B}_2^* = \overline{B}_{2L}^* = \overline{B}_{2R}^*$ and $\overline{B}_3^* = \overline{B}_{3L}^* = \overline{B}_{3R}^*$. The normal magnetic field $\overline{B}_1$ is not assumed to be constant but jumps only across the middle wave (modeling the linear degenerate "divergence wave" implied by (2.4)), and $\overline{B}_1$ is constant across the outer waves. The intermediate states are determined by local conservation across the two outermost waves and the middle wave resulting in,

$$s_\sigma \mathbf{W}_\sigma^* - \mathbf{F}_\sigma^* = s_\sigma \mathbf{W}_\sigma - \mathbf{F}_\sigma, \quad s_M \mathbf{W}_R^* - s_M \mathbf{W}_L^* = \mathbf{F}_R^* - \mathbf{F}_L^* + \mathbf{s}^{1,*} \tag{2.8}$$

where $\sigma \in \{L, R\}$ and

$$\mathbf{s}^{1,*} = \begin{pmatrix} 0 \\ -\frac{\overline{B}_{1R}^2 - \overline{B}_{1L}^2}{2}) \\ -\left(\overline{B}_2^*\right)\left(\overline{B}_{1R} - \overline{B}_{1L}\right) \\ -\left(\overline{B}_3^*\right)\left(\overline{B}_{1R} - \overline{B}_{1L}\right) \\ -\mathbf{u}^*\left(\overline{B}_{1R} - \overline{B}_{1L}\right) \\ -u_1^* \frac{\overline{B}_{1R}^2 - \overline{B}_{1L}^2}{2} - \left(u_2^* \overline{B}_2^* + u_3^* \overline{B}_3^*\right)\left(\overline{B}_{1R} - \overline{B}_{1L}\right) \end{pmatrix}. \tag{2.9}$$

This amounts to integrating the source $\mathbf{s}^1$ across the wave fan.

Applying the conservation relations, we obtain (check [5], section 3.1.2 for details) the following intermediate states,

$$\rho_\theta^* = \rho_\theta \frac{u_{1\theta} - s_\theta}{s_M - s_\theta}, \quad \pi_{1\theta}^* = \pi_{1\theta} + \rho_\theta(u_{1\theta} - s_\theta)(u_{1\theta} - s_M), \quad \theta \in \{L, R\},$$

$$s_M = u_1^* = \frac{\pi_{1R} - \pi_{1L} + \rho_R u_{1R}(u_{1R} - s_R) - \rho_L u_{1L}(u_{1L} - s_L)}{\rho_R(u_{1R} - s_R) - \rho_L(u_{1L} - s_L)},$$

$$u_\sigma^* = \frac{\zeta c_\sigma - \xi d_\sigma}{\alpha\zeta + \xi^2}, \quad \overline{B}_\sigma^* = \frac{-\alpha d_\sigma - \xi c_\sigma}{\alpha\zeta + \xi^2}, \quad \sigma \in \{2, 3\}$$

$$E_\theta^* = \frac{1}{s_M - s_\theta}\Big(E_\theta\,(u_{1\theta} - s_\theta) + \pi_{1\theta}u_{1\theta} - \pi_{1\theta}^* s_M + \frac{\overline{B}_{1\theta}^2}{2}\,(u_{1\theta} - s_M)$$
$$+ \left(\overline{B}_{1\theta}\right)\left(\overline{B}_{2\theta}u_{2\theta} + \overline{B}_{3\theta}u_{3\theta} - \overline{B}_{2\theta}^* u_{2\theta}^* - \overline{B}_{3\theta}^* u_{3\theta}^*\right)\Big), \quad \theta \in \{L, R\},$$

$$c_\sigma = \rho_R u_{\sigma R}\,(u_{1R} - s_R) - \rho_L u_{\sigma L}\,(u_{1L} - s_L) - \left(\overline{B}_{1R}\overline{B}_{\sigma R} - \overline{B}_{1L}\overline{B}_{\sigma L}\right),$$
$$d_\sigma = \overline{B}_{\sigma R}\,(u_{1R} - s_R) - \overline{B}_{\sigma L}\,(u_{1L} - s_L) - \left(\overline{B}_{1L}u_{\sigma L} - \overline{B}_{1R}u_{\sigma R}\right),$$
$$\alpha = \rho_R\,(u_{1R} - s_R) - \rho_L\,(u_{1L} - s_L), \quad \zeta = s_R - s_L, \quad \xi = \overline{B}_{1R} - \overline{B}_{1L}.$$

The intermediate fluxes are obtained in terms of the intermediate states by local conservation (2.8),

$$\mathbf{F}_L^* = \mathbf{F}_L + s_L(\mathbf{W}_L^* - \mathbf{W}_L), \quad \mathbf{F}_R^* = \mathbf{F}_R + s_R(\mathbf{W}_R^* - \mathbf{W}_R).$$

The discrete source term takes the form,

$$\mathbf{S}_i^{1,n} = \mathbf{s}_{i-1/2}^{1,*}\chi_{(s_{M,i-1/2}\ge 0)} + \mathbf{s}_{i+1/2}^{1,*}\chi_{(s_{M,i+1/2}<0)}, \tag{2.10}$$

where $\mathbf{s}_{i\pm 1/2}^{1,*}$ is defined in analogy to (2.9).

For the three dimensional form of the equations, the fluxes $\mathbf{G}, \mathbf{H}$ and the sources $\mathbf{S}^2$ and $\mathbf{S}^3$ can be defined analogously to obtain the following finite volume scheme:

$$\mathbf{W}_{i,j,k}^{n+1} = \mathbf{W}_{i,j,k}^{n} - \Delta t^n\left(\frac{\mathbf{F}_{i+1/2,j,k}^n - \mathbf{F}_{i-1/2,j,k}^n}{\Delta x} + \frac{\mathbf{G}_{i,j+1/2,k}^n - \mathbf{G}_{i,j-1/2,k}^n}{\Delta y}\right)$$
$$- \Delta t^n\left(\frac{\mathbf{H}_{i,j,k+1/2}^n - \mathbf{H}_{i,j,k-1/2}^n}{\Delta z} - \mathbf{S}_{i,j,k}^{1,n} - \mathbf{S}_{i,j,k}^{2,n} - \mathbf{S}_{i,j,k}^{3,n}\right), \tag{2.11}$$

for the three dimensional version of the MHD equations with Godunov-Powell source term (2.3).

The above scheme (2.11) is first-order accurate in both space and time. Second-order of accuracy in space is obtained by using non-oscillatory piecewise linear reconstructions using minmod, ENO and WENO limiters. Second-order accuracy in time is obtained using strong stability preserving Runge-Kutta methods. Details of how these procedures are adapted for the MHD equations are described in [4].

2.2 Numerical experiment: isothermal blast wave

This standard test case easily exhibits spurious behavior if the equations are not properly discretized. The equation of state is isothermal in this case, i.e, $p = \rho$, the computational domain is $[0,1] \times [0,1]$ and the initial data are

$$\{\mathbf{u}, B_1, B_{2,3}\}_{t=0} = \{\mathbf{0}, \frac{5}{\sqrt{\pi}}, 0\}, \rho_0 = \begin{cases} 100 & \text{if } |(x,y) - (0.5, 0.5)| \leq 0.05, \\ 0 & \text{otherwise.} \end{cases}$$

In order to illustrate the necessity of the *upwinded* Godunov-Powell source term, we also considered the central source discretisation of [9] and schemes with the Godunov-Powell source set to zero. The results with three schemes: no source, central source and upwind source (2.11), with a second-order discretization, are presented in Figure 2.1. They clearly illustrate the necessity of using the Godunov-Powell source term as well as discretizing it in the correct manner.

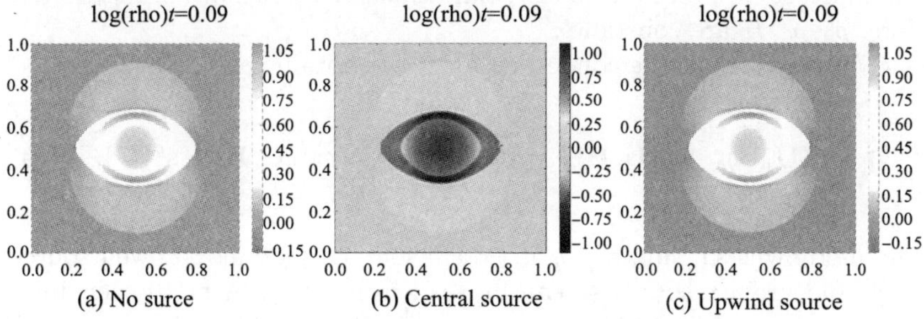

Figure 2.1 Density in the isothermal blast wave problem.

3 Schemes for stratified MHD

We model the outer solar atmosphere in terms of the following modified *stratified MHD equations*:

$$\begin{aligned}
&\rho_t + \operatorname{div}(\rho \mathbf{u}) = 0, \\
&(\rho \mathbf{u})_t + \operatorname{div}\left(\rho \mathbf{u} \otimes \mathbf{u} + \left(p + \frac{|\mathbf{B}|^2}{2} + \widetilde{\mathbf{B}} \cdot \mathbf{B}\right) I - \mathbf{B} \otimes \mathbf{B} - \widetilde{\mathbf{B}} \otimes \mathbf{B} - \mathbf{B} \otimes \widetilde{\mathbf{B}}\right) \\
&\qquad = -\left(\mathbf{B} + \widetilde{\mathbf{B}}\right)(\operatorname{div}\mathbf{B}) - \rho g \mathbf{e}_3, \\
&\mathbf{B}_t + \operatorname{div}\left(\mathbf{u} \otimes \mathbf{B} - \mathbf{B} \otimes \mathbf{u} + \mathbf{u} \otimes \widetilde{\mathbf{B}} - \widetilde{\mathbf{B}} \otimes \mathbf{u}\right) = -\mathbf{u}(\operatorname{div}\mathbf{B}), \\
&E_t + \operatorname{div}\left(\left(E + p + \frac{1}{2}|\mathbf{B}|^2 + \mathbf{B} \cdot \widetilde{\mathbf{B}}\right)\mathbf{u} - (\mathbf{u} \cdot \mathbf{B})\mathbf{B} - \left(\mathbf{u} \cdot \widetilde{\mathbf{B}}\right)\mathbf{B}\right) \\
&\qquad = -(\mathbf{u} \cdot \mathbf{B})(\operatorname{div}\mathbf{B}) - \rho g (\mathbf{u} \cdot \mathbf{e}_3),
\end{aligned} \tag{3.1}$$

Here, we assume that there exist a *potential* magnetic field $\widetilde{\mathbf{B}}$ satisfying the following assumptions,

$$\widetilde{\mathbf{B}}_t = 0, \quad \mathrm{div}(\widetilde{\mathbf{B}}) = 0, \text{ and } \mathrm{Curl}(\widetilde{\mathbf{B}}) = 0. \tag{3.2}$$

and solve for the perturbation $\mathbf{B} = \overline{\mathbf{B}} - \widetilde{\mathbf{B}}$.

Waves in the solar atmosphere are modeled as perturbations of the steady state:

$$\mathbf{u} \equiv \mathbf{0}, \quad \mathbf{B} \equiv 0 \quad \rho(z) = \frac{\rho_0 T_0}{T(z)} e^{-\frac{\alpha(z)}{H}}, \quad p(z) = p_0 e^{-\frac{\alpha(z)}{H}}. \tag{3.3}$$

Here,

$$\alpha(x,y,z) = \alpha(z) = \int_0^z \frac{1}{T(s)} ds, \tag{3.4}$$

with $T = T(z)$ being a given temperature distribution. An example for temperature distributions in the solar atmosphere is shown in Figure 1.1. Here, ρ_0, p_0, H are constants.

Writing (3.1) in the condensed form, we obtain the balance law:

$$\mathbf{U}_t + f(\mathbf{U}, \widetilde{\mathbf{B}})_x + g(\mathbf{U}, \widetilde{\mathbf{B}})_y + h(\mathbf{U}, \widetilde{\mathbf{B}})_z = \sum_{l=1}^{3} \mathbf{s}^l(\mathbf{U}, \widetilde{\mathbf{B}}) + \mathbf{s}^g(\mathbf{U}), \tag{3.5}$$

The unknowns $\mathbf{U}$, fluxes f, g, h, co-efficient $\widetilde{\mathbf{B}}$, Godunov-Powell source $\mathbf{s}^{1,2,3}$ and gravity source $\mathbf{s}^g$ can be read from (3.1). A first-order finite volume scheme for (3.5) is given by,

$$\begin{aligned}
&\mathbf{W}^{n+1}_{i,j,k} \\
=&\mathbf{W}^n_{i,j,k} - \Delta t^n \left(\frac{\mathbf{F}^n_{i+1/2,j,k} - \mathbf{F}^n_{i-1/2,j,k}}{\Delta x} + \frac{\mathbf{G}^n_{i,j+1/2,k} - \mathbf{G}^n_{i,j-1/2,k}}{\Delta y} \right) \\
&- \Delta t^n \left(\frac{\mathbf{H}^n_{i,j,k+1/2} - \mathbf{H}^n_{i,j,k-1/2}}{\Delta z} - \mathbf{S}^{1,n}_{i,j,k} - \mathbf{S}^{2,n}_{i,j,k} - \mathbf{S}^{3,n}_{i,j,k} - \mathbf{S}^{g,n}_{i,j,k} \right).
\end{aligned} \tag{3.6}$$

The fluxes $\mathbf{F}, \mathbf{G}$ and sources $\mathbf{S}^{1,2}$ in the x- and y-directions are computed using the procedure outlined in section 2.

3.1 Fluxes and sources in the z-direction

The numerical flux $\mathbf{H}$ and discrete Godunov-Powell source term $\mathbf{S}^3$ in (3.6) are described in terms of the following Riemann problem,

$$\mathbf{W}_t + h(\mathbf{W}, \widetilde{\mathbf{B}}_m)_z = \mathbf{s}^3(\mathbf{W}, \widetilde{\mathbf{B}}_m, \mathbf{W}_z), \quad \mathbf{W}(z,0) = \begin{cases} \mathbf{W}_T & z < 0, \\ \mathbf{W}_B & z > 0. \end{cases}$$

The natural way to specify initial data $\mathbf{W}_{T,B}$ in the above problem is to use the states $\mathbf{W}_B = \mathbf{W}^n_{i,j,k}$ and $\mathbf{W}_T = \mathbf{W}^n_{i,j,k+1}$. However, this approach leads to a scheme that does not preserve discrete versions of the interesting steady states (3.3). Therefore we must design suitable fluxes in order to design well-balanced schemes.

3.1.1 Local hydrostatic pressure reconstructions

Instead of just using the cell averages below and above the interface as data in the Riemann problem, we follow [5, 6] and utilize the special structure of the steady states (3.3) to perform a local hydrostatic reconstruction of the pressure inside the cell and use the local primitive variables,

$$\mathbf{V}_B = \{\rho^n_{i.j,k}, \mathbf{u}^n_{i,j,k}, \mathbf{B}^n_{i,j,k}, p^{n,-}_{i,j,k+1/2}\},$$
$$\mathbf{V}_T = \{\rho^n_{i,j,k+1}, \mathbf{u}^n_{i,j,k+1}, \mathbf{B}^n_{i,j,k+1}, p^{n,+}_{i,j+1/2}\},$$

where the reconstructed pressure is given in terms of extrapolated cell averages by first defining the local temperature,

$$T^n_{i,j,k} = \frac{p^n_{i,j,k}}{gH\rho^n_{i,j,k}}. \tag{3.7}$$

The piecewise constant temperature defines the scaling function α by (3.4). We can compute the differences in α and use it to define the reconstructed local pressure,

$$p^{n,-}_{i,j,k+1/2} = p^n_{i,j,k} e^{\frac{-\Delta z}{2HT^n_{i,j,k}}}, \quad p^{n,+}_{i,j,k+1/2} = p^n_{i,j,k+1} e^{\frac{\Delta z}{2HT^n_{i,j,k+1}}}. \tag{3.8}$$

The data $\mathbf{W}_B$ and $\mathbf{W}_T$ (in terms of conservative variables) are easily obtained from the primitive variables $\mathbf{V}_B$, $\mathbf{V}_T$. The coefficient $\widetilde{\mathbf{B}}_m$ is given by the average,

$$\widetilde{\mathbf{B}}_m = \widetilde{\mathbf{B}}_{i,j,k+1/2} = \frac{\widetilde{\mathbf{B}}_{i,j,k} + \widetilde{\mathbf{B}}_{i,j,k+1}}{2}. \tag{3.9}$$

The HLL three wave solver of the previous section is easily obtained for the z-direction by repeating the approach of describing the solver in the x- direction.

We discretize the gravitational source term as

$$\mathbf{S}^{g,n}_{i,j,k} = \left\{0, 0, \frac{p^{n,-}_{i,j,k+1/2} - p^{n,+}_{i,j,k-1/2}}{\Delta z}, 0, 0, 0, 0, -\rho^n_{i,j,k}(u^n_3)_{i,j,k}\, g\right\}. \tag{3.10}$$

where $p^{n,-}_{i,j,k+1/2}, p^{n,+}_{i,j,k-1/2}$ are defined in (3.8).

Periodic boundary conditions are used in the x and y directions and we use the following *non-reflecting* Neumann type boundary conditions in the vertical direction:

$$\mathbf{W}^n_{i,j,0} = \mathbf{W}^n_{i,j,1} e^{\frac{\Delta z}{T_{i,j,1} H}}, \quad \mathbf{W}^n_{i,j,N_z+1} = \mathbf{W}^n_{i,j,N_z} e^{-\frac{\Delta z}{T_{i,j,N_z} H}}. \tag{3.11}$$

The resulting scheme satisfies the following properties.

Theorem 3.1. *Consider the scheme* (3.6) *approximating the system* (3.5). *This scheme has the following properties,*

(i) *The scheme* (3.6) *is consistent with* (3.5), *and it is first order accurate in both space and time (for smooth solutions).*

(ii) *The scheme* (3.6) *is* well-balanced *and preserves discrete versions of the steady states* (3.3),

The proof of the above theorem is presented in [6].

Remark 3.2. The finite volume scheme (3.6) is first-order accurate. For second-order accuracy, we need to employ locally hydrostatic piecewise linear non-oscillatory reconstructions as designed in [5, 6].

3.2 Numerical experiments

3.2.1 Waves in a two dimensional model solar atmosphere

We consider a synthetic potential magnetic field $\widetilde{\mathbf{B}}$ given in [4] and plot the velocity (in the direction of the magnetic field) in two cases: first, with an isothermal atmosphere (modeling the lower chromosphere) and then with a realistic temperature distribution similar to the one shown in Figure 1.1. Both are plotted in Figure 3.1 and computed with the second-order version of the well-balanced scheme (3.6). Both results show the effectiveness of the schemes in resolving complex physical phenomena, in particular, the focusing of waves due to the magnetic field and the wave acceleration at the transition layer.

3.3 Waves in a realistic solar chromosphere

Finally, we consider a background magnetic field measured from the MDI instrument on SOHO and bottom velocity measurements, also from SOHO and present the velocity (in the direction of magnetic field) in Figure 3.2. The results demonstrate how the well-balanced schemes are robust and effective in simulating realistic scenarios in the solar atmosphere.

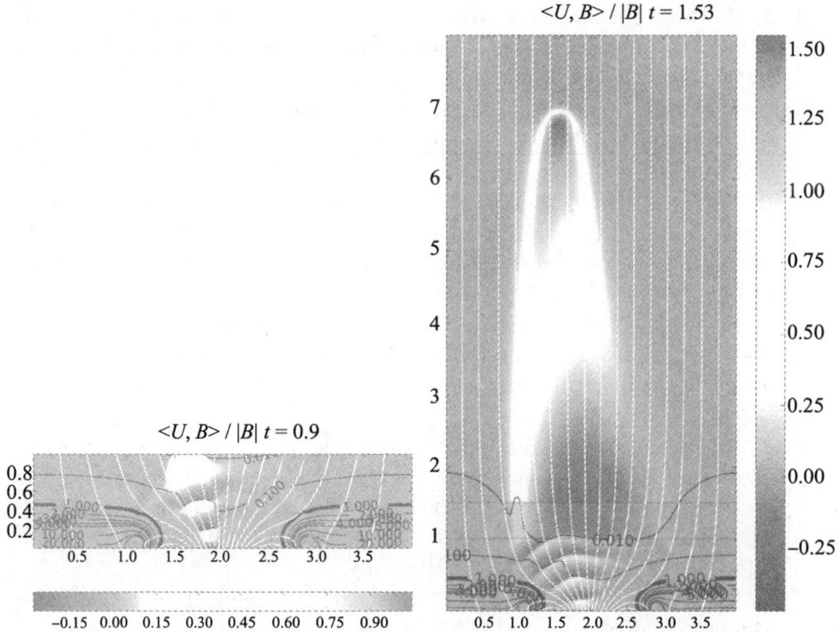

(a) Isothermal (b) Realistic

Figure 3.1 Velocity in the direction of the magnetic field, computed with a second-order well-balanced scheme for both the lower chromosphere and the chromosphere, corona, transition region combination.

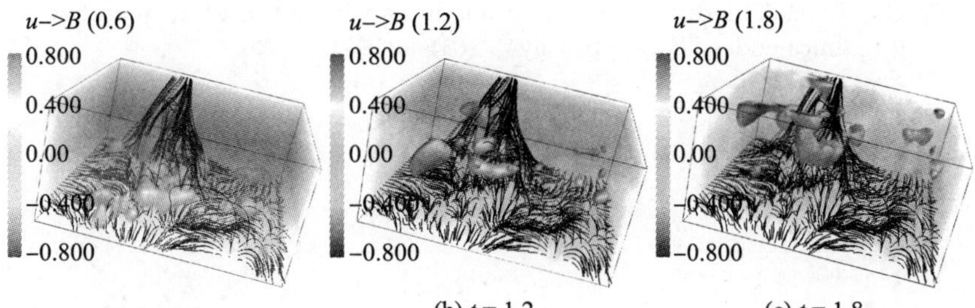

(a) $t = 0.6$ (b) $t = 1.2$ (c) $t = 1.8$

Figure 3.2 Velocity in the direction of the magnetic field, computed with a second-order well-balanced scheme for the lower chromosphere with realistic magnetic field and velocity forcing.

References

[1] T. J. Bogdan, et al, Waves in the magnetized solar atmosphere II: Waves from localized sources in magnetic flux concentrations. Astrophys. Jl, 599: 626–660, 2003.

[2] J. U. Brackbill and D. C. Barnes, The effect of nonzero div$B$ on the numerical solution of the magnetohydrodynamic equations. J. Comp. Phys., 35: 426–430, 1980.

[3] F. Fuchs, S. Mishra and N. H. Risebro, Splitting based finite volume schemes for the ideal MHD equations. J. Comp. Phys., 228 (3): 641–660, 2009.

[4] F. Fuchs, A. D. McMurry, S. Mishra, N. H. Risebro and K. Waagan, Approximate Riemann solver based high-order finite volume schemes for the Godunov-Powell form of ideal MHD equations in multi-dimensions. Comm. Comput. Phys., to appear.

[5] F. Fuchs, A. D. McMurry, S. Mishra, N. H. Risebro and K. Waagan, Well-balanced high-order finite volume methods for simulating wave propagation in stratified magneto-atmospheres. J. Comput. Phys., to appear.

[6] F. Fuchs, A. D. McMurry, S. Mishra and K. Waagan, Well-balanced high-order finite volume methods for simulating wave propagation in three-dimensional non-isothermal stratified magneto-atmospheres. Preprint 2010, available from ftp://ftp.sam.math.ethz.ch/pub/sam-reports/reports/reports2010/2010-27.pdf

[7] K. F. Gurski, An HLLC-type approximate Riemann solver for ideal Magneto-hydro dynamics. SIAM. J. Sci. Comp., 25(6): 2165–2187, 2004.

[8] R. J. LeVeque, Finite Volume Methods for Hyperbolic Problems. Cambridge university press, Cambridge, 2002.

[9] K. G. Powell, P. L. Roe. T. J. Linde, T . I. Gombosi and D . L. De zeeuw, A solution adaptive upwind scheme for ideal MHD. J. Comp. Phys., 154(2): 284–309, 1999.

[10] G. Toth, The div$B = 0$ constraint in shock capturing magnetohydrodynamics codes. J. Comp. Phys., 161: 605–652, 2000.

Maximum Principle of Central Schemes with k-monotone Fluxes*

Orhan Mehmetoglu
Department of Mathematics, Texas A&M University
College Station, Texas, USA
E-mail: omehmet@math.tamu.edu

Bojan Popov
Department of Mathematics, Texas A&M University
College Station, Texas, USA
E-mail: popov@math.tamu.edu

Abstract

The Nessyahu-Tadmor (NT) scheme is a simple yet robust second order non-oscillatory scheme, which relies on a piecewise linear reconstruction. A typical reconstruction choice is based on the standard minmod limiter. This choice gives a maximum principle for the scheme. However, it reduces the reconstruction to first order at local extrema. In this paper we show that a maximum principle is still valid for second order schemes when a new MAPR-like limiter is used, see [2]. To prove this result we require that the flux is k-monotone. We also show that the maximum principle implies the usual TVD bound.

1 Introduction

This paper is concerned with scalar hyperbolic conservation laws

$$\begin{cases} u_t + f(u)_x = 0, & (x,t) \in \mathbb{R} \times (0,\infty), \\ u(x,0) = u^0(x), & x \in \mathbb{R}, \end{cases} \tag{1.1}$$

where f is a given flux function. There have recently been numerous advances in the development of new numerical algorithms. The key idea of non-oscillatory second order numerical methods is to use a piecewise

*This material is based upon work supported by the National Science Foundation grant DMS-0811041.

linear reconstruction based on a slope limiter such as minmod, UNO, and superbee, see for instance [3, 9]. An unfortunate requirement for maximum principle is that we need to restrict the slope to zero at local extrema in order to be able to prove a maximum principle. That is, we reduce our local reconstruction to first order at all local extrema.

In this paper, we present an alternative approach in which we use a class of new limiters, inspired by the MAPR limiter, see [2]. The advantage of our limiters is that we do not need to restrict the slope of the reconstruction to zero at local extrema. However, using a second order reconstruction at local extrema requires a new approach when trying to prove a maximum principle.

The paper is organized as follows. In section 2, we briefly summarize the Nessyahu-Tadmor (NT) staggered central scheme. In section 3, we prove the maximum principle for a conservation law with any k-monotone flux for $k \geq 2$. The class of k-monotone functions include all strictly convex functions (for $k = 2$) and also any polynomial flux of a fixed degree k, see [5] or Section 3 for a definition of k-monotonicity. This result generalizes our previous result for convex fluxes from [8]. Finally, we show that the usual TVD bound follows from the maximum principle.

2 Non-oscillatory central schemes

In this section we give a brief overview of second order non-oscillatory staggered central difference approximations to the scalar conservation law

$$u_t + f(u)_x = 0. \tag{2.1}$$

We restrict our attention to the one-dimensional staggered NT scheme [9], which was the motivation for the construction of many other central staggered schemes, see for example [4, 7, 2, 1, 6]. Let $v(x,t)$ be an approximate solution to (2.1) and assume that the space mesh Δx, and the time mesh Δt are uniform. Let $x_j := j\Delta x$, $j \in \mathbb{Z}$, $I_j := \{s : |s - x_j| < \frac{\Delta x}{2}\}$, $t^n := n\Delta t$, $n \in \mathbb{N}$, and

$$v_j^n := \frac{1}{\Delta x} \int_{I_j} v(x, t^n)\, dx \tag{2.2}$$

be the average of $v(x,t^n)$ over I_j. Next, assume that $v(\cdot, t^n)$ is a piecewise linear function of the form

$$v(x,t^n) = \sum L_j(x,t^n) := \sum \left(v_j^n + (x - x_j)\frac{1}{\Delta x} v_j' \right) \chi_j(x), \tag{2.3}$$

where χ_j is the characteristic function over I_j and $\frac{1}{\Delta x}v'_j$ is the numerical derivative of $v(x = x_j, t^n)$ which is yet to be specified. We proceed by integrating (2.3) over $I_{j+\frac{1}{2}} \times (t^n, t^{n+1})$ which yields

$$
\begin{aligned}
v_{j+\frac{1}{2}}(t^{n+1}) =& \frac{1}{\Delta x}\left(\int_{x_j}^{x_{j+\frac{1}{2}}} L_j(x,t^n)dx + \int_{x_{j+\frac{1}{2}}}^{x_{j+1}} L_{j+1}(x,t^n)dx\right) \\
&- \frac{1}{\Delta x}\left(\int_{t^n}^{t^{n+1}} f(v(x_{j+1},\tau))d\tau - \int_{t^n}^{t^{n+1}} f(v(x_j,\tau))d\tau\right).
\end{aligned} \tag{2.4}
$$

The first two integrands on the right of (2.4), $L_j(x,t^n)$ and $L_{j+1}(x,t^n)$, can be integrated exactly and if the CFL condition (with mesh ratio $\lambda := \frac{\Delta t}{\Delta x}$)

$$
\lambda \max_{x_j \le x \le x_{j+1}} |f'(v(x,t))| \le \frac{1}{2}, \quad j \in \mathbb{Z} \tag{2.5}
$$

is satisfied, then the last two integrands on the right of (2.4), $f(v(x_j,\tau))$ and $f(v(x_{j+1},\tau))$, can be integrated approximately by the midpoint rule at the expense of $O(\Delta t)^3$ local truncation error. Thus we arrive at

$$
v_{j+\frac{1}{2}}^{n+1} = \frac{1}{2}(v_j^n + v_{j+1}^n) + \frac{1}{8}(v'_j - v'_{j+1}) - \lambda\left(f(v_{j+1}^{n+\frac{1}{2}}) - f(v_j^{n+\frac{1}{2}})\right). \tag{2.6}
$$

The approximate midvalue, $v_j^{n+\frac{1}{2}}$, satisfying second order accuracy requirement can be chosen as

$$
v_j^{n+\frac{1}{2}} = v_j^n - \frac{1}{2}\lambda f'_j, \tag{2.7}
$$

owing to Taylor expansion and (2.1). Here, $\frac{1}{\Delta x}f'_j$ stands for an approximate numerical derivative of the flux $f(v(x = x_j, t^n))$, which is yet to be determined. Although there are many different recipes to construct v'_j and f'_j, in this paper the following are considered to be the approximations of the numerical derivatives

$$
v'_j = \mathrm{m}(v_{j+1}^n - v_j^n, v_j^n - v_{j-1}^n), \tag{2.8}
$$

$$
f'_j = f'(v_j^n)v'_j, \tag{2.9}
$$

where $\mathrm{m}(\cdot,\cdot)$ is the modified minmod limiter given by

$$
\mathrm{m}(a,b) := \begin{cases} \mathrm{sgn}\ (a)\ \min(|a|,|b|), & ab \ge 0, \\ \sigma \min(|a|,|b|), & ab < 0, \end{cases} \tag{2.10}
$$

with $\sigma \in \mathbb{R}, |\sigma| \le 1$.

Remark 2.1. Note that the choice

$$\sigma = \operatorname{sgn}(s), \quad \text{where} \quad s = \begin{cases} a, & |a| \le |b| \\ b, & |b| \le |a| \end{cases} \tag{2.11}$$

recovers the MAPR limiter introduced in [2].

Using the approximate slopes (2.8) and the approximate flux derivatives (2.9), we end up with a family of central differencing schemes in the predictor-corrector form

$$v_j^{n+\frac{1}{2}} = v_j^n - \frac{1}{2}\lambda f_j', \tag{2.12}$$

$$v_{j+\frac{1}{2}}^{n+1} = \frac{1}{2}(v_j^n + v_{j+1}^n) + \frac{1}{8}(v_j' - v_{j+1}') - \lambda \left(f(v_{j+1}^{n+\frac{1}{2}}) - f(v_j^{n+\frac{1}{2}}) \right). \tag{2.13}$$

Remark 2.2. Note that (2.10) allows the predicted values $\left\{v_j^{n+\frac{1}{2}}\right\}_j$ in (2.12) to violate a strong maximum principle. That is, the minimum (maximum) of the sequence $\left\{v_j^{n+\frac{1}{2}}\right\}_j$ could be smaller (bigger) than that of $\left\{v_j^n\right\}_j$. This presents two new challenges: (i) bounding the minimum/maximum of the sequence $\left\{v_{j+\frac{1}{2}}^{n+1}\right\}_j$; (ii) bounding the local speed of propagation because in (2.13) we have $f(v_{j+1}^{n+\frac{1}{2}}) - f(v_j^{n+\frac{1}{2}}) = f'(\xi)\left(v_{j+1}^{n+\frac{1}{2}} - v_j^{n+\frac{1}{2}}\right)$ for some ξ between $v_{j+1}^{n+\frac{1}{2}}$ and $v_j^{n+\frac{1}{2}}$. Then ξ could (and sometimes will) be out of the range of $\left\{v_j^n\right\}_j$.

3 Maximum principle

In this section we will prove a maximum principle for k-monotone flux functions, see Definition 3.1, which is the main result of this paper, see Theorem 3.4. We will also show that this type of maximum principle implies the usual TVD bound as described in [9], see Theorem 3.7. We start with the following definition and properties of k-monotone functions, see [5, 10, 11].

Definition 3.1. A function $f : [a,b] \to \mathbb{R}$ is said to be k-monotone, $k \ge 1$, on $[a,b]$ if and only if for all choices of $(k+1)$ distinct nodes $x_0, \ldots, x_k$ in $[a,b]$ the inequality

$$[x_0, \ldots, x_k]f \ge 0 \tag{3.1}$$

holds, where $[x_0, \ldots, x_k]f := \sum_{j=0}^{k}(f(x_j)/w'(x_j))$ denotes the kth divided difference of f at $x_0, \ldots, x_k$ and $w(x) = \prod_{j=0}^{k}(x - x_j)$. f is said to be k-strictly monotone if the inequality (3.1) is strict.

Remark 3.2. If $f \in C^k[a,b]$, then f is k-strictly monotone if and only if there exists $\gamma_1 \in \mathbb{R}$ such that $0 < \gamma_1 \le f^{(k)}(x),\ x \in [a,b]$.

Example 1. Some well known examples of k-monotone fluxes are

1. Polynomial fluxes: $f(u) = \sum_{i=0}^{k} a_i u^i \in \mathrm{P}_k$, $a_k > 0$. A special case is the Burgers' flux, $f(u) = \frac{u^2}{2}$, which is 2-strictly monotone, i.e., strictly convex.
2. Buckley-Leverett flux: $f(u) = u^2/(u^2 + a(1-u)^2)$, $a > 0$, which is 1-monotone for $0 \le u \le 1$.

Definition 3.3. The range of a function $g : \mathbb{R} \to \mathbb{R}$ is defined to be the interval

$$R(g) := [\operatorname{essinf}_{x\in\mathbb{R}} g(x), \operatorname{esssup}_{x\in\mathbb{R}} g(x)].$$

The following theorem is the main result of this section.

Theorem 3.4. *Let v_j' be chosen as in (2.8) and $f_j' = f'(v_j^n)v_j'$. If f satisfies the following properties*

1. *f or $(-f)$ is $(m+2)$-strictly monotone on $R(u^0)$,*
2. *$f \in C^{m+2}(\mathbb{R})$ and there exists a constant $\gamma_2 \in \mathbb{R}$ such that $\left|f^{(m+2)}(x)\right| \le \gamma_2$, for all $x \in \mathbb{R}$,*

then the NT-scheme described by (2.12)–(2.13) satisfies the maximum principle

$$\min(v_j^n, v_{j+1}^n) \le v_{j+\frac{1}{2}}^{n+1} \le \max(v_j^n, v_{j+1}^n), \tag{3.2}$$

under the CFL condition

$$\lambda \max_{w\in R(u^0)} |f'(w)| \le \beta, \tag{3.3}$$

where β is a fixed constant which depends only on γ_1, γ_2 and m, see (3.15).

Remark 3.5. Note that the CFL condition (3.3) depends only on the maximum initial speed and smoothness properties of the flux. The CFL condition, see (3.15), amounts to solving a quadratic inequality similar to the one in [9]. For example, for the Burgers' equation, (3.3) is valid with $\beta = \frac{1}{8}$, see [9] for details.

Proof. It suffices to prove the theorem for the case when f is $(m+2)$-strictly monotone on $R(u^0)$. By remark 3.2 and the second property of f, we have that there exist $\gamma_1, \gamma_2 \in \mathbb{R}$ such that

$$\begin{aligned} f^{(m+2)}(x) &\le \gamma_2, \quad \forall x \in \mathbb{R}, \\ 0 < \gamma_1 \le f^{(m+2)}&(x), \quad \forall x \in R(u^0). \end{aligned} \tag{3.4}$$

We recall the definition of the open Newton-Cotes quadrature formula. For any given interval $[a, b]$, we define $h = (b-a)/(m+2)$ and introduce the nodes $x_i = a + (i+1)h$, where $i = -1, 0, \ldots, m+1$. Note that each x_i, $1 \le i \le m$, is a convex combination of a and b, that is

$$x_i = \theta a + (1-\theta)b, \quad \theta \in \left\{\frac{1}{m+2}, \frac{2}{m+2}, \ldots, \frac{m+1}{m+2}\right\}.$$

Next, we define

$$\ell_i(x) = \prod_{\substack{j=0 \\ j \neq i}}^{m} \frac{x - x_i}{x_i - x_j}, \quad L_m(x) = \sum_{i=0}^{m} f'(x_i)\ell_i(x), \quad \text{and} \quad A_i = \int_a^b \ell_i(x)dx,$$

and estimate

$$\begin{aligned}
\left|\int_a^b f'(x)dx\right| &\le \left|\frac{1}{(m+1)!}\int_a^b f^{(m+2)}(\xi_x)\prod_{i=0}^{m}(x-x_i)dx\right| + \left|\sum_{i=0}^{m} A_i f'(x_i)\right| \\
&\le \left|\frac{\gamma_2}{(m+1)!}\int_a^b \prod_{i=0}^{m}(x-x_i)dx\right| + \sum_{i=0}^{m}|A_i|\,|f'(x_i)| \\
&\le c_m\gamma_2|b-a|^{m+2} + \sum_{i=0}^{m}|A_i|\,|f'(x_i)| \\
&\le c_m|b-a|\left(\gamma_2|b-a|^{m+1} + \max_{0\le i\le m}|f'(x_i)|\right),
\end{aligned} \tag{3.5}$$

where c_m is a constant depending only on m, whose value may change at each occurrence. With this convention, let $x_{\frac{1}{2}} = x_0 + \frac{h}{2}$ and observe that

$$\begin{aligned}
\left|f'(x_{\frac{1}{2}}) - L_m(x_{\frac{1}{2}})\right| &= \left|\frac{f^{(m+2)}(\xi)}{(m+1)!}\right| \frac{h}{2}\prod_{i=1}^{m}(2i-1)\frac{h}{2} \\
&= c_m\left|f^{(m+2)}(\xi)\right|\,|b-a|^{m+1},
\end{aligned} \tag{3.6}$$

for some $\xi \in [x_0, x_m]$. From (3.4) and (3.6) it follows that

$$\begin{aligned}
|b-a|^{m+1} &\le \frac{c_m}{\gamma_1}\left|f'(x_{\frac{1}{2}}) - L_m(x_{\frac{1}{2}})\right| \\
&\le \frac{c_m}{\gamma_1}\max_{0\le i\le m}|f'(x_i)|.
\end{aligned} \tag{3.7}$$

Together (3.5) and (3.7) imply,

$$\left|\int_a^b f'(x)dx\right| \le c_m\left(\frac{\gamma_2}{\gamma_1} + 1\right)\max_{0\le i\le m}|f'(x_i)|\,|b-a|. \tag{3.8}$$

Let $a = v_j^{n+\frac{1}{2}}$ and $b = v_{j+1}^{n+\frac{1}{2}}$ in the above calculations and introduce the notation $v_{j+\frac{1}{2}}^{\theta} := \theta v_j^{n+\frac{1}{2}} + (1-\theta)v_{j+1}^{n+\frac{1}{2}}$. Then

$$v_{j+\frac{1}{2}}^{\theta} = \theta v_j^n + (1-\theta)v_{j+1}^n - \frac{\lambda}{2}\left(\theta f'(v_j^n)v_j' + (1-\theta)f'(v_{j+1}^n)v_{j+1}'\right). \quad (3.9)$$

Note that,

$$\begin{aligned} v_{j+\frac{1}{2}}^{\theta} &\geq \min(v_j^n, v_{j+1}^n) + \left(\theta - \max(\theta, 1-\theta)\lambda \left\|f'\right\|_{L^\infty(R(v^n))}\right)\left|v_{j+1}^n - v_j^n\right| \\ &\geq \min(v_j^n, v_{j+1}^n) + (\theta - \max(\theta, 1-\theta)\beta)\left|v_{j+1}^n - v_j^n\right|, \\ v_{j+\frac{1}{2}}^{\theta} &\leq \max(v_j^n, v_{j+1}^n) - \left(\theta - \max(\theta, 1-\theta)\lambda \left\|f'\right\|_{L^\infty(R(v^n))}\right)\left|v_{j+1}^n - v_j^n\right| \\ &\leq \max(v_j^n, v_{j+1}^n) - (\theta - \max(\theta, 1-\theta)\beta)\left|v_{j+1}^n - v_j^n\right|. \end{aligned} \quad (3.10)$$

Under the CFL condition $\beta \leq \frac{1}{m+1}$ the above inequalities imply

$$\min(v_j^n, v_{j+1}^n) \leq v_{j+\frac{1}{2}}^{\theta} \leq \max(v_j^n, v_{j+1}^n). \quad (3.11)$$

Therefore, we conclude that $\min(v_j^n, v_{j+1}^n) \leq x_i \leq \max(v_j^n, v_{j+1}^n)$, for $i = 0, \ldots, m$. Next we rewrite (3.8) with $a = v_j^{n+\frac{1}{2}}$ and $b = v_{j+1}^{n+\frac{1}{2}}$,

$$\begin{aligned} \left|f(v_{j+1}^{n+\frac{1}{2}}) - f(v_j^{n+\frac{1}{2}}))\right| &= \left|\int_a^b f'(x)dx\right| \\ &\leq c_m\left(\frac{\gamma_2}{\gamma_1} + 1\right)\max_{0\leq i\leq m}\left|f'(x_i)\right|\left|v_{j+1}^{n+\frac{1}{2}} - v_j^{n+\frac{1}{2}}\right| \\ &\leq c_m\left(\frac{\gamma_2}{\gamma_1} + 1\right)(1+\beta)\left\|f'\right\|_{L^\infty(R(v^n))}\left|v_{j+1}^n - v_j^n\right|. \end{aligned} \quad (3.12)$$

Using this result in (2.13) yields the estimates

$$\begin{aligned} v_{j+\frac{1}{2}}^{n+1} &\leq \frac{1}{2}(v_j^n + v_{j+1}^n) + \frac{1}{8}\left|v_j' - v_{j+1}'\right| + \lambda\left|f(v_{j+1}^{n+\frac{1}{2}}) - f(v_j^{n+\frac{1}{2}})\right| \\ &\leq \frac{1}{2}(v_j^n + v_{j+1}^n) + \left(\frac{1}{4} + \beta(1+\beta)c_m\left(\frac{\gamma_2}{\gamma_1} + 1\right)\right)\left|v_{j+1}^n - v_j^n\right|, \end{aligned} \quad (3.13)$$

and

$$\begin{aligned} v_{j+\frac{1}{2}}^{n+1} &\geq \frac{1}{2}(v_j^n + v_{j+1}^n) - \frac{1}{8}\left|v_j' - v_{j+1}'\right| - \lambda\left|f(v_{j+1}^{n+\frac{1}{2}}) - f(v_j^{n+\frac{1}{2}})\right| \\ &\geq \frac{1}{2}(v_j^n + v_{j+1}^n) - \left(\frac{1}{4} + \beta(1+\beta)c_m\left(\frac{\gamma_2}{\gamma_1} + 1\right)\right)\left|v_{j+1}^n - v_j^n\right|. \end{aligned} \quad (3.14)$$

Hence, under the CFL condition

$$\beta \leq \frac{1}{m+1} \quad \text{and} \quad \beta(1+\beta)c_m\left(\frac{\gamma_2}{\gamma_1}+1\right) \leq \frac{1}{4}, \tag{3.15}$$

we have

$$\min(v_j^n, v_{j+1}^n) \leq v_{j+\frac{1}{2}}^{n+1} \leq \max(v_j^n, v_{j+1}^n). \tag{3.16}$$

□

We end this section by showing that the maximum principle (3.2) implies the usual TVD bound, see [9]. We will need the following two lemmas to prove this result.

Lemma 3.6. *Given any non-constant sequence $\{x_i\}_{i=m}^n$ (i.e. not all x_i's are equal) with $n > m$, there exist piecewise monotone subsequences $\{x_{i_j}\}_{j=1}^k$ (with $k \geq 2$) such that if $\{x_{i_j}, x_{i_j+1}, \ldots, x_{i_{j+1}}\}$ is non-decreasing, then $\{x_{i_{j+1}}, x_{i_{j+1}+1}, \ldots, x_{i_{j+2}}\}$ is non-increasing, and vice versa, with the property that $\left|x_{i_j} - x_{i_{j+1}}\right| > 0$ for all j.*

Proof. We will construct a subsequence $S_k := \{x_{i_j}\}_{j=1}^k$ of $\{x_i\}_{i=m}^n$ that satisfies the conditions in the lemma. Let $S = \{\bar{x}_{i_s}\}_{s=1}^l$ denote the set of all local extremas of $\{x_i\}_{i=m}^n$ such that all the $\bar{x}_{i_s} \in S$ excluding the endpoints, namely x_n and x_m, satisfy

$$|\bar{x}_{i_s} - \bar{x}_{i_s-1}| + |\bar{x}_{i_s} - \bar{x}_{i_s+1}| > 0, \tag{3.17}$$

$$\operatorname{sgn}(\bar{x}_{i_s} - \bar{x}_{i_s-1}) \neq \operatorname{sgn}(\bar{x}_{i_{s+1}+1} - \bar{x}_{i_{s+1}}). \tag{3.18}$$

Observe that $\bar{x}_{i_1} = x_m$ and $\bar{x}_{i_l} = x_n$. In order to construct $S_k \subset S$ we remove $\bar{x}_{i_{s+1}}$ from S if $\bar{x}_{i_s} = \bar{x}_{i_{s+1}}$ for $1 \leq s \leq l-2$, and we remove $\bar{x}_{i_{l-1}}$ from S if $\bar{x}_{i_{l-1}} = \bar{x}_{i_l}$. Finally, we reindex the remaining elements of S to get $S_k = \{x_{i_j}\}_{j=1}^k$. □

Lemma 3.7. *Given sequences $\{x_j\}_{j=m}^n$ and $\{y_j\}_{j=m}^{n-1}$ with $m \leq n-1$, satisfying*

$$\min(x_j, x_{j+1}) \leq y_j \leq \max(x_j, x_{j+1}) \ , \ m \leq j \leq n-1, \tag{3.19}$$

we have

$$\sum_{j=m}^{n-2} |y_{j+1} - y_j| \leq \sum_{j=m}^{n-1} |x_{j+1} - x_j| . \tag{3.20}$$

Proof. Consider the subsequence $\{x_{i_j}\}_{j=1}^k$ of the sequence $\{x_i\}_{i=m}^n$ described in Lemma 3.6. For any given j with $2 \leq j \leq k-1$, assume without loss of generality that $\{x_{i_{j-1}}, x_{i_{j-1}+1}, \ldots, x_{i_j}\}$ is non-decreasing. That is, $\{y_{i_{j-1}}, y_{i_{j-1}+1}, \ldots, y_{i_j-1}\}$ is non-decreasing, and

$\{x_{i_j}, x_{i_j+1}, \ldots, x_{i_{j+1}}\}$ and $\{y_{i_j}, y_{i_j+1}, \ldots, y_{i_{j+1}-1}\}$ are non-increasing. There are two possible cases.

Case 1 : $y_{i_j-1} \leq y_{i_j}$. We consider the sequences $\{y_{i_{j-1}}, y_{i_{j-1}+1}, \ldots, y_{i_j}\}$ and $\{y_{i_j+1}, \ldots, y_{i_{j+1}-1}\}$, which are non-decreasing and non-increasing, respectively.

$$\begin{aligned}\sum_{k=i_{j-1}}^{i_{j+1}-2} |y_{k+1} - y_k| &= \sum_{k=i_{j-1}}^{i_j-1} (y_{k+1} - y_k) + \sum_{k=i_j}^{i_{j+1}-2} (y_k - y_{k+1}) \\ &= 2y_{i_j} - y_{i_{j-1}} - y_{i_{j+1}-1} \leq 2x_{i_j} - x_{i_{j-1}} - x_{i_{j+1}} \\ &= \sum_{k=i_{j-1}}^{i_j-1} (x_{k+1} - x_k) + \sum_{k=i_j}^{i_{j+1}-1} (x_k - x_{k+1}) \\ &= \sum_{k=i_{j-1}}^{i_{j+1}-1} |x_{k+1} - x_k| . \end{aligned} \tag{3.21}$$

Case 2 : $y_{i_j-1} \geq y_{i_j}$. The proof is analogous to Case 1.

In both cases we get $\mathrm{Var}(\{y_k\}_{k=i_{j-1}}^{i_{j+1}-1}) \leq \mathrm{Var}(\{x_k\}_{k=i_{j-1}}^{i_{j+1}})$. Similar arguments apply for the case when $\{x_{i_{j-1}}, x_{i_{j-1}+1}, \ldots, x_{i_j}\}$ is non-increasing. Therefore,

$$\mathrm{Var}\left(\{y_k\}_{k=i_{j-1}}^{i_{j+1}-1}\right) \leq \mathrm{Var}\left(\{x_k\}_{k=i_{j-1}}^{i_{j+1}}\right), \text{ for } 2 \leq j \leq k-1, \tag{3.22}$$

which implies

$$\mathrm{Var}\{y_j\}_{j=m}^{n-1} = \sum_{j=m}^{n-2} |y_{j+1} - y_j| \leq \sum_{j=m}^{n-1} |x_{j+1} - x_j| = \mathrm{Var}\{x_j\}_{j=m}^{n}. \tag{3.23}$$

□

Remark 3.8. Lemma 3.7 applied to the sequences $\{v_j^n\}_j$ and $\{v_{j+\frac{1}{2}}^{n+1}\}_j$ gives that our maximum principle (Theorem 3.4) implies the usual TVD bound for the NT scheme, see Theorem 3.2 in [9].

4 Optimal choice for σ

In this section we try to find an "optimal" σ for the MAPR-like limiters defined in (2.10). The goal is to make the l_∞-norm of the staggered averages when the MAPR-like limiter is used bigger than the l_∞-norm of the staggered averages when standard minmod limiter is used for the same initial data. The following lemma motivates our "optimal" choice of σ, see (4.1).

Lemma 4.1. *Let $\{v_j^n\}_j$ be a sequence with isolated local extrema. Let $\left\{v_{j+\frac{1}{2}}^{n+1}\right\}_j$ and $\left\{\bar{v}_{j+\frac{1}{2}}^{n+1}\right\}_j$ denote the staggered averages for standard MAPR-like limiter and minmod limiter, respectively. If the parameter σ in the MAPR-like limiter (2.10) is chosen as*

$$\sigma = \operatorname{sgn}\left(\bar{v}_{j+\frac{1}{2}}^{n+1} - \bar{v}_{j-\frac{1}{2}}^{n+1}\right), \tag{4.1}$$

then we have

$$\left\|\left\{\bar{v}_{j+\frac{1}{2}}^{n+1}\right\}_j\right\|_{l_\infty} \le \left\|\left\{v_{j+\frac{1}{2}}^{n+1}\right\}_j\right\|_{l_\infty}. \tag{4.2}$$

Proof. We will only consider the case of isolated local maxima with the case of isolated local minima being analogous. If the sequence $\left\{\bar{v}_{i+\frac{1}{2}}^{n+1}\right\}_i$ assumes its local maximum at $\bar{v}_{j+\frac{1}{2}}^{n+1}$, then we need to choose σ such that the difference $\Delta v_{j+\frac{1}{2}}^{n+1} := v_{j+\frac{1}{2}}^{n+1} - \bar{v}_{j+\frac{1}{2}}^{n+1}$ is positive. Let us rewrite the formula (2.13) for the staggered averages,

$$\begin{aligned} v_{j+\frac{1}{2}}^{n+1} &= \frac{1}{2}(v_j^n + v_{j+1}^n) + \frac{1}{8}(v_j' - v_{j+1}') - \lambda\left(f(v_{j+1}^{n+\frac{1}{2}}) - f(v_j^{n+\frac{1}{2}})\right), \\ \bar{v}_{j+\frac{1}{2}}^{n+1} &= \frac{1}{2}(v_j^n + v_{j+1}^n) + \frac{1}{8}(\bar{v}_j' - \bar{v}_{j+1}') - \lambda\left(f(\bar{v}_{j+1}^{n+\frac{1}{2}}) - f(\bar{v}_j^{n+\frac{1}{2}})\right). \end{aligned} \tag{4.3}$$

Subtracting the above equations and using $\bar{v}_j' = 0$, and $\bar{v}_{j+1}' = v_{j+1}'$, we obtain

$$\begin{aligned} \Delta v_{j+\frac{1}{2}}^{n+1} &= \frac{1}{8}v_j' + \lambda\left(f(v_j^{n+\frac{1}{2}}) - f(\bar{v}_j^{n+\frac{1}{2}})\right) \\ &= \frac{1}{2}\left(\frac{1}{4} - \lambda^2 f'(\xi_j^{n+\frac{1}{2}})f'(v_j^n)\right)v_j', \end{aligned} \tag{4.4}$$

where $\min(\bar{v}_j^{n+\frac{1}{2}}, v_j^{n+\frac{1}{2}}) \le \xi_j^{n+\frac{1}{2}} \le \max(\bar{v}_j^{n+\frac{1}{2}}, v_j^{n+\frac{1}{2}})$. By our maximum principle for MAPR-like limiters, see Theorem 3.4, we conclude that $\left|\lambda^2 f'(\xi_j^{n+\frac{1}{2}})f'(v_j^n)\right| \le \frac{1}{4}$. Thus, for $\Delta v_{j+\frac{1}{2}}^{n+1} > 0$ we need $v_j' > 0$.

If the sequence $\left\{\bar{v}_{i+\frac{1}{2}}^{n+1}\right\}_i$ assumes its local maximum at $\bar{v}_{j-\frac{1}{2}}^{n+1}$, we need to choose σ such that the difference $\Delta v_{j-\frac{1}{2}}^{n+1}$ is positive. By similar arguments we get

$$\Delta v_{j-\frac{1}{2}}^{n+1} = \frac{1}{2}\left(\lambda^2 f'(\xi_j^{n+\frac{1}{2}})f'(v_j^n) - \frac{1}{4}\right)v_j'. \tag{4.5}$$

Hence, for $\Delta v_{j-\frac{1}{2}}^{n+1} > 0$ we need $v_j' < 0$. This motivates our MAPR-like limiter choice which takes σ such that we maximize the l_∞ norm of the staggered averages in one time step. □

Remark 4.2. The inequality (4.2) is strict if the l_∞- norm of the sequence $\left\{v_j^n\right\}_j$ is achieved in one cell only. That is, if $|v_i^n| = \| \left\{v_j^n\right\}_j \|_{l_\infty}$, $|v_i^n - v_{i-1}^n| > 0$ and $|v_i^n - v_{i+1}^n| > 0$, then

$$\left\| \left\{ \bar{v}_{j+\frac{1}{2}}^{n+1} \right\}_j \right\|_{l_\infty} < \left\| \left\{ v_{j+\frac{1}{2}}^{n+1} \right\}_j \right\|_{l_\infty}.$$

Remark 4.3. Note that the choice of σ in (4.1) is not just a function of cell averages (like a standard slope reconstruction) but also depends on the flux in a nonlinear way.

Our numerical experiments suggest that the above choice of σ always gives bigger l_∞-norms of the numerical solutions, but at this point it is not feasible to prove this result for multiple time steps.

References

[1] P. Arminjon and M.-C. Viallon, Généralisation du schéma de Nessyahu-Tadmor pour une équation hyperbolique à deux dimensions d'espace. C. R. Acad. Sci. Paris Sér. I Math., 320(1): 85–88, 1995.

[2] I. Christov and B. Popov, New non-oscillatory central schemes on unstructured triangulations for hyperbolic systems of conservation laws. J. Comput. Phys., 227(11): 5736–5757, 2008.

[3] A. Harten and S. Osher, Uniformly high-order accurate nonoscillatory schemes. I. SIAM J. Numer. Anal., 24(2): 279–309, 1987.

[4] G.-S. Jiang and E. Tadmor, Nonoscillatory central schemes for multidimensional hyperbolic conservation laws. SIAM J. Sci. Comput., 19(6): 1892–1917 (electronic), 1998.

[5] K. A. Kopotun, Approximation of k-monotone functions. J. Approx. Theory, 94(3): 481–493, 1998.

[6] A. Kurganov, S. Noelle and G. Petrova, Semidiscrete central-upwind schemes for hyperbolic conservation laws and Hamilton-Jacobi equations. SIAM J. Sci. Comput., 23(3): 707–740 (electronic), 2001.

[7] A. Kurganov and E. Tadmor, New high-resolution central schemes for nonlinear conservation laws and convection-diffusion equations. J. Comput. Phys., 160(1): 241–282, 2000.

[8] O. Mehmetoglu and B. Popov, Maximum principle and convergence of central schemes based on slope limiters. Math. Comp., 2010. In review.

[9] H. Nessyahu and E. Tadmor, Nonoscillatory central differencing for hyperbolic conservation laws. J. Comput. Phys., 87(2): 408–463, 1990.

[10] J. E. Pecarić, F. Proschan and Y. L. Tong, Convex Functions, Partial Orderings, and Statistical Applications. Mathematics in Science and Engineering, Vol. 187. Academic Press Inc., Boston, MA, 1992.

[11] A. W. Roberts and D. E. Varberg, Convex Functions. Pure and Applied Mathematics, Vol. 57. Academic Press [A subsidiary of Harcourt Brace Jovanovich, Publishers], New York-London, 1973.

Darcy's Law in One-dimensional Isentropic Porous Medium Flow*

Ronghua Pan
School of Mathematics, Georgia Institute of Technology
Atlanta, GA 30332, USA
E-mail: panrh@math.gatech.edu

Abstract

We study the asymptotic behavior of compressible isentropic flow through porous medium with general L^∞ initial data. The model system is the compressible Euler equation with frictional damping. As $t \to \infty$, the density is conjectured to obey to the well-known porous medium equation and the momentum is expected to be formulated by Darcy's law. Recent progress gives a definte answer to this conjecture for one-dimenstional isentropic flow through the important entropy dissipation principle.

1 Introduction

This note is to survey the recent progress of the program toward the mathematical justification of Darcy law as long time asymptotic limit for compressible isentropic porous medium flow, modeled by the following Cauchy problem of compressible Euler equation with frictional damping,

$$\begin{cases} \rho_t + (\rho u)_x = 0, \\ (\rho u)_t + (\rho u^2 + p(\rho))_x = -\alpha \rho u, \\ \rho(x,0) = \rho_0(x), \ u(x,0) = u_0(x). \end{cases} \tag{1.1}$$

Here ρ, u and $p = \kappa\rho^\gamma$, $\kappa = \frac{(\gamma-1)^2}{4\gamma}(1 < \gamma < 3)$ denotes density, velocity, momentum and pressure, respectively. $\alpha > 0$ is a given positive constant modeling frictional force induced by the medium. We also use momentum $m =: \rho u$ in what follows for convenience. For simplicity, we assume $\alpha = \kappa = \frac{(\gamma-1)^2}{4\gamma}$. Such choice of constants κ and α is purely

*Pan's research is supported by National Science Foundation under grant DMS-0807406.

for convenience, which simplifies the form of the entropy functions we employed below.

Roughly speaking, because of the external frictional force, the inertial terms in the momentum equation decay to zero faster than other terms so that the pressure gradient force is balanced by the frictional force, which was stated as Darcy law. Therefore, one expects that, as $t \to \infty$, the density is approximated by certain solution of the well-known porous medium equation and the momentum is formulated by Darcy law, as observed in experiments. Therefore, time asymptotically, the system (1.1) is conjectured to be equivalent to the following decoupled system

$$\begin{cases} \bar{\rho}_t = (\bar{\rho}^\gamma)_{xx}, & \textbf{Porous Medium Equation}, \\ \bar{m} = -(\bar{\rho}^\gamma)_x, & \textbf{Darcy Law}. \end{cases} \tag{1.2}$$

Mathematical study of system (1) dated back to 1970s. Following the pioneer work of Nishida [29], many contributions have been made for this problem. During the early stage of the development, major attentions were paid on the case away from vacuum, system (1) can be transferred to the damped p-system by changing to the Lagrangian coordinates; see [34]. The frictional damping prevents the breaking of waves with small amplitude, leading to the global existence of smooth solutions when initial data is small and smooth [29]. However, waves break down in finite time when the initial derivatives of initial data exceed certain threshold [37]. Therefore, nice theory were developed mainly in the class of small smooth solution away from vacuum. The asymptotic limit has been well justified in such case by Hsiao and Liu in [11] and [12], and further improved by many mathematicians for small smooth or piecewise smooth solutions away from vacuum based on the energy estimates for derivatives; see [10], [13], [14], [15], [16], [28], [30], [31], [32], and [36]. Recently, Dafermos and Pan [7] constructed the global BV solutions to damped p-system and proved the conjecture with sharp decay rates in L^2. In these results, the solutions of damped p-system were shown to converge to the self-similar solutions of the corresponding porous medium equations constructed in [33] since the end-states of the initial density are away from vacuum.

Recently, huge efforts have been put to tackle the large solutions. This is based on the global existence of weak solutions in L^∞ was established by the method of compensated compactness in [8], [9], [23], [38] and [17]. We remark that in this case, the vacuum is allowed in the solution. When a vacuum occurs in the solution, the difficulty of the problem greatly increased mainly due to the interaction of nonlinear convection, lower order dissipation of damping and the resonance due to vacuum.

Definition 1.1. *For any $T > 0$, the bounded measurable functions*

$(\rho, m)(x,t) \in L^\infty(\mathbf{R} \times [0,T])$ *are called entropy solutions of (1), if*

$$\begin{cases} \rho_t + m_x = 0, \\ m_t + \left(\dfrac{m^2}{\rho} + \kappa\rho^\gamma\right)_x + \alpha m = 0, \\ \eta_t + q_x + \alpha\eta_m m \leq 0 \end{cases} \tag{1.3}$$

hold in the sense of distributions, where (η, q) *is any weak convex entropy–flux pair* $(\eta(\rho,m), q(\rho,m))$ *satisfying*

$$\nabla q = \nabla\eta\nabla f, \quad f = \left(m, \frac{m^2}{\rho} + \kappa\rho^\gamma\right)^t, \quad \eta(0,0) = 0. \tag{1.4}$$

Since the L^∞ weak solution does not have any degree of regularity, the methods for the case away from vacuum are not applicable here. Earlier attempts were made by Huang and Pan [17], where the authors followed the rescaling argument due to Serre and Hsiao [35] and obtained the first justification to the conjecture for vacuum case. In [18], Huang and Pan started to develope a new technique based on the conservation of mass and entropy dissipation analysis to attack this conjecture. They showed that the L^∞ weak entropy solutions with vacuum, selected by the physical entropy-flux pairs, converge strongly in $L^p(R)$ ($p \geq p_0$ for some $p_0 \geq 2$) with decay rates to the similarity solution of the porous medium equation determined uniquely by the end-states and the mass distribution of the initial data provided that the end-states are away from vacuum. This approach seems remarkable since it does not need smallness assumptions on the solutions. Inspired by this result, Huang, Marcati and Pan [19] further studied this problem with finite total mass, where the asymptotic profile is the celebrated Barenblatt's solution of the porous medium equation. These results, valid for some physical flows, depend only on the physcial mechanical energy.

In order to give a further definite answer to the problem, Huang, Pan and Wang devoted further efforts in [20] and [21] to explore the better application of the rich entropy of the isentropic Euler system. In these two papers, they proved that the conjecture is true for most physical flows, that is any entropy L^∞ weak solutions.

In section 2, the self-similar solution and the Barenblatt solution of porous medium equation are reviewed. In section 3, some results on the entropy were collected, and an invariant region result is proved for L^∞ entropy solutions of (1). Finally, in section 4, the main results were presented along with an outline of the proof. In case when one of the end states is away from vacuum, we prove that any L^∞ weak entropy solutions converges to the corresponding self-similar solution of porous medium equation. When the initial total mass is finite, we prove that any L^∞ weak entropy solutions converges to the Barenblatt solution, with a surprising L^1 decay results.

2 Self-similar solutions and Barenblatt solutions

The PME is strictly parabolic if $\rho > 0$ and degenerate if $\rho = 0$. This equation has been well-understood, see [1] and [3].

2.1 Self-similar solutions

Consider the following Cauchy problem for porous medium equation(PME):

$$\begin{cases} \rho_t = (\rho^\gamma)_{xx}, \quad \gamma \geq 1, \quad t > 0, \\ \rho(x,0) = \rho_0(x) = \rho_-\chi(x<0) + \rho_+\chi(x>0), \end{cases} \tag{2.1}$$

where, χ is the characteristic function and $\rho_\pm$ are non-negative constants. In [17], Huang and Pan proved the uniqueness of weak solutions with L^∞ initial data for PME. Since the initial data is self-similar, the uniqueness theorem of [17] implies that ρ itself is self-similar, i.e., $\rho(x,t) = \rho_s(z)$ $(z = \frac{x}{\sqrt{t}})$, which satisfies

$$\begin{cases} (\gamma\bar{\rho}^{\gamma-1}\bar{\rho}_z)_z + \dfrac{1}{2}z\bar{\rho}_z = 0, \\ \bar{\rho}(-\infty) = \rho_-, \quad \bar{\rho}(+\infty) = \rho_+. \end{cases} \tag{2.2}$$

The following two lemmas provide some some properties on ρ_s.

Lemma 2.1. *If $\rho_- \geq \rho_+ > 0$ then (5) has a unique solution $\rho_s(z) \in C^2$ satisfying the following:*

(1) $\rho_+ \leq \rho_s(z) \leq \rho_-$ is monotone decreasing on $\mathbf{R}$.

(2) $(|(\rho_s)_x|, |(\rho_s)_t|) \leq C(t_1)(t^{-\frac{1}{2}}, t^{-1})$ for any $t \geq t_1 > 0$.

Lemma 2.2. *If $\rho_+ = 0$ and $\rho_- > 0$, then there is one and only one solution $\rho_s(z)$ to (5). Furthermore, the follows hold.*

- *(1) $0 \leq \rho_s(z) \leq \rho_-$ is continuous and monotone decreasing on $\mathbf{R}$.*
- *(2) ${\rho_s}^\gamma(z)$ is smooth on $\mathbf{R}$.*
- *(3) There is a number $b > 0$, such that $\rho_s(z) > 0$ if $z < b$ and $\rho_s(z) = 0$ if $z \geq b$.*
- *(4) $\rho_s(z)$ is smooth if $z < b$.*
- *(5) $({\rho_s}^\gamma)'(z) \to 0$ as $z \to b - 0$.*
- *(6) $(|\partial_x({\rho_s}^{\gamma-1})|, |\partial_t({\rho_s}^{\gamma-1})|) \leq C(t_1)(t^{-\frac{1}{2}}, t^{-1})$ for any $t \geq t_1 > 0$.*

2.2 Barenblatt solutions

According to [19], the solutions to (1) with finite total mass should converge in large time to the fundamental solutions of the porous media equation, i.e., the Barenblatt's solutions [2].

Consider

$$\begin{cases} \bar{\rho}_t = (\bar{\rho}^\gamma)_{xx}, \\ \bar{\rho}(-1,x) = M\delta(x), \quad M > 0, \end{cases} \tag{2.3}$$

which admits a unique solution (c.f. [1], [2]) given below

$$\rho_b(x,t) = (t+1)^{-\frac{1}{\gamma+1}}\{(A - B\xi^2)_+\}^{\frac{1}{\gamma-1}}. \tag{2.4}$$

Here $\xi = x(t+1)^{-\frac{1}{\gamma+1}}$, $(f)_+ = \max\{0, f\}$, $B = \frac{\gamma-1}{2\gamma(\gamma+1)}$ and A is determined by

$$2A^{\frac{\gamma+1}{2(\gamma-1)}}B^{-\frac{1}{2}}\int_0^{\frac{\pi}{2}} (\cos\theta)^{\frac{\gamma+1}{\gamma-1}}\, d\theta = M. \tag{2.5}$$

The following two lemmas summarize some of the properties of ρ_b.

Lemma 2.3. *If $M > 0$ is finite, then there is one and only one solution $\rho_b(x,t)$ to (11). Furthermore, the follows hold.*

- *$\rho_b(x,t)$ is continuous on $\mathbf{R}$.*
- *There is a number $b = \sqrt{\frac{A}{B}} > 0$, such that $\rho_b(x,t) > 0$ if $|x| < bt^{\frac{1}{\gamma+1}}$ and $\rho_b(x,t) = 0$ if $|x| \geq bt^{\frac{1}{\gamma+1}}$.*
- *$\rho_b(x,t)$ is smooth if $|x| < bt^{\frac{1}{\gamma+1}}$.*

In terms of the explicit form of ρ_b, it is easy to check the following estimates.

Lemma 2.4. *For ρ_b defined in (2.4) and $t > 0$, it holds that*

$$\begin{cases} |\rho_b| \leq C(1+t)^{-\frac{1}{\gamma+1}}, \\ |({\rho_b}^{\gamma-1})_x| \leq C(1+t)^{-\frac{\gamma}{\gamma+1}}, \quad |({\rho_b}^{\gamma-1})_t| \leq C(1+t)^{-\frac{2\gamma}{\gamma+1}}, \\ |({\rho_b}^{\gamma})_x| \leq C(1+t)^{-1}, \quad |({\rho_b}^{\gamma})_t| \leq C(1+t)^{-\frac{2\gamma+1}{\gamma+1}}. \end{cases} \tag{2.6}$$

3 Entropy and Invariant Region

First of all, we recall some results on the entropies available for (1). According to [22], all weak entropies of (1) are given by the following formula:

$$\eta(\rho, u) = \int g(\xi)\chi(\xi; \rho, u)\, d\xi = \rho \int_{-1}^{1} g(u + z\rho^\theta)(1 - z^2)^\lambda\, dz,$$
$$q(\rho, m) = \int g(\xi)(\theta\xi + (1-\theta)u)\chi(\xi; \rho, u)\, d\xi \tag{3.1}$$
$$= \rho \int_{-1}^{1} g(u + z\rho^\theta)(u + \theta z\rho^\theta)(1 - z^2)^\lambda\, dz$$

where $\theta = \frac{\gamma-1}{2}$, $\lambda = \frac{3-\gamma}{2(\gamma-1)}$ and $g(\xi)$ is any smooth function of ξ and

$$\chi(\xi; \rho, u) = (\rho^{\gamma-1} - (\xi - u)^2)_+^\lambda. \tag{3.2}$$

This remarkable formula can be derived from the entropy equation (4) utilizing the kinetic formulation or by fundamental solution of linear wave equation. We remark that when $g(\xi) = 1$, $\eta(\rho, m) = \rho$; when $g(\xi) = \xi$, $\eta(\rho, m) = m$ and when $g(\xi) = \frac{1}{2}\xi^2$, then

$$\eta_e = \frac{m^2}{2\rho} + \frac{\kappa}{\gamma - 1}\rho^\gamma$$

is the mechanical energy.

As the convexity of entropy function is crucial in the definition of admissible weak solutions, the characterization of convexity of entropy functions is important. In our case, the following lemma provides full details in this direction.

Lemma 3.1 (Lions-Perthame-Tadmor, [22]). *Weak entropy $\eta(\rho, m)$ defined in (3.1) is convex with respect to ρ and m if and only if $g(\xi)$ is a convex function.*

Another important entropy in our context is the one measures $L^{\gamma+1}$ norm in density. Choosing $g(\xi) = |\xi|^{\frac{2\gamma}{\gamma-1}}$, this entropy reads as

$$\tilde{\eta} = \rho \int_{-1}^{1} |u + z\rho^\theta|^{\frac{2\gamma}{\gamma-1}}(1 - z^2)^\lambda dz. \tag{3.3}$$

Recall that $B(p, q)$ the Beta function is defined by

$$B(p, q) = \int_0^1 x^{p-1}(1 - x)^{q-1} dx.$$

It is proved in [20] that

Lemma 3.2. *For entropy $\tilde{\eta}$ defined in (3.3), it holds that*

$$\tilde{\eta} = C_1\rho^{\gamma+1} + C_2 m^2 + A(\rho, m), \tag{3.4}$$

where

$$C_1=\frac{1}{2}B\left(\frac{\gamma+1}{2(\gamma-1)},\frac{\gamma+1}{2(\gamma-1)}\right),\quad C_2=\frac{\gamma(\gamma+1)}{(\gamma-1)^2}B\left(\frac{\gamma+1}{2(\gamma-1)},\frac{\gamma+1}{2(\gamma-1)}\right).$$

Furthermore, $A(\rho,m)\geq 0$, $A_m m\geq 0$, *and*

$$|A|\leq C\rho|u|^3(|u|^{\frac{3-\gamma}{\gamma-1}}+\rho^{1-\theta}).$$

As a direct application of entropy analysis, we will present an invariant region theorem for L^∞ weak entropy solutions of (1). Choosing $g(\xi)=g_k(\xi)=e^{k\xi^2}$ in (3.1), for positive parameter $k>0$, the corresponding entropy $\eta_k\geq 0$ is clearly convex. Using this sequence of entropies, [20] proved the following invariant region theorem, which confirms that any L^∞ weak entropy solutions to (1) will stay inside the physical region

$$0\leq\rho(x,t)\leq C,\quad |m(x,t)|\leq C\rho(x,t),$$

if the initial data does so. More precisely, the theorem read as

Theorem 3.1 (Huang, Pan, Wang, [20]). *Suppose* $(\rho_0,u_0)(x)\in L^\infty(\mathbf{R})$ *such that*

$$0\leq\rho_0(x)\leq C,\quad |m_0(x)|\leq C\rho_0(x).$$

Let $(\rho,u)\in L^\infty(R\times[0,T])$ *be an* L^∞ *weak entropy solution of the system (1) with* $\gamma>1$. *Then* (ρ,m) *satisfies*

$$0\leq\rho(x,t)\leq C,\quad |m(x,t)|\leq C\rho(x,t),\tag{3.5}$$

where the constant C *depends solely on the initial data.*

4 Large time behavior

In this section, we will report the main results obtained in [20] and [21] for the large time asymptotic behavior of general L^∞ weak entropy solutions of (1). For this purpose, we suppose that (ρ,m) is a L^∞ weak entropy solution of (1) such that

$$0\leq\rho_0(x)\leq C,\quad |m_0(x)|\leq C\rho_0(x).\tag{4.1}$$

For the long time profile, we will also denote by ρ_s the self-similar solution of PME, by ρ_b the Barenblatt solution of PME. Due to the limitation of the length, we will only outline the proof for the case with finite total mass.

4.1 Infinite total mass

In this case, one assumes that

$$\lim_{x\to\pm\infty}(\rho_0(x), m_0(x)) = (\rho_\pm, m_\pm). \tag{4.2}$$

When the initial end stats of density has a non-vacuum state, say $\rho_- > 0$, the results in [21] confirms that (ρ, m) converges to the corresponding self-similar profiles ρ_s in large time. With the procedure introduced in [11] and [12], one could assume that $m_- = m_+ = 0$ and

$$\int_{-\infty}^{\infty} (\rho_0(x) - \rho_s(x,0))\, dx = 0. \tag{4.3}$$

A convenient variable y is thus introduced

$$y = -\int_{-\infty}^{x} (\rho - \rho_s)(r,t)\, dr. \tag{4.4}$$

The following two theorems are proved in [21].

Theorem 4.1. *Assume $1 < \gamma \le 2$, $\rho_- > \rho_+ = 0$ and $m_\pm = 0$. Suppose (ρ, m) is a L^∞ weak entropy solution of (1) satisfies (4.1) and (4.2). Let ρ_s be the selfsimilar solution of (1.3) with $\rho_s(\pm\infty) = \rho_\pm$ and $\bar{m} = -P(\rho_s)_x$. If $y \in H^1(\mathbf{R})$, then there exists positive constant C such that, for any $\varepsilon > 0$,*

$$\|\rho - \rho_s\|_{L^{\gamma+1}}^{\gamma+1} \le C(1+t)^{-\frac{1}{4}+\varepsilon}. \tag{4.5}$$

Theorem 4.2. *Assume $1 < \gamma < 3$, $\rho_- \ge \rho_+ > 0$, and $m_- = m_+ = 0$. Suppose (ρ, m) is a L^∞ weak entropy solution of (1) satisfies (4.1) and (4.2). Let ρ_s be the selfsimilar solution of (1.3) with $\rho_s(\pm\infty) = \rho_\pm$ and $\bar{m} = -P(\rho_s)_x$. If $y_0(x) \in H^1$, then there exist constants $C > 0$ and $0 < \beta < \frac{1}{2}$ independent of time such that*

$$\|(\rho(x,t) - \rho_s(x,t), m(x,t) - \bar{m}(x,t))\|_{L^p} \le C(1+t)^{-\beta/p}, \quad 2 \le p < \infty. \tag{4.6}$$

4.2 Finite total mass

In this case, one assumes that

$$\int_{-\infty}^{+\infty} \rho(x,t)\, dx = \int_{-\infty}^{+\infty} \rho_b(x,t)\, dx = \int_{-\infty}^{+\infty} \rho_0(x)\, dx = M < \infty. \tag{4.7}$$

The following Theorem, proved in [20], shows that $\rho(x,t)$ converges to the corresponding Barenblatt solution ρ_b with the same total mass. The decay rates in both energy norm ($L^{\gamma+1}$) and mass norm (L^1) are obtained.

Theorem 4.3. *Suppose $\rho_0(x) \in L^1(\mathbf{R}) \cap L^\infty(\mathbf{R})$, $u_0(x) \in L^\infty(\mathbf{R})$ and*

$$M = \int_{-\infty}^{\infty} \rho_0(x)\, dx > 0.$$

Let $1 < \gamma < 3$ and (ρ, m) be an L^∞ entropy solution of the Cauchy problem (1). Let ρ_b be the Barenblatt's solution of porous medium equation (2) with mass M and $\bar{m} = -(\rho_b{}^\gamma)_x$. Define

$$y = -\int_{-\infty}^{x} (\rho - \rho_b)(r,t)\, dr.$$

If $y(x,0) \in L^2(\mathbf{R})$, then for any $\varepsilon > 0$ and $t > 0$,

$$\begin{aligned} &\|(\rho - \rho_b)(\cdot,t)\|_{L^{\gamma+1}}^{\gamma+1} \le C(1+t)^{-1+\frac{1}{2(\gamma+1)}+\varepsilon}, \\ &\|(\rho - \rho_b)(\cdot,t)\|_{L^1} \le C(1+t)^{-\frac{1}{4(\gamma+1)}+\varepsilon}. \end{aligned} \tag{4.8}$$

4.3 Outline of the proof to Theorem 4.3

In order to control the nonlinearity and singularity near vacuum, the following lemma is proved in [20].

Lemma 4.1. *If $0 \le \rho, \bar\rho \le C$, there are two constants $c_1 > 0$ and $c_2 > 0$ such that*

$$\begin{cases} c_1(\rho^{\gamma-1} + \bar\rho^{\gamma-1})(\rho - \bar\rho)^2 \le \rho^{\gamma+1} - \bar\rho^{\gamma+1} - (\gamma+1)\bar\rho^\gamma(\rho - \bar\rho) \\ \qquad\qquad\qquad\qquad \le c_2(\rho^{\gamma-1} + \bar\rho^{\gamma-1})(\rho - \bar\rho)^2 \\ c_1(\rho^{\gamma-1} + \bar\rho^{\gamma-1})(\rho - \bar\rho)^2 \le (\rho^\gamma - \bar\rho^\gamma)(\rho - \bar\rho) \le c_2(\rho^{\gamma-1} + \bar\rho^{\gamma-1})(\rho - \bar\rho)^2. \end{cases} \tag{4.9}$$

Using the notion y, the momentum equation is equivalent to

$$y_{tt} + (\frac{m^2}{\rho})_x + \kappa(\rho^\gamma - \bar\rho^\gamma)_x + \kappa y_t = -\bar{m}_t. \tag{4.10}$$

By standard energy method, we first obtain the following estimate

Lemma 4.2. *Let the conditions of Theorem 4.3 be satisfied, it holds*

$$\begin{aligned} &\int_{-\infty}^{+\infty} (y_t y + \frac{\kappa}{2} y^2)\, dx + \int_0^t \int_{-\infty}^{+\infty} \kappa(\rho^\gamma - \bar\rho^\gamma)(\rho - \bar\rho)\, dx d\tau \\ &\le C + \int_0^t \int_{-\infty}^{+\infty} y_t^2\, dx d\tau + \int_0^t \int_{-\infty}^{+\infty} \frac{m^2}{\rho} y_x\, dx d\tau. \end{aligned} \tag{4.11}$$

In the second step, we use mechanical energy

$$\eta_e = \frac{m^2}{2\rho} + \frac{\kappa}{\gamma - 1}\rho^\gamma,$$

and q_e the corresponding flux in the entropy inequality,

$$\eta_{et} + q_{ex} + \kappa\frac{m^2}{\rho} \leq 0, \tag{4.12}$$

to prove the following estimate

Lemma 4.3. *Under the conditions of Theorem 4.3, for any $t > 0$, it holds that*

$$\begin{aligned}\int_{-\infty}^{+\infty} (\rho^\gamma + y_t^2 + \frac{m^2}{\rho} + y^2)\, dx + \int_0^t \int_{-\infty}^{+\infty} (\frac{m^2}{\rho} + y_t^2)\, dx d\tau \\ + \int_0^t \int_{-\infty}^{+\infty} (\rho^\gamma - {\rho_b}^\gamma)(\rho - \rho_b)\, dx d\tau \leq C.\end{aligned} \tag{4.13}$$

In the third step, we employ the entropy-entropy flux pair $(\tilde{\eta}, \tilde{q})$, which induces

$$\eta_* = \tilde{\eta} - C_1{\rho_b}^{\gamma+1} - C_1(\gamma+1){\rho_b}^\gamma(\rho - \rho_b). \tag{4.14}$$

The entropy inequality implies

$$\eta_{*t} + (C_1{\rho_b}^{\gamma+1} + C_1(\gamma+1){\rho_b}^\gamma(\rho-\rho_b))_t + \tilde{q}_x + 2\kappa C_2 m^2 + \kappa A_m m \leq 0, \tag{4.15}$$

which further implies that

$$\eta_{*t} - 2\kappa C_2({\rho_b}_x^\gamma y)_t + 2\kappa C_2(m-\bar{m})^2 + 4\kappa C_2({\rho_b}^\gamma)_t(\rho-\rho_b) + \kappa A_m m + (\cdots)_x \leq 0. \tag{4.16}$$

For any small positive constant $\varepsilon > 0$, we define

$$\mu(\varepsilon) = 1 - \frac{1}{2(\gamma+1)} - \varepsilon.$$

Multiplying (4.16) by $(1+t)^{\mu(\varepsilon)}$, integrating the result on $\mathbf{R} \times [0, t]$ and using Lemma 4.4, we shall reach, after some computation, that

$$\int_{-\infty}^{+\infty} \eta_*\, dx \leq C(1+t)^{-\mu(\varepsilon)}. \tag{4.17}$$

Therefore we proved the following lemma,

Lemma 4.4. *Under the conditions of Theorem 4.3, it holds for any $t > 0$ that*

$$\begin{aligned}&\|(m-\bar{m})(\cdot,t)\|_{L^2}^2 + \|(\rho-\rho_b)(\cdot,t)\|_{L^{\gamma+1}}^{\gamma+1} \\ &+ \int_{-\infty}^{+\infty} (\rho^{\gamma-1}+{\rho_b}^{\gamma-1})(\rho-\rho_b)^2\, dx \le C(1+t)^{\frac{1}{2(\gamma+1)}+\varepsilon-1}, \\ &\int_0^t (1+\tau)^{1-\frac{1}{2(\gamma+1)}-\varepsilon}\|(m-\bar{m})(\cdot,\tau)\|_{L^2}^2\, d\tau \le C,\end{aligned} \tag{4.18}$$

for any positive constant ε.

This crucial sharper decay rates in $L^{\gamma+1}$, together with the following key observation will lead to the decay in L^1.

Lemma 4.5. *If $\rho \ge 0$ and $\bar{\rho} \ge 0$ have the same total mass M, then for any $t > 0$,*

$$\int_{-\infty}^{+\infty} |\rho-\bar{\rho}|(x,t)dx \le 2\int_{\bar{\rho}>0} |\rho-\bar{\rho}|(x,t)dx. \tag{4.19}$$

Now, it is straightforward to combine Lemma 4.4 and Lemma 4.5 to prove the following decay estimates.

Lemma 4.6. *Assume the conditions in Theorem 1.2 are satisfied, then*

$$\|\rho-\rho_b\|_{L^1} \le C(1+t)^{-\frac{1}{4(\gamma+1)}+\varepsilon}, \ \forall t > 0,$$

for any $\varepsilon > 0$.

We thus conclude the proof of Theorem 4.3 from Lemmas 4.4 and 4.6.

References

[1] D. G. Aronson, The porous media equations. In: Nonlinear Diffusion Problem, Lecture Notes in Math., Vol. 1224 (A. Fasano, M. Primicerio, Eds), Springer-Verlag, Berlin, 1986.

[2] G. I. Barenblatt, On one class of the one-dimensional problem of non-stationary filtration of a gas in a porous medium. Prikl. Mat. i Mekh. 17 (1953), 739–742.

[3] J. A. Carrillo and G. Toscani, Asymptotic L^1-decay of solutions of the Porous Medium Equation to self-similarity. Indiana U. Math. J. 49 (2000), 113–142.

[4] G. Chen and H. Frid, Divergence-measure fields and hyperbolic conservation laws. Arch. Rat. Mech. Anal. 147 (1999), 89–118.

[5] K. Chueh, C. Conley and J. Smoller, Positively invariant regions for systems of nonlinear diffusion equations. Indiana U. Math. J. 26 (1977), 373–392.

[6] C. M. Dafermos, A system of hyperbolic conservation laws with frictional damping. Z. Angew. Math. Phys. 46 (1995), 294–307.

[7] C. M. Dafermos and R. Pan, Global BV solutions for the p-system with frictional damping. SIAM J. Math. Anal. 41, No. 3 (2009), 1190–1205.

[8] X. Ding, G. Chen and P. Luo, Convergence of the fractional step Lax-Friedrichs and Godunov scheme for isentropic system of gas dynamics. Commun. Math. Phys. 121 (1989), 63–84.

[9] R. Diperna, Convergence of viscosity method for isentropic gas dynamics. Comm. Math. Phys. 91 (1983), 1–30.

[10] L. Hsiao, Quasilinear Hyperbolic Systems and Dissipative Mechanisms. World Scientific, 1997.

[11] L. Hsiao and T. Liu, Convergence to nonlinear diffusion waves for solutions of a system of hyperbolic conservation laws with damping. Comm. Math. Phys. 143 (1992), 599–605.

[12] L. Hsiao and T. Liu, Nonlinear diffusive phenomena of nonlinear hyperbolic systems. Chin. Ann. of Math. 14B, 4 (1993), 465–480.

[13] L. Hsiao and T. Luo, Nonlinear diffusive phenomena of entropy weak solutions for a system of quasilinear hyperbolic conservation laws with damping. Q. Appl. Math. 56, No.1 (1998), 173–198.

[14] L. Hsiao and R. Pan, The damped p-system with boundary effects. Contemporary Mathematics, 255 (2000), 109–123.

[15] L. Hsiao and S. Tang, Construction and qualitative behavior of solutions for a system of nonlinear hyperbolic conservation laws with damping. Q. Appl. Math. LIII, NO. 3 (1995), 487–505.

[16] L. Hsiao and S. Tang, Construction and qualitative behavior of solutions of perturbated Riemann problem for the system of one-dimensional isentropic flow with damping. J. Differential Equations, 123, No. 2 (1995), 480–503.

[17] F. Huang and R. Pan, Asymptotic behavior of the solutions to the damped compressible Euler equations with vacuum. J. Differential Equations 220 (2006), 207–233.

[18] F. Huang and R. Pan, Convergence rate for compressible Euler equations with damping and vacuum. Arch. Rat. Mech. Anal. 166 (2003), 359–376.

[19] F. Huang, P. Marcati and R. Pan, Convergence to Barenblatt solution for the compressible Euler equations with damping and vacuum. Arch. Rat. Mech. Anal. 176 (2005), 1–24.

[20] F. Huang, R. Pan and Z. Wang, L^1 convergence to the Barenblatt solution for compressible Euler equations with damping. Arch. Rat. Mech. Anal. in press.

[21] F. Huang, R. Pan and Z. Wang, Large time behavior of solutions for compressible Euler equations with damping and vacuum, preprint, 2010.

[22] P. L. Lions, B. Perthame and E. Tadmor, Kinetic formulation of the isentropic gas dynamics and p-systems. Comm. Math. Phys. 163 (1994), 169–172.

[23] P. L. Lions, B. Perthame and P. Souganidis, Existence and stability of entropy solutions for the hyperbolic systems of isentropic gas dynamics in Eulerian and Lagrangian coordinates. Commun. Pure Appl. Math. 49 (1996), 599–638.

[24] T. Liu, Compressible flow with damping and vacuum. Japan J. Appl. Math. 13, No. 1 (1996), 25–32.

[25] M. Luskin and B. Temple, The existence of a global weak solution to the nonlinear water-hammar problem. Comm. Pure Appl. Math. 35 (1982), 697–735.

[26] T. Liu and T. Yang, Compressible Euler equations with vacuum. J. Differential Equations 140 (1997), 223–237.

[27] T. Liu and T. Yang, Compressible flow with vacuum and physical singularity. Methods Appl. Anal. 7 (2000), 495–509.

[28] T. Luo and T. Yang, Interaction of elementary waves for compressible Euler equations with frictional damping. J. Differential Equations 161 (2000), 42–86.

[29] T. Nishida, Nonlinear hyperbolic equations and related topics in fluid dynamics. Publ. Math. D'Orsay (1978), 46–53.

[30] K. Nishihara, Convergence rates to nonlinear diffusion waves for solutions of system of hyperbolic conservation laws with damping. J. Differential Equations 131 (1996), 171–188.

[31] K. Nishihara, W. Wang and T. Yang, L_p-convergence rate to nonlinear diffusion waves for p-system with damping. J. Differential Equations 161 (2000), 191–218.

[32] K. Nishihara and T. Yang, Boundary effect on asymptotic behavior of solutions to the p-system with damping. J. Differential Equations 156 (1999), 439–458.

[33] L. A. Peletier and C. J. Van Duyn, A class of similary solutions of the nonlinear diffusion equations. Nonlinear Analysis, TMA 1 (1977), 223–233.

[34] J. A. Smoller, Shock Waves and Reaction-diffusion Equations. Springer-Verlag, 1980.

[35] D. Serre and L. Xiao, Asymptotic behavior of large weak entropy solutions of the damped p-system. J. P. Diff. Equa. 10 (1997), 355–368.

[36] H. Zhao, Convergence to strong nonlinear diffusion waves for solutions of p-system with damping. J. Differential Equations 174 (2001), 200–236.

[37] Y. Zheng, Global smooth solutions to the adiabatic gas dynamics system with dissipation terms. Chinese Ann. of Math. 17A (1996), 155–162.

[38] C. J. Zhu, Convergence of viscosity solutions for the system Of nonlinear elasticity. J. Math. Anal. Appl. 209 (1997), 585–604.

[39] C. J. Zhu, Convergence rates to nonlinear diffusion waves for weak entropy solutions to p-system with damping. Sci. China Ser. A 46 (2003), 562–575.

Diffusive Expansion for Solutions of the Boltzmann Equation in the Whole Space*

Huijiang Zhao
School of Mathematics and Statistics, Wuhan University
Wuhan 430072, China
E-mail: hhjjzhao@hotmail.com

Abstract

This paper is concerned with global-in-time validity of the diffusive expansion for solutions of the Boltzmann equation in the whole space.

1 The problem

This paper is concerned with the diffusive expansion for solutions of the rescaled Boltzmann equation

$$\epsilon\partial_t F^\epsilon + v\cdot\nabla_x F^\epsilon = \frac{1}{\epsilon}Q(F^\epsilon,F^\epsilon),\quad x\in\Omega\subset\mathbf{R}^N,\ v\in\mathbf{R}^N,\ t>0 \tag{1.1}$$

with prescribed initial data

$$F^\epsilon(t,x,v)|_{t=0} = F_0^\epsilon(x,v),\quad x\in\Omega\subset\mathbf{R}^N,\ v\in\mathbf{R}^N. \tag{1.2}$$

Here Ω is a subset of $\mathbf{R}^N$ and $N\geq 3$ is the space dimension. Although the physical space is three dimensional, we consider in this paper the general space dimension $N\geq 3$ to show how the space dimension plays in the analysis.

Our main purpose is to justify the global-in-time validity of the diffusive expansion

$$F^\epsilon(t,x,v) = \mu + \sqrt{\mu}\left\{\sum_{j=1}^{n-1}\epsilon^j f_j(t,x,v) + \epsilon^n f_n^\epsilon(t,x,v)\right\} \tag{1.3}$$

for a solution $F^\epsilon(t,x,v)$ of the rescaled Boltzmann equation (1.1) for all $x\in\Omega$, $v\in\mathbf{R}^N$, and $t\geq 0$ with initial data $F_0^\epsilon(x,v)$ satisfying the initial

*The author is partly supported by NSFC (No. 10871151, No. 10925103) and "the Fundamental Research Funds for the Central Universities".

expansion

$$F_0^\epsilon(x,v) = \mu + \sqrt{\mu}\left\{\sum_{j=1}^{n-1}\epsilon^j f_j(0,x,v) + \epsilon^n f_n^\epsilon(0,x,v)\right\}, \quad x\in\Omega,\ v\in\mathbf{R}^N. \tag{1.4}$$

Here $\mu(v) = (2\pi)^{-\frac{N}{2}}\exp\left(-\frac{|v|^2}{2}\right)$ is a normalized global Maxwellian and $Q(f,g)$ is the collision operator taking the following form

$$\begin{aligned} Q(f,g) = &\int_{\mathbf{R}^N}\int_{\mathbf{S}^{N-1}}|v-u|^\gamma f(u')g(v')B(\theta)dud\omega \\ &-\int_{\mathbf{R}^N}\int_{\mathbf{S}^{N-1}}|v-u|^\gamma f(u)g(v)B(\theta)dud\omega, \end{aligned} \tag{1.5}$$

where $u' = u - [(u-v)\cdot\omega]\omega$, $v' = v + [(u-v)\cdot\omega]\omega$, γ and $B(\theta)$ are assumed to satisfy

$$-N < \gamma \le 1, \quad 0 \le B(\theta) \le c|\cos\theta|, \quad \cos\theta = \frac{(u-v)\cdot\omega}{|u-v|}. \tag{1.6}$$

Clearly, this is satisfied by the hard sphere gas with $\gamma = 1$ and by the inverse power law potential case under the Grad's cutoff assumption with $\gamma = 1 - \frac{2(N-1)}{s}$ satisfying $s > \frac{2(N-1)}{N+1}$. The case $0 \le \gamma \le 1$ is called hard potential, while the case $-N < \gamma < 0$ is called soft potential. Throughout this paper, we will only discuss the case of hard potential.

Recall that the linearized Boltzmann collision operator L satisfies (cf. [7])

$$Lg := -\frac{1}{\sqrt{\mu}}\{Q(\mu,\sqrt{\mu}g) + Q(\sqrt{\mu}g, g)\} = \nu(v)g - Kg,$$

where $\nu(v)$ is called the collision frequency which is given by

$$\nu(v) = \int_{\mathbf{R}^N}|v-u|^\gamma\mu(u)B(\theta)dud\omega = c\int_{\mathbf{R}^N}|v-u|^\gamma\mu(u)du \tag{1.7}$$

for some constant $c > 0$ and, as in [7], we can define the (non-symmetric) bilinear form $\Gamma[g_1,g_2]$ as

$$\Gamma[g_1,g_2] = \mu^{-\frac{1}{2}}(v)Q\left[\mu^{-\frac{1}{2}}g_1, \mu^{-\frac{1}{2}}g_2\right]. \tag{1.8}$$

In recent years the study on limiting process leading from the rescaled Boltzmann equation (1.1) of the classical kinetic theory of gases to the Navier-Stokes equations of incompressible fluids has been observed by many mathematicians and a lot of good results have been obtained (a complete literature in this direction is beyond the scope of this paper, however, we want to mention [2, 3, 4, 5, 6, 9, 10, 12] and the references cited therein). To go directly to the main points of the present paper, in what follows we only review some former results which are closely related to our main results:

- This limit was first discussed by Y. Sone in [12] in the steady case on the basis of formal asymptotic expansions and later by C. Bardos, F. Golse, and C. D. Levermore in [2, 3] in the time-dependent case by a systematic moment-closure method. The first complete mathematical proof of this limit is due to C. Bardos and S. Ukai [4] in the case of small initial data leading to smooth solutions; K. Asano [1] studied independently the same limit for short times;
- The rigorous proofs of convergence of DiPerna-Lions' renormalized solutions of (1.1) to the Leray-Hopf's weak solutions of the incompressible Navier-Stokes-Fourier equations was given by F. Golse and L. Saint-Raymond in [6];
- The problem on the validity of the diffusive expansion (1.3) was firstly studied by Y. Guo in the celebrated paper [9]. For the normalized global Maxwellian μ defined above, Y. Guo established the global-in-time validity of the diffusive expansion (1.3) for the solutions to the rescaled Boltzmann equation (1.1) inside a periodic box in three dimensions, i.e., $\Omega = \mathbf{T}^N$ and $N = 3$. For a solution to the rescaled Boltzmann equation (1.1), such an expansion yields a set of dissipative new partial differential equations such as the incompressible Navier-Stoke-Fourier system etc. Moreover the uniform estimate on the remainders is also obtained which guarantees the global-in-time validity of such an expansion up to any order.

Even so, since the analysis in [9] relies heavily on the Poincare inequality which does not hold in the whole space, i.e. for the case of $\Omega = \mathbf{R}^N$, the problem on whether the global validity of the diffusive expansion (1.3) holds for the solutions of the rescaled Boltzmann equation (1.1) with given initial data (1.2) in the whole space $\Omega = \mathbf{R}^N$ or not, to the best of our knowledge, seems to be unknown and this manuscript is devoted to this problem.

Roughly speaking, as in [9], we will be concentrated on the following two types of problems

- The first is to determine the coefficients $f_1(t,x,v), \cdots, f_{n-1}(t,x,v)$ in the diffusive expansion (1.3) by their initial data;
- The second is to prove boundedness of the remainder $f_n^\epsilon(t,x,v)$ in terms of its initial value, uniformly in ϵ and t. Such an estimate guarantees the validity of the diffusive expansion (1.3) in the large.

Recall that the linearized operator L is nonnegative and for any fixed (t,x), $\mathcal{N}(L)$, the null space of L, is generated by $\left[\sqrt{\mu}, v\sqrt{\mu}, v^2\sqrt{\mu}\right]$. Thus for any function $g(t,x,v)$, we can decompose it as follows

$$g = \mathbf{P}g + (\mathbf{I}-\mathbf{P})g,$$

where $\mathbf{P}g$ is the L^2_v-projection of g on $\mathcal{N}(L)$ for given (t,x) and we can further denote $\mathbf{P}g$ by

$$\mathbf{P}g = \left\{\rho_g(t,x) + v\cdot u_g(t,x) + \left(\frac{|v|^2}{2} - \frac{N}{2}\right)\theta_g(t,x)\right\}\sqrt{\mu}.$$

Here $\rho_g(t,x), u_g(t,x)$, and $\theta_g(t,x)$ represents the density, velocity, and temperature fluctuation physically respectively. For the velocity field $u_g(t,x)$, we further define its divergence-free part as $\mathbf{P}_0 u_g$ and it is easy to see that $\nabla\cdot\mathbf{P}_0 u_g \equiv 0$. Here

$$\begin{cases} \mathbf{P}_0 u = \mathcal{F}^{-1}[p_0(\xi)\mathcal{F}[u](\xi)], \\ p_0(\xi) = \left(\delta_{j,k} - \dfrac{\xi_j\xi_k}{|\xi|^2}\right)_{N\times N}, \end{cases}$$

and $\mathcal{F}$, $\mathcal{F}^{-1}$ denote the Fourier transform and inverse Fourier transform respectively.

Now we show how to determine $f_1(t,x.v),\cdots,f_{n-1}(t,x.v)$ in (1.3). For this purpose, we define $[\rho_m(t,x), u_m(t,x), \theta_m(t,x)]$ to be the corresponding hydrodynamic field of the mth-coefficient $f_m(t,x,v)$. Since $f_1(t,x,v)$ contains only the macroscopic (or fluid) component, its velocity fluctuation $u_1(t,x)$ is incompressible

$$\nabla\cdot u_1 \equiv 0 \quad \text{or} \quad u_1 = \mathbf{P}_0 u_1 \tag{1.9}$$

and its density and temperature fluctuation $\rho_1(t,x), \theta_1(t,x)$ satisfy the Boussinesq relation

$$\rho_1 + \theta_1 \equiv 0. \tag{1.10}$$

Moreover $[u_1(x), \theta_1(x)]$ satisfies the celebrated incompressible Navier-Stokes-Fourier equations

$$\partial_t u_1 + u_1\cdot\nabla u_1 + \nabla p_1 = \eta\triangle u_1, \tag{1.11}$$

$$\partial_t\theta_1 + u_1\cdot\nabla\theta_1 = \kappa\triangle\theta_1. \tag{1.12}$$

Here $p_1(t,x)$ is the pressure, η the viscosity and κ the heat conductivity.

As to $f_m(t,x,v)$ for $2\le m\le n$, its microscopic component of $f_m(t,x,v)$ is determined by

$$(\mathbf{I}-\mathbf{P})f_m = L^{-1}\left\{-\partial_t f_{m-2} - v\cdot\nabla_x f_{m-1} + \sum_{\substack{i+j=m\\ i,j\ge 1}}\Gamma(f_i,f_j)\right\}, \tag{1.13}$$

$(\mathbf{I}-\mathbf{P}_0)u_m$ can be obtained from the mth-order incompressibility condition

$$\nabla\cdot(\mathbf{I}-\mathbf{P}_0)u_m=-\partial_t\rho_{m-1}, \tag{1.14}$$

and ρ_m, $\mathbf{P}_0u_m,\theta_m$ can be determined by solving the mth-order Boussinesq relation

$$\begin{aligned}\rho_m+\theta_m=&\triangle^{-1}\nabla\cdot\{-u_1\cdot\nabla\{\mathbf{P}_0u_{m-1}\}-\mathbf{P}_0u_{m-1}\cdot\nabla u_1+R^u_{m-1}\}\\&+\frac{2-\delta_{1,(m-1)}}{N}u_{m-1}\cdot u_1,\end{aligned} \tag{1.15}$$

the mth-order linear Navier-Stokes-Fourier system for $[\mathbf{P}_0u_m,\theta_m]$

$$(\partial_t+u_1\cdot\nabla-\eta\triangle)\mathbf{P}_0u_m+\mathbf{P}_0u_m\cdot\nabla u_1+\nabla p_m=R^u_m, \tag{1.16}$$
$$(\partial_t+u_1\cdot\nabla-\kappa\triangle)\theta_m+\mathbf{P}_0u_m\cdot\nabla\theta_1=R^\theta_m \tag{1.17}$$

with prescribed initial data

$$(\mathbf{P}_0u_m(0,x),\theta_m(0,x))=\left(u^0_m(x),\theta^0_m(x)\right). \tag{1.18}$$

Here

$$\left\{\begin{aligned}R^u_m=&\left\langle v\cdot\nabla_xL^{-1}\left\{\partial_t\{\mathbf{I}-\mathbf{P}\}f_{m-1}-\sum_{\substack{i+j=m\\i,j>1}}\Gamma(f_i,f_j)\right\},v\sqrt{\mu}\right\rangle\\&+\langle v\cdot\nabla_xL^{-1}\{\mathbf{I}-\mathbf{P}\}(v\cdot\nabla_x(\mathbf{I}-\mathbf{P})f_m),v\sqrt{\mu}\rangle\\&-\langle v\cdot\nabla_xL^{-1}\{\Gamma(f_1,\{\mathbf{I}-\mathbf{P}\}f_m)+\Gamma(\{I-P\}f_m,f_1)\},v\sqrt{\mu}\rangle\\&-(\partial_t+u_1\cdot\nabla-\eta\triangle)\{\mathbf{I}-\mathbf{P}_0\}u_m-\{\mathbf{I}-\mathbf{P}_0\}u_m\cdot\nabla u_1\\&-(\nabla\cdot\{\mathbf{I}-\mathbf{P}_0\}u_m)u_1+\frac{N-2}{N}\nabla(\nabla\cdot\{\mathbf{I}-\mathbf{P}_0\}u_m),\\R^\theta_m=&\left\langle v\cdot\nabla_xL^{-1}\left\{\partial_t\{\mathbf{I}-\mathbf{P}\}f_{m-1}-\sum_{\substack{i+j=m\\i,j>1}}\Gamma(f_i,f_j)\right\},\frac{|v|^2\sqrt{\mu}}{N+2}\right\rangle\\&+\left\langle v\cdot\nabla_xL^{-1}\{\mathbf{I}-\mathbf{P}\}(v\cdot\nabla_x(\mathbf{I}-\mathbf{P})f_m),\frac{|v|^2\sqrt{\mu}}{N+2}\right\rangle\\&-\left\langle v\cdot\nabla_xL^{-1}\{\Gamma(f_1,\{\mathbf{I}-\mathbf{P}\}f_m)+\Gamma(\{\mathbf{I}-\mathbf{P}\}f_m,f_1)\},\frac{|v|^2\sqrt{\mu}}{N+2}\right\rangle\\&-\frac{2}{N+2}\partial_t\{\rho_m+\theta_m\}-(\nabla\cdot\{\mathbf{I}-\mathbf{P}_0\}u_m)\theta_1-\{\mathbf{I}-\mathbf{P}_0\}u_m\cdot\nabla\theta_1,\\p_m=&\rho_{m+1}+\theta_{m+1}-\frac{2-\delta_{1,m}}{N}(u_m\cdot u_1),\end{aligned}\right.$$

and $\delta_{i,j}$ is the Kronecker symbol.

It is easy to see that R^u_m and θ^θ_m depend only on $f_j(t,x,v)$ for $j\le m-1$ since $(\mathbf{I}-\mathbf{P})f_m(t,x,v)$, $(\mathbf{I}-\mathbf{P}_0)u_m(t,x)$, as well as $\rho_m(t,x)+\theta_m(t,x)$ have been determined by (1.13), (1.14), and (1.15).

For $m=n+1,n+2$, $f_m(t,x,v)$ are given by

$$f_m = L^{-1}\left\{-\partial_t f_{m-2} - v\cdot\nabla_x f_{m-1} + \sum_{i+j=m,i,j\ge 1}\Gamma(f_i,f_j)\right\}. \quad (1.19)$$

Notice that $f_{n+1}(t,x,v)$ and $f_{n+2}(t,x,v)$ contain only the non-fluid component.

Finally, the remainder $f^\epsilon_n(t,x,v)$ solves

$$\begin{aligned}
&\epsilon^2\partial_t f^\epsilon_n + \epsilon v\cdot\nabla_x f^\epsilon_n + Lf^\epsilon_n\\
&= \left\{-\partial_t f_{n-2} - v\cdot\nabla_x f_{n-1} + \sum_{\substack{i+j=n\\ i,j\ge 1}}\Gamma(f_i,f_j)\right\}\\
&\quad +\epsilon\left\{-\partial_t f_{n-1} + \sum_{\substack{i+j=n+1\\ 1\le i,j\le n-1}}\Gamma(f_i,f_j)\right\} + \epsilon^n\Gamma(f^\epsilon_n,f^\epsilon_n)\\
&\quad +\sum_{i=1}^{n-1}\epsilon^i\left\{\Gamma(f^\epsilon_n,f_i)+\Gamma(f_i,f^\epsilon_n)\right\}\\
&\quad +\sum_{i+j\ge n+2}\epsilon^{i+j-n}\left\{\Gamma(f_i,f_j)+\Gamma(f_j,f_i)\right\}. \qquad (1.20)
\end{aligned}$$

Before concluding this section, we collect some notations which are used throughout this paper.

For notational simplicity, we use $\langle\cdot,\cdot\rangle$ to denote the L^2_v-inner product in $\mathbf{R}^N_v$ with its L^2 norm given by $|\cdot|$ and $(\cdot,\cdot)$ stands for L^2-inner product either in $\mathbf{R}^N_x\times\mathbf{R}^N_v$ or in $\mathbf{R}^N_x$ with corresponding L^2 norm denoted by $\|\cdot\|$. L^p is the usual Lebesgue space with its norm denoted by $\|\cdot\|_p$ and H^s represent for the usual Sobolev space.

It is natural to define the following weighted L^2 norm to characterize the dissipation rate

$$|g|^2_\nu = \int_{\mathbf{R}^N} g^2(v)\nu(v)dv, \quad \|g\|^2_\nu = \int_{\mathbf{R}^N\times\mathbf{R}^N} g^2(x,v)\nu(v)dxdv,$$

and a weight function $w(v)$ is defined as $w(v)=\left(1+|v|^2\right)^{\frac12}$.

To be consistent with the hydrodynamic equations, we define $\partial^\beta_\alpha = \partial^{\alpha_1}_{x_1}\cdots\partial^{\alpha_N}_{x_N}\partial^{\beta_1}_{v_1}\cdots\partial^{\beta_N}_{v_N}$ and $\partial_\tau=\partial^{\tau_0}_t\partial^{\tau_1}_{x_1}\cdots\partial^{\tau_N}_{x_N}$. Here $\alpha=[\alpha_1,\cdots,\alpha_N]$ is related to the space derivatives, while $\beta = [\beta_1, \cdots, \beta_N]$ is related to the velocity derivatives, and $\tau = [\tau_0, \tau_1, \cdots, \tau_N]$ is related to the space-time derivatives.

For the sake of capturing the structure of the rescaled Boltzmann equation (1.1) and getting the uniform estimates on the remainder term, we introduce the instant energy functionals $\mathcal{E}_{s,l}$ and the dissipate rate $\mathcal{D}_{s,l}$ as follows for $s \geq N+1, l \geq 0$:

$$\mathcal{E}_{s,l}(g) \sim \sum_{|\alpha| \leq s} \|\partial_\alpha g\|^2 + \sum_{|\alpha|+|\beta| \leq s} \left\|w^l \partial_\alpha^\beta g\right\|^2$$

and

$$\mathcal{D}_{s,l}(g) = \sum_{0<|\alpha| \leq s} \|\partial_\alpha \mathbf{P} g\|^2 + \sum_{|\alpha| \leq s} \frac{1}{\epsilon^2} \|\partial_\alpha (\mathbf{I} - \mathbf{P}) g\|_\nu^2 + \frac{1}{\epsilon^2} \sum_{|\alpha|+|\beta| \leq s} \left\|w^l \partial_\alpha^\beta (\mathbf{I} - \mathbf{P}) g\right\|_\nu^2.$$

Furthermore in order to get the estimates on the diffusive coefficients in (1.3), we define the following norms and semi-norms:

$$\widetilde{\mathcal{E}}_{s,l}(g) \sim \sum_{|\tau| \leq s} \|\partial_\tau \mathbf{P} g\|^2 + \sum_{|\tau| \leq s} \left\|w^l \partial_\tau^\beta (\mathbf{I} - \mathbf{P}) g\right\|^2$$

and

$$\widetilde{\mathcal{D}}_{s,l}(g) = \sum_{0<|\tau| \leq s} \|\partial_\tau \mathbf{P} g\|^2 + \sum_{|\tau| \leq s} \left\|w^l \partial_\tau^\beta (\mathbf{I} - \mathbf{P}) g\right\|_\nu^2.$$

Moreover, for $T \geq t \geq 0$, we define

$$Z_{s,l}(g)(T) = \|g\|_{Z_{s,l}} = \sup_{0 \leq t \leq T} \widetilde{\mathcal{E}}_{s,l}(g)(t) + \int_0^T \widetilde{\mathcal{D}}_{s,l}(g)(t) dt.$$

And we denote $\widetilde{\mathcal{E}}_{s,0} = \widetilde{\mathcal{E}}_s$, $\widetilde{\mathcal{D}}_{s,0} = \widetilde{\mathcal{D}}_s$, $Z_{s,0} = Z_s$.

If g is independent of v, we can define similar norms or semi-norms as in the following:

$$\widetilde{\mathcal{E}}_s(g) \sim \sum_{|\tau| \leq s} \|\partial_\tau g\|^2$$

and

$$\widetilde{\mathcal{D}}_s(g) = \sum_{0<|\tau| \leq s+1} \|\partial_\tau g\|^2.$$

Furthermore, for $T \geq t \geq 0$,

$$Z_s(g)(T) = \|g\|_{Z_s} = \sup_{0 \leq t \leq T} \widetilde{\mathcal{E}}_s(g)(t) + \int_0^T \widetilde{\mathcal{D}}_s(g)(t) dt.$$

Remark 1. Some remarks concerning the instant energy and dissipation rate are listed below

i) For the case of hard sphere model, it is unnecessary to introduce the weight function $w(v)$, while for hard potential case with $0 \geq \gamma < 1$ the weight function $w(v)$ must be introduced since it is impossible to control v-derivatives ∂^β of the linear streaming term $v_j \partial^j f$ in term of the weaker term $\|\partial^\beta f\|_\nu$.

ii) Noticing that, unlike the corresponding problem inside the periodic box discussed in [9], the zeroth-order term of macroscopic part $\mathbf{P}g$ does't appear in $\mathcal{D}_{s,l}(g)$ and consequently we can not compare the instant energy with the dissipation rate, which cause some difficulties to perform the energy estimates for the remainder and to study the large time behaviors of the solutions. It is the very reason why we can not deduce a decay estimate in our main result.

iii) There is no $\frac{1}{\epsilon^2}$ in front of hydrodynamic component $\mathbf{P}g$ in the dissipate rate $\mathcal{D}_{s,l}(g)$, since only the microscopic component $(\mathbf{I} - \mathbf{P})g$ should vanish as $\epsilon \to 0$.

2 Main results

Now we turn to state our main results. The first result is on how to determine the coefficients $f_1(t,x,v), \cdots, f_m(t,x,v)$ in the diffusive approximation (1.3).

Theorem 1. Let $M_0 > 0$, $s_r \geq N+1, 2 \leq r \leq m$, given m divergence-free, vector-valued functions $\left[u_1^0(x), \cdots, u_m^0(x)\right]$ and m scalar functions $\left[\theta_1^0(x), \cdots, \theta_m^0(x)\right]$ such that

$$\begin{cases} \|u_1^0\|_{H^{2s_1}} + \|\theta_1^0\|_{H^{2s_1}} \leq M_0, & s_1 \geq N+m, \\ u_r^0(x) \in H^{2s_r+2(m-r)}\left(\mathbf{R}^N\right), & \theta_r^0(x) \in H^{2s_r+2(m-r)}\left(\mathbf{R}^N\right). \end{cases} \tag{2.1}$$

Then if we assume further that $N > 2m-4$ $(m > 2)$ and M_0 is sufficiently small, there exist unique functions $f_1(t,x,v), \cdots, f_m(t,x,v)$ such that

$$\mathbf{P}_0 u_r(0,x) = u_r^0(x), \quad \theta_r(0,x) = \theta_r^0(x),$$

and $f_1(t,x,v)$ satisfies (1.9), (1.10), (1.11), and (1.12), $f_r(t,x,v)$ satisfies (1.13), (1.14), (1.15), (1.16), (1.17) and (1.18) for $2 \leq r \leq m$.

Moreover, for $1 \leq r \leq m$, $s_r \geq N+1$, and each $l \geq 0$, there exists a polynomial U_{r,s_r} with $U_{r,s_r}(0) = 0$ such that

$$\|f_r(t,x,v)\|_{Z_{s_r,l}} \leq U_{r,s_r}\left(\sum_{1\leq j\leq r} \left\|\left(u_j^0, \theta_j^0\right)\right\|_{H^{2s_r+2(r-j)}}\right). \tag{2.2}$$

In addition, if $N > 4k + 2r - 2|\alpha| - 4$ $(k = 1, 2, \cdots)$, $k' \geq k$, we have

$$\begin{aligned} &Z_{s_{2k+r}}\left(\triangle^{-k}\partial_x^\alpha\partial_t^{k'}(\rho_r, u_r, \theta_r)\right) \\ \leq &U_{r,s_{2k+r}}\left(\sum_{i=1}^{r}\|(u_i^0, \theta_i^0)\|_{H^{2s_{2k+r}+2k'+2(r-i)}}\right). \end{aligned} \tag{2.3}$$

Remark 2. For the case of $N = 3$, we can only determine the coefficients $f_1(t, x, v)$ and $f_2(t, x, v)$. In order to deduce the desired estimates on the higher order coefficients $f_r(t, x, v)$ for $r \geq 3$ by exploiting the L^p-L^q-estimate on the Riesz potential, we need to ask that N, the spatial dimension, is suitable large, i.e. $N > 2r - 4$. Since we have to determine the $f_n(t, x, v)$ and to get the estimates on $\triangle^{-1}\nabla\partial_t(\rho_n, u_n, \theta_n)$ in order to obtain the uniform estimates on $f_n^\epsilon(t, x, v)$, we thus require that $N > 2n - 2$.

Remark 3. Notice that there is a loss of derivatives in this theorem, in order to construct $f_r(t, x, v)$ for $r \geq 2$, more regularity assumption should be imposed on $u_1^0(x), \theta_1^0(x)$. From (2.2), we can see that $s_1 \geq N + n + 2$.

Remark 4. Interestingly, the coefficients $f_1(t, x, v), \cdots, f_m(t, x, v)$ are determined uniquely by the divergence-free part of the velocity field only and by the temperature field of the initial distribution function $f_r(0, x, v)$. Except for $[u_1^0(x), \theta_1^0(x)]$, $[u_r^0(x), \theta_r^0(x)]$ can have arbitrarily large amplitude for $r \geq 2$.

Now we turn to the problem on the uniform estimates on the remainder $f_n^\epsilon(t, x, v)$. For results in this direction, we have

Theorem 2. Let $s \geq N+1$, $N > 2n-2$, given $f_1(t, x, v), \cdots, f_{n+2}(t, x, v)$ as constructed above, and assume

$$F^\epsilon(0, x, v) = \mu + \sqrt{\mu}\left\{\sum_{j=1}^{n-1}\epsilon^j f_j(0, x, v) + \epsilon^n f_n^\epsilon(0, x, v)\right\} \geq 0.$$

If there is a sufficiently small positive constant $M_1 > 0$ such that

$$\mathcal{E}_{s,l}(f_n^\epsilon - f_n)(0) \leq M_1,$$

then (1.1) has a unique global smooth solution $F^\epsilon(t, x, v)$ which satisfies

$$F^\epsilon(t, x, v) = \mu + \sqrt{\mu}\left\{\sum_{j=1}^{n-1}\epsilon^j f_j(t, x, v) + \epsilon^n f_n^\epsilon(t, x, v)\right\} \geq 0$$

and

$$\sup_{0\leq t\leq\infty}\mathcal{E}_{s,l}(f_n^\epsilon - f_n)(t) \leq e^{C\mathcal{U}_n}\left(\mathcal{E}_{s,l}(f_n^\epsilon - f_n)(0) + \epsilon^2\mathcal{U}_n\right). \tag{2.4}$$

Here $C > 0$ is a positive constant independent of t and x,

$$\mathcal{U}_n = U_{n,s}\left(\sum_{1\leq j\leq n}\left\|\left(u_j^0,\theta_j^0\right)\right\|_{H^{2s+2(n-j)+4}}\right),$$

and as in Theorem 1, $U_{n,s}(z)$ is a polynomial satisfying $U_{n,s}(0) = 0$.

Now we outline the main ideas used in this paper. The derivation of the Navier-Stokes-Fourier system (1.9), (1.10), (1.11), (1.12), (1.14), (1.15), (1.16), and (1.17) from the rescaled Boltzmann equation (1.1) follows essentially the same way as in [9] for the case of $N = 3$ provided that $f_1(t,x,v), \cdots, f_{n-1}(t,x,v)$ and $f_n^\epsilon(t,x,v)$ satisfy certain a priori estimates. To construct the diffusive coefficients $f_1(t,x,v), \cdots, f_n(t,x,v)$ rigorously, since (ρ_1, u_1, θ_1) solves the incompressible Navier-Stokes-Fourier equations (1.9), (1.10), (1.11) and (1.12), from the well-established results concerning the global smooth solvability of small amplitude solutions to the incompressible Navier-Stokes-Fourier equations (1.9), (1.10), (1.11), (1.12) (cf. [14] and the references cited therein), we can therefore determine $f_1(t,x,v)$ easily in Theorem 1. But for $m \geq 2$, $u_m(t,x)$ is no longer a divergence-free vector, i.e.

$$\nabla \cdot (\mathbf{I} - \mathbf{P}_0)u_m = -\partial_t \rho_{m-1} \neq 0$$

for $m \geq 2$. Thus to deduce the desired estimates on $(\mathbf{I} - \mathbf{P}_0)u_m(t,x)$ in the energy space, one had to deal with the problem on the gradient estimate on solutions to the Poisson equation. For the periodic initial-boundary value problem considered in [9], such an estimate is easy to deduce since the Poincare inequality holds. For the corresponding problem in the whole space discussed in this paper, since the Poincare type inequality does not hold any longer, it is difficult to get the desired L^2-estimates on $(\mathbf{I} - \mathbf{P}_0)u_m(t,x)$ for $m \geq 2$. To overcome this difficulty, we use the iteration argument to replace the term $\partial_t \rho_{m-1}$ $(m \geq 2)$ by the summation of some nonlinear terms and some terms with higher regularity. Consequently, we have to deal with the operator $\triangle^{-k}\partial^\alpha$ $(k > 0)$ which has high singularity and such a problem can be overcome by applying the L^p-L^q-estimate on the Riesz potential, which seems to be an effective tool to avoid the singularities. But as a compensation, the above analysis requires that the spatial dimension N should be large enough. It is worth to pointing out that, for $m \geq 3$, it was to deduce certain desired estimates on $p_m(t,x)$ by the L^p-L^q-estimate on the Riesz potential that we need to assume that $N > 2n - 2$.

Our analysis is also based on a nonlinear energy method for the Boltzmann equation developed recently in [8, 10] with a slight modification. The most important analytical difficulty lies in deriving the uniform estimates on $f_n^\epsilon(t,x,v)$ independent of ϵ and time variable t. For this purpose, the positivity of the collision operator L plays an essential role. We note, however, that due to the singular behavior of the time derivative in our problem, the positivity estimate for purely spatial derivative is invoked. For the periodic initial-boundary value problem discussed in [9], macroscopic equations and local conservation laws are proven to be the key tools to build the crucial positivity of the collision operator L with respect to the solution of the Boltzmann equation but not just the microscopic component only. But for the corresponding Cauchy problem, it seems that we can not hope such a strong positivity estimate on L holds since the Poincare's inequality does not hold any longer. In fact for the space-time L^2-estimates on the macroscopic component $\mathbf{P}f_n^\epsilon$ of $f_n^\epsilon(t,x,v)$, only its derivative but not $\mathbf{P}f_n^\epsilon$ itself, can be estimated from the macroscopic equations. Consequently, the corresponding dissipation rate $\mathcal{D}_{s,l}(f_n^\epsilon)$ is weaker than that in the periodic box. In such a case, to use the energy method to deduce certain estimates on the remainder $f_n^\epsilon(t,x,v)$, we need to estimate the nonlinear term $\Gamma(f_n^\epsilon, f_n^\epsilon)$ carefully via such a weaker $\mathcal{D}_{s,l}(f_n^\epsilon)$ which does not contain the space-time L^2-norm of $\mathbf{P}f_n^\epsilon$. Motivated by [8, 15], we split $f_n^\epsilon(t,x,v)$ as $f_n^\epsilon = \mathbf{P}f_n^\epsilon + (\mathbf{I}-\mathbf{P})f_n^\epsilon$ and then try to bound the nonlinear term $\Gamma(f_n^\epsilon, f_n^\epsilon)$ by the $L^{\frac{2N}{N-2}}$-norm of $(a_\epsilon(t,x), b_\epsilon(t,x), c_\epsilon(t,x))$ and some other estimates on $(\mathbf{I}-\mathbf{P})f_n^\epsilon$. Here we have used the following projection

$$\mathbf{P}f_n^\epsilon = \{a_\epsilon(t,x) + b_\epsilon(t,x)\cdot v + c_\epsilon(t,x)|v|^2\}\sqrt{\mu}.$$

Note that Sobolev's inequality tells us that

$$\|(a_\epsilon, b_\epsilon, c_\epsilon)\|_{L^{\frac{2N}{N-2}}(\mathbf{R}^N)} \leq \|\nabla_x(a_\epsilon, b_\epsilon, c_\epsilon)\|_{L^2(\mathbf{R}^N)},$$

and finally the nonlinear term can be suitably controlled by the entropy dissipation $\mathcal{D}_{s,l}(f_n^\epsilon)$.

References

[1] K. Asano, On the global solutions of the initial-boundary value problem for the Boltzmann equaiton with an external force. Transp. Theory Stat. Phys. 16 (1987), no. 2–4, 735–761.

[2] C. Bardos, F. Golse and C. D. Levermore, Fluid dynamic limits of the kinetic equation. I. Formal derivation. J. Stat. Phys. 63 (1991), no. 1–2, 323–344.

[3] C. Bardos, F. Golse and C. D. Levermore, Fluid dynamic limits of the kinetic equation. II. Convergence proofs for the Boltzmann equation. Commun. Pure Appl. Math. 46 (1993), no. 5, 667–753.

[4] C. Bardos and S. Ukai, The classical incompressible Navier-Stokes limit of the Boltzmann equation. Math. Models Methods Appl. Sci. 1 (1991), no. 2, 235–257.

[5] A. De Masi, R. Esposito and J. L. Lebowitz, Incompressible Navier-Stokes and Euler limits of the Boltzmann equation. Commun. Pure Appl. Math. 42 (1989), no. 8, 1189–1214.

[6] F. Golse and L. Saint-Raymond, The Navier-Stokes limit of the Boltzmann equation for bounded collision kernels. Invent. Math. 155 (2004), no. 1, 81–161.

[7] H. Grad, Asymptotic Theory of the Boltzmann Equation. II. Rarefied Gas Dynamics, Vol. I, 26–59, Academic Press, New York, 1963.

[8] Y. Guo, The Boltzmann equation in the whole space. Indiana. Univ. Math. J. 53 (2004), no. 4, 1081–1094.

[9] Y. Guo, Boltzmann diffusive limit beyond the Navier-Stokes approximation. Commun. Pure Appl. Math. 55 (2006), no. 9, 626–687.

[10] T.-P. Liu, T. Yang and S.-H. Yu, Energy method for the Boltzmann equation. Physica D 188 (2004), no. 3–4, 178–192.

[11] S.-Q. Liu and H.-J. Zhao, Diffusive expansion for solutions of the Boltzmann equation in the whole space. J. Differential Equations, in press.

[12] Y. Sone, Asymptotic Theory of Flow of a Rarefied Gas over a Smooth Boundary II. In: Rarefied Gas Dynamics. Vol. II, 737–749, ed. by D. Dini. Pisa: Editrice Tecnico Scientifica, 1971.

[13] E. M. Stein, Singular Integrals and Differentiability Properties of Functions. Princeton University Press, Princeton, New Jersey, 1979.

[14] R. Teman, Navier-Stokes Equations: Theory and Numerical Analysis. AMS Chelsea Publishing, American Mathematical Society, Providence, Rhode Island, 2001.

[15] T. Yang and H.-J. Zhao, A new energy method for the Boltzmann equation. J. Math. Phys. 47 (2006), no. 5, 053301.

Contributed Talks

A Nonlocal Conservation Law from a Model of Granular Flow

Debora Amadori
Department of Pure & Applied Mathematics, University of L'Aquila
L'Aquila, I-67010 Italy
E-mail: amadori@univaq.it

Wen Shen
Department of Mathematics, Penn State University
University Park, PA 16802 USA
E-mail: shen_w@math.psu.edu

Abstract

In this paper we study the well-posedness for a scalar conservation law in which the flux term is non-local in space.

This equation represents a reduced model for slow erosion in granular flow ([1, 6]) and describes roughly the evolution of a profile of stationary matter, under the effect of a thin moving layer of granular matter on the top of it.

We show that the present equation admits weak solutions existing globally in time and prove their stability w.r.t the initial data. These properties are related to the assumption on the erosion flux. Different assumptions may lead to significantly different behaviors, see [9].

1 Introduction

We consider the scalar integro-differential equation

$$q_t + (K(q)\, f(q))_x = 0, \tag{1.1}$$

where

$$K(q(t,\cdot))(x) = \exp\left\{ \int_x^0 f(q(t,\xi))\, d\xi \right\}, \tag{1.2}$$

for $(t,x) \in [0,T] \times \mathbb{R}_-$ and

$$\begin{cases} f : (-1, +\infty) \to \mathbb{R}, \quad f(0) = 0, \\ f' > 0, \quad f'' < 0 \\ \lim_{q \to -1} f(q) = -\infty, \qquad \lim_{q \to +\infty} \dfrac{f(q)}{q} = 0. \end{cases} \tag{1.3}$$

We are interested in the initial-boundary value problem for (1.1) on the domain $\mathbb{R}_+ \times \mathbb{R}_-$, with initial condition

$$q(0,x) = q_0(x) > -1, \quad x < 0 \tag{1.4}$$

and no boundary condition at $x = 0$.

This equation arises as a *slow erosion limit* in a model of granular flow, studied in [1]. In more detail, let h be the height of the moving layer, u the height of the standing layer and $p = u_x$ be its slope. Assume that $p > 0$. The granular flow in one space dimension is described by the following 2×2 system of balance laws (originally proposed in [6])

$$\begin{cases} h_t - (hp)_x = (p-1)h, \\ p_t + \big((p-1)h\big)_x = 0. \end{cases} \tag{1.5}$$

As the moving layer becomes very thin, i.e., as $\|h\|_{\mathbf{L}^\infty} \to 0$, we proved in [1] that the solution for the slope p in (1.5) provides the weak solution of the following scalar integro-differential equation

$$p_\mu + \left(\frac{p-1}{p} \cdot \exp \int_x^0 \frac{p(\mu,y)-1}{p(\mu,y)}\, dy\right)_x = 0.$$

Here, the new time variable $\mu(t) = \int_0^t p(\tau,0)h(\tau,0)\,d\tau$ accounts for the total mass of granular material being poured from above. Recognizing $p = 1$ as the equilibrium slope, we introduce $q \doteq p-1$. Using the variable t in place of μ, we obtain the equation (1.1) with

$$f(q) = \frac{q}{q+1}$$

that clearly satisfies (1.3).

In this paper we establish the well-posedness result for equation (1.1). In order to define entropy weak solutions, we recall some results in [1]. From Theorem 2 in [1], for h sufficiently small, we proved that the solution for p of (1.5) satisfies the following bounds for all $t \geq 0$, uniformly in h:

$$p(t,x) \geq p_0 > 0, \quad \text{TV } p(\cdot) \leq M, \quad \|p-1\|_{\mathbf{L}^1(\mathbb{R}_-)} \leq M. \tag{1.6}$$

This leads to the state of the solutions $q = p - 1$ for (1.1), on a domain $[0,T] \times \mathbb{R}_-$, for a given $T > 0$. To be precise, let C_0, p_0 be some positive constants and define $\mathcal{D}_{C_0,p_0}$ as

$$\mathcal{D}_{C_0,p_0} = \left\{ q(x) : \inf_{x<0} q(x) + 1 \geq p_0,\ \text{TV } q \leq C_0,\ \|q\|_{\mathbf{L}^1(\mathbb{R}_-)} \leq C_0 \right\}. \tag{1.7}$$

Note that for any $q \in \mathcal{D}$, the *local* term $f(q)$ of the flux satisfies $Kf' > 0$, i.e., the characteristic speed of equation (1.1) is positive. Therefore no boundary condition is posed at $x = 0$.

We now state a natural definition of an entropy weak solution of (1.1)–(1.4) on a domain $[0,T] \times \mathbb{R}_-$, with $T > 0$.

Definition 1.1. For some $C_0 > 0$ and $p_0 > 0$, let $q_0 \in \mathcal{D}_{C_0,p_0}$ (see (1.7)). We say that q provides an **entropy weak solution** to (1.1)–(1.2) on $[0,T] \times \mathbb{R}_-$ with initial condition (1.4), if the following conditions are satisfied:

(H1) $q : [0,T] \to \mathcal{D}_{C_1,p_1}$ for some $C_1 > 0$, $p_1 > 0$; the map $[0,T] \ni t \mapsto q(t)$ is Lipschitz in $\mathbf{L}^1(\mathbb{R}_-)$.

(H2) q is a weak entropy solution of the scalar conservation law

$$\begin{cases} q_t + (k(t,x)\, f(q))_x = 0, \\ q(0,x) = q_0(x) \end{cases} \tag{1.8}$$

where the coefficient k is defined by

$$k(t,x) = K(q(t,\cdot))(x) = \exp\left\{\int_x^0 f(q(t,\xi))\, d\xi\right\}. \tag{1.9}$$

The main result of the paper is the uniqueness and well-posedness of the solution for (1.1), as stated in the following Theorem.

Theorem 1.2. *Let C_0, p_0 be given positive constants. Then for any initial data $q_0 \in \mathcal{D}_{C_0,p_0}$ there exists a solution $q(t,x)$ to the initial-boundary value problem* (1.1) *on $x < 0$, $t \geq 0$ with the following properties:*

(i) *for all $t \geq 0$, $\inf_{x<0} q(t,x) + 1 \geq p_0 > 0$;*

(ii) *$q \in \mathbf{L}^\infty([0,+\infty) \times \mathbb{R}_-)$;*

(iii) *for all $t \geq 0$, $q(t,\cdot) \in \mathbf{L}^1(\mathbb{R}_-) \cap BV(\mathbb{R}_-)$.*

Moreover, consider two solutions $q_0(t,\cdot)$, $q_1(t,\cdot)$ of the integro-differential equation (1.1)*, corresponding to the initial data $\bar{q}_0$, $\bar{q}_1 \in \mathcal{D}_{C_0,p_0}$ respectively. Then for any $T > 0$ there exists $L = L(T, C_0, p_0) > 0$ such that, for $t \in [0,T]$*

$$\|q_0(t,\cdot) - q_1(t,\cdot)\|_{\mathbf{L}^1(\mathbb{R}_-)} \leq e^{Lt}\, \|\bar{q}_0 - \bar{q}_1\|_{\mathbf{L}^1(\mathbb{R}_-)}. \tag{1.10}$$

Recalling that $q = p - 1 = u_x - 1$, the solution q established by Theorem 1.2 allows us to reconstruct the profile u of the standing layer:

$$u(t,x) - x = \int_{-\infty}^x q(t,y)\, dy. \tag{1.11}$$

Moreover an equation for u is deduced as follows. From (1.2) we have

$$K_x = -Kf(q(t,x)),$$

so that equation (1.1) can be rewritten as $q_t - K_{xx} = 0$.

By integrating in space, using (1.11) and noticing that $K_x(q(t,\cdot)) \in \mathbf{L}^1(\mathbb{R}_-)$, we arrive at

$$u_t - K_x = u_t + Kf(u_x - 1) = 0.$$

This is a nonlocal Hamilton-Jacobi equation for the unknown u. An analysis for this type of equation has been carried out in [9] for a different class of functions f, that allows for singularities on the standing profile.

More precisely, the main difference between Theorem 1.2 and the results in [9] is due to the different behavior of f at $+\infty$. Indeed a slower increase of f as

$$\frac{f(q)}{q} \to 0 \quad \text{as } q \to +\infty, \tag{1.12}$$

assumed in (1.3), yields to globally bounded solutions. On the other hand, the faster behavior assumed in [9] requires that the above limit is positive and leads to blow-up of the solutions, that renders more convenient to study the equation in terms of the antiderivative u.

To be more specific, for both problems it is possible to derive an apriori, global bound on the quantity

$$\frac{q(t,x)}{f(q(t,x))}.$$

Thanks to (1.12), the function $q/f(q)$ diverges as $q \to +\infty$. From this one deduces that $q(t,x)$ must be globally bounded.

Other problems involving a nonlocal term in the flux have been studied in [5, 3, 4]. In the rest of the paper we review the proof of Theorem 1.2, whose details can be found in [2].

2 Well-posedness for the equation with a given $k(t,x)$

Towards the proof of Theorem 1.2, we study the equation with a given local coefficient $k(t,x)$

$$u_t + (k(t,x)f(u))_x = 0, \quad x \le 0, \quad t \ge 0 \tag{2.1}$$

where $k(t,x)$ satisfies the following assumptions, for some given $T > 0$:

(K1) $k(t,x) \in \mathbf{L}^\infty([0,T] \times \mathbb{R}_-)$, is Lipschitz continuous and $\inf_{t,x} k > 0$;

(K2) $\mathrm{TV}\, k(t,\cdot)$, $\mathrm{TV}\, k_x(t,\cdot)$ are bounded uniformly in time;

(K3) $[0,T] \ni t \to k_x(t,\cdot) \in \mathbf{L}^1(\mathbb{R}_-)$ is Lipschitz continuous.

The local existence and well-posedness for (2.1) is stated in the following Theorem.

Theorem 2.1. *Let f satisfy* (1.3) *and k satisfy* **(K1)**–**(K3)**. *Let $C_0 > 0$, $p_0 > 0$ be given. Then there exist two positive constants $C_1 > C_0$, $p_1 < p_0$ and an operator $P_t : [0,T] \times \mathcal{D}_{C_0,p_0} \to \mathcal{D}_{C_1,p_1}$ such that:*

1) *for all $\bar{u}_0$, $\bar{u}_1 \in \mathcal{D}_{C_0,p_0}$ one has*

$$\|P_t(\bar{u}_0) - P_t(\bar{u}_1)\|_{\mathbf{L}^1(\mathbb{R}_-)} \leq \|\bar{u}_0 - \bar{u}_1\|_{\mathbf{L}^1(\mathbb{R}_-)}; \tag{2.2}$$

2) *the function $u(t,x) = P_t(\bar{u}_0)$ is a weak entropy solution of* (2.1) *with initial data $u(0,\cdot) = \bar{u}_0 \in \mathcal{D}_{C_0,p_0}$.*

Moreover, the solutions of (2.1) depend continuously in $\mathbf{L}^1$ on the coefficient k (see also [7]).

Theorem 2.2. *Let k, $\tilde{k}$ satisfy the assumptions* **(K1)**–**(K3)**. *Let u, $\tilde{u}$ be the solutions of the conservation laws*

$$u_t + \Big(k(t,x) f(u)\Big)_x = 0, \quad x \leq 0, \quad t \geq 0,$$
$$u_t + \Big(\tilde{k}(t,x) f(u)\Big)_x = 0, \quad x \leq 0, \quad t \geq 0,$$

respectively, with the same initial data $\bar{u}_0 \in \mathcal{D}_{C_0,p_0}$, on the time interval $[0,T]$. Then, the following estimate holds

$$\frac{1}{t}\|u(t,\cdot) - \tilde{u}(t,\cdot)\|_{\mathbf{L}^1(\mathbb{R}_-)} \leq \hat{C}_1 \|k - \tilde{k}\|_{\mathbf{L}^\infty([0,t]\times\mathbb{R}_-)} + \hat{C}_2 \sup_{t\in[0,T]} \mathrm{TV}\left(k(t,\cdot) - \tilde{k}(t,\cdot)\right)$$

where $\hat{C}_1$, $\hat{C}_2$ depend on $\sup_t \mathrm{TV}\, u(t,\cdot)$, $\sup_t \mathrm{TV}\, \tilde{u}(t,\cdot)$ and on $\sup|f|$, $\sup|f'|$ taken over the range of the solutions.

3 Sketch of the proof of Theorem 1.2

The existence of solutions for (1.1), in the case of $f(q) = q/(q+1)$, is a consequence of the slow erosion limit studied in [1]. In the more general assumptions (1.3) for f, a proof can be obtained through a time step approximation of (1.1). Below we give some formal arguments for smooth solutions, while rigorous analysis can be found in [2].

(a) *Bound on the* $\mathbf{L}^1$ *norm.* Because $\text{sign}(q) = \text{sign}(f(q))$, the conservation law (1.1) implies that, for all $t \geq 0$, we have

$$\|q(t,\cdot)\|_{L^1(\mathbb{R}_-)} \leq \|q(0,\cdot)\|_{L^1(\mathbb{R}_-)} - \int_0^t |f(q(\tau,0))|\, d\tau. \tag{3.1}$$

(b) *Lower bound on* q. Along a characteristic curve $t \to x(t)$ we have

$$\begin{cases} x'(t) = k(t,x)f'(q), \\ q'(t) = -k_x(t,x)f(q) = kf(q)^2 \geq 0. \end{cases} \tag{3.2}$$

Then the solution q is non-decreasing along any characteristics, and therefore $\inf_x q(t,x) \geq \inf q_0(x) \geq p_0 - 1 > -1$ for all $t \geq 0$.

(c) *Bounds on* f, f', k. As a consequence of (a), (b) we immediately deduce apriori bounds on $\|f(q(t,\cdot))\|_\infty$, $\|f(q(t,\cdot))\|_{\mathbf{L}^1}$, $\|f'(q(t,\cdot))\|_\infty$, $\|k(t,\cdot)\|_\infty$, $\text{TV}\,(k)$ and the characteristic speed kf'.

(d) *Upper bound on* q. The physical meaning of the function $f(q)$ is the amount of erosion per unit length in x. Then the function $q/f(q)$ is a measurement for the change in height per unit erosion. Therefore, the function

$$\phi(t,x) \doteq \int_{-\infty}^x q(t,y)\, dy + \frac{q}{f(q)}$$

would remain constant along characteristics. In fact, we can easily check that

$$\dot\phi(t,x(t)) = \phi_t + \dot x \phi_x = [qf'(q) - f(q)]k + \frac{f(q) - qf'(q)}{f^2(q)} f^2(q)k = 0.$$

For any $q(0) > 0$, for any time $t > 0$, along the characteristic we have

$$\frac{q(t)}{f(q(t))} = \frac{q(0)}{f(q(0))} + \int_{-\infty}^{x(0)} q(0,y)\, dy - \int_{-\infty}^{x(t)} q(t,y)\, dy \leq M \tag{3.3}$$

which is bounded thanks to (a). Recalling the last assumption in (1.3), the function $q \to q/f(q)$ is monotone increasing and approaches $+\infty$ as $q \to +\infty$. Therefore (3.3) provides an upper bound for q for all $t > 0$.

(e) *BV bound.* For smooth solutions, the equation (1.1) gives

$$(q_x)_t + (kf'(q)q_x)_x = (kf^2(q))_x = -f^3(q)k + 2f(q)f'(q)q_x k.$$

Formally, the $\mathbf{L}^1$ norm of q_x provides the total variation for q. One has

$$\begin{aligned}\frac{d}{dt}\text{TV}\,(q(t,\cdot)) \leq \|k\|_\infty \cdot \|f(q(t,\cdot))\|_\infty^2 \cdot \|f(q(t,\cdot))\|_{\mathbf{L}^1} \\ +2\|k(t,\cdot)\|_\infty \cdot \|f(q(t,\cdot)\|_\infty \cdot \text{TV}\,(f(q(t,\cdot))).\end{aligned}$$

Then the total variation of q can grow exponentially in t, but remains bounded for finite t.

Finally, to see that the flow generated by equation (1.1) is Lipschitz continuous, we consider two solutions $q_0(t,\cdot)$, $q_1(t,\cdot)$ of the integro-differential equation (1.1) with initial data

$$q_0(0,x) = \bar{q}_0(x), \quad q_1(0,x) = \bar{q}_1(x) \quad x < 0,$$

and satisfying (1.7) for $t \in [0,T]$. We have the following estimate

$$\|q_0(t,\cdot) - q_1(t,\cdot)\|_{\mathbf{L}^1(\mathbb{R}_-)} \leq \|\bar{q}_0 - \bar{q}_1\|_{\mathbf{L}^1(\mathbb{R}_-)} + L\int_0^t \|q_0(s,\cdot) - q_1(s,\cdot)\|_{\mathbf{L}^1(\mathbb{R}_-)}\, ds \qquad (3.4)$$

for a suitable constant L. By Gronwall lemma, this yields (1.10), hence the Lipschitz continuous dependence of solutions of (1.1) on the initial data.

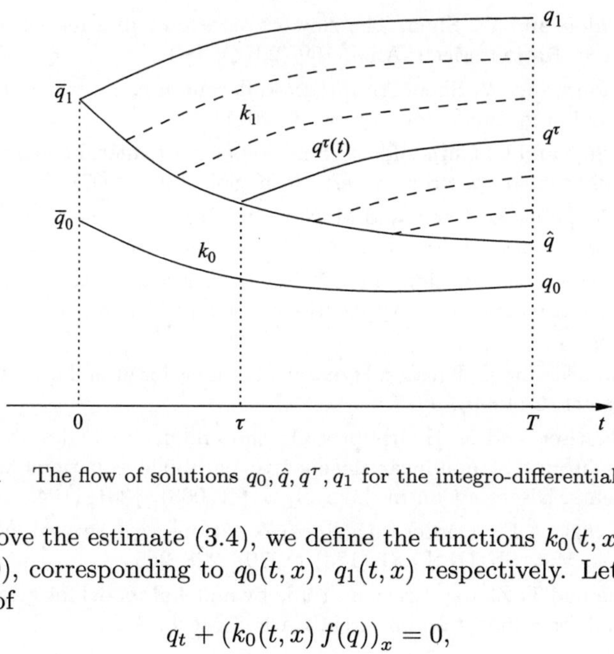

Figure 3.1 The flow of solutions $q_0, \hat{q}, q^\tau, q_1$ for the integro-differential equation.

To prove the estimate (3.4), we define the functions $k_0(t,x)$, $k_1(t,x)$ as in (1.9), corresponding to $q_0(t,x)$, $q_1(t,x)$ respectively. Let $\hat{q}$ be the solution of

$$q_t + (k_0(t,x)\, f(q))_x = 0,$$

with initial data $\hat{q}(0,x) = \bar{q}_1(x)$ (see Figure 3.1). Since q_0 and $\hat{q}$ satisfy the *same* equation (with coefficient k_0), we have

$$\|q_0(t,\cdot) - \hat{q}(t,\cdot)\|_{\mathbf{L}^1(\mathbb{R}_-)} \leq \|\bar{q}_0 - \bar{q}_1\|_{\mathbf{L}^1(\mathbb{R}_-)} \quad \forall t \in [0,T]. \qquad (3.5)$$

Now we evaluate $\|\hat{q}(t,\cdot) - q_1(t,\cdot)\|_{\mathbf{L}^1(\mathbb{R}_-)}$ as follows:

$$\|\hat{q}(T,\cdot) - q_1(T,\cdot)\|_{\mathbf{L}^1(\mathbb{R}_-)} \leq \int_0^T E(\tau)\, d\tau, \qquad (3.6)$$

where

$$E(\tau) \doteq \limsup_{h\to 0+} \frac{\|q^\tau(\tau+h,\cdot) - \hat{q}(\tau+h,\cdot)\|_{\mathbf{L}^1}}{h}$$

and $q^\tau(t,\cdot)$ is the solution, for $t \geq \tau$, of

$$q_t + (k_1(t,x)\, f(q))_x = 0, \quad q(\tau,x) = \hat{q}(\tau,x).$$

By Theorems 2.1 and 2.2 we get the following estimate for E:

$$E(\tau) \leq M\,\mathrm{TV}\,\{k_0(\tau,\cdot) - k_1(\tau,\cdot)\} \leq L \cdot \|q_0(\tau,\cdot) - q_1(\tau,\cdot)\|_{\mathbf{L}^1}$$

for some suitable constants M, L. By inserting this estimate into (3.6) and using (3.5) one finally obtains (3.4).

References

[1] D. Amadori and W. Shen, The slow erosion limit in a model of granular flow. Arch. Ration. Mech. Anal. 199 (2011), 1–31.

[2] D. Amadori and W. Shen, An integro-differential conservation law arising in a model of granular flow. Preprint (2011).

[3] G. Q. Chen and C. Christoforou, Solutions for a nonlocal conservation law with fading memory. Proc. Amer. Math. Soc. 135, (2007), 3905–3915.

[4] R. M. Colombo, M. Herty and M. Mercier, Control of the continuity equation with a non local flow. ESAIM COCV, to appear.

[5] C. M. Dafermos, Solutions in L^∞ for a conservation law with memory. Analyse Mathematique et Applications. Gauthier-Villars, Montrouge, 117–128, 1988.

[6] K. P. Hadeler and C. Kuttler, Dynamical models for granular matter. Granular Matter. 2 (1999), 9–18.

[7] K. H. Karlsen and N. H. Risebro, On the uniqueness and stability of entropy solutions of nonlinear degenerate parabolic equations with rough coefficients. Discrete Contin. Dyn. Syst. 9 (2003), 1081–1104.

[8] S. N. Kružkov, First order quasilinear equations with several independent variables. Mat. Sb. (N.S.) 81 (123) (1970), 228–255.

[9] W. Shen and T. Zhang, Erosion profile by a global model for granular flow. Accepted for publication on Arch. Ration. Mech. Anal.

Computing Time-periodic Solutions of Nonlinear Systems of Partial Differential Equations*

David M. Ambrose
Department of Mathematics, Drexel University
Philadelphia, PA 19104 USA
E-mail: ambrose@math.drexel.edu

Jon Wilkening
Department of Mathematics, University of California
Berkeley, CA 94720 USA
E-mail: wilken@math.berkeley.edu

Abstract

We describe an efficient numerical method for the computation of time-periodic solutions of nonlinear systems of partial differential equations. The strategy is to minimize a functional that measures how far solutions of the initial value problem are from being time-periodic by solving an adjoint problem to compute the gradient of the functional. We discuss applications to two different problems in interfacial fluid dynamics, the Benjamin-Ono equation and the vortex sheet with surface tension.

1 Introduction

Many evolutionary systems of partial differential equations possess special solutions such as traveling waves or time-periodic solutions. It can be useful to understand the set of all such solutions, but it can frequently be difficult to make either analytic or numerical studies of these solutions. In this paper, we will describe a numerical method for the computation of families of time-periodic solutions of nonlinear systems of partial

*The first author gratefully acknowledges support from the National Science Foundation through grants DMS-0926378, DMS-1008387, and DMS-1016267. The second author gratefully acknowledges support from the National Science Foundation through grant DMS-0955078 and from the Director, Office of Science, Computational and Technology Research, US Department of Energy under Contract DEAC02-05CH11231.

differential equations which the authors have recently developed. The solutions we study are taken to also be spatially periodic.

The main idea of this numerical method is to define a functional (depending on the initial data and the presumed period) which is zero if the solution of the system is time-periodic with this period, and which is positive otherwise. This functional is then minimized; in order to do this, we need to be able to compute the variational derivative of the functional with respect to the initial data and the presumed period.

There are two primary types of alternative methods. Orthogonal collocation methods, such as AUTO [6], can be efficient when computing traveling solutions, but can be difficult to use for the much larger problem of finding time-periodic solutions. There are also shooting and multi-shooting methods [12], which require computing the variation for each of the unknowns individually (the unknowns being the discretized values of the initial data, or its leading Fourier modes). We believe that the present method is in many cases more efficient than these alternatives.

We have applied the method to two problems from interfacial fluid dynamics, finding time-periodic solutions of the Benjamin-Ono equation [1, 2], and finding time-periodic symmetric vortex sheets with surface tension [3, 4].

2 Formulation and numerical method

Consider the solution $u(x,t)$ of the initial value problem

$$\partial_t u = F(u), \quad u(x,0) = u_0(x). \tag{2.1}$$

No assumptions are made about the operator F; it can include nonlinear, nonlocal, and unbounded operators. We take the spatial domain to be 2π, with periodic boundary conditions, which is to say, $u(x+2\pi,t) = u(x,t)$. We consider the following functional:

$$G(u_0,T) = \frac{1}{2}\int_0^{2\pi} (u(x,T) - u_0(x))^2 \, dx.$$

Clearly, a solution u of (2.1) with initial data u_0 is time-periodic with period T if and only if $G(u_0,T) = 0$.

We seek to minimize G numerically. To do this, we use the BFGS minimization algorithm [10]. This requires computing the derivative of G with respect to each of u_0 and T. It is clear how to compute the derivative with respect to T; we now describe how to calculate the derivative with respect to u_0. We remark that in applications, we may instead choose to minimize a different functional depending on the situation. In any case, computing the derivative with respect to the initial data will use the same ideas as the present calculation.

To begin, we compute the variation of G as follows:

$$\dot{G} = \frac{d}{d\varepsilon}\bigg|_{\varepsilon=0} G(u_0 + \varepsilon \dot{u}_0, T) = \int_0^{2\pi} (u(x,T) - u_0(x))(\dot{u}(x,T) - \dot{u}_0(x))\, dx.$$

We seek to write this as $\dot{G} = \langle \frac{\delta G}{\delta u_0}, \dot{u}_0 \rangle$. To this end, we define an auxiliary quantity $Q(x,s)$, where $Q(x,0) = u(x,T) - u_0(x)$ and $s = T - t$. We see then that $\dot{G} = \langle Q(\cdot,0), \dot{u}(\cdot,T) - \dot{u}_0 \rangle$. We will define $Q(x,t)$ for positive s so that $\langle Q(\cdot,s), \dot{u}(\cdot,T-s)\rangle$ is constant.

Notice that since $u_t = F(u)$, we have $\dot{u}_t = (DF(u))\dot{u}$, where DF is the linearization of the operator F. Now, differentiating our inner product with respect to s, we have

$$\langle Q_s(\cdot,s), \dot{u}(\cdot,T-s)\rangle - \langle Q(\cdot,s), (DF(u(\cdot,T-s)))\dot{u}(\cdot,T-s)\rangle = 0.$$

Taking the adjoint of DF, we se that this equality can be accomplished by setting $Q_s = (DF(u(\cdot,T-s))^* Q$. We then have

$$\frac{\delta G}{\delta u_0}(x) = Q(x,T) - Q(x,0).$$

3 Results: Benjamin-Ono

The Benjamin-Ono equation is

$$u_t - Hu_{xx} + uu_x = 0,$$

where H is the Hilbert transform and subscripts denote differentiation. This is a well-known dispersive PDE which arises as a long-wave model in interfacial fluid dynamics. For $\beta \in (-1,1)$, there are the stationary solutions

$$u(x;\beta) = \frac{1-3\beta^2}{1-\beta^2} + \frac{4\beta(\cos(x)-\beta)}{1+\beta^2-2\beta\cos(x)}.$$

In one period, these solutions $u(x;\beta)$ have one hump; by suitably rescaling, we have N-hump solutions, given by $u(x;\beta,N) = Nu(Nx;\beta)$. Furthermore, if $u(x)$ is any such stationary solution of the Benjamin-Ono equation, then $u(x-ct)+c$ is a traveling wave solution. These are known to be all of the periodic traveling wave solutions of Benjamin-Ono [5].

Linearizing about the stationary solutions, we find the linearized evolution $v_t = iBAv$, where $A = H\partial_x - u$ and $B = \frac{1}{i}\partial_x$. We are able to solve for the eigenvalues and eigenvectors of the operator BA, and these give formulas for time-periodic solutions of the linearized equation. We then look for time-periodic solutions of the full problem, using the period and initial data for the time-periodic solutions of the linearized equations to determine the initial guess in the BFGS algorithm when minimizing G.

Our findings in [1] and [2] are that there are indeed bifurcations from stationary or traveling solutions of Benjamin-Ono, at all of the periods predicted by the linear theory. In fact, for each bifurcation, we have a four-parameter family of solutions emanating from the trivially time-periodic (i.e., stationary or traveling) solution. We furthermore find that every such four-parameter family emanating from a trivially time-periodic solution coincides with a four-parameter family emanating from a different trivially time-periodic solution; that is to say, we find that these stationary and traveling solutions are connected to each other through continua of nontrivial time-periodic solutions. One such nontrivial time-periodic solution is shown in Figure 3.1.

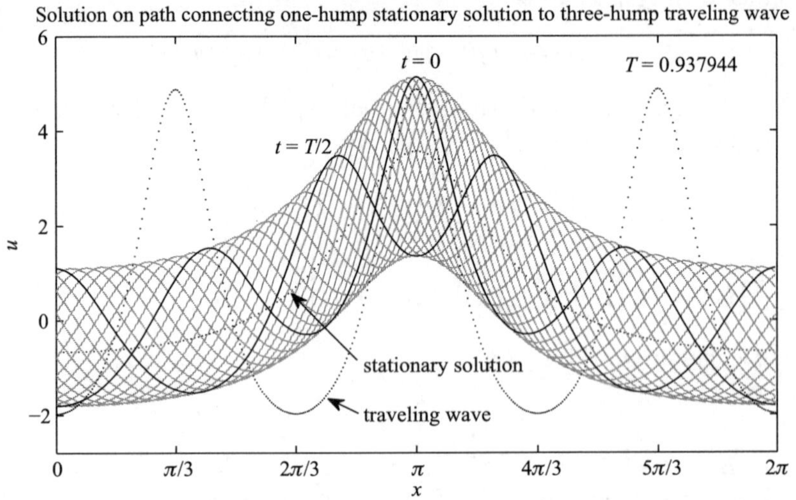

Figure 3.1 A time-periodic solution of the Benjamin-Ono equation. This solution belongs to a continuum of solutions which connect a stationary solution with one hump to a traveling solution with three humps; these trivially time-periodic solutions are shown with dotted curves.

For the Benjamin-Ono equation, several other authors have studied time-periodic solutions, finding explicit formulas for these solutions [9, 11]. It is in fact possible to use these explicit formulas to prove that the bifurcations we find in [1, 2] do in fact occur; in one case, this is proved in [2]. This is carried further by the second author in [13], and additionally, bifurcations which are not predicted by the linear theory are also shown to occur. The existence of explicit formulas for the time-periodic solutions of the Benjamin-Ono equation provides an important validation of our numerical method; we can confirm that every solution we have computed for this equation does in fact correspond to a time-periodic solution of the equation, and is not just a solution of the initial value problem which comes very close to returning to its initial state.

4 Results: vortex sheets

The vortex sheet with surface tension is the interface between two irrotational, immiscible fluids satisfying the incompressible Euler equations. Across the interface, the normal component of the velocity is continuous while the tangential velocity may jump. This implies that the vorticity is not identically zero, but is instead a Dirac measure supported on the interface. We describe the interface as a parameterized curve at each time, $(x(\alpha,t),y(\alpha,t))$, horizontally periodic with period 2π:

$$x(\alpha+2\pi,t)=x(\alpha,t)+2\pi, \qquad y(\alpha+2\pi,t)=y(\alpha,t).$$

For this problem, an efficient numerical method for solution of the initial value problem was developed by Hou, Lowengrub, and Shelley [7, 8]. We use this same method for the numerical solution of the initial value problem when solving for time-periodic solutions with our algorithm.

The method of [7, 8] uses a natural geometric formulation of the problem. Rather than evolving (x,y), the tangent angle $\theta=\tan^{-1}(y_\alpha/x_\alpha)$ and the arclength element $s_\alpha=(x_\alpha^2+y_\alpha^2)^{1/2}$ are evolved instead. Furthermore, the parameterization is taken to be a normalized arclength parameterization: if $L(t)$ is the length of one period of the interface at time t, then we insist on the condition $s_\alpha(\alpha,t)=L(t)/2\pi$. This condition can be maintained by properly choosing the tangential velocity of the interface; by contrast, the normal velocity is determined by the fluid dynamics. For a given 2π-periodic curve with corresponding tangent angle θ, the length L can be determined from θ. Therefore, the position of the interface can be determined from θ and the location of one point.

If we let $\hat{\mathbf{t}}$ and $\hat{\mathbf{n}}$ be unit tangent and normal vectors at each point of the interface at each time, then the velocity of the interface can be written $(x,y)_t=U\hat{\mathbf{n}}+V\hat{\mathbf{t}}$. It is sometimes convienient to denote the position of the interface as $z=x+iy$. The normal velocity is $U=\mathbf{W}\cdot\hat{\mathbf{n}}$, where $\mathbf{W}=(W_1,W_2)$ is the Birkhoff-Rott integral defined as

$$W_1-iW_2=\frac{1}{4\pi i}\mathrm{PV}\int_0^{2\pi}\gamma(\alpha')\cot\left(\frac{1}{2}(z(\alpha)-z(\alpha'))\right)\,d\alpha.$$

The quantity γ is the vortex sheet strength and measures the jump in tangential velocity across the interface.

We have the following evolution equations:

$$\theta_t=\frac{U_\alpha+V\theta_\alpha}{s_\alpha}, \quad \gamma_t=\tau\frac{\theta_{\alpha\alpha}}{s_\alpha}+\left(\frac{\gamma(V-\mathbf{W}\cdot\hat{\mathbf{t}})}{s_\alpha}\right)_\alpha.$$

The constant τ here is the coefficient of surface tension, which is positive. With surface tension present, the system is dispersive and is well-posed.

If we were to attempt to minimize the functional G described above, we would need to compute solutions of the initial value problem from time $t = 0$ until time $t = T$, and we would then need to solve the adjoint problem backward in time over this same interval. Instead of doing this, we utilize symmetry and only perform one-fourth of the computations. We do this differently in the cases where the mean of γ is zero or non-zero. In the case that the mean of γ is zero, we take initial data $\theta_0(\alpha) = 0$, and we seek solutions for which

$$G_1(\gamma_0, T) = \int_0^{2\pi} \gamma^2\left(\alpha, \frac{T}{4}\right)\, d\alpha = 0.$$

In the case that the mean of γ is non-zero, we again use initial data $\theta_0(\alpha) = 0$, and we seek solutions for which

$$G_2(\gamma_0, T) = \int_0^{2\pi} \left(\gamma\left(\alpha, \frac{T}{4}\right) - \gamma\left(\alpha - \pi, \frac{T}{4}\right)\right)^2 + \left(\theta\left(\alpha, \frac{T}{4}\right) + \theta\left(\alpha - \pi, \frac{T}{4}\right)\right)^2 d\alpha = 0. \quad (4.1)$$

We can argue that zeros of G_1 or G_2 lead to time-periodic solutions with period T. We now give the argument in the second case.

Consider γ_0 and T such that $G_2(\gamma_0, T) = 0$. We can then solve the initial value problem with data θ_0 and γ_0 to find θ and γ for $0 \le t \le \frac{T}{4}$, and we have $\theta\left(\alpha, \frac{T}{4}\right) = -\theta\left(\alpha - \pi, \frac{T}{4}\right)$ and $\gamma\left(\alpha, \frac{T}{4}\right) = \gamma\left(\alpha - \pi, \frac{T}{4}\right)$. If we define $\Theta(\alpha, t) = -\theta\left(\alpha - \pi, \frac{T}{4} - t\right)$ and $\Gamma(\alpha, t) = \gamma\left(\alpha - \pi, \frac{T}{4} - t\right)$, then Θ and Γ are also solutions of the vortex sheet evolution problem. Furthermore, $\Theta(\alpha, 0) = \theta\left(\alpha, \frac{T}{4}\right)$ and $\Gamma(\alpha, 0) = \gamma\left(\alpha, \frac{T}{4}\right)$. Therefore, Θ and Γ continue θ and γ. That is, for $\frac{T}{4} \le t \le \frac{T}{2}$, we can take $\theta(\alpha, t) = \Theta\left(\alpha, t - \frac{T}{4}\right) = -\theta\left(\alpha - \pi, \frac{T}{2} - t\right)$ and $\gamma(\alpha, t) = \Gamma\left(\alpha, t - \frac{T}{4}\right) = \gamma\left(\alpha - \pi, \frac{T}{2} - t\right)$. Similarly, we can continue the solution to $\frac{T}{2} \le t \le T$ by making the definitions $\theta(\alpha, t) = \theta\left(\alpha - \pi, t - \frac{T}{2}\right)$ and $\gamma(\alpha, t) = \gamma\left(\alpha - \pi, t - \frac{T}{2}\right)$. We see then that $\theta(\alpha, T) = \theta\left(\alpha - \pi, \frac{T}{2}\right) = -\theta(\alpha, 0) = \theta(\alpha, 0)$. Similarly, we have $\gamma(\alpha, T) = \gamma\left(\alpha - \pi, \frac{T}{2}\right) = \gamma(\alpha, 0)$. So, we have indeed found a time-periodic solution.

Using the BFGS minimization algorithm requires computing $\frac{\delta G_1}{\delta \gamma_0}$ or $\frac{\delta G_2}{\delta \gamma_0}$ as appropriate. As in the calculation of $\frac{\delta G}{\delta u_0}$ in Section 2, these variational derivatives can be found in terms of the solution of the adjoint PDE for the linearization of the equations of motion. Thus, we linearize the evolution equations about an arbitrary state, calculating the operator $DF(u)$ (using the notation of Section 2, with $u = (\theta, \gamma)$ here), and then taking its adjoint $(DF(u))^*$. This is significantly complicated by the presence of nonlinear, nonlocal singular integral operators, such as the Birkhoff-Rott integral. Details of the calculation can be found in [3].

In [3] and [4], we compute bifurcations from the flat equilibrium $\theta = 0$ in the case of zero and nonzero mean vortex sheet strength, respectively. In both cases, we predict bifurcation from the flat vortex sheet using linear theory, and we then find numerically that the expected bifurcation does occur. We are thus able to compute many time-periodic solutions, continuing the bifurcation curves as far as possible. Two such time-periodic solutions are shown in Figures 4.1 and 4.2.

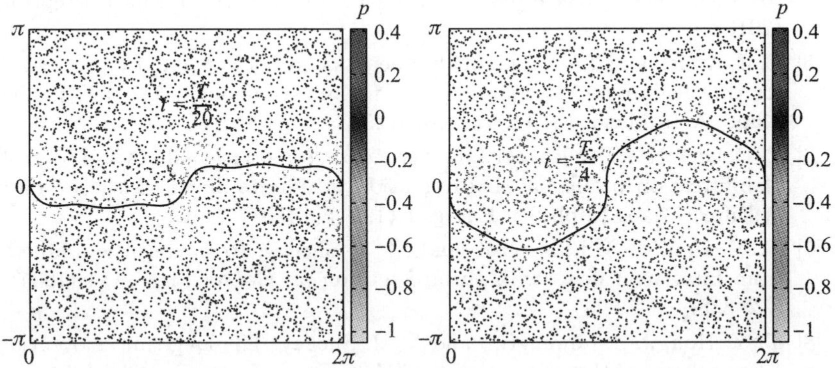

Figure 4.1 Two profiles of a time-periodic vortex sheet with surface tension with $\tau = 1$ and zero mean vortex sheet strength. The particles in the fluid are passive tracers added for visualization.

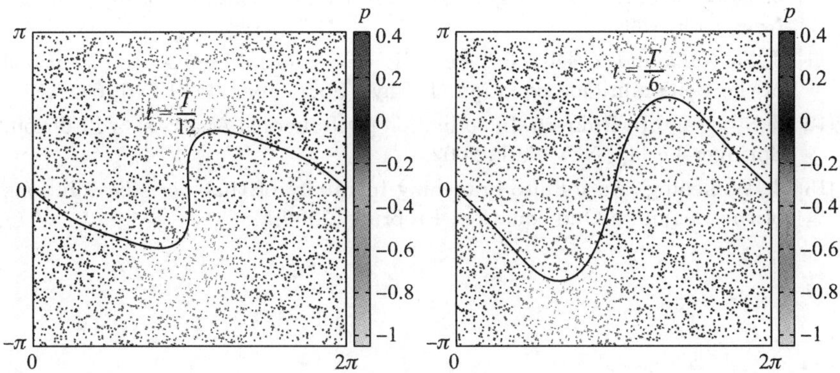

Figure 4.2 Two profiles of a time-periodic vortex sheet with surface tension with $\tau = 1$ and nonzero mean vortex sheet strength. Notice that the solution exhibits less symmetry than the solution in Figure 4.1.

References

[1] D. M. Ambrose and J. Wilkening, Global paths of time-periodic solutions of the Benjamin-Ono equation connecting pairs of traveling waves. Comm.

Appl. Math. and Computational Sci. 4 (2009), 177–215.

[2] D. M. Ambrose and J. Wilkening, Computation of time-periodic solutions of the Benjmain-Ono equation. J. Nonlinear Sci. 20 (2010), 277–308.

[3] D. M. Ambrose and J. Wilkening, Computation of symmetric, time-periodic solutions of the vortex sheet with surface tension. Proc. Nat. Acad. Sci. USA 107 (2010), 3361–3366.

[4] D. M. Ambrose and J. Wilkening, Dependence of time-periodic vortex sheets with surface tension on mean vortex sheet strength. Preprint (2010).

[5] C. J. Amick and J. F. Toland, Uniqueness and related analytic properties for the Benjamin-Ono equation — a nonlinear Neumann problem in the plane. Acta Math. 167 (1991), 107–126.

[6] E. Doedel, H. B. Keller and J. P. Kernévez, Numerical analysis and control of bifurcation problems. II. Bifurcation in infinite dimensions. Internat. J. Bifur. Chaos Appl. Sci. Engrg. 1 (1991), 745–772.

[7] T. Y. Hou, J. S. Lowengrub and M. J. Shelley, Removing the stiffness from interfacial flows with surface tension. J. Comp. Phys. 114 (1994), 312–338.

[8] T. Y. Hou, J. S. Lowengrub and M. J. Shelley, The long-time motion of vortex sheets with surface tension. Phys. Fluids 9 (1997), 1933–1954.

[9] Y. Matsuno, New representations of multiperiodic and multisoliton solutions for a class of nonlocal soliton equations. J. Phys. Soc. Japan 73 (2004), 3285–3293.

[10] J. Nocedal and S. J. Wright, Numerical Optimization. Springer-Verlag, New York, 1999.

[11] J. Satsuma and Y. Ishimori, Periodic wave and rational soliton solutions of the Benjamin-Ono equation. J. Phys. Soc. Japan 46 (1979), 681–687.

[12] J. Stoer and R. Bulirsch, Introduction to Numerical Analysis. 3rd Edition. Springer-Verlag, New York, 2002.

[13] J. Wilkening, An infinite branching hierarchy of time-periodic solutions of the Benjamin-Ono equation. Preprint (2008).

Boundary Value Problem for the Linearized Euler-Korteweg System on the Half Line

Corentin Audiard
Université Claude Bernard Lyon 1
43 *Bvd du* 11 *novembre 1918*
69622 Villeurbanne Cedex, France
E-mail: audiard@math.univ-lyon1.fr

Abstract

We deal here with the initial boundary value problem for the Euler-Korteweg equations in dimension 1 by developping a generalization of Kreiss symmetrizer. To do so, we first introduce a suitable condition for the boundary data (the so called "generalized Kreiss-Lopatinskiĭ condition"), and show that it implies well-posedness for the pure boundary value problem. The well-posedness of the initial boundary value problem is obtained by the mean of sharp trace estimates for the Cauchy problem and application of the principle of superposition.

1 Introduction

The Euler-Korteweg system (in any dimension) consists of the Euler equations with an additional "stress" term:

$$\begin{cases} \partial_t \rho + \operatorname{div}(\rho u) = 0, \\ \partial_t u + (u \cdot \nabla)u + \nabla g(\rho) = \nabla\left(K(\rho)\Delta\rho + \dfrac{1}{2}K'(\rho)|\nabla\rho|^2\right), \end{cases} \tag{1.1}$$

where ρ represents the density, $u \in \mathbb{R}^d$ is the velocity of a compressible fluid. The applications $g: \rho \to g(\rho)$ (chemical potential) and $K: \rho \to K(\rho)$ (capillarity) are given smooth functions. The additional Korteweg tensor is intended to take into account capillarity effects. It is named after the work of Korteweg [5] in the XIXth century, see [11] for the derivation of the modern form of the equations.

This is a quasilinear third order system and it is known to admit local in time smooth solutions [2] and global smooth solutions such as traveling waves [3] in the whole space $\mathbb{R}^d$. A key point for the Cauchy problem analysis is the use of an additional dependent variable $v = \nabla\zeta$, where $\nabla\zeta = \sqrt{\frac{K'(\rho)}{\rho}}\nabla\rho$: the extended system in the variables (ζ, u, v) takes the form of a transport equation with source term for ζ coupled with a degenerate Schrödinger equation for $z = u+iv$. A priori estimates without loss of derivatives can then be obtained by integration by parts, suitable choices of weight functions and commutator estimates.

From now on, we focus on the case $d = 1$. The extended system is

$$\begin{cases} \partial_t\zeta + u\partial_x a(\zeta)\partial_x u = 0, \\ \partial_t u + u\partial_x u - \partial_x\left(\dfrac{1}{2}|v|^2\right) - \partial_x(a(\zeta)\partial_x v) = -g'(\zeta)v, \\ \partial_t v + \partial_x(u\,v) + \partial_x(a(\zeta)\partial_x u) = 0, \end{cases} \tag{1.2}$$

and we are concerned with the linearized initial boundary value problems (IBVP) for the extended system, which are more relevant than the Cauchy problem from a physical point of view. The linearized system reads

$$\begin{cases} \partial_t\zeta + \underline{u}\partial_x\zeta + \underline{a}\partial_x u = 0, \\ \partial_t u + \underline{u}\partial_x u - \partial_x(\underline{v}\,v) - \partial_x(\underline{a}\partial_x v) = -\underline{g}'v, \\ \partial_t v + \partial_x(\underline{u}\,v) + \partial_x(\underline{a}\,\partial_x u) = 0. \end{cases} \tag{1.3}$$

The analysis of a constant coefficient linear problem in the half line $\{x > 0\}$ is classically tackled by Fourier-Laplace transform in time, which transforms the PDE system into an ODE of the form $\partial_x\widehat{U} + G\widehat{U} = \widehat{U}_0$. The matrix G depends on $\tau \in \mathbb{C}$, dual variable to t, and $\widehat{U_0}$ is the Fourier-Laplace transform of the initial condition for the unknown U. Our approach is inspired by the strategy of Métivier and Zumbrun [9] for hyperbolic-parabolic IBVP, which consists in constructing quasi-homogeneous generalized Kreiss symmetrizers to obtain a priori estimates. A noticeable novelty is the fact that the real part of G's eigenvalues cancels when $\mathrm{Im}(\tau) \to \infty$. The careful study of these eigenvalues leads to *a priori* estimates with a gain of regularity of the trace. This sharp smoothing of the trace for solutions of dispersive equations was described for the first time by Kenig, Ponce and Vega [4], in the framework of the Cauchy problem for the Airy equation.

Our paper is organized as follows:

1. We first describe the *stable subspace* of G and derive a generalized Kreiss-Lopatinskiĭ condition (i.e. characterization of convenient boundary conditions) for the linearized system,
2. We construct generalized Kreiss symmetrized for the boundary value problem with convenient boundary conditions,
3. We deduce from the analysis of the pure boundary value problem and the Cauchy problem well-posedness results.

2 Kreiss-Lopatinskiĭ condition for the linearized Euler-Korteweg system

After a Fourier-Laplace transform in time of (1.3) we get the system:

$$\begin{cases} \underline{u}\partial_x\widehat{\zeta} = -(\tau\widehat{\zeta} + \underline{a}\,\partial_x\widehat{u}), \\ \partial_x\widehat{u} = \partial_x\widehat{u}, \\ \underline{a}\partial_x^2\widehat{u} = -\big(\tau\widehat{v} + \partial_x(\underline{u}\widehat{v})\big), \\ \partial_x\widehat{v} = \partial_x\widehat{v}, \\ \underline{a}\partial_x^2\widehat{v} = \tau\widehat{u} + \underline{u}\partial_x\widehat{u} + \underline{g'}\widehat{v} - \partial_x\underline{v}\widehat{v}, \end{cases} \tag{2.1}$$

or equivalently $\partial_x U = GU$, with

$$G = \begin{pmatrix} \frac{-\tau}{\underline{u}} & 0 & \frac{-\underline{a}}{\underline{u}} & 0 & 0 \\ 0 & 0 & 1 & 0 & 0 \\ 0 & 0 & 0 & \frac{-\tau}{\underline{a}} & \frac{-\underline{u}}{\underline{a}} \\ 0 & 0 & 0 & 0 & 1 \\ 0 & \frac{\tau}{\underline{a}} & \frac{\underline{u}}{\underline{a}} & \frac{\underline{g'}}{\underline{a}} & \frac{-\underline{v}}{\underline{a}} \end{pmatrix}, \quad U = \begin{pmatrix} \widehat{\zeta} \\ \widehat{u} \\ \partial_x\widehat{u} \\ \widehat{v} \\ \partial_x\widehat{v} \end{pmatrix}. \tag{2.2}$$

After a galilean change of frame we may assume $\underline{v} = 0$. The restriction of the extended problem to the new variables $V = (\sqrt{|\tau|}\widehat{u}, \sqrt{|\tau|}\widehat{v}, \partial_x\widehat{u}, \partial_x\widehat{v})$ reads:

$$\partial_x V = \sqrt{|\tau|}HV,\ H = \begin{pmatrix} 0 & 0 & 1 & 0 \\ 0 & 0 & 0 & 1 \\ 0 & \frac{-\tau}{\underline{a}|\tau|} & 0 & \frac{-\underline{u}}{\underline{a}\sqrt{|\tau|}} \\ \frac{\tau}{\underline{a}|\tau|} & \frac{\underline{g'}}{\underline{a}|\tau|} & \frac{\underline{u}}{\underline{a}\sqrt{|\tau|}} & 0 \end{pmatrix}. \tag{2.3}$$

The characteristic polynomial of $\sqrt{|\tau|}H$ is

$$\chi_H(X) = \frac{1}{\underline{a}^2}\left(\left(\frac{\tau}{|\tau|} + X\frac{\underline{u}}{\sqrt{|\tau|}}\right)^2 - X^2\left(\frac{\underline{g'}}{\underline{a}|\tau|} - X^2\right)\right). \tag{2.4}$$

Note that H is nearly homogeneous of order 0 in the sense that $H(\tau)$ admits a radial limit when $|\tau| \to \infty$.

We take boundary conditions of the form $FV = \widehat{\varphi}$, with F a smooth matrix homogeneous in τ of maximal rank. We define the stable subspace of H

$$E^-(H(\tau)) := \{V \in \mathbb{C}^4 :\ \sup_{x\geq 0}|e^{xH}V| < \infty\}.$$

If the problem

$$\begin{cases} \partial_x V = \sqrt{|\tau|}HV + \widehat{f}, & x \geq 0, \\ FV(0) = \widehat{\varphi}, \end{cases} \tag{2.5}$$

admits a unique bounded solution for some $\tau \in \{\mathrm{Re}(\tau) > 0\}$ and any $\widehat{f} = 0$, then F must be an isomorphism $E^-(H(\tau)) \to \mathrm{Im}(F)$.

Definition 2.1. The generalized Kreiss-Lopatinskiĭ condition is satisfied by the operator F if there exists $\Gamma > 0$ such that for any $\tau \in \{\mathrm{Re}(z) \geq \Gamma\}$, $F(\tau)$ is an isomorphism $E^-(G(\tau)) \to \mathrm{Im}(F)$. It satisfies the uniform Kreiss-Lopatinskiĭ condition (KLU) if moreover F and its inverse are bounded uniformly in τ.

The following proposition describes E^-.

Proposition 2.2. *Let $\tau = \gamma + i\delta$. If $\gamma > 0$, H has two eigenvalues of positive real part (counted with multiplicity), two of negative real part, and no purely imaginary roots. The stable subspace is the subspace associated to the eigenvalues of negative real part.*

Moreover, if we denote by (ω_1^+, ω_2^+), resp. (ω_1^-, ω_2^-) the roots of positive real part, resp. negative, of $P(\cdot\,;\tau)$, then two of them have their real part vanishing when $\delta \longrightarrow +\infty$, while the other two have their real part vanishing when $\delta \longrightarrow -\infty$.

More precisely, we may choose the numbering of eigenvalues in such a way that:

$$\begin{aligned} Re(\omega_1^{\pm}) &\sim \pm\frac{\gamma}{2\sqrt{\delta}}, & \delta &\longrightarrow +\infty, \\ Re(\omega_2^{\pm}) &\longrightarrow \pm\infty, & \delta &\longrightarrow +\infty, \\ Re(\omega_1^{\pm}) &\longrightarrow \pm\infty, & \delta &\longrightarrow -\infty, \\ Re(\omega_2^{\pm}) &\sim \pm\frac{\gamma}{2\sqrt{|\delta|}}, & \delta &\longrightarrow -\infty. \end{aligned}$$

The proposition entails that the number of boundary condition (i.e. the rank of F) is two. Thus a natural and simple candidate for the boundary operator F is the operator corresponding to Dirichlet boundary conditions

$$F = \begin{pmatrix} 1\,0\,0\,0 \\ 0\,1\,0\,0 \end{pmatrix}. \tag{2.6}$$

Using Proposition 2.2 and an explicit computation of eigenvectors, it can be checked that these conditions do satisfy (KLU). For simplicity we shall use the Dirichlet conditions in the rest of this text, although every result is true for any boundary condition satisfying (KLU).

3 Symmetrizers and well-posedness results

Definition 3.1. Let $\Gamma > 0$ be fixed. A (generalized) symmetrizer S for (2.5) is a $d \times d$ self adjoint operator, depending smoothly on τ for $\gamma := Re(\tau) \geq \Gamma$, uniformly bounded and satisfying:

$$\exists\, \alpha(\tau) > 0 : \ \operatorname{Re}(S(\tau)\sqrt{|\tau|}H) \geq \gamma\alpha(\tau)I_d, \qquad (3.1)$$

$$\exists\, C(\tau) > 0, \exists\, \beta(\tau) > 0 : \ S(\tau,\eta) \geq \beta(\tau)I_d - C(\tau)F^*F. \qquad (3.2)$$

Proposition 3.2. *If there is a symmetrizer S satisfying* (3.1), (3.2), *any smooth solution of* (2.5) *satisfies:*

$$\alpha(\tau)\frac{\gamma}{2}\|\widehat{U}\|^2 + \beta(\tau)|\widehat{U}(0)|^2 \leq C(\tau)|\widehat{\varphi}|^2 + \frac{\|S\|_\infty^2}{2\gamma\alpha(\tau)}\|f\|^2, \qquad (3.3)$$

where $|\cdot|$ is the usual euclidean norm in $\mathbb{C}^{2d+3}$, and $\|\cdot\|$ is the L^2 norm for functions of $x_d \in \mathbb{R}^+$.

Proof. Taking the inner product of $S\partial_x\widehat{U} = S(\sqrt{|\tau|}H\widehat{U} + f)$ with $\overline{\widehat{U}}$, integrating on $\mathbb{R}$ and using that S is hermitian gives:

$$\int_0^\infty \widehat{U}^* \cdot S\sqrt{|\tau|}H\widehat{U}\,dx + \widehat{U}^*(0)\cdot S\widehat{U}(0) = -\int_0^\infty \widehat{U}^* \cdot Sf \qquad (3.4)$$

for $\widehat{U}$ smooth enough and vanishing at ∞.

Using (3.1), (3.2) we obtain:

$$\alpha\gamma\|\widehat{U}\|^2 + \beta|\widehat{U}(0)|^2 \leq C|\widehat{\varphi}|^2 + \|S\|_\infty\|f\|\,\|\widehat{U}\| \qquad (3.5)$$

hence (3.3) by Young's inequality. □

Proposition 3.3. *For $\Gamma > 0$ large enough there exists a symmetrizer for* (2.5) *such that:*

- *The applications C, β are constants independent of τ,*
- *The application α has the form $\alpha = \dfrac{c}{\sqrt{|\tau|}}$, with c a constant independent of τ.*

Sketch of proof. In order to use the 'quasi-homogeneity' of H, we use the variables $(\widehat{\tau},\varepsilon) = (\frac{\tau}{|\tau|}, \frac{1}{\sqrt{|\tau|}}) \in \mathbb{S}^+ \times]0,\infty[$, with $\mathbb{S}^+ := \{\widehat{\tau} \in \mathbb{C} : \ |\tau| = 1,\ \operatorname{Re}(\widehat{\tau}) \geq 0\}$. The matrix $H(\tau)$ is (abusively) written $H(\widehat{\tau},\varepsilon)$. It is easy to see that

$$\lim_{\varepsilon\to 0} H(\widehat{\tau},\varepsilon) = \begin{pmatrix} 0 & 0 & 1 & 0 \\ 0 & 0 & 0 & 1 \\ 0 & \dfrac{-\widehat{\tau}}{\underline{a}} & 0 & 0 \\ \dfrac{\widehat{\tau}}{\underline{a}} & 0 & 0 & 0 \end{pmatrix} =: H_0(\widehat{\tau}),$$

whose structure is very simple. In particular, it is easily checked that H_0 has four distinct eigenvalues for any $\widehat{\tau} \in \mathbb{S}^+$, and thus can be smoothly diagonalized as

$$P^{-1}HP = \begin{pmatrix} \lambda_1^+ & & & \\ & \lambda_2^+ & & \\ & & \lambda_1^- & \\ & & & \lambda_2^- \end{pmatrix}, \quad \pm\mathrm{Re}(\lambda_j^\pm) \geq 0.$$

To satisfy (3.1) $\mathrm{Re}(SH) \geq \alpha$ it is natural to choose

$$S = (P^{-1})^* \begin{pmatrix} \kappa & & & \\ & \kappa & & \\ & & -1 & \\ & & & -1 \end{pmatrix} P^{-1}$$

and after some elementary algebra, it may be checked that for κ large enough the condition (3.2) is satisfied too. This construction is only valid for $\varepsilon = 0$ but, using some perturbation arguments, we may obtain a symmetrizer $S(\widehat{\tau}, \varepsilon)$ defined on $\mathbb{S}^+ \times [0, \varepsilon_0]$ for ε_0 small enough. Using finally the bijection $\mathbb{S}^+ \times [0, \varepsilon_0] \to \{\tau \in \mathbb{C},\ \mathrm{Re}(\tau) \geq 0,\ |\tau| \geq 1/\varepsilon_0^2\}$ we obtain a symmetrizer defined on $\{\mathrm{Re}(\tau) \geq 1/\varepsilon_0^2\}$. □

Propositions 3.2, 3.3 give the *a priori* estimate.

$$\frac{\gamma\sqrt{|\tau|}}{2} \|(\widehat{u}, \widehat{v})\|^2 + \frac{\gamma}{2\sqrt{|\tau|}} \|\partial_x(\widehat{u}, \widehat{v})\|^2 + \beta|\tau|\,|(\widehat{u}, \widehat{v})(0)|^2 + \beta|\partial_x(\widehat{u}, \widehat{v})(0)|^2 \leq C|\tau|\,|\widehat{\varphi}|^2 + \frac{|\sqrt{\tau}|}{2\gamma} \|0, f\|^2. \tag{3.6}$$

In order to apply the Plancherel theorem, we introduce the weighted Sobolev spaces

$$H_\gamma^s(\mathbb{R}) = \left\{ f : \|f\|_{s,\gamma} = \left(\int |\widehat{e^{-\gamma \cdot} f}(\delta)|^2 (\gamma^2 + \delta^2)^s d\delta \right)^{\frac{1}{2}} < \infty \right\}. \tag{3.7}$$

These norms appear naturally because of the coefficient $\sqrt{|\tau|}$ in our inequalities. Note in particular that $\sqrt{|\tau|} = (\gamma^2 + \delta^2)^{\frac{1}{4}}$.

By integration in $Im(\tau)$ and use of the Plancherel theorem, the previous estimate gives the *a priori* estimate for smooth solutions of (2.5) with Dirichlet boundary conditions.

$$\frac{\gamma}{2} \|(u, v)\|^2_{L_x^2(H_{\gamma,t}^{1/2})} + \frac{\gamma}{2} \|\partial_x(u, v)\|_{L^2_{x,t}} + \beta|(u, v)(0)|_{H_{\gamma,t}^{3/4}} + \beta|\partial_x(u, v)(0)|_{H_{\gamma,t}^{1/4}} \leq C\,|\varphi|^2_{H_{\gamma,t}^{3/4}} + \frac{1}{2\gamma} \|f\|_{L_x^2(H_{\gamma,t}^{1/2})}. \tag{3.8}$$

(we used the notation $|\cdot|$ for the norm of functions defined on the boundary whereas $\|\cdot\|$ are norms on the half space.)

This *a priori* estimate can be used to obtain the well-posedness of the pure boundary value problem (zero initial data). Moreover, dispersive estimates for the Cauchy problem allow us (by the superposition principle) to state the following well posedness theorem for the initial boundary value problem.

Theorem 3.4. *For any $T > 0$, there exists $C(T)$ such that the problem*

$$\begin{cases} \partial_t u + \underline{u}\partial_x u - \underline{a}\partial_x^2 v = -\underline{g}' v, \\ \partial_t v + \underline{u}\partial_x v + \underline{a}\partial_x^2 u = 0, \qquad (x,t) \in \mathbb{R}^+ \times [0,T], \\ (u,v)|_{t=0} = (u_0, v_0) \in H^1(\mathbb{R}^+),\, x \in \mathbb{R}^+, \\ (u,v)|(0) = \varphi \in H^{3/4}([0,T]), \quad t \in [0,T], \end{cases} \tag{3.9}$$

admits a unique solution in $L^2([0,T]; H^1(\mathbb{R}^+)) \cap L^2(\mathbb{R}^+; H^{1/2}([0,T]))$ under the compatibility condition $\varphi(0) = (u_0(0), v_0(0))$.
Moreover, this solution satisfies the estimate:

$$\begin{aligned} &\|(u,v)\|_{L^2(\mathbb{R}^+;H^{1/2}([0,T]))} + \|\partial_x(u,v)\|_{L^2(\mathbb{R}^+\times[0,T])} + |(u,v)(0)|_{H^{3/4}([0,T])} \\ &\quad + |\partial_x(u,v)(0)|_{H^{1/4}([0,T])} \le C(T)(\|(u_0,v_0)\|_{H^1(\mathbb{R}^+)} + |\varphi|_{H^{3/4}([0,T])}). \end{aligned} \tag{3.10}$$

The reader may refer to [1] for a full proof, where are also exposed well-posedness results for data with higher regularity.

Conclusion: The construction of Kreiss symmetrizers for the extended system has allowed us to obtain a priori estimates and well-posedness results. It would be of very high interest to know how far the hyperbolic (homogeneous) theory may be adapted to the fully nonlinear, dispersive problem. A first step would be to use a quasi-homogeneous pseudo/para-differential calculus to treat the case of variable coefficients (as was done by Majda [7]), and eventually obtain at least local or for small data well-posedness results similar to Mokrane [10] or Métivier [8]. The formal algebraic construction is actually expected to be a straightforward adaptation of what was done in this article. It is the passage from a Kreiss symmetrizer to a functional symmetrizer that involves very technical difficulties that do not arise in the case of hyperbolic systems.

References

[1] C. Audiard, Kreiss symmetrizer and boundary conditions for the Euler-Korteweg system in a half space. J. Differential Equations, 249(3): 599–620, 2010.

[2] S. Benzoni-Gavage, R. Danchin and S. Descombes, On the well-posedness for the euler-korteweg model in several space dimensions. Indiana Univ. Math. J. 56: 1499–1579, 2007.

[3] S. Benzoni-Gavage, R. Danchin, S. Descombes and D. Jamet, Stability issues in the euler korteweg model. AMS, 220: 103–127, 2007.

[4] C. E. Kenig, Gustavo Ponce and Luis Vega, Well-posedness of the initial value problem for the Korteweg-de Vries equation. J. Amer. Math. Soc. 4(2): 323–347, 1991.

[5] D. J. Korteweg, Sur la forme que prennent les équations des mouvement des fluides si l'on tient compte des forces capillaires par des variations de densit. Arch. Néer. Sci. Exactes Sér. II, 6: 1–24, 1901.

[6] H.-O. Kreiss, Initial boundary value problems for hyperbolic systems. Comm. Pure Appl. Math. 23: 277–298, 1970.

[7] A. Majda, The stability of multidimensional shock fronts. Mem. Amer. Math. Soc. 41(275): iv+95, 1983.

[8] G. Métivier, Stability of multidimensional shocks. In: Advances in the Theory of Shock Waves, Progr. Nonlinear Differential Equations Appl., 47: 25–103. Birkhäuser Boston, Boston, MA, 2001.

[9] G. Métivier and K. Zumbrun, Large viscous boundary layers for non-characteristic nonlinear hyperbolic problems. Mem. Amer. Math. Soc. 175(826): vi+107, 2005.

[10] A Mokrane, Problèmes mixtes hyperboliques non linéaires. PhD thesis, Université de Rennes I, 1987.

[11] C. Truesdell and W. Noll, The Non-linear Field Theories of Mechanics. Third edition. Edited and with a preface by Stuart S. Antman. Springer-Verlag, Berlin, 2004.

A Simple Model for Cavitation with Non-condensable Gases

Mathieu Bachmann, Siegfried Müller
Institut für Geometrie und Praktische Mathematik, RWTH Aachen
52056 Aachen, Germany
E-mail: bachmann@igpm.rwth-aachen.de, mueller@igpm.rwth-aachen.de

Philippe Helluy, Hélène Mathis
IRMA, Université de Strasbourg
7 rue René Descartes, 67084 Strasbourg, France
E-mail: helluy@math.unistra.fr, mathis@math.unistra.fr

Abstract

In this paper, we propose a numerical method to model the dynamical behavior of a spherical bubble of vapor and air inside water. The air is assumed to be miscible with the vapor. Each phase is described by a stiffened gas law and the mixture pressure law is recovered by an entropy maximization process. We use an adaptive finite volume solver in order to solve the Euler equations. Numerical experiments are presented that validate our approach.

1 Introduction

In the present work, we focus on the modeling of the collapse and rebound of a single gas-filled spherical bubble in a liquid. This topic has been subject of numerous numerical and experimental investigations, see for instance [5], [3]. Here, we consider three-phase mixtures, i.e., a bubble of vapor *and* air is immersed in liquid water. Since the air is a non-condensable gas, it will act as an elastic material in the rebound. Stiffened gas laws are used to depict the thermodynamical behavior of each phase. The mixture pressure law is obtained from an entropy maximization procedure [2], where we assume thermodynamical equilibrium in the mixture. This procedure allows us to compute the mass fraction of vapor in the spirit of [1]. Once it is done, all the other quantities can be computed. In particular, the whole procedure determines the mixture pressure as a function of the internal energy, the specific volume and the mass fraction of air. Note that it defines in a unique manner the mass and volume fractions of *all* the three phases.

The outline is as follows. In Section 2, we describe the mixture pressure law construction. This pressure is then coupled with the fluid model introduced in Section 3. Finally some numerical results are presented in Section 4.

2 Construction of the mixture pressure law

We consider a mixture of three phases (1), (2) and (3) representing the vapor, the liquid and the non-condensable gas (air), respectively. The liquid is not miscible with the two others, while the vapor and the gas are miscible. We only admit phase transition between vapor (1) and liquid (2).

Each phase obeys a stiffened gas Equation Of State (EOS), where the entropy function is defined as

$$s_i = C_i \ln((e_i - Q_i - \pi_i \tau_i)\tau_i^{\gamma_i - 1}) + s_i^0, \text{ for } i = 1, 2, 3. \tag{2.1}$$

In this formula, C_i is the specific heat at constant volume, Q_i is the heat of formation, π_i is the reference pressure and s_i^0 the reference entropy. The specific energy is noted e_i and the specific volume τ_i is $1/\rho_i$.

Let us recall some basic computations related to the stiffened gas EOS. The temperature T_i and the pressure p_i are given by

$$\frac{1}{T_i} = \frac{\partial s_i}{\partial e_i}, \quad \frac{p_i}{T_i} = \frac{\partial s_i}{\partial \tau_i}.$$

Inserting (2.1) this implies the following relations

$$C_i T_i = e_i - Q_i - \pi_i \tau_i, \tag{2.2}$$

$$p_i + \pi_i = (\gamma_i - 1)\rho_i C_i T_i. \tag{2.3}$$

The chemical potential $\mu_i = e_i + p_i \tau_i - T_i s_i$ thus is

$$\mu_i = Q_i + \gamma_i C_i T_i - T_i (C_i \ln(C_i T_i \tau_i^{\gamma_i - 1}) + s_0^i).$$

We now consider a mixture of the three phases. We denote by V_i the volume, M_i the mass and E_i the energy of each phase. We wish to compute the pressure law for a mixture of volume V, mass M and energy E. In other words, we want to compute the pressure from the knowledge of $\tau = V/M$, $e = E/M$ and the mass fraction $\varphi_3 = M_3/M$ of the non-condensable gas, i.e., $p = p(\tau, e, \varphi_3)$. We introduce the volume fraction $\alpha_i = V_i/V$, the mass fraction $\varphi_i = M_i/M$ and the energy fraction $\zeta_i = E_i/E$ of phase (i). With the fractions, we can express the partial intensive quantities from the global ones

$$\tau_i = \frac{\alpha_i}{\varphi_i}\tau, \quad e_i = \frac{\zeta_i}{\varphi_i} e. \tag{2.4}$$

Because phases (1) and (3) are miscible and phase (3) is immiscible with the other phases, the volumes are subject to the constraints

$$V_1 = V_3, \quad V_1 + V_2 = V. \tag{2.5}$$

The conservation of mass and energy read

$$M = M_1 + M_2 + M_3, \quad E = E_1 + E_2 + E_3.$$

From these constraints, we deduce

$$\alpha_1 = \alpha_3, \qquad \alpha_1 + \alpha_2 = 1, \tag{2.6}$$

$$\varphi_1 + \varphi_2 + \varphi_3 = 1, \quad \zeta_1 + \zeta_2 + \zeta_3 = 1. \tag{2.7}$$

At this point, the pressure would depend on τ, e, φ_3 but also on $\alpha_1, \varphi_1, \zeta_1, \zeta_2$. We have to eliminate the four supplementary fractions. The computation proceeds in three steps, which consist in setting the temperature, the pressure and the chemical equilibrium.

Temperature equilibrium. Setting $T = T_1 = T_2 = T_3$, we eliminate the energy fractions (ζ_1, ζ_2). Using relations (2.2) and (2.4), we obtain

$$\sum_{i=1}^{3} \varphi_i C_i T = e - \sum_{i=1}^{3} \varphi_i Q_i - \sum_{i=1}^{3} \alpha_i \pi_i \tau.$$

We then define

$$C = \sum_{i=1}^{3} \varphi_i C_i, \quad Q = \sum_{i=1}^{3} \varphi_i Q_i, \quad \pi = \sum_{i=1}^{3} \alpha_i \pi_i, \tag{2.8}$$

in such a way that

$$CT = e - Q - \pi\tau. \tag{2.9}$$

Analogously, we derive

$$p + \pi = \sum_{i=1}^{3} \alpha_i (\gamma_i - 1) \frac{\varphi_i}{\alpha_i} \rho C_i T.$$

Next we introduce

$$\gamma = \frac{\sum_{i=1}^{3} \varphi_i \gamma_i C_i}{\sum_{i=1}^{3} \varphi_i C_i}, \tag{2.10}$$

in such a way that

$$p + \pi = (\gamma - 1)\rho C T. \tag{2.11}$$

We observe that assuming the temperature equilibrium leads again to a stiffened gas EOS

$$p = (\gamma - 1)\rho(e - Q) - \gamma\pi,$$

where the mixture parameters are defined by (2.8). This way, the mixture parameters do not depend on the energy fractions.

Pressure equilibrium. The gas pressures satisfy Dalton's law $\tilde{p} = p_1 + p_3$. The non-miscibility of the gases implies that $p = \tilde{p} = p_2 = p_1 + p_3$. Using relation (2.11), we get

$$\left((\gamma_1 - 1)\frac{\varphi_1}{\alpha_1}\frac{C_1}{C} + (\gamma_3 - 1)\frac{\varphi_3}{\alpha_1}\frac{C_3}{C}\right)\rho CT - \pi_1 - \pi_3 = \\ (\gamma_2 - 1)\frac{1 - \varphi_1 - \varphi_3}{1 - \alpha_1}\frac{C_2}{C}\rho CT - \pi_2, \quad (2.12)$$

where we deduce from (2.9) together with (2.6) and (2.8)

$$CT = e - Q - (\alpha_1(\pi_1 + \pi_3 - \pi_2) + \pi_2)\tau. \quad (2.13)$$

If we multiply (2.12) by $\alpha_1(1 - \alpha_1)$, we obtain a second order equation in α_1. It is easy to check that this equation admits one single root in $(0, 1)$ denoted by $\alpha_1(\varphi_1)$.

Chemical equilibrium. From the chemical equilibrium, we compute φ_1. The first step is to express the chemical potential as a function of τ, e and φ_1, i.e.,

$$\mu_i = Q_i + C_i T \times \left(\gamma_i - \frac{s_i^0}{C_i} - \ln\left(C_i T \left(\frac{\alpha_i}{\varphi_i}\tau\right)^{\gamma_i - 1}\right)\right),$$

where the mixture parameters C, Q, π are given by (2.8) and the temperature $T = T(\varphi_1)$ is computed from (2.13). Since φ_3 is assumed to be constant, the mass fraction for the liquid is only a function of φ_1 according to (2.7), i.e., $\varphi_2 = 1 - \varphi_1 - \varphi_3$. Therefore, the mass fractions of the liquid and the vapor are bounded by $[0, 1 - \varphi_3]$. Then we determine φ_1 in order to compute the pressure law. Going back to chemistry theory, we apply the following entropy maximization procedure

- if $\mu_1(1 - \varphi_3) < \mu_2(1 - \varphi_3)$ and $T(1 - \varphi_3) \geq 0$, the mass fraction of the liquid $\varphi_2 = 0$ corresponds to a stable case, thus $\varphi_1 = 1 - \varphi_3$, i.e., the mixture contains no liquid but vapor only;
- else if $\mu_1(0) > \mu_2(0)$ and $T(0) \geq 0$, the mass fraction of the liquid $\varphi_2 = 1 - \varphi_3$ corresponds to a stable case, thus $\varphi_1 = 0$, i.e., the mixture contains no vapor but liquid only;
- else the mixture is necessarily at saturation and we have to compute the mass fraction $\varphi_1 \in [\varphi_{min}, \varphi_{max}]$ such that $\mu_1(\varphi_1) = \mu_2(\varphi_1)$ where $\varphi_{\min} := \min\{\varphi \in [0, 1] : T(\varphi) \geq 0\}$ and $\varphi_{\max} := \max\{\varphi \in [0, 1] : T(\varphi) \geq 0\}$. This gives a mass fraction such that $0 < \varphi_1 < 1 - \varphi_3$, i.e., the mixture is at saturation and contains vapor and liquid.

Note that the case $T(1-\varphi_3) < 0$ and $T(0) < 0$ never should occur. Once the mass fraction φ_1 is computed, it is possible to compute all the other quantities and thus the mixture pressure $p = p(\tau, e, \varphi_3)$ according to (2.11).

3 Compressible fluid model

Once the pressure law is defined, we consider the three-phase mixture as a single inviscid compressible medium described by the density ρ, the pressure p, the velocity u and the internal energy e, all depending on the space variable x and the time t. The balances of mass, momentum and energy lead to the Euler equations

$$\begin{cases} \partial_t(\rho) + \partial_x(\rho u) = 0, \\ \partial_t(\rho u) + \partial_x(\rho u^2 + p) = 0, \\ \partial_t(\rho E) + \partial_x((\rho E + p)u) = 0, \end{cases} \tag{3.1}$$

where $E = e + u^2/2$ is the total energy. In principle, the amount of *air* should be evolved by the homogeneous transport equation

$$\partial_t \varphi_3 + u \partial_x \varphi_3 = 0. \tag{3.2}$$

In order to avoid the pressure oscillation phenomenon, described for instance in [6], we consider a uniformly constant fraction of air φ_3. Finally, the system (3.1) is closed by the pressure law $p = p(\tau, e, \varphi_3)$ according to the procedure described in Section 2.

4 Numerical results

The first test case consists of a Riemann problem which presents two rarefaction waves moving in opposite directions in water. The value of φ_3 is varying in the range $[10^{-9}, 10^{-1}]$. The computational domain is $[0, 1]$ and the simulation time is 0.2 ms. The computations are performed by an adaptive multiresolution finite volume solver, cf. [4]. The coarse grid consists of $N_0 = 20$ cells and is locally refined by at most L refinement levels. L varies between 5 and 15 and the CFL number is 0.8. The initial conditions are $\rho_R = \rho_L = 1000 \text{ kg} \cdot \text{m}^{-3}$, $u_R = -u_L = 100 \text{ m} \cdot \text{s}^{-1}$ and $e_R = e_L = 1.32 \times 10^5$ J for the right and left state, respectively. The physical parameters are listed in Table 1.

First of all, we observe in Figure 4.1a that the rarefaction waves are smeared for $L = 5, 7$ whereas the smearing reduces significantly for $L \geq 10$. In Figure 4.1b, we note that the density drops significantly in the cavitation zone around $x = 0.5$ m with decreasing φ_3. When the amount of air is close to zero, the density is almost zero in the cavitation zone. In particular, the higher the amount of air, the less vapor is in the cavitation zone. Indeed the air occupies the area, which limits the vaporization process as can be observed in Figure 4.1d. Thus from (2.6), it can be deduced that with increasing φ_3, the volume fraction of liquid also increases. Hence, the density is higher in the cavitation zone. Note that the initial pressure changes as φ_3 varies, see Figure 4.1c. However, this change in pressure does not affect the conserved quantities. In Figure 4.1c, we observe in the pressure profile discontinuities in the cavitation region when $\varphi_3 = 10^{-9}$. These are probably split waves due to the phase transition across the saturation line.

Table 1 Physical parameters.

	unit	vapor (1)	liquid (2)	air (3)
γ		1.3	3	1.4
π	Pa	0	8533×10^5	0
C_v	J/kg K	1614.38	1400	719.28
Q	J/kg	1892000	-0.1148×10^7	-0.26979×10^6
s_i^0	J/K	-14684	1460	-8725

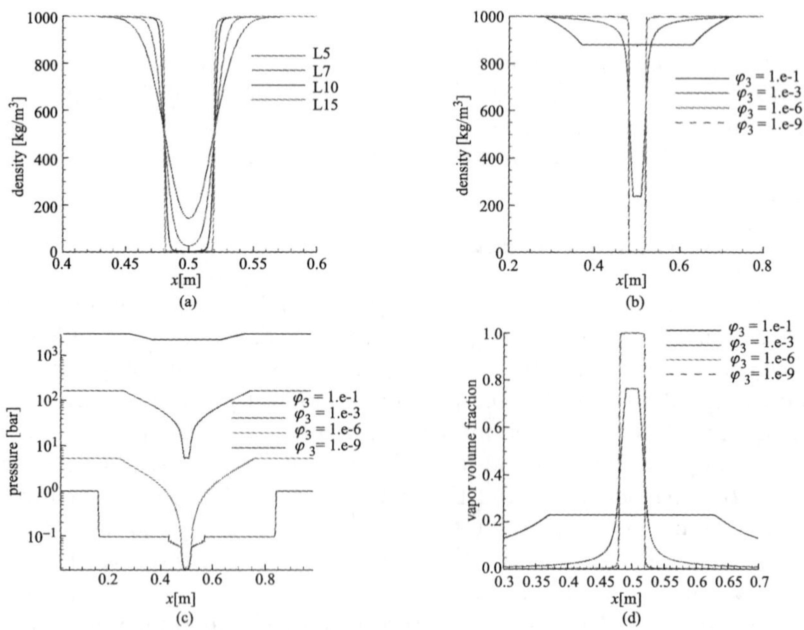

Figure 4.1 Test case 1: (4.1a) density profiles for $\varphi_3 = 10^{-6}$ and different discretizations. (4.1b) density, (4.1c) pressure, vapor volume fraction for different values of φ_3 and $L = 15$.

The second test case is a collapse of a spherical vapor bubble of radius 0.7469 mm. The vapor state is determined by $\rho_V = 9.57 \times 10^{-4}$ kg $\cdot$ m^{-3}, $u_V = 0$ m $\cdot$ s^{-1} and $e_V = 1.90795263 \times 10^6$ J whereas the ambient liquid state is $\rho_L = 1000$ kg $\cdot$ m^{-3}, $u_L = 0$ m $\cdot$ s^{-1} and $e_L = 1.32 \times 10^5$ J, see [5]. Again, the physical parameters are those in Table 1. We can observe in Figure 4.2a that φ_3 has an important effect on the collapse time. The higher the amount of air, the faster the collapse is. This is due to the fact that the pressure increases in the liquid when the amount of air increases but keeping the same conservative variables. In order to gain inside in the dynamics of the bubble collapse we present in Figure 4.2b a space-time plot of the density gradient magnitude for the results corresponding to $\varphi_3 = 10^{-2}$. When the shock (s) focuses in the center, pressure and temperature increase significantly such that condensation occurs in the bubble center. As the bubble is collapsing, the region of pure vapor, indicated by the isoline $\alpha_3 = 0.99$, reduces. Thus the reflected shock (rs) interacts with the liquid at $t = 0.85$ µs and $x = 0.27$ mm and it splits in a transmitted shock (ts) in the liquid and in a reflected shock (rrs) moving towards the bubble center. From the comparison with experiments, see [5], we finally conclude for the configuration at hand that a reasonable value for φ_3 to investigate bubble oscillations is around 10^{-3}.

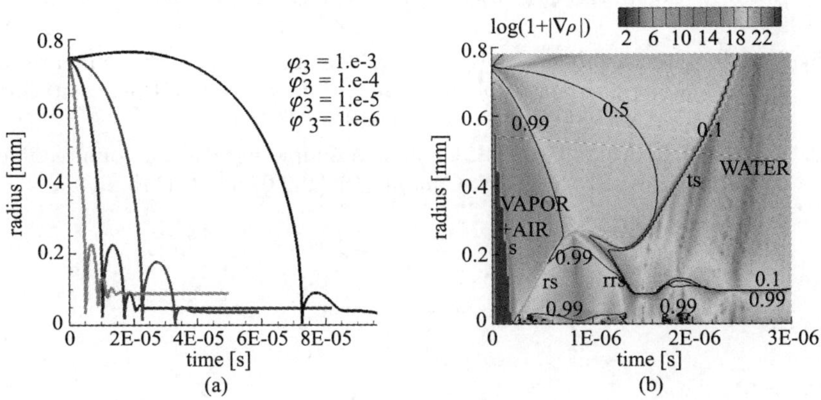

Figure 4.2 Test case 2: (4.2a) isoline $\alpha_3 = 0.5$ for computations with $\varphi_3 = 10^{-3}, 10^{-4}, 10^{-5}$ and 10^{-6}. (4.2b) density gradient magnitude and isolines of α_3 in case of $\varphi_3 = 10^{-2}$.

5 Conclusions

In this paper we have proposed a cavitation model with a fixed amount of non-condensed gas. Assuming thermodynamical equilibrium in the mixture, we have shown that the mixture pressure law can be computed as a stiffened gas law. This allows us to use a classical finite volume scheme with a relaxation method. The numerical results illustrate a cavitation behavior depending on the amount of air.

In a forthcoming work we will investigate whether the pressure jumps that can be observed in the cavitation region of test case 1, are caused by the saturation line of the mixture.

Another step is to use a non-constant mass fraction of air. Numerical difficulties may arise. In particular, pressure oscillations can appear at the interface between vapor and water, like those observed in many works about two-phase flows, see for instance [6] and [1].

References

[1] T. Barberon and P. Helluy, Finite volume simulation for cavitating flows. Comput. Fluids 34 (2005), no. 7, 832–858.

[2] H. B. Callen, Thermodynamics and an Introduction to Thermostatistics. Wiley, New York, 1985.

[3] W. Lauterborn and T. Kurz, Physics of bubble oscillations. Rep. Prog. Phys. 73 (2010) 106501.

[4] S. Müller, Adaptive Multiscale Schemes for Conservation Laws. In: Lecture Notes in Computational Science and Engineering, volume 27. Springer, 2003.

[5] S. Müller, M. Bachmann, D. Kröninger, T. Kurz and P. Helluy, Comparison and validation of compressible flow simulations of laser-induced cavitation bubbles. Comput. Fluids 38 (2009), no. 9, 1850–1862.

[6] R. Saurel, R. Abgrall and M. Choulli, A simple method for compressible multifluid flows. SIAM J. Sci. Comput. 21 (1999), no. 3, 1115–1145.

Adjoint IMEX-based Schemes for the Numerical Solution of Optimal Control Problems Governed by Conservation Laws*

Mapundi K. Banda
University of the Witwatersrand, Faculty of Science
WITS 2050, South Africa
E-mail: mapundi.banda@wits.ac.za

Michael Herty
RWTH Aachen University
D-52056 Aachen, Germany
E-mail: herty@mathc.rwth-aachen.de

Abstract

Starting from relaxation schemes for hyperbolic conservation laws we derive continuous and discrete schemes for optimization problems subject to nonlinear, scalar hyperbolic conservation laws. We discuss properties of first-order discrete schemes and give some numerical results with non-smooth desired states.

1 Introduction

We are concerned with numerical approaches for optimization problems governed by hyperbolic conservation laws. As a prototype, we consider the problem

$$\min_{u_0} \mathbf{J}(u(\cdot,T),u_d) \text{ subject to } u_t + f(u)_x = 0,\; u(x,t=0) = u_0(x).$$

It is known that in general the semi-group generated by a conservation law is non-differentiable in L^1 even in the scalar, one-dimensional case. A differential structure on general BV-solutions for hyperbolic conservation laws in one space dimension has been introduced and discussed in [5, 10, 11, 12, 19]. Based on the derived calculus first-order optimality

*This work has been supported by DFG SPP1253 HE5386/8-1, DAAD D/08/11076 and DAAD 50727872.

conditions for systems have been given in [13]. Theoretical discussion on the resulting non-conservative equations can be found in [6, 7, 8]. Numerical results in the scalar, one-dimensional case with distributed control are also presented in [18, 19, 20, 15]. The results presented here are discussed in more detail in [2]. We give here more explanations and numerical results in this communication.

2 Relaxation of the optimization problem

In [14] a relaxation system as an approximation to a general nonlinear conservation law was introduced. Since then relaxation methods have been investigated by many authors, e.g., [1, 4, 3, 9, 16, 17]. A different discretization of relaxation schemes has been introduced in [1] . Therein, estimates on the discrete scheme as well as convergence results for first- and second-order spatial dimension have been given. Here, we use relaxation in the context of the optimization problem given above.

The hyperbolic relaxation for

$$u_t + f(u)_x = 0, \quad u(x, t = 0) = u_0(x) \tag{2.1}$$

is given by

$$\begin{aligned} u_t + v_x &= 0, & u(x, t = 0) &= u_0(x), & (2.2)\\ v_t + a^2 u_x &= -\frac{1}{\epsilon}\left(v - f(u)\right), & v(x, t = 0) &= f(u_0(x)), & (2.3)\end{aligned}$$

where a^2 satisfies the sub-characteristic condition

$$a^2 - f'(u)^2 \geq 0.$$

For positive ϵ, it is known that a solution (u, v) yields an approximation to the solution (2.1) in the following sense. For ϵ sufficiently small, we have that up to second-order in ϵ

$$u_t + f(u)_x = \epsilon \left(\left(a^2 - f'(u)^2\right) u_x\right)_x. \tag{2.4}$$

We now consider the problem

$$\min_{u_0} \mathbf{J}(u(\cdot, T), u_d) \text{ subject to (2.2)–(2.3).}$$

We obtain by formal Lagrange calculus the equations for the multipliers p and q as

$$\begin{aligned} -p_t - a^2 q_x &= \frac{q}{\epsilon} f'(u), & p(x, t = T) &= p_T(x), & (2.5)\\ -q_t - p_x &= -\frac{q}{\epsilon}, & q(x, t = T) &= q_T(x). & (2.6)\end{aligned}$$

An expansion in terms of ϵ yields the solution (p, q) which solves up to second-order in ϵ (see [2]):

$$-p_t - f'(u)p_x = \epsilon a^2 p_{xx}. \tag{2.7}$$

Since the equation is solved backwards in time, we obtain a numerical diffusion without further constraints on the numerical scheme. In the following paragraphs, we apply a similar framework to the discretizations and show that we obtain suitable discretizations of (2.5), (2.6) and (2.7), respectively.

Our motivation is to pursue the same idea on the discrete level. To this end, we consider a first–order space and time discretization (IMEX): Assume an equally spaced grid with grid spacing $\Delta x = x_{i+\frac{1}{2}} - x_{i-\frac{1}{2}}$ where $x_{i+\frac{1}{2}} = i\Delta x + \frac{1}{2}\Delta x$, and a uniform time-step $\Delta t = t_{n+1} - t_n$. Let $w^n_{i+\frac{1}{2}} = w(x_{i+\frac{1}{2}}, t_n)$ and

$$D_x w_i = \frac{w_{i+\frac{1}{2}} - w_{i-\frac{1}{2}}}{\Delta x}.$$

The variables $v \pm au$ are transported with velocities $\pm a$. Hence, a first–order upwind scheme is given by

$$(v + au)_{i+\frac{1}{2}} = v_i + au_i, \; (v - au)_{i+\frac{1}{2}} = v_{i+1} - au_{i+1},$$

or equivalently,

$$u_{i+\frac{1}{2}} = \frac{u_i + u_{i+1}}{2} - \frac{v_{i+1} - v_i}{2a}, \; v_{i+\frac{1}{2}} = \frac{v_i + v_{i+1}}{2} - a\frac{u_{i+1} - u_i}{2},$$

and an implicit-explicit scheme (IMEX) for the time-discretization for (2.2) and (2.3) is therefore given by

$$\begin{gathered} u_i^* = u_i^n, \; v_i^* = v_i^n - \frac{\Delta t}{\epsilon}\left(v_i^* - f(u_i^*)\right), \\ u_i^1 = u_i^* - \Delta t D_x v_i^*, \; v_i^1 = v_i^* - \Delta t a^2 D_x u_i^*, \\ u_i^{n+1} = u_i^1, \; v_i^{n+1} = v_i^1. \end{gathered}$$

Hence, on the discrete level we obtain the adjoint scheme to the previous equations given by

$$\begin{gathered} p_i^1 = p_i^{n+1}, \; q_i^1 = q_i^{n+1}, \\ p_i^* = p_i^1 + \frac{\Delta t}{\epsilon} q_i^* f'(u_i^*) - \Delta t a^2 D_x^* q_i^1, \; q_i^* = q_i^1 - \frac{\Delta t}{\epsilon} q_i^* - \Delta t D_x^* p_i^1, \\ p_i^n = p_i^*, \; q_i^n = q_i^*. \end{gathered}$$

Since $D_x^* w_i = \frac{1}{\Delta x}(w_{i+\frac{1}{2}} - w_{i-\frac{1}{2}})$, we obtain in a compact form

$$q_i^* = q_i^{n+1} - \Delta t D_x^* p_i^{n+1}, \; q_i^n = q_i^* - \frac{\Delta t}{\epsilon} q_i^n, \tag{2.8}$$

$$p_i^* = p_i^{n+1} - \Delta t a^2 D_x^* q_i^{n+1}, \; p_i^n = p_i^* + \frac{\Delta t}{\epsilon} q_i^n f'(u_i^n). \tag{2.9}$$

For more details we refer to [2]. The following results can be shown [2, Proposition 2.2]

Proposition 2.1. *Let $\epsilon > 0$. The discrete adjoint equations (2.8) and (2.9) are a first-order discretization in time and space of the continuous adjoint equations (2.5) and (2.6). The time discretization is a (backwards) explicit Euler scheme in the transport and (backwards) implicit Euler step in the stiff part. In space we obtain a first–order Upwind scheme for the variables $p \pm aq$.*

Next, we explain how the numerical results for problems of the type

$$\min_{u_0} \mathbf{J}(u(\cdot, T), u_d) \text{ subject to } u_t + f(u)_x = 0, u(x,0) = u_0(x) \tag{2.10}$$

can be obtained. Based on the previous discussion the numerical schemes are devised by solving the relaxation approximation (2.11) *instead of* (2.10).

$$\begin{aligned} &\min_{u_0} \mathbf{J}(u(\cdot, T), u_d) \text{ subject to} \\ &u_t + v_x = 0, u(x,0) = u_0(x), \\ v_t + a^2 u_x = -&\frac{1}{\epsilon}(v - f(u)), \; v(x,0) = f(u_0(x)). \end{aligned} \tag{2.11}$$

We consider functionals $\mathbf{J}$ of the integral type on the spatial domain $I \subset \mathbb{R}$

$$\mathbf{J}(u(\cdot, T), u_d) := \int_I J(u(x,T), u_d(x))dx \tag{2.12}$$

for some desired state u_d. In this context, the control variable is the initial profile, u_0. For numerical solutions of (2.11), we discretize the terminal data to (2.11) as

$$p_{N,i} = J_u(u_{i,N}, u_{d,i}), \; q_{N,i} = 0. \tag{2.13}$$

and the discrete gradient of the reduced cost functional as

$$\nabla_{u_{0,i}} \tilde{J} = p_{0,i} + q_{0,i} f'(u_{0,i}). \tag{2.14}$$

Finally, we present some results on Burger's equation and different parameters $\epsilon > 0$. Other numerical results and further discussion can be found in [2]. We will use the solution at time $t = 2$ to Burger's equation $f(u) = \frac{1}{2}u^2$ with initial data $u_0(x) = \frac{1}{2}+\sin(x)$ as desired states u_d in the optimization problem (2.11). We use the first-order IMEX discretization for the numerical solution on an equidistant grid with $N_x = 100$ grid points in space. The stopping criteria for the gradient descent method is either 500 optimization steps or a functional value less than $1.e-8$. The relaxation rate is taken to be $\epsilon \in \{1.e-4, 1.e-8, 1.e-16\}$ and the corresponding results are presented in Figure 2.1. The state $u(x,t)$

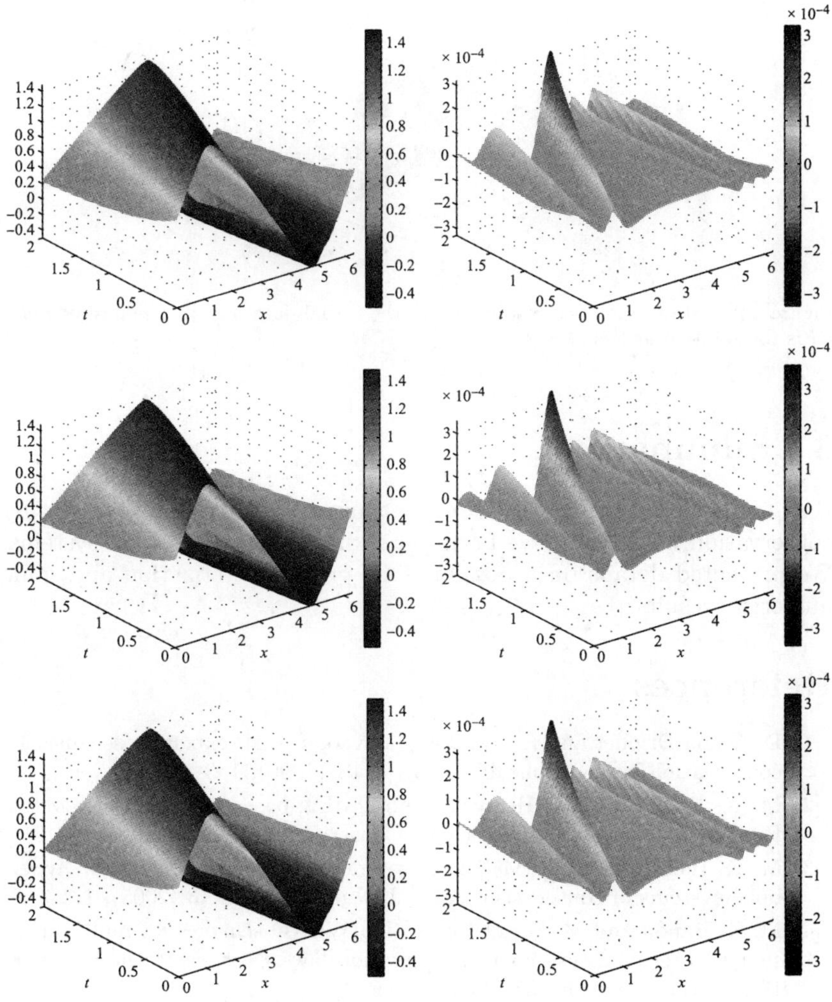

Figure 2.1 State $u(x,t)$ and adjoint $p(x,t)$ for the solution to the optimization problem (2.11) and with the values of ϵ being $\{1.e-4, 1.e-8, 1.e-16\}$ from top to bottom.

corresponding to the initial data u_0 is depicted in Figure 2.2. We present the state $u(x,t)$ as well as the adjoint $p(x,t)$ after optimization in Figure 2.1 for different values of ϵ. We observe that despite the different values of ϵ, we obtain similar results in the numerical experiments. We recover, outside the area confined by the extremal backwards characteristics of the shock the initial data u_0. Inside the confined area we observe a linear decay of the initial data for small values and a smoother transition for the high values of ϵ.

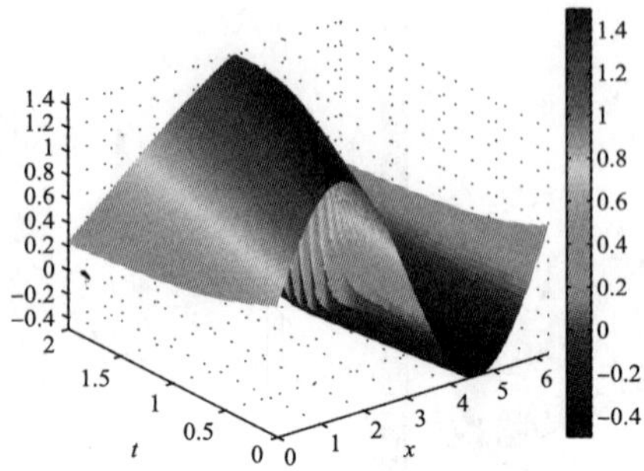

Figure 2.2 State $u(x,t)$ corresponding to the initial data $u_0(x)$. The desired state u_d is the solution at time $t = 2$.

3 Summary

Using recent results, we discussed optimization problems governed by conservation laws that have been approximated by relaxation systems. We presented details on a possible first–order discretization and some numerical results.

References

[1] D. Areba-Driollet and R. Natalini, Convergence of relaxation schemes for conservation laws. Applicable Analysis 61 (1996), 163–193.

[2] M. K. Banda and M. Herty, Adjoint IMEX-based schemes for Control Problems Governed by Hyperbolic Conservation Laws. preprint (2010).

[3] M. K. Banda and M. Seaïd, Higher-order relaxation schemes for hyperbolic systems of conservation laws. J. Numer. Math. 13 (2005), 171–196.

[4] M. K. Banda and M. Seaïd, Relaxation WENO schemes for multidimensional hyperbolic systems of conservation laws. Numer. Methods Partial Differential Equations 23 (2007), 1211–1234.

[5] S. Bianchini, On the shift differentiability of the flow generated by a hyperbolic system of conservation laws. Discrete Contin. Dynam. Systems 6 (2000), 329–350.

[6] F. Bouchut and F. James, One-dimensional transport equations with discontinuous coefficients. Nonlinear Anal. 32 (1998), 891–933.

[7] F. Bouchut and F. James, Differentiability with respect to initial data for a scalar conservation law. In: Hyperbolic Problems: Theory, Numerics, Applications, vol. I (Zürich, 1998), vol. 129 of Internat. Ser. Numer. Math., Birkhäuser, Basel, 1999, 113–118.

[8] ——, Duality solutions for pressureless gases, monotone scalar conservation laws, and uniqueness, Comm. Partial Differential Equations 24 (1999), 2173–2189.

[9] Y. Brenier, R. Natalini and M. Puel, On a relaxation approximation of the incompressible Navier-Stokes equations. Proceedings of the American Mathematical Society 132 (2003), 1021.

[10] A. Bressan and G. Guerra, Shift-differentiability of the flow generated by a conservation law. Discrete Contin. Dynam. Systems 3 (1997), 35–58.

[11] A. Bressan and M. Lewicka, Shift differentials of maps in BV spaces. In: Nonlinear theory of generalized functions (Vienna, 1997), vol. 401 of Chapman & Hall/CRC Res. Notes Math., Chapman & Hall/CRC, Boca Raton, FL, 1999, 47–61.

[12] A. Bressan and A. Marson, A variational calculus for discontinuous solutions to conservation laws. Communications Partial Differential Equations, 20 (1995), 1491–1552.

[13] A. Bressan and W. Shen, Optimality conditions for solutions to hyperbolic balance laws, control methods in PDE-dynamical systems. Contemp. Math. 426 (2007), 129–152.

[14] S. Jin and Z. Xin, The relaxation schemes for systems of conservation laws in arbitrary space dimensions. Comm. Pure Appl. Math. 48 (1995), 235–276.

[15] Z. Liu and A. Sandu, On the properties of discrete adjoints of numerical methods for the advection equation. Int. J. for Num. Meth. in Fluids 56 (2008), 769–803.

[16] R. Natalini, Convergence to equilibrium for the relaxation approximations of conservation laws. Comm. Pure Appl. Math. 49 (1996), 795–823.

[17] R. Natalini and A. Terracina, Convergence of a relaxation approximation to a boundary value problem for conservation laws. Comm. Partial Differential Equations, 26 (2001), 1235–1252.

[18] S. Ulbrich, Optimal Control of Nonlinear Hyperbolic Conservation Laws with Source Terms. Technische Universitaet Muenchen, 2001.

[19] S. Ulbrich, A sensitivity and adjoint calculus for discontinuous solutions of hyperbolic conservation laws with source terms. SIAM J. Control Optim. 41 (2002), 740.

[20] ——, Adjoint-based derivative computations for the optimal control of discontinuous solutions of hyperbolic conservation laws. Systems & Control Letters 3 (2003), 309.

A Two-dimensional Model of Pedestrian Flow Generating Pattern Formation*

Stefan Berres
Departamento de Ciencias Matemática y Física
Universidad Católica de Temuco, Temuco, Chile
E-mail: sberres@uct.cl

Ricardo Ruiz-Baier
Modeling and Scientific Computing, MATHICSE
École Polytechnique Fédérale de Lausanne
CH-1015 Lausanne, Switzerland
E-mail: ricardo.ruiz@epfl.ch

Hartmut Schwandt
Institut für Mathematik, Technische Universität Berlin
Straße des 17. Juni 136, D-10623 Berlin, Germany
E-mail: schwandt@math.tu-berlin.de

Elmer M. Tory
Department of Mathematics and Computer Science
Mount Allison University, Sackville, NB E4L 1G6, Canada
E-mail: elmer@sherpasoftware.ca

Abstract

The pedestrian flow at a crossing is modeled by a two-dimensional system of conservation laws with a location-dependent flux function. Different populations moving in different directions are represented by different phases. Numerical simulations show that the hyperbolic-elliptic degenerate type of equation-system leads to spatial segregation if the initial data are chosen inside the elliptic region of the phase space.

*The first author is supported by Conicyt (Chile) through Fondecyt project #11080253. The work of the second author is supported by the European Research Council Advanced Grant "Mathcard, Mathematical Modelling and Simulation of the Cardiovascular System", Project ERC-2008-AdG 227058. The third author is supported by the Deutsche Forschungsgemeinschaft (German Research Foundation) through project SCHW548/5-1.

1 Introduction

The idea of considering traffic flow in the context of continuum mechanics was first formulated in the Lighthill-Whitham-Richards (LWR) model for traffic flow [9, 10], where traffic dynamics are described by the scalar one-dimensional conservation law

$$u_t + f(u)_x = 0. \tag{1.1}$$

The unknown variable $u = u(x, t)$ denotes a locally averaged car-density assigned at each time t and each position x. The flux function

$$f(u) = auV(u),$$

encapsulates the model assumptions about the average velocity. Here, a is a constant that corresponds to the maximum velocity and $V = V(u)$ is a normalized density-dependent velocity function. A natural assumption is that the velocity function V is monotonically decreasing with density. For car-traffic, this formalizes the observation that, due to mutual blocking on a road, the average speed decreases as the number of cars increases. In one extreme case, the maximum speed is attained when there are no other cars on the road. In the other extreme case, there is no movement when the road is completely jammed. The meaning of the velocity function is analogous for pedestrian traffic. These considerations lead to a monotonically decreasing velocity function

$$V(0) = 1, \quad V(1) = 0, \quad V'(u) < 0 \quad \text{for} \quad u \in (0, 1), \tag{1.2}$$

for example $V(u) = 1 - u$.

One generalization of the scalar LWR model is to extend it to multiphase traffic flow. The first formulation of a two-phase description of traffic flow was the model of Bick and Newell [5]

$$u_t + (u(1 - u - \beta v))_x = 0, \quad v_t + (-v(1 - \beta u - v))_x = 0, \tag{1.3}$$

with $\beta > 0$, which describes the flow in opposite directions that are mutually affected. In more general multiphase traffic-flow models, the unknown variables represent several phases corresponding to different lanes or car types [1]. A special property of the Bick-Newell model (1.3) is that it contains an elliptic region, i.e., a region where the Jacobian matrix of the flux function has non-real eigenvalues.

Another generalization of the one-dimensional scalar equation (1.1) consists of an extension to two space dimensions. In the subsequent sections, a two-dimensional model for pedestrian flow is presented for typical benchmark settings. We particularly focus on the effect of the elliptic region on the solution structure. This contribution is a short version of [4], where a more general model framework, details of the numerical method, and simulations of various model variants can be found.

2 Two-species pedestrian model

The two-dimensional pedestrian-flow model describing the spatial interaction of two species of pedestrians is formulated as the system of two convection-diffusion equations

$$u_t + \operatorname{div} F(u,v) = \varepsilon\Delta u, \quad v_t + \operatorname{div} G(u,v) = \varepsilon\Delta v, \tag{2.1}$$

in $\Omega_T = \Omega \times (0,T)$, where u, v denote the two distinct pedestrian species in the domain $\Omega = [-1,1]^2$. The parabolic terms $\varepsilon\Delta u, \varepsilon\Delta v$ account for hydrodynamic diffusion that is included for completeness. However, this diffusion is assumed to be small (if not negligible) and the equations are dominated by convection. Convection-dominated equations are known for their special need for a careful discretization.

In generalizing the LWR model to two space dimensions, the main issue is to determine the directions of the flow. Pedestrian flow in two space dimensions follows the same physical principles as in its reduction to one space dimension. The main difference is that the *total flux* of a species, which describes the overall movement, is distributed to the directions. The flux functions are specified as

$$F(u,v) = a_u u V(u,v)\boldsymbol{d}_u^{\mathrm{T}}, \quad G(u,v) = a_v v V(u,v)\boldsymbol{d}_v^{\mathrm{T}}, \tag{2.2}$$

where the directions $\boldsymbol{d}_u = (d_u^x, d_u^y)$ and $\boldsymbol{d}_v = (d_v^x, d_v^y)$ denote the preferred directions of species u, v, respectively. The flux components split the total flux, which is the flux in the direction of principal movement, into directions. For simplicity we choose the parameters $a_u = a_v = 1$.

In the case of countercurrent flow, where $\boldsymbol{d}_u = (0,1)$, $\boldsymbol{d}_v = (0,-1)$, the system (2.1), (2.2) can be specified as

$$u_t + \big(u(1-u-v)\big)_y = \varepsilon\Delta u, \quad v_t - \big(v(1-u-v)\big)_y = \varepsilon\Delta v. \tag{2.3}$$

In the case of perpendicular flow, where $\boldsymbol{d}_u = (1,0)$, $\boldsymbol{d}_v = (0,1)$, the system (2.1), (2.2) can be specified as

$$u_t + \big(u(1-u-v)\big)_x = \varepsilon\Delta u, \quad v_t + \big(v(1-u-v)\big)_y = \varepsilon\Delta v. \tag{2.4}$$

3 Numerical approximation

For the numerical simulation of the two-dimensional system (2.1), a fully adaptive multiresolution scheme is employed. Space-adaptive methods represent an adequate tool for accurately capturing the wave fronts at a reduced computational cost. The concept of multiresolution methods in the numerical solution of PDEs goes back to the pioneering work of Harten [8]. The multiresolution setting permits a navigation between different resolution levels via certain transfer operators. In particular, projection and prediction operators are defined for coarsening and refining, respectively. Thereby, the discrete representation of the approximate solution on a given fine grid is equivalent to the projected solution on a coarse grid plus a series of detail-coefficients. The data compression is controlled by a thresholding procedure, which preserves only the relevant information stored in the original data. The flexibility of multiresolution methods also permits the straightforward combination with other numerical strategies such as ENO schemes or time-adaptive methods (see e.g. [6]) and a dynamic storage and access to data to construct fully adaptive methods.

4 Numerical examples

In the numerical examples, we take the coefficients $a_u = a_v = 1$. The flow characteristics are mainly described by the boundary conditions and the superimposed flow directions. The solution behaviour of the relevant equation depends on the position of the elliptic region in the phase space and, in particular, whether the initial data are chosen inside or outside the elliptic region.

4.1 Example 1: countercurrent flow

The domain of interest is $\Omega = [-1, 1]^2$, and the initial conditions are

$$u(\boldsymbol{x}, t = 0) = u_0 + \eta_u(\boldsymbol{x}), \quad v(\boldsymbol{x}, t = 0) = v_0 + \eta_v(\boldsymbol{x}), \quad \text{for} \quad \boldsymbol{x} \in \Omega, \tag{4.1}$$

where η_u, η_v are uniformly distributed random noise with variations of 10% and 1.5% for u and v, respectively. The flow is countercurrent: $\boldsymbol{d}_u = (0, 1)$, $\boldsymbol{d}_v = (0, -1)$. In Example 1A, the initial data $\boldsymbol{u}_0 = (0.4, 0.35)^{\mathrm{T}}$ are taken inside the elliptic region, whereas in Example 1B, the initial data $\boldsymbol{u}_0 = (0.2, 0.2)^{\mathrm{T}}$ are taken outside. As one can see in Figure 4.1, a strong spatial phase-separation is confirmed for initial data inside the elliptic region. This separation leads to channel formation, where the phases separate vertically and move horizontally through channels. Moreover, clusters with a predominant species, moving in the directions of that species, are formed by the smearing effect of the parabolic term. (Here that term is weighted by a constant $\varepsilon = 1.5 \times 10^{-3}$.) With respect to the boundary condition, the default setting is a zero-flux Neumann boundary condition, which prescribes that there is no flux through the boundary and corresponds to an impermeable wall. For initial data outside the elliptic region, as used in Figure 4.2, we observe a countercurrent flow of overlying layers without

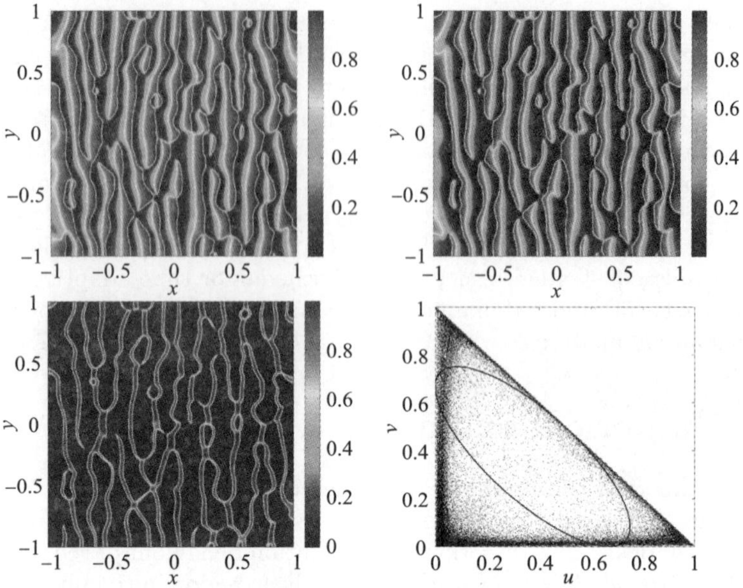

Figure 4.1 Example 1A. Species' densities u, v, void fraction $(1 - u - v)$ at time $t = 2$, and corresponding phase diagram where the elliptic region is depicted.

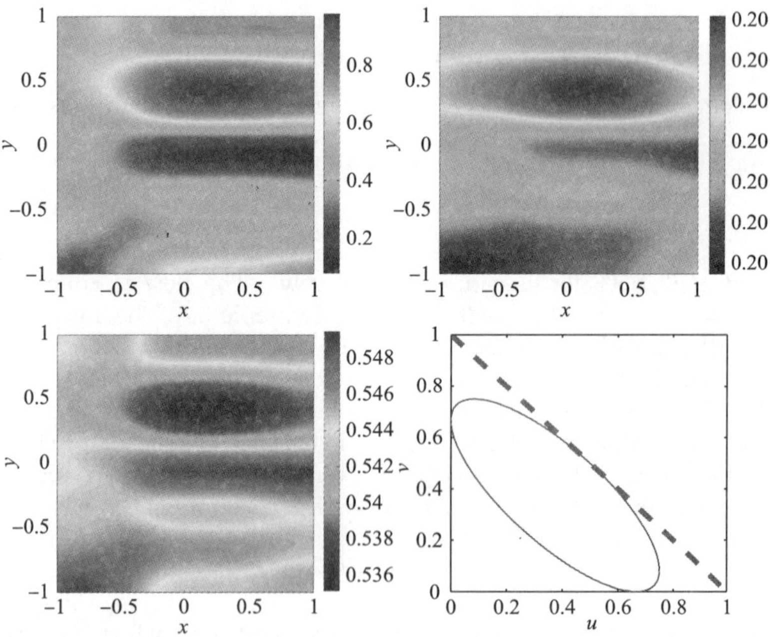

Figure 4.2 Example 1B. Species' densities u, v, void fraction $(1 - u - v)$ at time $t = 5$, and corresponding phase diagram where the elliptic region is depicted.

producing a channel structure. Furthermore, the variation of the density values is low and they remain outside the elliptic region, near the initial state.

For the multiresolution setting, $L = 10$ resolution levels are used with a reference tolerance of $\varepsilon_{\text{ref}} = 1.25 \times 10^{-2}$ for Example 1A, and $\varepsilon_{\text{ref}} = 5.0 \times 10^{-4}$ for Example 1B.

4.2 Example 2: perpendicular flow

In Example 2, a crossing with rectangular flow directions is considered. The directions are perpendicular: $\boldsymbol{d}_u = (1, 0)$, $\boldsymbol{d}_v = (0, 1)$. For this example, the diffusion term is reduced further, i.e. $\varepsilon = 1 \times 10^{-3}$ and, as in Example 1A, zero-flux boundary conditions and homogeneous initial data (4.1) with $\boldsymbol{u}_0 = (0.4, 0.35)^{\mathrm{T}}$ are taken inside the elliptic region. A small channeling effect is also observed in the corresponding Figure 4.3. It is weaker than in Examples 1A and 1B since the flow is not in opposite but rather in perpendicular directions. We observe that both species follow an intermediate direction.

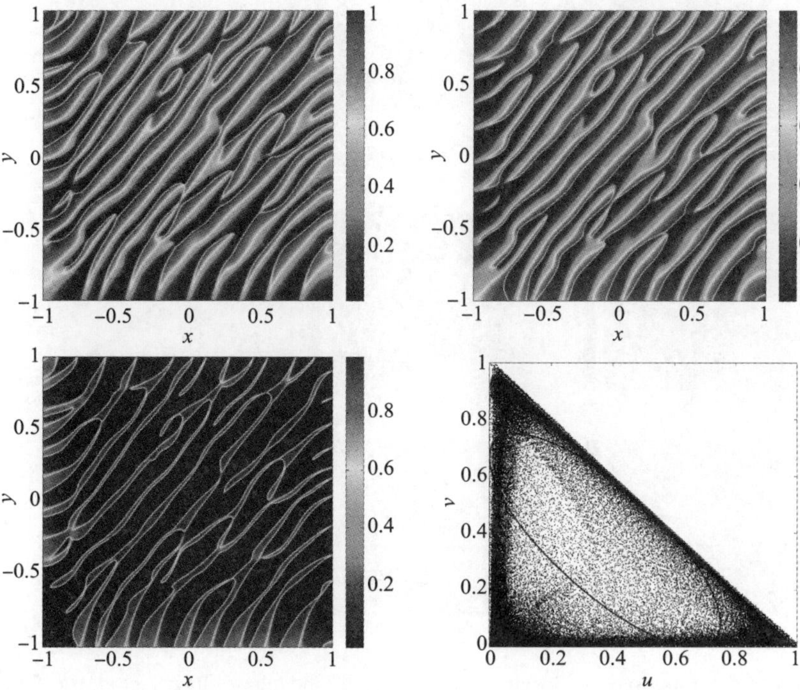

Figure 4.3 Example 2. Species' densities u, v, velocity modulus $V(u, v)$ at time $t = 2.5$, and corresponding phase diagram where the elliptic region is depicted.

4.3 Example 3: flow with different velocity functions

In Example 3, the same flow as in Example 2 is considered, but with different velocity functions. These functions describe real and hypothesized forces during the interactions between pedestrians. The velocity function $V(u,v) = 1 - u - v$ considers a slow-down that is proportional to $u + v$. By the variant $V(u,v) = 1 - \max(u,v)$, a superposition of different pedestrians types would be allowed. However, in this case, the desired pattern formation cannot be obtained; see Figure 4.4. Another desirable feature of the velocity function is that the mutual blocking at almost equal concentrations ($u \approx v$) would be higher than otherwise. A desirable property for $V(u,v)$ is to be convex satisfying

$$V(0,0) = 1, \quad V(1,0) = 0, \quad V(0,1) = 0. \tag{4.2}$$

Different flow patterns corresponding to variants of the velocity function can be seen in Figure 4.4.

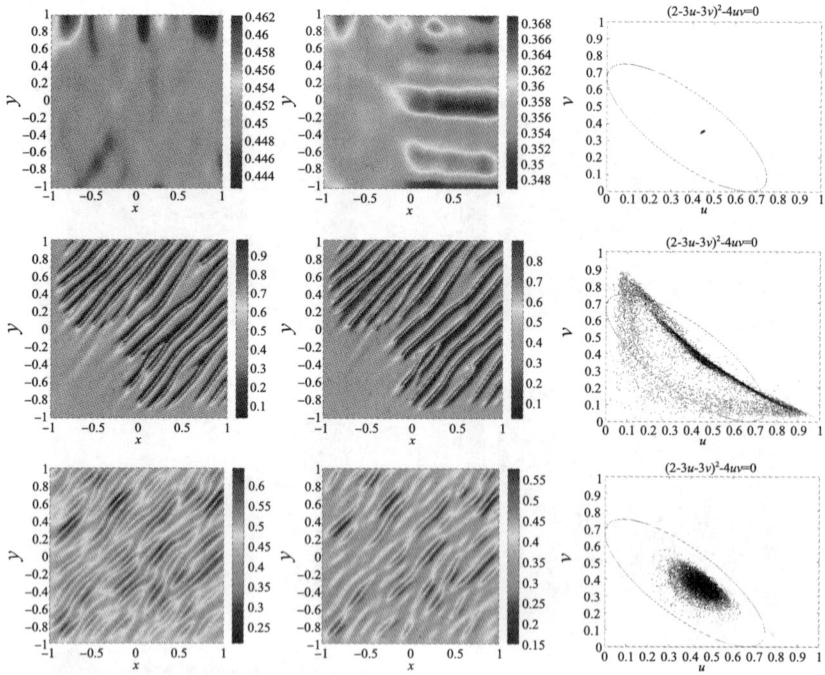

Figure 4.4 Example 3. Species' densities u, v, and phase diagrams at the final time $t = 2.5$ for different moduli $V(u,v)$. Top: $V(u,v) = 1 - \max(u,v)$, middle: $V(u,v) = 1 - u - v - uv$, and bottom: $V(u,v) = (1-u)(1-v)(1-u-v)$.

5 Conclusion

A model for pedestrian flow in two space dimensions considering two pedestrian types is proposed. Pattern formations emerge when the perturbed initial data are chosen inside the elliptic region. This pattern formation correpsonds to the oscillation waves that have been observed and studied in one-dimensional equations [3, 7].

References

[1] S. Benzoni-Gavage and R. M. Colombo, An n-populations model for traffic flow. Eur. J. Appl. Math. 14 (2003), 587–612.

[2] S. Berres, R. Bürger, K. H. Karlsen and E. M. Tory, Strongly degenerate parabolic-hyperbolic systems modeling polydisperse sedimentation with compression. SIAM J. Appl. Math. 64 (2003), 41–80.

[3] S. Berres, R. Bürger and A. Kozakevicius, Numerical approximation of oscillatory solutions of hyperbolic-elliptic systems of conservation laws by multiresolution schemes. Adv. Appl. Math. Mech. 1 (2009), 581–614.

[4] S. Berres, R. Ruiz-Baier, H. Schwandt and E. M. Tory, A multiresolution finite-volume method for pedestrian flow at a crossing. in preparation.

[5] J. H. Bick and G. F. Newell, A continuum model for two-directional traffic flow. Quart. Appl. Math. 18 (1960), 191–204.

[6] R. Bürger, R. Ruiz-Baier, K. Schneider and M. Sepúlveda, Fully adaptive multiresolution schemes for strongly degenerate parabolic equations in one space dimension. M2AN Math. Model. Numer. Anal. 42 (2008), 535–563.

[7] H. Frid and I-S. Liu, Oscillation waves in Riemann problems for phase transitions. Quart. Appl. Math. 56 (1998), 115–135.

[8] A. Harten, Multiresolution representation of data: A general framework. SIAM J. Numer. Anal. 33 (1996), 1205–1256.

[9] M. J. Lighthill and G. B. Whitham, On kinematic waves. II. A theory of traffic flow on long crowded roads. Proc. Roy. Soc. London Ser A 229 (1955), 317–345.

[10] P. I. Richards, Shock waves on the highway. Oper. Res. 4 (1956), 42–51.

Well-posedness and Travelling Wave Analysis for a Strongly Degenerate Parabolic Aggregation Equation*

Fernando Betancourt, Raimund Bürger
Departamento de Ingeniería Matemática, Universidad de Concepción
Casilla 160-C, Concepción, Chile
E-mail: fbetanco@ing-mat.udec.cl, rburger@ing-mat.udec.cl

Kenneth H. Karlsen
Centre of Mathematics for Applications (CMA), University of Oslo
P.O. Box 1053, Blindern, N-0316 Oslo, Norway
E-mail: kennethk@math.uio.no

Abstract

The aggregation of the individuals of a population is modeled by a 1-d strongly degenerate parabolic equation whose convective flux depends on the total mass to one side of the given position. The aggregation mechanism is balanced by a strongly degenerate diffusion term accounting for dispersal. Solutions of this nonlocal problem are usually discontinuous and are here defined as weak solutions. Existence of the unique weak solution follows from the convergence of a difference scheme. We obtain finite speed of propagation under additional assumptions on the coefficients. Numerical examples are presented.

1 Introduction

We consider the initial value problem for a strongly degenerate convection-diffusion equation of the form

$$u_t + \left(\Phi'\left(\int_{-\infty}^{x} u(y,t)\,\mathrm{d}y\right)u\right)_x = A(u)_{xx}, \quad x \in \mathbb{R}, \quad 0 < t \leq T, \tag{1.1}$$

$$u(x,0) = u_0(x) \geq 0, \quad x \in \mathbb{R}, \quad u_0 \in (L^1 \cap L^\infty)(\mathbb{R}), \tag{1.2}$$

*FB acknowledges support by CONICYT fellowship. RB acknowledges support by Fondecyt project 1090456, BASAL project CMM, Universidad de Chile and Centro de Investigación en Ingeniería Matemática (CI²MA), Universidad de Concepción.

$$A(u) := \int_0^u a(s)\,\mathrm{d}s, \quad \text{where } a(u) \geq 0 \text{ for } u \in \mathbb{R}. \tag{1.3}$$

The models (1.1), (1.2) were studied by a series of authors, see e.g. [1, 3, 7], all of which assumed that $a(u) = 0$ at most at isolated values of u. In [2] we study (1.1), (1.2) under the more general assumption that $a(u) = 0$ on bounded u-intervals on which (1.1) reduces to a first-order conservation law with non-local flux. We herein summarize the analysis of [2] and, generalizing the analysis of [7] to the strongly degenerate case, show that (1.1) admits travelling wave solutions with finite speed of propagation if

$$\Phi'' < 0; \quad a(u) = 0 \quad \text{for } u \leq u_{\mathrm{c}},\ u_{\mathrm{c}} \geq 0; \quad a(u) > 0 \quad \text{for } u > u_{\mathrm{c}}, \tag{1.4}$$

i.e., (1.1) degenerates only on the interval $[0, u_{\mathrm{c}}]$. See Section 5 for an example of a function that satisfies (1.4). However, we do not view (1.4) as essential; numerical experiments suggest that travelling waves possibly also exist in certain cases where Φ has several extrema.

We always assume that $A(u) \to \infty$ as $u \to \infty$, i.e. $a(\cdot)$ vanishes only on bounded intervals. We also need that $\Phi \in C^2(\mathbb{R})$, and that Φ has exactly one maximum, i.e. there exsists a number $v^* > 0$ such that $\Phi'(v^*) = 0$, $\Phi'(v) > 0$ for $v < v^*$, and $\Phi'(v) < 0$ for $v > v^*$. This assumption is introduced in [2] to facilitate some of the steps of the analysis; it is, however, not essential. Moreover we require that $u_0 \geq 0$, u_0 has compact support and $u_0 \in \mathrm{BV}(\mathbb{R})$. A typical example is $\Phi(v) = -(1-v)^2$ with $v^* = 1$; if $\|u_0\|_{L^1(\mathbb{R})} =: C_0 = 2$, then the left half of the total mass will move to the right and the right half to the left, eventually forming a single "herd" of individuals, see Section 5.

The key observation is that if u is an L^1 solution of (1.1), (1.2), then

$$v(x,t) := \int_{-\infty}^{x} u(\xi, t)\,\mathrm{d}\xi, \quad t \in (0, T] \tag{1.5}$$

is a solution of the following local initial value problem, which is more amenable to well-posedness and numerical analysis:

$$v_t + \Phi(v)_x = A(v_x)_x, \quad x \in \mathbb{R}, \quad t \in (0, T], \tag{1.6}$$

$$v(x,0) = v_0(x), \quad x \in \mathbb{R}, \quad v_0(x) := \int_{-\infty}^{x} u_0(\xi)\,\mathrm{d}\xi. \tag{1.7}$$

Since (1.1) is strongly degenerate parabolic, its solutions are usually discontinuous and need to be defined as weak solutions. For the present model, weak solutions are also entropy solutions (see [2]), which allows us to apply an available L^1 stability and uniqueness theory [6].

The novelty of [2] is the formulation and convergence proof of a finite difference scheme for (1.1), (1.2). The scheme is based on a monotone difference scheme for (1.6), (1.7), which is a special case of the schemes formulated and analyzed in [4]. It is proved in [2] that the numerical approximations converge to the unique weak solution of (1.1), (1.2). An appealing feature is that the primitive (1.5) never needs to be calculated explicitly (except for the computation of v_0).

2 Definitions and uniqueness

Definition 2.1. A measurable function u is a *weak solution* of (1.1), (1.2) if it satisfies:

1. $u \in L^\infty(\Pi_T) \cap L^1(\Pi_T) \cap L^\infty(0,T;BV(\mathbb{R}))$, and $A(u) \in L^2(0,T;H^1(\mathbb{R}))$, where $\Pi_T := \mathbb{R} \times (0,T)$.
2. The initial condition (1.2) holds in the sense

$$\lim_{t\downarrow 0} \int_{\mathbb{R}} \big|u(x,t) - u_0(x)\big| \,\mathrm{d}x = 0. \tag{2.1}$$

3. The following equality is satisfied for all test functions $\phi \in C_0^\infty(\Pi_T)$, where $v(x,t)$ is given by (1.5):

$$\iint_{\Pi_T} \Big\{ u\big(\phi_t + \Phi'(v)\phi_x\big) + A(u)\phi_{xx} \Big\} \,\mathrm{d}x\,\mathrm{d}t = 0. \tag{2.2}$$

Definition 2.2. A measurable, non-negative function u is an *entropy solution* of (1.1), (1.2) if it satisfies (1) and (2) of Def. 2.1 and if for all $0 \le \varphi \in C_0^\infty(\Pi_T)$, the following entropy inequality is satisfied:

$$\forall k \in \mathbb{R}: \iint_{\Pi_T} \Big\{ |u-k|\big(\varphi_t + \Phi'(v)\varphi_x\big) - \operatorname{sgn}(u-k)uk\Phi''(v)\varphi + \big|A(u) - A(k)\big|\varphi_{xx} \Big\} \,\mathrm{d}x\,\mathrm{d}t \ge 0 \tag{2.3}$$

It is standard to check that an entropy solution of (1.1), (1.2) is, in particular, a weak solution. Moreover, conversely, it can be shown that any weak solution is an entropy solution [2].

Standard results from the theory of strongly degenerate parabolic equations state that a discontinuity from u^- to u^+ is possible only if $A(u)$ is flat for $u \in \mathcal{I}(u^-,u^+) := (\min\{u^-,u^+\}, \max\{u^-,u^+\})$. In that case, the Rankine-Hugoniot condition, which follows from (2.2), reads

$$s = \Phi'\big(v(x_0,t_0)\big) - [(A(u)_x)^+ - (A(u)_x)^-]/(u^+ - u^-), \tag{2.4}$$

where we have used that v is continuous.

The uniqueness of weak solutions is an immediate consequence of a result proved in [6] regarding continuous dependence of entropy solutions with respect to the flux function. More precisely, we have:

Theorem 2.3. *There exists at most one weak solution of* (1.1), (1.2). *Moreover, if u and $\bar{u}$ are weak solutions with initial data u_0 and $\bar{u}_0$, respectively, then there exists a constant $C = C(\max|\Phi'|)$ such that $\|u(\cdot,t) - \bar{u}(\cdot,t)\|_{L^1(\mathbb{R})} \leq C\|u_0 - \bar{u}_0\|_{L^1(\mathbb{R})}$ for all $t \in (0,T]$.*

3 Convergence analysis

We define $U^n := \{u^n_{j+1/2}\}_{j\in\mathbb{Z}}$, $V^n := \{v^n_j\}_{j\in\mathbb{Z}}$, $x_j := j\Delta x$ for $j \in \mathbb{Z}$, and $t_n := n\Delta t$ for $n = 0,\dots,N$, where $\Delta t := T/N$, $N \in \mathbb{N}$. We denote by $u^n_{j+1/2}$ the cell average over $I_j := [x_j, x_{j+1}]$ at time t_n. We also define $\lambda := \Delta t/\Delta x$ and $\mu := \Delta t/\Delta x^2$. Moreover, using the discrete integral operator $\mathcal{S}_{\Delta x}$ and the discrete differential operator $\mathcal{S}^{-1}_{\Delta x}$, the numerical scheme for (1.1), (1.2) can be written as

$$U^{n+1} = \left[\mathcal{S}^{-1}_{\Delta x} \circ \mathcal{H} \circ \mathcal{S}_{\Delta x}\right]U^n, \quad n = 0,\dots,N-1. \tag{3.1}$$

The basic idea is to utilize a standard scheme of the form $V^{n+1} = \mathcal{H}(V^n)$, $n = 0,\dots,N-1$ for the local PDE (1.6), which is given by

$$v^{n+1}_j = v^n_j - \lambda\Delta_+\left[h\left(v^n_{j-1}, v^n_j\right) - A\left(\Delta_- v^n_j/\Delta x\right)\right], \tag{3.2}$$

where $h(\cdot,\cdot)$ is the Engquist-Osher flux, and we start from $v^0_j := v_0(x_j)$ for $j \in \mathbb{Z}$. Note that (3.1) implies that $U^n = [\mathcal{S}^{-1}_{\Delta x} \circ \mathcal{H}^n \circ \mathcal{S}_{\Delta x}]U^0$. We assume that Δt and Δx satisfy the CFL stability condition

$$\mathrm{CFL}_\varepsilon := 2\lambda \max_{u\in\mathbb{R}}\left|\Phi'(u)\right| + 4\mu \max_{u\in\mathbb{R}} a(u) \leq 1 - \varepsilon, \quad \varepsilon \in (0,1). \tag{3.3}$$

The u-scheme results from $u^n_{j+1/2} := (\mathcal{S}^{-1}_{\Delta x}(V^n))_{j+1/2} = (v^n_{j+1} - v^n_j)/\Delta x$, and can in closed form be written as

$$u^{n+1}_{j+1/2} = u^n_{j+1/2} - \mu\Delta^2_+ h\left(v^n_{j-1}, v^n_j\right) + \mu\Delta^2 A\left(u^n_{j+1/2}\right). \tag{3.4}$$

Theorem 3.1 ([2]). *Assume that Δx and Δt satisfy* (3.3), *and that the numerical solution $\{u^n_{j+1/2}\}$ generated by* (3.4) *is properly converted to a function u^Δ defined on Π_T, with $u^\Delta(x,t) = u^n_{j+1/2}$ if $(x,t) \in I_j \times [t_n, t_{n+1})$. Then u^Δ converges in the strong topology of $L^1(\Pi_T)$ to a weak solution of* (1.1), (1.2) *as $\Delta = (\Delta x, \Delta t) \downarrow 0$.*

Sketch of proof. Theorem 3.1 is proved in [2] by a series of lemmas, which we now summarize, and where $C_0, C_1, C_2, \dots$ are always constants that are independent of Δ. One first shows that the v-scheme (3.2) is monotone, which implies that the numerical solution $\{v^n_j\}$ satisfies

$$0 \leq v^n_j \leq C_0, \quad v^n_j \leq v^n_{j+1} \quad \text{for all } j \in \mathbb{Z},\ n = 1,\dots,N, \tag{3.5}$$

$$\mathrm{TV}(V^n) = \sum_{j\in\mathbb{Z}} \left|v^n_{j+1} - v^n_j\right| = C_0 := \int_{\mathbb{R}} u_0(x)\,\mathrm{d}x. \tag{3.6}$$

Next, one proves that the numerical solution $\{v_j^n\}$ satisfies

$$\sum_{j\in\mathbb{Z}} |v_j^{n+1} - v_j^n| \leq C_1\lambda, \quad |\Delta_+ v_j^n/\Delta x| \leq C_3. \tag{3.7}$$

(The proof of the second bound relies on $A(u) \to \infty$ as $u \to \infty$.) Equivalently, the solution $\{u_{j+1/2}^n\}$ generated by (3.4) satisfies the uniform L^∞ bound $|u_{j+1/2}^n| \leq C_3$. One can then prove that

$$\mathrm{TV}\big(A(U^n)\big) = \sum_{j\in\mathbb{Z}} \big|\Delta_+ A\big(u_{j-1/2}^n\big)\big| \leq C_4,$$

but this bound implies a uniform bound on $\mathrm{TV}(U^n)$ only in the special case that $A(\cdot)$ is strictly increasing, and $a(u)$ vanishes at most at isolated values of u. In the strongly degenerate case one can, however, prove the weaker (time-dependent) total variation bound

$$\mathrm{TV}(U^n) = \sum_{j\in\mathbb{Z}} |u_{j+1/2}^n - u_{j-1/2}^n| \leq \big(C_5 + \mathrm{TV}(U^0)\big)\exp(C_6 T). \tag{3.8}$$

The proof of (3.8) requires an analysis of the numerical solution near values v_j^n that include v^*, the extremum of Φ at which $\Phi'(v^*) = 0$. To simplify the discussion, in [2] (and herein) it is assumed that there is just one maximum of that kind; the same arguments can, however, be applied when Φ has several extrema since in view of (3.5) and the second bound of (3.7), the corresponding numerical solution values will be assumed at separated spatial positions. Next, one shows that the solution $\{u_{j+1/2}^n\}$ generated by the u-scheme (3.4) is L^1 Hölder continuous in time, i.e.,

$$\sum_{j\in\mathbb{Z}} |u_{j+1/2}^m - u_{j+1/2}^n|\Delta x \leq C_8\sqrt{\Delta t|m-n|}, \quad m,n \in \mathbb{N}_0, \tag{3.9}$$

and that the discrete version of the integrated diffusion coefficient $A(u)$ has the required regularity:

$$\sum_{n=1}^{N}\sum_{j\in\mathbb{Z}} \big(\Delta_- A\big(u_{j+1/2}^n\big)/\Delta x\big)^2 \Delta t\Delta x \leq C_9, \tag{3.10}$$

$$\sum_{j\in\mathbb{Z}} \big|A\big(u_{j+1/2}^m\big) - A\big(u_{j+1/2}^n\big)\big|^2 \Delta x \leq C_{10}|m-n|\Delta t. \tag{3.11}$$

The previous estimates permit to use standard compactness arguments to deduce that u^Δ converges a.e. on Π_T, as $\Delta \downarrow 0$, along a subsequence to a limit function u having the regularity requested in Definition 2.2. Similarly, $A(u^\Delta)$ converges to $A(u)$ strongly in $L^2_{\mathrm{loc}}(\Pi_T)$, and $A(u)$ has the requested regularity properties. Satisfaction of (2.1) follows from (3.9), while that of (2.2) is achieved by a Lax-Wendroff argument. □

4 Finite speed of propagation

If Φ and A satisfy (1.4), then the left portion (for which $v < v^*$) of the total "population" (mass) moves to the right and the right portion moves to the left. One then expects that one "herd" of bounded spatial extension forms which moves to the left, right, or is stationary depending on whether the initially left-moving or right-moving individuals outnumber, or are equal to, those moving initially in opposite direction. This is indeed the case, as will be shown in the following characterization of the process of aggregation-dispersion by a travelling wave analysis. We obtain finite speed of propagation.

Lemma 4.1. *Assume that* Φ *and* A *satisfy* (1.4), *and let* $u(x,t)$ *be a solution of* (1.1), (1.2). *Then there exist constants* α *and* β *such that for* $t \leq T$, $u(x,t) = 0$ *outside of* $\alpha \leq x - st \leq \beta$, *where* $s = \sigma(0, C_0) := (\Phi(C_0) - \Phi(0))/C_0$.

Proof. We consider the following regularized equation

$$\partial_t v_\varepsilon + \partial_x \Phi(v_\varepsilon) = \partial_x A_\varepsilon(\partial_x v_\varepsilon), \quad x \in \mathbb{R}, \quad t \in (0,T], \tag{4.1}$$

$$v_\varepsilon(-\infty) = 0 \quad \text{and} \quad v_\varepsilon(+\infty) = C_0 \tag{4.2}$$

(see (3.6)), where $A_\varepsilon(u) = A(u) + \varepsilon u$. We seek a travelling wave solution $v_\varepsilon(x,t) = w_\varepsilon(x - st)$ of (4.1), (4.2). Hence, $w_\varepsilon(\xi)$ (with $\xi := x - st$) is a solution of the following problem, where $' = \mathrm{d}/\mathrm{d}\xi$:

$$-s w_\varepsilon' + \Phi(w_\varepsilon)' = \left(A_\varepsilon(w_\varepsilon')\right)', \tag{4.3}$$

$$w_\varepsilon(-\infty) = 0 \quad \text{and} \quad w_\varepsilon(+\infty) = C_0. \tag{4.4}$$

Integrating (4.3) we find

$$-s w_\varepsilon + \Phi(w_\varepsilon) - A_\varepsilon(w_\varepsilon') \equiv \overline{w} \quad (= \text{const.}).$$

Applying (4.4) we get $\overline{w} = \Phi(0)$ and $s = \sigma(0, C_0)$. By using $\Phi'' < 0$ we observe that $A_\varepsilon(w_\varepsilon') = \Phi(w_\varepsilon) - [\Phi(0) + \sigma(0, C_0) w_\varepsilon] \geq 0$ for $0 \leq w_\varepsilon \leq C_0$. Since $A_\varepsilon(\cdot)$ is strictly increasing we can write

$$w_\varepsilon' = A_\varepsilon^{-1}\left(\Phi(w_\varepsilon) - \left[\Phi(0) + \sigma(0, C_0) w_\varepsilon\right]\right) \geq 0.$$

Hence, w_ε is determined by the relation

$$\xi_2 - \xi_1 = \int_{w_\varepsilon(\xi_1)}^{w_\varepsilon(\xi_2)} \frac{\mathrm{d}\tilde{w}_\varepsilon}{A_\varepsilon^{-1}(\gamma(\tilde{w}_\varepsilon))} \quad \text{for } \xi_1, \xi_2 \in \mathbb{R},$$

$$\text{where} \quad \gamma(\tilde{w}_\varepsilon) := \Phi(\tilde{w}_\varepsilon) - [\Phi(0) + \sigma(0, C_0)\tilde{w}_\varepsilon].$$

Thus we have constructed a solution w_ε for the problems (4.3), (4.4). The above integral gives a well-defined function w_ε since $A_\varepsilon^{-1}(\gamma(w_\varepsilon)) > 0$ for $w_\varepsilon \in (0, C_0)$ and $w'_\varepsilon > 0$. Now, we study the limit $\varepsilon \to 0$. We note that

$$A_\varepsilon : [0, u_c) \to [0, \varepsilon u_c) \Rightarrow A_\varepsilon^{-1} : [0, \varepsilon u_c) \to [0, u_c) \quad \text{for } w_\varepsilon < u_c,$$
$$A_\varepsilon : [u_c, \infty) \to [\varepsilon u_c, \infty) \Rightarrow A_\varepsilon^{-1} : [\varepsilon u_c, \infty) \to [u_c, \infty) \quad \text{for } u_c \le w_\varepsilon.$$

Let ξ_1, ξ_2 be fixed. We may assume that $\varepsilon u_c < w_\varepsilon(\xi_1) < w_\varepsilon(\xi_2)$ (otherwise we could choose a smaller ε using that $w'_\varepsilon > 0$ in $(0, C_0)$). Thus

$$\int_{w_\varepsilon(\xi_1)}^{w_\varepsilon(\xi_2)} \frac{d\tilde{w}_\varepsilon}{A_\varepsilon^{-1}(\gamma(\tilde{w}_\varepsilon))} \le \frac{C_0}{u_c}. \tag{4.5}$$

Moreover, $A_\varepsilon^{-1} \overset{\varepsilon \to 0}{\longrightarrow} A_0^{-1}$ pointwise, where $A_0^{-1} : [0, +\infty) \to [u_c, +\infty)$ is the inverse function of A restricted to $[u_c, +\infty)$. Using the Lebesgue theorem we obtain passing to the limit $\varepsilon \downarrow 0$

$$\xi_2 - \xi_1 = \int_{w(\xi_1)}^{w(\xi_2)} \frac{d\tilde{w}}{A_0^{-1}(\gamma(\tilde{w}))} \quad \text{for } \xi_1, \xi_2 \in \mathbb{R}. \tag{4.6}$$

Since the right-hand side of (4.6) is bounded, there exist two constants α y β such that $w(\xi) = 0$ for $\xi \le \alpha$ and $w(\xi) = C_0$ for $\xi \ge \beta$ and $w'(\xi) > 0$ for $\alpha < \xi < \beta$. Let $W(\xi) := w'(\xi)$, which is the desired function and satisfies $W(\xi) = 0$ outside (α, β) and $W(\xi) > 0$ on (α, β).

Let u be the solution of (1.1), (1.2), and let $v(x,t)$ and $v_0(x)$ be defined by (1.5) and (1.7), respectively. Since u_0 is assumed to have compact support, there exist constants α_1 and β_1 such that $v_0(x) = 0$ for $x \le \alpha_1$ and $v_0(x) = C_0$ for $x \ge \beta_1$. On the other hand, from results in [2] we have that v is a solution of (1.6), (1.7). From the above discussion we can construct two travelling wave solutions $w_1(x - st)$ and $w_2(x - st)$ of (1.6), (1.7) with the same speed $s = \sigma(0, C_0)$ which satisfy $w_1(x) \le v_0(x) \le w_2(x)$, $w_2(x - st) = 0$ for $x - st \le \alpha$, and $w_1(x - st) = C_0$ for $x - st \ge \beta$, where we used that (1.1) and (1.6) are invariant under translation of x. From [2] we know that the solution v of (1.6), (1.7) is monotone (as a function of x, for each fixed t), and we find $w_1(x - st) \le v(x,t) \le w_2(x - st)$ on Π_T. Therefore $v(x,t) = 0$ if $x - st \le \alpha$ and $v(x,t) = C_0$ if $x - st \ge \beta$, which implies $u(x,t) = 0$ outside $\alpha \le x - st \le \beta$ for $t \le T$. □

5 Numerical examples

We use the degenerating diffusion coefficient $A(u) = 0$ for $0 \le u \le 10$ and $A(u) = 0.1(u - 10)$ for $u > 10$. The left plot of Figure 5.1 shows the

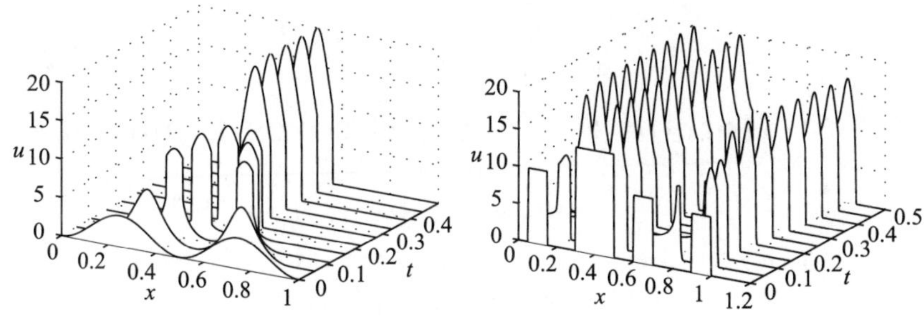

Figure 5.1 Numerical examples ($\Delta x = 0.001$ and $\mu = 0.2$).

numerical solution of (1.1), (1.2) for $\Phi(v) = -(1-v)^2$ and the initial datum is $u_0(x) = 2(1 - \cos(4\pi x))$. Since $C_0 = 2$ and $\Phi(C_0) - \Phi(0) = 0$, we expect the formation of a stationary travelling wave according to Sect. 4. We observe that all individuals aggregate into one group. Moreover, the initially smooth profile becomes discontinuous after finite time.

In the right plot we use a function Φ with several extrema given by $\Phi(v) = -0.5(\cos(v\pi) + 1)$ along with the initial datum $u_0(x) = 10$ for $x \in [0.05, 0.15]$, $u_0(x) = 9$ for $x \in [0.6, 0.7]$, $u_0(x) = 14$ for $x \in [0.3, 0.5]$, $u_0(x) = 8$ for $x \in [0.9, 1]$, and $u_0(x) = 0$ otherwise. We observe the formation of three groups, but the third moves to the right "looking for more" mass since it is not a full state, cf. the conditions of Lemma 4.1.

References

[1] W. Alt, Degenerate diffusion equations with drift functionals modelling aggregation. Nonlin. Anal. TMA 9: 811–836, 1985.

[2] F. Betancourt, R. Bürger and K. H. Karlsen, A strongly degenerate parabolic aggregation equation. Commun. Math. Sci. 9: 711–742, 2011.

[3] J. I. Diaz, T. Nagai and S. I. Shmarev, On the interfaces in a nonlocal quasilinear degenerate equation arising in population dynamics. Japan J. Indust. Appl. Math. 13: 385–415, 1996.

[4] S. Evje and K. H. Karlsen, Discrete approximations of *BV* solutions to doubly degenerate parabolic equations. Numer. Math. 86: 377–417, 2000.

[5] K. H. Karlsen and N. H. Risebro, Convergence of finite difference schemes for viscous and inviscid conservation laws with rough coefficients. M2AN Math. Model. Numer. Anal. 35: 239–269, 2001.

[6] K. H. Karlsen and N. H. Risebro, On the uniqueness and stability of entropy solutions for nonlinear degenerate parabolic equations with rough coefficients. Discr. Contin. Dyn. Syst. 9: 1081–1104, 2003.

[7] T. Nagai and M. Mimura, Asymptotic behavior for a nonlinear degenerate diffusion equation in population dynamics. SIAM J. Appl. Math. 43: 449–464, 1983.

CDG Method for Navier-Stokes Equations

Slavko Brdar
Section of Applied Mathematics, University of Freiburg
Freiburg i. Br. 79104, Germany
E-mail: slavko@mathematik-uni-freiburg.de

Andreas Dedner
Section of Applied Mathematics, University of Freiburg
Freiburg i. Br. 79104, Germany
E-mail: dedner@mathematik-uni-freiburg.de

Robert Klöfkorn
Section of Applied Mathematics, University of Freiburg
Freiburg i. Br. 79104, Germany
E-mail: robertk@mathematik-uni-freiburg.de

Abstract

In this work we formulate the Compact Discontinuous Galerkin (CDG) method of Peraire and Persson [1] for advection-difussion problems and prove its stability for the linear heat equation. We compare the method numerically with three other well-established methods: the Interior Penalty (IP), the Bassy-Rebay (BR2), the Local Discontinuous Galerkin (LDG) method. In contrast to the LDG method, the primal formulation of the CDG method only involves the direct neighbors of a cell, thus resulting in a minimal stencil. The IP method also has this compactness property, but unlike CDG, it requires penalization of the jumps in the numerical solution across grid interfaces.

1 Introduction

The CDG method uses the Discontinuous Galerkin (DG) approach to discretize diffusion fluxes for advection-diffusion problems. It was derived and analysed for the first time by Peraire and Persson in 2008 for elliptic problems, see [1].

In general the CDG method inherits all the properties of the DG method: it is highly efficient on parallel computers; higher accuracy is achieved by choosing higher polynomial degree locally in each cell;

the implementation on unstructured grid is straightforward. At the same time, it has the potential to overcome some shortcomings of other DG methods, e.g., of the Interior Penalty (IP), the Local Discontinuous Galerkin (LDG) and the Bassy-Rebay 2 (BR2) method.

The IP method was proposed by Douglas and Dupont in 1975 in [4] and revised by Hartmann and Houston in 2008 for Navier-Stokes equations in [2]. In comparison to CDG the IP method, due to stability reasons, can not be used without the penalization of the jump of numerical solution across the grid interfaces (see, for example, [2, 3]). In comparison to LDG the CDG method is *compact*, in the sense that updating the degrees of freedom (DOF) in one time step in one cell requires only data from the directly neighboring cells. This leads to a better sparsity pattern and allows for a more efficient parallelization of the method. The BR2 method is also a compact DG method. It is proposed by Bassy and Rebay in [7] and unlike IP it doesn't require any penalization of the numerical solution across grid interfaces.

The paper is organized as follows. In Sec. 2 we formulate the CDG method for advection-diffusion problem. In Sec. 2.1 we give stability results and finally, in Sec. 3 we show some numerical experiments.

2 Primal formulation

We consider a general non-linear scalar advection-diffusion equation. A system of advection-diffusion equations can be treated by applying the same algorithm to each equation in the system.

$$\partial_t u + \nabla \cdot (F(u) - A(u, \nabla u)) = 0 \quad \text{in } \Omega \times [0, T] \tag{2.1}$$

equipped with Dirichlet boundary data g_D on $\Gamma = \partial\Omega$. For scalar case we have $u : \mathbb{R}^d \times [0, T] \to \mathbb{R}$, $F : \mathbb{R}^d \to \mathbb{R}^d$, $A : \mathbb{R} \times \mathbb{R}^d \to \mathbb{R}^d$ and A is linear w.r.t. second argument. We start by rewriting (2.1) as a system of first order PDEs

$$\partial_t u + \nabla \cdot (F(u) - A(u, \boldsymbol{\sigma})) = 0, \tag{2.2a}$$

$$\boldsymbol{\sigma} = \nabla u. \tag{2.2b}$$

Next, we employ a finite polygonal grid $\mathcal{T}_h = \{K\}$ which partitions Ω to define the following function spaces for some $l, k \in \mathbb{N}$

$$V_h^l = \{\boldsymbol{v} \in L^\infty(\Omega, \mathbb{R}^l) | \boldsymbol{v}|_K \in [\mathcal{P}_k(K)]^l\}.$$

Multiplication of (2.2a) with an arbitrary $\phi \in V_h^1$ and (2.2b) by $\boldsymbol{\psi} \in V_h^d$, followed by integration (and one additional partial integration for (2.2b)) over Ω yields *the flux formulation* of the DG method

$$\begin{aligned}\int_\Omega \partial_t u &= \int_\Omega (F(u) - A(u,\boldsymbol{\sigma}))\cdot\nabla\varphi - \sum_{e\in\Gamma}\int_e (\hat{F}(u) - \hat{A}(u,\boldsymbol{\sigma}))\cdot[\![\varphi]\!]_e,\\ \int_\Omega \boldsymbol{\psi}\cdot\boldsymbol{\sigma} &= \int_\Omega \boldsymbol{\psi}\cdot\nabla u - \sum_{e\in\Gamma_0}\int_e \{\!\{\boldsymbol{\psi}\}\!\}_e\cdot[\![u]\!]_e - \sum_{e\in\Gamma}\int_e \{u-\hat{u}\}_e[\boldsymbol{\psi}]_e,\end{aligned} \tag{2.3}$$

where one still needs to specify the diffusion fluxes $(\hat{u}, \hat{A})$ and replace the advection flux $\hat{F}$ with one of the available Riemann solvers (Local Lax-Friedrichs, Roe, HLLC, Godunov, etc.).

In (2.3) we used the following notation: Γ is $\bigcup_{K\in\mathcal{T}_h}\partial K$, whereas $\Gamma_0 = \bigcup_{K\in\mathcal{T}_h}(\partial K\setminus\partial\Omega)$. For $e = K^+\cap K^-$ and Hausdorff measure $\mathcal{H}_{d-1}(e) > 0$ we define operators $[\cdot]_e$, $\{\cdot\}_e$ acting on $\phi\in V_h^1$ and $[\![\cdot]\!]_e$, $\{\!\{\cdot\}\!\}_e$ on $\boldsymbol{\tau}\in V_h^d$. When $e\cap\partial\Omega = 0$ we let

$$\begin{aligned}[\phi]_e &= \phi^-\boldsymbol{n}^- + \phi^+\boldsymbol{n}^+, \quad [\![\boldsymbol{\tau}]\!]_e = \boldsymbol{\tau}^-\cdot\boldsymbol{n}^- + \boldsymbol{\tau}^+\cdot\boldsymbol{n}^+,\\ \{\phi\}_e &= \phi^- + \phi^+, \quad \{\!\{\boldsymbol{\tau}\}\!\}_e = \boldsymbol{\tau}^- + \boldsymbol{\tau}^+,\end{aligned}$$

otherwise if $e\subset\partial\Omega$

$$\begin{aligned}[\phi]_e &= (\phi - g'_D)\boldsymbol{n}, \quad [\![\boldsymbol{\tau}]\!]_e = \boldsymbol{\tau}\cdot\boldsymbol{n},\\ \{\phi\}_e &= \phi, \quad \{\!\{\boldsymbol{\tau}\}\!\}_e = \boldsymbol{\tau}.\end{aligned}$$

$\boldsymbol{n}^\pm$ is outer normal on $K^\pm\in\mathcal{T}_h$ and $\boldsymbol{n}$ on Ω. g'_D is g_D only when ϕ is numerical solution u, otherwise 0.

In the following we derive *the primal formulation* of the LDG method. To that end, we introduce *the lifting operators* $\boldsymbol{r}_e : [L^2(\Gamma)]^d \to V_h^d$ and $\boldsymbol{l}_e : L^2(\Gamma_0)\to V_h^d$ as

$$\int_\Omega \boldsymbol{r}_e(\boldsymbol{\eta})\cdot\boldsymbol{\tau} = -\int_e \boldsymbol{\eta}\cdot\{\!\{\boldsymbol{\tau}\}\!\}_e, \quad \int_\Omega \boldsymbol{l}_e(\phi)\cdot\boldsymbol{\tau} = -\int_e \phi\cdot\{\boldsymbol{\tau}\}_e.$$

We extend $\boldsymbol{l}_e$ for $e\subset\partial\Omega$ by setting $l_e(\cdot)\equiv 0$. Notice that second equation in (2.3) now becomes

$$\boldsymbol{\sigma} = \nabla u + \sum_{e\in\Gamma_0}\boldsymbol{r}_e([\![u]\!]_e) + \boldsymbol{l}_e([u-\hat{u}]_e).$$

In case of LDG the diffusion fluxes are

$$\begin{aligned} e\in\Gamma_0 :\quad & \hat{A}(u,\boldsymbol{\sigma}) = \{\!\{A(u,\boldsymbol{\sigma})\}\!\}_e + \boldsymbol{\beta}_e[A(u,\boldsymbol{\sigma})]_e - \xi[\![u]\!]_e/|e|,\\ & \hat{u} = \{u\}_e - \boldsymbol{\beta}_e\cdot[\![u]\!]_e,\\ e\notin\Gamma_0 :\quad & \hat{A}(u,\boldsymbol{\sigma}) = A(u,\boldsymbol{\sigma}) - \xi(u-g_D)/|e|, \quad \hat{u} = g_D.\end{aligned}$$

where *the switch* is chosen like $\boldsymbol{\beta}_e = \frac{1}{2}\mathrm{sgn}(\boldsymbol{w}\cdot\boldsymbol{n})\boldsymbol{n}$, with *the upwind* vector $\boldsymbol{w}$ prescribed a-priori and $\boldsymbol{n}$ is any unit vector normal to e. $\xi \geq 0$ is *the penalty coefficient.* Therefore, if we neglect the advection fluxes (for shorter notation only) the primal formulation of the LDG method can be written as

$$\int_\Omega \varphi\partial_t u + \int_\Omega A(u,\nabla u)\cdot\nabla\varphi + \sum_{e\in\Gamma}\sum_{f\in\Gamma}\int_\Omega \boldsymbol{L}_e(\varphi)\cdot A(u,\boldsymbol{L}_f(u))$$

$$-\sum_{e\in\Gamma}\int_e \left(\{\!\{A^T(u,\nabla\varphi)\}\!\}_e\cdot[\![u]\!]_e + \{\!\{A^T(u,\nabla u)\}\!\}_e\cdot[\![\varphi]\!]_e - \frac{\xi}{|e|}[\![u]\!]_e\cdot[\![\varphi]\!]_e\right)$$

$$-\sum_{e\in\Gamma_0}\boldsymbol{\beta}_e\cdot\left([A(u,\nabla u)]_e[\![\varphi]\!]_e + [A(u,\nabla\varphi)]_e[\![u]\!]_e\right) = 0,$$

where $\boldsymbol{L}_e(\phi) = \boldsymbol{r}_e([\![\phi]\!]_e) + \boldsymbol{l}_e(\boldsymbol{\beta}_e\cdot[\![\phi]\!]_e)$.

Note that the product of the lifting operators $\boldsymbol{L}_e(\varphi)\cdot A(u,\boldsymbol{L}_f(u))$, where $e\neq f$, connects the DOFs in non-neighboring elements. The idea of Peraire and Persson in [1], and similarly, of Brezzi and collaborators in [5] is to remove such terms and replace the double sum in the upper equation with

$$\sum_{e\in\Gamma}\int_\Omega \boldsymbol{L}_e(\varphi)\cdot A(u,\boldsymbol{L}_e(u)). \tag{2.4}$$

In this way we obtain the CDG method. For stability reasons we will multiply this term with a coefficient $\chi\geq 0$.

2.1 Stability analysis

In this section we want to give L^2-stability conditions for the CDG method in the case of the linear heat equations. The L^2-stability is defined as $\frac{d}{dt}\int_\Omega u^2\,dt \leq 0$.

Lemma 2.1 (L^2-stability). *The CDG method is L^2-stable in case of linear heat equation if one of the two following conditions is fulfilled*

- $\chi \geq \mathcal{N}(\mathcal{T}_h)$, *and* $\xi\geq 0$,
- $\chi\geq 0$ *and* ξ *large enough,*

where $\mathcal{N}(\mathcal{T}_h)$ is maximal number of the outflow faces e ($\boldsymbol{n}_{K,e}\cdot\boldsymbol{w} > 0$) that one grid element K in $\mathcal{T}_h$ can have.

Proof. Notice first that $\mathrm{supp}\,\boldsymbol{L}_e(u) = K$ if $\boldsymbol{n}_{K,e}\cdot\boldsymbol{w} < 0$ and $e\subset\partial K$, and denote from now on $\boldsymbol{L}_e(u)$, simply with $\boldsymbol{L}_e$. Since $\boldsymbol{L}_e|_K\equiv 0$ when e is an inflow face for K ($\boldsymbol{n}_{K,e}\cdot\boldsymbol{w} < 0$), we deduce the following inequality

$$\|\sum_{e\in\Gamma}\boldsymbol{L}_e\|^2_\Omega \leq \mathcal{N}(\mathcal{T}_h)\sum_{e\in\Gamma}\|\boldsymbol{L}_e\|^2_\Omega. \tag{2.5}$$

If we now rewrite the CDG method in the form

$$\int_\Omega \varphi \partial_t u + B(u,\varphi) = 0 \tag{2.6}$$

where $\varphi \in V^1$, and replacing φ with u we get

$$B(u,u) = \|\nabla u + \sum_{e\in\Gamma} \boldsymbol{L}_e\|_\Omega^2 - \|\sum_{e\in\Gamma} \boldsymbol{L}_e\|_\Omega^2 + \chi \sum_{e\in\Gamma} \|\boldsymbol{L}_e\|_\Omega^2 + \frac{\xi}{|e|}\|[\![u]\!]_e\|_e^2. \tag{2.7}$$

From (2.6) it is sufficient to show that $B(u,u) \geq 0$ in order to show L^2-stability. Because of (2.5) we can choose $\chi = \mathcal{N}(\mathcal{T}_h)$ in (2.7) to get $B(u,u) \geq 0$ what implies the L^2-stability and proves the first part of the lemma.

The second part is a simple consequence of the lemma of Brezzi and collaborators in [5] which states that

$$\|\boldsymbol{r}_e([\![u]\!]_e)\|_\Omega \leq \frac{C_2}{\sqrt{|e|}}\|[\![u]\!]_e\|_e, \tag{2.8}$$

for some constant $C_2 > 0$ where h_e is volume of grid interface e; and the fact that

$$\|\sum_{e\in\Gamma} \boldsymbol{L}_e\|_\Omega^2 \leq 4\mathcal{N}(\mathcal{T}_h) \sum_{e\in\Gamma} \|\boldsymbol{r}_e([\![u]\!]_e)\|_\Omega^2, \tag{2.9}$$

which is a consequence of (2.5). From (2.7) we analyse possible negative terms

$$-\|\sum_{e\in\Gamma} \boldsymbol{L}_e\|_\Omega^2 + \sum_{e\in\Gamma} \xi\|[\![u]\!]_e\|_e^2 \geq \sum_{e\in\Gamma} \frac{1}{|e|}(-4\mathcal{N}(\mathcal{T}_h)C_2^2 + \xi)\|[\![u]\!]_e\|_e^2 \tag{2.10}$$

and choose $\xi > 4\mathcal{N}(\mathcal{T}_h)C_2^2$ to assure positivity of $B(u,u)$. Thus, the proof of the second part is finished.

3 Numerical results

In this section we compare IP, LDG and CDG method on unstructured triangular grids for solving the linear heat equation and the Navier-Stokes equations. It is interesting to point out that we have performed tests on structured rectangular grids and came out with similar results. For both equations we have set analytical solutions to compare the convergence rate of the methods. For Navier-Stokes equations we have chosen the analytical solution proposed in [6]. The CDG method in all numerical test cases is used with $\chi = 1$.

The LDG method is implemented in the flux formulation, meaning that one spatial evaluation of the LDG operator requires two iterations through grid, first one to compute $\boldsymbol{\sigma}$ and the second for u. In each grid iteration one DG computation is applied. In this way computation of lifting operators is avoided. Nevertheless, the LDG spatial operator only becomes more expensive to compute than other mention methods when one solves system of PDEs. This is depicted in Figure 3.1 on the example of the linear heat equation and Navier-Stokes system, where explicit RK method with time-stepping is used. However, in practical application, for example using the Navier-Stokes system, the usage of explicit time integrators is inefficient, as the time step restriction is severe. The usage of implicit, or semi-implicit methods is a requirement in such cases. Because of this reason we compare methods with implicit-explicit Runge-Kutta (IMEX-RK) time integrators for different orders.

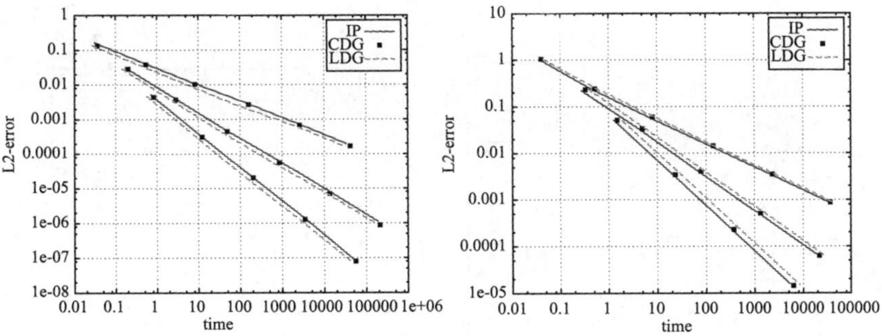

Figure 3.1 Different methods for the linear heat (left) and Navier-Stokes equations on unstructured triangular grid using explicit RK with fixed time stepping.

In Table 1 we compare the methods using Runge-Kutta (RK) time integration with fixed time stepping. In this way we avoid the influence of the time integration method in measuring the cost of the spatial op-

Table 1 Different methods for Navier-Stokes equations on unstructured triangular grid using explicit RK with fixed time stepping. 22, 88, ... are number of elements in grids. p is polynomial degree for DG method.

p	Method	352	1408		5632		22528	
	IP	5.96e-2	1.47e-2	2.02	3.67e-3	2.01	9.16e-4	2.00
1	LDG	6.22e-2	1.49e-2	2.06	3.57e-3	2.06	8.72e-4	2.04
	CDG	5.94e-2	1.44e-2	2.03	3.53e-3	2.03	8.78e-4	2.01
p		88	352		1408		5632	
	IP	3.36e-2	4.25e-3	2.98	5.35e-4	2.99	6.68e-5	3.00
2	LDG	3.46e-2	4.25e-3	3.03	5.21e-4	3.03	6.45e-5	3.01
	CDG	3.33e-2	4.08e-3	3.03	5.07e-4	3.01	6.30e-5	3.01
p		22	88		352		1408	
	IP	5.24e-2	3.67e-3	3.84	2.45e-4	3.91	1.56e-5	3.97
3	LDG	5.04e-2	3.60e-3	3.81	2.35e-4	3.94	1.48e-5	3.99
	CDG	5.12e-2	3.47e-3	3.88	2.29e-4	3.93	1.44e-5	3.99

erator. One can see that all method perform equally well, with IP being slightly less accurate. In Figure 3.1 we see that the IP method with explicit RK method would be the first choice in terms of efficiency for solving Navier-Stokes system, whereas LDG the last choice. The results are quite different when one uses IMEX-RK time-stepping.

In Figure 3.2 we couple our methods with IMEX-RK. When we choose the IP penalty coefficient to be 10 for $k = 1$ and 20 for $k = 2, 3$ we get the following. All methods for the Navier-Stokes equations on unstructured triangular grid converge optimally. But the CPU time of CDG in case of 4th order methods is 40% better than the one of LDG and IP; 55% than BR2; whereas accuracy of CDG and LDG is 14% better than IP; 18% than BR2. Notice that cheaper-to-evaluate IP spatial

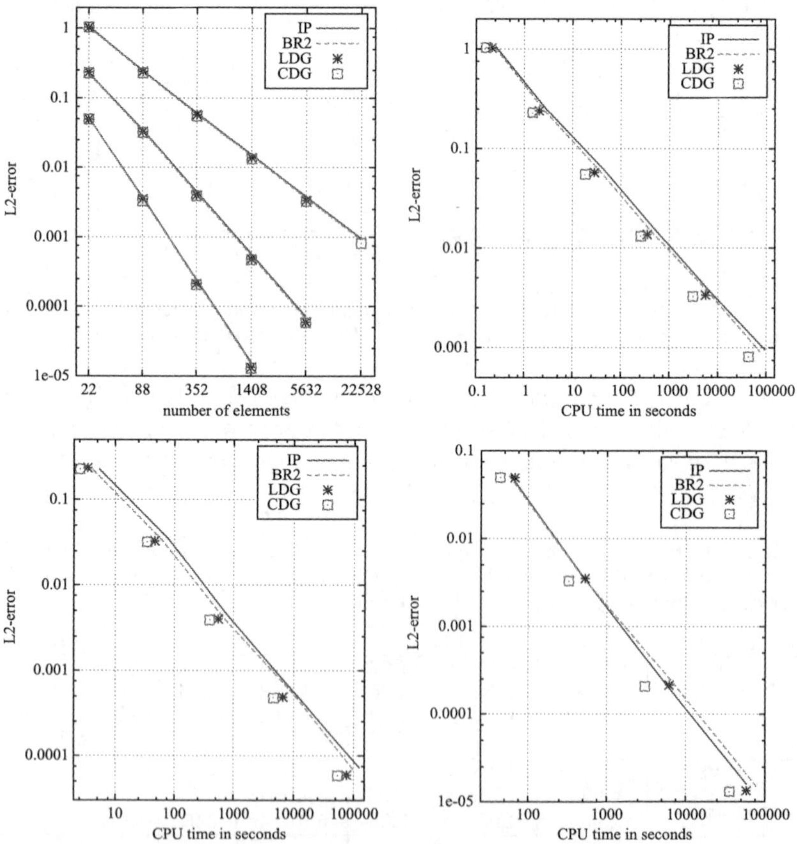

Figure 3.2 Navier-Stokes. Unstructured triangular grid. *Top left:* Convergence rate (EOC) *Top right:* efficiency of the 2nd order methods *Bottom left:* efficiency of the 3rd order methods *Bottom right:* efficiency of the 4th order methods

operator doesn't lead to faster method in terms of CPU time, the reason being the badly conditioned linear systems for the IP method, induced by large penalty coefficients. Comparing the iterations of the GMRES solver we get that for the 4th order methods, IP has 100% iterations of the GMRES solver more than CDG.

4 Conclusions

In this paper, we have formulated a higher order CDG method of Perraire and Persson for advection-diffusion problems and in case of linear heat equation we have proven the stability under certain conditions. The CDG method was computationally verified for the linear heat equation and the Navier-Stokes equations using known analytical solutions. The expected order of convergence was achieved. In comparisons for the Navier-Stokes system using the IMEX-RK as time integrator we conclude that CDG with $\chi = 1$ is more efficient than other methods.

References

[1] J. Peraire and P. O. Persson, The compact discontinuous galerkin method for elliptic problems. SIAM J. Sci. Comput. 30 (2008), no. 4, 806–824.

[2] R. Hartmann and P. Houston, An optimal order penalty discontinuous galerkin discretization of the compressible navier-stokes equations. J. Comput. Phys. 22 (2008), 9670–9685.

[3] H. Liu and J. Yan, The direct discontinuous galerkin (DDG) method for diffusion with interface corrections. SIAM J. Numer. Anal. 47 (2009), no. 1, 675–698.

[4] J. Douglas and T. Dupont, Interior Penalty Procedures for Elliptic and Parabolic Galerkin Methods. Lecture Notes in Phys. 58, Springer-Verlag, Berlin 1976.

[5] F. Brezzi, G. Manzini, D. Marini, P. Pietra and A. Russo, Discontinuous galerkin approximations for elliptic problems. Numer. Methods Partial Dffer. Equations. 16 (2000), no. 4: 365–378.

[6] G. Gassner, F. Lörcher and C.-D. Munz, A discontinuous galerkin scheme based on a space-time expansion ii. Viscous Fow Equations in Multidimensions. J. Sci. Comp. 34 (2008), no. 3, 260–286.

[7] F. Bassi and S. Rebay, Numerical evaluation of two discontinuous galerkin methods for the compressible Navier-Stokes equations. Int. J. Numer. Meth. Fluids. 40 (2002), 197–207.

IFCP Riemann Solver for the Pitman-Le System*

Manuel J. Castro
Dpto. Análisis Matemático. Universidad de Málaga
Málaga, 29071, Spain
E-mail: castro@anamat.cie.uma.es

Enrique D. Fernández-Nieto
Dpto. Matemática Aplicada. E.T.S. Arquitectura
Universidad de Sevilla. Sevilla, 41012, Spain
E-mail: edofer@us.es

José M. González-Vida
Dpto. Matemática Aplicada. E.T.S. Ing. Telecomunicación
Universidad de Málaga. Málaga, 29071, Spain
E-mail: jgv@uma.es

Abstract

In this work we present a first order numerical scheme for solving the Pitman-Le model. It is constructed by using a suitable decomposition of a Roe matrix (see [14]) by means of a parabolic viscosity matrix (see [5]), that captures information of the intermediate fields and we named it as IFCP (Intermediate Field Capturing Parabola). It is linearly L^{∞}-stable and well-balanced with respect to the water at rest solution. A numerical experiment is presented to compare the behaviour of this new scheme with Roe and FORCE methods.

1 Introduction

Let us consider the two-fluid flow model of Pitman and Le [13] under the formulation given by Pelanti et al. in [12], where the friction terms

*This research has been partially supported by the Spanish Government Research projects MTM2006-01275, MTM2006-08075, MTM2009-07719, MTM2009-11923 and P06-RNM-01594. The numerical computations have been performed at the Laboratory of Numerical Methods of the University of Málaga.

are neglected

$$
\begin{cases}
\dfrac{\partial h_f}{\partial t} + \dfrac{\partial q_f}{\partial x} = 0, \\
\dfrac{\partial q_f}{\partial t} + \dfrac{\partial}{\partial x}\left(\dfrac{q_f^2}{h_f} + \dfrac{g}{2}h_f^2\right) + gh_f\dfrac{\partial h_s}{\partial x} = -gh_f\dfrac{db}{dx}, \\
\dfrac{\partial h_s}{\partial t} + \dfrac{\partial q_s}{\partial x} = 0, \\
\dfrac{\partial q_s}{\partial t} + \dfrac{\partial}{\partial x}\left(\dfrac{q_s^2}{h_s} + \dfrac{g}{2}h_s^2 + g\dfrac{1-r}{2}h_sh_f\right) + rgh_s\dfrac{\partial h_f}{\partial x} = -gh_s\dfrac{db}{dx}.
\end{cases}
\tag{1.1}
$$

In these equations, index s (f respectively) makes reference to the solid (fluid respectively) phase. The granular fluid is assumed to occupy a straight channel with constant rectangular cross-section and constant width. The coordinate x refers to the axis of the channel, t is time, and g is the acceleration due to gravity. $b(x)$ represents the fixed bottom topography and r is the ratio of densities between the solid and fluid phase. The unknowns h_s and h_f are related to the total height of the granular fluid h and the solid fraction ψ by $h_s = \psi h$ and $h_f = (1-\psi)h$. The unknowns q_s and q_f represent the mass-flow of each phase.

System (1.1) can be written under the form

$$
w_t + F(w)_x + B(w)\cdot w_x = S(w)b_x, \tag{1.2}
$$

by setting

$$
w(x,t) = \begin{bmatrix} h_f(x,t) \\ q_f(x,t) \\ h_s(x,t) \\ q_s(x,t) \end{bmatrix}, \quad F(w) = \begin{bmatrix} q_f \\ \dfrac{q_f^2}{h_f} + \dfrac{g}{2}h_f^2 \\ q_s \\ \dfrac{q_s^2}{h_s} + \dfrac{g}{2}h_s^2 + g\dfrac{1-r}{2}h_fh_s \end{bmatrix},
$$

$$
B(w) = \begin{bmatrix} 0 & 0 & 0 & 0 \\ 0 & 0 & gh_f & 0 \\ 0 & 0 & 0 & 0 \\ grh_s & 0 & 0 & 0 \end{bmatrix}, \quad S(w) = \begin{bmatrix} 0 \\ -gh_f \\ 0 \\ -gh_s \end{bmatrix}.
$$

As in the case of two-layer shallow-water system (see [1] for example), simple expression of the eigenvalues of matrix $A(w)$ cannot be obtained and the hyperbolicity of system (1.1) is not always ensured. In [12] some bounds of the eigenvalues of $A(w)$ are obtained and conditions ensuring the hyperbolicity of system (1.1) are derived.

By adding to (1.2) the equation

$$
b_t = 0, \tag{1.3}
$$

the system can be rewritten in the form:

$$W_t + \mathcal{A}(W) \cdot W_x = 0, \tag{1.4}$$

where W is the augmented vector

$$W = \begin{bmatrix} w \\ b \end{bmatrix} \in \Omega = \mathcal{O} \times \mathbb{R},$$

and $\mathcal{A}(W)$ is the 5×5 matrix whose block structure is given by:

$$\mathcal{A}(W) = \left[\begin{array}{c|c} A(w) & -S(w) \\ \hline 0 & 0 \end{array}\right],$$

where

$$A(w) = J(w) + B(w), \quad \text{being } J(w) = \frac{\partial F}{\partial w}(w).$$

Solutions of (1.4) may develop discontinuities and, due to the non-divergence form of the equations, the notion of weak solution in the sense of distributions cannot be used. The theory introduced by Dal Maso, LeFloch, and Murat [4] is followed here to define weak solutions of (1.4). This theory allows one to define the nonconservative product $\mathcal{A}(W) \cdot W_x$ as a bounded measure provided a family of Lipschitz continuous paths $\Phi : [0,1] \times \Omega \times \Omega \to \Omega$ is prescribed, which must satisfy certain natural regularity conditions. Here, the family of straight segments is considered:

$$\Phi(s; W_L, W_R) = W_L + s(W_R - W_L).$$

The outline of the paper is the following: the numerical scheme is introduced in Section 2. A numerical example to compare the behaviour of this new scheme with Roe and FORCE methods is performed in Section 3, and finally some conclusions are drawn.

2 Numerical scheme

In [6] authors introduce a first order numerical scheme, named IFCP for solving the two-layer shallow-water system. IFCP numerical scheme is constructed by using a suitable decomposition of a Roe matrix of system (1.1) by means of a parabolic viscosity matrix (see [5]), that captures information of the intermediate fields. IFCP numerical scheme can be written as follows

$$w_i^{n+1} = w_i^n - \frac{\Delta t}{\Delta x}\left(D_{i-1/2}^{+} + D_{i+1/2}^{-}\right), \tag{2.1}$$

being

$$D^{\pm}_{i+1/2} = \frac{1}{2}(F(w^n_{i+1}) - F(w^n_i) + B_{i+1/2}(w^n_{i+1} - w^n_i) - S_{i+1/2}(b_{i+1} - b_i)$$
$$\pm Q_{i+1/2}(w^n_{i+1} - w^n_i - A^{-1}_{i+1/2}S_{i+1/2}(b_{i+1} - b_i))), \qquad (2.2)$$

and $B_{i+1/2}$ is a matrix satisfying:

$$B_{i+1/2} \cdot (w_{i+1} - w_i) = \int_0^1 B(\Phi(s; W_i, W_{i+1}))\frac{\partial \Phi_{[1,\cdots,4]}}{\partial s}(s; W_i, W_{i+1})\, ds; \qquad (2.3)$$

$S_{i+1/2}$ is a vector satisfying:

$$S_{i+1/2}(b_{i+1} - b_i) = \int_0^1 S(\Phi(s; W_i, W_{i+1}))\frac{\partial \Phi_5}{\partial s}(s; W_i, W_{i+1})\, ds; \qquad (2.4)$$

$A_{i+1/2} = J_{i+1/2} + B_{i+1/2}$, being $J_{i+1/2}$ a Roe linearization of the Jacobian of he flux F in the usual sense:

$$J_{i+1/2} \cdot (w_{i+1} - w_i) = F(w_{i+1}) - F(w_i); \qquad (2.5)$$

and $Q_{i+1/2}$ is a viscosity matrix.

Remark 2.1. Note that the numerical scheme depends on the choice of the family of path Φ. In fact, this scheme is a path-conservative numerical scheme in the sense introduced by Parés in [9]. In [2] and [10] it has been proved that, in general, the numerical solutions provided by a path-conservative numerical scheme converge to functions which solve a perturbed system in which an error source-term appears on the right-hand side. The appearance of this source term, which is a measure supported on the discontinuities, has been first observed in [7] when a scalar conservation law is discretized by means of a nonconservative numerical method. Nevertheless, in certain special situations the convergence error vanishes for finite difference methods: this is the case for systems of balance laws (see [8]). Moreover for more general problems, even when the convergence error is present, it may be only noticeable for very fine meshes, for discontinuities of large amplitude, and/or for large-time simulations: see [2], [10] for details.

The key point is the definition of the matrix $Q_{i+1/2}$, that in the case of the IFCP is defined by:

$$Q_{i+1/2} = \alpha_0 Id + \alpha_1 A_{i+1/2} + \alpha_2 A^2_{i+1/2}, \qquad (2.6)$$

where α_j, $j = 0, 1, 2$ are defined by:

$$\begin{aligned}
\alpha_0 &= \delta_1 \lambda_{4,i+1/2}\chi_{int} + \delta_{4,i+1/2}\lambda_{1,i+1/2}\chi_{int} + \delta_{int}\lambda_{1,i+1/2}\lambda_{4,i+1/2},\\
\alpha_1 &= -\lambda_1(\delta_4 + \delta_{int}) - \lambda_{4,i+1/2}(\delta_1 + \delta_{int}) - \chi_{int}(\delta_{1,i+1/2} + \delta_{4,i+1/2}), \qquad (2.7)\\
\alpha_2 &= \delta_1 + \delta_4 + \delta_{int},
\end{aligned}$$

being

$$\delta_1 = \frac{|\lambda_{1,i+1/2}|}{(\lambda_{1,i+1/2} - \lambda_{4,i+1/2})(\lambda_{1,i+1/2} - \chi_{int})},$$
$$\delta_4 = \frac{|\lambda_{4,i+1/2}|}{(\lambda_{4,i+1/2} - \lambda_{1,i+1/2})(\lambda_{4,i+1/2} - \chi_{int})},$$
$$\delta_{int} = \frac{|\chi_{int}|}{(\chi_{int} - \lambda_{1,i+1/2})(\chi_{int} - \lambda_{4,i+1/2})},$$

where $\lambda_{1,i+1/2} < \lambda_{2,i+1/2} < \lambda_{3,i+1/2} < \lambda_{4,i+1/2}$ are the eigenvalues of matrix $A_{i+1/2}$ and

$$\chi_{int} = \mathcal{S}_{ext} \max(|\lambda_2|, |\lambda_3|), \tag{2.8}$$

with

$$\mathcal{S}_{ext} = \begin{cases} \mathrm{sgn}(\lambda_1 + \lambda_4), & \text{if } (\lambda_1 + \lambda_4) \neq 0, \\ 1, & \text{otherwise.} \end{cases} \tag{2.9}$$

It can be proved that the numerical scheme previously defined is linearly L^∞ stable under the usual CFL condition

$$\max\{|\lambda_{l,i+1/2}|, 1 \leq l \leq 4,\ i \in \mathbb{Z}\} \frac{\Delta t}{\Delta x} = CFL \leq 1. \tag{2.10}$$

Remark 2.2. Note that the coefficients α_i are defined in terms of the eigenvalues of the matrix $A_{i+1/2}$. Here, we use the bounds derived in [12] to estimate the wave speeds $\lambda_{l,i+1/2}$, $l = 1, \cdots, 4$.

Remark 2.3. Note that if $\alpha_0 = (1-\omega)\frac{\Delta x}{\Delta t}$, $\alpha_1 = 0$ and $\alpha_2 = \omega \frac{\Delta t}{\Delta x}$, then the numerical scheme (2.1)–(2.2) coincides with the family introduced in [3]. This family contains, as particular cases, a well-balanced extension of the Lax-Friedrichs ($\omega = 0$), Lax-Wendroff ($\omega = 1$), FORCE ($\omega = \frac{1}{2}$), and GFORCE ($\omega = \frac{1}{1+CFL}$) methods.

Remark 2.4. Notice that in the definition of (2.2) the term

$$C = Q_{i+1/2} A^{-1}_{i+1/2} S_{i+1/2}(b_{i+1} - b_i),$$

that can be interpreted as the upwinding part of the source term discretization, makes no sense if one of the eigenvalues of $A_{i+1/2}$ vanishes. In this case the problem is said to be resonant. Resonant problems exhibit an additional difficulty, as weak solutions may not be uniquely determined by their initial data, and the limits of the numerical solutions may depend both on the family of paths and the numerical scheme itself. The analysis of this difficulty is beyond the scope of this article. Here, we follow the strategy described in [3] to get rid of this difficulty and to obtain well-balanced numerical schemes for a given set of stationary solutions.

3 Numerical example

In this section, we compare the IFCP numerical scheme introduced here with a Roe scheme for system (1.1) that is constructed following [11], and GFORCE method.

An experiment that is an extension of LeVeque's classical test for one-layer shallow-water system with bottom topography is performed, that is, the behaviour of small perturbation of steady state solution corresponding to water at rest over a non flat bottom topography is studied. This numerical test is taken from [12]. The computational

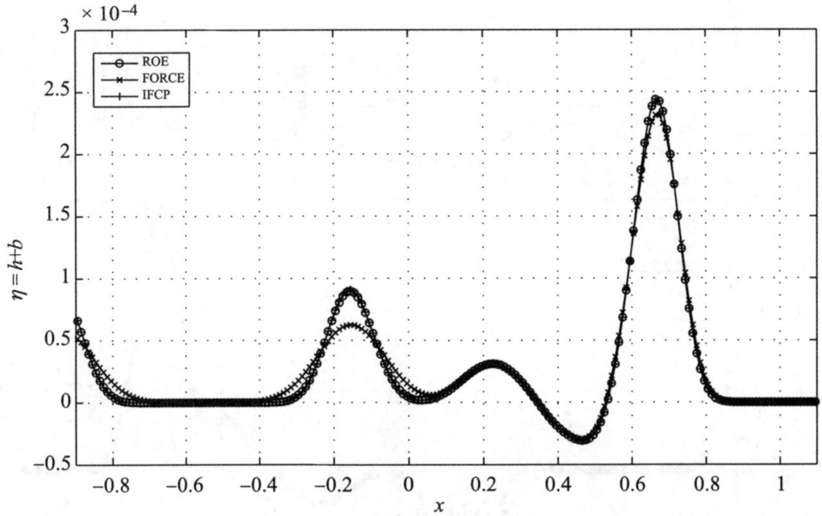

(a) Free surface $\eta = h+b$.

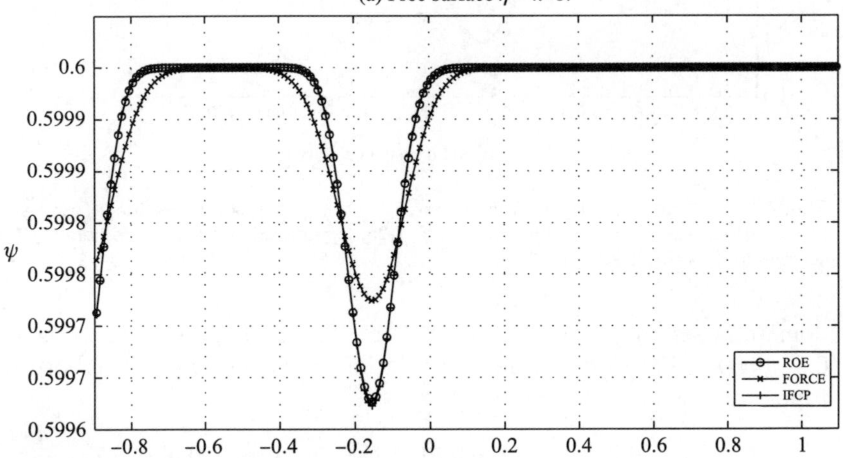

(b) Solid fraction ψ.

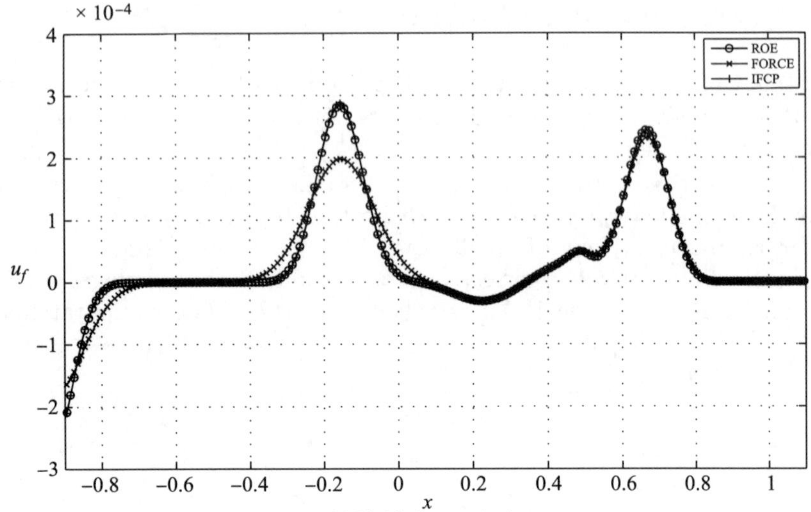

(c) Fluid phase velocity u_f.

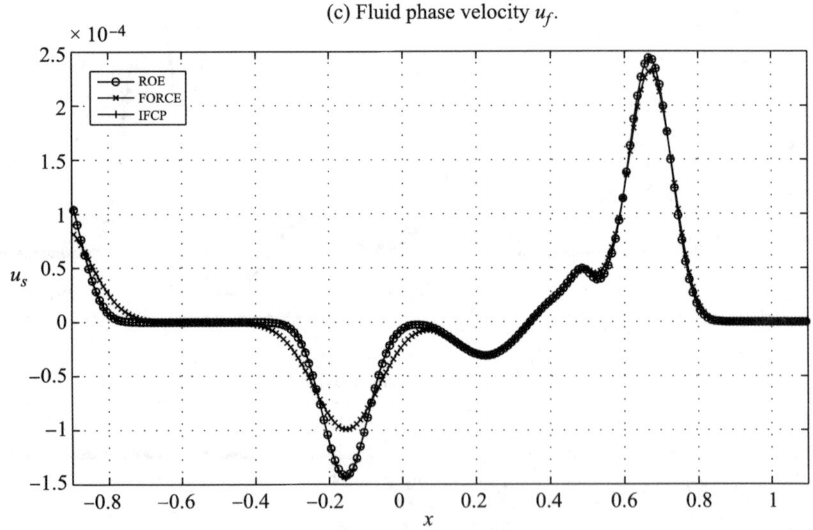

(d) Solid phase velocity u_s.

Figure 3.1 Numerical test: Free surface $\eta = h + b$, the solid fraction ψ, and the phase velocities u_f, and u_s at time $t = 1.25$, computed with IFCP scheme, Roe scheme and FORCE scheme.

domain is set to $[-0.9, 1.0]$ and $g = 1$. The CFL parameter is set to 0.9, $r = 0.5$. Free flow boundary conditions are considered. The bottom topography is given by the function

$$b(x) = \begin{cases} 0.25(\cos(\pi(x - 0.5)/0.1) + 1), & \text{if } |x - 0.5| < 0.1, \\ 0, & \text{otherwise.} \end{cases}$$

As initial condition we set $u_s(x,0) = u_f(x,0) = 0$ and

$$h(x,0) = \begin{cases} 1 + 10^{-3}, & \text{if } -0.6 < x < -0.5, \\ 1 - b(x), & \text{otherwise}, \end{cases}$$

$$\psi(x,0) = \begin{cases} 0.6 - 10^{-3}, & \text{if } -0.6 < x < -0.5, \\ 0.6, & \text{otherwise}. \end{cases}$$

Figure 3.1 shows the results for the free surface $\eta = h + b$, the solid fraction ψ, and the phase velocities u_f, and u_s at time $t = 1.25$, computed with ICFP scheme, Roe scheme and FORCE scheme using $\Delta x = 0.01$. A reference solution computed with Roe scheme for $\Delta x = 1/3200$ is also obtained. The initial perturbation is split into four waves, two going to the right and two to the left. The two waves going to the left have gone out and only the two going to the right are present in the domain. It can be seen that the right-going external wave has just passed over the obstacle at the bottom, and it has been partially reflected. It is well known that numerical schemes that are not well-balanced for the steady state at rest may produce in this kind of numerical test unphysical disturbances. All the numerical schemes considered here are well-balanced and they do not produce spurious oscillations. Note that Roe and IFCP give similar solutions, while FORCE is more diffusive. Concerning the CPU time, Roe is about 3 times more expensive than IFCP and FORCE.

4 Conclusions

In this work a first order numerical scheme for solving the Pitman-Le system has been introduced. IFCP is a method constructed by a suitable decomposition of a Roe matrix $A_{i+1/2}$, whose viscosity matrix is computed by a linear combination of the identity matrix, $A_{i+1/2}$ and $A^2_{i+1/2}$ and whose coefficients are given in terms of the eigenvalues of $A_{i+1/2}$. The resulting numerical scheme is linearly L_∞-stable and well-balanced for the water at rest solution. The numerical tests performed and, in particular, the one presented in this paper, show that the results obtained with IFCP are close to those obtained with ROE scheme and much better than FORCE scheme.

References

[1] M. J. Castro, J. Macías and C. Parés, A Q-Scheme for a class of systems of coupled conservation laws with source term. Application to a two-layer 1-D shallow water system. Math. Mod. Num. Anal. 35(1)(2001), 107–127.

[2] M. J. Castro, P. G. LeFloch, M. L. Muñoz and C. Parés. Why many theories of shock waves are necessary: Convergence error in formally path-consistent schemes. Jour. Comp. Phys. 3227(2008), 8107–8129.

[3] M. J. Castro, A. Pardo, C. Parés and E. F. Toro, On some fast well-balanced first order solvers for nonconservative systems. Math. Comp. 79(2010), 1427–1472.

[4] G. Dal Maso, P. G. LeFloch and F. Murat, Definition and weak stability of nonconservative products. J. Math. Pures Appl. 74(1995), 483–548.

[5] P. Degond, P. F. Peyrard, G. Russo and Ph. Villedieu, Polynomial upwind schemes for hyperbolic systems. C. R. Acad. Sci. Paris. 1328(1999), 479–483.

[6] E. D. Fernández-Nieto, M. J. Castro and C. Parés, IFCP Riemann solver for the two-layer shallow-water system. Submitted J. Sci. Comput. (2010).

[7] T. Y. Hou and P. G. LeFloch, Why nonconservative schemes converge to wrong solutions: error analysis. Math. of Comput. 62(1994), 497–530.

[8] M. L. Muñoz and C. Parés, On the convergence and well-balanced property of path-conservative numerical schemes for systems of balance laws. Submitted to Jour. Sci. Comp. (2010).

[9] C. Parés, Numerical methods for nonconservative hyperbolic systems: a theoretical framework. SIAM J. Num. Anal. 44(1)(2006), 300–321.

[10] C. Parés and M. L. Muñoz Ruíz, On some difficulties of the numerical approximation of nonconservative hyperbolic systems. Boletín SEMA. 47(2009), 23–52.

[11] C. Parés and M. J. Castro, On the well-balance property of Roe's method for nonconservative hyperbolic systems. Applications to Shallow-Water Systems. M2AN. Vol 38(5)(2004), 821–852 .

[12] M. Pelanti, F. Bouchut and A. Mangeney, A Roe-type scheme for two-phase shallow granular flows over variable topography. M2AN. 42(2008), 851–885.

[13] E. B. Pitman and L. Le, A two-fluid model for avalanche and debris flows. Phil. Trans. R. Soc. A. 363(2005), 1573–1601.

[14] I. Toumi, A weak formulation of Roe approximate Riemann solver. J. Comp. Phys. 102(2)(1992), 360–373.

On the Hyperbolicity of Two- and Three-Layer Shallow Water Equations

M. Castro
Universidad de Málaga

J. T. Frings
RWTH Aachen
E-mail: frings@igpm.rwth-aachen.de

S. Noelle
RWTH Aachen

C. Parés
Universidad de Málaga

G. Puppo
Politecnico di Torino

Abstract

The two-layer shallow water system looses hyperbolicity if the magnitude of the shear velocity is above a certain threshold, essentially determined by the density difference between the two layers. Introducing an additional third layer might recover hyperbolicity in regions of strong shear. We demonstrate that this adaptive two/three-layer approach can cure some of the shortcomings of the two-layer model.

1 Introduction

In this paper we consider stratified flows. An example which has recently found attention in applied mathematics is the superimposition of the Mediterranean and the Atlantic in the Strait of Gibraltar, e.g. [4]. This layering persists as long as the velocity difference (shear velocity) is not too large. Under this condition the flow may be modeled by two coupled shallow water models, yielding a non-conservative hyperbolic system in one or two space dimensions which may be solved efficiently and accurately by well-balanced finite volume schemes, e.g. [3].

In regions of strong gradients in the bottom topography, however, the lower layer Mediterranean water may accelerate, causing a strong shear and hence a Kelvin-Helmholtz instability. Mathematically, the hyperbolicity of the two-layer model is lost, and usually the finite volume solvers break down or produce questionable results. The obvious, but expensive, solution would be to switch to the full 2D or 3D Navier-Stokes equations and apply the respective free surface incompressible solvers.

A challenging question is whether one can model such local instabilities without giving up the depth-averaging everywhere. Several authors have suggested such models, e.g. LeVeque and Kim [6], Castro and Parés [5], Bouchut and Morales [2] and Audusse [1].

Here we explore an adaptive two/three layer model. We introduce an intermediate third layer to regain hyperbolicity in case it is lost.

2 Multilayer flows

The multilayer shallow water system The flow of $i = 1, \ldots, N$ layers of water, ordered from top to bottom, is governed by

$$\begin{aligned} &\partial_t h_i + \partial_x q_i = 0, \\ &\partial_t q_i + \partial_x \left(q_i^2/h_i + g/2\, h_i^2\right) = -g h_i \partial_x \left(\sum_{j=1}^{i-1} r_{ji} h_j + \sum_{j=i+1}^{N} h_j + b\right) \end{aligned} \tag{2.1}$$

with densities $\rho_i < \rho_{i+1}$ and $r_{ij} := \frac{\rho_i}{\rho_j}$ denoting the density ratios. The coordinate x refers to the spatial position, t is the time, and g is gravity. The unknowns $q_i(x,t)$ and $h_i(x,t)$ represent respectively the discharge and the height, i.e. thickness of the i-th layer at the section of coordinate x at time t. The bottom topography is given by the function b.

The system (2.1) can be written in quasi-linear form

$$w_t + \mathcal{A}(w) w_x = S(w) b_x, \tag{2.2}$$

cf. [3]. The eigenvalues of matrix $\mathcal{A}(w)$ may become complex even for physically relevant data, corresponding to the development of shear instabilities. We will call a multilayer state w *hyperbolic*, if the matrix $\mathcal{A}(w)$ is diagonalizable and all its eigenvalues are real.

Loss of hyperbolicity In the two-layer case, the matrix $\mathcal{A}$ becomes a 4×4 matrix. An approximation formula for the eigenvalues in the limit of almost constant water level, almost equal densities and vanishing velocity difference is given in [7]. The authors distinguish two pairs of eigenvalues, the internal and the external ones:

$$\lambda_{ext}^{\pm} = \frac{u_1 h_1 + u_2 h_2}{h_1 + h_2} \pm \left(g(h_1 + h_2)\right)^{\frac{1}{2}}, \tag{2.3}$$

$$\lambda_{int}^{\pm} = \frac{u_1 h_2 + u_2 h_1}{h_1 + h_2} \pm \left[g' \frac{h_1 h_2}{h_1 + h_2} \left(1 - \frac{(u_1 - u_2)^2}{g'(h_1 + h_2)}\right)\right]^{\frac{1}{2}}, \tag{2.4}$$

where $g' := (1 - r_{12})g$ is a small constant in this limit. The external eigenvalues $\lambda^{\pm}_{ext}$ correspond to the eigenvalues of a single-layer shallow water system with total water height $H := h_1 + h_2$ and an averaged total velocity $u = \frac{u_1 h_1 + u_2 h_2}{h_1 + h_2}$. These eigenvalues are therefore distinct and real, thus they are not important regarding the hyperbolicity. If

$$\kappa := \frac{(u_1 - u_2)^2}{g'(h_1 + h_2)} > 1, \tag{2.5}$$

the approximative internal eigenvalues become complex. In [5] it is shown that (2.5) is a good approximative criterion for the loss of hyperbolicity when r_{12} is close to 1. The ratio κ can be interpreted as the balance between the stabilizing influence of the difference in density and the destabilizing one of the velocity shear.

We quantify this loss of hyperbolicity by the indicator δ as follows:

Definition 1. The hyperbolicity indicator δ for a state U is given by

$$\delta(U) := |\Re(\lambda^{+}_{int}(U)) - \Re(\lambda^{-}_{int}(U))| - |\Im(\lambda^{+}_{int}(U)) - \Im(\lambda^{-}_{int}(U))|, \tag{2.6}$$

where $\Re$ denotes the real part and $\Im$ the imaginary part.

This indicator is nonnegative if the eigenvalues are real and negative if they are complex conjugate. It is extended to the multilayer case by applying it to each pair of internal eigenvalues, then taking the minimum.

3 The three-layer adaptation

The breakdown of hyperbolicity corresponds to a Kelvin-Helmholtz instability and the onset of mixing [7]. Our interest is to model this situation in the context of shallow water models. We introduce an intermediate layer around the original interface with constant values for velocity and density attached to it, assuming that all the possible turbulence and mixing is contained within this layer and does not affect the horizontal velocities of the other layers in a substantial way.

A strategy for choosing the intermediate layer Given a two-layer state $[(\rho_1, h_1, u_1), (\rho_2, h_2, u_2)]$ we are now looking at a way to find a corresponding three layer state $[(\tilde{\rho}_1, \tilde{h}_1, \tilde{u}_1), (\rho_m, h_m, u_m), (\tilde{\rho}_2, \tilde{h}_2, \tilde{u}_2)]$ such that the discharge (Q), mass (M) and total water height (H) are conserved:

$$\sum_i \rho_i h_i u_i = \sum_i \tilde{\rho}_i \tilde{h}_i \tilde{u}_i + \rho_m h_m u_m, \tag{Q}$$

$$\sum_i \rho_i h_i = \sum \tilde{\rho}_i \tilde{h}_i + \rho_m h_m, \tag{M}$$

$$\sum_i h_i = \sum_i \tilde{h}_i + h_m, \tag{H}$$

In the upper and lower layer the densities and the velocities are set to

$$\tilde{u}_1 = u_1, \qquad \tilde{u}_2 = u_2, \qquad \tilde{\rho}_1 = \rho_1, \qquad \tilde{\rho}_2 = \rho_2. \tag{C.1}$$

Next, among several possible choices we pick $\rho_m = \frac{1}{2}(\rho_1 + \rho_2)$ in the intermediate layer, thus enforcing constant density throughout the layer, and consider h_m a free parameter. This gives

$$\tilde{h}_1 = h_1 - \tfrac{1}{2}h_m, \qquad \tilde{h}_2 = h_2 - \tfrac{1}{2}h_m, \qquad u_m = \tfrac{\rho_1 u_1 + \rho_2 u_2}{\rho_1 + \rho_2}. \tag{C.2}$$

The density ρ_m is independent of x and t. Thus even for multiple adaption steps the three-layer context persists, simplifying the construction of an adaptive solver.

Hyperbolic domains We test the hyperbolicity of 10.000 randomly chosen two-layer test states and compare with those of the corresponding three-layer states gained by the transformation formula (C). The ratio $r_{12} = 0.98$ is kept fix.

The hyperbolicity is tested by inserting the test states into the matrix $\mathcal{A}$ in (2.2) and then numerically calculating the eigenvalues. In case of all real, distinct eigenvalues we plot a blue dot in the left graph in Figure 3.1, otherwise we plot a red dot in the right graph. The axes are $(u_2 - u_1)^2$ and $h_1 + h_2$. With this choice, the hyperbolic and the non-hyperbolic region are approximately[1] separated by the straight line $\kappa = 1$.

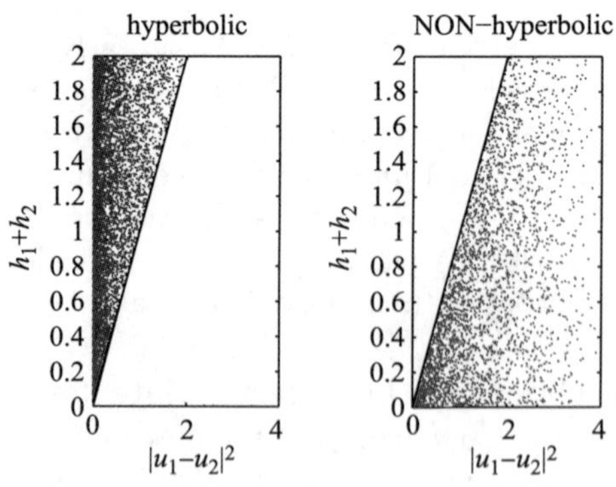

Figure 3.1 Hyperbolic domain for 2-layer test cases, $r_{12} = 0.98$.

To check whether we gain hyperbolicity with our adaptation strategy, we try to find any intermediate layer height for the given test states

[1] Note that there are indeed some hyperbolic states with $\kappa > 1$ close to the approximative boundary. Also, there is a second region of hyperbolicity for very large shear velocities, but this is not in the focus of our work.

that gives a hyperbolic state with three positive layer heights. The intermediate layer height h_m^{opt} is chosen to maximize the indicator δ among all states with positive layer heights. Again we plot dots indicating the hyperbolicity, this time of the three-layer system, but at the location given by the values of the associated original two-layer system. Thus we can observe from the plots the change in hyperbolicity triggered by the two/three-layer adaptation. We can indeed notice a gain in hyperbolicity but a clear picture only emerges if the data are ordered by the ratio $\Theta := \frac{h_2}{H}$ with $H := h_1 + h_2$ the total water height. The plots on the left in Figures 3.2, 3.3 and 3.4 show the gain for different values of Θ. Note that there is a small area in Figure 3.2 where the hyperbolicity is lost when transferring from two to three layers. This does not pose a problem as those states are hyperbolic for the two-layer model.

Choice of intermediate layer height We fix Θ and randomly generate a set with 10000 two-layer test states with $h_2/H = \Theta$. For each test state, using the transformation rules (C), we search for a three-layer state with all positive layer heights that maximizes δ. This gives an intermediate height h_m^{opt}. Now plotting $(\kappa, h_m^{opt}/H)$ we found interesting structures as shown by the exemplary results in the plots on the right of Figures 3.2, 3.3 and 3.4. Θ is fixed for each plot, but varies from plot to plot. Considering several values of Θ, the points $(\kappa, h_m^{opt}/H)$ seem to form a surface $h_m^{opt}(\kappa, \Theta)$. Note that for each Θ, there is a value $\kappa_{\max}$ such that for $\kappa > \kappa_{\max}$ the third layer cannot cure the hyperbolicity loss, at least not without deteriorating to a two-layer state again.

There are three additional alleged curves in these plots. The two sets of pairs $(\kappa, h_m^{\min/\max})$ are defined as follows: define $\mathcal{H}^{opt}$ for a given κ as the largest interval of hyperbolic points containing h_m^{opt}. This interval exists as the eigenvalues depend smoothly on the state (and thus on h_m). Then $\mathcal{H}^{opt} =: [h_m^{\min}, h_m^{\max}]$. The remaining curve is given by a heuristic approach for the intermediate layer height.

Heuristic intermediate layer height As a result of the structure revealed by the Figures 3.2, 3.3 and 3.4 we can define a heuristic method to determine h_m. We fit curves to the point values of h^{opt} given for different fixed values of Θ and then use interpolation between the discrete curves to get a function $\tilde{h}_m^{heu}$ depending on Θ and κ. The resulting function h_m^{heu} is defined only for $1 < \kappa < \kappa_{\max}$, i.e. for those values of κ for which hyperbolicity could be regained. h_m^{heu} is shown in the figures mentioned above as solid lines lying within the optimal intervals $\mathcal{H}^{opt}$ found for the tested states. In the interval $[(\kappa_{\max} + 1)/2, \kappa_{\max})$ it is defined as a least-squares linear fit to the optimal heights. In the interval $[1, (\kappa_{\max} + 1)/2)$ the definition is $h_m^{heu}(\kappa, \Theta) := c_\Theta(1 - \kappa)^{\alpha_\Theta}$,

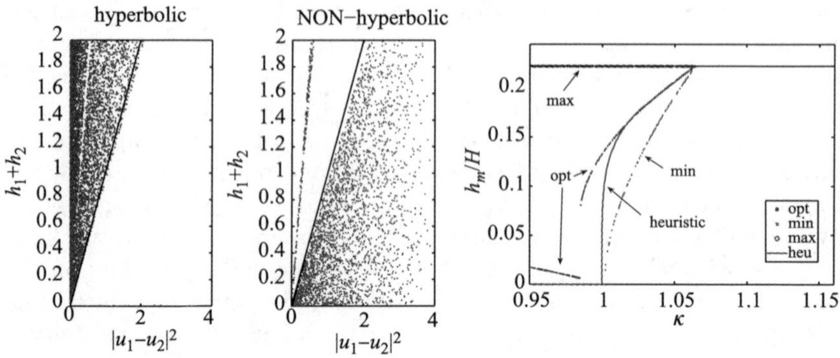

Figure 3.2 Left: Hyperbolic domain for 2-3-layer adaptation, $\Theta = 0.111$; Right: h_m^{opt}, $\mathcal{H}^{opt}$ and heuristic $\tilde{h}_m^{heu}$.

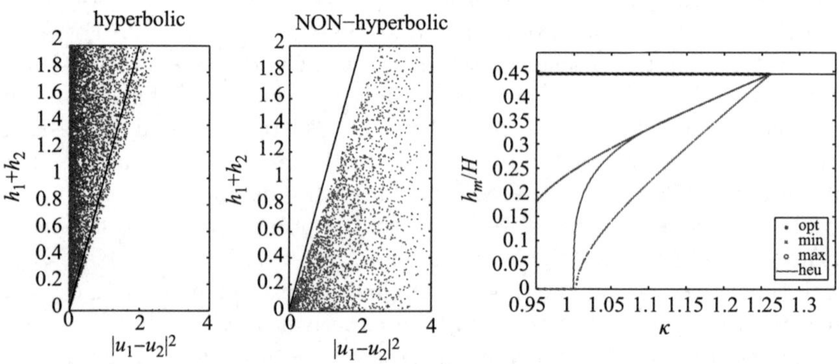

Figure 3.3 Left: Hyperbolic domain for 2-3-layer adaptation, $\Theta = 0.222$; Right: h_m^{opt}, $\mathcal{H}^{opt}$ and heuristic $\tilde{h}_m^{heu}$.

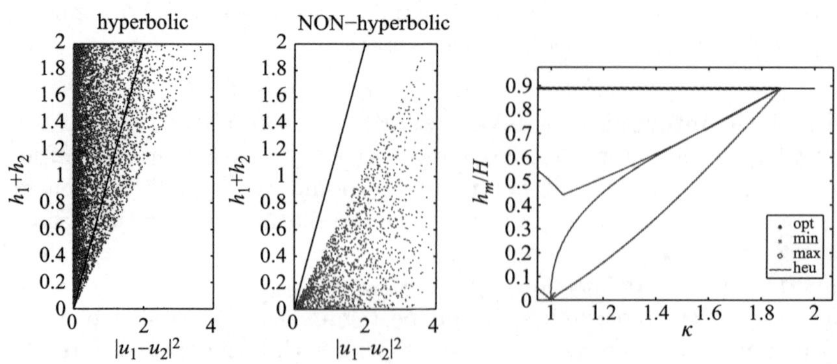

Figure 3.4 Left: Hyperbolic domain for 2-3-layer adaptation, $\Theta = 0.444$; Right: h_m^{opt}, $\mathcal{H}^{opt}$ and heuristic $\tilde{h}_m^{heu}$.

where $c_\Theta, \alpha_\Theta > 0$ are chosen in such a way that the linear fit in $\kappa_{max}/2$ is matched with $\mathcal{C}_1$ smoothness. For $\kappa = 1$ it connects to the corresponding two-layer hyperbolic state.

Loss of hyperbolicity for a two-layer tidal flow over a hump We present the details of a tidal flow over a hump as shown in Figure 3.5. The initial data are set to:

$$\eta := h_1 + h_2 + b = 2, \qquad h_2 = 1 + 0.45 \cdot \mathrm{atan}(1.5(x + 0.75)) - b,$$
$$q_1 = 0.001, \qquad q_2 = -0.001, \qquad b = e^{-x^2},$$

with $r = 0.98$ and $g = 9.812$. The spatial domain is $I = (-3, 3)$ subdivided into 20 cells and CFL-Number is cfl $= 0.8$. Absorbing boundary conditions are applied.

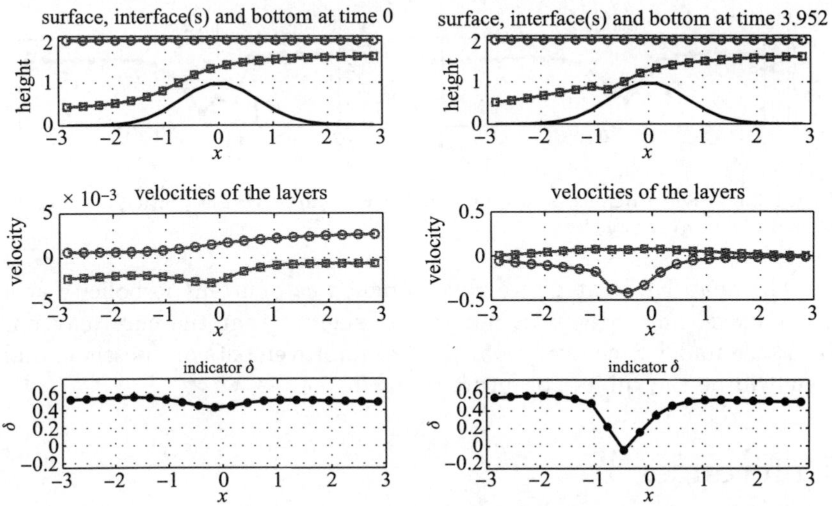

Figure 3.5 Left: Initial state; Right: Situation close to loss of hyperbolicity.

This test problem is an idealized version of a natural two-layer flow being found e.g. at the Strait of Gibraltar, where the denser Mediterranean Sea forms the lower layer and "spills" over a hump while being moved by tidal forces. Our goal is to design an efficient numerical solver that retains most of the physical features of the real flow.

The simulation runs up until time $t = 3.9524$, where the indicator shows that hyperbolicity is close to being lost, see Figure 3.5. At this time an adaptation step is carried out using our heuristic approach to determine the intermediate height. The simulation then continues to run for some time steps until time $t = 4.1689$ when the three-layer system loses hyperbolicity in the same cell that was marked critical before the adaptation step, see Figure 3.6. This is caused mainly by the transport of the intermediate layer. Mass is flowing away from the formerly critical cell and the intermediate layer height drops below h_m^{min} for that state.

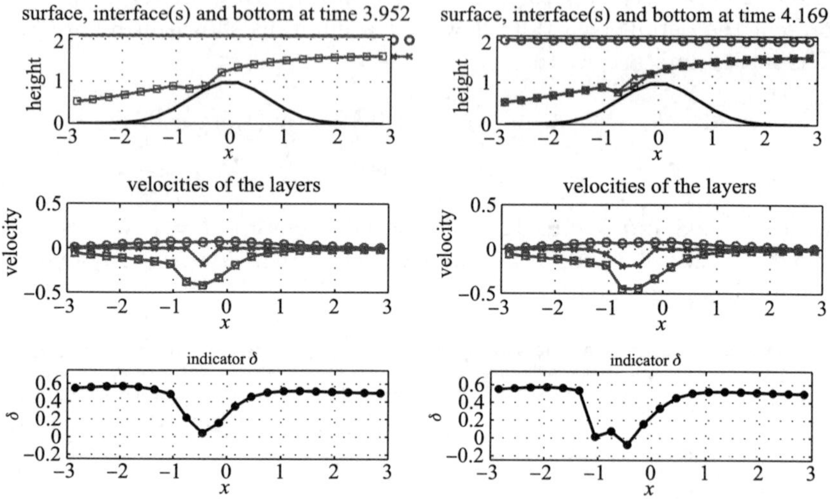

Figure 3.6 Left: Refined state, hyperbolicity regained; Right: Situation close to second loss of hyperbolicity.

The adaptive strategy we have designed can cure hyperbolicity at a given fixed time. However, our results suggest that the cure may not be stable under time evolution. Several improvements are possible, and they will be the subject of future research.

References

[1] E. Audusse, A multilayer Saint-Venant model: derivation and numerical validation. Discrete Contin. Dyn. Syst. Ser. B 5(2)(2005), 189–214.

[2] F. Bouchut and T. Morales, An entropy satisfying scheme for two-layer shallow water equations with uncoupled treatment. M2AN Math. Model. Numer. Anal. 42(4)(2008), 683–698.

[3] M. Castro, J. M. Gallardo and C. Parés, High order finite volume schemes based on reconstruction of states for solving hyperbolic systems with nonconservative products. Applications to shallow-water systems. Math. Comp. 75(255) (2006), 1103–1134.

[4] M. Castro, J. Macías and C. Parés, A Q-scheme for a class of systems of coupled conservation laws with source term. Application to a two-layer 1-D shallow water system. M2AN Math. Model. Numer. Anal. 35(1)(2001), 107–127.

[5] M. J. Castro Daz, E. D. Fernndez-Nieto, J. M. Gonzlez-Vida and C. Pares, Numerical treatment of the loss of hyperbolicity of the two-layer shallow-water system. Submitted to Journal of Scientific Computing.

[6] J. Kim and R. J. LeVeque, Two-layer Shallow Water System and Its Applications. Proceedings of the Twelth International Conference on Hyperbolic Problems. Maryland 2008.

[7] J. B. Schijf and J. C. Schonfeld, Theoretical considerations on the motion of salt and fresh water. In: Proceedings of the Minn. Int. Hydraulic Conv., Joint Meeting IAHR Hydro. Div. ASCE. (September 1953), (1953), 321–333.

Computing Material Fronts with a Lagrange-Projection Approach*

Christophe Chalons
CEA-Saclay, DEN/DANS/DM2S/SFME/LETR
F-91191 Gif-sur-Yvette, France
E-mail: chalons@math.jussieu.fr

Frédéric Coquel
CNRS & Centre de Mathématiques Appliquées, U.M.R. 7641
Ecole Polytechnique, Route de Saclay, 91128 Palaiseau Cedex, France
E-mail: frederic.coquel@cmap.polytechnique.fr

Abstract

This paper reports investigations on the computation of material fronts in multi-fluid models using a Lagrange-Projection approach. Various forms of the Projection step are considered. Particular attention is paid to minimization of conservation errors.

1 Introduction

It is well-known that standard conservative discretizations of gas dynamics equations in Eulerian coordinates generally develop non physical pressure oscillations near contact discontinuities, and more generally near material fronts in multi-component flows. Several cures based on a local non conservative modification have been proposed. Let us quote for instance the hybrid algorithm derived by Karni in [10] and the Two-Flux method proposed by Abgrall and Karni in [2] for multi-fluid flows. See also [1], [3] and the references therein.

We investigate here a Lagrange-Projection type method to get rid of pressure oscillations. The basic motivation lies in the fact that oscillations do not exist in Lagrangian computations. It is then possible to clearly determine which operation in the projection step sparks off

*Research of the first author was partially supported by the French Agence Nationale de la Recherche, contract ANR-08-JCJC-0132-01. Research of the authors was partially supported by the Ministère de la Recherche under grant ERTint 1058 entitled *Simulation Avancée du Transport des Hydrocarbures*.

pressure oscillations. As in [10], [2], a non conservative correction is proposed. It is based on a local pressure averaging and a random sampling strategy on the mass fraction in order to strictly preserve isolated material fronts and get a statistical conservation property. Numerical results are proposed and compared with the Two-Flux method [2].

2 The model under consideration

We consider a nonlinear partial differential equations model governing the flow of two species Σ_1 and Σ_2 separated by a material interface. For instance, we focus on two perfect gases and we set $p_i(\rho_i, e_i) = (\gamma_i - 1)\rho_i e_i$, $\gamma_i = C_{p,i}/C_{v,i}$ and $T_i(\rho_i, e_i) = e_i/C_{v,i}$ where p_i, ρ_i, e_i, $\gamma_i > 1$, T_i, $C_{p,i} > 0$, $C_{v,i} > 0$ respectively denote the pressure, the density, the internal energy, the adiabatic coefficient, the temperature and the specific heats of Σ_i, $i = 1, 2$. The mixture density is given by $\rho = \rho_1 + \rho_2$ and we adopt a Dalton's law for the mixture pressure $p = p_1(\rho_1, e_1) + p_2(\rho_2, e_2)$. We assume in addition that the two species evolve according to the same velocity u and are at thermal equilibrium, that is $T = T_1(\rho_1, e_1) = T_2(\rho_2, e_2)$. The mixture internal and total energies e and E are defined by $\rho e = \rho_1 e_1 + \rho_2 e_2$ and $\rho E = \frac{1}{2}\rho u^2 + \rho e$. Then, introducing the mass fraction $Y = \rho_1/\rho$, straightforward manipulations yield $\rho e = \rho C_v T$ with $C_v = C_v(Y) = Y C_{v,1} + (1 - Y)C_{v,2}$ and

$$p = (\gamma - 1)\rho e \quad \text{with} \quad \gamma = \gamma(Y) = \frac{Y C_{p,1} + (1 - Y)C_{p,2}}{Y C_{v,1} + (1 - Y)C_{v,2}} > 1.$$

In one-space dimension, the model under consideration writes

$$\begin{cases} \partial_t \rho + \partial_x(\rho u) = 0, \\ \partial_t(\rho u) + \partial_x(\rho u^2 + p) = 0, \\ \partial_t(\rho E) + \partial_x(\rho E + p)u = 0, \\ \partial_t \rho Y + \partial_x(\rho Y u) = 0, \end{cases} \tag{2.1}$$

and for the sake of conciseness, we set

$$\partial_t \mathbf{u}(x, t) + \partial_x \mathbf{f}(\mathbf{u}(x, t)) = 0. \tag{2.2}$$

The flux function $\mathbf{f}$ finds a natural definition with respect to the conservative unknowns $\mathbf{u} = (\rho, \rho u, \rho E, \rho Y)$. Let us mention that (2.2) is hyperbolic with eigenvalues $\lambda_0(\mathbf{u}) = u$ and $\lambda_\pm(\mathbf{u}) = u \pm c(\mathbf{u})$, $c(\mathbf{u}) = \sqrt{\gamma p/\rho}$, provided that $\rho > 0$, $0 \leq Y \leq 1$ and $p > 0$. The characteristic field associated with λ_0 is linearly degenerate, leading to *contact discontinuities* or *material fronts*. The two extreme fields are genuinely nonlinear.

3 Numerical schemes

This section is devoted to the discretization of (2.2). As already stated, a specific attention must be paid to the contact discontinuities to avoid pressure oscillations. With this in mind, we first revisit the "Two-Flux Method" proposed by Abgrall and Karni [2] and then present a new numerical procedure based on a Lagrangian approach and a random sampling strategy. Comparisons will be proposed in section 4.

Let us introduce a time step $\Delta t > 0$ and a space step $\Delta x > 0$ that we assume to be constant for simplicity. We set $\lambda = \Delta t/\Delta x$ and define the mesh interfaces $x_{j+1/2} = j\Delta x$ for $j \in \mathbb{Z}$, and the intermediate times $t^n = n\Delta t$ for $n \in \mathbb{N}$. In the sequel, $\mathbf{u}_j^n$ denotes the approximate value of $\mathbf{u}$ at time t^n and on the cell $\mathcal{C}_j = [x_{j-1/2}, x_{j+1/2}[$. For $n = 0$ and $j \in \mathbb{Z}$, we set $\mathbf{u}_j^0 = \frac{1}{\Delta x}\int_{x_{j-1/2}}^{x_{j+1/2}} \mathbf{u}_0(x)dx$ where $\mathbf{u}_0(x)$ is the initial condition.

3.1 The Two-Flux method revisited

Aim of this section is to review the Two-Flux method proposed by Abgrall and Karni [2]. Let us first recall that pressure oscillations do not systematically appear in *single-fluid* computations. Abgrall and Karni [2] then propose to replace any conservative *multi-fluid* strategy by a non conservative approach based on the definition of two *single-fluid* numerical fluxes at each interface. We first recall the algorithm in details and then suggest a slight modification in order to lessen the conservation errors. This strategy will be used as a reference to assess the validity of the Lagrangian strategies proposed in the next subsection.

The original algorithm. Let us consider a two-point numerical flux function $\mathbf{g}$ consistent with $\mathbf{f}$. The Two-Flux method proposes to update the sequence $(\mathbf{u}_j^n)_{j\in\mathbb{Z}}$ into two steps, under an usual 1/2 CFL restriction.

First step : evolution of ρ, ρu and p ($t^n \to t^{n+1-}$)
Let us first define $\mathbf{v} = (\rho, \rho u, p, Y)$ and the one-to-one mapping $\mathbf{u} \to \mathbf{v} = \mathbf{v}(\mathbf{u})$ thanks to the thermodynamics closures. Two interfacial numerical fluxes $\mathbf{g}_{j+1/2,L}$ and $\mathbf{g}_{j+1/2,R}$ are then defined by

$$\mathbf{g}_{j+1/2,L} = \mathbf{g}(\mathbf{u}_j^n, \overline{\mathbf{u}}_{j+1,L}^n), \quad \mathbf{g}_{j+1/2,R} = \mathbf{g}(\overline{\mathbf{u}}_{j,R}^n, \mathbf{u}_{j+1}^n), \tag{3.1}$$

where $\overline{\mathbf{u}}_{j+1,L}^n = \mathbf{u}(\overline{\mathbf{v}}_{j+1,L}^n)$ with $\overline{\mathbf{v}}_{j+1,L}^n = (\rho_{j+1}^n, (\rho u)_{j+1}^n, p_{j+1}^n, Y_j^n)$, and $\overline{\mathbf{u}}_{j,R}^n = \mathbf{u}(\overline{\mathbf{v}}_{j,R}^n)$ with $\overline{\mathbf{v}}_{j,R}^n = (\rho_j^n, (\rho u)_j^n, p_j^n, Y_{j+1}^n)$. In some sense, the mass fraction Y is then assumed to be the same on both side of each interface since $Y = Y_j^n$ is used for the computation of $\mathbf{g}_{j+1/2,L}$ and $Y = Y_{j+1}^n$ for $\mathbf{g}_{j+1/2,R}$. At last, we use $\mathbf{g}_{j+1/2,L}$, respectively $\mathbf{g}_{j+1/2,R}$, to update the conservative unknowns ρ, ρu and ρE on the cell j, resp.

$(j+1)$. With clear notations, we get for all $j \in \mathbb{Z}$

$$\begin{aligned} \rho_j^{n+1-} &= \rho_j^n - \lambda(\mathbf{g}_{j+1/2,L}^{\rho} - \mathbf{g}_{j-1/2,R}^{\rho}), \\ (\rho u)_j^{n+1-} &= (\rho u)_j^n - \lambda(\mathbf{g}_{j+1/2,L}^{\rho u} - \mathbf{g}_{j-1/2,R}^{\rho u}), \\ (\rho E)_j^{n+1-} &= (\rho E)_j^n - \lambda(\mathbf{g}_{j+1/2,L}^{\rho E} - \mathbf{g}_{j-1/2,R}^{\rho E}). \end{aligned} \tag{3.2}$$

Let us note that $\rho_1 = \rho Y$ is not concerned with (3.2). We simply set $(\rho_1)_j^{n+1-} = \rho_j^{n+1-} \times Y_j^n$ (Y is naturally kept constant in this step) and then $p_j^{n+1-} = p(\mathbf{u}_j^{n+1-})$ with $\mathbf{u}_j^{n+1-} = (\rho, \rho u, \rho E, \rho_1)_j^{n+1-}$.

Second step : evolution of ρ_1 ($t^{n+1-} \to t^{n+1}$)
In this step, ρ_1 is evolved in a conservative way using the numerical flux function g while the values of ρ, ρu and p are kept unchanged. Again with clear notations, the vector $\mathbf{v}_j^{n+1} = (\rho_j^{n+1}, (\rho u)_j^{n+1}, p_j^{n+1}, Y_j^{n+1})$ is then defined $\rho_j^{n+1} = \rho_j^{n+1-}$, $(\rho u)_j^{n+1} = (\rho u)_j^{n+1-}$, $p_j^{n+1} = p_j^{n+1-}$ and $Y_j^{n+1} = (\rho_1)_j^{n+1}/\rho_j^{n+1}$, where

$$(\rho_1)_j^{n+1} = (\rho_1)_j^n - \lambda(\mathbf{g}_{j+1/2}^{\rho Y} - \mathbf{g}_{j-1/2}^{\rho Y}), \tag{3.3}$$

with $\mathbf{g}_{j+1/2} = \mathbf{g}(\mathbf{u}_j^n, \mathbf{u}_{j+1}^n)$. At last, we set $\mathbf{u}_j^{n+1} = \mathbf{u}(\mathbf{v}_j^{n+1})$.

It is worth noticing that the Two-Flux method is not conservative on ρ and ρu since the fluxes $\mathbf{g}_{j+1/2,L}$ and $\mathbf{g}_{j+1/2,R}$ defined at each interface are different as soon as $Y_j^n \neq Y_{j+1}^n$. It is not conservative on ρE either, but by construction the Two-Flux method is conservative on ρ_1.

The associated quasi-conservative algorithm. It is actually clear from [10], [2] and the references therein that in standard conservative discretizations of (2.2), only the update formula of the total energy ρE is responsible for the pressure oscillations. We are then tempted to propose a quasi-conservative variant of the Two-Flux method such that only the total energy is treated in a non conservative form. For all $j \in \mathbb{Z}$, we simply replace (3.2) by

$$\begin{aligned} \rho_j^{n+1-} &= \rho_j^n - \lambda(\mathbf{g}_{j+1/2}^{\rho} - \mathbf{g}_{j-1/2}^{\rho}), \\ (\rho u)_j^{n+1-} &= (\rho u)_j^n - \lambda(\mathbf{g}_{j+1/2}^{\rho u} - \mathbf{g}_{j-1/2}^{\rho u}), \\ (\rho E)_j^{n+1-} &= (\rho E)_j^n - \lambda(\mathbf{g}_{j+1/2,L}^{\rho E} - \mathbf{g}_{j-1/2,R}^{\rho E}), \end{aligned} \tag{3.4}$$

where again $\mathbf{g}_{j+1/2} = \mathbf{g}(\mathbf{u}_j^n, \mathbf{u}_{j+1}^n)$. The second step is unchanged.

3.2 The Lagrange-Projection approach with random sampling

In this section, we propose a Lagrangian approach for approximating the solutions of (2.2). The general idea is to first solve this system in Lagrangian coordinates, and then to come back to an Eulerian description of the flow with a projection step. Under its classical conservative form, the Lagrange-Projection method generates spurious oscillations near the material fronts. In order to remove these oscillations, we propose to adapt the projection step (only). We begin with a description of the Lagrangian step and then recall, for the sake of clarity, the usual conservative projection step (see for instance [8]). Again, an usual 1/2 CFL restriction is used.

The Lagrangian step ($t^n \to t^{n+1-}$)

In this step, (2.2) in written in Lagrangian coordinates and solved by an acoustic scheme (see for instance [7]), which gives

$$\begin{aligned} \tau_j^{n+1-} &= \tau_j^n - \lambda\tau_j^n(u_{j+1/2}^n - u_{j-1/2}^n), \\ u_j^{n+1-} &= u_j^n - \lambda\tau_j^n(p_{j+1/2}^n - p_{j-1/2}^n), \\ E_j^{n+1-} &= E_j^n - \lambda\tau_j^n((pu)_{j+1/2}^n - (pu)_{j-1/2}^n), \\ Y_j^{n+1-} &= Y_j^n, \end{aligned} \tag{3.5}$$

where the velocity and the pressure at interfaces are defined by

$$\begin{cases} u_{j+1/2}^n = \dfrac{1}{2}(u_j^n + u_{j+1}^n) + \dfrac{1}{2(\rho c)_{j+1/2}^n}(p_j^n - p_{j+1}^n), \\ p_{j+1/2}^n = \dfrac{1}{2}(p_j^n + p_{j+1}^n) + \dfrac{(\rho c)_{j+1/2}^n}{2}(u_j^n - u_{j+1}^n). \end{cases} \tag{3.6}$$

The proposed local approximation $(\rho c)_{j+1/2}^n$ of the Lagrangian sound speed is $(\rho c)_{j+1/2}^n = \max((\rho c)_j^n, (\rho c)_{j+1}^n)$ but other definitions may be found for instance in [11]. In this step, the grid points $x_{j+1/2}$ move at velocity $u_{j+1/2}^n$ so that $\rho_j^{n+1-} = 1/\tau_j^{n+1-}$, $(\rho u)_j^{n+1-} = \rho_j^{n+1-} \times u_j^{n+1-}$, $(\rho E)_j^{n+1-} = \rho_j^{n+1-} \times E_j^{n+1-}$ and $(\rho_1)_j^{n+1-} = \rho_j^{n+1-} \times Y_j^{n+1-}$ define approximate values of $\mathbf{u}$ on a Lagrangian grid with mesh interfaces $x_{j+1/2}^* = x_{j+1/2} + u_{j+1/2}^n \Delta t$.

The usual projection step ($t^{n+1-} \to t^{n+1}$)

Aim of this step is to project the solution obtained at the end of the first step on the Eulerian grid defined by the mesh interfaces $x_{j+1/2}$. *Usually*, the choice is made to project the conservative vector $\mathbf{u}$ in order to obtain a conservative Lagrange-Projection scheme (see again [8]). More precisely, such a choice writes

$$\varphi_j^{n+1} = \frac{1}{\Delta x}\int_{x_{j-1/2}}^{x_{j+1/2}} \varphi^{n+1-}(x)dx \quad \text{with} \quad \varphi = \rho, \rho u, \rho E, \rho_1 = \rho Y, \tag{3.7}$$

or, with $\Delta x_j^* = x_{j+1/2}^* - x_{j-1/2}^*$ and $\varepsilon(j,n) = \begin{cases} -1/2 \text{ if } u_{j+1/2}^n > 0, \\ 1/2 \text{ if } u_{j+1/2}^n < 0, \end{cases}$

$$\varphi_j^{n+1} = \frac{1}{\Delta x}\{\Delta x_j^* \varphi_j^{n+1-} - \Delta t(u_{j+1/2}^n \varphi_{j+1/2+\varepsilon(j,n)}^{n+1-} - u_{j-1/2}^n \varphi_{j-1/2+\varepsilon(j-1,n)}^{n+1-})\}.$$

What is wrong with this scheme? The Lagrange-Projection approach allows to precisely reveal the operation that makes the material fronts necessarily damaged by pressure oscillations. Let us indeed consider an isolated material front with uniform velocity and pressure profiles : $u(x,0) = u^0 > 0$ and $p(x,0) = p^0$, while $Y(x,0) = 1$ if $x < 0$ and $Y(x,0) = 0$ if $x > 0$. The density is also set to be uniform for simplicity: $\rho(x,0) = \rho^0$. We first observe that this profile is clearly preserved in the Lagrangian step since by (3.6) we have $u_{j+1/2}^0 = u^0$ and $p_{j+1/2}^0 = p^0$ by (3.6). Then, the projection procedure (3.7) gives $\rho_j^1 = \rho^0$ and $(\rho u)_j^1 = \rho^0 u^0$ so that $u_j^1 = (\rho u)_j^1/\rho_j^1 = u^0$. After the first time iteration, the velocity profile is then still free of spurious oscillations. At last, using the property that this velocity is constant and positive, and focusing for instance on the cell of index $j = 1$, (3.7) gives for $\varphi = \rho E, \rho_1$

$$(\rho E)_1^1 = (\rho E)_1^0 - u^0\lambda\big((\rho E)_1^0 - (\rho E)_0^0\big) \quad \text{and} \quad Y_1^1 = u^0\lambda.$$

The pressure is then given after easy calculations by

$$p_1^1 = p^0 \times (\gamma(Y_1^1) - 1) \times \left(\frac{1}{\gamma_2 - 1}(1 - u^0\lambda) + u^0\lambda\frac{1}{\gamma_1 - 1}\right).$$

At this stage, there is no reason for p_1^1 to equal p^0 From the very first time iteration, a pressure oscillation is then created. As an immediate consequence, the velocity profile will not remain uniform in the next time iteration and the numerical solution is damaged for good.

It is then clear that the way the pressure is updated in the projection step (only) is responsible for the spurious oscillations in an usual conservative Lagrange-Projection scheme. We propose to modify this step. As for the Two-Flux method, the idea is to give up the conservation property in order to maintain uniform the pressure p across material fronts. Let us emphasize that the Lagrangian step is unchanged.

The quasi-conservative p-projection step ($t^{n+1-} \to t^{n+1}$)
First of all, ρ, ρu and ρ_1 still evolve according to (3.7) so that the algorithm remains conservative on these variables. We will keep on using (3.7) for $\varphi = \rho E$ only for j not in a subset $\mathbb{Z}_p^\varepsilon$ of $\mathbb{Z}$ defined below. On the contrary, the pressure p (instead of ρE) is averaged for $j \in \mathbb{Z}_p^\varepsilon$:

$$p_j^{n+1} = \frac{1}{\Delta x}\int_{x_{j-1/2}}^{x_{j+1/2}} p^{n+1-}(x)dx. \tag{3.8}$$

For $j \in \mathbb{Z}_p^\varepsilon$, we then set $\mathbf{v}_j^{n+1} = (\rho_j^{n+1}, (\rho u)_j^{n+1}, p_j^{n+1}, Y_j^{n+1})$ with $Y_j^{n+1} = (\rho_1)_j^{n+1}/\rho_j^{n+1}$, and $\mathbf{u}_j^{n+1} = \mathbf{u}(\mathbf{v}_j^{n+1})$.

Definition of $\mathbb{Z}_p^\varepsilon$. Up to our knowledge, the idea of averaging the pressure p in a Lagrange-Projection strategy first appeared in [3]. This way to proceed is clearly sufficient to remove the pressure oscillations near the material fronts if $\mathbb{Z}_p^\varepsilon = \mathbb{Z}$. However, averaging the pressure p instead of the total energy ρE for all $j \in \mathbb{Z}$ gives a non conservative scheme that is expected to provide discontinuous solutions violating the Rankine-Hugoniot conditions, see for instance Hou and LeFloch [9] (note however that here, the Lagrangian system associated with (2.1) is actually treated in a *conservative* form, while the pressure averaging takes place in the projection step only). This was confirmed in practice when considering solutions involving shocks with large amplitude. Then, in order to lessen the conservation errors, we propose to localize the non conservative treatment around the contact discontinuities setting $\mathbb{Z}_p^\varepsilon = \{j \in \mathbb{Z}, \max(|Y_j^n - Y_{j-1}^n|, |Y_{j+1}^n - Y_j^n|) > \epsilon\}$ for a given $\epsilon > 0$. Following Karni [10], we will use $\epsilon = 0.05$ in practice.

The quasi-conservative p-projection step with sampling ($t^{n+1-} \to t^{n+1}$)
The quasi-conservative p-projection step will be seen in the next section to properly compute large amplitude shock propagations. Localizing the averaging process of p nevertheless prevents the method from keeping strictly uniform the velocity and pressure profiles of an isolated material front, see **Test A** below. Indeed, note that since $\mathbb{Z}_p^\varepsilon$ is generally a *strict* subset of $\mathbb{Z}$ due to the numerical diffusion on Y (*i.e.* $\mathbb{Z}_p^\varepsilon \subsetneq \mathbb{Z}$), an usual conservative treatment is still used on ρE as soon as j not in $\mathbb{Z}_p^\varepsilon$. This is sufficient to create pressure oscillations. In order to cure this problem, we propose to get rid of the numerical diffusion on Y so as to enforce the non conservative treatment (3.8) across an isolated material front. This objective is achieved when replacing the conservative updating formula (3.7) for ρ_1 with random sampling strategy applied to Y (see also [4] and [5] for similar ideas). More precisely, we consider an equidistributed random sequence (a_n) in $(0,1)$ (following Collela [6], we take in practice the celebrated van der Corput sequence), define $x_j^* = x_{j-1/2} + a_{n+1}\Delta t$ for all $j \in \mathbb{Z}$ and set

$$\begin{cases} Y_j^{n+1} = Y_{j-1}^{n+1-} \text{ if } x_j^* \le x_{j-1/2}^*, \\ Y_j^{n+1} = Y_j^{n+1-} \text{ if } x_{j-1/2}^* \le x_j^* \le x_{j+1/2}^*, \\ Y_j^{n+1} = Y_{j+1}^{n+1-} \text{ if } x_j^* \ge x_{j+1/2}^*. \end{cases}$$

Then, we set $(\rho_1)_j^{n+1} = \rho_j^{n+1} Y_j^{n+1}$ so that the conservation of ρ_1 now holds only statistically.

4 Numerical results

We propose two numerical experiments with $\gamma_1 = 1.4$ and $\gamma_2 = 1.6$ associated with a Riemann initial data. The left and right vectors $\mathbf{v}$ are denoted $\mathbf{v}_L$ and $\mathbf{v}_R$ and the initial discontinuity is at $x = 0.5$. In the first simulation (**Test A**), we consider the propagation of an isolated material interface with $\mathbf{v}_L = (1, 1, 1, 1)$ and $\mathbf{v}_R = (0.1, 0.1, 1, 0)$. We take $\Delta x = 0.005$ and plot the solutions at time $t = 0.15$. The second simulation (**Test B**) develops a strong shock due to a large initial pressure ratio. More precisely, we choose $\mathbf{v}_L = (1, 0, 500, 1)$ and $\mathbf{v}_R = (1, 0, 0.2, 0)$. We

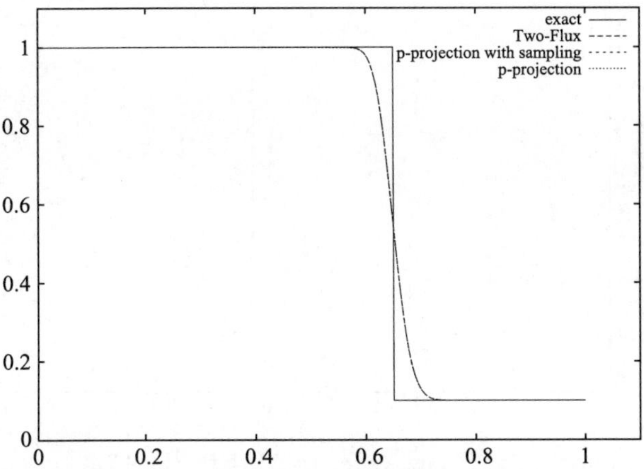

Figure 4.1 ρ (**Test A**)

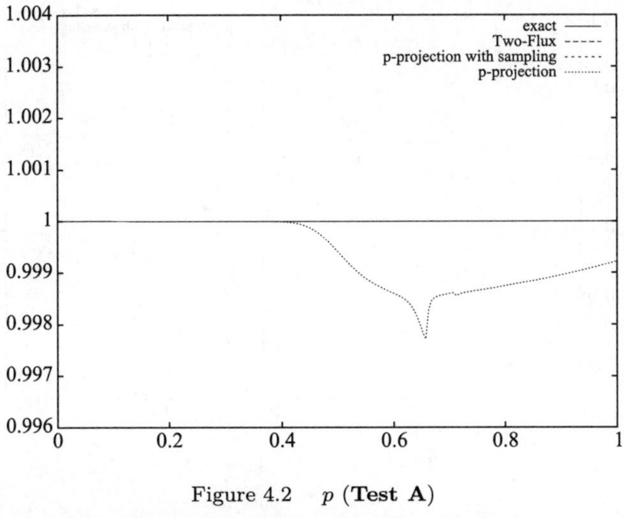

Figure 4.2 p (**Test A**)

take $\Delta x = 0.00125$ and plot the solutions at time $t = 0.008$.

We observe that the Two-Flux method and the Lagrangian methods are in agreement with the exact solutions and give similar results. As expected, note that the Lagrangian approach without sampling does not strictly maintain uniform the pressure profile for **Test A**. Note also that the mass fraction Y is sharp when a random sampling is used. At last, the relative conservation error on ρE (see for instance [5] for more details) for the Lagrangian approach with random sampling is actually less important and swings around 0.2% only.

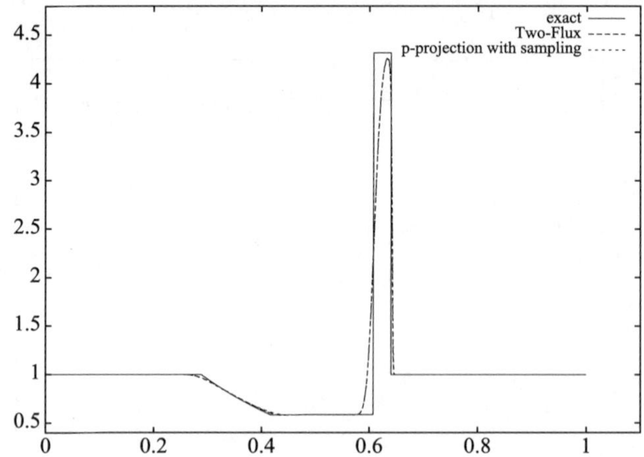

Figure 4.3 ρ (**Test B**)

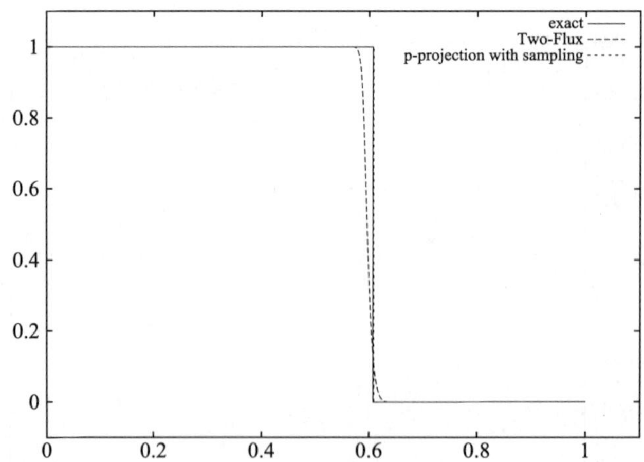

Figure 4.4 Y (**Test B**)

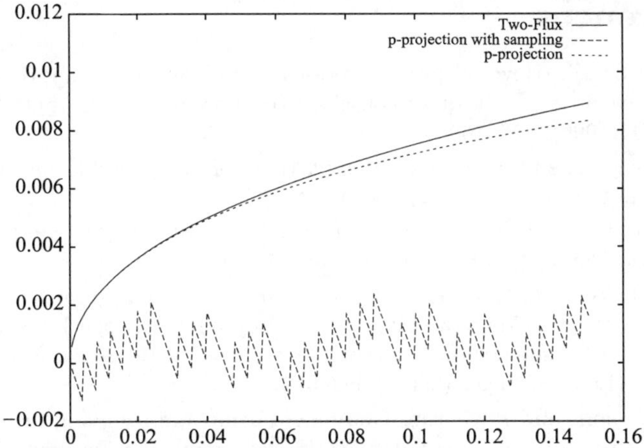

Figure 4.5 Conservation errors (**A**)

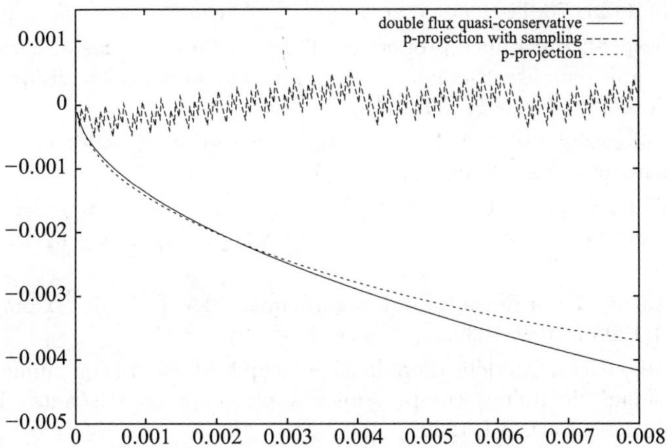

Figure 4.6 Conservation errors (**B**)

5 Concluding remarks

We have investigated a Lagrange-Projection approach for computing material fronts in multi-fluid models. We get similar results to the Two-Flux method [2] with less important conservation errors on ρE. Let us mention that other strategies, like for instance the one consisting in a local random sampling of $\mathbf{u}$ (instead of Y only) in the Lagrangian step, have been investigated. The results are not reported here.

Acknowledgements. The authors are grateful for helpful discussions and exchanges with P. Helluy and F. Lagoutière.

References

[1] R. Abgrall, How to prevent pressure oscillations in multicomponent flow calculations: a quasi-conservative approach. J. Comput. Phys. vol 125(1)(1996), 150–160.

[2] R. Abgrall and S. Karni, Computations of compressible multifluids. J. Comput. Phys. vol 169(2)(2001), 594–623.

[3] T. Barberon, P. Helluy and S. Rouy, Practical computation of axisymmetrical multifluid flows. IJFV, vol 1(1), 2004.

[4] C. Chalons and F. Coquel, Capturing infinitely sharp discrete shock profiles with the Godunov scheme, in: Proceedings of the Eleventh International Conference on Hyperbolic Problems. S. Benzoni-Gavage and D. Serre (Eds), Springer (2008), 363–370 .

[5] C. Chalons and P. Goatin, Godunov scheme and sampling technique for computing phase transitions in traffic flow modeling, Interfaces and Free Boundaries, vol 10, no. 2(2008), 197–221.

[6] P. Collela, Glimm's method for gas dynamics. SIAM J. Sci. Stat. Comput. vol 3(1982), 76–110.

[7] B. Després, Invariance properties of Lagrangian systems of conservation laws, approximate Riemann solvers and the entropy condition, Numer. Math. vol 89(1)(2001), 99–134.

[8] E. Godlewsky and P. A. Raviart, Numerical Approximation of Hyperbolic Systems of Conservation Laws, Springer (1995).

[9] T. Y. Hou and P. G. LeFloch, Why nonconservative schemes converge to wrong solutions: error analysis. Math. of Comp. vol 62(206)(1994), 497–530.

[10] S. Karni, Hybrid multifluid algorithms. SIAM J. Sci. Comput. vol 17(5)(1996), 1019–1039.

[11] F. Lagoutière, Modélisation mathématique et résolution numérique de problèmes de fluides compressibles à plusieurs constituants. Thèse de l'Université Paris VI, (2000).

A Numerical Method for a Shallow Water Model with Dispersive Source Terms*

Florent Chazel
Université de Toulouse, UPS/INSA
31077 Toulouse, France
E-mail: florent.chazel@math.univ-toulouse.fr

David Lannes
DMA, Ecole Normale Supérieure
75005 Paris, France
E-mail: David.Lannes@ens.fr

Fabien Marche
Université Montpellier 2, I3M
34090 Montpellier, France
E-mail: Fabien.Marche@math.univ-montp2.fr

Abstract

The fully nonlinear and weakly dispersive Green-Naghdi model for shallow water waves of large amplitude is studied. A hybrid finite volume and finite difference splitting approach is proposed. Numerical validations are then performed in one horizontal dimension.

1 Introduction

In an incompressible, homogeneous, inviscid fluid, the propagation of surface waves is governed by the Euler equations with nonlinear boundary conditions at the surface and at the bottom. Simpler models have been derived to describe the behavior of the solution in some specific physical regimes.

Of particular interest in coastal oceanography is the *shallow-water* regime, which corresponds to the configuration where the wave length λ of the flow is large compared to the typical depth h_0 (i.e. when $\mu =$

*The authors are partly supported by the ANR MathOcean.

$\frac{h_0^2}{\lambda^2} \ll 1$). When the typical amplitude a of the wave is small (i.e. when $\varepsilon = \frac{a}{h_0} = O(\mu)$), an approximation of order $O(\mu^2)$ of the free surface Euler equations is furnished by the Green and Naghdi equations [8]. We refer to [9] for more details; note also that a rigorous mathematical justification of these models has been given in [1].

These equations provide a correct description of the waves up to the breaking point. But there are theoretical and numerical difficulties when we try to apply these equations in coastal oceanography, including wave breaking and a description of the shoreline. Therefore the aim of this work is to develop a model and a numerical scheme that describes correctly both phenomena. More precisely, we want to

♯ 1 Provide a good description of the dispersive effects,

♯ 2 Take into account wave breaking in a simple way,

♯ 3 Allow the possibility of vanishing depth.

In the following, we will only focus on the construction of a numerical schemes to handle difficulties ♯ 1 and ♯ 3, while we refer to [4] for a study of the numerical solutions developed for difficulty ♯ 2. The strategy adopted here starts from a new Green-Naghdi model (see §2), for which a robust and accurate numerical scheme is proposed (see §3). Finally, numerical validations are shown.

2 The physical model

Throughout this paper, we denote by $\zeta(t, X)$ the elevation of the surface with respect to its rest state, and by $-h_0 + b(X)$ a parametrization of the bottom, where h_0 is a reference depth (see Figure 2.1). Here X stands for the horizontal variables ($X = (x, y)$ for 2D surface waves, and $X = x$ for 1D surface waves), and t is the time variable; we also denote by z the vertical variable.

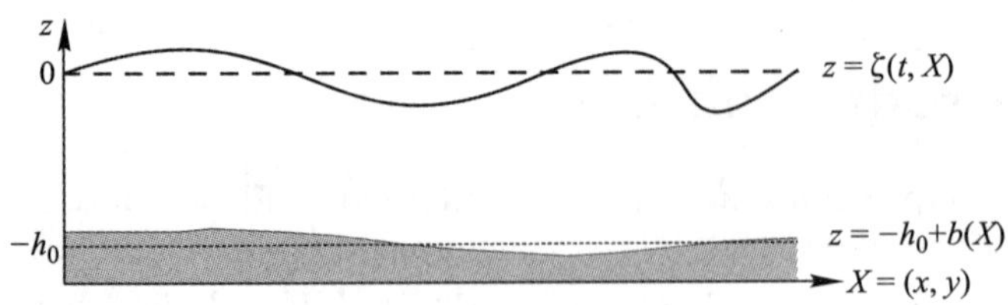

Figure 2.1 Sketch of the domain

If U_{hor} denotes the horizontal component of the velocity field in the fluid domain, we then define V as

$$V(t, X) = \frac{1}{h} \int_{-h_0+b}^{\zeta} U_{hor}(t, X, z) dz,$$

where $h := h_0 + \zeta - b$ is the water depth. We thus have $V = (u, v) \in \mathbb{R}^2$ for 2D surface waves, and $V = u \in \mathbb{R}$ for 1D surface waves.

It is shown in [3] and [6] that the usual Green-Naghdi equations [8] are asymptotically equivalent, with a precision up to terms of order $O(\mu^2)$ to the following system:

$$\begin{cases} \partial_t h + \nabla \cdot (hV) = 0, \\ \partial_t (hV) + \nabla \cdot (hV \otimes V) + (I + h\mathcal{T}\frac{1}{h})^{-1}[gh\nabla\zeta + h\mathcal{Q}_1(V)] = 0, \end{cases} \tag{2.1}$$

where the operators $\mathcal{T}$ and $\mathcal{Q}_1$ are defined as follows:

$$\mathcal{T}[h,b]W = \mathcal{R}_1[h,b](\nabla \cdot W) + \mathcal{R}_2[h,b](\nabla b \cdot W), \tag{2.2}$$

$$\mathcal{Q}_1[h,b](V) = -2\mathcal{R}_1(\partial_1 V \cdot \partial_2 V^{\perp} + (\nabla \cdot V)^2) + \mathcal{R}_2(V \cdot (V \cdot \nabla)\nabla b), \tag{2.3}$$

$$\mathcal{R}_1[h,b]w = -\frac{1}{3h}\nabla(h^3 w) - \frac{h}{2} w \nabla b, \tag{2.4}$$

$$\mathcal{R}_2[h,b]w = \frac{1}{2h}\nabla(h^2 w) + w\nabla b. \tag{2.5}$$

Looking at the linearization of (2.1) around the rest state $h = h_0$, $V = 0$, and flat bottom $b = 0$, one derives the dispersion relation associated to (2.1). It is found by looking for plane wave solutions of the form $(\underline{h}, \underline{hV})e^{i(\mathbf{k}\cdot x - \omega t)}$ to the linearized equations, and consists of two branches parametrized by $\omega_{\alpha,\pm}(\cdot)$,

$$\omega_{\alpha,\pm}(\mathbf{k}) = \pm|\mathbf{k}|\sqrt{gh_0}\sqrt{\frac{1}{1 + (|\mathbf{k}|h_0)^2/3}}. \tag{2.6}$$

3 Numerical methods

This new formulation is well-suited for a splitting approach separating the hyperbolic and the dispersive part of the equations (2.1).

3.1 The splitting scheme

We decompose the solution operator $S(\cdot)$ associated to (2.1) at each time step by the second order splitting scheme

$$S(\delta_t) = S_1(\delta_t/2)S_2(\delta_t)S_1(\delta_t/2), \tag{3.1}$$

with:

- $S_1(t)$ is the solution operator associated to NSWE

$$\begin{cases} \partial_t h + \nabla \cdot (hV) = 0, \\ \partial_t (hV) + \nabla\left(\frac{1}{2}gh^2\right) + \nabla \cdot (hV \otimes V) = -gh\nabla b. \end{cases} \tag{3.2}$$

- $S_2(t)$ is the solution operator associated to the remaining (dispersive) part of the equations,

$$\begin{cases} \partial_t h = 0, \\ \partial_t(hV) - \dfrac{1}{\alpha} gh\nabla\zeta + \left(I + \alpha hT\dfrac{1}{h}\right)^{-1}\left[\dfrac{1}{\alpha} gh\nabla\zeta + h\mathcal{Q}_1(V)\right] = 0. \end{cases} \tag{3.3}$$

From this point, *we only consider* 1D *surface waves.*

3.2 Spatial discretization of the hyperbolic component $S_1(\cdot)$

When specified in one space dimension, the system under consideration reads as follows:

$$\begin{cases} \partial_t h + \partial_x(hu) = 0, \\ \partial_t(hu) + \partial_x(hu^2 + gh^2/2) = -gh\partial_x b. \end{cases} \tag{3.4}$$

For the sake of simplicity in the notations, it is convenient to rewrite the system (3.4) in the following condensed form:

$$\partial_t \mathbf{w} + \partial_x \mathbf{f}(\mathbf{w}) = \mathbf{S}(\mathbf{w}, b), \tag{3.5}$$

with

$$\mathbf{w} = \begin{pmatrix} h \\ hu \end{pmatrix}, \quad \mathbf{f}(\mathbf{w}) = \begin{pmatrix} hu \\ hu^2 + \dfrac{g}{2}h^2 \end{pmatrix} \quad \text{and} \quad \mathbf{S}(\mathbf{w}) = \begin{pmatrix} 0 \\ -gh\partial_x b \end{pmatrix}, \tag{3.6}$$

where $\mathbf{w} : \mathbb{R} \times \mathbb{R}^+ \to \Omega$ is the state vector in conservative variables and $\mathbf{f}(\mathbf{w}) : \Omega \to \mathbb{R}^2$ stands for the flux function, with Ω the set of the admissible states. Considering numerical approximations of system (3.5), we seek a numerical scheme that provides stable simulations of the processes occurring in surf and swash areas, with a precise control of the spurious effects induced by numerical dissipation and dispersion. Moreover, the scheme should be able to handle the complex interactions between waves and topography, including the preservation of motionless steady states:

$$u = 0, \quad h + b = \text{cste}.$$

The homogeneous nonlinear shallow water equations (NSWE) associated with (3.4), given by

$$\partial_t \mathbf{w} + \partial_x \mathbf{f}(\mathbf{w}) = 0, \tag{3.7}$$

is known to be hyperbolic over Ω. The spatial discretization of the homogeneous system (3.7) can be recast under the following classical semi-discrete finite-volume formalism:

$$\frac{d}{dt}\bar{\mathbf{w}}_i(t) + \frac{1}{\delta_x}\left(\tilde{\mathbf{f}}\left(\bar{\mathbf{w}}_i, \bar{\mathbf{w}}_{i+1}\right) - \tilde{\mathbf{f}}\left(\bar{\mathbf{w}}_{i-1}, \bar{\mathbf{w}}_i\right)\right) = 0$$

where $\tilde{\mathbf{f}}$ is a numerical flux function based on a conservative flux consistent with the homogeneous NSWE. For the numerical validations shown in §4, we use the numerical flux issued from the relaxation approach introduced in [2]. To achieve a second order accuracy, we use the modified ENO scheme detailed in [5] for instance. The robustness of the resulting second order scheme is shown in [2]. We finally introduce a well-balanced discretization of the topography source term, using the *hydrostatic reconstruction* [5].

3.3 Spatial discretization of the dispersive component $S_2(\cdot)$

The system corresponding to the operator $S_2(\cdot)$ writes in one dimension

$$\begin{cases} \partial_t h = 0, \\ \partial_t(hu) - \dfrac{1}{\alpha}gh\partial_x\zeta + (1 + \alpha h\mathcal{T}\dfrac{1}{h})^{-1}\left[\dfrac{1}{\alpha}gh\partial_x\zeta + h\mathcal{Q}_1(u)\right] = 0 \end{cases} \tag{3.8}$$

where the operators $\mathcal{T}$ and $\mathcal{Q}_1$ are explicitly given by

$$\mathcal{T}w = -\frac{h^2}{3}\partial_x^2 w - h\partial_x h\partial_x w + \left(\partial_x\zeta\partial_x b + \frac{h}{2}\partial_x^2 b\right)w, \tag{3.9}$$

and

$$\mathcal{Q}_1(u) = 2h\partial_x\left(h + \frac{b}{2}\right)(\partial_x u)^2 + \frac{4}{3}h^2\partial_x u\partial_x^2 u + h\partial_x^2 bu\partial_x u + \left(\partial_x\zeta\partial_x^2 b + \frac{h}{2}\partial_x^3 b\right)u^2. \tag{3.10}$$

The system (3.8) is solved at each time step using a classical finite-difference technique. Details can be found in [3]

3.4 Time discretization

As far as time discretization is concerned, we choose to use explicit methods. The systems corresponding to S_1 and S_2 are integrated in time using a classical fourth-order Runge-Kutta scheme.

4 Numerical validation

In this section we focus on two test cases. The first one is devoted to assess that the balance between non-linear and dispersive effects is preserved by our splitting approach. In the second test, we show that the use of a finite-volume scheme for the hyperbolic part of the equations leads to an accurate description of the shoreline motions.

4.1 Propagation of a solitary wave

It is known that for horizontal bottoms, the Green-Naghdi model admits exact solitary wave solutions [3]. These solutions can be used as a validation tool for our present numerical scheme. We successively consider the propagation of two solitary waves of different relative amplitude a/h_0, on a 30 m long domain with a constant depth $h_0 = 0.5$ m. The considered relative amplitudes are $a/h_0 = 0.05$ for the first solitary wave, and $a/h_0 = 0.2$ for the second one. Periodic conditions are imposed on each boundary.

The numerical solution is computed for several time steps δ_t and cell sizes δ_x, over a sufficient duration $T = 3$ s. Starting with $\delta_x = 1$ m and $\delta_t = \delta_x/\sqrt{gh_0} = 0.45$ s, we successively divide the time step by two, while keeping the CFL equal to 1. For each computation and each discrete time $t_n = n\delta_t$, the relative errors E^n_ζ and E^n_u on the free surface elevation and the averaged velocity are computed using the discrete L^∞ norm $||.||_\infty$:

$$E^n_\zeta = \frac{||h_{num} - h_{sol}||_\infty}{||h_{sol} - h_0||_\infty} \quad ; \quad E^n_u = \frac{||u_{num} - u_{sol}||_\infty}{||u_{sol}||_\infty}$$

where (h_{num}, u_{num}) are the numerical solutions and (ζ_{sol}, u_{sol}) denotes the analytical ones.

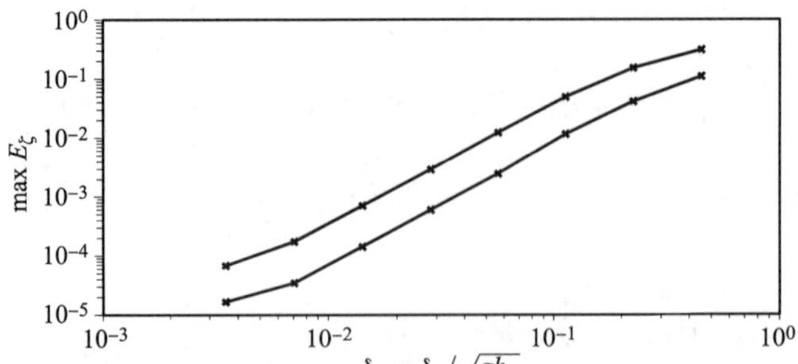

Figure 4.1 Propagation of a solitary wave over a flat bottom, maximum of the relative error on the free surface elevation. Case $a/h_0 = 0.05$ in black, case $a/h_0 = 0.2$ in grey.

Results are gathered in Figure 4.1. The convergence of our numerical scheme is clearly demonstrated. Computing a linear regression on all points yields a slope equal to 1.91 for the first case and 1.83 for the second one.

4.2 Run-up and run-down of a breaking solitary wave over a planar beach

This test is based on experiments carried out by Synolakis [10] for an incident solitary wave of relative amplitude $a_0/h_0 = 0.28$, which propa-

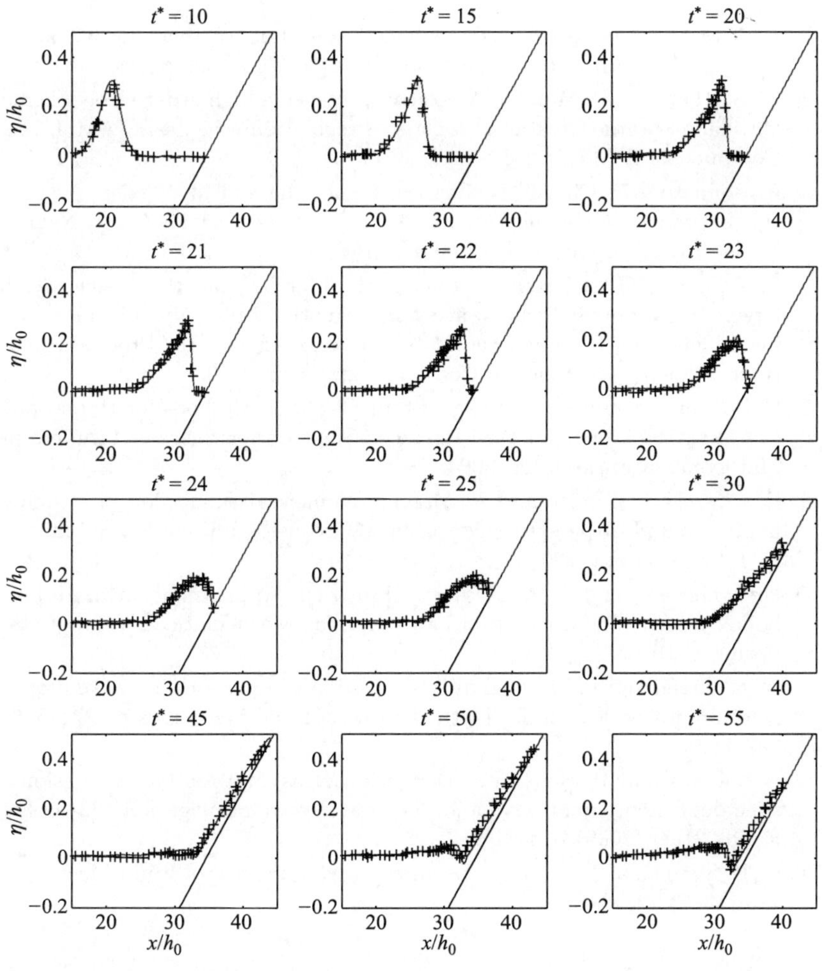

Figure 4.2 Comparisons of model predictions (—) and experimental snapshots (+) for a breaking solitary wave, on a 1 : 19.85 constant slope beach, with $t^* = t(g/h_0)^{1/2}$.

gates and breaks over a planar beach with a slope of 1:19.85. Free surface elevations at different times are available thanks to video measurements. The still water level in the horizontal part of the beach is $h_0 = 0.3\,\mathrm{m}$. The simulations are performed using the cell size $\delta_x = 0.08\,\mathrm{m}$ and $\delta_t = 0.02\,\mathrm{s}$.

The comparison between measured and computed waves is presented in Figure 4.2. It shows a good agreement between model predictions and laboratory data and illustrates the ability of our model to reproduce shoaling, breaking, run-up and run-down, as well as the formation and breaking of the backwash bore.

References

[1] B. Alvarez-Samaniego and D. Lannes, Large time existence for 3D water waves and asymptotics. Invent. Math. 171(3) (2008) 485–541.

[2] C. Berthon and F. Marche, A positive preserving high order vfroe scheme for shallow water equations: a class of relaxation schemes. SIAM J. Sci. Comput. 30(5) (2008) 2587–2612.

[3] P. Bonneton, F. Chazel, D. Lannes, F. Marche and M. Tissier, A splitting approach for the fully nonlinear and weakly dispersive Green-Naghdi model. In correction for J. Comput. Phys.

[4] P. Bonneton, F. Chazel, D. Lannes, F. Marche and M. Tissier, Serre Green-Naghdi modelling of wave transformation breaking and run-up using a high-order finite-volume finite-difference scheme. In: Proceedings of International Conference in Coastal Engineering 2010.

[5] F. Bouchut, Nonlinear stability of Finite Volume Methods for Hyperbolic Conservation Laws and Well-Balanced Schemes for Sources. Frintiers in Mathematics, Birkhauser, 2004.

[6] F. Chazel, D. Lannes and F. Marche, Numerical simulation of strongly nonlinear and dispersive waves using a Green-Naghdi model. Submitted to J. Sci. Comput. (2010).

[7] E. Godlewski and P.-A. Raviart, Numerical Approximation of Hyperbolic Systems of Conservation Laws. Applied Mathematical Sciences 118, Springer, 1996.

[8] A. E. Green and P. M. Naghdi, A derivation of equations for wave propagation in water of variable depth. Journal of Fluid Mechanics 78(2) (1976) 237–246.

[9] D. Lannes and P. Bonneton, Derivation of asymptotic two-dimensional time-dependent equations for surface water wave propagation. Physics of Fluids 21(1) (2009) 016601.

[10] C. E. Synolakis, The runup of solitary waves. Journal of Fluid Mechanics 185 (1987) 523–545.